Agroforestry-Based Ecosystem Services

Agroforestry-Based Ecosystem Services

Editor

Meine van Noordwijk

MDPI • Basel • Beijing • Wuhan • Barcelona • Belgrade • Manchester • Tokyo • Cluj • Tianjin

Editor
Meine van Noordwijk
Plant Production Systems
Wageningen University and
Research
Wageningen
The Netherlands

Editorial Office
MDPI
St. Alban-Anlage 66
4052 Basel, Switzerland

This is a reprint of articles from the Special Issue published online in the open access journal *Land* (ISSN 2073-445X) (available at: www.mdpi.com/journal/land/special_issues/agroforestry_ES).

For citation purposes, cite each article independently as indicated on the article page online and as indicated below:

LastName, A.A.; LastName, B.B.; LastName, C.C. Article Title. *Journal Name* **Year**, *Volume Number*, Page Range.

ISBN 978-3-0365-1742-1 (Hbk)
ISBN 978-3-0365-1741-4 (PDF)

Contents

About the Editor

Meine van Noordwijk

Meine van Noordw (Ph.D. Wageningen University, 1987) served, until mid-2017, as Chief Scientist at World Agroforestry (ICRAF), based in Bogor (Indonesia), and since that time as Distinguished Senior Fellow. He is also a Professor of Agroforestry (em.) at Wageningen University (The Netherlands), and part of the agroforestry research group at Brawijaya University (Malang, Indonesia). Trained as biologist-ecologist he has experience with systems analysis and modeling of systems that range from single roots in the soil, via tree-soil-crop interactions, to the understanding of water, biodiversity, and greenhouse gasses at landscapes scales, and the conflicts that arise over multiple claims to landscape functions, to the institutional translation of ecosystem service concepts in a comprehensive approach to land use policy. His work experience includes Southeast Asia, East Africa and The Netherlands.

Preface to "Agroforestry-Based Ecosystem Services"

With 21 papers and an editorial overview published, more than a hundred authors involved, and contributions from tropical Asia, Africa, and Latin America, the response to this Special Issue exceeded initial expectations. The collection of papers is not a comprehensive, encyclopedic treatment of the topic, but it provides little gemstones of case studies on a good cross-section of the social and ecological issues that are part of the current debate. The collection can serve as teaching material for interdisciplinary courses on land-use science and may provide inspiration (and specific hypotheses) for further research by graduate students or researchers interested in use-oriented science.

Although the call for contributions was not explicit on the geographical focus of the contributed papers, the resulting focus on (sub)tropical case studies probably reflects the historical development of agroforestry science, where attention to temperate regions only took off after two decades of tropical research. Temperate-zone agroforestry is now a flourishing field of research, so future compilations may become more geographically balanced.

Acknowledgements are due to the positive responses of all authors in their initial submissions and subsequent revisions, the many (mostly anonymous) reviewers, other members of the *Land* editorial board, and the professional services by the MDPI publisher.

Meine van Noordwijk
Editor

land

Editorial

Agroforestry-Based Ecosystem Services

Meine van Noordwijk [1,2,3]

1   World Agroforestry (ICRAF), Bogor 16155, Indonesia; M.vannoordwijk@cgiar.org
2   Plant Production Systems, Wageningen University & Research, 6700 AK Wageningen, The Netherlands
3   Study Group Agroforestry, Faculty of Agriculture, Brawijaya University, Jl Veteran, Malang 65145, Indonesia

1. Introduction

Agroforestry, land use at the agriculture-forestry interface that implies the presence of trees on farms and/or farmers in forests, has a history that may be as old as agriculture, but as an overarching label and topic of formal scientific analysis, it is in its fifth decade. The trees as such, and the agroforestry system they are part of, provide direct benefits to the farmer (land manager), often through a combination of marketable goods, subsistence needs of the farm household, buffering climate variability, and protecting soil and water resources. However, it also provides benefits to those sharing the same landscape, the same watershed, biome, or even planet Earth, the latter especially as part of the global climate and biodiversity conservation discourses. These external benefits are generally discussed under the heading 'ecosystem services' (Figure 1), and are the topic of the collection of papers in this Special Issue.

Citation: van Noordwijk, M. Agroforestry-Based Ecosystem Services. *Land* **2021**, *10*, 770. https://doi.org/10.3390/land10080770

Received: 12 July 2021
Accepted: 20 July 2021
Published: 22 July 2021

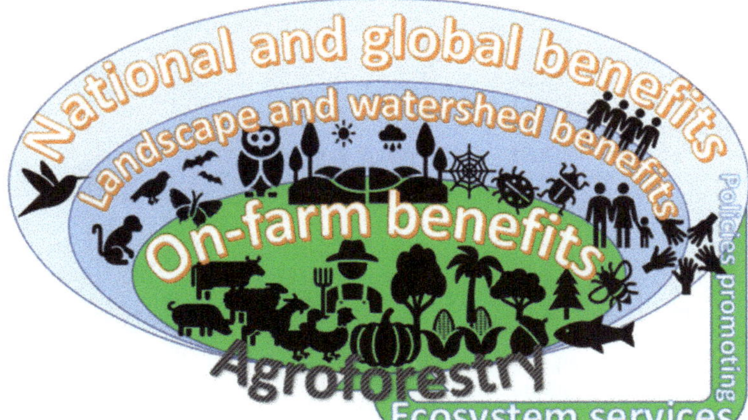

Figure 1. Agroforestry-based ecosystem services are human benefits achieved beyond the farm scale from the way trees on farm interact with soil and water, carbon storage, and biodiversity and of the cultural/relational aspects of landscapes with partial tree cover.

For this Special Issue, we invited case studies or synthesis papers that achieve the following:

1.   Quantify change in ecosystem services in forest–agriculture interface landscapes and relate such to stakeholder concerns and farmer/manager decisions;
2.   Analyze efforts to increase the feedback from external stakeholders to land use decisions (including 'agroforestry') within landscapes; and/or
3.   Describe and analyze efforts to transcend an existing forestry versus agriculture dichotomy in land use policies.

2. Quantifying Ecosystem Services

Of the twelve papers in the first category (Table 1), seven have a biophysical/ecological focus and five a social one. For the current discussion, four groups of ecosystem services are important and distinct: provisioning and economic services, soil and water conservation, carbon-related roles in the global climate discourse, and biodiversity-related services. Many papers discuss more than one type of service.

Table 1. Papers quantifying change in ecosystem services related to farmer/manager decisions.

Title and Reference	Provisioning, Economic	Soil & Water	Carbon-Related	Biodiversity-Related
Biophysical focus:				
Traditional pollarding practices for dimorphic ash tree (*Fraxinus dimorpha*) support soil fertility in the Moroccan High Atlas [1]	Traditional tree management	Mycorrhiza		Local tree species
Soil organic matter, and mitigation of and adaptation to climate change in cocoa–based agroforestry systems [2]		Corg, water holding capacity	Carbon stocks	Comparing land use systems
Earthworm diversity, forest conversion, and agro-forestry in Quang Nam province, Viet Nam [3]		LU impacts on 'soil engineers'		Earthworm diversity
Assessing context-specific factors to increase tree survival for scaling ecosystem restoration efforts in East Africa [4]	Farmer's technology tested	Tree seedling management		
Tree roots anchoring and binding soil: reducing landslide risk in Indonesian agroforestry [5]		Slope stabilization		Functional tree diversity
Infiltration-friendly agroforestry land uses on volcanic slopes in the Rejoso watershed, East Java, Indonesia. [6]		Infiltration and erosion control		Comparing land use systems
Groundwater-extracting rice production in the Rejoso watershed (Indonesia) reducing urban water availability: characterization and intervention priorities [7]	Rice production using ground-water from AF hillslopes	Water balance effects of lowland water use		
Focus on farmers/managers:				
Fruit tree-based agroforestry systems for smallholder farmers in northwest Vietnam—a quantitative and qualitative assessment [8]	Farmer knowledge, economic analysis			
Local knowledge about ecosystem services provided by trees in coffee agroforestry practices in Northwest Vietnam [9]	Farmer knowledge, preferences	Soil & water protection		Functional tree diversity
Gendered species preferences link tree diversity and carbon stocks in cacao agroforest in Southeast Sulawesi, Indonesia [10]	Gendered tree preferences		Carbon stocks	Functional tree diversity
Agroforestry innovation through planned farmer behavior: trimming in pine–coffee system [11]	Constraints to farmer tree management			
Gendered migration and agroforestry in Indonesia: livelihoods, labor, know-how, networks [12]	Human migration~ AF knowledge			

In agropastoral parklands of the Moroccan High Atlas, traditional practices of managing the local dimorphic ash tree (*Fraxinus dimorpha*), endemic to North Africa, involve pollarding [1]. Compared to soil without trees, the soil under trees was shown to be enriched with phosphorus, nitrogen, carbon, and mycorrhizal spore density, suggesting soil fertility benefits generated by sound agroforest management. Belowground roles of agroforestry in climate change mitigation (C storage) and adaptation (reduced vulnerability to drought) are less obvious than easy-to-measure aspects aboveground. Comparisons between cocoa agroforestry systems, intermediate in properties between remnant forest, and cocoa monocultures [2] showed that increased soil organic matter content can support about a week's worth of evapotranspiration without rain, assisting in climate change adaptation. Agroforestry systems in Viet Nam were also found to be intermediate between natural forest and cropland in terms of earthworm diversity [3]. Out of a total of 25 different earthworm species, 21 were found in natural forests, 15 in agroforestry, 14 in planted forests, and 7 in annual croplands and home gardens. A cosmopolitan species, *Pontoscolex corethrurus*, dominated habitats with intensive anthropogenic activities but was rare in natural forests. The study concluded that protection of the remaining natural forests is urgent, while the promotion of a tree-based farming system such as agroforestry can reconcile earthworm conservation and local livelihoods.

Especially in drylands, tree planting is not enough to get trees growing, and survival rates depend on management. A survey in Kenya and Ethiopia [4] showed low tree survival especially for tree species preferred by local communities, but also local knowledge on options to increase seedling survival in local context. Soil quality ranking was positively correlated with tree survival in Ethiopia, regardless of species assessed, while in Kenya the presence of soil erosion on a farm had a negative effect on seedling survival. The need for watering, manuring, and protection from grazing varied with context and tree species.

Biophysical properties at the plot level influence ecological relationships at the landscape scale. Once established, tree root systems can stabilize hillslopes and riverbanks, reducing landslide risk, but comparisons across a wide range of tree species are rare as research methods are laborious. Observing proximal (close to the tree stem) woody roots and applying fractal allometry hypotheses, a comparison [5] of 55 tree species in Indonesia found differences in the ratio of stem to root cross-sectional area (relative amount of roots) as well as relative distribution in topsoil ('Soil-Root Binding') and subsoil ('Root Anchoring') that both contribute to soil stabilization. The study concluded that a mix of tree species with deep roots and grasses with intense fine roots provides the highest hillslope and riverbank stability.

Rapid infiltration of rainfall into soils is important to reduce surface runoff and erosion on mountain slopes exposed to high-intensity rainfall. The degree of tree cover that is needed to achieve a desirable level of infiltration, however, depends on further characteristics of soil, tree species, and zone-dependent rainfall properties. A study in Indonesia [6] found that for midstream conditions only a tree canopy cover of >80% qualified as "infiltration-friendly" land use, but erosion rates were relatively low for a tree canopy cover in the range of 20–80%. In the upstream watershed, a tree canopy cover > 55% was associated with the infiltration rates needed, as soil erosion per unit overland flow was high. The tree canopy characteristics required for infiltration-friendly land use clearly varied over short distances with soil type and rainfall intensity showing that generic rules, such as a 30% forest cover requirement, cannot be the basis for local resource management using agroforestry concepts. A study of the water balance and the impacts of lowland water use in the same watershed through uncontrolled flow of Artesian wells found that equal attention is needed for the upstream and downstream parts of the watershed if urban water supply is to be secured [7].

The second group of papers described in Table 1 has farmer knowledge, choices, and preferences as their primary focus. Agroforestry practices with fruit trees can be more profitable than sole-crop cultivation within a few years, as data for Viet Nam and two local fruit tree species show [8]. After seven years agroforestry systems with longan

(*Dimocarpus longan*) and son tra (*Docynia indica*) had generated 2.4-times higher average annual income than sole maize, the main comparator in the area. Farmers also reported that agroforestry enhanced ecosystem services by controlling surface runoff and erosion, increasing soil fertility and improving resilience to extreme weather, indicating a win-win in economic and environmental terms, if the initial investment hurdle can be overcome.

Farmer knowledge of ecosystem services provided by trees in coffee agroforestry practices in Northwest Vietnam was found to differ between three indigenous groups surveyed [9]. Most farmers were aware of the benefits of trees for soil improvement, shelter (from wind and frost), and the provision of shade and mulch. In contrast, farmers had limited knowledge of the impact of trees on coffee quality and other interactions amongst trees and coffee. The farmers' selection of tree species to combine with coffee was highly influenced by economic benefits provided, especially by intercropped fruit trees, which was influenced by market access, determined by the proximity of farms to a main road.

Surveys in Southeast Sulawesi of tree diversity on cocoa agroforestry systems [10] showed that gendered preferences for trees partly diverge and may contribute to overall diversity. Male farmers selected timber and fruit tree species with economic benefits as shade trees, while female farmers preferred production for household needs (fruit trees and vegetables). Tree portfolios reflected the preferences of both genders. In agroforestry, tree diversity was found to be proportional to differences in carbon stock, with an intermediate position between the concave relationship in forest decline and the convex one in reforestation responses.

In a specific form of agroforestry in which forest authorities own the land and trees and farmers are allowed to intercrop, ecological management choices reflect the social relations of power. In a setting where farmers have contracts permitting coffee cultivation under pine trees, experiments tested canopy trimming to improve light for coffee production while maintaining tree density [11]. Exploring planned farmer behavior brought 'path dependency' and 'lack of trust' to the forefront as issues to be understood before agroforestry innovation can contribute in the triangle of farmers, forest authorities, and empirical science.

A wider policy perspective is needed to understand the gendered decisions to migrate into or away from areas where agroforestry is practiced in Indonesia [12]. Most of the decision making that the research revealed was linked to perceived poverty, natural resource and land competition, and emergencies, such as natural disasters or increased human conflicts. Movements of a temporary labor force and/or migrants who might still return to their landscape of origin, do contribute to the spread of agroforestry knowledge (and germplasm) between areas of higher and lower human population density, contributing to ecosystem service awareness.

3. Co-Investment in Ecosystem Service Provision by External Stakeholders

Five papers considered the logical next step after recognizing that ecosystem services depend on farmers' land management: co-investment (Table 2).

Table 2. Papers addressing efforts to increase the feedback from external stakeholders to land use decisions (including 'agroforestry') within landscapes.

Title and Reference	Ecosystem Services			
	Provisioning, Economic	Soil & Water-Related	Carbon-Related	Biodiversity-Related
Effects of agroforestry and other sustainable practices in the Kenya Agricultural Carbon Project (KACP) [13]	Project impact study		Co-investment in carbon stocks	
Discounted cash flow and capital budgeting analysis of silvopastoral systems in the Amazonas region of Peru [14]	Farm economic analysis			

Table 2. Cont.

Title and Reference	Ecosystem Services			
	Provisioning, Economic	Soil & Water-Related	Carbon-Related	Biodiversity-Related
Carbon storage potential of silvopastoral systems of Colombia [15]	Options for national climate policy		Potential increase in silvopastoral tree density	
Enhancing Vietnam's nationally determined contribution with mitigation targets for agroforestry: a technical and economic estimate [16]	Options for national climate policy		Potential increase in various AF systems	

Support for sustainable agricultural land management can also be motivated by global climate change concerns, with externally funded advisory services. An evaluation of the Kenya Agricultural Carbon Project (KACP) [13] concluded that farmers benefitted from increased maize yields after participation in the program that supported terraces and agroforestry practices. Study results, showing that the KACP farms had higher food self-sufficiency and tended to have higher monetary savings than control farms, suggest that, apart from the carbon stock gains that generated the co-investment by external stakeholders, local benefits were clear.

A study of the economic consequences of silvopastoral systems in the Peruvian Amazon [14] found that these were above those for either planted forests or conventional cattle-pasture systems. Benefits could be substantially higher—at least for the farmers pioneering such activities—where farmers generated added-value through on-site retail stands and direct links to customers.

Nine Latin American countries plan to use silvopastoral practices—incorporating trees into grazing lands—to mitigate climate change, but the cumulative potential of scaling up silvopastoral systems at national levels is not well quantified [15]. The range, 5 to 122 Mg ha^{-1}, of carbon stock values in Colombian grasslands in 2017 based on ecofloristic zones, suggests a potential for further increase. If all existing grasslands could be brought to the tree density of the current median or 75th percentile, silvopastoral systems could be a substantive part of nationally determined contributions (NDCs) and nationally appropriate mitigation actions (NAMAs) in Colombia and other Latin American countries with similar contexts.

The Nationally Determined Contributions (NDCs) of several non-Annex I countries already mention agroforestry but mostly without associated mitigation targets. The absence of reliable data, including on existing agroforestry practices and their carbon storage, partially constrains the target setting. A study for Viet Nam [16] tried to fill this gap by synthesizing above- and belowground vegetation and soil carbon for the close to 0.8 M ha of existing agroforestry systems identified. Estimates are that expansion to 0.9–2.4 M ha of agroforestry is technically and economically feasible, to offset the greenhouse gas emissions of the agriculture sector by 2015.

4. Addressing the Agricultural-Forestry Policy Interface

A final group of five papers (Table 3) considered the agricultural-forestry interface, in policy and institutional terms, as essential to the further development of agroforestry.

Table 3. Papers analyzing efforts to transcend the existing forestry versus agriculture dichotomy in land use.

Title and Reference	Ecosystem Services			
	Provisioning, Economic	Soil & Water-Related	Carbon-Related	Biodiversity-Related
Agroforestry as policy option for forest-zone oil palm (OP) production in Indonesia [17]	Smallholder OP as AF option	Concerns over OP disturbing hydrology	Concerns over C emissions	Concerns over OP causing deforestation
People-centric nature-based land restoration through agroforestry: a typology [18]	Range of restoration intensities	Subsoil recovery slow	Realistic expectations of tradeoffs	Realistic expectations of tradeoffs
Cost-benefit analysis of landscape restoration: a stocktake [19]	Efficient use of scarce public funds	Realistic expectations of benefits	Realistic expectations of tradeoffs	
Sustainable agroforestry landscape management: changing the game [20]	Social-ecological systems perspective	Need for shared understanding		
Agroforestry-based eco-system services: reconciling values of humans and nature in sustainable development [21]				Relational values (both + and −) of biodiversity are under-studied

With 15–20% of Indonesian oil palms located, without a legal basis and permits, within the forest zone ('Kawasan hutan'), international concerns regarding deforestation affect the totality of Indonesian palm oil exports [17]. Data analysis showed that 'Forest zone oil palm' (FZ-OP) is a substantive issue with substantial geographic variation in intensity within Indonesia that requires analysis and policy change. Responses will need to take the legal basis of the forest zone and its conversion into account, as well as the existing social stratification in oil palm production (large-scale, plasma and independent growers), and the various environmental consequences of forest conversion to FZ-OP depending on the location. Conditional acceptance of diversified smallholder plantings in 'agroforestry concessions' is one of the policy options to be considered.

In the decade of ecological restoration, just started, agroforestry can be a major 'people-centric nature-based' solution, if contextualized appropriately [18]. Restoration entails innovation to halt ongoing and reverse past degradation. Four intensities of land restoration can be distinguished in their interaction: R.I. Ecological intensification within a land use system, R.II. Recovery/regeneration, within a local social-ecological system, R.III. Reparation/recuperation, requiring a national policy context, and R.IV. Remediation, requiring international support and investment. Relevant interventions start from core values of human identity while addressing five potential bottlenecks: Rights, Know-how, Markets (inputs, outputs, credit), Local Ecosystem Services (including water, agrobiodiversity, micro/mesoclimate) and Teleconnections (global climate change, biodiversity).

With the increase in demand for landscape restoration and the limited resources available, there is need for economic analysis of landscape restoration to help prioritize investment of the resources [19]. However, cost-benefit analysis (CBA) seems limited and varied in its application as a commonly applied tool in the economic analysis of landscape restoration. Of the 2056 studies identified in a literature search, only 31 met the predefined criteria of rigor and relevance. About 60% of those focused on agroforestry, afforestation, reforestation, and assisted natural regeneration practices, but only 16% covered all cost categories, with opportunity costs being the least covered. Eighty-four percent apply direct use values, while only 16% captured the non-use values. The study thus suggests a strong need for improvements in both the quantity and quality of CBA to better inform planning, policies, and investments in landscape restoration.

While location-specific forms of agroforestry management can probably reduce problems in the forest–water–people nexus by balancing upstream and downstream interests, social and ecological finetuning is needed and requires a shared understanding of the underlying relations. 'Serious games' have been shown to contribute to such understanding but they so far (1) appear to be ad hoc and case-dependent, with poorly defined extrapolation domains, (2) require heavy research investment, (3) have untested cultural limitations, and (4) lack clarity on where and how they can be used in policy making. The final contribution to this special issue in this section [20] addresses ways to overcome these four challenges through a more systematic approach to game prototypes linked to a typology of forest–water–people nexus issues, in which agroforestry-based ecosystem services can be appreciated.

5. Re-Imagining Agroforestry-Based Ecosystem Services

Beyond directly responding to the three questions raised in the call for papers for the special issue, the concept of agroforestry-based ecosystem services itself evolved [21] as part of the broader debates on Sustainable Development Goals and the multifunctionality of land use, understood as a mosaic of forests, agroforestry, agriculture, and urban areas, at coarse or finely grained mosaic of interacting components. New perspectives that were introduced in [20] but elaborated in [21] include the balance between relational (two-way, reciprocal relations) and instrumental (goal-oriented, substitutable) value articulation on ecosystem functions relevant for human well-being at local, national, and global scales. Whereas the 'ecosystem services' language emphasized human benefits, part of which can also be substituted by technical means (potentially at higher cost) and are thus nice to have but not essential. The tone of the debate is changing, with the realization that tinkering with all non-human life on this planet is a huge risk to humanity. The recent "making peace with nature" report [22] urges for a coherent approach to climate change, loss of biodiversity, and pollution as part of reimagining and transforming the ways in which the values of humans and nature are reconciled. An ambitious vision of the way agroforestry can be part of the solution needs to connect local to global scales and vice versa.

Funding: This research received no external funding.

Acknowledgments: We acknowledge the positive response to the call for manuscripts, the many reviewers and MDPI editorial staff who facilitated the completion of the special issue.

Conflicts of Interest: The author declares no conflict of interest.

References

1. Fakhech, A.; Genin, D.; Ait-El-Mokhtar, M.; Outamamat, E.M.; M'Sou, S.; Alifriqui, M.; Meddich, A.; Hafidi, M. Traditional Pollarding Practices for Dimorphic Ash Tree (*Fraxinus dimorpha*) Support Soil Fertility in the Moroccan High Atlas. *Land* **2020**, *9*, 334. [CrossRef]
2. Gusli, S.; Sumeni, S.; Sabodin, R.; Muqfi, I.H.; Nur, M.; Hairiah, K.; Useng, D.; van Noordwijk, M. Soil Organic Matter, Mitigation of and Adaptation to Climate Change in Cocoa–Based Agroforestry Systems. *Land* **2020**, *9*, 323. [CrossRef]
3. Mulia, R.; Hoang, S.V.; Dinh, V.M.; Duong, N.B.T.; Nguyen, A.D.; Lam, D.H.; Hoang, D.T.; van Noordwijk, M. Earthworm diversity impacts of forest conversion and agroforestry in Quang Nam province, Viet Nam. *Land* **2021**, *10*, 36. [CrossRef]
4. Magaju, C.; Winowiecki, L.; Crossland, M.; Frija, A.; Ouerghemmi, H.; Hagazi, N.; Sola, P.; Ochenje, I.; Kiura, E.; Kuria, A.; et al. Assessing Context-Specific Factors to Increase Tree Survival for Scaling Ecosystem Restoration Efforts in East Africa. *Land* **2020**, *9*, 494. [CrossRef]
5. Hairiah, K.; Widianto, W.; Suprayogo, D.; Van Noordwijk, M. Tree Roots Anchoring and Binding Soil: Reducing Landslide Risk in Indonesian Agroforestry. *Land* **2020**, *9*, 256. [CrossRef]
6. Suprayogo, D.; van Noordwijk, M.; Hairiah, K.; Meilasari, N.; Rabbani, A.L.; Ishaq, R.M.; Widianto, W. Infiltration-Friendly Agroforestry Land Uses on Volcanic Slopes in the Rejoso Watershed, East Java, Indonesia. *Land* **2020**, *9*, 240. [CrossRef]
7. Khasanah, N.; Tanika, L.; Pratama, L.D.Y.; Leimona, B.; Prasetiyo, E.; Marulani, F.; Hendriatna, A.; Zulkarnain, M.T.; Toulier, A.; van Noordwijk, M. Groundwater-extracting rice production in the Rejoso watershed (Indonesia) reducing urban water availability: Characterization and intervention priorities. *Land* **2021**, *10*, 586. [CrossRef]
8. Do, V.H.; La, N.; Mulia, R.; Bergkvist, G.; Dahlin, A.S.; Nguyen, V.T.; Pham, H.T.; Öborn, I. Fruit Tree-Based Agroforestry Systems for Smallholder Farmers in Northwest Vietnam—A Quantitative and Qualitative Assessment. *Land* **2020**, *9*, 451. [CrossRef]
9. Nguyen, M.P.; Vaast, P.; Pagella, T.; Sinclair, F. Local Knowledge about Ecosystem Services Provided by Trees in Coffee Agroforestry Practices in Northwest Vietnam. *Land* **2020**, *9*, 486. [CrossRef]
10. Sari, R.R.; Saputra, D.D.; Hairiah, K.; Rozendaal, D.; Roshetko, J.M.; van Noordwijk, M. Gendered Species Preferences Link Tree Diversity and Carbon Stocks in Cacao Agroforest in Southeast Sulawesi, Indonesia. *Land* **2020**, *9*, 108. [CrossRef]

11. Cahyono, E.D.; Fairuzzana, S.; Willianto, D.; Pradesti, E.; McNamara, N.P.; Rowe, R.L.; Noordwijk, M.V. Agroforestry Innovation through Planned Farmer Behavior: Trimming in Pine–Coffee Systems. *Land* **2020**, *9*, 363. [CrossRef]

12. Mulyoutami, E.; Lusiana, B.; van Noordwijk, M. Gendered migration and agroforestry in Indonesia: Livelihoods, labor, know-how, networks. *Land* **2020**, *9*, 529. [CrossRef]

13. Nyberg, Y.; Musee, C.; Wachiye, E.; Jonsson, M.; Wetterlind, J.; Öborn, I. Effects of Agroforestry and Other Sustainable Practices in the Kenya Agricultural Carbon Project (KACP). *Land* **2020**, *9*, 389. [CrossRef]

14. Chizmar, S.; Castillo, M.; Pizarro, D.; Vasquez, H.; Bernal, W.; Rivera, R.; Sills, E.; Abt, R.; Parajuli, R.; Cubbage, F. A Discounted Cash Flow and Capital Budgeting Analysis of Silvopastoral Systems in the Amazonas Region of Peru. *Land* **2020**, *9*, 353. [CrossRef]

15. Aynekulu, E.; Suber, M.; van Noordwijk, M.; Arango, J.; Roshetko, J.M.; Rosenstock, T.S. Carbon Storage Potential of Silvopastoral Systems of Colombia. *Land* **2020**, *9*, 309. [CrossRef]

16. Mulia, R.; Nguyen, D.D.; Nguyen, M.P.; Steward, P.; Pham, V.T.; Le, H.A.; Rosenstock, T.; Simelton, E. Enhancing Vietnam's Nationally Determined Contribution with Mitigation Targets for Agroforestry: A Technical and Economic Estimate. *Land* **2020**, *9*, 528. [CrossRef]

17. Purwanto, E.; Santoso, H.; Jelsma, I.; Widayati, A.; Nugroho, H.Y.; van Noordwijk, M. Agroforestry as policy option for forest-zone oil palm production in Indonesia. *Land* **2020**, *9*, 531. [CrossRef]

18. Van Noordwijk, M.; Gitz, V.; Minang, P.A.; Dewi, S.; Leimona, B.; Duguma, L.; Pingault, N.; Meybeck, A. People-Centric Nature-Based Land Restoration through Agroforestry: A Typology. *Land* **2020**, *9*, 251. [CrossRef]

19. Wainaina, P.; Minang, P.A.; Gituku, E.; Duguma, L. Cost-Benefit Analysis of Landscape Restoration: A Stocktake. *Land* **2020**, *9*, 465. [CrossRef]

20. Van Noordwijk, M.; Speelman, E.; Hofstede, G.J.; Farida, A.; Abdurrahim, A.Y.; Miccolis, A.; Hakim, A.L.; Wamucii, C.N.; Lagneaux, E.; Andreotti, F.; et al. Sustainable Agroforestry Landscape Management: Changing the Game. *Land* **2020**, *9*, 243. [CrossRef]

21. Van Noordwijk, M. Agroforestry-Based ecosystem services: Reconciling values of humans and nature in sustainable development. *Land* **2021**, *10*, 699. [CrossRef]

22. United Nations Environment Programme. Making Peace with Nature: A Scientific Blueprint to Tackle the Climate, Biodiversity and Pollution Emergencies. Nairobi. 2021. Available online: https://www.unep.org/resources/making-peace-nature (accessed on 1 July 2021).

 *land*

Article

Traditional Pollarding Practices for Dimorphic Ash Tree (*Fraxinus dimorpha*) Support Soil Fertility in the Moroccan High Atlas

Abdessamad Fakhech [1], Didier Genin [2,*], Mohamed Ait-El-Mokhtar [3], El Mustapha Outamamat [1], Soufiane M'Sou [1], Mohamed Alifriqui [1], Abdelilah Meddich [3] and Mohamed Hafidi [1,4]

[1]   Laboratory of Microbiological Biotechnology, Agricultural Sciences & Environment (BioMAgE), Faculty of Sciences Semlalia, Cadi Ayyad University, Marrakech 40000, Morocco; abdessamad.fakhech@edu.uca.ma (A.F.); elmostapha.outamamat@ced.uca.ma (E.M.O.); soufiane.msou@ced.uca.ma (S.M.); Alifriqui@uca.ma (M.A.); Hafidi@uca.ma (M.H.)
[2]   IRD, Laboratory Population, Environment, Dévelopment, UMR151 AMU-IRD, 1331 Marseille, France
[3]   Laboratory of Agri-food,, Biotechnologies & Bioresource Valorisation (AGROBIOVAL), Faculté des Sciences Semlalia, Université Cadi Ayyad, Marrakech 40000, Morocco; mohamed.aitelmokhtar@ced.uca.ma (M.A.-E.-M.); a.meddich@uca.ma (A.M.)
[4]   AgroBioSciences Program, Mohammed IV Polytechnic University (UM6P), Benguerir 43150, Morocco
*   Correspondence: didier.genin@univ-amu.fr

Received: 4 September 2020; Accepted: 18 September 2020; Published: 21 September 2020

Abstract: Shaping and pollarding of dimorphic ash tree (*Fraxinus dimorpha*) are two traditional practices used by the local inhabitants in agropastoral parklands of the Moroccan High Atlas to secure their production systems and increase tree production and strength. This study focused on assessing the impact of these practices on soil quality. Abiotic parameters and mycorrhizal attributes of the samples of four soil types related to different ash tree morphotypes were assessed and compared. Rhizospheric soils (Rs) of three *F. dimorpha* morphotypes were sampled: trees regularly pollarded and shaped for stem anastomosis (An), regularly pollarded multistemmed trees (Na), and multistemmed trees belonging to a public forest under national forestry service management and sporadically illegally pollarded (Fo). The fourth soil was a non-Rs found in bare soils, which represented the control (Nr). Results showed a sizable difference between An soil properties and the other soil types ones, with significantly higher phosphorus (×6), nitrogen (×5), and carbon (×2) levels and higher mycorrhizal (×6) status than Nr soil, and showed 37% more mycorrhization intensity than Fo. Na showed intermediary levels between An and Fo. Fo had ×2 P, ×3 Total Kjeldahl Nitrogen (TKN), 58% more Total Organic Carbon (TOC) content, and twice the spore density compared with Nr. It is concluded that shaping and pollarding have a positive impact on the soil characteristics of the studied species and could make a useful contribution to sound agroforest management schemes.

Keywords: *Fraxinus dimorpha*; soil chemical characteristics; mycorrhizal attributes; traditional ecological knowledge; anastomosis; agroforest

---

1. Introduction

Traditional ecological knowledge (TEK) is defined as "a cumulative body of knowledge, practices and beliefs, handed down through generations by cultural transmission and evolving by adaptive processes, about the relationship between living beings with one another and with their forest environment" [1]. It refers to the knowledge developed by native or local people over generations through direct interactions with their environment. While TEK remains under-recognized, efforts to preserve it are increasing [2]. Over the last years, TEK has become an alternative approach to better understand and adapt to climate and biodiversity change [3,4]. Examples of how TEK provides insights into balanced forest management have been reported and discussed [5]. In Morocco, this approach was particularly documented by Genin et al. [6] in the forested landscapes of the Aït M'hamed region in Azilal province, situated in the Central High Atlas. A unique traditional practice relative to dimorphic ash tree stands was described, consisting of strict periodic tree pollarding and favoring tree stem coalescence and fusion in order to improve the productivity of foliar fodder [6,7].

The dimorphic ash tree, *Fraxinus dimorpha*, is a native tree growing spontaneously in some parts of the Moroccan High Atlas. It is called "imts" in the local Amazigh language, and it grows primarily on rocky inclines and valleys, at altitudes between 1200 and 2000 m. For the local inhabitants, it is a multifunctional tree, which constitutes a keystone species in the functioning of their agro-silvo-pastoral systems and livelihoods [8]. It supplies them with firewood, poles, and beams for houses and agricultural tools, spices, tinctorial, and medicinal products [6]. But its most essential use is as fodder, in times when pasture is scarce and insufficient. From August until November, *F. dimorpha* trees are relied upon for providing fodder for small stocks. For a single tree, branches are pollarded every 4 years to allow enough regrowth time. Sheep and goats graze directly on the ground chopped leafy branches [7].

F. dimorpha in this area is characterized by trunk heterogeneity with four reported morphotypes: (1) large anastomosed trunks located mainly in privately owned, cultivated, and/or grazed parklands; (2) multistemmed trunks with multiple 10–20 cm diameter stems located both in parklands and in the adjacent public forest; (3) single-stemmed trunks, which are the least frequently occurring morphotype; and (4) shrublike trees, which dominate the public forest (Figure 1) [6].

Figure 1. Four different *F. dimorpha* morphotypes: (**a**) anastomosed (background) and multistemmed (foreground) (image © Genin), (**b**) single-stemmed, (**c**) shrublike (images © Fakhech).

The most widespread morphotype within the privately owned lands is the anastomosed trunks. This morphotype is the fruit of one of the most remarkable features of traditional tree management of this region. In the local Amazigh language, this is called "Tahboucht", which translates roughly as "educating" or "mothering." It consists in building rock walls around small or overgrazed trees until they are out of the reach of sheep and goats. Only the best-developed and straightest stems are then attached together as close as possible to favor the process of "anastomosis," by which they can fuse and form a single large trunk. Anastomosis promotes anatomical changes, such as increased proportion of parenchymatous cells and cross section surface area [9]. This singular practice is viewed by Genin et al. [6] as an original and effective option for resource scarcity management in this region.

As perennial plants, woody trees and shrubs represent an important ecosystem stabilization component by ensuring fundamental ecosystem functions, such as soil organic matter improvement and mutual symbiosis [10,11], while producing biomass, services, and goods to support local populations. The case of *F. dimorpha* could help to develop a more global approach to ensure both the preservation and valorization of efficient TEK practices, and the conservation of ecosystemic functionality. Genin et al. [6] showed that the practice of trunk anastomosis allowed a 36% increase in foliage production after a 4-year cycle of exploitation, compared with non-anastomosed trees, and promoted the resilience and longevity of the trees. The effects of this practice on belowground properties have still to be investigated. Here we focused on mycorrhizal status since the arbuscular mycorrhizal fungi (AMF) are a major rhizospheric component of most plant roots [12].

The aim of this study was thus to investigate the response of some of *F. dimorpha* rhizospheric soil (Rs) characteristics found below different tree ports as a result of contrasted tree exploitation practices. We hypothesized that aerial improvement of *F. dimorpha* should be linked to an improvement of its belowground characteristics since the improvement of the tree's aerial parts requires a mobilization of all the rhizospheric components.

2. Material and Methods

2.1. Study Site and Sampling

Field sampling was conducted in Aït M'hamed rural commune, located in the Central High Atlas, Azilal province, Morocco (31°49′N; 6°35′W) (Figure 2). Dimorphic ash is the dominantly distributed spontaneous tree species in the landscape, forming tree parklands fully integrated within local agro-silvo-pastoral systems. Altitudes range from 1300 to 1700 m. The climate is mountain Mediterranean with annual rainfall between 450 and 600 mm. The mean minimum temperature is 5 °C, and the mean maximum temperature is 28 °C. Soils are relatively homogeneous, largely dominated by a sandy-loamy calcareous skeletal soil. Sampling was carried out in May on rhizospheric soil (Rs) found below three *F. dimorpha* morphotypes: anastomosed (An), multistemmed (Na) (both tree types located in occasionally cultivated parklands), and public forest (Fo) tree types. Ten 1 kg soil samples (Rs) were randomly collected from between 10 and 40 cm depth under the tree cover for each morphological type of trees, within a homogeneous edaphotopographic slope. Ten additional samples of non-Rs, situated 50 m away from any dimorphic ash tree cover, were also randomly collected as control (Nr). Root samples of *F. dimorpha* were collected for mycorrhizal colonization assessment for An, Na, and Fo trees.

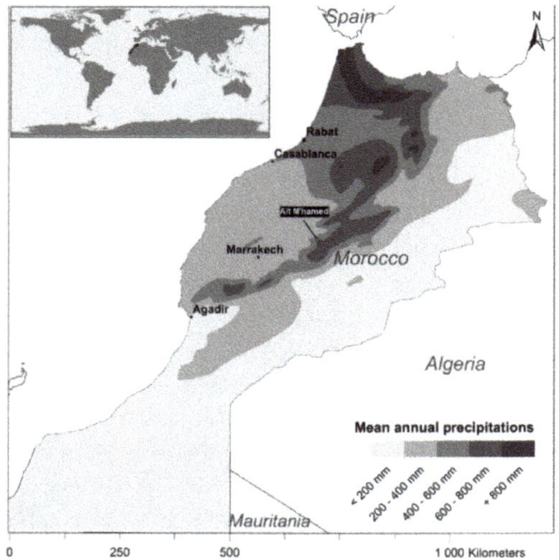

Figure 2. Localization of the study area.

2.2. Soil Chemical and Physical Analysis

Soil chemical and physical characteristics were determined using the following methods: Total Kjeldahl nitrogen (TKN) was determined using the Kjeldahl method [13], which consists in distilling the transformed organic nitrogen in a soil sample in boric acid (H_3BO_3) as ammonium (NH^+_4) and titrating it with dilute sulfuric acid. Available phosphorus (P) was determined using the Olsen method [14]; soil phosphoric acid was extracted using sodium bicarbonates ($NaHCO_3$), and the resulting proportional blue color of the phosphomolybdic complex was evaluated using spectrophotometry at 820 nm. Total organic carbon (TOC) was determined with the potassium dichromate method [13], soil sample organic matter was completely oxidized using potassium dichromate ($K_2Cr_2O_7$), and residual potassium dichromate was dosed with Mohr salt ((NH_4)$_2$Fe(SO_4)$_2$(H_2O)$_6$). Carbon-to-nitrogen (C/N) ratio was calculated using the last two parameters. Soil texture was determined using a Robinson pipette [13]; soil fine particles (sands, silt, and clay) were separated by diameter and gravity after different time periods and at different depths using the Stokes equation. Electrical conductivity and pH were measured using a conductometer (Basic 30) and a pH meter (Basic 20), respectively, for a 1:5 soil-to-water ratio.

2.3. Mycorrhizal Attributes

2.3.1. Root Mycorrhization

Arbuscular mycorrhizal fungi colonization assessment was performed according to the Phillips and Hayman [15] clearing and coloring method. Fine roots were cleared in 10% potassium hydroxide (KOH). They were then colored with 5 mL Trypan blue with repeated washing between and after. Next, samples were cut into 1 cm fragments, and 20 were placed on each slide, with three repetitions for each sample. Glycerol was added for conservation, as well as a cover slip before visualization under microscope. Mycorrhizal structures appear in dark blue. Relative mycorrhization intensity was measured by assigning a colonization index from 0 to 5: 0, no colonization; 1, colonization percentage less than 1%; 2, between 1% and 10%; 3, between 11% and 50%; 4, between 51% and 90%; 5, greater than 91% [16]:

$$M\% = \frac{(95 \times n5 + 70 \times n4 + 30 \times n3 + 5 \times n2 + n1)}{Tf} \tag{1}$$

where n5 is the fragments number that represents the colonization degree corresponding to index 5, the same for n4, n3, n2, and n1, respectively; and Tf is total fragments number.

2.3.2. Spore Enumeration

Arbuscular mycorrhizal fungi spores isolation was conducted using the Gerdemann and Nicolson method [17]: wet sieving and decanting, in combination with the Walker sucrose gradient centrifugation method [18]. Soil samples were passed through a set of sieves (800 and 50 μm) under running water. The trapped fraction in between was processed with two-step centrifugation at 3000 rpm, first with distilled water, then with two sucrose solutions (40% and 60%), with filtration being carried out on Whatman paper. Spores were observed and counted under a stereoscopic microscope at ×40.

2.4. Statistical Analysis

Results were tested using one-way analysis of variance (ANOVA) after dismissing heteroscedasticity and confirming normality using the Levene and Shapiro–Wilk tests, respectively [18]. Tukey's honest significant difference (HSD) was performed to describe significance. The Pearson product-moment coefficient was used to measure the linear correlation between the studied variables, for which p-values were approximated using the F-distributions. Values of p lower than 0.05 were considered significant. Principal component analysis (PCA) was also performed on all the variables to emphasize their variation and visualize how they evolve [19]. Statistical analyses were performed using the LibreOffice Calc v6.4.4 and R v4.0 software under a rolling ArchLabs distribution.

3. Results

3.1. Soil Chemical and Physical Analyses

Variability in Rs' abiotic variables is represented with box and whisker plots (Figure 3). An soil had significantly higher P (Figure 3a), TKN (Figure 3b), and TOC (Figure 3c) levels than the other soils (Table 1). Na had higher levels of the same parameters than Fo with no statistical significance. Nr had significantly the lowest levels of these parameters. The C/N ratio was 8 for An and Na and 9 for Fo. The Nr C/N ratio, on the other hand, was 19 and significantly higher than the other soil ratios (Figure 3d and Table 1). Texture did not change significantly for all the soil types and ranged from loam to loamy sand; pH was neutral, and Electrical Conductivity (EC) was around 3 μS/cm for all soils with no statistical difference. Thus these were left out of the box and whisker plots.

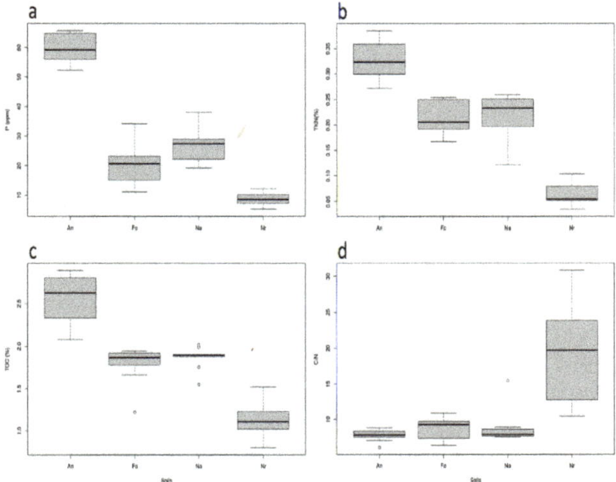

Figure 3. Variability of (**a**) available phosphorus (P), (**b**) total Kjeldahl nitrogen (TKN), (**c**) total organic carbon (TOC), and (**d**) carbon-to-nitrogen (C/N) ratios between the studied soils. The box and whisker diagrams include median value (dark rectangle), range of 50% of the samples (large rectangle), maximum and minimum values (cross bars), and outliers (circle).

Table 1. Analysis of variance and Tukey's honest significant difference (HSD) *p*-values of the studied variables and soil types.

	ANOVA *p*-Values		Tukey's HSD *p*-Values					
	F-Value	Pr (>F)	An–Na	An–Fo	An–Nr	Na–Fo	Na–Nr	Fo–Nr
M	321.9	$<2 \times 10^{-16}$ ***	3.13×10^{-5}	0	0	0.06	0	0
Sn	92.38	$<2 \times 10^{-16}$ ***	0	0	0	0.09	2.4×10^{-6}	0.003
P	142.4	$<2 \times 10^{-16}$ ***	0	0	0	0.07	1×10^{-7}	0.0001
TKN	95.32	$<2 \times 10^{-16}$ ***	4×10^{-7}	0	0	0.86	0	0
TOC	68.31	6.24×10^{-15} ***	4×10^{-7}	0	0	0.86	0	5×10^{-7}
C/N	20.36	6.9×10^{-8} ***	0.9	0.9	4×10^{-7}	0.9	2.6×10^{-6}	2.1×10^{-6}
pH	0.82	0.49	-	-	-	-	-	-
EC	0.65	0.59	-	-	-	-	-	-

Significance codes: ***, 0.001; P, available phosphorus; TKN, total Kjeldahl nitrogen; TOC, total organic carbon; M, relative mycorrhization intensity; Sn, spore number; EC, electrical conductivity. An, Anastomosed trees; Na: Non-anastomosed trees; Fo: Public forest; Nr: non-rhyzospheric soil.

3.2. Soil Mycorrhizal Attributes

An significantly had the highest spore density (Figure 4a) and mycorrhization intensity (Figure 4b), followed by Na, Fo, and Nr (Table 1). All root samples were mycorrhized and consequently had 100% mycorrhization frequency. The mycorrhized fragments number was equal to the total observed fragments number (M% = m%). That is the reason why only the relative mycorrhization intensity (M%) was used to evaluate the difference in colonization levels.

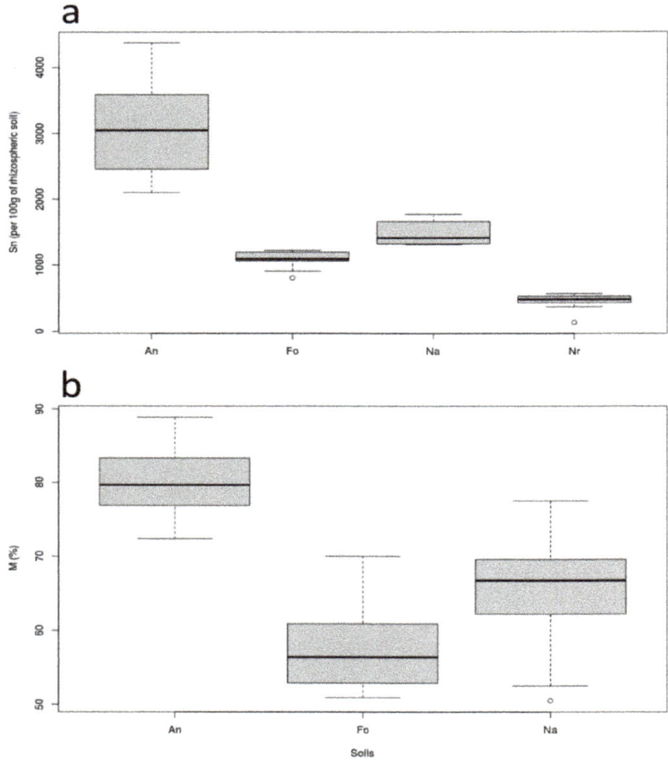

Figure 4. Variability of (**a**) spore number (Sn) and (**b**) mycorrhization intensity (M) between the studied soils. The box and whisker diagrams include median value (dark rectangle), range of 50% of the samples (large rectangle), maximum and minimum values (cross bars), and outliers (circle).

Compared with the control, An had six times higher P levels, five times higher TKN levels, and twice higher TOC levels. It also showed 37% higher mycorrhization intensity than Fo and six times higher spore density than Nr. Compared with Na, An had twice higher P levels, 47% higher TKN levels, 35% higher TOC levels, 22% higher M, and twice higher spore density. Compared with Nr, Na had three times higher P levels, three times higher TKN levels, and 65% times higher TOC levels. It also showed 12% higher mycorrhization intensity than Fo and three times higher spore density than Nr. Na and Fo did not differ much, and all differences were insignificant. Lastly, Fo had twice higher P content than Nr, three times higher TKN content, 58% higher TOC content, and twice higher spore density than Nr.

The studied variables' correlations were grouped by soil type in Tables 2–5. For An, a strong and significant positive correlation was found between P and M, P and Sn, M and Sn, Sn and TKN, and Sn and TOC and a moderate positive correlation between P and TKN (Table 2). For Na, a strong and significant positive correlation was found between P and M and between M and TKN, a moderate positive correlation between P and TKN and between M and TOC) and a strong negative one between TKN and C/N (Table 3). For Fo, a strong and significant positive correlation was noted between P and M and between M and Sn and a strong negative one between TKN and C/N) (Table 4). Lastly, the Nr soil type had one strong significant negative correlation between TKN and C/N (Table 5).

Figure 5a shows PCA compressed 56.5% of the data in the first component and 15.7% in the second. Figure 5b shows the biplot of the PCA with samples and variables represented in the plot. The PCA biplot shows the intimate co-evolution of P, TKN, and Sn, followed by M and TOC, highly driven by the An soil, while Nr had no influence. An showed a small overlap with Na and an even smaller one with Fo. Nr did not show any overlap with the other soil types.

Table 2. An soil parameters' correlation coefficients (left down) and corresponding *p*-values (right up); significant cases are highlighted.

	M	Sn	P	TKN	TOC	C/N	pH	EC
M		0.01	0.01	0.07	0.05	0.99	0.3	0.19
Sn	0.76		0.01	0.02	0	0.72	0.59	0.28
P	0.77	0.78		0.04	0.09	0.63	0.71	0.34
TKN	0.59	0.73	0.66		0.06	0.14	0.53	0.35
TOC	0.63	0.91	0.56	0.61		0.28	0.44	0.47
C/N	0.01	0.13	−0.17	−0.5	0.38		0.07	0.8
pH	−0.36	−0.2	0.14	0.22	−0.28	−0.59		0.91
EC	0.45	0.38	0.34	0.33	0.26	−0.09	0.04	

P, available phosphorus; TKN, total Kjeldahl nitrogen; TOC, total organic carbon; M, relative mycorrhization intensity; Sn, spore number; EC, electrical conductivity.

Table 3. Na soil parameters' correlation coefficients (left down) and corresponding *p*-values (right up); significant cases are highlighted.

	M	Sn	P	TKN	TOC	C/N	pH	EC
M		0.13	0	0.01	0.04	0.07	0.49	0.25
Sn	0.51		0.22	0.26	0.18	0.44	0.28	0.26
P	0.94	0.43		0.04	0.07	0.19	0.63	0.4
TKN	0.79	0.39	0.66		0.25	0	0.44	0.18
TOC	0.66	0.46	0.6	0.4		0.95	0.66	0.29
C/N	−0.59	−0.28	−0.46	−0.9	0.02		0.28	0.4
pH	0.25	0.38	0.17	0.28	−0.16	−0.38		0.95
EC	0.4	−0.39	0.3	0.46	0.37	−0.3	−0.02	

P, available phosphorus; TKN, total Kjeldahl nitrogen; TOC, total organic carbon; M, relative mycorrhization intensity; Sn, spore number; EC, electrical conductivity.

Table 4. Fo soil parameters' correlation coefficients (left down) and corresponding *p*-values (right up); significant cases are highlighted.

	M	Sn	P	TKN	TOC	C/N	pH	EC
M		0	0.01	0.53	0.13	0.69	0.53	0.89
Sn	0.82		0.29	0.62	0.2	0.61	0.58	0.99
P	0.77	0.37		0.2	0.23	0.65	0.15	0.73
TKN	0.23	0.18	0.44		0.4	0.03	0.84	0.52
TOC	0.51	0.44	0.42	0.3		0.17	0.24	0.09
C/N	0.14	0.18	−0.16	−0.69	0.47		0.61	0.04
pH	−0.22	0.2	−0.49	−0.08	−0.41	−0.18		0.86
EC	0.05	0	0.13	0.23	−0.57	−0.66	0.06	

P, available phosphorus; TKN, total Kjeldahl nitrogen; TOC, total organic carbon; M, relative mycorrhization intensity; Sn, spore number; EC, electrical conductivity.

Table 5. Nr soil parameters' correlation coefficients (left down) and corresponding *p*-values (right up); significant cases are highlighted.

	Sn	P	TKN	TOC	C/N	pH	EC
Sn		0.1	0.54	0.42	0.85	0.52	0.81
P	0.55		0.66	0.29	0.86	0.26	0.25
TKN	0.22	−0.16		0.92	0	0.92	0.23
TOC	0.29	−0.37	−0.04		0.14	0.33	0.41
C/N	0.07	0.06	−0.85	0.5		0.54	0.58
pH	−0.23	0.4	0.04	−0.35	−0.22		0.41
EC	−0.09	−0.4	0.41	0.3	−0.2	0.29	

P, available phosphorus; TKN, total Kjeldahl nitrogen; TOC, total organic carbon; Sn, spore number; EC, electrical conductivity.

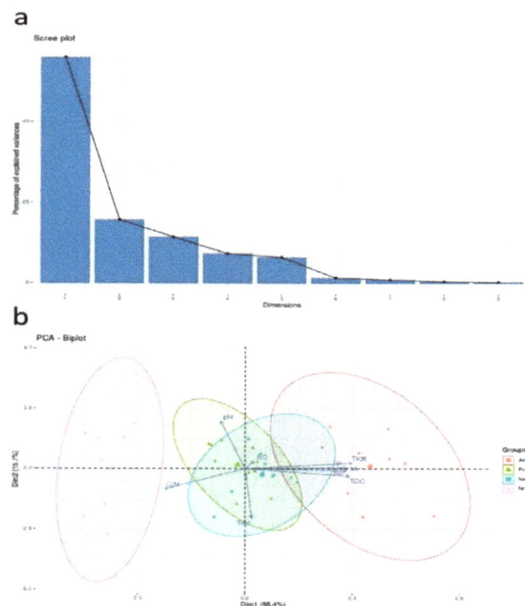

Figure 5. (**a**) Percentage of variances explained by each principal component; (**b**) PCA biplot of samples and variables, obtained from the analyzed variables for the different soil types. P, available phosphorus; TKN, total Kjeldahl nitrogen; TOC, total organic carbon; M, relative mycorrhization intensity; Sn, spore number; EC, electrical conductivity.

4. Discussion

The differences between the studied soil types are very considerable, considering that they are under the same climatic, topographic, and edaphic conditions. Since their only differences rely on whether or not they are located under specific tree morphotypes or not submitted to tree management (Fo and Nr), we consider that these differences are to be related to human intervention in tree shaping. Tree shaping seems to lead to an increase in biological activity and nutrient fluxes, and thus to differences in soil properties, as found in a Himalayan context [20]. For the An and Na soil types, the only difference is the presence/absence of anastomosis. Fo trees are subjected to higher but still moderate browsing [7]. This could explain the Fo soil's low recorded values for the studied variables and high C/N ratio compared with the other soil types. Heavy grazing has been proved to have a negative impact on soil fertility and AMF attributes [21–26]. In the present case, slight grazing could explain the relatively elevated N levels in soils [27,28], as well as the technique used to deliver forage material to flocks. Grazing in moderation was also recorded to have an increasing effect on AMF [29]. The only difference between An and Na samples was the shaping applied on trees. Some samples were no more than 10 m away from each other, but presented completely different soil profiles (Figure 1a). Most Na and Fo parameters were statistically similar and had almost the same profile. This suggests that the major explanation for the remarkably high An soil parameters is linked to the tree shaping status. The biomass produced by anastomosed trees is higher [6], which enriches the soil more than elsewhere, leading to increased soil biological activity.

The C/N ratio is used to characterize the patterns of change in organic matter in soil. It can represent an important indicator describing the organic matter state and its influence on soil microbial activity and identity and soil productivity [30,31]. Carbon is lost by mineralization faster than nitrogen, and the C/N ratio decreases over time. Low C/N ratios generally indicate that plants provide soil biological activity with necessary resources so that it releases, in return, the elements contained in the organic matter faster, thus helping to nourish the trees more actively [32]. In this study, An, Na, and Fo C/N ratios point to rapid decomposition of soil organic matter and release of N into the soil, which can also be an alternative explanation for the soils' elevated N levels. The higher Nr C/N ratio was still within the range, favoring good organic matter decomposition with little to no release of N into the soil [33].

Correlating the studied variables showed some similarities and clear differences between the behaviors of these variables in the different studied soil types. Starting with the similarities, all soil types had a significant positive correlation between P and M. An and Fo had one similarity as both had a strong and significant positive correlation between M and Sn. An and Na also had one similarity, a strong and significant positive correlation between P and TKN. The differences were that Nr showed no distinctive correlations between almost all of its parameters, and only An had a strong and significant positive correlation between Sn and TOC and between Sn and TKN. An had a distinctive new pattern never previously mentioned in the literature, where three variables (P, M, and Sn) positively correlated strongly with each other. To our knowledge, this is the first time such a pattern emerged. In the available literature, one or two of these variables commonly correlate negatively with the others, often being spore density and available phosphorus [34–37], although some studies sometimes showed positive correlations [38–40]. One study even showed that P can correlate with either M or Sn [41], but we could not find one that cites such case where the three variables simultaneously correlate positively with each other. This newly observed pattern could be interpreted as the result of an indirect effect of stems shaping on the discussed soil variables.

Arbuscular mycorrhizal fungi are a fungi group that forms mutualistic symbiosis with most land plants [42]. They are well known for increasing plant nutrient uptake and productivity [43]. Their impacts on a wide range of ecosystem processes and several species are very well documented [44–46]. In high concentrations, phosphorus is known to suppress AMF infection and spore density [47], whereas in low concentrations, AMF take on the task of P prospectors and make it more available to the host at the hyphosphere and mycorrhizosphere [48]. Arbuscular Mycorrhizal Fungi's positive correlation with TKN has also been reported [49,50], while different nitrogen responses have also been mentioned, ranging from positive correlation to P [51] to even opposing levels of P [52,53].

Principal Component analysiss confirmed the noted correlations and showed that Na and Fo shared more similarities with high variance overlap, while for Nr, the absence of variance overlap singled it out, not having any significant similarities with any of the other soil types. M and Sn co-evolution has also been noted in another ecosystem on a different tree species, *Juniperus phoenicea* [36]. It also showed that pH, EC, and texture evolved independently of the rest of the variables not having any drive effect on them.

Agroforestry is undergoing a revival of interest nowadays, particularly to manage soil fertility and in the context of ongoing climate change [54]. Pollarding is a secular technique found historically in many countries [55]. If it almost disappeared in many of them due to the eviction of trees in croplands, it is still alive in various developing countries and supports local livelihoods [20,56]. It is considered to be an efficient technique to delay tree ageing [57], as well as to modify the allocation of resources, especially in the short-term shift in shoot–root balance induced by pruning or pollarding, which can lead to altered root distribution and fine-root turnover enriching soil biota [58]. This knowledge has to be enriched, by both observation of traditional practices and experimentation, and could be useful to develop alternative schemes for agricultural production.

Finally, these results provide arguments in support of the intermediate disturbance hypothesis [59,60], which argues that biological diversity and ecological functioning are enhanced when environments are subjected to intermediate levels of disturbance. Trunk anastomosis, by enhancing tree productivity, seems to promote organic restitutions to the soil and favor its biological activity. These results have to be confirmed in other situations, particularly in zones where pollarding has been used for long time.

5. Conclusions

Traditional tree shaping by pollarding and favoring trunk anastomosis have a clear and distinct positive impact on the dimorphic ash tree's rhizospheric soil variables found in agroforestral systems in the Central Moroccan High Atlas, particularly on phosphorus, nitrogen, and carbon content, and on mycorrhizal activity, compared with uncultivated or poorly managed soils. This illustrates that, in some cases, human practices can indirectly improve the delivery of ecosystem services from spontaneous forested stands. Our results confirm the initial hypothesis that soil characteristics respond to *F. dimorpha* shaping, as such changes in the aerial parts of the tree require recruitment of all the belowground actors, both abiotic and biotic. They tend to show that soil biological activity can be boosted by modifying natural tree port, and by a management that allows higher organic restitutions. However, these results suggest that abiotic factors alone are insufficient to fully explain the spectacular improvement of leaf productivity on anastomosed tree, and physiological, histological, and biomolecular experiments have to be conducted to better understand the impact of shaping dimorphic ash tree at these different levels, and their mutual interactions.

Author Contributions: Conceptualization, A.F., D.G., M.A., M.H.; writing—original draft, A.F., D.G.; methodology, M.A.-E.-M., A.M., A.F., M.H.; investigation, A.F., E.M.O., S.M.; writing—review and editing, A.F., M.A.-E.-M., D.G., M.A., A.M., M.H. All authors have read and agreed to the published version of the manuscript.

Funding: This research was partly funded by Agence Nationale de la Recherche, France, within the Research Project Med-Inn-Local (ANR-12-TMED-0001-01), with further support from participating authors' institutions.

Conflicts of Interest: The authors declare no conflict of interest.

References

1. Parrotta, J.; Agnoletti, M. Traditional forest knowledge: Challenges and opportunities. *Forest Ecol. Manag.* **2007**, *249*, 1–4. [CrossRef]
2. Maffi, L.; Woodley, E. *Biocultural Diversity Conservation: A Global Sourcebook*; Routledge: London, UK, 2012.

3. De Freitas, J.G.; Bastos, M.R.; Dias, J.A. Traditional Ecological Knowledge as a Contribution to Climate Change Mitigation and Adaptation: The Case of the Portuguese Coastal Populations. In *Handbook of Climate Change Communication: Case Studies in Climate Change Communication*; Filho, W.L., Manolas, E., Azul, A.M., Azeiteiro, U.M., McGhie, H., Eds.; Springer International Publishing: Cham, Switzerland, 2018; pp. 257–269. [CrossRef]

4. Wyllie de Echeverria, V.R.; Thornton, T.F. Using traditional ecological knowledge to understand and adapt to climate and biodiversity change on the Pacific coast of North America. *Ambio* **2019**, *48*, 1447–1469. [CrossRef] [PubMed]

5. Parrotta, J.; Yeo-Chang, Y.; Camacho, L.D. Traditional knowledge for sustainable forest management and provision of ecosystem services. *Int. J. Biodivers. Sci. Ecosyst. Serv. Manag.* **2016**, *12*, 1–4. [CrossRef]

6. Genin, D.; M'Sou, S.; Ferradous, A.; Alifriqui, M. Another vision of sound tree and forest management: Insights from traditional ash shaping in the Moroccan Berber mountains. *For. Ecol. Manag.* **2018**, *429*, 180–188. [CrossRef]

7. Genin, D.; Crochot, C.; MSou, S.; Araba, A.; Alifriqui, M. Meadow up a tree: Feeding flocks with a native ash tree in the Moroccan mountains. *Pastoralism* **2016**, *6*. [CrossRef]

8. Genin, D.; Alifriqui, M. The agroforestry parklands with dimorphic ash trees of Aït M'Hamed. A well-guarded local specificity. In *The Emergence of Local Specificities in the Mediterranean Hinterlands*; Aderghal, M., Genin, D., Hanafi, A., Landel, P.A., Michon, G., Eds.; Laboratoire Population-Environnement-Développement: Marseille, France, 2019; pp. 30–51. ISBN 979-10-96763-08-5.

9. Schweingruber, F.H. *Wood Structure and Environment*; Springer Science & Business Media: Berlin/Heidelberg, Germany, 2007.

10. Howard, K.S.C.; Eldridge, D.J.; Soliveres, S. Positive effects of shrubs on plant species diversity do not change along a gradient in grazing pressure in an arid shrubland. *Basic Appl. Ecol.* **2012**, *13*, 159–168. [CrossRef]

11. Birhane, E.; Ahmed, S.; Hailemariam, M.; Negash, M.; Rannestad, M.M.; Norgrove, L. Carbon stock and woody species diversity in homegarden agroforestry along an elevation gradient in southern Ethiopia. *Agrofor. Syst.* **2020**, *94*, 1099–1110. [CrossRef]

12. Rillig, M.C.; Mummey, D.L. Mycorrhizas and soil structure. *New Phytol.* **2006**, *171*, 41–53. [CrossRef]

13. Schinner, F.; Öhlinger, R.; Kandeler, E.; Margesin, R. *Methods in Soil Biology*; Springer Science & Business Media: Berlin/Heidelberg, Germany, 2012.

14. Olsen, S.R. Estimation of available phosphorus in soils by extraction with sodium bicarbonate. *USDA Circular* **1954**, *24*. [CrossRef]

15. Phillips, J.M.; Hayman, D.S. Improved procedures for clearing roots and staining parasitic and vesicular-arbuscular mycorrhizal fungi for rapid assessment of infection. *Trans. Br. Mycol. Soc.* **1970**, *55*, 158. [CrossRef]

16. Trouvelot, A.; Kough, J.L.; Gianinazzi-Pearson, V. Measurement of the mycorrhization rate VA of a root system. In search for an estimation method with functional significance Presented at the Physiological and genetical aspects of mycorrhizae. In Proceedings of the 1st European Symposium on Mycorrhizae, Dijon, France, 1–5 July 1985; pp. 217–221.

17. Gerdemann, J.W.; Nicolson, T.H. Spores of mycorrhizal Endogone species extracted from soil by wet sieving and decanting. *Trans. Br. Mycol. Soc.* **1963**, *46*, 235–244. [CrossRef]

18. Scheffe, H. *The Analysis of Variance*; John Wiley & Sons, Inc.: Hoboken, NJ, USA, 1959; pp. 351–358.

19. Jolliffe, I.T. Principal Component Analysis. In *Series in Statistics*; Springer: New York, NY, USA, 2012. [CrossRef]

20. Singh, R.K.; Singh, A.; Garnett, S.T.; Zander, D.K.; Lobsang, D.K.; Tsering, D. Paisang (Quercus griffithii); a keystone tree species in sustainable agroecosystem management and livelihoods in Arunachal Pradesh, India. *Environ. Manag.* **2015**, *55*, 187–204. [CrossRef] [PubMed]

21. Walker, C.; Mize, C.W.; McNabb, H.S. Populations of endogonaceous fungi at two locations in central Iowa. *Can. J. Bot.* **1982**, *60*, 2518–2529. [CrossRef]

22. Liu, J.; Zhang, Q.; Li, Y.; Di, H.; Xu, J.; Li, J.; Pan, H. Effects of pasture management on soil fertility and microbial communities in the semi-arid grasslands of Inner Mongolia. *J. Soils Sediments* **2016**, *16*, 235–242. [CrossRef]

23. Guangyu, Z.; Zhuangsheng, T.; Lei, C.; Zhouping, S.; Lei, D. Overgrazing depresses soil carbon stock through changing plant diversity in temperate grassland of the Loess Plateau. *Plant Soil Environ.* **2018**, *64*, 1–6. [CrossRef]

24. Cai, Y.; Yan, Y.; Xu, D.; Xu, X.; Wang, C.; Wang, X.; Eldridge, D.J. The fertile island effect collapses under extreme overgrazing: Evidence from a shrub-encroached grassland. *Plant Soil* **2020**, *448*, 201–212. [CrossRef]

25. Faghihinia, M.; Zou, Y.; Chen, Z.; Bai, Y.; Li, W.; Marrs, R.; Staddon, P.L. The response of grassland mycorrhizal fungal abundance to a range of long-term grazing intensities. *Rhizosphere* **2020**, *13*, 100178. [CrossRef]

26. Yang, X.; Chen, J.; Shen, Y.; Dong, F.; Chen, J. Global negative effects of livestock grazing on arbuscular mycorrhizas: A meta-analysis. *Sci. Total Environ.* **2020**, *708*, 134553. [CrossRef]

27. Li, Y.; Wu, J.; Shen, J.; Liu, S.; Wang, C.; Chen, D.; Zhang, J. Soil microbial C:N ratio is a robust indicator of soil productivity for paddy fields. *Sci. Rep.* **2016**, *6*, 35266. [CrossRef]

28. Shen, H.; Dong, S.; Li, S.; Xiao, J.; Han, Y.; Yang, M.; Yeomans, J.C. Grazing enhances plant photosynthetic capacity by altering soil nitrogen in alpine grasslands on the Qinghai-Tibetan plateau. *Agric. Ecosyst. Environ.* **2019**, *280*, 161–168. [CrossRef]

29. Wan, X.; Huang, Z.; He, Z.; Yu, Z.; Wang, M.; Davis, M.R.; Yang, Y. Soil C:N ratio is the major determinant of soil microbial community structure in subtropical coniferous and broadleaf forest plantations. *Plant Soil* **2015**, *387*, 103–116. [CrossRef]

30. Eom, A.-H.; Wilson, G.W.T.; Hartnett, D.C. Effects of ungulate grazers on arbuscular mycorrhizal symbiosis and fungal community structure in tallgrass prairie. *Mycologia* **2001**, *93*, 233–242. [CrossRef]

31. Li, W.; Cao, W.; Wang, J.; Li, X.; Xu, C.; Shi, S. Effects of grazing regime on vegetation structure, productivity, soil quality, carbon and nitrogen storage of alpine meadow on the Qinghai-Tibetan Plateau. *Ecol. Eng.* **2017**, *98*, 123–133. [CrossRef]

32. Brust, G.E. Chapter 9-Management Strategies for Organic Vegetable Fertility. In *Safety and Practice for Organic Food*; Biswas, D., Micallef, S.A., Eds.; Academic Press: Cambridge, MA, USA, 2019; pp. 193–212. [CrossRef]

33. Zhang, S.; Zheng, Q.; Noll, L.; Hu, Y.; Wanek, W. Drivers of microbial nitrogen use efficiency and soil inorganic N processes at the continental scale. *Geophys. Res. Abstr.* **2019**, *21*, 1–4.

34. Khakpour, O.; Khara, J. Spore density and root colonization by arbuscular mycorrhizal fungi in some species in the northwest of Iran. *Int. Res. J. Appl. Basic Sci.* **2012**, *3*, 977–982.

35. Tamil Pradha, K.; Sivakumar, K. Effects of am fungi on spores density and root colonization of chilli (*Capsicum annum*) at different levels of rock phosphate. *Int. J. Curr. Res. Life Sci.* **2018**, *7*, 1415–1419.

36. Fakhech, A.; Ouahmane, L.; Hafidi, M. Seasonality of mycorrhizal attributes, soil phosphorus and nitrogen of Juniperus phoenicea and Retama monosperma boiss. in an Atlantic sand dunes forest. *J. Sustain. For.* **2019**, *38*, 1–17. [CrossRef]

37. Gao, D.; Pan, X.; Zhou, X.; Wei, Z.; Li, N.; Wu, F. Phosphorus fertilization and intercropping interactively affect tomato and potato onion growth and rhizosphere arbuscular mycorrhizal fungal community. *Arch. Agron. Soil Sci.* **2020**. [CrossRef]

38. Rubio, R.; Borie, F.; Schalchli, C.; Castillo, C.; Azcón, R. Occurrence and effect of arbuscular mycorrhizal propagules in wheat as affected by the source and amount of phosphorus fertilizer and fungal inoculation. *Appl. Soil Ecol.* **2003**, *23*, 245–255. [CrossRef]

39. Cuéllar, A.E.; Martinez, L.R.; Espinosa, R.R.; Cuéllar, E.E. The Inoculation with an Effecient AMF Strain Decreases the Phosphoric Fertilizer Requirements in Ipomea Batata (L.), Lam in Dry Period. *J. Chem. Environ. Biol. Eng.* **2019**, *3*, 13. [CrossRef]

40. Effendy, M.; Wijayani, B.W. Estimation of Available Phosphorus in Soil Using the Population of Arbuscular Mycorrhizal Fungi Spores. *J. Trop. Soils* **2019**, *16*, 225–232. [CrossRef]

41. Abdel-Fattah, G.M.; El-Dohlob, A.F.; El-Dohlob, G.M.; El-Haddad, S.M.; Hafez, S.A.; Rashad, Y.M. An ecological view of arbuscular mycorrhizal status in some Egyptian plants. *J. Environ. Sci.* **2010**, *37*, 123–136.

42. Brundrett, M.C.; Tedersoo, L. Evolutionary history of mycorrhizal symbioses and global host plant diversity. *New Phytol.* **2018**, *220*, 1108–1115. [CrossRef] [PubMed]

43. Smith, S.E.; Smith, F.A. Roles of Arbuscular Mycorrhizas in Plant Nutrition and Growth: New Paradigms from Cellular to Ecosystem Scales. *Annu. Rev. Plant Biol.* **2011**, *62*, 227–250. [CrossRef] [PubMed]

44. Powell, J.R.; Rillig, M.C. Biodiversity of arbuscular mycorrhizal fungi and ecosystem function. *New Phytol.* **2018**, *220*, 1059–1075. [CrossRef] [PubMed]

45. Fakhech, A.; Ouahmane, L.; Hafidi, M. Analysis of symbiotic microbial status of Atlantic sand dunes forest and its effects on Acacia gummifera and Retama monosperma (Fabaceae) to be used in reforestation. *J. For. Res.* **2019**. [CrossRef]

46. Ait-El-Mokhtar, M.; Fakhech, A.; Anli, M.; Ben-Laouane, R.; Boutasknit, A.; Wahbi, S.; Meddich, A. Infectivity of the palm groves arbuscular mycorrhizal fungi under arid and semi-arid climate and its edaphic determinants towards efficient ecological restoration. *Rhizosphere* **2020**, 100220. [CrossRef]

47. Tucker, M. Effects of Phosphorus on Mycorrhizal Ccolonization of Wetland Plants under Natural and Controlled Conditions. Master's Thesis, Wilfrid Laurier University, Waterloo, Belgium, 2020.

48. Linderman, R.G. The mycorrhizosphere phenomenon. In Proceedings of the International Symposium "Mycorrhiza for Plant Vitality" and the Joint Meeting of Working Groups 1–4 of COST Action 870, Hannover, Germany, 3–5 October 2007; pp. 341–355.

49. Nuccio, E.E.; Hodge, A.; Pett-Ridge, J.; Herman, D.J.; Weber, P.K.; Firestone, M.K. An arbuscular mycorrhizal fungus significantly modifies the soil bacterial community and nitrogen cycling during litter decomposition: AMF alters soil bacterial community and N cycling. *Environ. Microbiol.* **2013**, *15*, 1870–1881. [CrossRef]

50. Lin, C.; Wang, Y.; Liu, M.; Li, Q.; Xiao, W.; Song, X. Effects of nitrogen deposition and phosphorus addition on arbuscular mycorrhizal fungi of Chinese fir (Cunninghamia lanceolata). *Sci Rep.* **2020**, *10*, 12260. [CrossRef]

51. Liu, C.; Ravnskov, S.; Liu, F.; Rubæk, G.H.; Andersen, M.N. Arbuscular mycorrhizal fungi alleviate abiotic stresses in potato plants caused by low phosphorus and deficit irrigation/partial root-zone drying. *J. Agric. Sci.* **2018**, *156*, 46–58. [CrossRef]

52. Camenzind, T.; Homeier, J.; Dietrich, K.; Hempel, S.; Hertel, D.; Krohn, A.; Rillig, M.C. Opposing effects of nitrogen versus phosphorus additions on mycorrhizal fungal abundance along an elevational gradient in tropical montane forests. *Soil Biol. Biochem.* **2016**, *94*, 37–47. [CrossRef]

53. Xiao, D.; Che, R.; Liu, X.; Tan, Y.; Yang, R.; Zhang, W.; Wang, K. Arbuscular mycorrhizal fungi abundance was sensitive to nitrogen addition but diversity was sensitive to phosphorus addition in karst ecosystems. *Biol. Fertil. Soils* **2019**, *55*, 457–469. [CrossRef]

54. Jose, S. Agroforestry for ecosystem services and environmental benefits: An overview. *Agrofor. Syst.* **2009**, *76*, 1–10. [CrossRef]

55. Petit, S.; Watkins, C. Forgotten peasant practices: Tree pollarding in Great Britain. *Etudes Rural.* **2004**, *169*, 197–214. [CrossRef]

56. Genin, D.; Aumeerudy-Thomas, Y.; Balent, G.; Nasi, R. The multiple dimensionsof rural forests: Lessons from a comparative analysis. *Ecol. Soc.* **2013**, *18*, 27. [CrossRef]

57. Ferrini, F. Pollarding and its effects on tree physiology: A look to mature and senescent tree management in Italy. In *Pollard Trees in Europe, European Meeting on Pollard Trees*; Dumont, E., Ed.; Maison Botanique: Boursay, France, 2006; pp. 71–79.

58. Van Noordwijk, M.; Purnomosidhi, P. Root architecture in relation to tree-soil-crop interactions and shoot pruning in agroforestry. *Agrofor. Syst.* **1995**, *30*, 161–173. [CrossRef]

59. Connell, J. Diversity in tropical rain forests and coral reefs. *Science* **1978**, *199*, 1302–1310. [CrossRef]

60. Siebert, S.F.; Belsky, J.M. Historic livelihoods and land uses as ecological disturbances and their role in enhancing biodiversity: An example from Bhutan. *Biol. Conserv.* **2014**, *177*, 82–89. [CrossRef]

land

Article

Soil Organic Matter, Mitigation of and Adaptation to Climate Change in Cocoa–Based Agroforestry Systems

Sikstus Gusli [1,2,*], Sri Sumeni [1], Riyami Sabodin [1], Ikram Hadi Muqfi [1], Mustakim Nur [1], Kurniatun Hairiah [3], Daniel Useng [2,4] and Meine van Noordwijk [5,6]

1 Department of Soil Science, Faculty of Agriculture, Hasanuddin University, Makassar 9245, Indonesia; srisumeni31@gmail.com (S.S.); riyamial@gmail.com (R.S.); ikramhadimuqfi@gmail.com (I.H.M.); takimkure.20@gmail.com (M.N.)
2 Natural Resource Research and Development Center, Hasanuddin University, Makassar 9245, Indonesia; daniel.useng@agri.unhas.ac.id
3 Department of Soil Science, Faculty of Agriculture, Brawijaya University, Malang 65415, Indonesia; kurniatun_h@ub.ac.id
4 Department of Agricultural Engineering, Hasanuddin University, Makassar 9245, Indonesia
5 World Agroforestry Centre (ICRAF), Bogor 16001, Indonesia; m.vannoordwijk@cgiar.org
6 Plant Production Systems, Wageningen University, 6708 PB Wageningen, The Netherlands
* Correspondence: sikstusgusli@unhas.ac.id

Received: 11 August 2020; Accepted: 10 September 2020; Published: 14 September 2020

Abstract: Belowground roles of agroforestry in climate change mitigation (C storage) and adaptation (reduced vulnerability to drought) are less obvious than easy-to-measure aspects aboveground. Documentation on these roles is lacking. We quantified the organic C concentration (C_{org}) and soil physical properties in a mountainous landscape in Sulawesi (Indonesia) for five land cover types: secondary forest (SF), multistrata cocoa–based agroforestry (CAF) aged 4–5 years (CAF4), 10–12 years (CAF10), 17–34 years (CAF17), and multistrata (mixed fruit and timber) agroforest (MAF45) aged 45–68 years. With four replicate plots per cover type, we measured five pools of C-stock according to IPCC guidelines, soil bulk density (BD), macro porosity (MP), hydraulic conductivity (K_s), and available water capacity of the soil (AWC). The highest C-stock, in SF, was around 320 Mg ha^{-1}, the lowest, 74 Mg ha^{-1}, was in CAF4, with the older agroforestry systems being intermediate with 120 to 150 Mg ha^{-1}. Soil compaction after forest conversion led to increased BD and reduced MP, K_s, and AWC. Older agroforestry partly recovered buffering: AWC per m of rooted soil profile increased by 5.7 mm per unit (g kg^{-1}) increase of C_{org}. The restored AWC can support about a week's worth of evapotranspiration without rain, assisting in climate change adaptation.

Keywords: cocoa agroforestry; climate adaptation; soil restoration; soil organic carbon; soil macro-porosity; soil water availability; inceptisols

---

1. Introduction

Agroforestry, integrating trees in farms, implies reforestation (or at least re–treeing) within an agricultural framework of land tenure and practice [1,2]. Growing cocoa between various other trees as shade or companion trees to increase overall income from farming is one of the many forms of agroforestry [3]. Agroforestry can connect the climate change (CC) policy imperatives on mitigation —reducing net emissions of greenhouse gasses—and adaptation: reducing human vulnerability to increased climate variability and global warming trends [4,5]. These connections are made aboveground as increased tree cover in agricultural landscapes stores carbon [6,7] and modifies microclimates [8].

Moreover, belowground soil carbon storage will increase in agroforestry [9] while the soil water buffer function can be enhanced [10]. Yet, the belowground effects are slower (both in their degradation and their restoration phases [11]) and less visible than those aboveground and need to be teased out from a substantial background variability of soil diversity [12,13]. In accordance with the 'cascade' perspective on ecological structure and function, which shapes ecosystem services that represent the benefits people derive from functioning (agro)ecosystems as a basis for valuation and explicit decision making (Figure 1A, [14]), the chain of required evidence for belowground relevance of agroforestry for CC adaptation can be considered in three steps (Figure 1B):

A. Effects of trees as part of land cover change on soil organic matter (and soil carbon storage),
B. Relations between soil organic matter and soil physical properties relevant to crop vulnerability to climate variability, and
C. Deliberate farmer use of trees to reduce their own vulnerability (the adaptation line), potentially supported by global climate policies (if incentives will reach the farmer, then the dotted line).

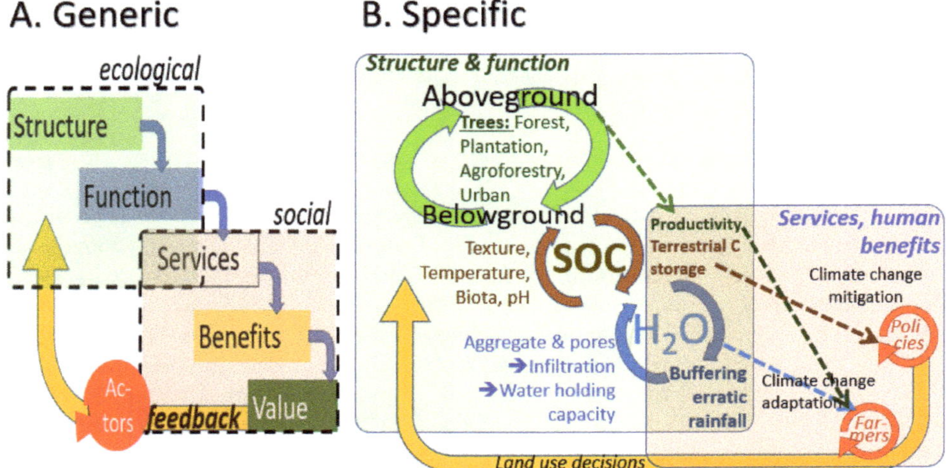

Figure 1. (**A**) Cascade representation of ecosystem structure and function, interacting with social systems via 'ecosystem services' [14]; (**B**) Relationship between the agroecosystem structure (observable), function (inferred), and human benefits (policy focus) in relation to the growth cycle of trees, the formation, and the decay of soil organic carbon (SOC), climate change, and the water (H_2O) cycle.

Soil organic matter needs to be understood in its central role in belowground C storage (mitigation) and its influence on soil water balance and hence on climate vulnerability (adaptation). Obtaining a full account of all inputs, both aboveground and belowground, and their associated rates of turnover and interaction with mineral soil particles in shaping soil physical properties is no simple task for even relatively simple forms of cocoa agroforestry. Stock changes over time can provide an indication of the net effects of interacting processes, as both plants [15] and farmers adapt to climate change [16]. Co-adaptation of people and trees is needed for resilience to climate change in the medium term [17].

Cocoa (*Theobroma cacao* L.) is grown across the humid tropics and is a major driver of tropical forest transitions [18]. Whether or not cocoa agroforestry increases or reduces vulnerability to uncertain rainfall is contested [19] and may depend on the tree species used, the soil, and climatic conditions. Within cocoa production, comparisons between more open and more shaded systems have revealed ecological, social, and economic variables [20–22], with farmer choices depending on location, markets, and evolving knowhow. According to global agriculture database data [23], Indonesia still is the third-largest global cocoa producer, while international cocoa organization data [24] suggests that it is the fifth-largest global cocoa producer. Indonesian smallholder cocoa production confronts multiple challenges on how to increase productivity as well as smallholder's prosperity, reduce pressure on forest conversion, and be adaptive and resilient to climate change [25]. Nearly a million hectares of forest was converted to cocoa farmland in Indonesia, especially in the island of Sulawesi [26], most notably in the 1980s [27] during the period of cocoa booming. Such conversion was intended to increase smallholder income, however it led to declining soil health by reducing soil biodiversity and the soil microbiome [28,29], compacting soils, reducing infiltration and soil water storage [30], depleting soil fertility [31], and reducing environmental services [32,33] such as carbon storage [34,35].

In the last twenty years, cocoa farmland management in Sulawesi has been dynamic, very much influenced by field experience and information gathered by returning migrant farmers from Malaysia [24]. In the early 1970s to 1980s, cocoa was grown as mixed cropping with food crops (mostly in flat land); much of the land was converted to monoculture (full sun) cocoa in the 1990s. The monoculture practice expanded rapidly including on sloping lands, in response to the information from other places that this system produces higher yields than cocoa grown under shade, especially in the short term [36]. Yet, farmers eventually learnt that, in the longer term, cocoa grown as a single crop suffered from pests and diseases, including the damaging vascular strike dieback, which can kill cocoa trees under heavy infestation [37]. Farmers also claimed that monoculture cocoa suffers more from extreme weather conditions (heavy rain or long dry periods), while monoculture cocoa only allows farmers to earn money in the months during the harvest periods. Learning from this experience, many cocoa farmers at present also grow other plants between cocoa trees, including horticultural crops, as shade trees for the cocoa and as extra income sources for the farmers. While the structure and functions of cocoa-based agroforestry practices are easily observed aboveground, their belowground roles are less obvious and remain poorly documented, including carbon storage dynamics in terms of the various agroforestry types and ages and its effects on climate change mitigation and adaptation, as well as changes of soil properties affecting vulnerability to drought.

In this study, we explored the relationships between A) Agroforestry land cover, aboveground and belowground carbon stocks, and B) Soil carbon and soil physical conditions, as steps towards assessing contributions to CC adaptation have occurred in the context of southern Sulawesi where farmers in the Polewali-Mandar District grow cocoa with three major systems: monoculture with minimum shade trees and both simple- and complex- (multilayers) agroforestry systems of varying ages. We tested the hypothesis that in cacao production systems, the farmer-level benefits from agroforestry through increased water storage, contributing to climate change adaptation, are parallel to global benefits of climate change mitigation through increased C storage. If such a synergy exists between local and global climate change agendas, then it is easier for climate policy to provide appropriate incentives.

2. Methods

2.1. Sampling Sites

We conducted the study in a mountainous landscape of Sulawesi, Indonesia with sample sites ranging from 93 to 575 m above sea level (Figure 2), on five land cover types (all were directly converted from primary forest according to local land use history) consisting of young smallholder cocoa agroforestry (CAF4, 4 to 5 years old), medium-aged smallholder cocoa agroforestry (CAF10, 10 to 12 years old), old smallholder cocoa agroforestry (CAF17, 17 to 34 years), multistrata agroforest (MAF45, 45 to 68 years after forest conversion) owned by smallholders, and remnant secondary tropical forest (SF). Information on the year planted or converted from forest was obtained directly from the farm owners who lived on and managed the farms. The sites of these agroforestry systems (CAF4, CAF10, CAF1,7 and MAF45) are scattered at elevations between 100 to 300 m above the sea level, with an 8 to 30% slope. The three sites of SFs are at >400 m above the sea level and 30 to over 40% slope. The average annual rainfall of the area is 2113 mm.

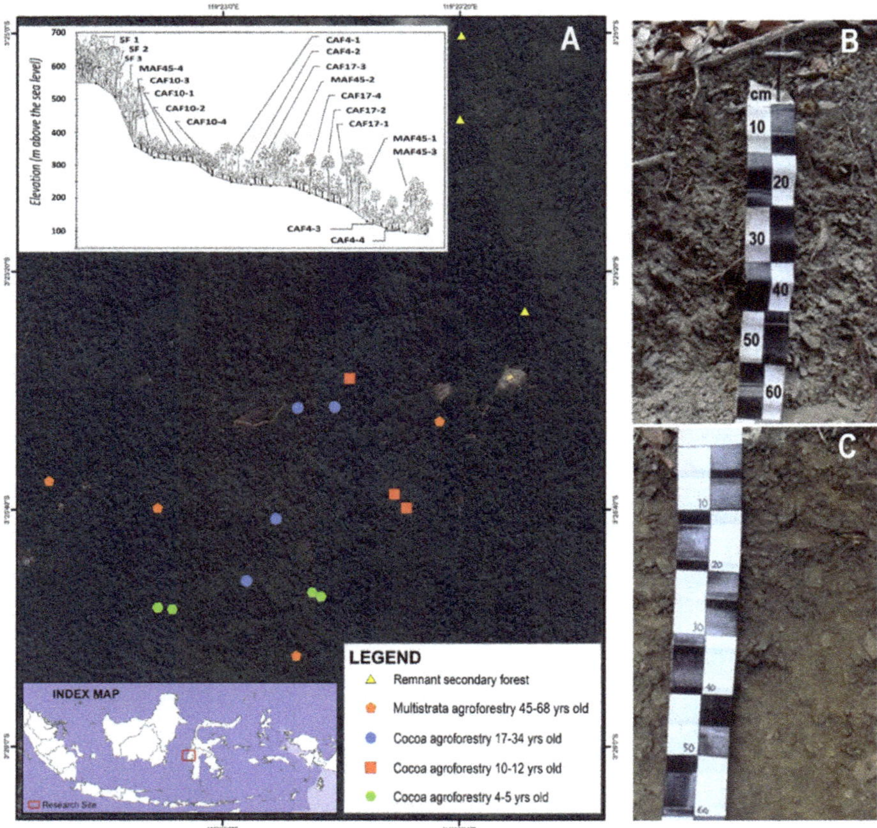

Figure 2. (**A**) The site distribution (satellite image) of remnant secondary forest (SF), multistrata agroforest that is 45–68 years old (MAF45), cocoa agroforestry that is 17–34 years old (CAF17), cocoa agroforestry that is 10–12 years old (CAF10), and cocoa agroforestry that is 4–5 years old (CAF4). The topo-sequence positions are shown to indicate slopes and elevations of the systems. The cocoa agroforestry sites are scattered randomly within 100 to 300 m above sea level (a.s.l.), generally 8 to 30% in slope, but the SFs are at >400 m a.s.l. with a slope from 30 to over 40%. CAF (of all ages) refers to cocoa farming practices where at least five other tree species (with shade and/or direct use functions) are used alongside cocoa trees as the main crop. MAF denotes a farming practice of multiple tree species, dominated by horticultural trees, cocoa is not the main crop. (**B**) A typical (common) soil profile of the study area for all land uses showing abundance of rock fragments, especially below a 20 or 30 cm depth. (**C**) A profile of less stony soil (uncommon).

The plots differed in vegetation structure and stratification, but all had permanent litter layers (Figure 3). We did not find remnants of natural (primary) forest in the neighborhood. The surface soil of all land-uses was well covered by mulch (litter layer). For each land-cover type, we selected four sites (replicates), managed by separate households; however, due to forest area limitations, only three sites were sampled in SF, in a single landscape location. Hence, across these five land-uses we evaluated 19 plots.

Figure 3. Typical look of the five land-use systems, namely remnant secondary forest (SF), multistrata agroforest that is 45–68 years old (MAF45), cocoa agroforestry that is 17–34 years old (CAF17), cocoa agroforestry that is 10–12 years old (CAF10), and cocoa agroforestry that is 4–5 years old (CAF4). Bottom right corner: SF being land–cleared using a chain saw and machete operated by the owner for a new mixed cocoa agroforestry (CAF), without burning.

At the soil surface, the rock fragments covered 10 to 20 percent of the soil surface, but beyond a 30 cm depth, rocks occupied nearly half of the soil volume. Soils in each site varied from coarse– to medium–textures (sandy loam, sandy clay loam, and loam), with an increase of clay content as the depth increases. To minimize confounding effects of soil texture and elevation on C-organic concentrations, we used reference values based on a pedotransfer function [9] as a basis of comparisons. The pedotransfer function we used accounts for texture (microaggregate C protection) and elevation (as proxy for temperature) effects on expected C-organic concentrations (with coefficients for silt, clay, pH, elevation, and specific soil types) [13]; this was reconfirmed in recent analysis [9], with an additional correction for sampling depths. The C_{ref} values can help in judging the validity of the 'chronosequence' interpretation of observed differences between land uses as being 'caused' by land use in an otherwise homogeneous background situation. Variations in texture (Table 1) stayed within the 'Loam' and 'Sandy Loam' classes, with the lowest C_{ref} value 9% below and the highest 14% above the mean for layer–specific results. This can be compared to C_{org} with a lowest value 25% below and a highest one 59% above the average. Variation in texture can thus not be ignored in the interpretation of plot–level C_{org} results in relation to land use. For part of the analysis, we relied on the C_{org}/C_{ref} ratio as a potentially more sensitive indicator of land use effects [9], as it has considered that the 'chronosequence' interpretation assumes that there is no variation in texture.

Table 1. Particle–size distribution (USDA classification), pH and derived reference organic carbon concentration (C-ref, responding to texture and pH) of the soil to a depth of 30 cm, as test of homogeneity of the sample sites used for the five land uses; values in brackets are standard errors of the means of four measurements; for C-ref and measured C-org, the values are also provided after normalization on the mean across land uses per soil layer.

Land Use *	Soil Depth (cm)	Particle-Size Distribution (g kg^{-1})			Soil Texture	pH (1N KCl)	C-Ref (g kg^{-1})	Normalized C-Ref Per Layer	C-org (g kg^{-1})	Normalized C-Org Per Layer
		Clay	Silt	Sand						
SF	0–10	128 (49)	176 (25)	674 (54)	Sandy loam	3.8 (0.2)	22.1 (1.5)	1.13	21.8 (0.3)	1.32
	10–20	192 (48)	175 (41)	610 (67)	Sandy loam	3.6 (0.3)	15.3 (1.3)	1.14	21.6 (0.5)	1.59
	20–30	118 (52)	294 (14)	571 (65)	Sandy loam	3.8 (0.5)	13.0 (1.5)	1.09	17.7 (3.2)	1.47
CAF4	0–10	82 (13)	309 (52)	606 (42)	Sandy loam	4.4 (0.3)	17.0 (1.0)	0.87	16.2 (0.6)	0.98
	10–20	96 (46)	323 (33)	567 (32)	Sandy loam	4.1 (0.3)	12.4 (0.7)	0.92	12.4 (1.3)	0.91
	20–30	86 (22)	375 (18)	527 (24)	Loam	3.7 (0.3)	11.4 (0.5)	0.96	10.4 (1.5)	0.87
CAF10	0–10	94 (19)	288 (57)	602 (43)	Sandy loam	4.1 (0.3)	19.8 (1.1)	1.01	15.1 (1.2)	0.92
	10–20	80 (25)	325 (27)	578 (39)	Sandy loam	3.9 (0.2)	13.2 (0.8)	0.98	11.4 (1.3)	0.84
	20–30	67 (19)	390 (37)	530 (34)	Loam	3.8 (0.2)	11.8 (0.4)	0.99	10.4 (1.4)	0.87
CAF17	0–10	93 (17)	341 (43)	554 (47)	Sandy loam	3.9 (0.4)	20.7 (1.6)	1.06	13.5 (1.1)	0.82
	10–20	138 (42)	319 (26)	525 (28)	Loam	3.4 (0.4)	14.2 (0.7)	1.05	10.2 (0.3)	0.75
	20–30	138 (31)	347 (44)	504 (39)	Loam	3.5 (0.4)	12.4 (0.9)	1.04	10.0 (0.6)	0.83
MAF45	0–10	119 (33)	254 (28)	607 (36)	Sandy loam	4.3 (0.4)	18.5 (1.1)	0.94	15.7 (2.0)	0.95
	10–20	131 (45)	267 (15)	583 (32)	Sandy loam	4.2 (0.3)	12.2 (0.5)	0.91	12.2 (2.2)	0.90
	20–30	124 (50)	313 (45)	540 (32)	Sandy loam	4.1 (0.3)	10.8 (0.4)	0.91	11.6 (2.6)	0.97

* SF = secondary forest, CAF = cocoa agroforestry, MAF—multistrata agroforest, with years since the plot establishment indicated.

2.2. Land Cover

Beyond cocoa, a range of other fruit tree species was part of local agroforestry systems, including langsat (*Lansium domesticum* Corr.), durian (*Durio zibethinus* Murr.), *rambutan* (*Nephelium lappaceum* L.), jackfruit (*Artocarpus heterophyllus* Lam.), cempedak (*Artocarpus integer* (Thunb.) Merr.), star fruit (*Averrhoa carambola* L.), and candlenut trees (*Aleurites moluccanus* L.). Timber trees include white teak (*Gmelina arborea* Roxb.), sandalwood (*Santalum album* L.), cananga (*Cananga odorata* (Lam.) Hook.), and *bitti* (*Vitex cofassus Reinw.*). Various bamboo species (*Bambusoideae sp.*) are also a source of construction material, with young shoots as a food source. Bananas (*Musa spp.*), papaya (*Carica papaya* L.), sugar palm (*Arenga pinnata* Merr.), and chili (*Capsicum annuum* L.) are present as readily marketable food sources. As the 'mother of cocoa', gliricidia (*Gliricidia sepium* Jack.) was common as nitrogen-fixing shade trees, as was leucaena (*Leucaena leucocephala* (Lam.) de Wit). Most of these trees were planted, although some sufficiently naturalized to spread spontaneously. Farmer management is based on selective retention of trees with use value, similar to what was documented for Southeast Sulawesi [38].

The vegetation found in the remnant forest (SF) was dominated by *Castanopsis acuminatissima* (Blume) A., *Symplocos polyandra* (Blanco) Brand., rattan (*Calamus deerratus* G.Mann & H.Wendl.), forest/wild pandan (*Pandanus amaryllifolius* Roxb.), bamboo (*Bambusa* spp), and some hardwood tree species (locally named *dao, bubun, nato*) including kluwak (*Pangium edule* Reinw.), trees that are locally called forest teak, *gattungan*, ebony (*Diospyros celebica* Bakh.), trees that are locally named *belating*, candlenut, coffee (*Coffea sp*), cananga trees, and various shrubs. All these species are part of Sulawesi's native flora.

2.3. C–stock Measurements

Within the selected sites, based on the field vegetation conditions that represent the land-use type, we set up four C-stock observation plots according to the RaCSA (Rapid Carbon Stock Appraisal) procedure outlined by [39] and used in a cocoa landscape context by [38]. While other C sampling protocols prefer circular plots to maximize plot homogeneity, the basic plot size in the RACSA protocol is a transect–like 40 m by 5 m to include variation within a plot; some adjustments to plot dimensions were made where 40 m × 5 m was not possible. Where we found trees with DBH greater than 30 cm as in the SF, plot size was expanded for such trees to 100 m by 20 m, as they tend to occur at lower frequency but have considerable impact on total biomass estimates [39]. On the other hand, in cocoa farms (CAF17, CAF4 and CAF10) with regular cocoa planting distance of 3 m × 4 m, the plot size was increased to be 20 m by 20 m. In these plots, we measured all trees with a stem diameter at breast height (DBH) of 5 to 30 cm. In every basic plot, we randomly selected six sub-plots of 50 cm by 50 cm, where we measured C–stock of understory (destructing sampling), non-woody necromass, and soil organic carbon (depth of 0–30 cm). For woody necromass, where present, we followed the protocol of [39].

2.4. Aboveground and Belowground Biomass

Tree biomass was estimated allometrically from the stem diameters, using allometric equations for Aboveground biomass (AGB in kg per tree) (Table 2), on the basis of stem diameter (D, in cm) at 1.3 m above the soil surface, with wood density ρ (in g cm^{-3}, obtained from the World Agroforestry database [http://db.worldagroforestry.org/wd, April 2014]. Wood density is used in allometric equations that are derived across many different tree species; as these equations are calibrated on a wider range of tree diameter data, equations that deviate from the simple power laws used for the other equations account for a larger share of the variation in the calibration datasets [40].

Table 2. Allometric equations used for aboveground biomass (AGB$_{est}$) based on stem diameter D.

Component	Allometric Equation for AGB$_{est}$	Reference
Cocoa	$0.1208\ D^{1.98}$	[41]
Generic trees, humid tropics (rainfall 1500–4000 mm yr^{-1})	$\rho \times \exp\,(-1.499 + 2.148\ \ln(D) + 0.207\ (\ln(D))^2 - 0.0281\ (\ln(D))^3)$	[40]
Banana	$0.030\ D^{2.13}$	[42]
Palm	$0.118\ D^{2.53}$	[43]

Tree basal area (m^2 ha^{-1}) was calculated by summing their cross–sectional area $\pi\ D^2/4$ of all sampled trees and scaling up from the sampled area, while converting from cm^2 to m^2. Formulas used to calculate woody necromass depend on the wood decomposition stage [39]. The decomposition stage was assessed in the field based on ease of inserting a metal rod into woody necromass. On recent tree falls, where the decomposition stage was low, the equations of Table 2 were used. However, where the necromass was highly decomposed, Equation (1) was used, and wood density was assumed to be 0.4 g cm^{-3} [39]:

$$N_D = \rho\ H\ D^2/40 \tag{1}$$

where N_D is dry necromass (kg).

Measurement of understory biomass was done on six randomly selected quadrants of 50 cm by 50 cm in every land–use. Understory vegetation in the quadrants was collected, weighed, then taken to laboratory for oven–drying at 80 °C for 48 h. Due to massive and bulky understory biomass samples, approximately 300 g sub–samples were taken from the field samples for dry weights. If the samples were less than 300 g, the whole biomass was oven–dried. The understory dry weight (UD) was calculated as the ratio of oven–dry weight to the sub–sample fresh weight multiplied by the total fresh weight of understory biomass sampled in the field. Understory dry biomass of each land-use is the average of the six replicates of the respective land–use. The litter collected was carefully cleaned to remove adhering soil particles, before being taken to a laboratory for oven–drying at 80 °C, for 48 h. Calculation of litter biomass follows the same procedure as that for understory biomass.

Belowground biomass was estimated from the aboveground–tree root biomass relationship. A default value of the canopy/root ratio of 4/1 for wet tropical forest trees was applied [44]. However, for cocoa trees, we estimated the value based on the 6.7/1 ratio developed by [45].

2.5. Soil Properties

On 19 sub–plots (four sub–plots each in CAF4, CAF10, CAF17 and MAF45, and 3 sub–plots in SF) we collected 57 of both disturbed and undisturbed soil samples at 0–10, 10–20, and 20–30 cm depths. In each sub–plot, we collected about 1 kg of disturbed soil samples per layer, mixed composite from six random spots within the same sub–plot for the measurements of soil organic carbon (C–org), soil texture, and soil carbon pools. The undisturbed soil samples (also collected from three soil depths at six random spots in each sub–plot) were used for bulk density, saturated hydraulic conductivity, water retention, soil porosity, macro–porosity and aeration, available water capacity (AWC), and soil structural quality measurements. C–org was determined according to the Walkley and Black procedure outlined by [46]; bulk density was determined in each land use using six large (20 cm long × 20 cm wide × 10 cm deep) core samples as well as six standard small (70 mm inner diameter, 50 mm length) ring samplers according to [47]; soil texture was determined using the pipet method outlined by [48] saturated hydraulic conductivity was measured in a laboratory at a constant head [49] using undisturbed samples with six replicates for each form of land use. Using the same undisturbed samples for saturated hydraulic conductivity and water retention at 0 and 5 kPa, matric suctions were determined using a hanging water column system [50,51]. Soil porosity was calculated from bulk density, assuming a particle density of mineral soils of 2.65 Mg m^{-3} [52]. Macro–porosity (effective diameter >60 μm) was calculated according to the soil pore (capillarity radius)—matric suction relationship formula [53] at 5 kPa (0.50 m). Available water capacity was calculated as the difference between water content at field capacity (−5 kPa matric potential) and at the wilting point (−1500 kPa), measured using a pressure chamber [54]. Soil carbon stocks were calculated from bulk density and C-org data [9], pH was measured in 1N KCl.

3. Results

3.1. Density and Basal Area

Land occupancy by trees, expressed in terms of tree basal area, of secondary forest was the most common form of land use (over 14 m^2 ha^{-1}); it was significantly reduced ($p < 0.05$) when land was converted to other land uses (Table 3). Ten–to–twelve years old cocoa agroforestry (CAF10) had the lowest tree occupancy (just over 5 m^2 ha^{-1}), but its difference with younger cocoa agroforestry systems (CAF4) was not statistically significant. Generally, the basal area increased as the cocoa–based agroforestry system aged.

Table 3. Tree basal area and plant density of remnant secondary forest (SF), multistrata agroforest that is 45–68 years old (MAF45), cocoa agroforestry that is 17–34 years old (CAF17), cocoa agroforestry that is 10–12 years old (CAF10), and cocoa agroforestry that is 4–5 years old (CAF4). The high plant density for CAF4 was mainly due to a high population of cocoa and shade trees introduced into the system. Shade trees are trees mainly planted to provide shade for cocoa, but several types also fix atmospheric nitrogen and thus improve soil fertility.

Type of Tree or Plant	SF	MAF45	CAF17	CAF10	CAF4
Basal area			$(m^2\ ha^{-1})$		
Timber trees	12.42	4.50	0.19	2.02	
Fruit trees		4.02	3.11	0.50	2.73
Cocoa		0.70	3.71	2.13	2.71
Shade trees			2.88	0.76	1.75
Palms		0.90			
Other trees	1.91	0.67			0.84
Total	14.33 [a]	10.78 [b]	9.88 [b]	5.41 [c]	8.04 [bc]
Plant density			$(trees\ ha^{-1})$		
Timber trees	498	79	5	5	
Fruit trees		251	64	28	175
Cocoa		44	263	188	650
Shade trees			225	87	469
Palms		5			
Bananas					75
Other trees	550	183			25
Total	1048 [a]	563 [b]	55 6 [b]	307 [b]	1394 [a]

a,b,c: Different letters along rows indicate significant differences ($p < 0.05$) according to Tukey's test.

Young cocoa agroforestry systems had similar tree density to secondary forests (over 1000 trees per hectare), which was two to four times that of other land uses (Table 3). The young cocoa–based agroforestry had a significantly higher tree density compared with old cocoa–agroforestry and multipurpose tree species agroforestry plots. While a 3 m × 3 m planting grid yields 1111 stems ha^{-1}, the tree population reduces as trees grow older.

3.2. Soil Properties

Converting natural forest to any land use under cocoa agroforestry systems, including multistrata agroforestry in Sulawesi, markedly reduced topsoil (0 to 30 cm depth) organic matter concentration. In the remnant secondary forest, soil organic carbon concentration at the 0–10 cm depth was 22 g kg^{-1}, categorized as 'high' according to [55] compared with only around 15 g kg^{-1} (moderate) for non-forest systems (Figure 4a). Variations among the agroforestry land uses (non–forest) were relatively small, within a 4 g kg^{-1} (0.4%) range, compared to the decline from forest to non–forest.

Surface soil bulk density values (Figure 4b) changed in the opposite direction from the pattern of soil organic matter (Figure 4a). Having the highest C-organic concentration, secondary forest was the least compacted. Consequently, inferred changes in the soil C stock in the top 30 cm of soil were small.

Converting forest to cocoa agroforestry led to soil compaction and therefore decreased total porosity (data not shown as it is the mirror of bulk density data); multistrata agroforestry mimics the forest condition closest (it is the least compacted).

Increased bulk density (reduced total porosity) resulted in the disappearance of part of the macropores when they are >60 μm (Figure 5a), which consequently reduced the soil's capacity to conduct water (Figure 5b) and affected soil water availability (Figure 5c). The reduction in macro-porosity caused by shifting a forest to CAF resulted in a decreased saturated hydraulic conductivity (Figure 5b). Cocoa farms which were recently converted from forest had the lowest hydraulic conductivity; apparently the hydraulic conductivity improved and became closer to that for forest as the land use was transformed to cocoa agroforestry, especially for old cocoa agroforestry and multistrata agroforest.

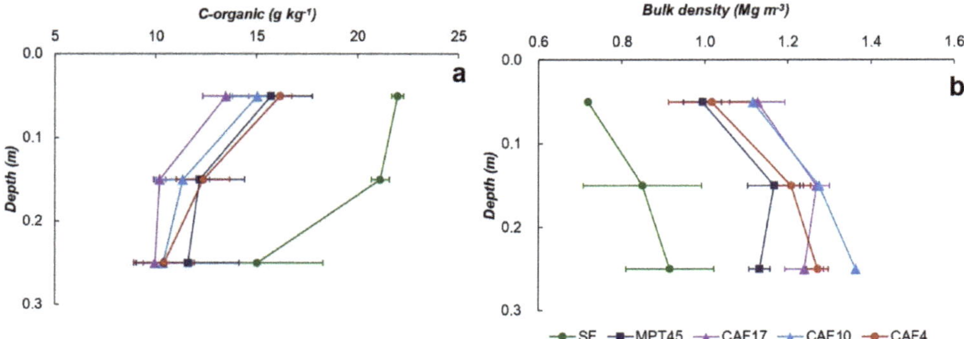

Figure 4. Surface soil (0–30 cm depth) soil organic carbon status (**a**) and bulk density (**b**) of secondary tropical forest (SF), multistrata agroforestry that is 45–68 years old (MAF45), cocoa agroforestry that is 17–34 years old (CAF17), cocoa agroforestry that is 10 years old (CAF10), and cocoa agroforestry that is 4–5 years old (CAF4). Bars are standard errors of the means.

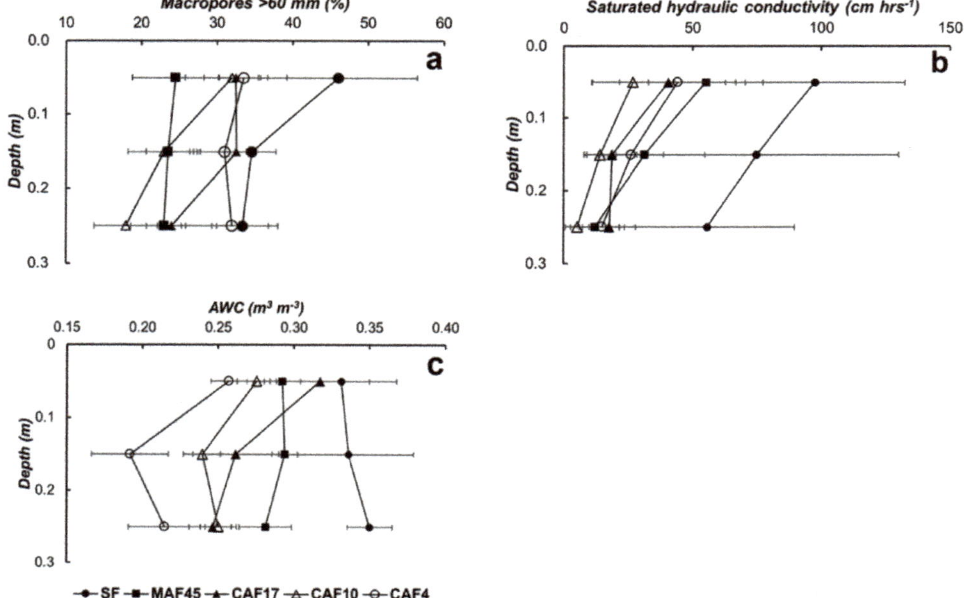

Figure 5. Changes of soil surface (0–30 cm depth) macro–porosity (**a**), saturated hydraulic conductivity (**b**) and available water capacity (**c**), resulted from shifting forest to cocoa farms and its restoration through aging cocoa agroforestry systems that are 4–5 years old (CAF4), 10–12 years old (CAF10), and 17–34 years old (CAF17), and through multistrata agroforest that is 45–68 years old (MAF45). Bars are standard errors of the means.

Converting forest to other land–uses also led to a change of soil water retention, which specifically reduces the available water capacity, AWC (Figure 5c). Young cocoa agroforestry led to the lowest hydraulic conductivity, but the figure improved closer toward that for forest as the land use was managed using aged cocoa agroforestry and multipurpose tree species.

3.3. C–stock

Converting tropical forest to cocoa–based agroforestry led to a substantial total C–stock decrease, from in total around 320 Mg ha^{-1} of secondary forest to about 75 to 150 Mg ha^{-1} of cocoa–based agroforestry (Figure 6). However, C–stock can be restored to approximately half of that found in the forest through aging cocoa agroforestry or practicing long–term multistrata agroforestry farming. While there was a large effect of land–use on the aboveground C–stock, no influence on the soil C–stock was evident (Figure 6), due to the inclusion of a higher soil dry weight in the 0–30 cm soil layers [9].

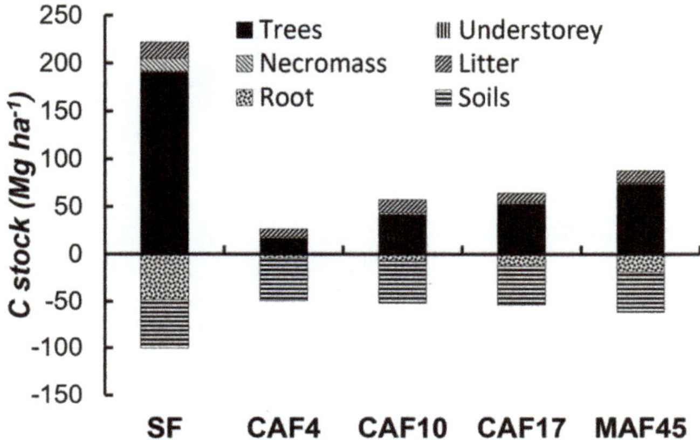

Figure 6. Total above– and below–ground carbon stock of remnant secondary forest (SF) compared with cocoa agroforestry that is 4–5 years old (CAF4), cocoa agroforestry that is 10–12 years old (CAF10), cocoa agroforestry that is 17–34 years old (CAF17), and multistrata agroforest that is 45–68 years old (MAF45). The figure demonstrates the total C–stock restoration from 4–5 (CAF4) to 17–34 years old cocoa agroforestry (CAF17) and to 45–68 years old multistrata agroforest (MAF45).

3.4. Available Water Capacity

Variable soil organic matter concentrations from different CAFs are associated with soil bulk density (Figure 7a), and hence with soil macro–pore reduction (Figure 7b) and soil hydraulic conductivity (Figure 7c). The decline of soil organic matter concentrations was correlated with soil macro–pore collapse and modified pore–size distribution, eventually affecting available water capacity (Figure 7d). Forest, having the highest organic matter concentrations, had the lowest bulk density, and highest macro–porosity, hydraulic conductivity, and plant water availability (Figure 8).

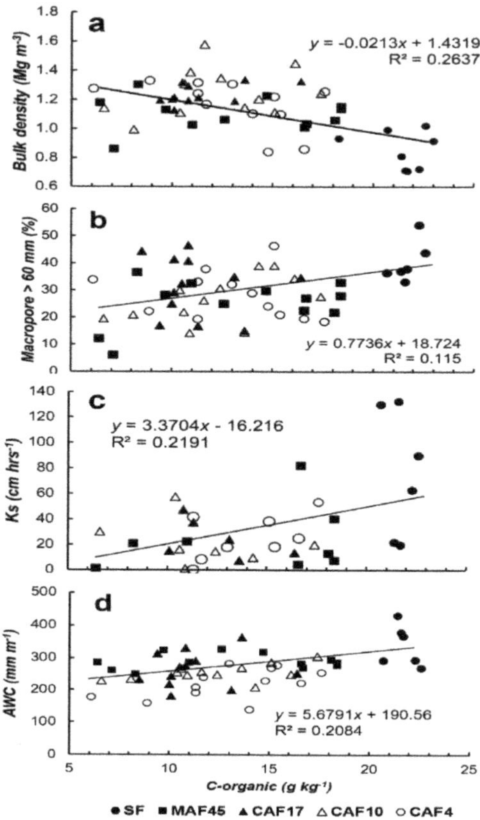

Figure 7. Surface soil bulk density (**a**), macro–porosity as percentage of total soil pores (**b**), saturated hydraulic conductivity – Ks (**c**) and available water capacity – AWC, expressed in mm of water per meter of soil (**d**) as related to soil organic matter measured at different cocoa–based agroforestry systems, i.e., secondary forest (SF), cocoa agroforestry that is 4–5 years old (CAF4), 10–12 years old (CAF10), and 17–34 years old (CAF17) and multistrata agroforest that is 45–68 years old (MAF45).

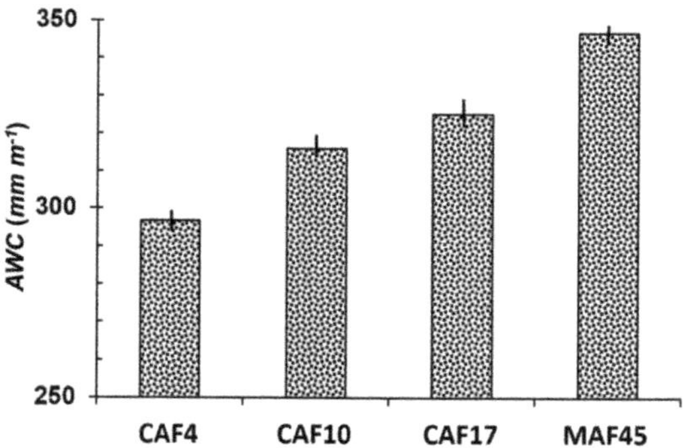

Figure 8. Available water capacity (AWC), expressed as mm of water per meter of soil of young (4–5 yrs.) cocoa agroforestry (CAF4) compared to aging (17 to 34 yrs.) cocoa–based agroforestry, and old (45–68 yrs.) multistrata agroforest (MAF45). Bars are standard errors of the means.

4. Discussion

Our results provided partial support for the hypothesis that in cacao production systems, the farmer–level benefits of agroforestry through increased water storage, contributing to climate change adaptation, are parallel to global benefits of climate change mitigation through increased C storage. However, the changes in soil properties are slow and gradual and only a partial recovery towards the conditions found in secondary forest was recorded, even for 45–year old mixed agroforest. This slow belowground change relative to aboveground appearance is in–line with restoration experience elsewhere [56].

Three steps are needed to assert a belowground contribution to climate change adaptation by farmers (Figure 1): A. an increase in soil organic matter content, B. a functional role of changed soil properties in the water balance reducing crop vulnerability to rainfall variability, and C. Deliberate use by farmers to reduce their vulnerability. Our data referred to steps A and B.

In the tradition of 'chronosequence' research where surveys of existing plot–level land cover are interpreted as a pseudo time–sequence, such as we did here, it is not easy, especially for soil properties, to separate cause–effect relations from background variability. While aboveground C stock data clearly differentiate the land cover types (Table 3), with only a partial recovery in older mixed agroforest of the C stocks lost when local forests were converted, soil changes are more subtle. As noted by [9] increased soil bulk density partly conceals reductions in C_{org} (expressed per unit dry soil) when carbon stocks in the top 30 cm of soil are calculated. Thus, specific effects of land cover on C_{org} and soil C stock are more clearly observed when C_{org}/C_{ref} ratios are considered rather than direct C_{org} observations.

As second step in the argument we similarly obtained clearer evidence when the plot–level C_{org} parameter, rather than land cover classes, was used as the basis for correlation with other soil parameters. At around 20% of the variance accounted for (Figure 7), other sources of variation clearly dominate. According to the regression equation of Figure 7d, increasing soil organic carbon by 1 g per kilogram of soil (0.1% C-organic improvement) would give an increase of available soil water of about 6 mm per meter of soil. Our data (Figures 4 and 7) reveal that an increase of C_{org} of 4 g kg^{-1} (0.4%) is feasible when land cover is transformed to the MAF 45 conditions in multistrata agroforestry available plant water, which can be around 9.6 mm per 0.4 m of soil (as most of the cocoa roots on these stony soils are concentrated in this layer). An increase of 0.4% in 45 years from a base of around 2% corresponds with a soil C increase of 4 ‰ y^{-1} that is desirable as part of global climate change policy [57], as the relative increase from 2.0 to 2.4 is 20% and $1.2^{1/45} = 1.004$, corresponding with a relative rate of increase of 4 ‰ y^{-1}.

As the average evapotranspiration rate for the humid tropics is 3–4 mm day^{-1} [58] and measurements in the dry seasons in Central Sulawesi indicated around 3.6 mm day^{-1} [59], this extra available water can support 50% of cocoa demand for water for any week without any rain in the growing season. In the study area, the average rainfall in the driest months is only 50 mm [60]. The additional AWC that can be expected with higher soil organic matter content, combines with the reduction of soil evaporation where a permanent litter layer is present [61,62]. Such effects are not dramatic but can at least partly offset increased demand for water by trees with active leaf canopies [63]. However, if shade trees are as shallowly rooted as those in a recent West African case study [64] were, then negative effects on cocoa survival are logical outcomes. Only with deep–rooted trees that have access to deeper soil layers and share up to 1 mm day^{-1} to the topsoil by nighttime hydraulic equilibration [65,66] can we expect substantive increases in dry season tolerance in agroforestry systems. A number of the cocoa companion trees may have the deep rooting pattern necessary for such effects, but location–specific observations will be needed to test that [67]. Shade trees in cocoa agroforestry not only improved soil carbon concentrations in Southeast Sulawesi, but also soil structure and aggregation [68], possibly facilitating cocoa root development, in line with our results (Figure 7b) on macroporosity, that will also increase infiltration into the profile and reduction of surface runoff [69]. The water–related functionality of C_{org} had coefficients of determination of around 20%, suggesting that other factors, such as details of soil structure, play a major role as well.

What matters to farmers is the overall effects on yield and economic performance of the plots (e.g., in terms of return to labor). Intercropped shade–tree species with cocoa had little or no effects on cocoa yield in earlier studies in Sulawesi [59,66]. Elsewhere the distance between shade trees and cocoa was found to be important [70]. Although cocoa yields maybe higher under monocultures system [71,72], returns derived from other agroforestry products compensated the difference in economic terms [73]. Part of the tree diversity on cocoa agroforestry in SE Sulawesi was recently attributed [39] to the fact that they harbor tree species prioritized by male farmers as well as those favored by female farmers. At the interface of aboveground and belowground effects and contributions to both mitigation and adaptation climate change agendas, cocoa agroforestry can indeed be 'climate–smart', contributing to resilience and policy synergy [74]. As detailed in a recent study in Southeast Sulawesi [75], the improvements of cocoa agroforestry relative to cocoa monoculture are modest steps in the direction of physically restoring forest soil conditions.

5. Conclusions

In terms of total system, C stock cocoa agroforestry could recover up to half of the stocks of secondary forest in the landscape, while conversion only left 23% of the original stock, essentially the soil component. Belowground losses after forest conversion but also restoration are slower than the changes aboveground.

Trees in agroforestry are not only relevant for C storage, but also as provider of farmer income, as modifier of microclimate and in support of soil C storage. In contrast to aboveground C storage, belowground increases in stock have functional relevance. Soil conditions connected to the capacity for CC adaptation, especially minimum soil compaction, greater macro–porosity, and hydraulic conductivity to allow high rain infiltration and higher available water capacity could be directly linked to higher C_{org} concentrations across the land covers sampled. An increase of $1 \mathrm{~g~kg}^{-1}$ (0.1%) of soil organic carbon increased available soil water capacity by 6% (vol/vol). Our data confirmed an increase in available water capacity, relevant for survival of week–long dry spells. As such, a contribution can be made to human adaptation to increased rainfall variability.

The water–related functionality of C_{org} had coefficients of determination of around 20%. Belowground consequences of tree cover in agroforestry are only quantifiable in large data sets but are likely as relevant as the more visible effects aboveground.

Our results confirmed that a shift from cocoa monoculture to forms of agroforestry has benefits for the farmer in terms of an increased soil water buffering and soil structure that, even though modest in effect size, contributes to tolerance of dry season conditions. These systems also contribute, in modest amounts, to increased storage of carbon in aboveground and belowground pools that ought to be reflected in the country's national C accounting system and be included in the Nationally Determined Contributions to the Paris Agreement on combatting global climate change. A shift from a focus on cocoa to be intensively managed as monoculture crop to one that is best grown in mixed agroforestry systems, along with other tree commodities will be appropriate.

Author Contributions: Conceptualization and methodology (S.G., K.H. and M.v.N.); investigation (S.S., R.S., I.H.M., M.N., D.U.); data curation (S.S., K.H., S.G., M.v.N.); writing—original draft preparation (S.G., K.H., M.v.N.); writing—review and editing (S.G., M.v.N.). All authors have read and agreed to the published version of the manuscript.

Funding: We thank Ford Foundation for funding the project (grant number 0155–1534) as an effort to enhance cocoa farmers' business performance in Polman, West Sulawesi.

Acknowledgments: We are grateful to all farmers for fieldwork supports during the research, especially Annas, Sahabuddin, and Abdullah, and their team coordinated through the farmer group. The land clearing photo in Figure 3 was kindly provided by Anas and Abdullah. We also thank Samsuar for preparing the site map. The publication was prepared with the support of Ministry of Research and Technology, High Education, Republic Indonesia through WCU collaborative research scheme between Brawijaya, Hasanuddin, and Sebelas Maret Universities.

Conflicts of Interest: The authors declare no conflict of interest. The funders had no role in the design of the study; in the collection, analyses, or interpretation of data; in the writing of the manuscript, or in the decision to publish the results.

References

1. Atangana, A.; Khasa, D.; Chang, S.; Degrande, A. *Tropical Agroforestry*; Springer Science: London, UK, 2013.
2. Van Noordwijk, M. *Sustainable Development through Trees on Farms: Agroforestry in its Fifth Decade*; World Agroforestry (ICRAF): Bogor, Indonesia, 2019.
3. Tscharntke, T.; Clough, Y.; Bhagwat, S.A.; Buchori, D.; Faust, H.; Hertel, D.; Hölscher, D.; Juhrbandt, J.; Kessler, M.; Perfecto, I.; et al. Multifunctional shade-tree management in tropical agroforestry landscapes—A review. *J. Appl. Ecol.* **2011**, *48*, 619–629. [CrossRef]
4. Duguma, L.A.; Minang, P.A.; Van Noordwijk, M. Climate change mitigation and adaptation in the land use sector: From complementarity to synergy. *Environ. Manag.* **2014**, *54*, 420–432. [CrossRef] [PubMed]
5. Harvey, C.A.; Chac, M.; Donatti, C.I.; Garen, E.; Hannah, L.; Andrade, A.; Bede, L.; Brown, D.; Calle, A.; Char, J. Climate-Smart Landscapes: Opportunities and Challenges for Integrating Adaptation and Mitigation in Tropical Agriculture. *Conserv. Lett.* **2014**, *7*, 77–90. [CrossRef]
6. Rosenstock, T.S.; Wilkes, A.; Jallo, C.; Namoi, N.; Bulusu, M.; Suber, M.; Mboi, D.; Mulia, R.; Simelton, E.; Richards, M.; et al. Making trees count: Measurement and reporting of agroforestry in UNFCCC national communications of non-Annex I countries. *Agric. Ecosyst. Environ.* **2019**, *284*, 106569. [CrossRef]
7. Zomer, R.J.; Neufeldt, H.; Xu, J.; Ahrends, A.; Bossio, D.; Trabucco, A.; Van Noordwijk, M.; Wang, M. Global Tree Cover and Biomass Carbon on Agricultural Land: The contribution of agroforestry to global and national carbon budgets. *Sci. Rep.* **2016**, *6*, 29987. [CrossRef]
8. Van Noordwijk, M.; Bayala, J.; Hairiah, K.; Lusiana, B.; Muthuri, C.; Khasanah, N.; Mulia, R. Agroforestry solutions for buffering climate variability and adapting to change. In *Climate Change Impact and Adaptation in Agricultural Systems*; CAB-International: Wallingford, UK, 2014; pp. 216–232.
9. Hairiah, K.; van Noordwijk, M.; Sasi, R.R.; Dwi, D.; Suprayogo, D.; Kurniawan, S.; Prayogo, C.; Gusli, S. Soil carbon stocks in Indonesian (agro) forest transitions: Compaction conceals lower carbon concentrations in standard accounting. *Agric. Ecosyst. Environ.* **2020**, *294*, 106879. [CrossRef]
10. Sanchez, P.A. *Properties and Management of Soils in the Tropics*; Cambridge University Press: Cambridge, UK, 2019. [CrossRef]
11. Wang, F.; Li, Z.; Ding, Y.; Sayer, E.J.; Li, Q.; Zou, B.; Mo, Q.; Li, Y.; Lu, X.; Tang, J.; et al. Tropical forest restoration: Fast resilience of plant biomass contrasts with slow recovery of stable soil C stocks. *Funct. Ecol.* **2017**, *31*, 2344–2355. [CrossRef]
12. Dollinger, J.; Jose, S. Agroforestry for soil health. *Agrofor. Syst.* **2018**, *92*, 213–219. [CrossRef]
13. Van Noordwijk, M.; Cerri, C.; Woomer, P.L.; Nugroho, K.; Bernoux, M. Soil carbon dynamics in the humid tropical forest zone. *Geoderma* **1997**, *79*, 187–225. [CrossRef]
14. Lusiana, B.; Kuyah, S.; Öborn, I.; van Noordwijk, M. Typology and metrics of ecosystem services and functions as the basis for payments, rewards and co-investment. In *Co-Investment in Ecosystem Services: Global Lessons from Payment and Incentive Schemes*; Namirembe, S., Leimona, B., van Noordwijk, M., Minang, P., Eds.; World Agroforestry Centre (ICRAF): Nairobi, Kenya, 2017.
15. Van Noordwijk, M.; Martikainen, P.; Bottner, P.; Cuevas, E.; Rouland, C.; Dhillion, S.S. Global change and root function. *Glob. Chang. Biol.* **1998**, *4*, 759–772. [CrossRef]
16. Van Noordwijk, M.; Bizard, V.; Wangpakapattanawong, P.; Tata, H.L.; Villamor, G.B.; Leimona, B. Tree cover transitions and food security in Southeast Asia. *Glob. Food Secur.* **2014**, *3*, 200–208. [CrossRef]
17. Van Noordwijk, M.; Hoang, M.H.; Neufeldt, H.; Oborn, I.; Yatich, T. *How Trees and People Can Co-Adapt to Climate Change: Reducing Vulnerability through Multifunctional Agroforestry Landscapes*; World Agroforestry Centre (ICRAF): Nairobi, Kenya, 2011.
18. Dewi, S.; Van Noordwijk, M.; Zulkarnain, M.T.; Dwiputra, A.; Hyman, G.; Prabhu, R.; Gitz, V.; Nasi, R. Tropical forest-transition landscapes: A portfolio for studying people, tree crops and agro-ecological change in context. *Int. J. Biodivers. Sci. Ecosyst. Serv. Manag.* **2017**, *13*, 312–329. [CrossRef]

19. Abdulai, I.; Vaast, P.; Hoffmann, M.P.; Asare, R.; Jassogne, L.; van Asten, P.; Rötter, R.P.; Graefe, S. Cocoa agroforestry is less resilient to sub-optimal and extreme climate than cocoa in full sun: Reply to Norgrove (2017). *Glob. Chang. Biol.* **2018**, *24*, e733–e740. [CrossRef] [PubMed]

20. Cerda, R.; Deheuvels, O.; Calvache, D. Contribution of cocoa agroforestry systems to family income and domestic consumption: Looking toward intensification. *Agrofor. Syst.* **2014**, *88*, 957–981. [CrossRef]

21. Somarriba, E.; Beer, J. Productivity of Theobroma cacao agroforestry systems with timber or legume service shade trees. *Agrofor. Syst.* **2011**, *81*, 109–121. [CrossRef]

22. Vaast, P.; Somarriba, E. Trade-offs between crop intensification and ecosystem services: The role of agroforestry in cocoa cultivation. *Agrofor. Syst.* **2014**, *88*, 947–956. [CrossRef]

23. World Atlas. Top 10 Cocoa Producing Countries—WorldAtlas [WWW Document]. Available online: https://www.worldatlas.com/articles/top-10-cocoa-producing-countries (accessed on 13 August 2019).

24. Tromba, A. What Stats Reveal about the Top 10 Cocoa Producing Countries [WWW Document]. ICCO. 2019. Available online: https://foodensity.com/cocoa-producing-countries/ (accessed on 13 August 2019).

25. Kroeger, A.; Koenig, S.; Thomson, A.; Streck, C. *Forest- and Climate-Smart Cocoa in Côte d'Ivoire and Ghana, Aligning Stakeholders to Support Smallholders in Deforestation-Free Cocoa*; World Bank: Washington, DC, USA, 2017.

26. PDSIP. *Outlook Kakao—Komoditas Pertanian Subsektor Perkebunan*; Kementerian Pertanian: Jakarta, Indonesia, 2016.

27. Ruf, F. The Sulawesi case: Deforestation, pre-cocoa and cocoa migrations. In *Beyond tropical Deforestation: From Tropical Deforestation to Forest Cover Dynamics and Forest Development*; Babin, D., Ed.; CIRAD: Montpellier, France, 2004; pp. 277–298.

28. Hartmann, M.; Niklaus, P.A.; Zimmermann, S.; Schmutz, S.; Kremer, J.; Abarenkov, K.; Lu, P. Resistance and resilience of the forest soil microbiome to logging-associated compaction. *ISME J.* **2013**, *8*, 226–244. [CrossRef]

29. Udawatta, R.P.; Rankoth, L.M.; Jose, S. Agroforestry and Biodiversity. *Sustainability* **2019**, *11*, 2879. [CrossRef]

30. Gijsman, A.J. *Deforestation and Land Use: Changes in Physical and Biological Soil Properties in Relation to Sustainability*; Agricultural University Wageningen, Department of Soil Science and Geology: Wageningen, The Netherlands, 1992.

31. Wang, Q.; Wang, S.; Yu, X. Decline of soil fertility during forest conversion of secondary forest to chinese fir plantations in subtropical China. *Land Degrad. Dev.* **2011**, *22*, 444–452. [CrossRef]

32. Mortimer, R.; Saj, S.; David, C. Supporting and regulating ecosystem services in cacao agroforestry systems. *Agrofor. Syst.* **2018**, *6*, 1639–1657. [CrossRef]

33. Rousseau, G.X.; Deheuvels, O.; Arias, I.R.; Somarriba, E. Indicating soil quality in cacao-based agroforestry systems and old-growth forests: The potential of soil macrofauna assemblage. *Ecol. Indic.* **2012**, *23*, 535–543. [CrossRef]

34. Fearnside, P.M. Global Warming and Tropical Land-Use Change: Greenhouse Gas Emission from Biomass Burning, Decomposition and Soils in Forest Conversion, Shifting Cultivation and Secondary Vegetation. *Clim. Chang.* **2000**, *46*, 115–158. [CrossRef]

35. Van Noordwijk, M.; Rahayu, S.; Hairiah, K.; Wulan, Y.C.; Farida, A.; Verbist, B. Carbon stock assessment for a forest-to-coffee conversion landscape in Sumber-Jaya (Lampung, Indonesia): From allometric equations to land use change analysis. *Sci. China Ser. C* **2002**, *45*, 75–86.

36. Schneider, B.M.; Andres, C.; Trujillo, G.; Alcon, F.; Amurrio, P.; Perez, E.; Weibel, F.; Milz, J. Cocoa and total system yields of organic and conventional agroforestry vs. monoculture systems in a long-term field trial in Bolivia. *Expl. Agric.* **2016**, *53*, 1–24. [CrossRef]

37. Guest, D.; Keane, P. Vascular-Streak Dieback: A New Encounter Disease of Cacao in Papua New Guinea and Southeast Asia Caused by the Obligate Basidiomycete Oncobasidium theobromae. *Phytopathology* **2007**, *97*, 1654–1657. [CrossRef]

38. Sari, R.R.; Saputra, D.D.; Hairiah, K.; Rozendaal, D.M.A.; Roshetko, J.; van Noordwijk, M. Gendered species preferences link tree diversity and carbon stocks in cacao agroforest in Southeast Sulawesi, Indonesia. *Land* **2020**, *9*, 108. [CrossRef]

39. Hairiah, K.; Dewi, S.; Agus, F.; Velarde, S.; Ekadinata, A.; Rahayu, S.; van Noordwijk, M. *Measuring Carbon Stocks: Across Land Use Systems: A Manual*; World Agroforestry Centre (ICRAF): Bogor, Indonesia, 2011.

40. Chave, J.; Andalo, C.; Brown, S.; Cairns, M.A.; Chambers, J.Q.; Eamus, D.; Fölster, H.; Fromard, F.; Higuchi, N.; Kira, T.; et al. Tree allometry and improved estimation of carbon stocks and balance in tropical forests. *Oecologia* **2005**, *145*, 87–99. [CrossRef]

41. Yuliasmara, F.; Wibawa, A.; Prawoto, A. Carbon stock in different ages and plantation systemof cocoa: Allometric approach. *Pelita Perkeb. Coffee Cocoa Res. J.* **2009**, *25*, 132. [CrossRef]

42. Arifin, J. *Estimasi Penyimpanan C Pada Berbagai Sistem Penggunaan Lahan di Kecamatan Ngantang*; Universitas Brawijaya: Malang, Indonesia, 2001.

43. Brown, S. *Estimating Biomass and Biomass Change of Tropical Forests: A Primer*; Food and Agriclture Organization: Rome, Italy, 1997.

44. Mokany, K.; Raison, R.J.; Prokushkin, A.S. Critical analysis of root: Shoot ratios in terrestrial biomes. *Glob. Chang. Biol.* **2006**, *12*, 84–96. [CrossRef]

45. Norgrove, L.; Hauser, S. Carbon stocks in shaded *Theobroma cacao* farms and adjacent secondary forests of similar age in Cameroon. *Trop. Ecol.* **2013**, *54*, 15–23.

46. Naelson, D.W.; Sommers, L.E. Carbon and Organic Matter. In *Methods of Soil Analysis, Part 3, Chemical Methods*; Page, A.L., Baker, D.E., Keeney, D.R., Eds.; American Society of Agronomy: Madison, WI, USA, 1996; pp. 1004–1005.

47. Blake, G.R.; Hartge, K.H. Bulk density. In *Methods of Soil Analysis: Part 1—Physical and Mineralogical Methods*; Klute, A., Ed.; Soil Science Society of America, American Society of Agronomy: Madison, WI, USA, 1986; pp. 363–375.

48. Gee, G.W.; Or, D. 2.4 Particle-size analysis. In *Methods of Soil Analysis: Part 4 Physical Methods*; Klute, A., Ed.; Soil Science Society of America, American Society of Agronomy: Madison, WI, USA, 2002; pp. 255–293.

49. Klute, A.; Dirksen, C. Hydraulic conductivity and diffusivity: Laboratory methods. In *Methods of Soil Analysis: Part 1—Physical and Mineralogical Methods*; Klute, A., Ed.; Soil Science Society of America, American Society of Agronomy: Madison, WI, USA, 1986; pp. 687–734.

50. Klute, A. Water retention: Laboratory methods. In *Methods of Soil Analysis: Part 1—Physical and Mineralogical Methods*; Klute, A., Ed.; Soil Science Society of America, American Society of Agronomy: Madison, WI, USA, 1986; pp. 635–662.

51. McIntyre, D.S. Sample preparation. In *Methods for Analysis of Irrigated Soils*; Loveday, J., Ed.; Commonwealth Agricultural Bureaux, Famham Royal: Bucks, UK, 1974; pp. 21–37.

52. Weil, R.; Brady, N. *The Nature and Properties of Soils*, 15th ed.; Prentice Hall: Upper Saddle River, NJ, USA, 2017.

53. Marshall, T.J.; Holmes, J.W. *Soil Physics*, 2nd ed.; Cambridge University Press: Cambridge, UK, 1988.

54. Cassel, D.K.; Nielsen, D.R. Field capacity and available water capacity. In *Methods of Soil Analysis: Part 1—Physical and Mineralogical Methods*; Klute, A., Ed.; Soil Science Society of America, American Society of Agronomy: Madison, WI, USA, 1986; pp. 901–926.

55. Ghaley, B.B.; Wösten, H.; Olesen, J.E.; Schelde, K.; Baby, S.; Karki, Y.K.; Børgesen, C.D.; Smith, P.; Yeluripati, J.; Ferrise, R.; et al. Simulation of Soil Organic Carbon Effects on Long-Term Winter Wheat (*Triticum aestivum*) Production Under Varying Fertilizer Inputs. *Front. Plant Sci.* **2018**, *9*, 1158. [CrossRef]

56. Van Noordwijk, M.; Gitz, V.; Minang, P.A.; Dewi, S.; Leimona, B.; Duguma, L.; Pingault, N.; Meybeck, A. People-Centric Nature-Based Land Restoration through Agroforestry: A Typology. *Land* **2020**, *9*, 251. [CrossRef]

57. Minasny, B.; Malone, B.P.; McBratney, A.B.; Angers, D.A.; Arrouays, D.; Chambers, A.; Chaplot, V.; Chen, Z.S.; Cheng, K.; Das, B.S.; et al. Soil carbon 4 per mille. *Geoderma* **2017**, *292*, 59–86. [CrossRef]

58. Juarez, R.I.N.; Hodnett, M.G.; Fu, R.; Goulden, M.L.; Randow, C. Control of Dry Season Evapotranspiration over the Amazonian Forest as Inferred from Observations at a Southern Amazon Forest Site. *J. Clim.* **2007**, *20*, 2827–2838. [CrossRef]

59. Olchev, A.; Ibrom, A.; Erasmi, S. Effects of land-use changes on evapotranspiration of tropical rain forest margin area in Central Sulawesi (Indonesia): Modelling study with a regional SVAT model. *Ecol. Model.* **2008**, *212*, 131–137. [CrossRef]

60. BPS Polman. *Kabupaten Polewali Mandar dalam Angka (Polewali Mandar Regency in Figures*; PEMDA Polman: Polewali, Indonesia, 2017.

61. Niether, W.; Schneidewind, U.; Armengot, L.; Adamtey, N.; Schneider, M.; Gerold, G. Spatial-temporal soil moisture dynamics under different cocoa production systems. *Catena* **2017**, *158*, 340–349. [CrossRef]

62. Jackson, N.A.; Wallace, J.S. Soil evaporation measurements in an agroforestry system in Kenya. *Agric. For. Meteorol.* **1999**, *94*, 203–215. [CrossRef]

63. Lin, B.B. Agroforestry management as an adaptive strategy against potential microclimate extremes in coffee agriculture. *Agric. For. Meteorol.* **2007**, *144*, 85–94. [CrossRef]

64. Abdulai, I.; Jassogne, L.; Graefe, S.; Asare, R.; Asten, P.; Van Vaast, P.; La, P. Characterization of cocoa production, income diversification and shade tree management along a climate gradient in Ghana. *PLoS ONE* **2018**, *13*, e0195777. [CrossRef] [PubMed]

65. Bayala, J.; Heng, L.K.; van Noordwijk, M.; Ouedraogo, S.J. Hydraulic redistribution study in two native tree species of agroforestry parklands of West African dry savanna. *Acta Oecol.* **2008**, *34*, 370–378. [CrossRef]

66. Van Noordwijk, M.; Lawson, G.; Hairiah, K.; Wilson, J. Root distribution of trees and crops: Competition and/or complementarity. In *Tree-Crop Interactions: Agroforestry in a Changing Climate*; Ong, C.K., Black, C.R., Wilson, J., Eds.; CAB International: Wallingford, UK, 2015; pp. 221–257. [CrossRef]

67. Hairiah, K.; Widianto, W.; Suprayogo, D.; van Noordwijk, M. Tree Roots Anchoring and Binding Soil: Reducing Landslide Risk in Indonesian Agroforestry. *Land* **2020**, *9*, 256. [CrossRef]

68. Wartenberg, A.C.; Blaser, W.J.; Roshetko, J.M.; van Noordwijk, M.; Six, J. Soil fertility and *Theobroma cacao* growth and productivity under commonly intercropped shade-tree species in Sulawesi, Indonesia. *Plant Soil* **2019**. [CrossRef]

69. Suprayogo, D.; van Noordwijk, M.; Hairiah, K.; Meilasari, N.; Rabbani, A.L.; Ishaq, R.M.; Widianto, W. Infiltration-Friendly Agroforestry Land Uses on Volcanic Slopes in the Rejoso Watershed, East Java, Indonesia. *Land* **2020**, *9*, 240. [CrossRef]

70. Koko, L.K.; Snoeck, D.; Lekadou, T.T.; Assiri, A.A. Cacao-fruit tree intercropping effects on cocoa yield, plant vigour and light interception in Côte d'Ivoire. *Agrofor. Syst.* **2013**, *87*, 1043–1052. [CrossRef]

71. Armengot, L.; Barbieri, P.; Andres, C.; Milz, J.; Schneider, M. Cacao agroforestry systems have higher return on labor compared to full-sun monocultures. *Agron. Sustain. Dev.* **2016**, *36*. [CrossRef]

72. Niether, W.; Schneidewind, U.; Fuchs, M.; Schneider, M.; Armengot, L. Below- and aboveground production in cocoa monocultures and agroforestry systems. *Sci. Total Environ.* **2019**, *657*, 558–567. [CrossRef] [PubMed]

73. Somarriba, E.; Cerda, R.; Orozco, L.; Cifuentes, M.; Dávila, H.; Espin, T.; Mavisoy, H.; Ávila, G.; Alvarado, E.; Poveda, V.; et al. Carbon stocks and cocoa yields in agroforestry systems of Central America. *Agric. Ecosyst. Environ.* **2013**, *173*, 46–57. [CrossRef]

74. Van Noordwijk, M. Agroforestry as nexus of sustainable development goals. *IOP Conf. Ser. Earth Environ. Sci.* **2020**, *449*, 012001. [CrossRef]

75. Saputra, D.D.; Sari, R.R.; Hairiah, K.; Roshetko, J.; Suprayogo, D.; van Noordwijk, M. Can cocoa agroforestry restore degraded soil structure following conversion from forest to agricultural use? *Agrofor. Syst.* **2020**, in press.

Article

Earthworm Diversity, Forest Conversion and Agroforestry in Quang Nam Province, Vietnam

Rachmat Mulia [1], Sam Van Hoang [2,*], Van Mai Dinh [3], Ngoc Bich Thi Duong [4], Anh Duc Nguyen [5,6], Dang Hai Lam [7], Duyen Thu Thi Hoang [8] and Meine van Noordwijk [9]

[1] World Agroforestry (ICRAF), Hanoi 100000, Vietnam; r.mulia@cgiar.org
[2] Department of Forest Plant, Vietnam National University of Forestry, Hanoi 100000, Vietnam
[3] Department of Soil, Vietnam National University of Forestry, Hanoi 100000, Vietnam; vandm@vnuf.edu.vn
[4] Department of Environmental Engineering, Vietnam National University of Forestry, Hanoi 100000, Vietnam; ngocdtb@vnuf.edu.vn
[5] Department of Soil Ecology, Institute of Ecology and Biological Resources, Vietnam Academy of Science and Technology, Hanoi 100000, Vietnam; ducanh@iebr.ac.vn
[6] Department of Ecology and Biological Resources, Graduate University of Technology and Science, Vietnam Academy of Science and Technology, Hanoi 100000, Vietnam
[7] Department of Biology, School of Education, Can Tho University, Can Tho 900000, Vietnam; Lhdang@ctu.edu.vn
[8] Climate Change and Development Program, Vietnam-Japan University, Vietnam National University, Hanoi 100000, Vietnam; htt.duyen@vju.ac.vn
[9] World Agroforestry (ICRAF), Bogor 16155, Indonesia; m.vannoordwijk@cgiar.org
* Correspondence: samhv@vnuf.edu.vn; Tel.: +84-983-33-78-98

check for **updates**

Citation: Mulia, R.; Hoang, S.V.; Dinh, V.M.; Duong, N.B.T.; Nguyen, A.D.; Lam, D.H.; Thi Hoang, D.T.; van Noordwijk, M. Earthworm Diversity, Forest Conversion and Agroforestry in Quang Nam Province, Vietnam. *Land* **2021**, *10*, 36. https://doi.org/10.3390/land10010036

Received: 30 October 2020
Accepted: 30 December 2020
Published: 4 January 2021

Publisher's Note: MDPI stays neutral with regard to jurisdictional claims in published maps and institutional affiliations.

Abstract: The conversion of natural forests to different land uses still occurs in various parts of Southeast Asia with poor records of impact on ecosystem services and biodiversity. We quantified such impacts on earthworm diversity in two communes of Quang Nam province, Vietnam. Both communes are situated within buffer zones of a nature reserve where remaining natural forests are under threat of continued conversion. We identified 25 different earthworm species, out of which 21 were found in natural forests, 15 in agroforestry, 14 in planted forests, and seven each in annual croplands and home gardens. Out of the six species that were omnipresent inhabitants of all observed habitats, *Pontoscolex corethrurus* largely dominated habitats with intensive anthropogenic activities but was rare in natural forests. Natural and regenerated forests had a much denser earthworm population in the top 10 cm of soil rather than in deeper soil layers. We conclude that the conversion of natural forests into different land uses has reduced earthworm diversity which can substantially affect soil health and ecosystem functions in the two communes. Protection of the remaining natural forests is urgent, while the promotion of a tree-based farming system such as agroforestry can reconcile earthworm conservation and local livelihoods.

Keywords: land-use change; belowground biodiversity; soil engineers; *Pontoscolex corethrurus*; natural habitats; planted forest

1. Introduction

The inclusion of 'planted forest' in forest statistics has been a major factor in the 'forest transition' concept: the reversal after long periods of decline towards a net gain in forest area [1,2]. Forest statistics can thus mask the ongoing loss of natural forest by larger gains in planted forest area, even though in terms of ecosystem services old-growth natural forests are very different from young planted forests [3]. Beyond the direct benefits that land-use change brings (or is expected to bring) to land users, a wide range of 'externalities' are generated, changes that do not play a role in the land users' decision making [4]. Where such externalities are negative for others in the landscape, corrective action is needed in the form of regulations or incentives; where they are positive, forms of incentives are appropriate [5].

Many of the ecosystem services that, as externalities of decision-making, are influenced by land use are controlled by soil conditions [6]. Beyond vegetation with its aboveground litter and plant root systems and turnover, microbes, fungi, and soil fauna are important for modifying the soil conditions that in turn shape soil structure (pores, aggregates) influencing water flows and storage and net greenhouse gas emissions that determine ecosystem services [7]. As part of the global loss of biodiversity, loss of belowground biodiversity and its impact on ecosystem services is still poorly understood [8,9]. Changes in the soil can be relatively slow to be noticed by soil biota, as the system is buffered in many aspects, but once undesirable changes take place, they are also hard to correct, especially if they reach soil layers below the surface [10], as thresholds are past where both roots and earthworms have difficulties penetrating [11]. Better knowledge of changes in belowground biodiversity in the context of 'forest transition' patterns of land-use change is urgent for many parts of the tropics, especially.

Over the past decades, Vietnam's forest cover has continued to expand from 28% in 1993 to 42% in 2018 [12]. Policy reforms and nation-wide forest protection and replanting programs have been the main drivers [13,14]. Forest expansion, including planted forests, has substantially contributed to economic growth, job creation, and poverty alleviation in the country [12]. However, despite the increasing trend, the conversion and degradation of natural forests still occur in different parts of the country. The existing natural forests in Vietnam are largely in poor condition or regenerating [12]. The forest cover rates overlook forest quality. Competing land uses, exploitation of forest resources, and weak forest governance are serious challenges for the country to protect the remaining and regenerating poor natural forests [12]. Massive conversion of natural forests still occurs for example in the Central Highlands and Central Annamites region of the country. The expansion of commercial crops and infrastructure development have been the main drivers of forest conversion in Central Highlands [15]. In the Central Annamites, considered as one of the largest continuous natural forest areas and key biodiversity landscapes in Asia, forest resources and wildlife are continuously under threat of intensive exploitation by surrounding forest-dependent populations [16,17].

Concerns over the impacts of forest conversions on ecosystem services and biodiversity have recently become more prominent in Vietnam [18,19]. Several sub-national and national policies to include the 2011–2020 National Biodiversity Strategy with vision to 2030 underline the need of taking up existing scientific- and evidence-based information and undertaking further research on such impacts while strengthening forest protection efforts. Several studies from other parts of the world have reported impacts of forest conversions on biodiversity including belowground biodiversity (e.g., [20,21]). Notwithstanding, accounts of such impacts in Vietnam are limited.

The diversity and activities of earthworms as one of the most important soil biotas influence the physical, chemical, and biological properties of soils [22,23]. Their physical movement creates soil pores which ease nutrient and water dynamic in the soils. As recyclers of organic materials, earthworms facilitate microorganisms such as fungi and bacteria to undertake further decomposing process [24]. Through these activities, earthworms influence soil health and the provision of ecosystem services [25]. The population and diversity of earthworm species vary across land habitats due to variation in soil moisture, soil temperature, soil properties, the abundance of surface litter, vegetation types, land use management, and human interventions [20,26–30]. Earthworms are thus sensitive to change in land use. In general, earthworm diversity is reduced in habitats with more intensive anthropogenic interventions [23,31].

Here, we describe the results of a study on earthworm diversity in natural forests and different land uses in two communes of Quang Nam province, South Central Coast region of Vietnam. The two communes are within the buffer zones of the Song Thanh Nature Reserve, one of the main reserves within the Central Annamites mountains, and parts of main watersheds in the region. In both communes, the conversion of natural forests into shifting cultivations and agricultural systems, such as agroforestry and annual croplands, has been taking place for decades. The purpose of the study was to compare the diversity of earthworm species among natural forest and different land uses to investigate the impact of forest conversion and to recommend necessary measures related to earthworm conservation to maintain soil health and ecosystem functions in the two buffer-zone communes.

2. Material and Methods

2.1. Study Sites

The study was conducted in Phuoc My commune of Phuoc Son district and Ta Bhing commune of Nam Giang district, Quang Nam province (15°35′0″ N–107°55′0″ E), South Central Coast of Vietnam (Figure 1a). The two communes are within the buffer zones of Song Thanh Nature Reserve and parts of main watersheds in South Central Coast region which supply water for several cities including Da Nang as the biggest city in Central Vietnam. For example, Phuoc My is part of Dak Mi watershed (Figure 1b), while Ta Bhing is part of Vu Gia-Thu Bon as a bigger watershed [32]. Phuoc My is situated at 223–446 m above sea level while Ta Bhing is approximately 100 m above sea level. Both communes are mountainous with flat areas concentrated at the feet of the mountains, along riverbanks.

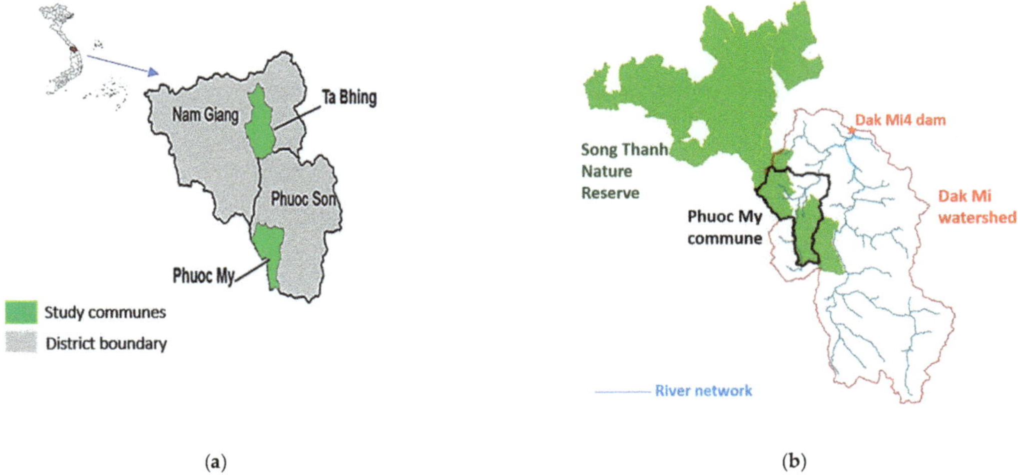

Figure 1. (**a**) Location of Phuoc My and Ta Bhing communes as study sites in Quang Nam province of Vietnam, (**b**) the boundary of Song Thanh Nature Reserve and Dak Mi watershed.

The two communes have a tropical monsoon climate. The annual rainfall is about 2650 mm in Phuoc My and 2300 mm in Ta Bhing [33]. In both communes, the rainy season usually peaks between September and November, and the dry season lasts from January until March. The highest temperature is usually recorded between May and July. The two communes have similar soil physical and chemical properties, dominated by sandy-loam and sandy-clay-loam textures [33].

Phuoc My has a smaller land size, but it has a larger area of rich natural forest than Ta Bhing in 2016. The total area of medium, poor, and regenerated natural forests accounts for 21% and 41.5% of the total land size in Phuoc My and Ta Bhing respectively. Land cover in the two communes has been classified into eleven types [16]. The area of each land cover in the two communes are described in Table A1.

2.2. Land-Use History

In both communes, massive conversion of natural forests took place more than 20 to 30 years ago [16]. Over the past two decades, natural forests have regenerated in some areas of the communes, thanks to enhanced protection efforts by the local authorities. Planted forests have become popular in the past decade, especially the short-rotation (3–4 year) of Acacia (mainly the hybrid *Acacia mangium x auriculiformis*) plantation for pulp and paper. Two types of agroforestry have developed in the two communes: a *taungya* or temporary agroforestry in which annual crops such as cassava or maize become intercrops in young planted forests, usually in the first 1–2 years after tree planting before tree canopy closure; and more permanent agroforestry in which timber tree species, such as *Melia azedarach*, *Vernicia montana*, or *Ficus racemosa*, or fruit trees such as *Dimocarpus longan*, were intercropped with annual crops such as peanut, beans, or cassava. In sloping lands, agroforestry practices with rows of trees and rows of pineapple to limit soil erosion also existed.

2.3. Sample Plots by Habitat

The earthworm's observations were carried out in seven land cover types (hereafter called habitats). Three represent habitats with less anthropogenic interventions namely natural forests, regenerated forests, and grasslands (hereafter referred to as 'natural habitats'); and four with more intensive anthropogenic interventions namely planted forests, agroforestry, mixed annual crops, and home gardens (hereafter referred to as 'human-disturbed habitats'). Among the four, the most frequent and intensive anthropogenic interventions occurred in mixed annual crops and home gardens.

In total, we took 28 and 53 sample plots from the different habitats in Phuoc My and Ta Bhing commune respectively. More sample plots were taken in natural and regenerated forests to anticipate a possibility of higher species diversity in those habitats. Although both communes have similarities in many of the factors that can influence earthworm diversity such as climate condition, soil properties, and landcover types, we took sample plots in both communes, but more samples plots in Ta Bhing than Phuoc My merely for time and resource efficiency. Most of the local assistants involved in the sampling activities in both communes were from Ta Bhing. The aim was to have replications from both communes, not to compare earthworm diversity between the two communes. The number of sample plot by habitat in the two communes are given in Table A2. Figure 2 illustrates the vegetation covers and types in the selected habitats.

Figure 2. *Cont.*

Figure 2. Vegetation cover and types in (**a**) natural forest, (**b**) regenerated forest, (**c**) agroforestry, (**d**) plantation of *Acacia mangium x auriculiformis*, (**e**) plantation of *Melia azedarach*, (**f**) plantation for *Machilus odoratissima* Nees, (**g**) home garden, (**h**) grasslands, and (**i**) upland annual crops.

2.4. Identification of Earthworm Species

Following [34], at each sample plot, earthworm specimens were collected from a 50 cm × 50 cm soil block of 30 cm depth (please see Figure 2e,h,i). A 50 cm × 50 cm supporting frame and a woody measure of 30 cm long were used to ensure identical surface areas and thicknesses of all soil blocks. The specimens were collected by carefully segregating the soils per 10 cm soil depth. All specimens were stored in cloth bags to keep them alive for further treatments in the laboratory.

To facilitate species identification, all earthworms were separated from soils by cleansing with water. They were subsequently immersed in liquid containing 4% formalin for 6–12 h. Thereafter, all earthworms were transferred to fresh liquid of 4% formalin for long-term storage and morphological studies. We investigated morphological features of collected earthworms, such as intestinal caeca (presence and shape), male pores (presence of copulatory pouches), genital markings (type, location, and number), spermathecal pores (location, number, and type), septa (thin, thick or absent), and additional characteristics such as prostomium (probilobous, epilobous or tanylobous), first dorsal pore (location), intestinal origin, last heart, ovaries, and testes. Earthworm samples were scrutinized using stereo microscope Motic DM143-FBGG-C (Motic Company, Hongkong) with drawing tube and camera attached to the monitor. All specimens were compared and identified using original papers as described in [35–43]. All specimens with different morphological characters compared to all known species were named using their genera and sp. 1, 2, and so on. For biomass measurement, we used paper tissues to remove all liquids (formalin, water) and cast from the preserved earthworms before weighting using the electronic weight Sartorius GM612.

2.5. Indicators for Earthworm Dominance and Diversity

We used seven indicators to assess the dominance, density, and diversity of earthworm species. Following [34], five indicators represent dominance and density, namely density and biomass per soil layer, quantity and biomass dominance, and occurrence frequency. Two indicators, the Shannon–Wiener index [44] and similarity index [45], measure the level of diversity of earthworm species within habitat and similarity in species composition among habitats, respectively. For the similarity index, we omitted the sub-species component in the calculation. The seven indicators are described in Table A3.

2.6. Aboveground Litter Biomass as A Covariate

In this study, we focused on the impact of conversion from natural to human-disturbed habitats on earthworm diversity. Notwithstanding, to consider the effect of aboveground plant litter biomass on earthworm density (e.g., [21,46,47]), we used the annual production rate of aboveground plant litter biomass (hereafter referred to as "annual litter production") (Table 1) as a covariate when comparing indicators among habitats. Due to a lack of data, we estimated the annual litter production from the annual production rate of aboveground plant biomass (hereafter referred to as "annual biomass production") of the habitats. We used a factor of 0.49 calculated from [48] for the biomass-litter conversion and assumed the same factor applies for all habitats. The biomass production data were obtained from [16] that estimated the standing aboveground plant biomass of all habitats in the two study communes, except NF and AF, under different ages using the Rapid Carbon Stock Appraisal approach [49]. The annual biomass production was calculated from the standing biomass and plot ages, assuming a constant biomass production across the year. We used an annual litter production of 9.3 ton ha^{-1} year^{-1} from [50] for NF and 2.4 $\pm$ 0.4 ton ha^{-1} year^{-1} calculated from [51] for AF. The earthworm and plant biomass assessment were not conducted using the same sample plots. Therefore, to create a variation of annual litter biomass among earthworm sample plots, we used random numbers from a uniform distribution to generate higher or lower litter biomass by a maximum of one standard error from the mean. We applied this for all habitats except for NF that lacks input data on the standard error of the mean.

Table 1. Annual production of aboveground plant litter biomass by habitat (source: calculated from [16]).

Habitat	Number of Sample Plots	Litter Production (ton ha^{-1} year^{-1})	
		Average	SE
Regenerated Forest (RF)	4	5.8	0.26
Grassland (GL)	18	0.7	0.05
Planted Forest (PF)	37	4.4	0.35
Upland Crops (UC)	11	2.7	0.45
Home Garden (HG)	8	3.4	0.87

2.7. Statistical Analysis

We used the IBM SPSS Statistics version 23 to test the difference in earthworm indicators among habitats using one-way analysis of covariance (ANCOVA). For the statistical test, we assumed an independency among sample plots. For all data and statistical analysis, we combined the earthworm data from the two study communes and the difference between the two communes was not investigated. All graphs were produced using the Microsoft Excel.

3. Results

3.1. Species Occupancy by Habitat

A total of 25 earthworm species from 3 families and 6 genera were found in the two study communes. Of these, 23 species belong to the *Megascolecidae* family and 17 species to the *Amynthas* genus. Only one genus and one species were found in each of the other two families, *Moniligastridae* and *Rhinodrilidae*. Seven out of the 25 species, namely *Pontoscolex corethrurus*, *Amynthas aspergillum*, *Amynthas divitopapillatus*, *Amynthas modiglianii*, *Amynthas* sp.1, *Metaphire houlleti*, and *Polypheretima taprobanae* were omnipresent. However, only *P. corethrurus* dominated the habitats, except in NF and RF. The average frequency of occurrence of this species in human-disturbed habitats was 87%. Based on the occurrence frequency, *P. corethrurus* is classified as rare species in NF and uncommon species in RF. In NF, only *Drawida beddardi* had occurrence frequency above 25%. The occurrence frequency of earthworm species by habitat is given in Table A4.

3.2. Earthworm Density and Biomass by Habitat

The NF tends to have a lower earthworm density compared to other habitats (Figure 3a). However, no significant difference (*p*-value > 0.05) was found among habitats due to large standard errors. In terms of biomass, the natural habitats had higher earthworm biomass compared to human-disturbed habitats (*p*-value < 0.001), likely because of the larger earthworm's physical size. The ratio between earthworm biomass and density was much lower in the human-disturbed habitats (*p*-value < 0.001) with an average of 0.73 $\pm$ 0.14 g per individual compared to 1.62 ± 0.23 g per individual in natural habitats. Related to quantity and biomass dominance, *P. corethrurus* was much more dominant in human-disturbed than natural habitats (*p*-value < 0.01) particularly due to extremely low dominance in NF (Figure 3b). The quantity and biomass dominance of all species by habitat is provided in Table A5.

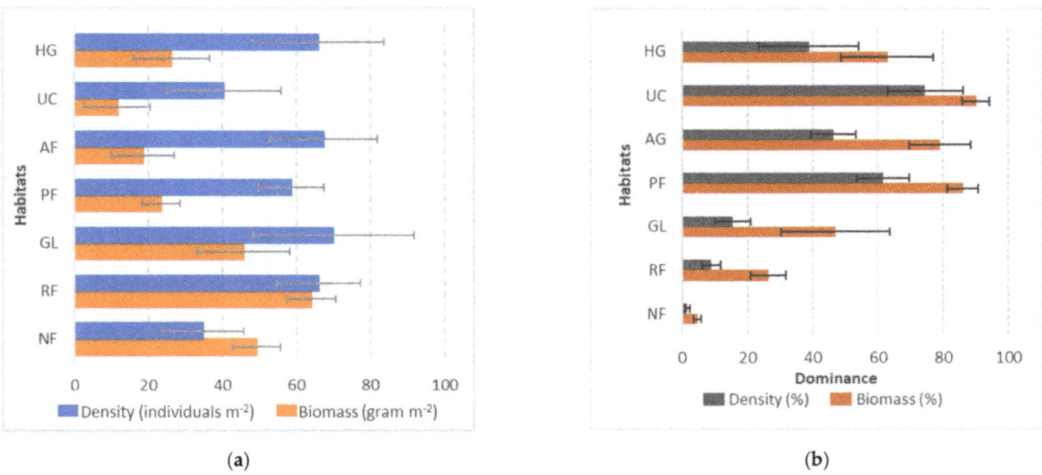

(a) (b)

Figure 3. (**a**) Earthworm density and biomass and (**b**) quantity and biomass dominance of *P. corethrurus*.

3.3. Earthworm Density and Biomass by Soil Depth

Both the natural and human-disturbed habitats generally had higher density of earthworm in the first 10 cm than deeper soil layers (*p*-value = 0.001) (Figure 4a). However, the decrease in the density by soil layer was more pronounced in natural habitats, especially in NF and RF, than human-disturbed habitats. In terms of biomass, both natural and human-disturbed habitats generally had comparable earthworm biomass in the first two soil layers, and lower biomass (*p*-value < 0.002) in the 20–30 cm soil depth (Figure 4b).

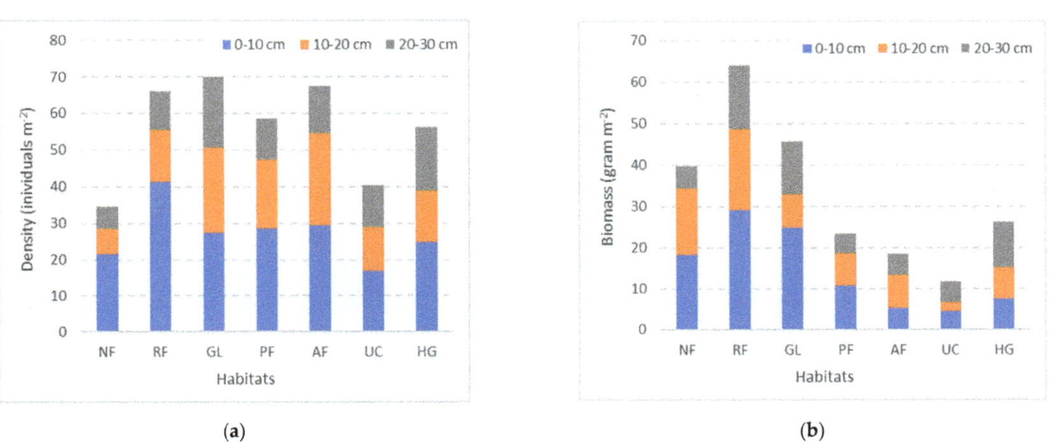

(a) (b)

Figure 4. Earthworm (**a**) density and (**b**) biomass by soil depth.

3.4. Species Diversity and Similarity Among Habitats

The natural habitats had a higher species diversity than the human-disturbed habitats (p-value < 0.001) (Figure 5a). The weak dominance of *P. corethrurus* especially in NF and RF was associated with higher species diversity. Among the human-disturbed habitats, the number of earthworm species found in UC and HG was equal, however, the Shannon-Wiener index of UC was much lower due to the strong dominance of few species.

The NF and RF had a similar earthworm species composition (Figure 5b). Among the four human-disturbed habitats, only UC and HG had a level of similarity. In general, the species composition in natural and human-disturbed habitats was different, except between RF and AF.

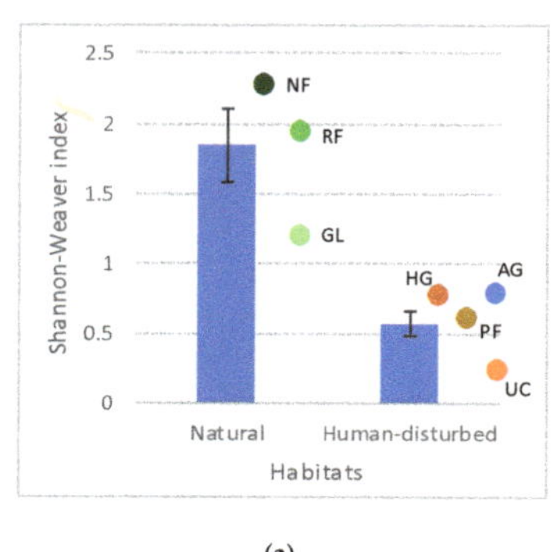

(a)

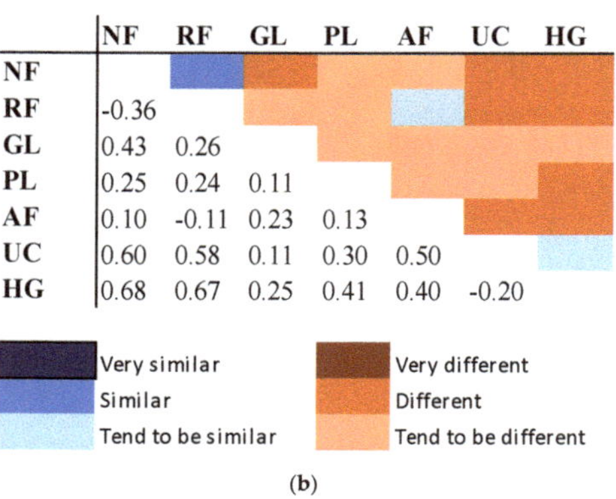

(b)

Figure 5. (**a**) Shannon-Wiener and (**b**) similarity index among habitats.

3.5. Variation among Types of Planted Forests

The Acacia plantation had more diverse earthworm species than Melia or Machilus plantation (Table 2). However, *P. corethrurus* was less dominant in Melia plantation because two other species, *Amynthas aspergillum,* and *Metapheretima tiencanhensis,* had an occurrence frequency above 25%. Also, earthworms in Melia plantation had a higher biomass and density ratio that reached 1.58 g per individual, compared to 0.3 and 0.5 g per individual in Acacia and Machilus plantations, respectively.

Table 2. Earthworm indicators comparing three types of planted forest.

Earthworm Indicators	Acacia	Melia	Machilus
No. of sample plots	15	5	4
No. of species	13	7	9
Frequency of occurrence of *P. corethrurus* (%)	97	72	100
No. of species with occurrence frequency $\geq 25\%$	1	3	2
Density (individuals m^{-2})	80	17	59
Biomass (gram m^{-2})	23	27	27
Shannon-Wiener index	0.48	0.95	0.5

4. Discussion

4.1. Earthworm Diversity in Agroforestry

In our study, among the human-disturbed habitats, higher earthworm diversity was found in tree-based farming systems like AF and PF. Other studies [52–57] reported similar evidence. Improved micro-climate, e.g., lower soil temperature and higher soil humidity, minimum plot management practices such as tillage, and a higher supply of organic matter from above- and below-ground litter all generate higher earthworm diversity in tree-based systems [58,59]. These factors also create contrast in earthworm diversity between areas nearby and far from the trees within AF, for example between tree row and crop alley in the alley-cropping system. In addition, minimal application of chemical inputs such as fertilizer and pesticide contribute to generating higher earthworm diversity within tree rows [58,60]. The variation in micro-climate and plot management practices also influence earthworm density. For example, in poplar AF in Canada, the average earthworm density within tree row was 182 individuals m^{-2}, compared to 117 and 95 individuals m^{-2} within the crop alley at two and six meters away from tree row, respectively [60]. In our study, some of the observed AFs were alley cropping, and others had scattered trees over the plots. In both cases, soil blocks for earthworm observation were taken relatively close to the trees, to avoid causing damage to annual crops. This likely explains the comparable earthworm diversity between AF and PF. AF can become a solution if larger areas of annual crops are needed without further converting NF or RF which otherwise will further decline the ecosystem health and functions in the two communes.

4.2. Potential Impact of P. corethrurus Dominance

The dominance of *P. corethrurus* in human-disturbed habitats has also been reported in other studies especially in tropical areas, e.g., see review in [61]. The species was exotic to Vietnam and initially inhabited hilly areas of the country [62,63]. Nowadays, it has a large distribution area including in the coastal areas of the country [64]. Compared to other exotic species which deliberately or inadvertently colonize new habitats, *P. corethrurus* has a high survival rate thanks to its tolerance to a range of biotic and abiotic environment [24,65]. Many other species either fail to survive or survive but not invasive [66].

P. corethrurus could bring favorable or unfavorable impacts to new habitats, although generally considered harmful to the environment and native species. As summarized in [61], as favorable impacts, the species could for example stimulate nutrient release in soil and enhance plant's resistance to phyto-parasitic nematodes. However, as unfavorable impacts, it can create soil compaction and alter biogeochemical processes to affect plants, native earthworms, and microbial communities. *P. corethrurus* can compact the soils through its feeding activity which accumulates small soil aggregates. In the absence of intervention from other soil biota, the small aggregates will progressively transform into larges aggregates and become compact [67,68]. In addition, layers of cast produced by this species on the soil surface, can turn into thick crust preventing water and air penetration to the soils [69]. The role of *P. corethrurus* in increasing soil bulk density and reducing soil porosity were reported in (e.g., [52,67,70,71]). Due to the unfavorable impacts, the presence and dominance of *P. corethrurus* have been considered as an indicator for soil health and level of disturbance to ecosystem and ecosystem services [72].

Our study shows the dominance of *P. corethrurus* in human-disturbed habitats namely PF, AF, UC, and HG which occupy about 17% and 20% of total land area in Phuoc My and Ta Bhing commune, respectively. These land uses potentially expand in the two communes driven by augmenting population and livelihood pressure. The two communes are therefore under a serious threat of further loss of earthworm diversity and stronger dominance of *P. corethrurus*, which can further reduce their ecosystem health and functions. Under the increasing dominance of the cosmopolitan species, the soil porosity and ecosystem functions in the two communes will mainly rely on tree roots. Therefore, tree-based farming systems such as AF are preferable to reconcile livelihood and provision of ecosystem services. Moreover, this land-use system generally has higher earthworm diversity than sole annual crop systems. Also, there is a need to select more suitable tree species for PF because of its popularity as land use in the two communes. Although Acacia is currently the most popular type of PF supported by local pulp and paper industries, Melia is worth to get further attention and research due to the lower dominance of *P. corethrurus* in the PF using this tree species. Based on the unpublished data as part of a study by [16], Melia in the communes likely had a superficial rooting system (Figure 6) which can create micro- and macropores in superficial soil layers through its coarse and fine roots. It has a root shallowness index of 0.83, higher than Acacia (0.69), and Machilus (0.63). The index measures the ratio of cross-sectional areas of horizontal and all proximal roots. Horizontal proximal roots are defined as those with an angle of less than 45 degrees relative to the soil surface. The number of replications in the root study was however limited, only involved three different trees, and worth for further investigation. Therefore, promoting this species in the communes can contribute to maintaining soil porosity with less dominance of *P. corethrurus*. Also, in terms of topographical, soil, and climate condition, Melia is suitable for the two study communes [16].

(a)

(b)

Figure 6. Distribution of proximal roots of *Melia azedarach* in (**a**) home garden and (**b**) forest plantation.

4.3. Caveats in Assessing Earthworm-Habitat Linkage

In this study, we could demonstrate that natural- and human-disturbed habitats have different levels of earthworm density and diversity. However, we could not do in-depth analysis on the effect of environmental factors to the differences among habitats due to lack of data such as on soil physical and chemical properties by sample plot. Several studies utilized the results of correlation analysis between earthworm and soil indicators for further purpose, for example considering earthworm density or diversity as proxy for levels of soil quality or pollution (e.g., [73–75]). Some of these studies also assessed the level of metal pollution in the soils and in the earthworm tissue or casts [74,76]. Further researches on earthworm in the two study communes are therefore necessary and should focus on assessing the impacts of soil and terrestrial conditions on the earthworm density and diversity. The terrestrial condition includes type and abundance of plant litter biomass. The researches can also associate the results of soil-earthworm analysis with the potential role of earthworm in the two study communes on watershed service.

4.4. Further Studies for the Unidentified Species

In our study, seven out of nine unidentified species belong to *Amynthas* genus. The seven species were mostly found in NF and RF with only *Amynthas* sp. 1 was an omnipresent inhabitant of all observed habitats. Also, *Amynthas* sp. 1 had a relatively high occurrence frequency in human-disturbed habitats such as AF and HG. Compared to the eleven identified species of *Amynthas* genus, three out of the eleven were omnipresent inhabitants, namely *Amynthas aspergillum*, *Amynthas divitopapillatus*, and *Amynthas modiglianii*, while the other eight were inhabitants of some habitats only, or one habitat such as the case for *Amynthas wui*. Therefore, earthworm species of *Amynthas* genus likely have different land colonization patterns or perhaps levels of resistance to the new habitat's condition.

The seven unidentified species have different external and internal morphological features compared to 115 earthworm species of *Amynthas* genus that have been identified from past studies across regions in Vietnam. To ascertain if all the seven species are new to the country, a comparison with four species of the genus *Amynthas* Kinberg, 1867 recently identified in the southeastern part of the country [77] is however necessary. Further studies can also compare morphological features of the unidentified species with those of *Amynthas* species identified in other countries. For example, earthworm species of the genus *Amynthas* Kinberg, 1867 of family Megascolecidae, the same family with *Amynthas* earthworms found in our study, had been known as common inhabitants of natural forests in the northern part of Laos [78]. Recently, three new earthworm species belonging to this genus were found in that part of the country [78]. Apart from morphological features, further studies should compare living habitat and land colonization pattern between the unidentified and known *Amynthas* species. Similar efforts are necessary for the unidentified species of other genus found in the current study, namely one each of *Metaphire* and *Polypheretima* genus of the Megascolecidae family.

5. Conclusions

The conversion of natural forests to different land uses in the two buffer-zone communes of Song Thanh Nature Reserve had reduced earthworm diversity which can substantially affect soil health and ecosystem functions in the two communes. Also, among the identified species, *P. corethrurus* was omnipresent and largely dominated human-disturbed habitats. The dominance of this species can bring unfavorable impacts such as soil compaction and continuous threat to native earthworm species. This will further affect soil porosity and related ecosystem functions and reduce the role of the two communes to the nature reserve and watershed of which the two communes are part of. The area of human-disturbed habitats in the two communes potentially expand driven by augmenting population and livelihood pressures. This will lead to further loss of earthworm diversity and stronger dominance of *P. corethrurus*. To avoid the further decline of earthworm diversity and rampant expansion of *P. corethrurus* in the two study communes, protection of the remaining natural and regenerated forests is urgent, and tree-based farming systems such as agroforestry should be promoted to reconcile earthworm conservation and local livelihoods. The results of this study enrich current limited knowledge on impacts of forest and land use conversion on earthworm density and diversity in Vietnam. Furthermore, the results can generate stronger concerns with respect to the degradation of belowground diversity in the country and call attention to the urgent need for strengthened forest protection efforts.

Author Contributions: Conceptualization, R.M. and M.v.N.; data curation, R.M., S.V.H., V.M.D., N.B.T.D., A.D.N., D.H.L., D.T.T.H., and M.v.N.; formal analysis, S.V.H., V.M.D., N.B.T.D., A.D.N., D.H.L., and D.T.T.H.; investigation, S.V.H., V.M.D., N.B.T.D., A.D.N., D.H.L., and D.T.T.H.; methodology, S.V.H., V.M.D., N.B.T.D., A.D.N., D.H.L., D.T.T.H., and M.v.N.; supervision, M.v.N.; writing—original draft, R.M., S.V.H., V.M.D., N.B.T.D., A.D.N., D.H.L., and D.T.T.H.; writing—review & editing, R.M., S.V.H., and M.v.N. All authors have read and agreed to the published version of the manuscript.

Funding: This research was funded by the United States Agency for International Development (USAID) to assess the livelihoods and ecological conditions of buffer-zone communes in Song Thanh Nature Reserve, Quang Nam province, and Phong Dien Nature Reserve, Thua Thien Hue province, grant number TF069018. The APC was funded by the same project.

Institutional Review Board Statement: Not applicable.

Informed Consent Statement: Not applicable.

Data Availability Statement: Data available in a publicly accessible repository.

Acknowledgments: We thank Le Tho Son, Nguyen Minh Quang, and Dao Van Chuong from the Vietnam National University of Forestry for their great contribution in the field inventory; Delia C. Catacutan, Nguyen Quang Tan, Do Trong Hoan, Nguyen Mai Phuong, Kurniatun Hairiah, and Pham

Thanh Van from the World Agroforestry (ICRAF) and Do Hoang Chung from Thai Nguyen University of Agriculture and Forestry, for their great help in field inventory and valuable inputs during data analysis and manuscript writing. We also thank valuable comments from three anonymous reviewers.

Conflicts of Interest: The authors declare no conflict of interest. The funders had no role in the design of the study; in the collection, analyses, or interpretation of data; in the writing of the manuscript, or in the decision to publish the results.

Appendix A

Table A1. Land cover types and areas in the study communes in 2016.

Land Cover Types	Phuoc My		Ta Bhing	
	Ha	**%**	**Ha**	**%**
Natural forest—rich	7788	61.4	5766	36.5
Logged over natural forest—medium	1447	11.4	5307	33.6
Logged over natural forest—poor	492	3.9	917	5.8
Regenerated natural forest	704	5.6	328	2.1
Planted forest	798	6.3	1687	10.7
Agroforestry	267	2.2	149	1.0
Home garden	28	0.2	56	0.4
Grasslands	641	5.1	774	4.9
Mixed annual crops	390	3.1	693	4.4
Paddy rice	63	0.5	52	0.3
Settlement, built up areas	59	0.5	89	0.6
Total (ha)	**12,677**		**15,818**	

Data source: [16].

Table A2. Selected habitats for earthworm observations and number of sample plots.

Habitat	Tree Cover (%)	No. of Sample Plots		Vegetation in Both Communes
		PM *	**TB ***	
Natural forest (NF)	>60	5	10	Fagaceae, Lauraceae, Meliaceae, Moraceae, Euphorbiaceae, Dipterocarpaceae, Sapindaceae
Regenerated forest (RF)	10–30	5	10	Bambosoideae such as *Bambusa natans*, *Dendrocalamus patellaris*, *Neohouzeauna dullooa*, and shrubs
Grassland (GL)	<10	1	3	Shrub and grass
Planted forest (PF)	Young: <30, mature: >30	9	15	In Phuoc My: Acacia (5 sample plots), *Machilus odoratissima Nees* (4); in Ta Bhing: Acacia (10), *Melia azedarach* (5). In both communes, Acacia variety is mostly the hybrid *Acacia mangium x auriculiformis*
Agroforestry (AF)	Young: <30, mature: >30	4	5	Acacia-based with cassava, banana, or herbal plants (4), Melia-based with banana or cassava (2), agroforestry with mixed tree species such as *Vernicia montana, Ficus racemose, Dimocarpus longan* and annual crops such as cassava, banana, vegetables (3)
Upland annual crops (UC)	<10	3	5	Key crops such as paddy rice, maize, and cassava
Home garden (HG)	>30	1	5	Mostly diverse vegetable, annual crop and fruit trees such as mango, jackfruit, and longan trees

* PM: Phuoc My commune, TB: Ta Bhing commune.

Table A3. Indicators for earthworm's dominance and diversity.

No	Indicators	Formula	Unit	Remark
1	Density per soil layer	$\dfrac{n_{ijk}}{N_{jk}} \times 4$	Individual m^{-2}	n_{ijk} = number of individual of species i in soil block k layer j N_{kj} = total number of individuals from all species in soil block k layer j
2	Biomass per soil layer	$\dfrac{b_{ijk}}{N_{jk}} \times 4$	g m^{-2}	b_{ijk} = total biomass of species i in soil block k layer j
3	Quantity dominance	$\dfrac{n_i}{N} \times 100\%$	%	n_i = number of individuals of species i, N = total number of individuals of all species in the habitat
4	Biomass dominance	$\dfrac{b_i}{B} \times 100\%$	%	b_i = total biomass of species i, B = total biomass of all species in the habitat
5	Occurrence frequency	$\dfrac{S_i}{S} \times 100\%$	%	s_i = number of sample plots having species i, S = total number of sample plots for the habitat. Range of values: >75% = very common species, >50–75% = common species, 25–50% = uncommon species, <25% = rare species
6	Shanon- Wiener index	$-\sum\left(\dfrac{n_i}{N} \times ln\left(\dfrac{n_i}{N}\right)\right)$	-	n_i = total of individuals of species i, N = total of individuals of all species in the habitat
7	Similarity index	$\dfrac{(2Rs + Rss)}{2+1}$ Where: $Rs = \dfrac{x+y-z}{x+y+z}$, and $Rss = \dfrac{x'+y'-z'}{x'+y'+z'}$	-	Rs = similarity level of species Rss = similarity of subspecies x (x'); y (y') = number of species (number of subspecies) found only in one habitat z(z') = number of species (number of subspecies) found in both habitats. Range of values: -1 to -0.7 = very similar, <-0.7 to -0.35 = similar, <-0.35 to 0 = tend to be similar, 0 to <0.35 = tend to be different, 0.35 to <0.7 = different, 0.7 to 1.0 = very different

Table A4. Occurrence frequency of earthworm species by habitat (codes as in Table A2).

Species	Occurrence Frequency by Habitat (%)							No. of Habitat *
	NF	RF	GL	PF	AF	UC	HG	
Family Rhinodrilidae *Pontoscolex corethrurus*	16	48	74	92	89	76	94	7
Family Moniligastridae *Drawida beddardi*	34	30	3.5		4.5			4
Family Megascolecidae *Amynthas alluxus*	7	3		3	2.5			4
Amynthas aspergillum	25	32	20	18	6	2	10	7
Amynthas cortices	9			2	2			3
Amynthas divitopapillatus	12	12	12	12	12	12	12	7
Amynthas exiguus austrinus	10			2				2
Amynthas exiguus chomontis		2	10	2	5			4
Amynthas falcipapillatus	4							1
Amynthas infantiloides	4	3						2
Amynthas modiglianii	13	34	30	6	8	7	10	7
Amynthas zoysiae			10		2			2
Amynthas wui		6						1
Amynthas sp.1	9	32	50	7	23	4	40	7
Amynthas sp.2	24	18		2	2.5			4
Amynthas sp.3	9	3		3	2			4
Amynthas sp.4	9	1						2
Amynthas sp.5	20	11		2				3

Table A4. *Cont.*

Species	Occurrence Frequency by Habitat (%)							No. of Habitat *
	NF	RF	GL	PF	AF	UC	HG	
Family Megascolecidae								
Amynthas sp.6	6	6						2
Amynthas sp.7	4	1						2
Metapheretima tiencanhensis	13	30	20	12	12			5
Metaphire houlleti	20	20	20	20	20	20	20	7
Metaphire sp.1				1				1
Polypheretima sp.1	4	14	30	4				4
Polypheretima taprobanae	20	32	17	8	9	2	14	7
Total number of species	21	20	12	14	15	7	7	

* Number of habitats where the species was found.

Table A5. Quantity (n′) and biomass (b′) dominance (%) of earthworm species by habitat and soil depth (NF: natural forest, RF: regenerated forest, GL: grasslands, PF: planted forest, AF: agroforestry, UC: upland annual crops, HG: home gardens).

Species	NF		RF		GL		PF		AF		UC		HG	
	n′	b′	n′	b′	n′	b′	n′	b′	n′	b′	n′	b′	n′	b′
Pontoscolex corethrurus														
0–10	1.7	1.2	35.7	13.6	48.6	13.8	91.1	51.8	90.1	61.8	85.4	70.6	88.4	65.8
10–20	6.7	2.2	30.7	7.3	47.4	19.0	85.6	76.6	86.8	43.4	95.6	92.6	44.3	10.5
20–30	3.2	0.5	12.9	6.0	45.4	13.7	82.0	56.2	60.7	33.9	89.6	60.8	56.5	40.5
Drawida beddardi														
0–10	2.3	0.9	11.7	3.7					0.4	1.1				
10–20	10.5	18.9	9.1	2.7	0.4	0.0								
20–30	4.7	1.4	9.1	1.1					2.2	0.4				
Amynthas alluxus														
0–10	0.4	0.0	1.2	0.3										
10–20			7.6	3.4			0.7	0.7						
20–30	6.3	2.3												
Amynthas aspergillum														
0–10	2.3	16.1	1.4	19.2	2.4	29.7	2.1	34.0					2.0	26.6
10–20	6.7	30.9	6.7	59.8	1.3	25.7	0.2	8.5	0.7	48.1			2.2	34.0
20–30	8.4	72.1	7.8	70.9	1.7	35.4	0.8	20.4	0.7	29.3	0.6	13.4	2.0	17.8
Amynthas corticis														
0–10	1.5	2.6												
10–20	1.4	1.6							0.3	1.1				
20–30	0.5	0.4												
Amynthas divitopapillatus														
0–10	0.8	1.3												
10–20	0.5	0.2												
20–30	0.5	0.2												
Amynthas exiguus austrinus														
0–10	10.5	3.5	0.5	0.1										
10–20	5.8	2.7			2.4	0.1	0.7	0.2	3.3	0.3				
20–30	18.8	1.5							18.2	2.8	0.6	0.1		
Amynthas exiguus chomontis														
0–10														
10–20					2.4	0.1			3.3	0.3				
20–30			1.2	0.0					18.2	2.8				

Table A5. *Cont.*

Species	NF		RF		GL		PF		AF		UC		HG	
	n'	b'	n'	b'	n'	b'	n'	b'	n'	b'	n'	b'	n'	b'
Amynthas falcipapillatus														
0–10	1.4	1.0												
10–20														
20–30														
Amynthas infantiloides														
0–10									0.8	5.8				
10–20	3.8	0.6	0.9	0.1					0.7	2.1				
20–30	9.4	0.7	1.4	0.0					1.1	4.1				
Amynthas modiglianii														
0–10	5.1	13.9	16.8	27.7	13.2	26.1	1.7	7.8	2.0	17.5	1.0	10.5	0.3	0.4
10–20	2.9	10.7	2.3	2.7										
20–30							5.1	10.8						
Amynthas zoysiae														
0–10					0.9	0.1								
10–20					4.8	0.8					1.2	1.0		
20–30									0.7	0.1				
Amynthas wui														
0–10			0.1	0.0										
10–20														
20–30														
Amynthas sp.1														
0–10	1.5	0.1	1.9	0.1	5.3	0.1	0.7	0.1	1.4	0.2			3.1	0.5
10–20	2.9	0.5	5.8	0.2	24.7	1.6	7.1	1.6	2.3	0.3	0.6	0.7	50.0	50.0
20–30	2.7	0.0	10.5	0.4	17.9	0.6	8.4	0.9	9.8	3.1			22.2	15.5
Amynthas sp.2														
0–10	10.5	0.8	2.7	0.2					0.4	0.1				
10–20	11.5	1.6	4.2	0.2										
20–30	3.1	0.1	1.7	0.1										
Amynthas sp.3														
0–10	1.3	0.1												
10–20	2.9	0.3	0.8	0.1										
20–30	3.7	0.1					1.0	4.2	4.5	4.5	2.1	0.8		
Amynthas sp.4														
0–10	2.4	2.3												
10–20	1.0	0.9												
20–30														
Amynthas sp.5														
0–10	19.7	0.9	1.0	0.1							9.1	1.2		
10–20	9.5	0.1	2.8	0.1			0.2	0.5			0.9	0.2		
20–30	13.3	0.4									0.6	13.4		
Amynthas sp.6														
0–10	1.4	0.2	0.5	0.0										
10–20	1.9	0.2												
20–30			0.6	0.0										
Amynthas sp.7														
0–10	0.6	0.6												
10–20	1.9	1.5	0.4	0.1										
20–30														

Table A5. *Cont.*

Species	NF		RF		GL		PF		AF		UC		HG	
	n′	b′	n′	b′	n′	b′	n′	b′	n′	b′	n′	b′	n′	b′
Metapheretima tiencanhensis														
0–10	0.3	1.5	4.2	6.8	6.1	2.2	0.9	1.9						
10–20			3.9	4.1			1.6	2.5	1.0	1.8				
20–30			1.2	3.0	16.7	21.5								
Metaphire houlleti														
0–10	7.4	10.4	0.3	0.4	3.2	4.5	0.5	1.2						
10–20	0.5	0.0					1.3	3.2						
20–30	1.1	0.2												
Metaphire sp1														
0–10							0.1	0.9						
10–20														
20–30														
Polypheretima sp.1														
0–10			0.5	1.0	10.5	14.7	0.3	1.4						
10–20	1.9	0.9	0.4	0.6	14.3	45.4								
20–30	3.1	0.5	1.4	0.3	16.7	28.0	1.0	3.8						
Polypheretima taprobanae														
0–10	3.9	8.3	4.8	6.8	9.1	8.7	0.5	0.3	0.6	1.8			3.1	3.3
10–20	4.3	1.2	4.5	4.0	4.4	5.4	2.3	3.4	0.7	1.7			1.8	2.9
20–30	1.1	0.6	3.0	3.6	1.7	0.8	0.2	0.7	1.1	12.5	0.6	0.8	1.2	1.4

References

1. Meyfroidt, P.; Lambin, E.F. Forest transition in Vietnam and displacement of deforestation abroad. *Proc. Natl. Acad. Sci. USA* **2009**, *106*, 16139–16144. [CrossRef] [PubMed]
2. Barbier, E.B.; Burgess, J.C.; Grainger, A. The forest transition: Towards a more comprehensive theoretical framework. *Land Use Policy* **2010**, *27*, 98–107. [CrossRef]
3. Meyfroidt, P.; Lambin, E.F. Global forest transition: Prospects for an end to deforestation. *Annu. Rev. Environ. Resour.* **2011**, *36*, 343–371. [CrossRef]
4. Tomich, T.P.; Thomas, D.E.; Van Noordwijk, M. Environmental services and land use change in Southeast Asia: From recognition to regulation or reward? In *Proceedings of the Agriculture, Ecosystems and Environment*; Elsevier: Amsterdam, The Netherlands, 2004; Volume 104, pp. 229–244.
5. Leimona, B.; van Noordwijk, M.; de Groot, R.; Leemans, R. Fairly efficient, efficiently fair: Lessons from designing and testing payment schemes for ecosystem services in Asia. *Ecosyst. Serv.* **2015**, *12*, 16–28. [CrossRef]
6. Bruijnzeel, L.A. Hydrological functions of tropical forests: Not seeing the soil for the trees? In *Proceedings of the Agriculture, Ecosystems and Environment*; Elsevier: Amsterdam, The Netherlands, 2004; Volume 104, pp. 185–228.
7. Wall, D.H.; Nielsen, U.N. Biodiversity and ecosystem services: Is it the same below ground? *Nat. Educ. Knowl.* **2012**, *3*, 8.
8. Delgado-Baquerizo, M.; Bardgett, R.D.; Vitousek, P.M.; Maestre, F.T.; Williams, M.A.; Eldridge, D.J.; Lambers, H.; Neuhauser, S.; Gallardo, A.; García-Velázquez, L.; et al. Changes in belowground biodiversity during ecosystem development. *Proc. Natl. Acad. Sci. USA* **2019**, *116*, 6891–6896. [CrossRef] [PubMed]
9. Wagg, C.; Bender, S.F.; Widmer, F.; Van Der Heijden, M.G.A. Soil biodiversity and soil community composition determine ecosystem multifunctionality. *Proc. Natl. Acad. Sci. USA* **2014**, *111*, 5266–5270. [CrossRef] [PubMed]
10. Greacen, E.; Sands, R. Compaction of forest soils. A review. *Aust. J. Soil Res.* **1980**, *18*, 163–189. [CrossRef]
11. Ruiz, S.; Schymanski, S.J.; Or, D. Mechanics and energetics of soil penetration by earthworms and plant roots: Higher rates cost more. *Vadose Zone J.* **2017**, *16*, vzj2017.01.0021. [CrossRef]
12. World Bank. *Country Forest Note Vietnam*; World Bank: Washington DC, USA, 2019.
13. To, X.P.; Tran, H.N. *Forest Land Allocation in the Context of Forestry Sector Restructuring: Opportunities for Forestry Development and Upland Livelihood Improvement*; Tropenbos International Vietnam: Hue City, Vietnam, 2014.
14. de Jong, W.; Do, D.S.; Trieu, V.H. *Forest Rehabilitation in Vietnam: Histories, Realities and Future*; Center for International Forestry Research (CIFOR): Jakarta, Indonesia, 2006.
15. Le, D.M.; Nguyen, M.H.; Tran, T. Deforestation in the Central Highlands: Implications for minority ethnic communities and a number of recommendations. In Proceedings of the 14th International Conference on Humanities and Social Sciences 2018 (IC-HUSO 2018), Khon Kaen, Thailand, 22–23 November 2018; pp. 491–507.

16. Catacutan, D.C.; Do, T.H.; Simelton, E.; Hanh, V.T.; Hoang, T.L.; Patton, I.; Hairiah, K.; Le, T.T.; van Noordwijk, M.; Nguyen, M.P.; et al. *Assessment of the Livelihoods and Ecological Conditions of Bufferzone Communes in Song Thanh National Reserve, Quang Nam Province, and Phong Dien Natural Reserve, Thua Thien Hue Province*; World Agroforestry (ICRAF): Hanoi, Vietnam, 2017.
17. Nguyen, H.T.; Yen, H.M. Song Thanh Nature Reserve. In *Evidence-Based Conservation. Lessons from the Lower Mekong*; Sunderland, T.C.H., Sayer, J., Hoang, M.H., Eds.; Center for International Forestry Research: Bogor, Indonesia, 2013; pp. 29–38.
18. FAO. *Vietnam Forestry Outlook Study*; FAO: Bangkok, Thailand, 2009.
19. MONRE. *Vietnam's Fifth National Report to the United Nations Convention on Biological Diversity (Reporting Period: 2009–2013)*; MONRE: Hanoi, Vietnam, 2014.
20. Bartz, M.L.C.; Brown, G.G.; da Rosa, M.G.; Filho, O.K.; James, S.W.; Decaëns, T.; Baretta, D. Earthworm richness in land-use systems in Santa Catarina, Brazil. *Appl. Soil Ecol.* **2014**, *83*, 59–70. [CrossRef]
21. Shakir, S.H.; Dindal, D.L. Density and biomass of earthworms in forest and herbaceous microecosystems in central New York, North America. *Soil Biol. Biochem.* **1997**, *29*, 275–285. [CrossRef]
22. Zhang, H.; Schrader, S. Earthworm effects on selected physical and chemical properties of soil aggregates. *Biol. Fertil. Soils* **1993**, *15*, 229–234. [CrossRef]
23. Blouin, M.; Hodson, M.E.; Delgado, E.A.; Baker, G.; Brussaard, L.; Butt, K.R.; Dai, J.; Dendooven, L.; Peres, G.; Tondoh, J.E.; et al. A review of earthworm impact on soil function and ecosystem services. *Eur. J. Soil Sci.* **2013**, *64*, 161–182. [CrossRef]
24. Lavelle, P.; Barois, I.; Cruz, I.; Fragoso, C.; Hernandez, A.; Pineda, A.; Rangel, P. Adaptive strategies of Pontoscolex corethrurus (Glossoscolecidae, Oligochaeta), a peregrine geophagous earthworm of the humid tropics. *Biol. Fertil. Soils* **1987**, *5*, 188–194. [CrossRef]
25. González, G.; Huang, C.Y.; Zou, X.; Rodríguez, C. Earthworm invasions in the tropics. In *Biological Invasions Belowground: Earthworms as Invasive Species*; Springer: Berlin/Heidelberg, Germany, 2006; pp. 47–56, ISBN 1402052820.
26. Bhadauria, T.; Ramakrishnan, P.S. Earthworm population dynamics and contribution to nutrient cycling during cropping and fallow phases of shifting agriculture (Jhum) in North-East India. *J. Appl. Ecol.* **1989**, *26*, 505. [CrossRef]
27. Nunes, D.H.; Pasini, A.; Benito, N.P.; Brown, G.G. Earthworm diversity in four land use systems in the region of Jaguapitã, Paraná State, Brazil. *Caribb. J. Sci.* **2006**, *42*, 331–338.
28. Singh, S.; Singh, J.; Vig, A.P. Effect of abiotic factors on the distribution of earthworms in different land use patterns. *J. Basic Appl. Zool.* **2016**, *74*, 41–50. [CrossRef]
29. Tao, Y.; Gu, W.; Chen, J.; Tao, J.; Xu, Y.J.; Zhang, H. The influence of land use practices on earthworm communities in saline agriculture soils of the west coast region of China's Bohai Bay. *Plant Soil Environ.* **2013**, *59*, 8–13. [CrossRef]
30. Rajkhowa, D.J.; Bhattacharyya, P.N.; Sarma, A.K.; Mahanta, K. Diversity and distribution of earthworms in different soil habitats of Assam, North-East India, an Indo-Burma biodiversity hotspot. *Proc. Natl. Acad. Sci. India Sect. B-Biol. Sci.* **2015**, *85*, 389–396. [CrossRef]
31. Dewi, W.S.; Senge, M. Earthworm diversity and ecosystem services under threat. *Rev. Agric. Sci.* **2015**, *3*, 25–35. [CrossRef]
32. ICEM. *Strategic Environmental Assessment of the Quang Nam Province Hydropower Plan for the Vu Gia-Thu Bon River Basin*; ICEM: Hanoi, Vietnam, 2008.
33. Mulia, R.; Khasanah, N.; Catacutan, D. Alternative forest plantation systems for Southcentral Coast of Vietnam: Projections of growth and production using the WaNuLCAS model. In *Towards Low-Emission Landscapes in Vietnam*; Mulia, R., Simelton, E., Eds.; World Agroforestry (ICRAF) Vietnam, World Agroforestry (ICRAF) Southeast Asia Regional Program: Hanoi, Vietnam, 2018; pp. 45–60.
34. Górny, M.; Grüm, L. *Methods in Soil Zoology*; Elsevier Science: Amsterdam, The Netherlands, 1993.
35. Gates, G.E. Burmese earthworms: An introduction to the systematics and biology of Megadrile Oligochaetes with special reference to Southeast Asia. *Trans. Am. Philos. Soc.* **1972**, *62*, 1. [CrossRef]
36. Sims, R.W.; Easton, E.G. A numerical revision of the earthworm genus Pheretima auct. (Megascolecidae: Oligochaeta) with the recognition of new genera and an appendix on the earthworms collected by the Royal Society North Borneo Expedition. *Biol. J. Linn. Soc.* **1972**, *4*, 169–268. [CrossRef]
37. Pham, T.H. The earthworm fauna of Quang Nam—Da Nang, Hanoi National University of Education. Ph.D. Thesis, Hanoi National University, Hanoi, Vietnam, 1995.
38. Nguyen, T.T.; Tran, B.T.T.; Nguyen, A.D. Earthworms of the "acaecate" Pheretima group in Vietnam (Oligochaeta: Megascolecidae), with description of a new species from the Mekong delta. *Zootaxa* **2014**, *3866*, 105–121. [CrossRef] [PubMed]
39. Thai, T. Earthworms of Vietnam (Systematic, Fauna, Distribution and Zoogeographic). Ph.D.Thesis, Lomonosov Moscow State University, Moscow, Russia, 1983.
40. Thai, T.B. New species of the genus Pheretima in Vietnam. *Zool. J.* **1984**, *63*, 1317–1327.
41. Thai, T.B. New species and subspecies of the genus Pheretima (Oligochaeta, Megascolecidae) from Vietnam. *Zool. J.* **1984**, *63*, 613–617.
42. Nguyen, T.T.; Nguyen, A.D.; Tran, B.T.T.; Blakemore, R.J. A comprehensive checklist of earthworm species and subspecies from Vietnam (Annelida: Clitellata: Oligochaeta: Almidae, Eudrilidae, Glossoscolecidae, Lumbricidae, Megascolecidae, Moniligastridae, Ocnerodrilidae, Octochaetidae). *Zootaxa* **2016**, *4140*, 1–92. [CrossRef]
43. Thai, T.B. Description of five new species of the acaecate earthworms of the genus Pheretima Kinberg in Vietnam and key to the species of acaecate Pheretima recorded from Indochinese area. *J. Biol.* **1996**, *18*, 1–6.

44. Shannon, C.E.; Weaver, W. *The Mathematical Theory of Communication*; University of Illinois Press: Urbana, IL, USA, 1949.
45. Stugren, B.; Radulescu, M. Metode matematice in zoogeografia regionala. *Studies Univ. Babes-Bolyai Biol.* **1961**, *18*, 8–24.
46. Gonzalez, G.; Zou, X. Plant and litter influences on earthworm abundance and community structure in a tropical wet forest. *Biotropica* **1999**, *31*, 486–493. [CrossRef]
47. Dechaine, J.; Ruan, H.; Sanchez-De Leon, Y.; Zou, X. Correlation between earthworms and plant litter decomposition in a tropical wet forest of Puerto Rico. *Pedobiologia (Jena)* **2005**, *49*, 601–607. [CrossRef]
48. Sangha, K.K.; Jalota, R.K.; Midmore, D.J. Litter production, decomposition and nutrient release in cleared and uncleared pasture systems of central Queensland, Australia. *J. Trop. Ecol.* **2006**, *22*, 177–189. [CrossRef]
49. Van Noordwijk, M.; Hairiah, K. Rapid carbon stock appraisal. In *Negotiation-Support Toolkit for Learning Landscapes*; van Noordwijk, M., Lusiana, B., Leimona, B., Dewi, S., Wulandari, D., Eds.; World Agroforestry (ICRAF), Southeast Asia Regional Program: Bogor, Indonesia, 2013.
50. Chakravarty, S.; Prakash, R.; Vineeta, N.A.; Pala, G.S. Litter production and decomposition in tropical forest. In *Handbook of Research on the Conservation and Restoration of Tropical Dry Forests*; Bhadouria, R., Ed.; IGI Global: Hershey, PA, USA, 2020; pp. 193–212.
51. Cardinael, R.; Umulisa, V.; Toudert, A.; Olivier, A.; Bockel, L.; Bernoux, M. Revisiting IPCC Tier 1 coefficients for soil organic and biomass carbon storage in agroforestry systems. *Environ. Res. Lett.* **2018**, *13*, 124020. [CrossRef]
52. Barros, E.; Curmi, P.; Hallaire, V.; Chauvel, A.; Lavelle, P. The role of macrofauna in the transformation and reversibility of soil structure of an oxisol in the process of forest to pasture conversion. *Geoderma* **2001**, *100*, 193–213. [CrossRef]
53. Hauser, S. Distribution and activity of earthworms and contribution to nutrient recycling in alley cropping. *Biol. Fertil. Soils* **1993**, *15*, 16–20. [CrossRef]
54. Hauser, S.; Asawalam, D.O.; Vanlauwe, B. Spatial and temporal gradients of earthworm casting activity in alley cropping systems. *Agrofor. Syst.* **1998**, *41*, 127–137. [CrossRef]
55. Marsden, C.; Martin-Chave, A.; Cortet, J.; Hedde, M.; Capowiez, Y. How agroforestry systems influence soil fauna and their functions-a review. *Plant Soil* **2020**, *453*, 29–44. [CrossRef]
56. Norgrove, L.; Csuzdi, C.; Forzi, F.; Canet, M.; Gounes, J. Shifts in soil faunal community structure in shaded cacao agroforests and consequences for ecosystem function in Central Africa. *Trop. Ecol.* **2009**, *50*, 71–78.
57. Sánchez-De León, Y.; De Melo, E.; Soto, G.; Johnson-Maynard, J.; Lugo-Pérez, J. Earthworm populations, microbial biomass and coffee production in different experimental agroforestry management systems in Costa Rica. *Caribb. J. Sci.* **2006**, *42*, 397–409.
58. Cardinael, R.; Hoeffner, K.; Chenu, C.; Chevallier, T.; Béral, C.; Dewisme, A.; Cluzeau, D. Spatial variation of earthworm communities and soil organic carbon in temperate agroforestry. *Biol. Fertil. Soils* **2019**, *55*, 171–183. [CrossRef]
59. Tian, G.; Olimah, J.A.; Adeoye, G.O.; Kang, B.T. Regeneration of earthworm populations in a degraded soil by natural and planted fallows under humid tropical conditions. *Soil Sci. Soc. Am. J.* **2000**, *64*, 222–228. [CrossRef]
60. Price, G.W.; Gordon, A.M. Spatial and temporal distribution of earthworms in a temperate intercropping system in southern Ontario, Canada. *Agrofor. Syst.* **1998**, *44*, 141–149. [CrossRef]
61. Taheri, S.; Pelosi, C.; Dupont, L. Harmful or useful? A case study of the exotic peregrine earthworm morphospecies Pontoscolex corethrurus. *Soil Biol. Biochem.* **2018**, *116*, 277–289. [CrossRef]
62. Nguyen, V.T. *The Earthworm Fauna of Binh Tri Thien Region*; Hanoi National University of Education: Hanoi, Vietnam, 1994.
63. Thai, T.B. Species diversity of earthworms in Vietnam. In *Proceedings of the National Workshop on the Basic Issues in Life Science*; Hanoi Science and Technics Publishing House: Hanoi, Vietnam, 2000; pp. 307–311.
64. Thai, T.B.; Huynh, T.K.H.; Nguyen, D.A. Remarks of earthworms on the islands in southern of Vietnam. In *Proceedings of the National Workshop on the Basic Issues in Life Science*; Hanoi Science and Technics Publishing House: Hanoi, Vietnam, 2004; pp. 757–760.
65. Fragoso, C.; Kanyonyo, J.; Moreno, A.; Senapati, B.K.; Blanchart, E.; Rodriguez, C. A survey of tropical earthworms: Taxonomy, biogeography and environmental plasticity. In *Earthworm Management in Tropical Agroecosystems*; Lavelle, P., Brussaard, L., Hendrix, P., Eds.; CAB International: Wallingford, UK, 1999; pp. 27–55.
66. Williamson, M.; Fitter, A. The varying success of invaders. *Ecology* **1996**, *77*, 1661–1666. [CrossRef]
67. Alegre, J.C.; Pashanasi, B.; Lavelle, P. Dynamics of soil physical properties in Amazonian agroecosystems inoculated with earthworms. *Soil Sci. Soc. Am. J.* **1996**, *60*, 1522–1529. [CrossRef]
68. Blanchart, E.; Lavelle, P.; Braudeau, E.; Le Bissonnais, Y.; Valentin, C. Regulation of soil structure by geophagous earthworm activities in humid savannas of Cote d'Ivoire. *Soil Biol. Biochem.* **1997**, *29*, 431–439. [CrossRef]
69. Chauvel, A.; Grimaldi, M.; Barros, E.; Blanchart, E.; Desjardins, T.; Sarrazin, M.; Lavelle, P. Pasture damage by an Amazonian earthworm. *Nature* **1999**, *398*, 32–33. [CrossRef]
70. Hallaire, V.; Curmi, P.; Duboisset, A.; Lavelle, P.; Pashanasi, B. Soil structure changes induced by the tropical earthworm Pontoscolex corethrurus and organic inputs in a Peruvian ultisol. *Eur. J. Soil Biol.* **2000**, *36*, 35–44. [CrossRef]
71. Sparovek, G.; Lambais, M.R.; Silva, Á.P.d.; Tormena, C.A. Earthworm (*Pontoscolex corethrurus*) and organic matter effects on the reclamation of an eroded oxisol. *Pedobiologia (Jena)* **1999**, *43*, 698–704.
72. Brown, G.G.; James, S.W.; Pasini, A.; Nunes, D.H.; Benito, N.P.; Martins, P.T.; Sautter, K. Exotic, peregrine, and invasive earthworms in Brazil: Diversity, distribution, and effects on soils and plants. *Caribb. J. Sci.* **2006**, *42*, 339–358.

73. Bamgbose, O.; Odukoya, O.; Arowolo, T. Earthworms as bio-indicators of metal pollution in dump sites of Abeokuta City, Nigeria. *Rev. Biol. Trop* **2000**, *48*, 229–234.
74. Suthar, S.; Singh, S.; Dhawan, S. Earthworms as bioindicator of metals (Zn, Fe, Mn, Cu, Pb and Cd) in soils: Is metal bioaccumulation affected by their ecological category? *Ecol. Eng.* **2008**, *32*, 99–107. [CrossRef]
75. Fründ, H.-C.; Graefe, U.; Tischer, S. Earthworms as bioindicators of soil quality. In *Biology of Earthworms*; Karaca, A., Ed.; Springer: Berli/Heidelberg, Germany, 2011; pp. 261–278.
76. Hirano, T.; Tamae, K. Earthworms and soil pollutants. *Sensors* **2011**, *11*, 11157–11167. [CrossRef] [PubMed]
77. Nguyen, T.T.; Tran, T.T.B.; Lam, H.D.; Nguyen, A.D. Four new species of Amynthas earthworms in southeastern Vietnam (Annelida, Oligochaeta, Megascolecidae). *Zootaxa* **2020**, *4790*, 277–290. [CrossRef] [PubMed]
78. Hong, Y. New earthworm species of Amynthas (Clitellata: Megascolecidae) from Nam Phouin National Protected Area, Laos. *J. Asia-Pac. Biodivers.* **2019**, *12*, 353–356. [CrossRef]

 *land*

Article

Assessing Context-Specific Factors to Increase Tree Survival for Scaling Ecosystem Restoration Efforts in East Africa

Christine Magaju [1,*], Leigh Ann Winowiecki [1], Mary Crossland [2], Aymen Frija [3], Hassen Ouerghemmi [3], Niguse Hagazi [4], Phosiso Sola [1], Ibrahim Ochenje [1], Esther Kiura [1], Anne Kuria [1], Jonathan Muriuki [1], Sammy Carsan [1], Kiros Hadgu [4], Enrico Bonaiuti [5] and Fergus Sinclair [1,2]

[1] World Agroforestry (ICRAF), UN Avenue, P.O. Box 30677, Nairobi 00100, Kenya; l.a.winowiecki@cgiar.org (L.A.W.); p.sola@cgiar.org (P.S.); ibraochenje@gmail.com (I.O.); muthoni.kiura@gmail.com (E.K.); a.kuria@cgiar.org (A.K.); j.muriuki@cgiar.org (J.M.); s.carsan@cgiar.org (S.C.); f.sinclair@cgiar.org (F.S.)

[2] School of Natural Sciences, Bangor University, Bangor, Gwynedd LL57 2DG, UK; afp43d@bangor.ac.uk

[3] International Center for Agricultural Research in the Dry Areas, ICARDA, Rue Hedi Karray, CP 2049 Ariana, Tunisia; a.frija@cgiar.org (A.F.); ouerghemmihassen@gmail.com (H.O.)

[4] World Agroforestry (ICRAF), Gurd Shola, P.O. Box 5689 Addis Ababa, Ethiopia; n.hagazi@cgiar.org (N.H.); k.hadgu@cgiar.org (K.H.)

[5] International Center for Agricultural Research in the Dry Areas, ICARDA, 6/106, Osiyo St, Tashkent 100084, Uzbekistan; E.Bonaiuti@cgiar.org

* Correspondence: c.magaju@cgiar.org

Received: 30 October 2020; Accepted: 27 November 2020; Published: 4 December 2020

Abstract: Increasing tree cover in agricultural lands can contribute to achieving global and national restoration goals, more so in the drylands where trees play a key role in enhancing both ecosystem and livelihood resilience of the communities that depend on them. Despite this, drylands are characterized by low tree survival especially for tree species preferred by local communities. We conducted a study in arid and semi-arid areas of Kenya and Ethiopia with 1773 households to assess how different tree planting and management practices influence seedling survival. Using on-farm planned comparisons, farmers experimented and compared tree survival under different planting and management practices as well as under varying socioeconomic and biophysical contexts in the two countries. Seedling survival was monitored at least six months after planting. Results show that watering, manure application, seedling protection by fencing and planting in a small hole (30 cm diameter and 45 cm depth) had a significant effect on tree seedling survival in Kenya, while in Ethiopia, mulching, watering and planting niche were significant to tree survival. Household socioeconomics and farms' biophysical characteristics such as farm size, education level of the household head, land tenure, age of the household head had significant effects on seedling survival in both Ethiopia and Kenya while presence of soil erosion on the farm had a significant effect in Kenya. Soil quality ranking was positively correlated with tree survival in Ethiopia, regardless of species assessed. Current findings have confirmed effects of context specific variables some involving intrahousehold socioeconomic status such education level of the household head, and farm size that influence survival.

Keywords: trees on farm; options by context; on-farm planned comparison; tree seedling survival

---

1. Introduction

Restoration and avoiding further degradation through tree (re)establishment can be a key pathway towards achieving the UN's Sustainable Development Goals if successful restoration efforts can reach larger numbers of farmers and hectares over the coming decade. In fact, several global initiatives and commitments have come up in recognition of the need to intensify restoration efforts. This includes the UN decade on ecosystem restoration whose goal is to prevent, halt and reverse degradation of ecosystems by acting as an accelerator for ongoing restoration efforts[1]; the Bonn Challenge whose goal is to restore 150 million hectares of degraded and deforested landscapes by 2020 and 350 million hectares by 2030[2]; and the African Forest Landscape Restoration Initiative (AFR100) in which countries have committed to restoring 100 million hectares of degraded land in Africa by 2030[3]. Ethiopia and Kenya have each committed to restore 15 and 5.1 million hectares, respectively, as a contribution to AFR100 and the Bonn challenge. In addition, Kenya, through its national strategy, aims at achieving and maintaining over 10% tree cover by 2022 [1].

Achieving the restoration goals outlined in international and national commitments will require promotion of context-specific land restoration options across diverse social, economic, and biophysical realities [2–4]. These options will also need to be scaled up and out across the scaling domains [5]. For the most part, farmers targeted by restoration efforts work in complex, heterogenous and dynamic systems and, as such, no one single restoration option can suit all [3,4,6,7]. Many factors, both socioeconomic and biophysical, can affect the suitability and performance of the restoration options at different scales [8]. These include household characteristics such as farm size, age of the household head as well as education level and labor availability, farming practices, land degradation status, the policy environment [5,9] as well as farmer values and preferences [5,6,10,11].

Therefore, to understand which options are suitable, we must be cognizant of the variation within and between farms, landscapes and communities for all the context variables that might be important [3]. We must also take into account the goals, values and preferences of the people living on the land [6,10,11]. It is also imperative to consider the impact of restoration options such as tree planting on gender dynamics within a household if we are to ensure women, men and the youth benefit from restoration [10,12–14]. This includes understanding the roles and responsibilities of men and women in managing natural resources, decision-making within restoration options for example, who decides what restoration options to take part in and where on the farm these options are implemented.

Tree (re)establishment on agricultural land through natural regeneration or direct planting is often considered a key approach to restoration in the drylands [15–17]. This is largely driven by the recognition of the vital role that trees play in enhancing ecosystem and livelihood resilience [18–20]. Trees, for example, provide goods and services such as food and fuel, enhancing soil health, enhancing biodiversity, opportunities for generating income and contributing to climate change mitigation and adaptation [21–23]. However, low survival rates of planted trees, especially species highly valued by farmers, is a major limitation in the drylands [19,20]. This is partly due to unreliable rainfall, high levels of land degradation resulting in low soil productivity, planting of ecologically unsuitable tree species, and poor tree seedling management practices [18,19,24].

Understanding what determines tree seedling survival is fundamental to successful tree planting initiatives. For example, which trees species are suitable for which ecological context, which tree species farmers prefer, and what management practices increase tree survival in the different agroecological and socioeconomic conditions. In this paper, we evaluate how tree seedling planting and management practices influence tree survival across various agroecological conditions and the farmer circumstances in Ethiopia and Kenya, as well as the effect of socioeconomic and farm characteristics on tree survival.

---

[1] https://www.decadeonrestoration.org/
[2] https://www.bonnchallenge.org/
[3] https://afr100.org/

2. Materials and Methods

The study was conducted within the context of a donor-funded project, 'Restoration of degraded land for food security and poverty reduction in East Africa and the Sahel: taking success in land restoration to scale'[4], (henceforth referred to as 'Restoration project'). The project's goal was to reduce food insecurity and improve livelihoods of poor people living in African drylands by restoring degraded land, and returning it to effective and sustainable tree, crop and livestock production, thereby increasing land profitability and landscape and livelihood resilience. To achieve this, the project employed a Research in Development (RinD) approach in which research activities are embedded within development activities [3]. Thus, the activities of the Restoration project were co-located with those of the Drylands Development Program[5], (henceforth referred to as 'DryDev'), an international development initiative.

A key aspect of RinD that was implemented by the restoration project, is the use of planned comparisons, an approach where farmers compare and test promising options, and variations thereof, in their fields across a varying range of ecological and socioeconomic conditions [3–5,24,25]. Ref. [26] defines planned comparisons as the systematic and deliberate comparison of options where options refer to what is being done differently to address a particular challenge [25]. The approach involves engagement of local communities including farmers, researchers, government extension agents, and development actors in identifying current challenges facing farmers, selecting and prioritizing the initial set of promising options to be compared, and in monitoring the performance of the options being compared. Involvement of local communities has an impact on the outcome on whether the options succeed or fail and, in most cases, local communities are responsible for long-term management of these options [27,28].

2.1. Site Description

The study was conducted across four woredas in Tigray and Oromia regions in Ethiopia and across six subcounties in Kitui, Machakos and Makueni counties in Kenya. Specifically, the study took place in Boset and Gursum woredas in the Oromia region of Ethiopia, Samre and Tsaeda Emba woredas in the Tigray region of Ethiopia. In Kenya, the study covered Kitui Rural and Mwingi East subcounties in Kitui, Mwala and Yatta subcounties in Machakos, and Mbooni East and Kibwezi East subcounties in Makueni counties in the Eastern drylands of Kenya (Figure 1).

All the sites in Ethiopia are classified as semi-arid with annual rainfall ranging from 400 mm to 800 mm which varies from year to year [29] and are located between 779 m and 1362 m above sea level. The sites are characterized by low vegetation cover, low soil fertility, high rates of soil erosion, and recurrent droughts [29,30]. The study area in Kenya is largely arid and semi-arid and is characterized by highly erratic and unreliable rainfall. Annual average rainfall varies across and within the three counties. For example, Machakos receives an average of 500 mm to 1250 mm per annum while Makueni receives 250–400 mm in the lower region and 800–900 mm in the higher regions. Annual average temperature also varies across the three counties. In Kitui, annual temperatures ranges from 14–32 °C, in Machakos from 14–32 °C and in Makueni from 20.2–24 °C [31–33].

Agriculture is the dominant land use in all the sites and is characterized by low input subsistence farming. Main crops grown across the sites in Ethiopia include *Eragrostis tef*, *Triticum*, *Zea mays* and *Sorghum bicolor*, while *Zea mays*, *Sorghum bicolor*, *Pennisetum glaucum*, and pulses such as *Phaseolus vulgaris* and *Vigna unguiculata* are commonly grown across the sites in Kenya. The majority of households also own livestock.

---

4   http://www.worldagroforestry.org/project/restoration-degraded-land-food-security-and-poverty-reduction-east-africa-and-sahel-taking

5   http://www.worldagroforestry.org/project/drylands-development-programe-drydev

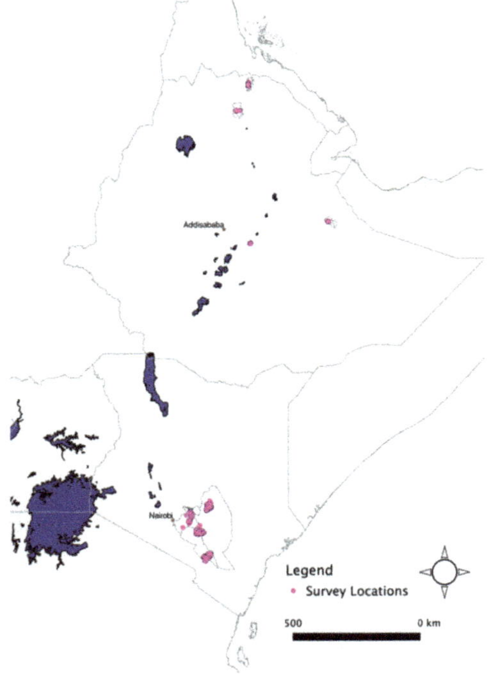

Figure 1. Map of the study locations in Ethiopia and Kenya.

2.2. Tree Planting Planned Comparison

The planned comparison on tree planting and management options was set up in response to a need by local communities in the study area to increase tree seedling survival. They identified tree planting and management as a learning priority during DryDev's community visioning and planning process [34]. Specifically, farmers identified low survival rates of planted trees as a key constraint to increasing tree cover and expressed interest in learning about planting and management methods that could increase establishment rates. Thus, the objective of the planned comparison on tree planting was to understand which planting and management practices can increase tree seedling survival rates for farmers across the study sites. More so, the practices that conferred the best chance of survival for the planted tree seedlings.

One thousand seven hundred and seventy-three households in both Kenya and Ethiopia volunteered to take part in the planned comparison during subsequent stakeholder and community engagement meetings. They compared the effect of different planting and management practices on seedling survival. Options compared included: tree species, planting hole sizes, planting with manure or without, physical protection of seedlings from livestock through spot fencing, and watering (Table 1). Water availability is a limiting factor for seedling survival in the study sites, thus the options compared were prioritized to increase the amount of water available to the seedlings after establishment. For example, digging a bigger planting hole to increase water capture thus increasing the rate of infiltration. This ensures that seedlings have sufficient water for initial growth and establishment [35]. Furthermore, browsing by livestock is one of the leading causes of seedling mortality.

Table 1. Options and contexts compared in the tree planting on-farm planned comparisons.

Options Compared	
Tree species	Categorical variable
Manure	Categorical variable with two levels, yes or no
Mulching	Categorical variable with two levels, yes or no (Mulch: a mix of grass, maize stalk and leaf litter is spread around the base of the seedling)
Hole size (diameter and depth)	Categorical variable with three levels (i) small hole (30 cm by 45 cm); (ii) big hole (60 cm by 45 cm); (iii) farmer usual hole size (diameter is not equal to 30 cm or 60 cm)
Watering	Categorical variable with two levels, yes or no
Watering frequency (Kenya)	Categorical variable with five levels (i) daily; (ii) every other day; (iii) weekly; (iv) bi-weekly; (v) monthly
Watering regime (Ethiopia)	Categorical variable with four levels (i) five litres every five days; (ii) five litres every ten days; (iii) three litres every five days; (iv) three litres every ten days
Fencing	Categorical variable with two levels, yes or no (physical protection of seedlings from livestock)
Pruning	Categorical variable with two levels, yes or no
Shading	Categorical variable with two levels, yes or no
Planting niche	Categorical variable with six levels, (i) external boundary; (ii) internal boundary; (iii) scattered in cropland; (iv) woodlot; (v) home compound; (vi) along terraces. (Planting niche: where on the farm the seedlings are planted)
Contexts Compared	
Age of household head	Continuous variable
Gender of household head	Categorical variable with two levels: (i) male; (ii) female
Education level of the household head	Categorical variable with four levels (i) no formal schooling; (ii) primary; (iii) secondary; (iv) tertiary (college/university/vocational)
Farm size (ha)	Continuous variable
Land tenure	Categorical variable with six levels: (i) private through purchase; (ii) private through customary inheritance; (iii) communal land; (iv) government; (v) settlement scheme; (vi) leasing
Erosion status	Categorical variable with two levels, yes or no. (Erosion status: Farmer experience with problems of soil erosion on the farm)
Soil quality ranking	Categorical variable with three levels (i) high (Good yields can be obtained without adding either chemical fertilizer or farmyard manure/compost); (ii) medium (Yields can be maximized with chemical fertilizer or farmyard manure/compost but fair yields can be obtained without); (iii) low (very little can grow without significant addition of chemical fertilizer or farmyard manure/compost). (Soil quality ranking: Farmer description of soil quality on the farm)

Farmers were trained on the various management options and on setting up the on-farm planned comparisons during several opportunities, during farmer workshops. This included practical training on how to set up the options being compared. Farmers had the choice of which option to implement and compare on their farms on condition of likeness. The number of options tested was subject to farmer interest and available resources for example if the farmer had access to manure or not. Farmers also had the choice of where on their farm to plant the seedlings (i.e., planting niche). Seedlings could be planted across all preferred niches provided variation in the farm characteristics was considered to ensure seedlings of a similar species were as homogenous as possible.

Implementing partners across the study sites distributed seedlings to farmers subsequent to training events. Tree seedlings were sourced from different nurseries across the study sites due to the large number required and the nature of the nursery enterprises in the study area. Seedlings were delivered potted in 10 cm by 15 cm potting bags and were on average, of good quality with variation depending on the nursery. The size of the seedlings varied depending on the nursery from which they were sourced. Farmers planted the seedlings within a week of receiving them.

In Kenya and Ethiopia, tree species selection was conducted through a consultative stakeholder engagement process. Twenty tree species were monitored in Ethiopia: *Acacia saligna, Azadirachta indica, Carica papaya, Casimiroa edulis, Citrus sinensis, Coffea arabica, Cordia africana, Faidherbia albida, Grevillea robusta, Jacaranda mimosifolia, Malus domestica, Mangifera indica, Melia volkensii, Moringa oleifera, Olea africana, Persea americana, Psidium guajava, Rhamnus prinodes and Vachellia seyal*. In Kenya, seven tree species were monitored: *Calliandra calothyrsus, Melia volkensii, Senna siamea, Mangifera indica, Carica papaya, Azadirachta indica* and *Moringa oleifera*. Farmers planted at least seven seedlings of each the selected species, applying the various management practices. Some species were planted exclusively in one woreda or county. For example, *Jacaranda mimosifolia* was only planted in Gursum Woreda while *Carica papaya* and *Melia volkensii* were exclusively planted in Boset Woreda, and in Kenya, *Calliandra calothyrsus* was only planted in Mwala, Machakos County.

2.3. Data Collection and Analysis

Data on tree seedling survival was collected from all households that received and planted tree seedlings through the project. Only tree seedlings planted as part of the on-farm planned comparisons were assessed. We assessed seedling survival under different planting and management practices at least six months after the seedlings were planted as it is widely accepted that the first six months after planting are most critical for survival as seedlings acclimatize and establish. This was done using a structured survey questionnaire administered through the Open Data Kit (ODK)[6] installed on mobile phones). Seedling survival was recorded as a categorical variable with two levels: yes or no represented by dummy variables 1 and 0. To understand men's and women's involvement in tree planting and management thereof, we collected information on the roles and responsibilities of men and women in decision-making within tree planting interventions.

We used a household survey to collect data on all farmers involved in the project. The survey included collection of basic socioeconomic and biophysical characteristics of the household and farm. Household demographic information collected included gender and age of household head, education level of the household head and household size. Farm characteristics such as farm size and ownership; soil erosion and control measures, trees on farm including the count, management and utility derived from the trees; crop and livestock production, climate change, including farmers' understanding, experience and response to climate change; and food security were also collected. We also conducted a tree inventory of the existing tree species within the household including data on species diversity within farms.

Statistical analysis on survival of planted trees under different planting and management options was conducted using the R software environment [36] including logistic regression to ascertain the probability of tree seedling survival under different management practices, and household socioeconomic and farm characteristics. Variation in the type and magnitude of the effect across the agroecological conditions represented in the study sites was also examined using a mixed-effect linear model. Since seedling survival varied highly across the tree species in Kenya, the effect of tree species was included as a constant in the model for Kenya. As not all practices were employed across the two countries, only those that were universally employed were compared across both countries in the results. For example, manure and variation in planting hole size was only assessed in study sites in Kenya while watering regime, disease, weed and pest control were only assessed in study sites in Ethiopia. Table 2 shows the number of observations in each option.

---

[6]   https://opendatakit.org/

Table 2. Number of tree seedlings under each option in Kenya and Ethiopia. Where available, a common name for each species is added in the brackets.

Country	Tree Species	Fencing	Manure	Mulch	Watering	Shade	Fertilizer
Ethiopia	*Acacia saligna*	14	-	4	12	-	-
	Azadiracta indica (Neem)	75	-	62	91	-	-
	Carica papaya (Papaya)	1	-			-	-
	Casimiroa edulis (White sapote)	5	-	13	19	-	-
	Citrus sinensis (Sweet orange)	172	-	102	155	-	-
	Coffea arabica (Arabica coffee)	36	-	35	55	-	-
	Cordia africana	1	-		1	-	-
	Faidherbia albida (Apple- ring acacia)	114	-	128	212	-	-
	Grevillea robusta (Silver oak)	450	-	222	346	-	-
	Jacaranda mimosifolia	6	-	12	11	-	-
	Malus domestica (Apple)	1	-		1	-	-
	Mangifera indica (Mango)	571	-	375	541	-	-
	Moringa oleifera (Moringa)	154	-	89	129	-	-
	Olea africana	2	-			-	-
	Persea americana (Avocado)	105	-	42	97	-	-
	Psidium guajava (Guava)	118	-	100	124	-	-
	Rhamnus prinoides	111	-	119	191	-	-
	Vachellia seyal (Shittah tree)	2	-	1	2	-	-
Kenya	*Azadirachta indica (Neem)*	899	1533	660	2436	82	8
	Calliandra calothyrsus	50	61	29	165	43	2
	Carica papaya (Pawpaw)	276	239	162	307	84	-
	Mangifera indica (Mango)	2157	4339	1658	6432	618	22
	Melia volkensii (Melia)	1113	1812	849	3221	335	-
	Moringa oleifera (Moringa)	167	489	113	883	41	9
	Senna siamea (Siamese senna)	662	1005	346	1808	59	25

3. Results

Data were collected from 173 households across the sites in Ethiopia in 2017 [37] and 1600 households in Kenya in 2018 [38]. Of these, 71% and 76% surveyed households in Kenya and Ethiopia were male-headed, respectively. Despite this, it was mostly women (i.e., their spouse) who registered to join the project and attended the training events. Median farm size was 2.02 Ha in Kenya and ranged from 0.05 Ha to 3.33 Ha while in Ethiopia, median farm size was 1 Ha and ranged from 0.1 Ha to 8 Ha. Household head age ranged from 18 years to 80 years in Ethiopia with a median of 40.5 years while in Kenya the age ranged from 23 years to 97 years with a median of 49 years.

3.1. Overview of Tree Seedling Survival

In Ethiopia, 4224 trees were monitored in 2017 and 17,520 trees were monitored in Kenya in 2018.

Overall, average seedling survival in the two countries varied across the species planted and agroecological conditions represented in the study sites. In Kenya, Kitui County had the highest average seedling survival at 53.4% while Machakos and Makueni counties had an average survival of 32.2% and 43.3%, respectively. The variation in average survival was also observed within each county. Highest variation was recorded in Makueni County with Mbooni East recording 66.7% survival compared to Kibwezi East which had survival rate of 33.8%. Comparatively, average seedling survival was high across all the woredas in Ethiopia. Tsaeda Emba and Boset had the highest average tree survival at 99% and 93% respectively while Gursum and Samre recorded 84% and 81% survival of all the trees planted respectively. Variation in average seedling survival was however observed within each woreda with highest variation recorded in Samre where Bara watershed recorded 95.8% seedling survival compared to Atami watershed which recorded 72.1% survival. We also found that different tree species performed better in some areas compared to others despite the broad similarities in agroecological conditions in Kenya (Figure 2) and Ethiopia (Table 3).

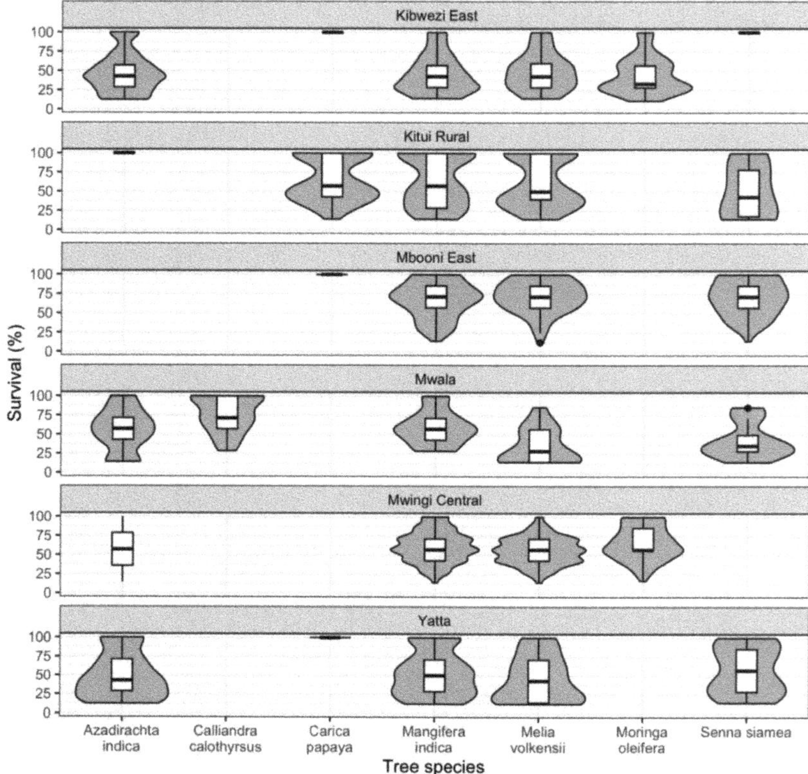

Figure 2. Boxplot of percentage tree seedling survival across the sites in Kenya (n = 7375 seedlings). The black horizontal line is the average percentage survival for each species and the gray shaded area shows the distribution of the data.

Table 3. Percentage seedling survival per species in each site in Ethiopia (number of trees monitored in each species is added in brackets).

Tree Species	Boset	Gursum	Samre	Tsaeda Emba
Acacia saligna	100 (1)	100 (5)		100 (8)
Azadiracta indica	66.7 (125)	85.5 (69)		
Carica papaya	100 (1)			
Casimiroa edulis				100 (19)
Citrus sinensis			93.8 (160)	
Coffea arabica				100 (59)
Cordia africana				100 (1)
Faidherbia albida	100 (1)			100 (215)
Grevillea robusta	93.6 (125)	71.3 (94)	87.7 (285)	
Jacaranda mimosifolia		91.7 (12)		
Malus domestica	100 (2)	100 (1)		
Mangifera indica	94.1 (169)	94.7 (95)	72.5 (280)	99.5 (204)
Melia volkensii	50 (2)			
Moringa oleifera			69.2 (156)	
Olea africana	100 (8)			
Persea americana	97.8 (45)	76.7 (103)		100 (1)
Psidium guajava		91.5 (94)		100 (60)
Rhamnus prinodes				99.5 (200)
Vachellia seyal				100 (2)

3.2. Effect of Tree Planting and Management Practices on Seedling Survival

Results show that despite influencing average survival, application of manure, watering regime, planting in a small hole relative to a big hole, and planting along the external boundary had a positive effect on seedling survival in Kenya (Table 4) while in Ethiopia, mulching, watering and physical protection were significant (Table 5). The type of effect (whether positive or negative) and magnitude of effect varied across the practices the sites where they were applied.

Overall, addition of manure increased seedling survival in Kenya by 12% with variation in the magnitude of effect across and within the counties. For example, we found that seedling survival in Kitui County increased by 8% with addition of manure while in Machakos and Makueni survival increased by 6% and 21% respectively. At subcounty level, Mbooni East in Makueni County had the greatest increase in seedling survival with addition of manure at 18.7%. There was no variation in the magnitude of the effect of manure at the species level although results of the model show that application of manure increased survival of all species by 10%. Addition of mulch had a positive effect on seedling survival and was significant in both Kenya and Ethiopia increasing survival by 5.8% and 5% respectively. Nevertheless, variation in the magnitude of the effect was observed across and within counties and woredas.

Watering the seedlings had a positive effect on tree survival and was significant in the two countries. In Kenya, survival increased by 35% when seedlings were watered with variation within and across the counties. Watering had the greatest effect on survival in Kitui (51% increase) and Makueni Counties (43% increase), which are considerably drier than Machakos County which recorded 7.5% increase in survival with watering. In Ethiopia, survival increased by 63% when the seedlings were watered. Farmers also compared four watering regimes in Ethiopia. They compared watering with five liters of water every five days, five liters every ten days, three liters every five days, and three liters every ten days. Our results show that the watering regime employed was not significant as long the seedlings were watered.

Planting in a small hole relative to a big hole had a negative effect on seedling survival in Kenya with variation across the counties. The greatest variation was observed in Machakos County where survival decreased by 21% when seedlings were planted in farmer hole size relative to a big hole. On the other hand, planting seedlings in a small hole relative to a big hole increased seedling survival by 12% in Makueni County and decreased survival by 2% and 8% in Machakos and Kitui Counties respectively. Variation in the effect of the size of planting hole was also observed across the tree species planted and was significant for *Calliandra calothyrsus*, *Mangifera indica* and *Melia volkensii*. For example, *Mangifera indica* had a 41.5% probability of survival when planted in a small hole compared to 45.6% when planted in a big hole and 39.3% when planted in normal-sized hole.

Physical protection of seedlings (fencing) had a significant effect on seedling survival in Kenya while in Ethiopia physical seedling protection was not significant. Overall, fencing increased seedling survival by 7.8% and 5% in Kenya and Ethiopia respectively. Variation in the type and magnitude of effect was observed across the counties and Woredas. For example, fencing increased survival by 9% and 15% in Machakos and Makeuni counties respectively and it decreased survival in Kitui County by 5%. On the other hand, survival increased by 8.5% in Boset, 5.3% in Gursum, 23.9% in Samre and by 0.4% in Tsaeda Emba when seedlings were fenced.

Finally, in Kenya, seedlings had a higher probability of survival when planted in woodlots and along the internal boundary and a lower probability of survival when planted along external boundaries, home compound and along terraces. In Ethiopia, seedlings had a higher probability of survival when planted along external boundaries and a lower probability of survival when planted scattered in cropland. All seedlings planted along terraces recorded 100% survival and, as such, were excluded from the regression model.

Table 4. Results of a logistic model on the probability of seedling survival under different planting and management practices on seedling survival in Kenya. Significance codes: 0 '***' 0.001 '**' 0.01 '*'.

Management Practice	n	Coefficient	Standard Error	Pr (>|z|)	Odds Ratio
Manure application	17,496				
No		−0.495	0.226	0.029 *	
Yes		0.466	0.033	$<2 \times 10^{-16}$ ***	1.641
Mulch application	17,496				
No		0.266	0.235	0.259	
Yes		0.233	0.041	1.34×10^{-8} ***	1.263
Watering	17,496				
No		−1.918	0.220	$<2 \times 10^{-16}$ ***	
Yes		1.838	0.069	$<2 \times 10^{-16}$ ***	6.284
Watering frequency	15,231				
Daily		0.249	0.239	0.296	1.283
Every other day		−0.063	0.069	0.363	0.939
Weekly		0.492	0.062	1.71×10^{-15} ***	0.611
Bi-weekly		−0.299	0.071	2.68×10^{-5} ***	0.741
Monthly		−0.416	0.143	0.004 **	0.66
Planting hole	17,496				
Big hole		−0.173	0.228	0.448	0.841
Small hole		−0.103	0.036	0.004 **	0.902
Farmer normal hole size		−0.046	0.071	0.52	0.955
Physical protection(fencing)	17,496				
No		−0.380	0.235	0.106	
Yes		0.405	0.037	$<2 \times 10^{-16}$ ***	1.5
Planting Niche	17,496				
Along terraces		−0.222	0.211	0.294	
External boundary		−0.271	0.076	0.000 ***	0.762
Home compound		−0.120	0.051	0.019 *	0.886
Internal boundary		−0.001	0.059	0.987	1.004
Scattered in cropland		−0.114	0.050	0.023 *	0.892
Woodlot		0.087	0.129	0.502	1.094

Table 5. Results of a logistic model on the probability of seedling survival under different planting and management practices on seedling survival in Ethiopia. Significance codes: 0 '***' 0.001 '**'.

Management Practice	n	Coefficient	Standard Error	Pr (>|z|)	Odds Ratio
Mulch application	2652				
No		2.414	0.812	0.295	
Yes		1.008	0.14	6.18×10^{-13} ***	2.739
Watering	2680				
No		−0.588	1.213	0.628	
Yes		5.021	0.366	$<2 \times 10^{-16}$ ***	151.608
Watering regime	2070				
Five liters every five days		3.447	1.028	0.001 ***	
Five liters every ten days		−0.004	0.277	0.990	0.996
Three liters every five days		−0.210	0.219	0.339	0.811
Three liters every ten days		−0.121	0.274	0.659	0.886
Physical protection (fencing)	2854				
No		2.202	0.842	0.009 **	
Yes		0.831	0.175	2×10^{-6} ***	2.296
Planting Niche	2695				
External boundary		3.065	0.844	0.000 ***	21.432
Home compound		0.591	0.391	0.131	1.806
Internal boundary		−0.538	0.359	0.134	1.713
Scattered in cropland		−0.623	0.422	0.140	1.864
Woodlot		−0.604	0.401	0.132	1.829

3.3. Options that Conferred the Best Chance of Survival for Planted Seedlings

To assess the planting and management options that conferred the best chance of survival, we calculated the survival rate of tree species planted using a combination of the options tested (Table 6).

Table 6. Description of options derived from a combination of those tested in Kenya.

Options	Planting Hole	Manure Quantity (kg)	Mulching	Watering	Observations
Option 1	Small*	2	No	No	32
Option 2	Small*	2	No	Yes	630
Option 3	Small*	2	Yes	No	26
Option 4	Small*	2	Yes	Yes	298
Option 5	Small*	0	Yes	No	34
Option 6	Small*	0	Yes	Yes	516
Option 7	Big**	4	No	No	19
Option 8	Big**	4	No	Yes	170
Option 9	Big**	4	Yes	No	7
Option 10	Big**	4	Yes	Yes	123
Option 11	Big**	0	Yes	No	20
Option 12	Big**	0	Yes	Yes	554
Option 0 (Farmer's practice)	Other***	Other	No/Yes	No/Yes	15,088
Total					17,517

Small* = 30 cm diameter × 45 cm depth; Big** = 60 cm diameter × 45 cm depth; Other*** = Common hole sizes.

In Kenya, seedlings planted in a big hole with 4 kg of manure and mulch had the highest survival at 47% while those planted in a small hole with 2 kg of manure and mulch had a survival of 36.1%. Variation in percentage survival was observed across the species planted (Figure 3). For example, *Azadirachta indica* recorded higher survival when planted in a small hole with addition of mulch and two kilograms of manure while *Moringa oleifera* recorded higher survival when planted in a big hole with watering and addition of mulch. In Ethiopia, almost all combinations of the options conferred at least 90% chance of survival for the planted seedlings. However, seedlings that were not watered but were applied mulch and controlled for weed recorded the least survival at 44%. This is consistent with the results of the statistical analysis where watering and watering regime were found to have a positive effect on survival of seedlings in Ethiopia.

3.4. Effect of the Household Socioeconomic Characteristics on Tree Survival

We assessed, using the household survey data, the effect of age and education level of the household head, land tenure, soil quality ranking (farmers ranked the quality of soil on their farm as either low, medium or high) and long-term experience with farms' soil erosion problems on seedling survival. In Kenya, our results showed that that there was a significant relationship between age and education level of the household head and seedling survival as well as between farm size, the type of land tenure and whether the farmer experienced soil erosion on their farm (Table 7). Variation was observed across the sites, for instance, in Mwingi Central, seedlings had a 59.3% probability of survival when farmers described their soil quality as high compared to 47% when soil quality was ranked as low. In Kitui Rural however, seedlings had a higher probability of survival when farmers ranked their soil quality as low suggesting that the low soil quality rating was overcome by the management strategy.

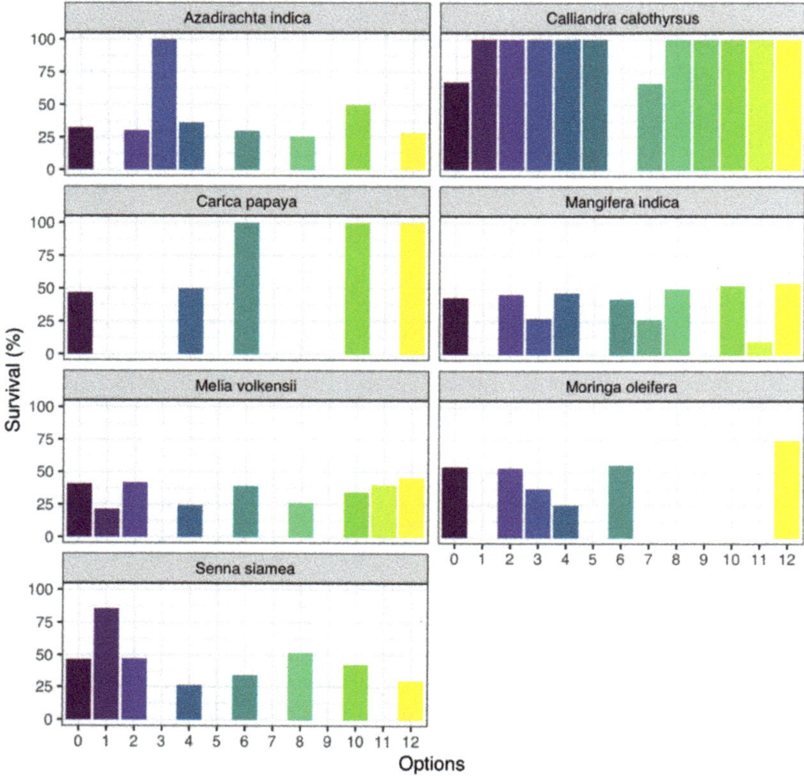

Figure 3. Performance of tree species against different combination of options that were tested in Kenya. Combinations presented here are described in Table 6.

Table 7. Results of the chi square test for a relationship between seedling survival and socioeconomic characteristics of a household and biophysical characteristics of the farm in Kenya. Significance levels: 1% '***' 5% '**'.

Variables	Seedling Survival (yes = 1)	n	Chi Square Value	*p* Value	Cramer's V	Relationship
Farm size		17,517	235.34 ***	0.000	0.12	weak
≤2 Ha	3145 (36%)					
>2 Ha	4239 (47%)					
Age category		17,517	50.95 ***	0.000	0.05	Very weak
<35 years	518 (40%)					
36–64 years	5833 (44%)					
>65 years	1033 (37%)					
Education level		15,723	59.84 ***	0.000	0.06	Very weak
No formal education	580 (40%)					
Primary education	4117 (44%)					
Secondary education	1484 (38%)					
Tertiary education	437 (49%)					
Land tenure		16,009	94.83 ***	0.000	0.08	Very weak
Title deed/	4841 (43%)					
Allotment/leasehold	1345 (35%)					
Others	501 (48%)					

Table 7. *Cont.*

Variables	Seedling Survival (yes = 1)	n	Chi Square Value	*p* Value	Cramer's V	Relationship
Soil quality		16,740	8.55 **	0.014	0.02	Very weak
High	1166 (44%)					
Low	1590 (42%)					
Medium	4262 (41%)					
Erosion		16,740	10.89 ***	0.001	0.03	Very weak
No	2372 (40%)					
Yes	4646 (43%)					

In Ethiopia, farm size and soil quality ranking had a significant relationship with seedling variation survival (Table 8). As soil quality ranking increased so did the likelihood of tree seedling survival (soil quality was ranked by the farmer as either low, medium or high).

Table 8. Results of the chi square test for a relationship between seedling survival and socioeconomic characteristics of a household and biophysical characteristics of the farm in Ethiopia. Significance levels: 1% '***' 5% '**' 10% '*'.

Variables	Seedling Survival (yes = 1)	n	Chi Square Value	*p* Value	Cramer's V	Relationship
Farm size		2854	22.72 ***	0.000	0.09	Very weak
<=2 Ha	980 (86%)					
>2 Ha	1562 (91%)					
Age category		2854	5.77 *	0.056	0.05	Very weak
<35 years	955 (88%)					
36–64 years	1234 (89%)					
>65 years	303 (93%)					
Education level		2710	9.58 **	0.022	0.06	Very weak
No formal education	1079 (87%)					
Primary education	1152 (91%)					
Secondary education	176 (86%)					
Tertiary education	7 (78%)					
Soil quality		2710	68.72 ***	0.000	0.16	weak
High	17 (44%)					
Low	516 (42%)					
Medium	1986 (41%)					

Variation in the effect of socioeconomic and biophysical variables was observed across the woredas. For example, increasing farm size decreased tree seedling survival in all woredas except Gursum which had increased tree survival with increased farm size.

3.5. Gendered Management and Use of Trees

In Kenya, analysis around intrahousehold decision-making over participation in the comparisons and division of labor in the management of trees was also conducted. The decision to get involved in the tree planting planned comparison was made by 94% of women respondents in female-headed households, 70% of women respondents in male-headed households and 70% of men respondents in female-headed and male-headed households respectively. A similar trend was observed in who was involved in deciding which tree species to plant and where on the farm to plant the trees.

We also wanted to understand the division of labor when managing the planted trees as, traditionally, men and women tend to have differing roles as well as areas of influence when it comes to tree planting [39,40]. Our results show that women-only labor was commonly used for adding mulch (44%), watering (42%), and adding manure (40%) compared to men-only labor, which was used for planting trees, fencing and pruning (Table 9).

Table 9. Gender of those involved in various management activities associated with tree planting.

	N *	Male Labor Only (%)	Male and Female Labor (%)	Female Labor Only (%)
Activities associated with tree planting				
Planting trees	1233	40	26	34
Adding manure	1064	34	26	40
Watering	1207	28	30	42
Fencing	668	50	16	33
Mulching	642	33	22	44
Pruning	613	52	16	32

* Differences in number of observations is due to "other" having been selected for those involved or households not being involved in certain activities

We also assessed if farmers were likely to invest in tree planting, which species they would consider for future planting among the surveyed households and how many seedlings they would plant. Results show that 74% of women respondents in female-headed households, 75% of women respondents in male-headed households, 77% of men respondents in female headed households and 75% of men respondents in male-headed households would consider planting additional trees in future. In terms of species consideration, *Mangifera indica* was the most considered species for future planting by 64% of all the surveyed households. Other species considered were *Melia volkensii* (24%), *Carica papaya* (21%), *Senna siamea* (20%) and *Azadirachta indica* (16%). Additional analysis revealed no major differences in the tree species desired for future planting between women and men respondents (Figure 4). We however found variation in the overall number of seedlings considered between female headed and male headed households. The maximum number of tree seedlings considered for future planting by respondents from male headed households (1000 seedlings) was twice that of respondents from female headed households (500 seedlings).

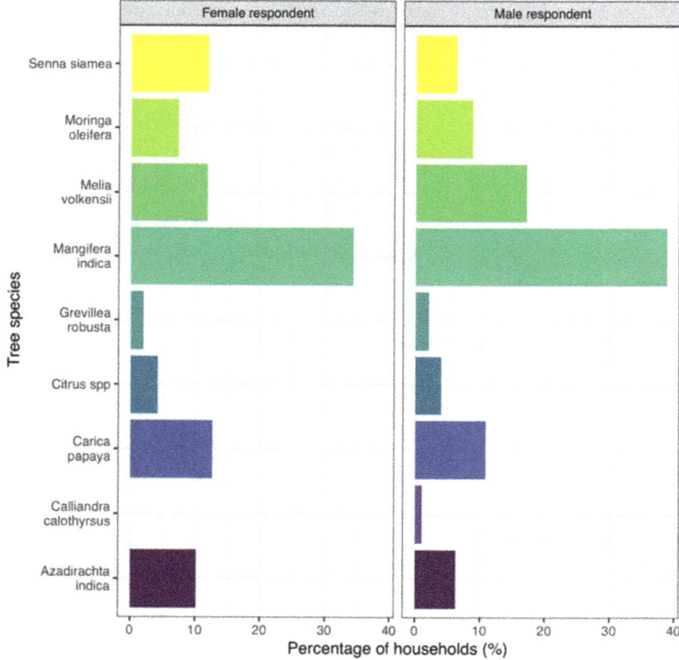

Figure 4. Species considered for future planting in female headed and male headed households by more than one percent of the surveyed households (n = 644 female respondents and 325 male respondents).

4. Discussion

The survival rate of tree seedlings in drylands is characteristically low partly due to unreliable rainfall, high levels of land degradation resulting in low soil productivity, planting of ecologically unsuitable tree species, and poor tree seedling management practices [18–20]. While planting and management practices employed can have a positive or negative effect on tree survival, the use of these practices is usually determined by knowledge, needs, perceptions and, availability and access to resources the communities implementing them [41]. The results show that planting and management practices such as watering and watering regime, manure application and seedling protection by fencing had a significant effect on tree seedling survival.

However, the type of effect (whether positive or negative) as well as the magnitude of this effect varied depending on the sites in which they were employed. Variation in average tree survival was also observed across tree species under the planting and management practices. For example, *Mangifera indica* recorded higher average survival when planted in a big hole in all the study sites in Kenya while *Melia volkensii* recorded higher survival when planted with smaller hole size (30 cm diameter by 45 cm depth). Water availability is a limiting factor to seedling survival especially in arid and semi-arid areas and the size of the planting hole can influence the water holding capacity. Bigger planting holes retain more water compared to smaller planting holes [35]. However, tree species have differing watering and management requirements. Additional discussion with farmers revealed that tree species such as *Melia volkensii* which are indigenous to arid and semi-arid areas are sensitive to waterlogging especially when young [42] and perform better when planted in smaller planting holes, while fruit trees such as *Mangifera indica* have a higher water requirement and perform better when planted in bigger planting holes.

That average seedling survival was higher when seedlings were physically protected from livestock across all the study sites in Kenya and Ethiopia was expected as browsing by livestock was reported as a key cause of seedling mortality across the study area. It is also consistent with findings by [43,44] who in their studies, found seedling survival rate was higher in the fenced plots compared to those not fenced. Farmers across the study sites used makeshift fences made up of twigs, old mosquito nets and old clothes. In some cases, seedlings whose tip had been previously browsed off, regenerated. The observed differences in the significance of seedling protection between Kenya and Ethiopia can partly be attributed to the varying grazing practices across the two countries. In Ethiopia, farmers practice communal grazing with designated grazing areas and as such is easier to protect seedlings planted. In contrast, grazing in Kenya is at the discretion of individual farmers with many letting their livestock roam in the homestead especially during the dry season when access to fodder is limited. This is not to mean that browsing by livestock is not a threat to seedling survival in Ethiopia. In fact, Ref. [45] found that browsing by livestock deterred farmers in Ethiopia from adopting high-value agroforestry.

Furthermore, seedling source plays a key role in the viability of planting material and is often overlooked in tree planting campaigns [28]. This is because seedling stage is considered the most sensitive stage in the lifecycle of trees and seedling care from early establishment is critical to survival [35,46]. Tree seedlings distributed to farmers were sourced from different nurseries across the study sites due to the large number of seedlings required and the structure of nursery enterprises in the area, that is, most nurseries are smallholder owned and run. As a result, the quality of seedlings varied depending on the nursery from which the seedlings were sourced.

Results also revealed a relationship between the household's socioeconomic characteristics and the survival of the planted seedlings in both Kenya and Ethiopia. The strength of that relationship was however either weak or very weak suggesting while farmer circumstances can influence the odds of survival of planted seedlings, they are not, on their own, determining factors for seedling survival. Instead, they interact with other factors for example the agroecological conditions prevalent in the locality. They also enhance or limit access to resources such as knowledge, manure and water, which were found to significantly increase the chances of survival for seedlings. Our study also found variation in the effect of both biophysical characteristics of the farm such as erosion status and soil quality and socioeconomic characteristic of the household at both national and local scale. For example, seedlings had a higher likelihood of survival when farmers described the soil quality as high across most of the study areas in the two countries. This was expected as high soil quality means that the soil has the requisite nutrients to support plant growth. However, in Kitui Rural within Kitui County, seedlings had a higher likelihood of survival when farmers described their soil quality as low. This suggests that farmers who considered their soil quality low were perhaps investing more in use of external inputs such as manure to improve their overall soil quality compared to farmers who considered the quality of their soil as high. This thus calls for quantitative analyses on seedling survival to be combined with in-depth qualitative work to understand the reason behind the variation across study sites.

Our results also revealed variation in survival among farmers within the same locality and farmer circumstances implying that some farmers have better experience growing and managing trees compared to their neighbors [22]. This can perhaps be explained by inherent local knowledge that has shaped how they grow and manage their trees necessitating the need for complementarity between scientific knowledge and local knowledge if good practices around tree growing that are suited to local agroecological conditions and farmer circumstance are to be identified and shared [22].

Our findings on the role of men and women in tree planting and management activities show that men-only labor was predominantly used for management activities such as pruning and fencing while women-only labor was mostly used for watering, adding manure and adding mulch. This was consistent with findings from [47], who found that traditionally women were discouraged from taking part in activities such as fencing and pruning as they were considered strenuous. The results were also consistent with previous studies that found that where both men and women were involved in tree management, women tended to be involved in the initial stages of tree establishment [13,39]. This suggests that existing customs and traditions influence the role that both men and women play in managing natural resources and as such tree planting.

We also found very little difference in tree species preferred for future planting between respondents from female-headed households and those from male-headed households in Kenya. However, there was variation in the total number of seedlings to be planted with respondents from male-headed households willing to plant twice the number of seedlings planted by respondents from female-headed households. This could partly be explained by differences in access to capital needed to purchase seedlings, labor availability and farm size between male-headed and female-headed households. These results are also consistent with the study by [48] who found that female-headed households tended to plant half the number of trees and/or shrubs planted by male-headed households and attributed the difference to variation in farm size. Tree species that are commonly found in households across the study sites were mostly preferred. Furthermore, species compared by farmers in the study were the top preferred species for future planting among the surveyed households. This suggests that farmers prioritize tree species whose survival they can assure based on their local knowledge and past experience with the tree species, species whose availability they are sure of, the value they attach to the tree species and species for which there is an already established market or demand [22].

5. Conclusions

Tree planting can have a positive effect on the environment as well as on the social-economic realities of farmers if seedlings are managed ensuring survival to maturity. However, to scale successful tree planting efforts, context-specific variables should be considered as they can impact the survival of planting seedlings either positively or negatively. Our study found that planting seedlings with manure, watering them and protecting them from physical harm significantly increased the probability of survival in Kenya while in Ethiopia, watering, the planting niche and protection from harm significantly increased the probability of survival. Our findings show that tree planting and management practices such as application of manure, mulch and fencing can significantly increase survival indicating the need to not only invest in such practices but also in continuous engagement and training with farmers and local community on how to implement the practices on their farms.

We also show that socioeconomic characteristics of the household as well as the biophysical characteristics of the farm can enhance or hinder tree seedling survival. That variation in both the type and magnitude of effect on seedling survival was recorded in most of the assessed socioeconomic variables across the study sites indicates the need to understand and be cognizant of the local agroecological conditions and the farmer circumstances in which these practices are employed. Results also show the need to consider the different role men and women play in tree planting including in the management and use of the planted trees as this can have an influence on practices that are feasible for different farmers. Finally, we show the need to consider the priorities and interests of farmers as well as the inherent local knowledge especially when it comes to the species preferred as well as the need for training on additional beneficial species for planting.

Author Contributions: Conceptualization of the work by F.S., L.A.W.; methodology by P.S., J.M., L.A.W., C.M., N.H., K.H., F.S.; formal analysis conducted by C.M., L.A.W., A.F., H.O., N.H., I.O., M.C.; data curation by A.F., H.O., E.B., E.K., C.M.; writing—original draft preparation, C.M., L.A.W., M.C., A.F., H.O., N.H., P.S., I.O., E.K., A.K., J.M., S.C., K.H., E.B., F.S., visualization, C.M., L.A.W., A.F., H.O., I.O.; supervision by L.A.W., F.S.; funding acquisition By F.S. All authors have read and agreed to the published version of the manuscript.

Funding: This work was funded by International Fund for Agricultural Development (IFAD) and the European Commission through the project on Restoration of degraded lands for food security and poverty reduction in East Africa and the Sahel: taking successes to scale, grant numbers: 2000000520 and 2000000976 and the CGIAR Research Programme on Forests, Trees and Agroforestry (FTA). The work also received funding support from the European Commission and IFAD through the project on Kenya Cereal Enhancement Programme—Climate Resilient Agricultural Livelihoods (KCEP-CRAL) window.

Acknowledgments: The authors acknowledge the collaboration of the Drylands Development Programme funded by the Ministry of Foreign Affairs (MoFA) of the Netherlands and its implementing partners in Kenya and Ethiopia, including World Vision Kenya, Netherlands Development Organization (SNV), Caritas-Kenya, Adventist Development and Relief Agency Kenya (ADRA), World Vision Ethiopia (WVE), Relief Society of Tigray (REST), Ethiopian Orthodox Church's Development and Inter-Church Aid Committee (EOC-DICAC). This project was also supported in part by the CGIAR Research Programme on Forests, Trees and Agroforestry (FTA), We would like to acknowledge the support of the enumerators and community facilitators including Angellinah Kimanzi, Caroline Mbuvi, Francisca Mutua, Felix Mbuvi, Mercy Mwea, Silas Muthuri, Stephen Maithya, and Sylvester Muendo.

Conflicts of Interest: The authors declare no conflict of interest and that the funders had no role in the design of the study; in the collection, analyses, or interpretation of data; in the writing of the manuscript, or in the decision to publish the results.

References

1.   Ministry of Environment and Forestry. *National Strategy for Achieving and Maintaining over 10% Tree Cover by 2022*; Ministry of Environment and Forestry: Nairobi, Kenya, 2019.

2.   Reubens, B.; Moeremans, C.; Poesen, J.; Nyssen, J.; TewoldeBerhan, S.; Franzel, S.; Deckers, J.; Orwa, C.; Muys, B. Tree species selection for land rehabilitation in Ethiopia: From fragmented knowledge to an integrated multi-criteria decision approach. *Agrofor. Syst.* **2011**, 303–330. [CrossRef]

3.   Coe, R.; Sinclair, F.; Barrios, E. Scaling up agroforestry requires research "in" rather than "for" development. *Curr. Opin. Environ. Sustain.* **2014**, *6*, 73–77. [CrossRef]

4.   Sinclair, F.; Coe, R. The Options by Context Approach: A Paradigm shift in agronomy. *Exp. Agric.* **2019**, *55*, 1–13. [CrossRef]

5.   Sinclair, F. System science at the scale of impact. In *Sustainable Intensification in Smallholder Agriculture: An Integrated Systems Research Approach*; Öborn, I., Vanlauwe, B., Phillips, M., Thomas, R., Brooijmans, W., Atta-krah, K., Eds.; Routledge: Abingdon, UK, 2017; pp. 43–57.

6.   Kuria, A.; Pagella, T.; Winowiecki, L.A.; Mitiku, H.; Sinclair, F.L. Local knowledge promotes design of adaptive land restoration options that deliver multiple ecosystem services at scale. **2019**. Manuscript in Preparation.

7.   Ojiem, J.O.; de Ridder, N.; Vanlauwe, B.; Giller, K.E. Socio-ecological niche: A conceptual framework for integration of legumes in smallholder farming systems. *Int. J. Agric. Sustain.* **2006**, *4*, 79–93. [CrossRef]

8.   Kindt, R.; Van Damme, P.; Simons, A.J. Tree Diversity in Western Kenya: Using Profiles to Characterise Richness and Evenness. *Biodivers. Conserv.* **2006**, *15*, 1253–1270. [CrossRef]

9. Kindt, R.; Simons, A.J.; Van Damme, P. Do Farm Characteristics Explain Differences in Tree Species Diversity among Western Kenyan Farms? *Agrofor. Syst.* **2004**, *63*, 63–74. [CrossRef]

10. Crossland, M.; Winowiecki, L.A.; Pagella, T.; Hadgu, K.; Sinclair, F. Implications of variation in local perception of degradation and restoration processes for implementing land degradation neutrality. *Environ. Dev.* **2018**, *28*, 42–54. [CrossRef]

11. Coe, R.; Njoloma, J.; Sinclair, F. Loading the dice in favour of the farmer: Reducing the risk of adopting agronomic innovations. *Exp. Agric.* **2016**, *55*, 67–83. [CrossRef]

12. Kiptot, E.; Franzel, S. Gender and agroforestry in Africa: A review of women's participation. *Agrofor. Syst.* **2011**, *84*, 35–58. [CrossRef]

13. Naz, F.; Catacutan, D. Gender roles, decision-making and challenges to agroforestry adoption in Northwest Vietnam. *Int. For.* **2015**, 22–32. [CrossRef]

14. Pehou, C.; Djoudi, H.; Vinceti, B.; Elias, M. Intersecting and dynamic gender rights to néré, a food tree species in Burkina Faso. *J. Rural Stud.* **2020**, *76*, 230–239. [CrossRef]

15. Bastin, J.; Finegold, Y.; Garcia, C.; Mollicone, D. The global tree restoration potential. *Science* **2019**, *365*, 76–79. [CrossRef] [PubMed]

16. Lohbeck, M.; Albers, P.; Boels, L.E.; Bongers, F.; Morel, S.; Sinclair, F.; Takoutsing, B.; Vågen, T.G.; Winowiecki, L.A.; Smith-Dumont, E. Drivers of farmer-managed natural regeneration in the Sahel. Lessons for restoration. *Sci. Rep.* **2020**, *10*, 15038. [CrossRef]

17. Brancalion, P.H.; Niamir, A.; Broadbent, E.; Crouzeilles, R.; Barros, F.S.; Zambrano, A.M.; Baccini, A.; Aronson, J.; Goetz, S.; Reid, J.; et al. Global restoration opportunities in tropical rainforest landscapes. *Sci. Adv.* **2019**, *5*, 1–12. [CrossRef]

18. De Leeuw, J.; Njenga, M.; Wagner, B.; Liyama, M. *Treesilience: An Assessment of the Resilience Provided by Trees in the Drylands of Eastern Africa*; De Leeuw, J., Njenga, M., Wagner, B., Liyama, M., Eds.; World Agroforestry Centre: Nairobi, Kenya, 2014; ISBN 9789290593522.

19. Ndegwa, G.; Iiyama, M.; Anhuf, D.; Nehren, U.; Schlüter, S. Tree establishment and management on farms in the drylands: Evaluation of different systems adopted by small-scale farmers in Mutomo District, Kenya. *Agrofor. Syst.* **2017**, *91*, 1043–1055. [CrossRef]

20. Syano, N.M.; Wasonga, O.V.; Nyangito, M.; Kironchi, G.; Egeru, A. Ecological and Socio-Economic Evaluation of Dryland Agroforestry Systems in East Africa. 2016. Available online: http://repository.ruforum.org (accessed on 19 January 2020).

21. Kuyah, S.; Öborn, I.; Jonsson, M.; Dahlin, A.S.; Barrios, E.; Muthuri, C.; Malmer, A.; Nyaga, J.; Magaju, C.; Namirembe, S.; et al. Trees in agricultural landscapes enhance provision of ecosystem services in Sub-Saharan Africa. *Int. J. Biodivers. Sci. Ecosyst. Serv. Manag.* **2016**, *12*, 255–273. [CrossRef]

22. Derero, A.; Coe, R.; Muthuri, C.; Hadgu, K.M.; Sinclair, F. Farmer-led approaches to increasing tree diversity in fields and farmed landscapes in Ethiopia. *Agrofor. Syst.* **2020**. [CrossRef]

23. Iiyama, M.; Mukuralinda, A.; Ndayambaje, J.; Musana, B.; Ndoli, A.; Mowo, J.; Garrity, D.; Ling, S.; Ruganzu, V. Tree-Based Ecosystem Approaches (TBEAs) as multi-functional land management strategies-evidence from Rwanda. *Sustainability* **2018**, *10*, 1360. [CrossRef]

24. Mganga, K.Z.; Nyariki, D.M.; Musimba, N.K.R.; Mwang'ombe, A.W. Indigenous Grasses for Rehabilitating Degraded African Drylands. In *Agriculture and Ecosystem Resilience in Sub Saharan Africa. Climate Change Management*; Bamutaze, Y., Kyamanywa, S., Singh, B., Nabanoga, G., Lal, R., Eds.; Springer: Cham, Switzerland, 2019. [CrossRef]

25. Coe, R.; Hughes, K.; Sola, P.; Sinclair, F. *Planned Comparisons Demystified*; ICRAF Working Paper No 263; Nairobi World Agroforestry Centre: Nairobi, Kenya, 2017. [CrossRef]

26. Hughes, K.; Oduol, J.; Coe, R.; Sinclair, F.; Peralta, A. *Guidelines for Identifying and Designing Planned Comparisons*; ICRAF: Nairobi, Kenya, 2016.

27. Franzel, S.; Sinja, J.; Simpson, B. *Farmer-to-Farmer Extension in Kenya: The Perspectives of Organizations Using the Approach*; ICRAF Working Paper No. 181; World Agroforestry Centre: Nairobi, Kenya, 2014. [CrossRef]

28. Duguma, L.; Minang, P.; Aynekulu, E.; Carsan, S.; Nzyoka, J.; Bah, A.; Jamnadass, R. *From Tree Planting to Tree Growing: Rethinking Ecosystem Restoration Through Trees*; ICRAF Working Paper No 304; World Agroforestry: Nairobi, Kenya, 2020. [CrossRef]

29. Hadgu, K.M.; Kooistra, L.; Rossing, W.A.H.; van Bruggen, A.H.C. Assessing the effect of Faidherbia albida based land use systems on barley yield at field and regional scale in the highlands of Tigray, Northern Ethiopia. *Food Secur.* **2009**, *1*, 337–350. [CrossRef]

30. Lemma, B.; Kleja, D.B.; Olsson, M.; Nilsson, I. Factors controlling soil organic carbon sequestration under exotic tree plantations: A case study using the CO2Fix model in southwestern Ethiopia. *For. Ecol. Manag.* **2007**, *252*, 124–131. [CrossRef]

31. County Government of Kitui. *Kitui County Intergrated Plan 2018–2022*; County Government of Kitui: Kitui, Kenya, 2018. Available online: https://www.cog.go.ke/downloads/category/106-county-integrated-development-plans-2018-2022# (accessed on 22 April 2020).

32. County Government of Machakos. Machakos County Intergrated Development Plan II (2018–2022). 2018. Available online: https://www.cog.go.ke/downloads/category/106-county-integrated-development-plans-2018-2022# (accessed on 22 April 2020).

33. County Government of Makueni. Makueni County Integrated Development Plan (Cidp) 2018–2022. 2018. Available online: https://www.cog.go.ke/downloads/category/106-county-integrated-development-plans-2018-2022# (accessed on 14 June 2019).

34. Sola, P.; Zerfu, E.; Coe, R.; Hughes, K. *Community Visioning and Action Planning: Guidelines for Integrating the Options by Context Approach*; ICRAF: Nairobi, Kenya, 2017.

35. Mwamburi, A.; Musyoki, J. *Improving Tree Survival in the Drylands of Kenya: A Guide for Farmers and Tree Growers in the Drylands*; Kenya Forestry Research Institute: Nairobi, Kenya, 2010.

36. R Core Team. R: A Language and Environment for Statistical Computing. Vienna, Austria, 2020. Available online: https://www.r-project.org (accessed on 8 January 2020).

37. Hagazi, N.; Hadgu, K.; Sitotaw, A.; Tofu, A.; Lavoll, V.; Winowiecki, L.; Magaju, C.; Nyaga, J.; Carsan, S.; Muriuki, J.; et al. Tree Planting Data 2017—Ethiopia 2019. MELDATA, V1. 2019. Available online: ahttps://hdl.handle.net/20.500.11766.1/FK2/O9LOGI (accessed on 11 March 2020).

38. Magaju, C.; Winowiecki, L.; Nyaga, J.; Ochenje, I.; Makui, P.; Kiura, E.; Crossland, M.; Valencia, A.M.; Kuria, A.; Carsan, S.; et al. Tree Planting Data 2018—Kenya 2019. MELDATA, V1. 2019. Available online: https://hdl.handle.net/20.500.11766.1/FK2/BLHHPR (accessed on 19 March 2020).

39. Kiptot, E.; Franzel, S.; Degrande, A. Gender, agroforestry and food security in Africa. *Curr. Opin. Environ. Sustain.* **2014**, *6*. [CrossRef]

40. Meijer, S.S.; Catacutan, D.; Ajayi, O.C.; Sileshi, G.W.; Nieuwenhuis, M. The role of knowledge, attitudes and perceptions in the uptake of agricultural and agroforestry innovations among smallholder farmers in sub-Saharan Africa. *Int. J. Agric. Sustain.* **2015**, *13*, 40–54. [CrossRef]

41. Maluki, J.M.; Kimiti, J.M.; Nguluu, S.; Musyoki, J.K. Adoption levels of agroforestry tree types and practices by smallholders in the semi-arid areas of Kenya: A case of Makueni County. *J. Agric. Ext. Rural Dev.* **2016**, *8*, 187–196. [CrossRef]

42. Kenya Forest Service. *A Guide to Growing Melia Volkensii in the Dryland Areas of Kenya*; Kenya Forest Service: Nairobi, Kenya, 2018.

43. Wassie, A.; Sterck, F.J.; Teketay, D.; Bongers, F. Effects of livestock exclusion on tree regeneration in church forests of Ethiopia. *For. Ecol. Manag.* **2009**, *257*, 765–772. [CrossRef]

44. Love, B.; Bork, E.; Spaner, D. Tree seedling establishment in living fences: A low-cost agroforestry management practice for the tropics. *Agrofor. Syst.* **2009**, *77*, 1–8. [CrossRef]

45. Iiyama, M.; Derero, A.; Kelemu, K.; Muthuri, C.; Kinuthia, R.; Ayenkulu, E.; Kiptot, E.; Hadgu, K.; Mowo, J.; Sinclair, F.L. Understanding patterns of tree adoption on farms in semi-arid and sub-humid Ethiopia. *Agrofor. Syst.* **2017**, *91*, 271–293. [CrossRef]

46. Bhadouria, R.; Singh, R.; Srivastava, P.; Raghubanshi, A.S. Understanding the ecology of tree-seedling growth in dry tropical environment: A management perspective. *Energy Ecol. Environ.* **2016**, *1*, 296–309. [CrossRef]

47. Crossland, M.; Valencia, A.M.P.; Pagella, T.; Magaju, C.; Kiura, E.; Winoweicki, L.; Sinclair, F. *Onto the Farm, into the Household: Intrahousehold Gender Relations Matter for Scaling Land Restoration Practices*, 2020; Manuscript Submitted for Publication.

48. Kiptot, E.; Franzel, S. Gender and agroforestry in Africa: Are women participating. *ICRAF Occas. Pap.* **2011**. [CrossRef]

Publisher's Note: MDPI stays neutral with regard to jurisdictional claims in published maps and institutional affiliations.

 *land*

Article

Tree Roots Anchoring and Binding Soil: Reducing Landslide Risk in Indonesian Agroforestry

Kurniatun Hairiah [1,*], Widianto Widianto [1], Didik Suprayogo [1] and Meine Van Noordwijk [1,2,3] 

[1] Faculty of Agriculture, Brawijaya University, Jl Veteran, Malang 65145, Indonesia; widianto@gmail.com (W.W.); suprayogo@ub.ac.id (D.S.); m.vannoordwijk@cgiar.org (M.V.N.)
[2] World Agroforestry (ICRAF), Bogor 16155, Indonesia
[3] Plant Production Systems, Wageningen University & Research, 6700 AK Wageningen, The Netherlands
* Correspondence: kurniatun_h@ub.ac.id or k.hairiah@cgiar.org

Received: 30 June 2020; Accepted: 30 July 2020; Published: 1 August 2020

Abstract: Tree root systems stabilize hillslopes and riverbanks, reducing landslide risk, but related data for the humid tropics are scarce. We tested fractal allometry hypotheses on differences in the vertical and horizontal distribution of roots of trees commonly found in agroforestry systems and on shear strength of soil in relation to root length density in the topsoil. Proximal roots of 685 trees (55 species; 4–20 cm stem diameter at breast height, dbh) were observed across six landscapes in Indonesia. The Index of Root Anchoring (IRA) and the Index of Root Binding (IRB) were calculated as $\Sigma D_v^2/dbh^2$ and as $\Sigma D_h^2/dbh^2$, respectively, where D_v and D_h are the diameters of vertical (angle > 45°) and horizontal (angle < 45°) proximal roots. High IRA values (>1.0) were observed in coffee and several common shade trees. Common fruit trees in coffee agroforestry had low medium values, indicating modest 'soil anchoring'. Where root length density (L_{rv}) in the topsoil is less than 10 km m^{-3} shear strength largely depends on texture; for L_{rv} > 10 shear strength was > 1.5 kg m^{-2} at the texture tested. In conclusion, a mix of tree species with deep roots and grasses with intense fine roots provides the highest hillslope and riverbank stability.

Keywords: coffee; fruit trees; index of root anchoring; slope stability; soil shear strength; root length density; root tensile strength

---

1. Introduction

Watersheds usually provide water but occasionally they generate mudflows. Intense seasonal precipitation during monsoons in mountainous watersheds can trigger landslides, debris flows, and flash floods. On steep slopes, landslides can destroy houses, villages, or any vegetation encountered in the downhill path. Soil and debris flow can also be a major contributor of sediment load in the river systems. Extreme rainfall events generate shallow slope failures by elevating pore pressures and decreasing effective stress, but numerous site-specific factors, such as preferential hydrologic flow paths, slope steepness, soil thickness, and existing plants root systems influence the potential for slope instability [1,2]. Agroforestry can reduce risks by having appropriate trees (species, age, diversity, management) at strategic locations at hillslope and landscape scale [3]. Higher plant diversity was found [4] to enhance soil stability in disturbed alpine ecosystems.

Shear strength is a term used in soil mechanics to describe the resistance against structural failure, or the magnitude of the shear stress that a soil can sustain before submitting to a sliding failure along a plane that is parallel to the direction of the force. The shear resistance of soil is a result of friction and interlocking of particles, and possibly cementation or bonding at particle contacts, strengthened by roots depending on their tensile strength. Tensile strength is a measure of the force required to pull something such as rope, wire, or a structural beam to the point where it breaks. Soil shear strength of rooted soil s_r exceeds that of unrooted soil s according to $s_r = s+1.2 \sum T_a$, where T_a it the tensile strength per unit cross-sectional area of all roots at the plane sheared. The factor 1.2 is an approximation derived for common slope angles [5]. A review of experimental methods and procedures to test root-strength and rooted-soil shear-strength behavior, focused on obtaining data to parameterize root reinforcement models. [6]. Common methods either determine the tensile strength of individual roots by loading the root in a pulling device until it breaks or determine the shear-strength of rooted soil in comparison to non-rooted soil in a Coulomb-type shear-box test. Laboratory shear-box tests encounter a difficulty in that the roots are not generally fixed or constrained at the base of the shear-box and are pulled rather than break.

The mechanics of 'failure' are broadly understood in terms of weight of the (wet) soil plus vegetation, slope, and critical soil shear strength. A higher soil clay content and lower water content lead to higher soil shear strength [7–9]. Intensifying soil management after forest conversion can lead to soil compaction meaning lower pore space and soil infiltration, enhancing shear strength and overland flow; along with a reduction of the weight of the vegetation such changes can reduce the chances of slope failure with time after forest conversion [10]. However, loss of 'root anchoring' associated with gradual decay of tree roots in the first few years after conversion will enhance the risk of landslides [11,12]. Conversion affecting only part of a hill slope, e.g., by road construction, can modify subsurface flows, leading to concentrated flows and locally enhanced landslide risk. The deepest and most destructive landslides, however, may be hardly influenced by vegetation. Still, a combination of deep-rooted trees for anchoring and shallow rooted grass (for stabilizing topsoil) is hypothesized to stabilize slopes prone to mass movement. At least four system levels are involved: the vegetation (species, stem diameters, spatial distribution of trees), the distribution of roots of an individual tree over the soil profile (shoot/root allometry, vertical root distribution), the effects of root length density on soil shear strength throughout the profile (it is the weakest plane on which the soil will shear), and the tensile strength of roots determining when individual roots will break. Most studies focus on only part of this range of scales.

Several studies on how woody species, mostly planted forests, influence slope stability have been reported in temperate areas such as Alaska [13,14], Canada [15], China [16–18], New Zealand [19], Australia [20], and the Mediterranean region [21,22]. A study in Australia [23] quantified increases in soil strength due to root reinforcement, on the basis of root strength, interface friction between the roots and the soil, and the distribution of roots within the soil for two common riparian species: river red gum (*Eucalyptus camaldulensis*) and swamp paperbark (*Melaleuca ericifolia*). They found that interspecies differences in the strength of living roots had less significance for bank reinforcement than interspecies differences in root distribution. A study in New Zealand found that shear strength contributions per tree did not differ between the native kanuka tree (*Kunzea ericoides*) and the species replacing it (*Pinus radiata*) on hillslopes prone to storm-generated land sliding, but higher stand density explained a slope protection advantage of the native vegetation [24]. A comparison between 9, 20, and 30-year-old stands of *Cryptomeria japonica* stands in China found that root density was highest in the 9-year-old stand, but that root tensile strength was lowest at this stage. As older stands had been thinned, however, the 9-year old stands contributed most to slope stability [25]. A study in Italy with three local shrub species found that these complement each other in different parts of the landscape in stabilizing steep hillslopes which are seasonally affected by storm-induced shallow landslides, by anchoring into the soil mantle. Root tensile force, at which a root breaks, increased with increasing root diameter, but tensile strength (per unit cross-sectional area) decreased with increasing root diameter following a power law curve [26]. Other studies found that when root distribution has a wide range of diameters, the root reinforcement results are controlled by large roots, which hold much more force than small roots [27].

Measurements from tropical vegetation, with higher tree diversity, on shear strength in relation to tree properties are scarce, however [28]. Interspecific variation in root strength in the tropics remains largely undescribed, with the exception of a study in southern Thailand [29] that compared roots of seven tree species in their effects on slope stabilization in the context of biotechnical slope protection.

The coarse root system of trees may differ in mixtures from that in monocultures [30], with shorter, more dense roots in focal trees growing with conspecific neighbors. A three-dimensional view on tree coarse roots is needed to verify this [31].

The mechanical effect of root systems in enhancing soil stability is based on at least three mechanisms [32]: (a) fine root systems in the surface layers bind soil particle strongly increasing cohesion [33]; through the more stable soil structure they reduce the entrainment of soil particles in overland flow of water; (b) the tensile strength of roots in the surface layers enhance shear strength and the risks that small blocks of soil break away on river banks, roadsides, gullies, or natural channels; (c) deep tree root systems support tree trunks and act as an anchor to the soil [34] resulting in a high root resistance to storm force and reducing the chance that larger soil masses are flushed away once channels are formed. Depending on the relative importance of those mechanisms, the choice of vegetation and its associated rooting pattern can influence slope stability. The susceptibility of cut slopes to land sliding can under certain circumstances be increased rather than reduced by the establishment of a vegetation cover, if the increase in infiltration rate offsets the mechanical benefits of soil reinforcement by roots.

Not all tree roots are the same. A study of 10 European perennials [35] found that root tensile strength–diameter relationships depend strongly on taxa. Root characteristics including root length density, diameter, lignin, and polyphenolic content are known to affect the soil cohesion [36]. Lignin content of wood is the primary determinant of strength, but also affects the rate of decomposition [18]. The role of polyphenolics and protein (linked to N content) is less clear in root strength, but these factors certainly influence the rate of decomposition and hence decay rate of strength of dead roots [37–39]. Root tensile strength, which is supposedly an intrinsic property that does not depend on the dimensions of the test specimen, can increase with decreasing root diameter, associated with a higher percentage of cellulose [40]. While trees differ in their spatial distribution of woody roots over the soil profile, their branching pattern generally adheres to fractal branching rules and allometric relations on the basis of proximal root diameters at the stem base can be used to predict total size of a root system, similar to the allometric equations that relate aboveground biomass to stem diameter [41,42].

Scaling up from a bundle of roots to tree stands [43] needs to consider counteracting processes. Tree roots not only influence tensile strength that reduces, but also infiltration and subsurface flows that contribute to landslide risk [44]. In the context of a broad evaluation of the role of agroforestry in maintaining or restoring watershed functions in the humid tropics [45], we tested hypotheses at four scales (individual) woody roots, rooted volumes of soil, tree root systems, and riparian vegetation:

1. The critical load where individual roots break is related to the root's lignin content,
2. The shear strength of a volume of soil increases proportionally to the number and strength of individual roots,
3. With a similar overall ratio of woody root to stem cross sectional area, there are consistent differences between tree species in root development in the topsoil and at depth, that contribute to differences in soil binding and soil anchoring, respectively,
4. Differences in the distribution of tree roots between species can be used to reduce landslide risks in the context of productive coffee agroforestry systems.

2. Materials and Methods

2.1. Location

The research was carried out in six sites: (a) Waybesai watershed, Sumberjaya (W. Lampung), (b) Ciherang and Cibadak sub-watershed, Sentul (West Java), (c) Kalikonto sub-watershed, Pujon and Ngantang-Malang (East java), (d) the UB-forest on the slopes of the Arjuna mountain (Malang), (e) Bangsri watershed Wajak-Malang (East java), (f) Upper Bedadung watershed, Jember, (East Java) (Figure 1 and Table 1). The measurements were performed in three steps: (a) individual root strength in relation to root properties; (b) root length density in relation to soil shear strength in the surface layer in Sumberjaya, West Lampung; (c) inventory of the potential of tree root systems as an anchor to maintain hillslope and river bank stability in all six sites.

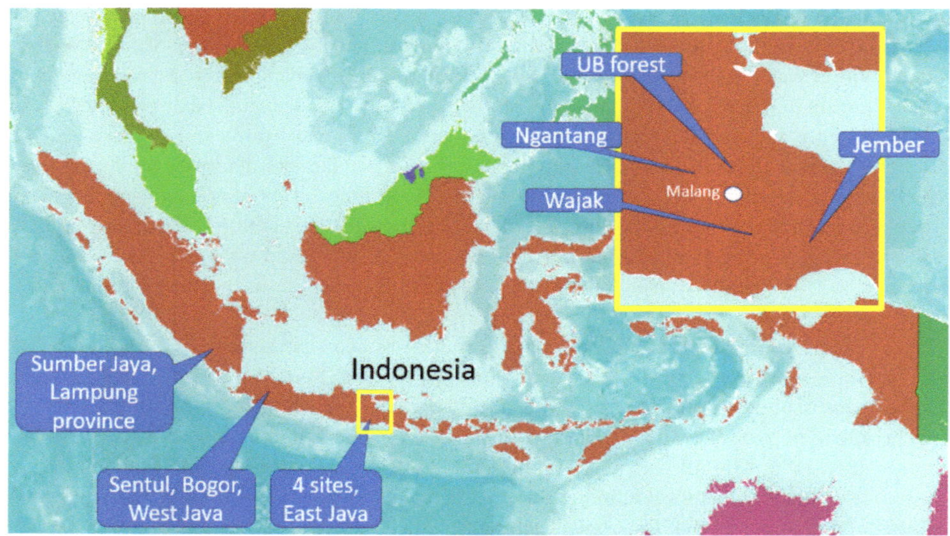

Figure 1. Location of six study sites in East Java, West Java, and Sumatra.

Table 1. Geoposition and characteristics of five study sites.

Watershed and Time of Study	Geoposition	Altitude, m above Sea Level	Annual Rainfall, mm	Soil Type	Main Tree Crops
Way Besai, Sumber Jaya, West Lampung, (2005–2006)	5°01′29.9″–5°02′34.2″ S, 104°25′46.5″–104°26′51.4″ E	700–1700	2500–2614	Inceptisol and Andisols	*Coffea, Gliricidia, Persea, Durio, Artocarpus, Aleurites*
Ciherang and Cibadak sub-watershed, Sentul Bogor, West Java (2007)	6°35′37″–6°35′29″ S, 106°54′58″–106°55′47″ E	1529	4000	Inceptisol	*Maesopsis, Pangium, Eugenia, Sandoricum*
Kali Konto sub-watershed (Pujon and Ngantang village), Malang, East Java (2006–2007)	7°46′03″–7°56′54″ S, 112°19′20″–112°29′55″ E.	1150	1797–3151	Inceptisol	*Coffea, Durio, Persea, Gliricidia, Pinus, Maesopsis, Toona, Agathis*
Kalisari sub-watershed (UB Forest), Malang, East Java (2019)	7°49′30″–7°51′36″ S, 112°34′38″–112°36′53″ E	900–1300	2005	Inceptisol, Andisol	*Pinus, Swietenia, Coffea, Gliricidia, Calliandra, Hibiscus, Parkia, Toona, Erythrina*
Bangsri sub-watershed, Wajak, Malang, East Java (2018)	8°7′29″–8°7′32″ S, 112°47′30″–112°48′30″ E	660–720	2220–2314	Inceptisol, Entisol	*Persea, Durio, Paraserianthes, Swietenia, Pinus, Michelia*
Upper Bedadung watershed, Jember, East Java (2006)	7°59′66″–8°33′56″ S, 113°19′–114°02′30″ E,	300–3000	2699	Inceptisol and Andisols	*Coffea, Leucaena, Swietenia, Hevea, Ochroma,*

2.2. Individual Root Strength in Relation to Root Properties

Root strength, the tensile force per unit area of root needed to break the root, was measured by clamping a weight onto a vertically hanging root and increasing the downward force until rupture. A tensile strength apparatus [46,47] was developed locally. Five tree species (all 5 years old) were selected for this study: mahogany (*Swietenia mahogani*), gmelina (*Gmelina arborea*), suren (*Toona sureni*), coffee (*Coffea canephora*), and bamboo (*Bambusa arundinacea*). Root samples were collected from three trees of each species from farmers' plots in Singosari village (Malang). Fresh roots with a length of about 25 cm and a diameter of close to 2 mm were placed in the tensile strength apparatus. The mean of six replicate root measurements of the force required to shear the root was used as indicator of root strength. A composite dry root sample of about 10 g from each species was used for analysis of chemical root characteristics: total carbon content C measured with the wet oxidation method of Walkley and Black [48], total nitrogen content (N) measured using the Kjeldahl distillation method, concentration of lignin by boiling plant sample with a sulfuric acid solution of Cetrimethyl Ammonium Bromide (CTAB) under controlled temperature [49], and polyphenolics by extracting plant tissues with methanol and subsequently colorimetrically measuring absorbance at a wavelength of 760 nm by the Folin–Denis method [50].

2.3. Role of Root Length Density in Soil Shear Strength

Roots add shear strength to soil when the root network penetrates a potential failure surface. The amount of tensile root force contributed to a potential slide mass should increase with increasing area of root intersection. Hence, we explored how soil shear strength was related to the root length density (L_{rv}) at different distances to tree trunks. Tree roots can increase the shear strength of shallow soils through mechanical reinforcement [9].

Measurement on tree root length density was done in two study sites, focused on coffee (*Coffea canephora*) and shade tree species commonly grown in coffee agroforestry system of each site. In Sumberjaya, root measurements thus included coffee (*Coffea canephora*), *Gliricidia sepium* ('kayu hujan'), *Maesopsis eminii* ('pohon afrika'), *Artocarpus heterophyllus* (jackfruit), and *Bambusa arundinacea* (bamboo), and mixed 'fallow vegetation' (dominated by *Trema orientalis* and *Melastoma*; local names are, respectively, 'anggrung' and 'harendong') and the edges of rice fields. In Malang, the measurement was performed on five tree species i.e., *Coffea canephora* (coffee) and *Bambusa arundinacea* (bamboo), *Swietenia mahogany* (mahogany), *Gmelina arborea* (gmelina), *Toona sureni* (suren). The trees selected for measurements were 5 years old, with three replications.

Soil and root samples were taken using two PVC rings of 5 cm height and 10 cm in diameter, temporarily connected, for measuring soil shear strength (Figure 2; [51]), and subsequently root length density. Soil and root samples were taken from the topsoil layer of 0–10 cm, at different distances to trees i.e., 50, 100, 150, 200 cm to obtain variation in root length density. For rice, samples were taken along a bund at 0–5 cm soil depth. Soil shear strength was measured by holding the lower and pulling the upper ring, recording the weight that had to be added to cause a break between the two rings. Soil texture and soil bulk density were measured in these same samples using the pipette and gravimetric technique. Subsequently, roots were separated from the soil by wet sieving. Roots of selected trees were separated from other roots of different crops. Root length density (L_{rv}) was measured using the line intersect method [52] on washed soil samples. Root samples were dried in the oven at 80 °C for 48 h to estimate their dry weight (D_{rv}). The GENSTAT 8.0 software [53] was used to carry out ANOVA (analysis of variance) including a *t*-test to separate means.

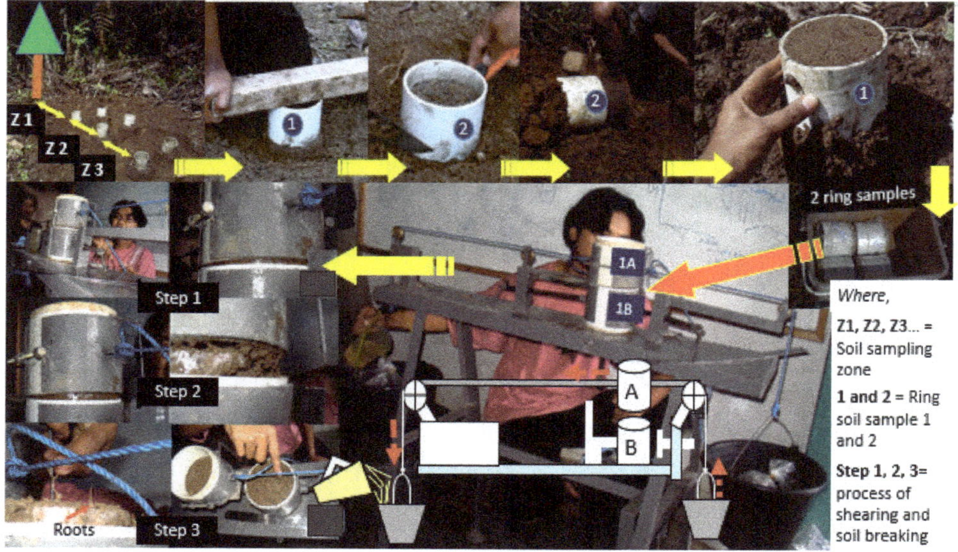

Figure 2. Taking soil samples in the study site in a double ring and measuring soil shear resistance by increasing the pull by adding water to a bucket until the upper ring moves.

2.4. Inventory of Potential Tree Root System as an Anchor to Maintain Slope and Riverbank Stability

At each of the six sites saplings (<5 cm dbh) and trees (>5 cm dbh) of common trees species were selected in multistrata coffee and other agroforestry systems, with a minimum age of 5 years (according to local farmer assessment). High economic value trees were prioritized for the observations and compared to bush fallow vegetation. Proximal roots (close to the tree stem) were exposed [41], with their diameter measured at 20–30 from the trunk, classified as either 'horizontal' or 'vertical' (using an angle of 45° as threshold; Figure 3). Two root indices were calculated i.e., Index of Root Anchoring (IRA) and Index of Root Binding of Soil (IRB). IRA was calculated as $\Sigma D_v^2/dbh^2$ where dbh is tree diameter at breast height (1.3 m height) and D_v is the diameter of vertical roots [54,55]; IRB was calculated as $\Sigma D_h^2/dbh^2$, where D_h is diameter of horizontal roots. Some vertical roots were sampled for measurement of wood density, lignin content, and polyphenolic content using the same method as described in Section 2.2.

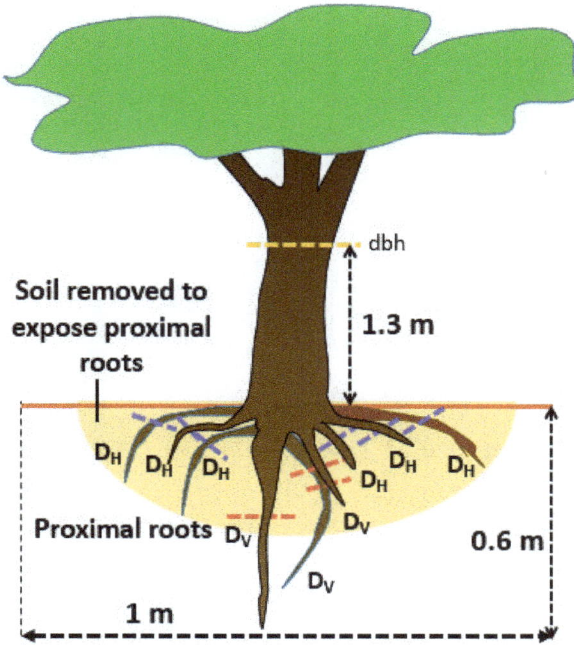

Figure 3. Schematic diagram of the distribution of proximal roots [41]; horizontal (H) roots descend at an angle of <45°, vertical (V) roots descend at an angle of >45°; D = root diameter; dbh = diameter at breast height.

3. Results

3.1. Individual Root Strength in Relation to Root Properties

Tree roots of approximately 2 mm in diameter (fine roots) were found to break when a weight of 2–15 kg was applied, which corresponds with 64–478 kPa (kg cm^{-2}) or 0.06–0.48 MPa of root tensile strength. There was considerable variation in lignin content of the roots (13–30%, with high contents in mahogany and coffee; Table 2) and this was associated with root tensile strength (Figure 4A). About 70% of the variation in root strength was associated with variation in lignin content. The ratio of (Lignin + Polyphenol)/N that generally associate with decomposition rate [38] had less predictive power on root strength than lignin content alone. A multiple regression shows that root strength was positively related with the N content, but negatively with the polyphenol content, with about 80% of variation accounted for (Figure 4B).

Table 2. Chemical characteristics of tree root system.

Litter Type	N (%)	L (%)	P (%)	L/N	P/N	(L + P)/N
Mahogany	0.6	29.2	26.4	45.0	40.7	85.7
Toona	1.2	18.9	7.7	16.4	6.7	23.1
Gmelina	1.3	13.5	8.1	10.1	6.0	16.1
Coffea	5.4	20.1	5.4	3.7	1.0	4.7
Bamboo	1.0	16.0	1.6	16.5	1.6	18.2

N: total nitrogen content, L: lignin content, P: polyphenol content.

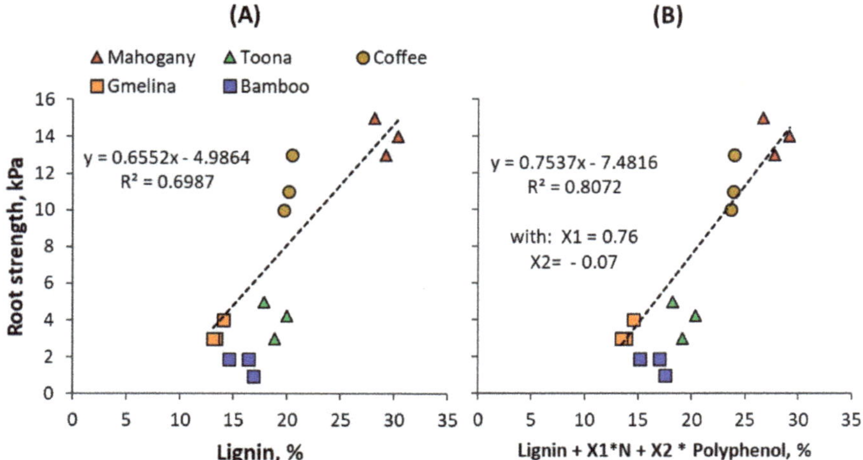

Figure 4. Regression of root tensile strength on (**A**) lignin content and (**B**) a linear combination of lignin content L, nitrogen content N, and polyphenol content P.

3.2. Tree Root Length Density (L_{rv})

Figure 5A shows root length density, L_{rv} (km m^{-3}) in the topsoil layer of 0–10 cm for different tree species with ages of 3, 5, and 7 years old. The average L_{rv} for coffee, *Gliricidia*, *Maesopsis*, and *Artocarpus* was 2.9, 1.4, 2.7, and 1.7 km m^{-3}, respectively. Statistical analysis revealed a significant ($p < 0.05$) interaction between tree species and age. Root length density of coffee was lower at 7 than at 5 years of age, probably affected by tree pruning regimes after the peak coffee harvest years around 5 years of age. The highest L_{rv} was found in 5-year-old coffee (6.5 km m^{-3}). At 7 years, *Maesopsis* and jackfruit had a L_{rv} of about 2.1 km m^{-3} significantly ($p < 0.05$) higher than that in 7-year old coffee and *Gliricidia*. The lowest L_{rv} was found in 3-year-old *Gliricidia* (1.2 km m^{-3}). In comparison to non-tree species i.e., bamboo and shrubs (about 3–5 years) and rice, the trees tested here had about 80 % lower L_{rv} than non-tree (average 10 km m^{-3}). Bamboo has the highest L_{rv}, with about 14.6 km m^{-3} followed by mixed shrub vegetation (10 km m^{-3}) and rice (5.6 km m^{-3}).

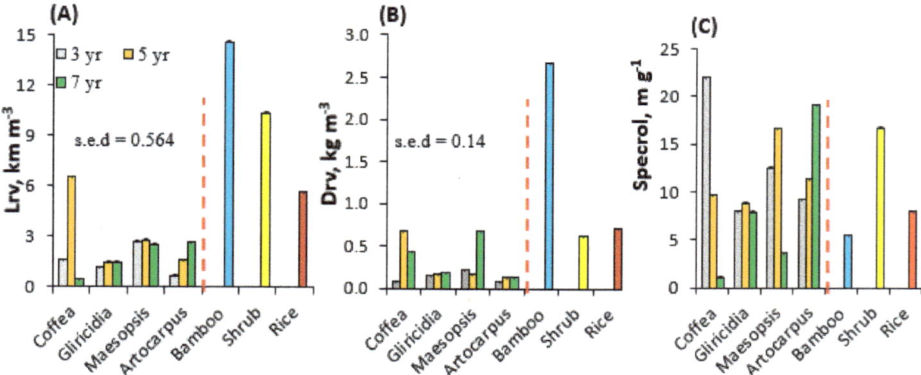

Figure 5. (**A**) Root length density (L_{rv}), (**B**) root dry weight (D_{rv}), and (**C**) specific root length (L_{rv}/D_{rv}) at the topsoil layer of 0–10 cm of different tree species and non-tree species (bamboo, shrub, and rice).

The highest root biomass and root weight per unit volume of soil, D_{rv}, was found in 5-year-old coffee (0.68 mg cm^{-3}); jackfruit had the lowest D_{rv} with an average of about 0.12 mg cm^{-3} (Figure 5B).

The specific root length (specrol = L_{rv}/D_{rv}, m g^{-1} of roots) differed between tree species; coffee had finer roots as shown by a high specrol value of about 22 m g^{-1}, but it declined rapidly with time to 9 m g^{-1} (Figure 5C); a similar case was also found in *Maesopsis*. Jackfruit roots, by contrast, increased in specrol with time up to 19 m g^{-1}, while the specrol of *Gliricidia* was not changing with time at about 8 m g^{-1}. Bamboo had the lowest specrol (5.5 m g^{-1}).

3.3. Soil Shear Strength

The average soil shear strength in our measurements was 3.37 MPa (Figure 6). There appeared to be ($p = 0.055$) a significant interaction between tree species and age on soil shear strength, accompanying a significant difference between tree species. The highest soil shear strength was found in the plot with bamboo (average of 4.82 MPa). Soil shear strength in the bund of a paddy rice field (2.38 MPa) was higher than that for all trees tested, except for coffee.

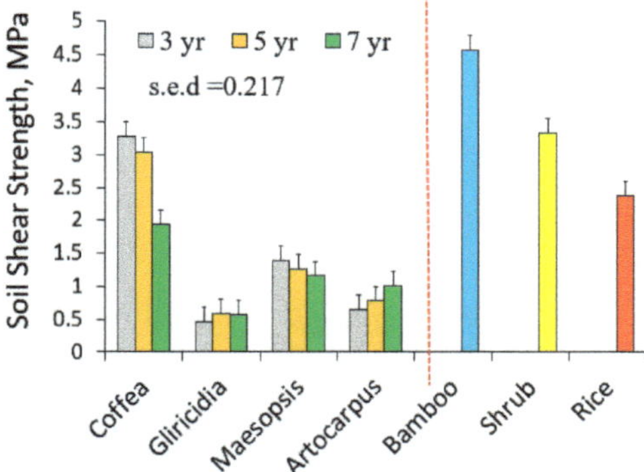

Figure 6. Soil shear strength (SS) at 5 cm depth for samples taken close to four species of trees and three types of non-trees.

3.4. Soil Shear Strength in Relation to Root Length Density in the Topsoil

The high shear strength close to bamboo stands was clearly correlated to the high root length density (Figure 7). In other trees and non-trees such relation was not clear, possibly due to the scarcity of soil samples with L_{rv} values above 10 km m^{-3}.

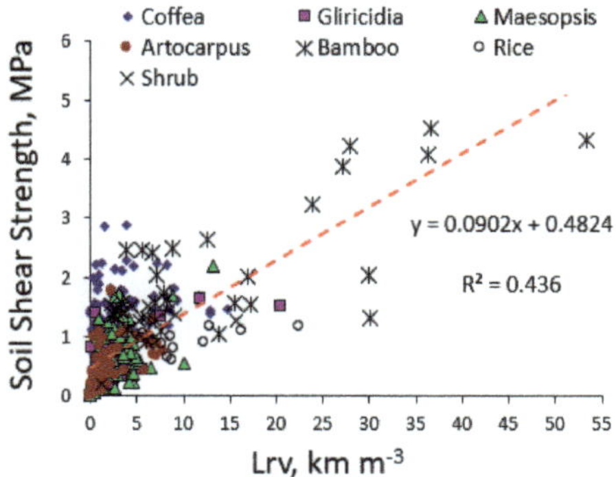

Figure 7. Regression of soil shear strength at 5 cm depth on root length density (L_{rv}) of various plant species in the layer 0–10 cm.

3.5. Allometry of Proximal Tree Root Systems

Results of the root survey of various tree species in six locations suggested difference among locations, with proximal root diameters relative to the stem diameters high in Bangsri and low in Ngantang (Figure 8). Our data allow a partial disentangling of site (e.g., soil or climate related), species, and age effects.

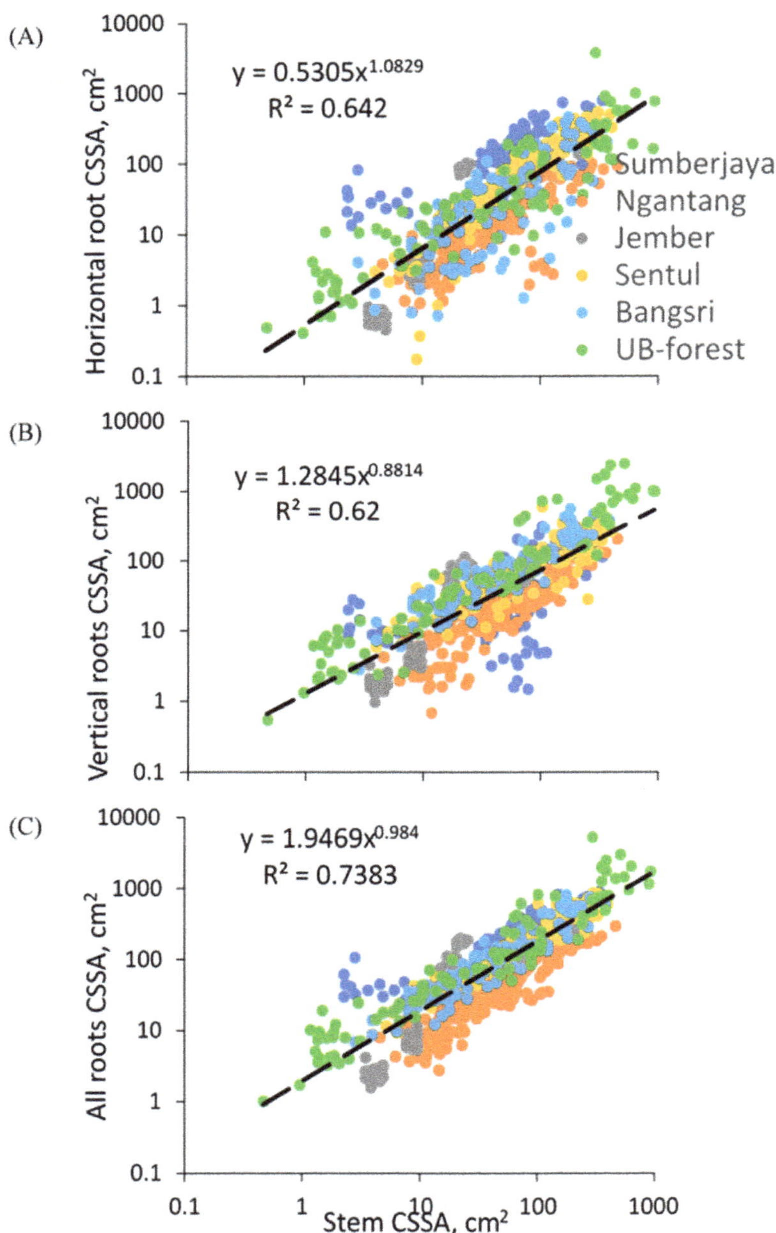

Figure 8. Relationship between cross-sectional area (CSSA) of tree stems and (**A**) horizontal roots ($D_h{}^2$), (**B**) vertical roots ($D_v{}^2$), and (**C**) total proximal roots for 685 tree samples of 60 tree species across five study sites.

Across all tree observations, variation in tree diameter accounted for 74% of variation in the sum of proximal root CSSA (Figure 8C), 62% of variation in vertical root CSSA (Figure 8B), and 64% of variation in horizontal root CSSA (Figure 8A). Average root-to-shoot ratio in terms of CSSA was around 2.39 (mean 2.21), average IRA 1.30 (median 1.11), and average IRB 1.09 (median 0.816).

Across all species the relative importance of vertical roots decreased with increasing stem diameter, and that of horizontal roots increased, with only a slight decline in root/shoot ratio (in terms of cross-sectional area) from 1.95 to 1.74 in the observed range of tree sizes (Table 3). The ratio of vertical to horizontal roots declined from 2.42 to 0.60, with lateral roots taking over from the tap root in most species.

Table 3. Average impact of increasing stem diameter (and cross-sectional area, CSSA) on vertical and horizontal proximal root CSSA, based on regression equations in Figure 8.

Stem CSSA, cm^2	Vertical (IRA)	Horizontal (IRB)	Root/Shoot CSSA	Vertical/Horizontal Ratio
1	1.28	0.53	1.95	2.42
10	0.98	0.64	1.88	1.52
100	0.74	0.78	1.81	0.96
1000	0.57	0.94	1.74	0.60

The ANOVA showed that tree species differed significantly ($p < 0.05$) in IRA (Index of Root Anchoring) and IRB (Index of Root Binding) values but given the overall relationship between stem diameter and both IRA and IRB values (Table 3), differences in the diameters at which tree species were observed could be (partially) responsible for this finding.

Appendix A lists the 50+ species to which the 685 measured trees belong with, a minimum of four, a median of 9.5, an average of 12.7, and a maximum of 58 observations per tree species (with robusta coffee as most studied tree). The mean diameters per species of the observed trees had a minimum of 1.5 cm, a median of 8.5 cm, an average of 8.0 cm, and a maximum of 19.9 cm. Across all tree species, there was no significant relationship (as evident from the low fraction of variance accounted for or R^2 value) between the average stem diameter at which a species was observed, and the average IRA (or IRB) value recorded (Figure 9). We can thus assume that differences in observed IRA and IRB values between species cannot be simply attributed to the differences in stem diameters for the trees observed.

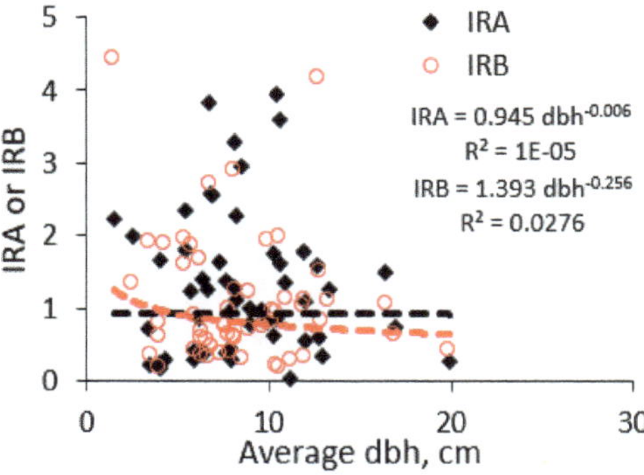

Figure 9. Relationship between average value of the IRA and IRB indices and average stem diameter at which a tree species was observed for the 50+ tree species in the survey.

Based on the observed IRA and IRB distributions at species level (rather than at the observed tree level, as the number of observations per species was uneven), the lower and upper quartile were 0.5 and 1.8 for IRA, and 0.5 and 1.5 for IRB, respectively (Figure 10). Using these thresholds, Table 4 shows a two-way classification of the tree species observed in the survey.

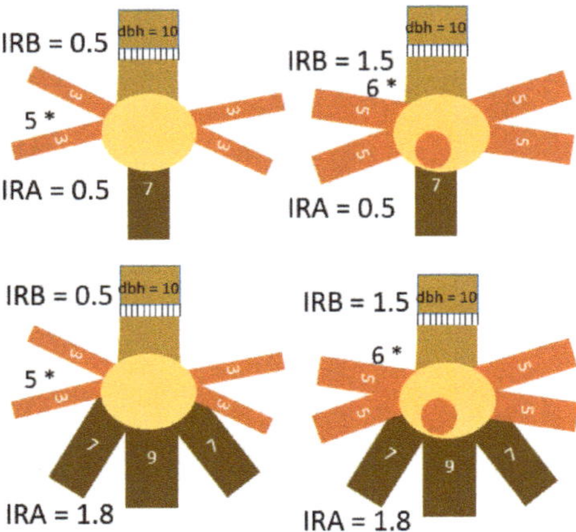

Figure 10. Examples of root-to-shoot relations at the four classification thresholds in Table 4. The IRB for a tree with stem diameter of 10 and 6 horizontal roots of 5 cm diameter is $6 * 5^2/10^2 = 1.5$, while that for a tree with 5 roots of 3.16 cm is $5 * 3.16^2/10^2 = 0.5$.

Table 4. Two-way classification of trees based on rooting indices IRA (Index of Root Anchoring) and IRB (Index of Root Binding soil particles) classified by lower and upper quartiles of their distributions

INDEX	IRB_Low (<0.5)	IRB_Medium (0.5–1.5)	IRB_High (>1.5)
IRA_Low <0.5	*Mangifera indica, Citrus × sinensis, Malus domestica, Cassia fistula, Agathis dammara, Ficus benjamina, Annona muricata*	*Spathodea campanulata, Pinus merkusii, Melia azedarach, Jatropha curcas, Ceiba pentandra, Trema orientalis*	*Litsea garcia*
IRA_Medium 0.5–1.8	*Eucalyptus deglupta, Cinnamommum burmannii*	*Parkia speciosa, Sandoricum koetjape, Hevea brasiliensis, Aleurites moluccana, Durio zibethinus, Erythrina subumbrans, Ochroma pyramidale, Syzygium polyanthum, Swietenia mahagoni, Persea americana, Calliandra calothyrsus, Pangium edule, Maesopsis eminii, Nephelium lappaceum, Psidium guajava, Toona sureni, Coffea canephora var. robinson, Artocarpus communis, Tectona grandis*	*Homalantus populneus, Artocarpus heterophyllus, Hibiscus tiliaceus, Pterocarpus indicus, Paraserianthes falcataria, Leucaena leucocephala, Artocarpus elasticus*
IRA_High >1.8	*Quercus lineata, Macarangga triloba, Syzygium aqueum, Croton argyratus, Ficus padana*	*Coffea canephora var. robusta, Piper aduncum, Gmelina arborea*	*Michellia alba, Coffea arabica, Gliricidia sepium, Senna spectabilis*

All nine cells of Table 4 had at least one tree species, with the largest number (obviously) in the mid-range values for both IRA and IRB. The seven species in the upper left cell (low IRA and low IRB) had a low overall root-to-shoot ratio, while the four species in the lower right cell (high IRA and high IRB) a high overall one. The other cells represent a partial tradeoff between vertical and horizontal roots (or IRA and IRB values, respectively).

The two main coffee species (arabica and robusta) had a relatively high IRA value (strong taproot system), with average or high lateral root development as well (relative to stem diameters). Most of the fruit trees commonly found in coffee agroforestry (such as *Artocarpus elasticus* (jackfruit), *Parkia speciosa* (petai), and *Durio zibethinus* (durian)) had intermediate values for both IRA and IRB. The same holds for common timber species such as *Maesopsis eminii* (pohon afrika), *Toona sureni* (suren), *Tectona grandis* (teak), *Swietenia mahagoni* (mahogany). However, their generally larger stem diameters, relative to coffee, still imply a larger absolute contribution to vertical root development in the plot as a whole. Legume trees commonly used as shade trees in current coffee agroforestry systems, *Gliricidia sepium* and *Senna spectabilis* (ramayana), provided a higher anchoring per unit stem diameter than the more traditional shade tree, *Erythrina subumbrams* (dadap).

A further comparison between pruned and unpruned robusta coffee in Sumberjaya (data not shown) found the highest IRA (7.7) for unpruned coffee, suggesting that unpruned coffee is a good species for anchoring riverbanks. This characteristic disappears with pruning, confirming earlier research where frequent pruning of other tree species also led to the formation of more roots in the surface layer [56].

4. Discussion

All four hypotheses were confirmed in this study. The break strength of woody roots across five species was positively related to lignin and nitrogen content and negatively to polyphenol content. A linear model explained 81% of the variation (Figure 4B). Lignin content alone accounted for 70% of the variation. Mahogany (*Swietenia mahogani*) and coffee (*Coffea canephora*) had the strongest roots, gmelina (*Gmelina arborea*) and suren (*Toona sureni*) the weakest, and giant bamboo (*Bambusa arundinacea*) had an intermediate root strength.

Soil shear strength depends on soil texture and soil water content, and we found a considerable spread in shear strength with low root length densities (Figure 7). The only species with sufficiently high root length densities to increase shear strength was bamboo. However, our test was based on the top 10 cm of soil, testing a possible break at 5 cm depth, which is unlikely to happen. The shallow landslides of most interest would slide at depths of 30 cm to 1 m. From our data it is unlikely that root length densities at such depth would be high enough to have a substantial effect on shear strength. However, as landslides are binary (yes or no) events, a small change in the threshold value may still be relevant. In terms of land use change, it is the change in root presence that matters, especially a decrease, as a preceding (forest) vegetation may have prevented landslides to occur that were otherwise 'due' (as the thickness and weight of soil layers exceeded what can be prevented from sliding by the soil properties as such). To judge absolute contributions to anchoring of the rooted soil, absolute tree size may be at least as important as the relative allocation over vertical and horizontal roots indicated by the IRA and IRB values.

The combination of root quality characteristics determining tensile strength according to our results in Figure 4 is different from that determining decomposition rate [39], although high lignin content is associated with strong and slowly decomposing roots. Tree species having specific rooting habits (high root strength and root diameter) can be used to control erosion and when linked with extreme flood probability can be used to indicate the risk of a storm likely to cause slope instability in the period between clear-felling and tree re-growth [19]. Our root tensile strength measurements of 64–478 kPa were lower than values reported in the literature, such as the 17.6 MPa for woody roots of 5.5 mm diameter measured for *Pinus radiata* [24]. More measurements on larger-diameter woody roots are still needed to clarify the difference between these results.

Trees with high IRA (Index of Root Anchoring) do not always have a low IRB (Index of Root Binding of soil particles) value or vice versa. Results in Table 4 can be interpreted as guide for suitability of trees as riverbank stabilizer. Coffee (especially without pruning) is potentially suitable for anchoring and soil surface holding at the riverbank, but it has a low root length density (Figure 5). Combination with other trees such as Gliricidia (high IRB as well as IRA) and other grasses is probably better to increase slope stability. High economic value trees (fruit and timber trees) in Sumberjaya (West Lampung) [57] are mostly grouped in the medium class (Table 4). Indeed, tree roots contribute to soil shear strength, but only in the upper 0.4 m of the soil. Most failures, however, occur at greater depths as anchorage by deeper roots was not effective or absent [58]. Evaluation of suitable trees for stabilizing riverbank should also consider the root strength differences [59].

Compared to other methods of observing tree root characteristics the proximal root method, based on fractal branching theory [42], allowed a much wider part of the field-level variation among sites, tree species, tree management, and tree size to be sampled, with an investment of 1–2 person-hours per tree for preparation and observation. As no roots are cut and soil can be returned to the original position the method is essentially 'non-destructive'. Among the limitations of the classification of proximal roots is that tree species with large horizontal roots may have 'sinker' roots further away from the stem that can have a major effect on strengthening the rooted soil as a whole. Root studies on Indonesian trees in the 1930s showed a wide range of rooting patterns along vertical and horizontal spread, including sinker roots [60].

A further caveat is that measurements in the dry season might be different; as all our measurements were during the rainy season. Landslide risk can be higher at the start of the rainy season (when roots are not yet fully active), as well as towards the end (when the whole soil system tends to be water-saturated).

The interest of farmers or watershed managers in deep roots for possible landslide risk reduction, may need to be balanced by considerations from the tree's perspective [61]. In terms of plant strategies, deep roots help first of all in water acquisition during dry (and sunny etc.) periods [62]. Horizontal roots provide nutrients and water during rainy seasons. It is tempting to interpret the root-to-sheet ratio in terms of cross-sectional area (and likely xylem vessel transport capacity) as a fail-safe strategy to be able to support the leaf canopy on either surface or deeper soil water, depending on conditions. Any impact on slope stabilization is at best a 'co-benefit' for the tree, not a major selection force.

Conceptually the different scales (root, tree, and vegetation) can be combined to understand landslide risk at landscape level and to target interventions to reduce risk, but a fully quantitative model that includes all relations of dynamic tree root system growth and decay along with the probability of rainfall intensities and slope geo-hydrology is not feasible at this stage. While most practices that try to reduce erosion target an increase of water infiltration [45], such increase can effectively increase landslide risk unless it is accompanied by deep and strong tree root systems. The collapse of terraces on steep slopes can be interpreted this way. Increased risk of landslides after reduction in tree cover is likely to peak once tree root system decay has reduced soil shear strength and before water infiltration into the soil profile has been reduced. Gradual replacement of trees on hill slopes, as practiced in the 'sisipan' management style of agroforests [63], will have lower risk than a clear felling-replanting. Maintaining and promoting mixed tree vegetation will reduce (but not necessarily minimize) risk, while providing other benefits to local livelihoods. Based on this study, we suggest a mix of tree species with deep roots and grasses with intense and strong fine roots will provide the highest riverbank and hillslope stability, especially if drastic tree pruning or clearing-replanting events can be avoided.

5. Conclusions

Our investigations confirmed that woody roots play a role in holding soil together and resisting mass movement in landslides. The critical load where individual roots break is related to the root's lignin content (its woodiness). The shear strength of a volume of soil over any plane increases proportionally with the number and strength of individual roots. With a similar overall ratio of woody root to stem cross sectional area, there are consistent differences between tree species in root development in the topsoil and at depth, that contribute to differences in soil binding and soil anchoring, respectively. Differences in the distribution of tree roots between species can be used to reduce landslide risks in the context of productive coffee agroforestry systems. Tree root architecture varies with species, location, and age. On average the cross-sectional area of proximal roots is twice that of the stem, with only a small reduction in the ratio with increasing stem diameter. The fraction of proximal roots in the vertical and horizontal categories varies more between species than the root-to-shoot ratio of cross-sectional areas. Using the Index of Root Anchoring (based on vertical roots) and the Index of Soil Binding (based on horizontal roots), tree species in the upper and lower quartile of each indicator were identified. However, the strong location effect that was observed implies that site-specific observations will be needed for selecting trees that are locally deep-rooted, taking current observations only as first indication. A mixed vegetation is the simplest way of ensuring that both soil binding and soil anchoring functions are provided.

Author Contributions: Conceptualization, K.H. and M.V.N; Investigation, K.H., W.W. and D.S.; Methodology, K.H., W.W., D.S. and M.V.N.; Validation, D.S.; Writing—original draft, K.H. and M.V.N.; Writing—review and editing, K.H. and M.V.N. All authors have read and agreed to the published version of the manuscript.

Funding: The Sumberjaya case study has been funded by the Australian Center for International Agricultural Research (ACIAR) through the World Agroforestry Center (ICRAF)-Southeast Asia, while the case study in Malang was funded by the Indonesian Ministry of Culture and Education (DIKTI, A2 Program). CIFOR through TroFCCA project supported our activity in Ciherang sub-watershed (Sentul, Bogor), UNDP-KLHK through CCCD Project in Bangsri sub watershed (Malang), and the work in UB Forest was supported by Rector of Brawijaya University through collaborative research UB-CEH(UK).

Acknowledgments: Ari Santosa, Mohadi Nurhada, Veronika Kurniasari, Emanuel Ardian Kristanto, Astrid Ferera, Arya Putra Widyatmadya, and Tsarwah Al Sausan offered valuable contributions during their field work as undergraduate students of University of Brawijaya, Malang. We appreciated the valuable inputs of Rachmat Mulia, Fahmuddin Agus, and Jianchu Xu to earlier drafts.

Conflicts of Interest: The authors declare no conflict of interest. The funders had no role in the design of the study; in the collection, analyses, or interpretation of data; in the writing of the manuscript, or in the decision to publish the results.

Appendix A

Table A1. Root observations across six sites, with number of trees observed (N), average dbh, IRA, and IRB, and quartiles in IRA and IRB distribution.

Species	N	Dbh Avg	IRA Avg	IRB Avg	Quartile	Quartile
Mangifera indica	6	11.2	0.04	0.29	IRA Q1	IRB Q1
Citrus × sinensis	6	4.0	0.18	0.19	IRA Q1	IRB Q1
Malus domestica	6	3.5	0.23	0.36	IRA Q1	IRB Q1
Cassia fistula	6	19.9	0.26	0.42	IRA Q1	IRB Q1
Agathis dammara	6	5.9	0.29	0.43	IRA Q1	IRB Q1
Ficus benjamina	6	7.9	0.29	0.41	IRA Q1	IRB Q1
Annona muricata	6	6.3	0.37	0.35	IRA Q1	IRB Q1
Spathodea campanulata	6	4.0	0.25	0.61	IRA Q1	IRB Q23
Pinus merkusii	11	12.9	0.34	0.83	IRA Q1	IRB Q23
Melia azedarach	11	7.6	0.40	0.75	IRA Q1	IRB Q23
Jatropha curcas	6	6.3	0.41	0.68	IRA Q1	IRB Q23
Ceiba pentandra	6	5.9	0.41	0.89	IRA Q1	IRB Q23
Trema orientalis	10	7.8	0.47	0.64	IRA Q1	IRB Q23
Litsea garcia	5	4.3	0.30	1.88	IRA Q1	IRB Q4
Eucalyptus deglupta	6	12.0	0.56	0.33	IRA Q23	IRB Q1
Cinnamommum burmannii	10	7.3	1.65	0.38	IRA Q23	IRB Q1
Parkia speciosa	19	12.8	0.59	1.50	IRA Q23	IRB Q23
Sandoricum koetjape	20	10.2	0.62	0.96	IRA Q23	IRB Q23

Table A1. *Cont.*

Species	N	Dbh Avg	IRA Avg	IRB Avg	Quartile	Quartile
Hevea brasiliensis	10	16.9	0.75	0.64	IRA Q23	IRB Q23
Aleurites moluccana	15	8.9	0.80	0.71	IRA Q23	IRB Q23
Durio zibethinus	15	10.4	0.87	0.95	IRA Q23	IRB Q23
Erythrina subumbrans	20	7.8	0.95	0.98	IRA Q23	IRB Q23
Ochroma pyramidale	7	9.6	0.99	0.76	IRA Q23	IRB Q23
Syzygium polyanthum	5	8.9	1.00	1.218	IRA Q23	IRB Q23
Swietenia mahagoni	45	12.0	1.09	1.11	IRA Q23	IRB Q23
Persea americana	21	8.2	1.13	1.26	IRA Q23	IRB Q23
Calliandra calothyrsus	15	6.6	1.26	0.57	IRA Q23	IRB Q23
Pangium edule	23	13.3	1.27	1.11	IRA Q23	IRB Q23
Maesopsis eminii	31	10.9	1.36	1.14	IRA Q23	IRB Q23
Nephelium lappaceum	32	7.6	1.37	0.77	IRA Q23	IRB Q23
Psidium guajava	10	6.3	1.40	0.58	IRA Q23	IRB Q23
Toona sureni	25	16.4	1.50	1.07	IRA Q23	IRB Q23
Coffea canephora var. *robinson*	16	4.0	1.66	0.80	IRA Q23	IRB Q23
Artocarpus communis	4	10.2	1.76	0.83	IRA Q23	IRB Q23
Tectona grandis	9	11.9	1.79	1.03	IRA Q23	IRB Q23
Homalantus populneus	5	3.4	0.73	1.90	IRA Q23	IRB Q4
Artocarpus heterophyllus	35	9.9	0.82	1.93	IRA Q23	IRB Q4
Hibiscus tiliaceus	20	6.2	0.86	1.68	IRA Q23	IRB Q4
Pterocarpus indicus	5	5.7	1.25	1.86	IRA Q23	IRB Q4
Paraserianthes falcataria	15	8.1	1.28	2.91	IRA Q23	IRB Q4
Leucaena leucocephala	15	12.7	1.59	4.17	IRA Q23	IRB Q4
Artocarpus elasticus	4	10.6	1.62	1.99	IRA Q23	IRB Q4
Quercus lineata	4	6.9	2.57	0.48	IRA Q4	IRB Q1
Macarangga triloba	4	8.5	2.96	0.30	IRA Q4	IRB Q1
Syzygium aqueum	4	10.6	3.60	0.19	IRA Q4	IRB Q1
Croton argyratus	4	6.7	3.84	0.34	IRA Q4	IRB Q1
Ficus padana	4	10.4	3.96	0.22	IRA Q4	IRB Q1
Coffea canephora var. *robusta*	58	2.5	1.99	1.35	IRA Q4	IRB Q23
Piper aduncum	4	8.2	2.28	0.58	IRA Q4	IRB Q23
Gmelina arborea	4	8.1	3.28	0.62	IRA Q4	IRB Q23
Michellia alba	5	5.4	1.81	1.96	IRA Q4	IRB Q4
Coffea arabica	10	1.5	2.23	4.43	IRA Q4	IRB Q4
Gliricidia sepium	14	5.4	2.35	1.60	IRA Q4	IRB Q4
Acacia auriculiformis	16	6.8	2.58	2.72	IRA Q4	IRB Q4

References

1. Sidle, R.C.; Dhakal, A.S. Recent advances in the spatial and temporal modeling of shallow landslides. In Proceedings of the 2003 MODSIM Conference, Townsville, Australia, 14–17 July 2003; pp. 602–607.
2. Stokes, A.; Norris, J.E.; Van Beek, L.P.H.; Bogaard, T.; Cammeraat, E.; Mickovski, S.B.; Jenner, A.; Di Iorio, A.; Fourcaud, T. *How vegetation reinforces soil on slopes. In Slope Stability and Erosion Control: Ecotechnological Solutions*; Springer: Dordrecht, The Netherlands, 2008; pp. 65–118.
3. Van Noordwijk, M.; Hairiah, K.; Lasco, R.; Tata, H.L. How can agroforestry be part of disaster risk management. In *Sustainable Development through Trees on Farms: Agroforestry in Its Fifth Decade*; Van Noordwijk, M., Ed.; World Agroforestry (ICRAF): Bogor, Indonesia, 2019; pp. 215–229.
4. Pohl, M.; Alig, D.; Körner, C.; Rixen, C. Higher plant diversity enhances soil stability in disturbed alpine ecosystems. *Plant Soil* **2009**, *324*, 91–102. [CrossRef]
5. Alam, S.; Banjara, A.; Wang, J.; Patterson, W.B.; Baral, S. Novel Approach in Sampling and Tensile Strength Evaluation of Roots to Enhance Soil for Preventing Erosion. *Open J. Soil Sci.* **2018**, *8*, 330. [CrossRef]
6. Giadrossich, F.; Schwarz, M.; Cohen, D.; Cislaghi, A.; Vergani, C.; Hubble, T.; Phillips, C.; Stokes, A. Methods to measure the mechanical behaviour of tree roots: A review. *Ecol. Eng.* **2017**, *109*, 256–271. [CrossRef]
7. Iverson, R.M. Landslide triggering by rain infiltration. *Water Resour. Res.* **2000**, *36*, 1897–1910. [CrossRef]

8.   Micheli, E.R.; Kirchner, J.W. Effects of wet meadow riparian vegetation on streambank erosion. 2. Measurements of vegetated bank strength and consequences for failure mechanics. *Earth Surf. Process Landf.* **2002**, *27*, 687–697. [CrossRef]

9.   Roering, J.J.; Schmidt, K.M.; Stock, J.D.; Dietrich, W.E.; Montgomery, D.R. Shallow landsliding, root reinforcement, and the spatial distribution of trees in the Oregon Coast range. *Can. Geotech.* **2003**, *40*, 237–253. [CrossRef]

10.   Wu, T.H.; Swanston, D.N. Risks of landslides in shallow soils and its relation to clearcutting in Southeastern Alaska. *For. Sci.* **1980**, *26*, 495–510. [CrossRef]

11.   O'Loughlin, C.L.; Watson, A.J. Root wood strength deterioration in radiate pine after clearfelling. *N. Z. J. For. Sci.* **1979**, *9*, 284–293.

12.   Wilkinson, P.L.; Anderson, M.G.; Lloyd, D.M. An integrated hydrological model for rain-induced landslide prediction. *Earth Surf. Processes Landf.* **2002**, *27*, 1285–1297. [CrossRef]

13.   Ziemer, R.R.; Swantson, D.N. *Root Strength Changes after Logging in Southeast Alaska*; USDA Forest Service Research Note PNW; Department of Agriculture, Forest Service, Pacific Northwest Forest and Range Experiment Station: Portland, OR, USA, 1997; Volume 306.

14.   Abe, K.; Ziemer, R.R. *Effect of Tree Roots on Shallow-Seated Land Slides*; USDA Forest Service Gen. Tech. Rep. PSW-GT130; International Union of Forestry Research: Vienna, Austria, 1991; pp. 11–20.

15.   Schiechtl, H.M. *Bioengineering for Land Reclamation and Conservation*; University of Alberta Press: Edmonton, AL, Canada, 1980.

16.   Shen, P.; Zhang, L.M.; Chen, H.X.; Gao, L. Role of vegetation restoration in mitigating hillslope erosion and debris flows. *Eng. Geol.* **2017**, *216*, 122–133. [CrossRef]

17.   Wang, B.; Zhang, G.H. Quantifying the binding and bonding effects of plant roots on soil detachment by overland flow in 10 typical grasslands on the Loess Plateau. *Soil Sci. Soc. Am. J.* **2017**, *81*, 1567–1576. [CrossRef]

18.   Zhu, J.; Wang, Y.; Wang, Y.; Mao, Z.; Langendoen, E.J. How does root biodegradation after plant felling change root reinforcement to soil? *Plant Soil* **2020**, *446*, 211–227. [CrossRef]

19.   Watson, A.; Phillips, C.; Marden, M. Root strength, growth, and rates of decay: Root reinforcement changes of two tree species and their contribution to slope stability. *Plant Soil* **1999**, *217*, 39–47. [CrossRef]

20.   Hubble, T.C.T.; Docker, B.B.; Rutherfurd, I.D. The role of riparian trees in maintaining riverbank stability: A review of Australian experience and practice. *Ecol. Eng.* **2010**, *36*, 292–304. [CrossRef]

21.   Bischetti, G.B.; Chiaradia, E.A.; Simonato, T.; Speziali, B.; Vitali, B.; Vullo, P.; Zocco, A. Root Strength and Root Area Ratio of Forest Species in Lombardy (Northern Italy). *Plant Soil* **2005**, *278*, 11–22. [CrossRef]

22.   De Baets, S.; Poesen, J.; Reubens, B.; Wemans, K.; De Baerdemaeker, J.; Muys, B. Root tensile strength and root distribution of typical Mediterranean plant species and their contribution to soil shear strength. *Plant Soil* **2008**, *305*, 207–226. [CrossRef]

23.   Abernethy, B.; Rutherfurd, I.D. The distribution and strength of riparian tree roots in relation to riverbank reinforcement. *Hydrol. Processes* **2001**, *15*, 63–79. [CrossRef]

24.   Ekanayake, J.C.; Marden, M.; Watson, A.J.; Rowan, D. Tree roots and slope stability: A comparison between Pinus radiata and kanuka. *N. Z. J. For. Sci.* **1997**, *27*, 216–233.

25.   Genet, M.; Kokutse, N.; Stokes, A.; Fourcaud, T.; Cai, X.; Ji, J.; Mickovski, S. Root reinforcement in plantations of *Cryptomeria japonica* D. Don: Effect of tree age and stand structure on slope stability. *For. Ecol. Manag.* **2008**, *256*, 1517–1526. [CrossRef]

26.   Tosi, M. Root tensile strength relationships and their slope stability implications of three shrub species in the Northern Apennines (Italy). *Geomorphology* **2007**, *87*, 268–283. [CrossRef]

27.   Giadrossich, F.; Cohen, D.; Schwarz, M.; Ganga, A.; Marrosu, R.; Pirastru, M.; Capra, G.F. Large roots dominate the contribution of trees to slope stability. *Earth Surf. Processes Landf.* **2019**, *44*, 1602–1609. [CrossRef]

28.   Collison, A.J.C.; Anderson, M.G. Using a combined slope hydrology/stability model to identify suitable conditions for landslide prevention by vegetation in the humid tropics. *Earth Surf. Processes Landf.* **1996**, *21*, 737–747. [CrossRef]

29.   Nilaweera, N.S.; Nutalaya, P. Role of tree roots in slope stabilisation. *Bull. Eng. Geol. Environ.* **1999**, *57*, 337–342. [CrossRef]

30. Madsen, C.; Potvin, C.; Hall, J.; Sinacore, K.; Turner, B.L.; Schnabel, F. Coarse root architecture: Neighbourhood and abiotic environmental effects on five tropical tree species growing in mixtures and monocultures. *For. Ecol. Manag.* **2020**, *460*, 117851. [CrossRef]

31. Danjon, F.; Reubens, B. Assessing and analyzing 3D architecture of woody root systems, a review of methods and applications in tree and soil stability, resource acquisition and allocation. *Plant Soil* **2008**, *303*, 1–34. [CrossRef]

32. Collison, A.; Pollen, N. The effects of riparian buffer strips on streambank stability: Root reinforcement, soil strength and growth rates. In *Roots and Soil Management: Interaction between Roots and the Soil*; Zobel, R.W., Wright, S.F., Eds.; American Society of Agronomy: Madison, WI, USA, 2005; Volume 48, pp. 15–56.

33. Reubens, B.; Poesen, J.; Danjon, F.; Geudens, G.; Muys, B. The role of fine and coarse roots in shallow slope stability and soil erosion control with a focus on root system architecture: A review. *Trees* **2007**, *21*, 385–402. [CrossRef]

34. Stokes, A.; Atger, C.; Bengough, A.G.; Fourcaud, T.; Sidle, R.C. Desirable plant root traits for protecting natural and engineered slopes against landslides. *Plant Soil* **2009**, *324*, 1–30. [CrossRef]

35. Boldrin, D.; Leung, A.K.; Bengough, A.G. Root biomechanical properties during establishment of woody perennials. *Ecol. Eng.* **2017**, *109*, 196–206. [CrossRef]

36. Chen, H.; Harmo, M.E.; Griffiths, R.P. Decomposition and nitrogen release from decomposing woody roots in coniferous forests of the Pacific Northwest: A chronosequence approach. *Can. J. For. Res.* **1999**, *31*, 246–260. [CrossRef]

37. Schroth, G. Decomposition and nutrient supply from biomass. In *Trees, Crops and Soil Fertility: Concepts and Research Methods*; Schroth, G., Sinclair, F.L., Eds.; CABI Publishing: Wallingford, UK, 2003; pp. 131–150.

38. Hairiah, K.; Sulistyani, H.; Suprayogo, D.; Widianto Purnomosidhi, P.; Widodo, R.H.; Van Noordwijk, M. Litter layer residence time in forest and coffee agroforestry systems in Sumberjaya, West Lampung. *For. Ecol. Manag.* **2006**, *224*, 45–57. [CrossRef]

39. Palm, C.A.; Sanchez, P.A. Nitrogen release from some tropical legumes as affected by lignin and polyphenol contents. *Soil Biol. Biochem.* **1991**, *23*, 83–88. [CrossRef]

40. Genet, M.; Stokes, A.; Salin, F.; Mickovski, S.B.; Fourcaud, T.; Dumail, F.; van Beek, R. The Influence of Cellulose Content on Tensile Strength in Tree Roots. *Plant Soil* **2005**, *278*, 1–9. [CrossRef]

41. Van Noordwijk, M.; Purnomosidhi, P. Root architecture in relation to tree-soil- crop interactions and shoot pruning in agroforestry. *Agrofor. Syst.* **1996**, *30*, 161–173. [CrossRef]

42. Van Noordwijk, M.; Mulia, R. Functional branch analysis as tool for scaling above- and belowground trees for their additive and non-additive properties. *Ecol. Model.* **2002**, *149*, 41–51. [CrossRef]

43. Schwarz, M.; Lehmann, P.; Or, D. Quantifying lateral root reinforcement in steep slopes—From a bundle of roots to tree stands. *Earth Surf. Processes Landf. J. Br. Geomorphol. Res. Group* **2010**, *35*, 354–367. [CrossRef]

44. Nespoulous, J.; Merino-Martin, L.; Monnier, Y.; Bouchet, D.C.; Ramel, M.; Dombey, R.; Viennois, G.; Mao, Z.; Zhang, J.L.; Cao, K.F.; et al. Tropical forest structure and understorey determine subsurface flow through biopores formed by plant roots. *Catena* **2019**, *181*, 104061. [CrossRef]

45. Van Noordwijk, M.; Agus, F.; Verbist, B.; Hairiah, K.; Tomich, T.P. Managing Watershed Services in Ecoagriculture Landscapes. In *Farming with Nature: The Science and Practice of Ecoagriculture*; Scherr, S.J., McNeely, J.A., Eds.; Island Press: Washington, DC, USA, 200; pp. 191–212.

46. Burroughs, E.R.; Thomas, B.R. *Declining Root Strength in Douglas-Fir after Felling as a Factor in Slope Stability*; US Department Agriculture Forest Services Research Paper INT-190; Department of Agriculture, Forest Service, Intermountain Forest and Range Experiment Station: Ogden, UT, USA, 1977.

47. Ziemer, R.R. An apparatus to measure the crosscut shearing strength of roots. *Can. J. For. Res.* **1978**, *8*, 142–144. [CrossRef]

48. Walkley, A.; Black, I.A. An examination of the Degtjareff method for determining soil organic matter, and a proposed modification of the chromic acid titration method. *Soil Sci.* **1934**, *37*, 29–38. [CrossRef]

49. Goering, H.K.; Van Soest, P.J. *Forage Fiber Analysis*; Agricultural Handbook no 379; USDA: Washington, DC, USA, 1970; p. 20.

50. Anderson, J.M.; Ingram, J.S.I. *Tropical Soil Biology and Fertility: A Handbook of Methods*; CAB. International: Wallingford, UK, 1989; p. 171.

51. Rose, C. *An Introduction to the Environmental Physics of Soils, Water and Watershed*; Cambridge University Press: Cambridge, UK, 2004; p. 65.

52. Tennant, D. A test of a modified line intersect method of estimating root length. *J. Ecol.* **1975**, *63*, 995–1001. [CrossRef]

53. Payne, R.W.; Lane, P.W.; Ainsley, A.E.; Bicknell, K.E.; Digby, P.G.N.; Harding, S.A.; Leech, P.K.; Simpson, H.R.; Todd, A.D.; Verrier, P.J.; et al. *Genstat 5 Reference Manual*; Clarendon Press: Oxford, UK, 1987.

54. Van Noordwijk, M. Functional branch analysis to derive allometric equations. In *Modelling Global Change Impacts on the Soil Environment: Report of Training Workshop 5–13 May 1998*; IC-SEA Report No., 6; Murdiyarso, D., van Noordwijk, M., Suyamto, D.A., Eds.; BIOTROP-GCTE: Bogor, Indonesia; ICSEA: Lisbon, Portugal, 1999; pp. 77–80.

55. Akinnifesi, F.K.; Rowe, E.C.; Livesley, S.J.; Kwesiga, F.R.; Van Lauwe, B.; Alegre, J.C. Tree root architecture. In *Below-Ground Interactions in Tropical Agroecosystems. Concepts and Models with Multiple Plant Components*; Van Noordwijk, M., Cadisch, G., Ong, C.K., Eds.; CABI Publishing: Wallingford, UK, 2004; pp. 61–82.

56. Hairiah, K.; Widianto, W.; Utami, S.R.; Sunaryo, D.S.; Sitompul, S.M.; Lusiana, B.; van Noordwijk, M.; Cadish, G. *Pengelolaan Tanah Masam Secara Biologi: Refleksi Pengelaman dari Lampung Utara*; World Agroforestry Centre (ICRAF): Bogor, Indonesia, 2000; p. 184.

57. Budidarsono, S.; Wijaya, K. Praktek Konservasi dalam Budidaya Kopi Robusta dan Keuntungan Petani. *Agrivita* **2004**, *26*, 107–117.

58. Cammeraat, E.; van Beek, R.; Kooijman, A. Vegetation succession and its consequences for slope stability in SE Spain. *Plant Soil* **2005**, *278*, 135–147. [CrossRef]

59. Stokes, A.; Mattheck, C. Variation of wood strength in tree roots. *J. Exp. Bot.* **1996**, *47*, 693–699. [CrossRef]

60. Coster, C. Wortelstudien in de tropen. V. Gebergtehoutsoorten [Root studies in the Tropics. V. Tree species of the mountain region]. *Tectona* **1935**, *28*, 861–878.

61. Van Noordwijk, M.; Lawson, G.; Hairiah, K.; Wilson, J. *Root Distribution of Trees and Crops: Competition and/or Complementarity. Tree-Crop Interactions: Agroforestry in a Changing Climate*; CABI: Wallingford, UK, 2015; pp. 221–257.

62. Van Noordwijk, M.; Martikainen, P.; Bottner, P.; Cuevas, E.; Rouland, C.; Dhillion, S.S. Global change and root function. *Glob. Change Biol.* **1998**, *4*, 759–772. [CrossRef]

63. Wibawa, G.; Hendratno, S.; van Noordwijk, M. Permanent smallholder rubber agroforestry systems in Sumatra, Indonesia. In *Slash and Burn: The Search for Alternatives*; Palm, C.A., Vosti, S.A., Sanchez, P.A., Ericksen, P.J., Juo, A.S.R., Eds.; Columbia University Press: New York, NY, USA, 2005; pp. 222–232.

Article

Infiltration-Friendly Agroforestry Land Uses on Volcanic Slopes in the Rejoso Watershed, East Java, Indonesia

Didik Suprayogo [1],*, Meine van Noordwijk [1,2,3] , Kurniatun Hairiah [1], Nabilla Meilasari [1], Abdul Lathif Rabbani [1], Rizki Maulana Ishaq [1] and Widianto Widianto [1]

[1] Soil Science Department, Faculty of Agriculture, Brawijaya University, Jl. Veteran no 1, Malang 65145, Indonesia; m.vannoordwijk@cgiar.org (M.v.N.); K.HAIRIAH@CGIAR.ORG (K.H.); nabilla.meilasari@gmail.com (N.M.); agroforestry@ub.ac.id (A.L.R.); rizky.maulana17@gmail.com (R.M.I.); widianto@gmail.com (W.W.)
[2] World Agroforestry Centre, ICRAF, Indonesia Office, Bogor 16001, Indonesia
[3] Plant Production Systems, Wageningen University, 6708 PB Wageningen, The Netherland
*   Correspondence: Suprayogo@ub.ac.id or Suprayogo09@yahoo.com

Received: 9 June 2020; Accepted: 14 July 2020; Published: 23 July 2020

Abstract: Forest conversion to agriculture can induce the loss of hydrologic functions linked to infiltration. Infiltration-friendly agroforestry land uses minimize this loss. Our assessment of forest-derived land uses in the Rejoso Watershed on the slopes of the Bromo volcano in East Java (Indonesia) focused on two zones, upstream (above 800 m a.s.l.; Andisols) and midstream (400–800 m a.s.l.; Inceptisols) of the Rejoso River, feeding aquifers that support lowland rice areas and drinking water supply to nearby cities. We quantified throughfall, infiltration, and erosion in three replications per land use category, with 6–13% of rainfall with intensities of 51–100 mm day^{-1}. Throughfall varied from 65 to 100%, with a zone-dependent intercept but common 3% increase in canopy retention per 10% increase in canopy cover. In the upstream watershed, a tree canopy cover > 55% was associated with the infiltration rates needed, as soil erosion per unit overland flow was high. Midstream, only a tree canopy cover of > 80% qualified as "infiltration-friendly" land use, due to higher rainfall in this zone, but erosion rates were relatively low for a tree canopy cover in the range of 20–80%. The tree canopy characteristics required for infiltration-friendly land use clearly vary over short distances with soil type and rainfall intensity.

Keywords: entrainment; erosion; forest conversion; overland flow; soil macroporosity; throughfall; water balance

---

1. Introduction

Water access for all, the Sustainable Development Goal 6 of the Agenda 2030 agreed by the United Nations [1], not only refers to drinking water and sanitation. It requires the protection of "infiltration-friendly" land uses in upland watersheds as a source of clean water [2]. Sufficient groundwater recharge is important to the sustainable management of groundwater resources to maintain streamflow throughout the year, as well as to feed springs [3,4]. While much of the public discourse is in terms of forest versus agriculture, thresholds for specific soil and climate regimes are needed within the intermediate agroforestry spectrum of land uses [5]. Thresholds to critical hydrological functions are likely dependent on local context but need to be understood to guide natural resource management in the challenging trade-offs between local and external priorities [6]. Hydrological functions, and their sensitivity to climate change, can be characterized by a number of metrics [7]. Much of the literature on forest hydrology is concerned with reductions in annual water yield due to increased canopy interception and/or tree water use by fast-growing forest stands [8], without a distinction between (fast) overland and (slower, infiltration-dependent) subsurface flow pathways. A recent review [9] found that the recovery of annual river flow with the age of planted forest is an exception rather than a rule. However, the recovery of infiltration with tree cover can increase dry season flows [10] without increasing annual water yield. Changes in streamflow regime will reflect both changes in evapotranspiration (ET) and in infiltration after the change in land use under given climatic conditions [11,12]. On the other hand, a high tree (*Acacia auriculiformis*) canopy, without an understory and permanent litter layer, was associated with high erosion rates due to high-impact drips from the leaves [13]. Empirical data and process-level understanding is needed of these diverse and partly contradictory effects of tree cover, especially in human-managed land uses.

Agroforestry systems with high canopy densities can, if a permanent litter layer is present, maintain high infiltration rates and can positively impact on hydrologic functions through: (1) a green canopy cover at the tree and understory level, (2) land surface roughness, (3) litter at the soil surface, and (4) water uptake by trees and other vegetation [14,15]. Five aspects that hydrologically differentiate natural forest from open-field agriculture, with intermediate functionality for managed forest, plantations, agroforestry, and trees outside forest [16,17], are: (1) the leaf area index (LAI) that allows photosynthesis when stomata are open and transpiring, and that, along with leaf morphology and rainfall intensity, determines canopy interception, retention, and subsequent evaporation, (2) the surface litter that prevents crusting and supports infiltration [18] while reducing soil evaporation and reduces the entrainment of soil particles if overland flow still occurs, (3) the soil macroporosity that governs infiltration and allows for the aeration of deeper soil layers between rainfall events while recovering at a decadal time scale after reforestation [19,20], (4) the root systems that govern water extraction from deeper soil layers, in conjunction with the phenology of the aboveground canopy [21], and (5) possible influences on rainfall events [22,23]. Each of these five aspects has its own dynamic (time constants) and dependency on the type of trees and their management, challenging the definition of hydrologically adequate land use choices. Rather than prescribing, independent of soil types and slope, the type and quantity of tree cover that is needed, as tends to happen in forest zoning, it may help if limits to infiltration-friendly land use (focused on the third function) can be operationalized in a local context. In terms of watershed hydrology, infiltration-friendly land uses can be interpreted as any land use that allows high rates of water infiltration so that surface runoff is a small (to be defined in local context) fraction of rainfall and the watershed functions of flow buffering and erosion control are secured (to specified standards). River flow in watersheds that provide perfect buffering might theoretically be constant every day, but in practice, a "flow persistence" metric of about 0.85 is hard to surpass [24]. Flow buffering is essential for climate resilience [25] and high flow persistence metrics are desirable, as they directly relate to peak flow transmission [26].

The discussion on forests and watershed functions in Indonesia became based on specific theories about underlying mechanisms and measurements in the 1930s [27–30], but at the policy level, a generic dichotomy between "forest" and "non-forest" conditions was maintained. The Indonesian spatial planning law prescribes that 30% forest cover is needed in all local government entities to secure hydrological forest functions [31]. As the 30% norm originated in studies of flooding risk in relation to spring snow melt in Switzerland [32], a more nuanced and process-based understanding is needed to underpin effective policies on desirable forest cover, especially in densely populated Java, where agroforests are common. In Southeast Asia, 8.5% of the global human population lives on 3.0% of the land area. With 7.9% of the global agricultural land base, the region has 14.7% and 28.9% of such land with at least 10% and 30% tree cover, respectively, and is the world's primary home of "agroforests" [33].

On densely populated Java, volcanic slopes are home to large numbers of farmers, while also serving as sources of water for lowland agriculture and the rapidly growing cities. The shrinking area of state-managed forests is no longer able to secure the required watershed functions, but at least part of the agroforestry managed by farmers can meet the required hydrological functions [34].

In the Rejoso Watershed (Pasuruan, East Java, Indonesia), numerous stakeholders depend on the watershed functions of densely populated mountain slopes to meet their water demand. These include local communities, farmers using water irrigation, the Regional Water Company, and bottled water industries. A major infrastructure is planned to bring water to Surabaya and the surrounding urban centers. However, the quantity and quality of the water at the source of the pipe have been decreasing over the past 10 years, putting the infrastructure investment at risk. Decreasing water resources are likely due to land use changes in the recharging area of the Rejoso Watershed on the northern slopes of the Bromo-Semeru volcanic mountain range, and/or decreased pressure on artesian wells across the land due to increased extraction for paddy rice fields. Among the hydrologic functions, infiltration is critical, as water travels through the subsoil to artesian wells at the foot of the volcano, in addition to surface rivers.

This research in the Rejoso Watershed using runoff and soil erosion plot-scale studies under natural rainfall [35] within the locally relevant range of land cover types was thus designed to assess which land uses can maintain infiltration rates under local peak rainfall intensity and restrict soil erosion to acceptable levels. The specific questions were:

1. Which existing land uses limit infiltration below the required rates at peak rainfall events?
2. Which factors that are directly observable, such as tree cover, litter layer thickness, or surface roughness, can be used to define thresholds for "infiltration-friendly land use"?
3. Do the answers to questions 1 and 2 need to be differentiated between the upper and middle watershed, with current vegetable production and agroforestry as dominant land uses, respectively?

2. Materials and Methods

2.1. Study Area

The Rejoso Watershed, is located on the northern slope of Mount Bromo, covering 16 sub-districts in Pasuruan District, East Java Province, Indonesia. The Rejoso Watershed is located between 7°37′13.35″ and 7°55′18.63″ South, and between 112°48′32.51″ to 113°55′55″ East (Figure 1).

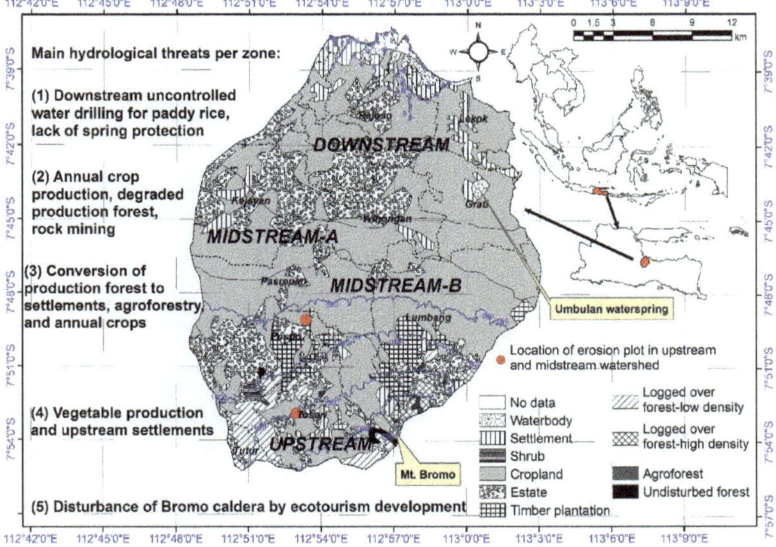

Figure 1. A. Position of the Rejoso Watershed in East Java as part of the Indonesian archipelago in Southeast Asia. B. The Rejoso Watershed from upstream (at the bottom) to sea level and land uses considered to be a hydrological threat; purple indicates open soil, green tree cover (base map in [36]).

The Rejoso Watershed covers an area of 634 km^2 with a hydrologic (watershed) length of the main channel of about 22 km. This study was conducted in two locations, namely in the upstream (above 800 m a.s.l.) and midstream (400–800 m a.s.l.) sections, with the dominant vegetation (land cover) selected for each location (Figure 1).

Climatic conditions that influence hydrology and erosion are largely determined by the influence of the northwest and southwest monsoons. The northwest monsoon, picking up large amounts of moisture over the Indian Ocean, brings in most of the annual precipitation in the area, and predominates during the period from November through April. Although there is considerable variation in the amount and distribution of rainfall from year to year, most places in the watershed receive about 91% of the rainfall during the November–May wet season (monthly rainfall > 100 mm) in the upper stream and about 91% of the rainfall during the November–April wet season in the midstream (Figure 2). Due to topographic influences, there is considerable spatial variation in annual precipitation as well, ranging from 1655 mm to 3675 mm, with extreme yearly rainfall in 2010, with an annual precipitation of 5298 mm over 24 years of rainfall measurements (1990–2013) in the upper stream compared with an annual precipitation ranging from 1020 mm to 2603 mm in the midstream. The May to October period is considered the dry season. Then the southeast monsoon predominates, bringing much smaller amounts of precipitation due to the lower atmospheric moisture caused by lower temperatures in the Southern Hemisphere at this time of the year. The annual precipitation in the upstream (average annual precipitation = 2488 mm) is higher than the in midstream (average annual precipitation = 1632 mm). Over 24 years of measurements, the maximum daily rainfall in the upper stream and midstream ranged from 80 mm day^{-1} to 200 mm day^{-1} and 60 mm day^{-1} to 320 mm day^{-1}, respectively. Based on Schmidt–Ferguson climate classification, the upper stream and midstream are considered rather wet (C) and average (D), respectively.

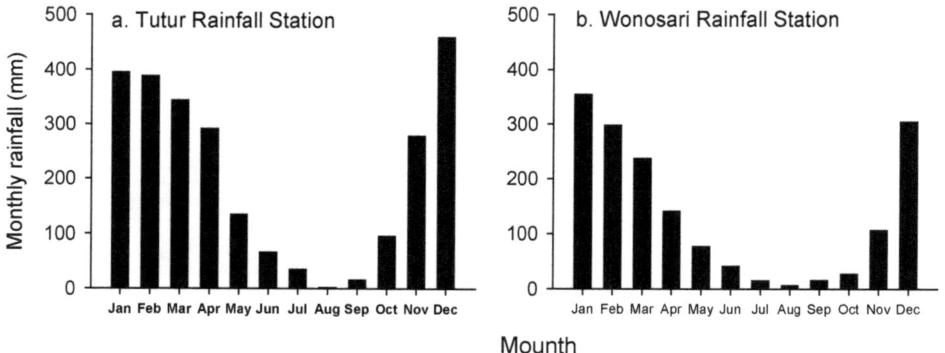

Figure 2. Monthly rainfall distribution in the Rejoso Watershed from the average of 24 years of rain events (1990–2013) in the (**a**) upstream (Tutur Rainfall Station) and (**b**) midstream (Wonosari Rainfall Station).

The Rejoso Watershed consists of four types of soil, namely: Andisols, Inceptisols, Alfisols, and Entisols. Andisols are mainly found on the upper slopes of the volcano. Andisols have a distinct black to very dark brown surface horizon rich in organic matter, which usually overlies a brown to dark yellowish-brown subsoil. The clay fraction is dominated by allophane. Andisols are highly permeable, porous with low bulk density, have high water-holding capacity, and a crumb structure. The most common texture is sandy loam. These soils have high inherent fertility and are highly erodible only when seriously disturbed. The middle and some lower volcanic slopes, consisting of easily weatherable permeable tuffs and ash deposits, give rise to deep soils—Inceptisols and Alfisols. Inceptisols have only limited horizon differentiation. Their texture ranges from deep friable clays to clay loams. Alfisols are soils which have an accumulation of clay in the subsoil. Their texture ranges from loam to clay loam in the topsoil and clay loam to clay in the subsoil. Both soils have moderate to high inherent fertility but are highly susceptible to erosion. The fourth group, Entisols, are soils that lack horizon development and are found on volcanic sands, ashes, and tuffs. Entisols occur on recent and sub-recent lahars of the Bromo volcano. Entisols with a coarse texture are extremely erodible and have very low water-holding capacities. Permanent vegetative cover, and especially diversified tree crops and agroforestry or forestry, are the most suitable land utilization types to prevent erosion.

2.2. Land Cover Types Compared

In important research traditions associated with the Universal Soil Loss Equation (USLE) [37], the quantification of erosion requires a "bare soil" reference, expressing the degree of protection provided by vegetation relative to this "control". Fortunately, bare soil is rare in this landscape, and it would be considered an extreme, rather than a standard agricultural point of reference. Artificially clearing land to allow such treatment to be measured would give results that are hard to be interpreted, as soil changes after clearing would lead to a time-dependence of the results, rather than being an unambiguous point of reference. By referring to the more process-based Rose equation [38], separating overland flow as a transport medium and "entrainment" as a soil characteristic relative to the energy-dependent transport capacity of such flow, we do not depend on the USLE framework (that infers that soil loss is universal, but does not account for its counterpart process, sedimentation [39]) but can focus on existing land covers and associated land uses in the landscape.

In both the upstream and midstream parts of the catchment, four dominant land use systems were assessed (Table 1), spatially replicated in three separate measurement plots. Upstream land uses included old and young pine plantations (production forest) and highland vegetable crops with variations in tree canopy cover in the landscape on steep (30–60%) to very steep (>60%) land with imperfect ridge terraces. Midstream land uses included production forest, multistrata coffee-based agroforestry, clove-based agroforestry, and several mixed agroforestry types with variations in tree canopy cover in the landscape on moderately steep (15–30%) and steep (30–60%) land with bench terraces sloping outward.

Table 1. Land use, vegetation, soil conservation measure, and slope of measurement plots.

Code	Land Use	Vegetation (the Average Height of Trees)	Terracing	Slope (Plot Level, %)
		Upstream Rejoso Watershed		
UT1	Old production forest	Pine (*Pinus merkusii*) (34 m) + grass	None	35–40
UT2	Young production forest	Pine (11 m) + grass	None	50–60
UT3	Agroforestry	Strip cemara (*Casuarina junghuniana*) (13 m) + cabbage	None	40–50
UT4	Arable land	Banana, maize, carrot	None	40–50
		Midstream Rejoso Watershed		
MT1	Old production forest	Mixed pine (28 m) or mahogany (*Swietenia macrophylla*) (12 m), banana, salak (*Salacca zalacca*), taro (*Colocasia esculenta*), elephant grass (*Miscanthus giganteus*).	Bench terrace sloping outward	3–8
MT2	Agroforestry	Coffee-based (2 m) mix with durian (*Durio zibethinus*) (10 m), mahogany (9 m), *Leucaena leucocephala* (8 m), *Paraserianthes falcataria* (11 m), *Albizia saman* (11 m), dadap (*Erythrina variegata*) (11 m), banana	Bench terrace sloping outward	3–8
MT3	Agroforestry	Clove (*Syzygium aromaticum*) (8 m), banana	Bench terrace sloping outward	3–8
MT4	Agroforestry	Mango (*Mangifera indica*) (10 m), durian (10 m), *Randu kapuk* (*Ceiba pentandra*) (11 m), maize, cassava, groundnut	Bench terrace sloping outward	3–8

2.3. Quantifying Terms of Water and Soil Balance

2.3.1. Overview

As forest and tree cover can influence various steps in the chain from rainfall to streamflow and erosion, we aimed to quantify (1) the direct effect of tree canopies on the retention of part of the rainfall (followed by direct evaporation), versus the fraction reaching the soil surface by throughfall and stemflow, (2) the partitioning of the latter into infiltration and overland flow, (3) the entrainment of soil particles into this overland flow. Within the time and resources available we did not assess (4) the seepage of groundwater beyond the root zone and access by vegetation, (5) the pathways and release of groundwater into streams, (6) the routing of overland flow into streams, or (7) the in-field (or riparian filter zone) sedimentation of entrained soil particles, beyond the scale of the measurement plots. However, we did characterize the vegetation and soil characteristics that influence the various processes.

2.3.2. Rainfall and Throughfall

Rain gauges, outside of the direct influence of tree canopies, were installed in four observation locations (with adjacent erosion plots) upstream and four observation locations midstream of the Rejoso Watershed. In each runoff plot, throughfall was measured with five replications. The throughfall gauges below the tree canopy had a horizontally placed 30 cm diameter funnel 120 cm above the soil surface and a collector bottle with a volume of 1.5 dm^3 placed with bamboo as a support. Throughfall and rainfall were collected every day for two months of the rainy season, from March to May 2017. Attempts were made to also quantify stemflow, but due to technical problems with the method used, no reliable data were obtained, and the results are not shown here.

2.3.3. Water Infiltration and Soil Erosion Measurement

Water infiltration was quantified in each land cover type via its complement, surface runoff, and expressed in the runoff/throughfall ratio. As the throughfall for quantified infiltration was measured below the tree canopy, the amounts were the net of canopy retention and possible stemflow. Surface runoff was measured in 6 m × 2 m plots protected from surface run-on, with two drums at the lower end to collect surface runoff and sediment concentrations for soil erosion measurements (Figure 3).

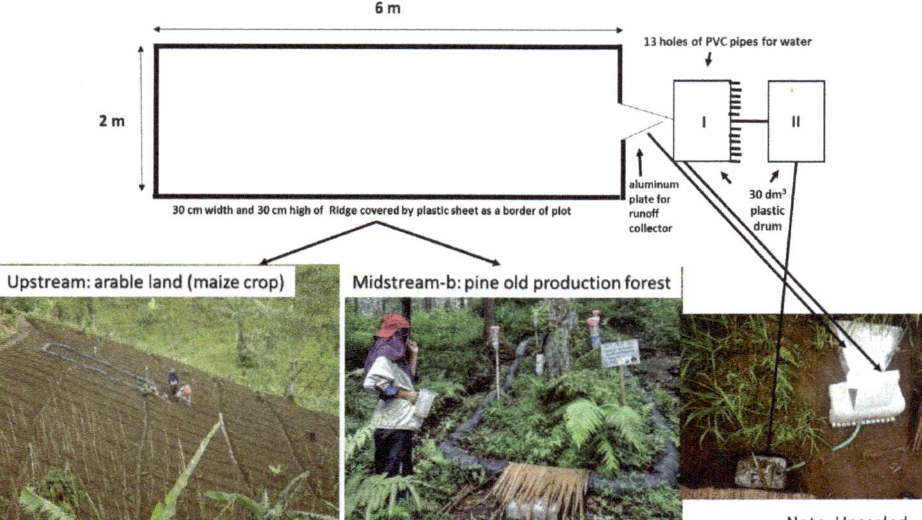

Figure 3. Runoff and soil erosion plot design.

In each plot, the water flow was collected into two collection drums with a capacity of 30 dm^3. The first drum had a divider system channeling into 13 channels (PVC pipes) with equal diameters and level positions, with one connected with a second drum. The volume of water flowing from each pipe was measured to calibrate the water volume proportion entering into the second drum. The potential capacity of the runoff collector thus was (30 dm^3 * 13) + 30 dm^3 = 420 dm^3 for 12 m^2 or 35 mm. We did nor encounter situations where the second tank overflowed. Runoff samples at each plot were collected every day and the rain that occurred during the measurement period was measured by measuring the water depth in each drum. The amount of runoff in each rain event was calculated using Equations (1) and (2):

$$R_t = V_{d-I} + (13 * V_{d-II}) \tag{1}$$

$$V_d = 1000 * (D * L * W) \tag{2}$$

where R_t is total runoff (dm^3), V_d is the water volume in drums I and II (dm^3), L = length and W = width of drum (cm), and D is the water depth in each drum (cm). The total runoff was then divided by the area of the plot (2 m × 6 m) to convert to mm. Data could be compared to a classification developed elsewhere [40] that indicates a runoff coefficient of 0.14 as adequate for Andisols, and 0.20 for Inceptisols.

Soil erosion in each rain event was determined by collecting 1 dm^3 of runoff sediment in each drum. The sample was filtered with "newsprint" and dried in an oven with a temperature of 105 °C to get the weight of the sediment (S). In earlier studies, we found that effluent from this readily available filter material had a negligible sediment concentration [41]. Erosion (E) in each rain event was calculated using Equation (3):

$$E = ((V_{d-1} * S) + (13 * (V_{d-2} * S))) * \left(\frac{10^{-2}}{A}\right) \tag{3}$$

where E is soil erosion (Mg ha^{-1}), S is sediment (g dm^{-3}), and A is the area of the plot (m^2).

2.4. Determination of Soil Properties

Three bulk mineral soil samples were collected from each layer of soil of 0–10 cm, 10–20 cm, 20–30 cm, 30–40, and 40–50 cm for soil texture analysis and each layer of soil of 0–10 cm, 10–20 cm, and 20–30 cm for soil bulk density, particle density, total soil porosity, and soil organic matter content. Particle size distribution (particles < 2 mm) was determined with the Bouyoucos densimeter method [42] after H_2O_2 pre-treatment and after samples had been dispersed in 5% sodium hexametaphosphate and 5% dispersing solution. Bulk density (oven dry weight per unit volume) was measured for a block-sized sample (20 cm × 20 cm × 10 cm = 4000 cm^3) collected in field moisture conditions (modified from [43]). Particle density was measured by the pycnometer method. Total soil porosity ($\emptyset$), the percentage of the total soil volume that is not filled by solid (soil) particles [44], was calculated from bulk density data and particle density using Equation (4):

$$\emptyset = \left(1 - \frac{\rho_b}{\rho_p}\right) \times 100\% \tag{4}$$

where $\emptyset$ is porosity (%), ρ_b is bulk density (g cm^{-3}), and ρ_p is particle density (g cm^{-3}).

Soil organic carbon (SOC) was determined by dichromate oxidation [45]. Soil infiltration was measured by the standard double-ring infiltrometer test [46]. The double-ring infiltrometer as often constructed from a thin-walled steel pipe with inner and outer cylinder diameters of 20 and 30 cm, respectively.

The soil macro-porosity was measured using the methylene blue method, by looking at the blue distribution pattern of the methylene blue solution in the soil profile. The methylene blue solution (70 g methylene blue per 200 L of water) was gradually poured into the ground, which had been bound by a metal frame measuring 100 cm × 50 cm × 30 cm (Figure 3) and left for 36 h until the methylene blue solution soaked into the soil. Methylene blue will pass through soil macropores but be absorbed by micropores and soil surfaces. After all the methylene blue solution had disappeared from the soil surface, the top 5 cm of soil was removed from a 100 cm × 100 cm sample area and infiltration patterns were recorded, before a further 10 cm of soil was removed for a second map (at 15 cm below the soil surface), and a further 10 cm of soil for a third map (25 cm below soil surface). For mapping blue patches in each horizontal plane, a transparent sheet of plastic was placed on the surface and all visible blue patches were mapped with marker pens. Blue distribution patterns, redrawn on tracing paper, were photocopied for analysis of the black-and-white pattern of the fraction of soil involved in macropore flow with the IDRISI computer program.

2.5. Other Plot Characteristics

2.5.1. Canopy Cover

The canopy cover can be defined as the percentage of tree canopy area occupied by the vertical projection of tree crowns [47]. The percentage of canopy cover is measured by scathing the shadow of sunshine at ground level using 10 m × 10 m sheets of white paper. The canopy projection when the sun was overhead was drawn to scale on white paper in each of the four quadrants of the 20 m × 20 m plots, after which the shaded areas were cut out and weighed separately. Canopy cover was calculated according to Equation (5):

$$\%\text{Canopy Cover} = \frac{\text{W Canopy}}{\text{W Total}} \times 100 \tag{5}$$

where %Canopy Cover is the percentage of tree canopy cover, W Canopy is the paper weight representing canopy cover and W Total is the paper weight representing the total area of observation, respectively.

2.5.2. Understory and Litter

Understory vegetation and litter were measured according to the rapid carbon stock appraisal protocol [48], using 50 cm × 50 cm samples for fresh weight, with subsamples dried for dry weight determination.

2.5.3. Land Surface Roughness

Surface roughness was measured in each plot as the standard deviation of elevation measured every 30 cm along a thread (thin rope) installed 30 cm from the surface vertically, horizontally, and diagonally over the erosion plot [49]. The measurement of the difference in elevation was set to a pixel size of 30 cm × 30 cm. Each plot was divided into six pixels for a 2 m plot width and 20 pixels for a 6 m plot length, so there were 120 pixels (N). Pixels were made on a flat plane 30 cm from the ground point of reference with a thin rope. In each center, the pixel was measured vertically parallel to the thin rope towards the surface of the ground with a ruler. The results of the measurements of height differences in each pixel were used to calculate Ra with the equation:

$$\text{Ra} = \frac{1}{N} \sum_{n=1}^{N} | h_n | \tag{6}$$

where N = Number of pixels in the patch and h_n = difference of elevation between the nth pixel in the patch and the mean value.

2.6. Data Analysis

To answer the first research question, the null hypothesis was that within the forest to open field agriculture continuum of any observed difference in soil hydrological functions could be due to random variation. To see if that null hypothesis could be rejected, we examined differences in soil infiltration, runoff coefficient, and soil erosion between the dominant land uses in the upstream and midstream with Fisher's Least Significant Differences (LSD) test. Fisher's LSD test, which establishes differences between groups defined for independent samples, was used for hypothesis testing, given that the data met the requirements for normality and the homogeneity of variances. A probability level of 0.05 was set for rejecting the null hypothesis of no difference in tests of statistical significance. We used the GenStat 15th edition software for Fisher's LSD tests. The soil infiltration, runoff coefficient, and soil erosion were then compared with the soil infiltration category [50], existing infiltration adequacy standards, and acceptable soil erosion rates. An acceptability threshold, below which soil erosion is less than an "agriculturally permissible" rate (E_{apr}, Mg ha^{-1} year^{-1}), was derived as:

$$E_{apr} = \left(\frac{\text{Depth of soil} * \text{Factor of soil depth}}{\text{Time horizon}} \right) * \text{Soil bulk density} \tag{7}$$

Both Andisols and Inceptisols are deep (beyond 120 cm soil depth) and have a soil depth factor of 1.0. We chose 400 years as a time horizon. Given the average soil bulk density of Andisols (0.83 g cm^{-3}) and Inceptisols (0.99 g cm^{-3}), we obtained E_{apr} Andisol = 24.9 Mg ha^{-1} year^{-1} and E_{apr} Inceptisol 29.7 Mg ha^{-1} year^{-1}.

For the second research question, we tested a number of plot scale characteristics as possible indicators of "infiltration-friendly" plot characteristics: tree canopy cover, understory vegetation, litter necromass, and land surface roughness. Linear regression relationships between the surface runoff/rainfall ratio or soil erosion and the amount of rainfall, tree canopy cover, understory, litter, and land surface roughness were determined using SigmaPlot version 10.0. While a search for "explanatory" factors might have explored multiple regression, our focus was single indicators that could be used as proxies for follow-up discussions with farmers about adjusting land use.

The third research question required the analysis of data for the first two research questions, with the expectation that any thresholds for acceptable hydrological disturbance could be zone-specific, given variations in rainfall, soil type, and the specific characteristics of land use and vegetation.

3. Results

3.1. Rainfall and Throughfall

Within the measurement period, 31 rainy days were recorded (Figure 4). Rainfall variation between the upstream and midstream observation plots was relatively high, with an average of 520 mm (range 476–556 mm among 12 rain gauge measurements), and an average of 666 mm (range 541–840 mm among 12 rain gauge measurements), respectively. In the upstream and midstream areas, 71% and 57% of the rainy days had < 20 mm day^{-1} ("light rain"), 24% and 31% had "moderate" rainfall (21–50 mm day^{-1}) and 6% and 13% "heavy" rain (51–100 mm day^{-1}), respectively; none had "very heavy rain" (>100 mm day^{-1}). Such rain conditions indicate that the rain erosivity in the midstream is higher than that of the upstream.

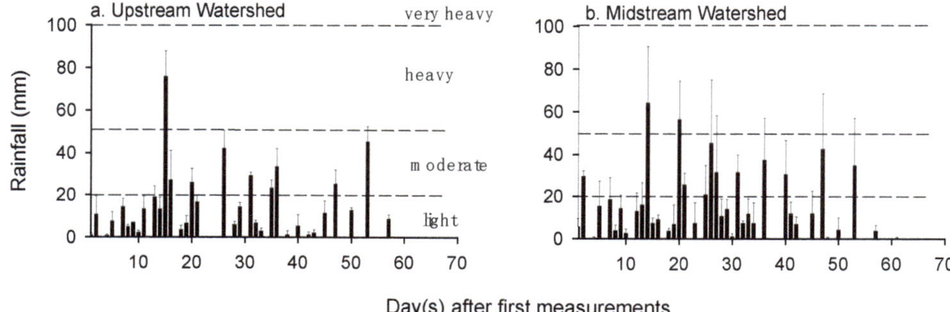

Figure 4. Distribution of rainfall during observation starting on March 03, 2017 in the Rejoso Watershed.

In the upstream area, the old production forest obtained a throughfall/rainfall ratio of 0.73 (standard deviation (SD) = 0.05), while for open-field agriculture it was 0.94 (SD = 0.5) (Figure 5a). For young production forests of *Casuarina junghuniana*-based agroforestry, the throughfall/rainfall ratio was 0.83 (SD = 0.05). In the midstream, the throughfall/rainfall ratio in agroforestry systems with tree canopies of 87%, 75%, and 52% were 0.81 (SD = 0.07) 0.81 (SD = 0.07), and 0.1 (SD = 06), respectively (Figure 5b). For agroforestry with low cover (26%), the throughfall/rainfall ratio was 0.96 (SD = 0.01).

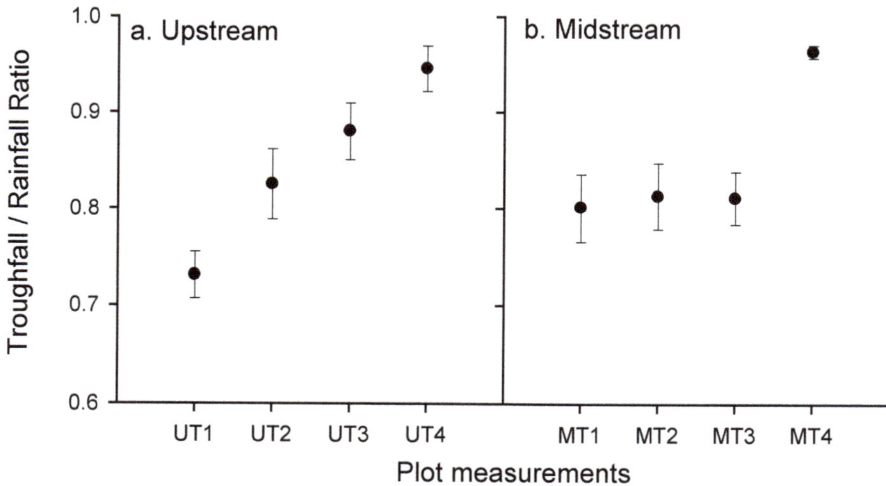

Figure 5. The throughfall/rainfall ratio variability in measured runoff plots (**a**) upstream, (**b**) midstream.

3.2. Soil Properties

The Andisols in the upstream area had a 40–60% silt fraction in all soil measured layers; the Inceptisols had a higher clay fraction (Appendix A). The upstream area had a lower bulk density and higher soil porosity, with a lower clay content than the midstream area (Table 2). The soil organic carbon content varied from 0.65 to 2.12%.

Table 2. Bulk density, particle density, soil porosity, macro-porosity, and organic C of runoff plots.

Location Code at Soil Depth (cm)	Bulk Density (g cm^{-3}) *			Particle Density (g cm^{-3}) *			Soil Porosity (%) *			Soil Macro- Porosity (%)			C$_{org}$ (%) *		
	0–10	10–20	20–30	0–10	10–20	20–30	0–10	10–20	20–30	0–10	10–20	20–30	0–10	10–20	20–30
Upstream Rejoso Watershed: Andisols															
UT1	0.87a	0.81a	0.83a	2.16a	2.23a	2.31a	60a	63a	64c	8.0b	5.2b	0.9a	2.05bc	1.61c	1.79b
UT2	0.85a	0.86a	0.82a	2.27a	2.30a	2.33a	63a	63a	65c	5.1ab	1.5a	0.3a	2.46c	1.56bc	1.78b
UT3	0.81a	0.84a	0.85a	2.14a	2.12a	2.28a	62a	60a	63b	4.7ab	2.1ab	1.4a	1.17a	0.58a	0.71a
UT4	0.84a	0.88a	0.84a	2.28a	2.29a	2.08a	63a	62a	60a	3.0a	0.3a	0.1a	1.35ab	1.06ab	0.92a
LSD	0.07	0.13	0.12	0.17	0.21	0.38	4	5	1	3.52	3.4	1.8	0.85	0.50	0.50
Midstream Rejoso Watershed: Inceptisols															
MT1	0.83a	0.85a	0.83a	2.20a	2.28a	2.20a	62c	63a	62b	13.6ab	7.0bc	2.5c	1.73a	1.87a	1.65b
MT2	0.96b	0.91a	0.91a	2.42b	2.38a	2.21a	60bc	62a	59ab	16.1b	8.3c	1.8bc	2.22a	1.59a	1.84b
MT3	1.03bc	0.96a	0.94ab	2.38b	2.36a	2.40a	57ab	59a	61b	11.7a	3.4ab	0.9ab	2.19a	1.61a	1.01a
MT4	1.09c	1.04a	1.04b	2.38b	2.33a	2.33a	54a	55a	55a	11.4a	0.8a	0a	1.71a	1.36a	1.12a
LSD	0.10	0.24	0.11	0.15	0.17	0.22	4	10	4	4.0	3.9	1.0	0.84	0.54	0.41

* The same letter indicates no statistically significant differences between locations with Fisher's Least Significant Differences (LSD) test ($p < 0.05$).

3.3. Land Characteristics Related to Runoff and Soil Erosion

Production forests in the upstream area had a lower tree canopy cover than those midstream but higher than those in agroforestry systems (Table 3). Agroforestry in the upstream area had a very low tree canopy cover because trees were planted only along field edges. Midstream agroforestry gardens ranged from high (75%) to low (26%) canopy cover. Understory vegetation was more prominent upstream than midstream. Litter layer necromass and land surface roughness were generally aligned with tree canopy cover.

Table 3. Canopy cover, understory vegetation, litter necromass, and soil roughness of the sample plots.

Code	Land Cover	Tree Canopy Cover (%) *	Understory Vegetation (Mg ha^{-1}) *	Litter (Mg ha^{-1}) *	Soil Roughness (%) *
Upstream Rejoso Watershed					
UT1	Old production forest	55b	10.1b	9.2b	8.5a
UT2	Young production forest	40b	10.5b	2.0a	7.0a
UT3	Agroforestry	4a	10.1b	2.1a	9.5a
UT4	Arable land	0a	3.7a	0.3a	7.7a
LSD		15	5.6	3.7	4.6
Midstream Rejoso Watershed					
MT1	Old Production Forest	87c	2.5a	9.8b	7.6b
MT2	Agroforestry	75c	2.5a	4.8a	5.4ab
MT3	Agroforestry	52b	2.1a	5.2a	2.8a
MT4	Agroforestry	26a	1.3a	3.5a	2.0a
LSD		14	2.6	2.4	4.5

* The same letter indicates no statistically significant differences between locations with Fisher's LSD test ($p < 0.05$).

3.4. Runoff and Soil Erosion

Decreasing tree canopy cover in agroforestry systems significantly increased the surface runoff/rainfall ratio or the surface runoff/throughfall ratio (Table 4). In these results, the relationship between surface runoff with rainfall or throughfall was in line, and the ratio of surface runoff/rainfall was further used. The ratio of surface runoff/rainfall is also known as the runoff coefficient.

Table 4. Rainfall, runoff, ratio runoff/rainfall, and soil erosion in the runoff plots in each land cover type.

Code	Land Cover	Rainfall (mm)	Runoff (mm) *	Runoff/ Rainfall Ratio *	Runoff/ Throughfall Ratio *	Soil Erosion (Mg ha^{-1}) *
Upstream Rejoso Watershed						
UT1	Old production forest	555	14.3a	0.03a	0.04a	5.86a
UT2	Young production forest	492	13.2a	0.03a	0.03a	1.47a
UT3	Agroforestry	476	203.3b	0.43b	0.56c	120.98b
UT4	Arable land	556	225.7b	0.41b	0.43b	163.22b
LSD			46.3	0.09	0.11	87
Midstream Rejoso Watershed						
MT1	Old Production Forest	616	80.2a	0.13a	0.16a	3.07a
MT2	Agroforestry	841	316.3c	0.38b	0.48b	2.88a
MT3	Agroforestry	616	228.8b	0.37b	0.46b	6.63ab
MT4	Agroforestry	541	344.9c	0.64c	0.66c	10.33b
LSD			86.6	0.12	0.12	4.22

* The same letter indicates no statistically significant differences between locations with Fisher's LSD test ($p < 0.05$).

Infiltration rates in the andisols of the upper watershed were all above 45 mm hour^{-1} (Figure 6a). In the midstream area, forest plots had a high infiltration rate, but in the agroforestry systems infiltration rates were low and the apparent declined with decreasing tree cover was not statistically significant (Figure 6b).

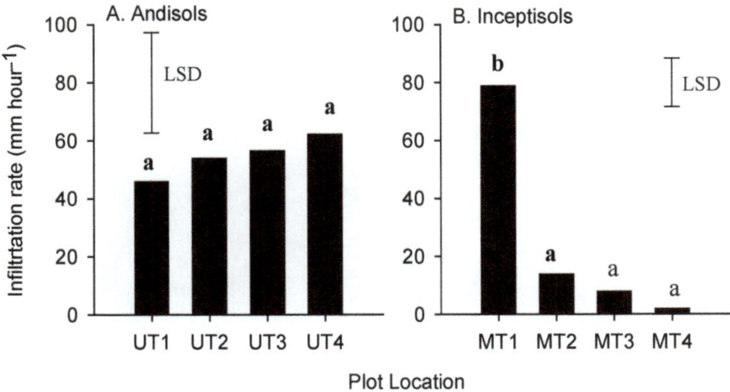

Figure 6. Soil infiltration rate measured using a double-ring infiltrometer (n = 6).

In the upstream area, with decreasing tree canopy cover, the surface runoff/rainfall ratio increased 16-fold compared to production forest (Figure 7a). In the midstream area, agroforestry systems with a tree canopy cover > 80% were still able to support low surface runoff (Figure 7b). With a tree canopy cover of < 80%, surface runoff increased rapidly on days with moderate rainfall (20–50 mm day^{-1}) (Figure 7b).

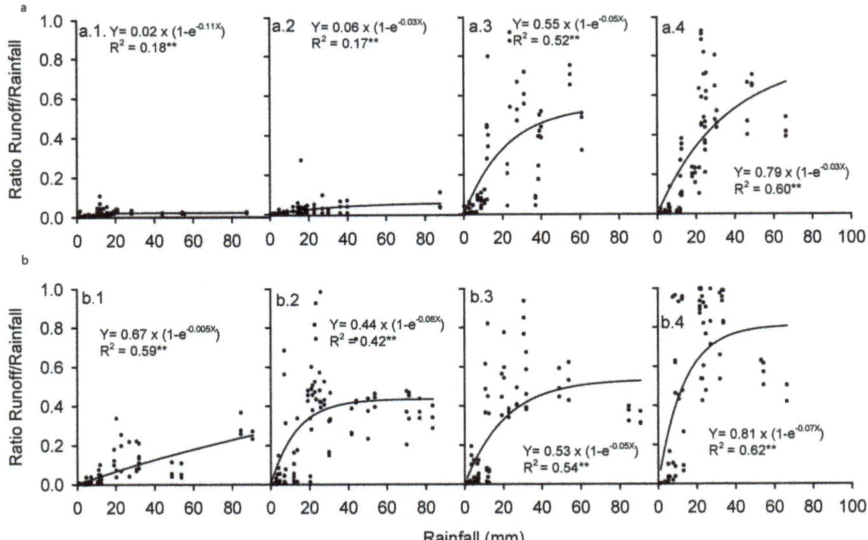

Figure 7. The relationship between surface runoff/rainfall ratio and the amount of rainfall in production forest and agroforestry systems in (**a**) the upstream Rejoso Watershed, under (a.1) 55% canopy cover of pine-based old production forest, (a.2) 40% canopy cover of pine-based young production forest, (a.3) 5% canopy cover of *Casuarina*-based agroforestry with cabbage crop, (a.4) 0% tree canopy cover of arable land (maize crop); (**b**) the midstream Rejoso Watershed under (b.1) 87 % canopy cover of pine/mahogany-based old production forest, (b.2) 75% canopy cover of coffee-based agroforestry, (b.3) 52% canopy cover of clove-based agroforestry, (b.4) 26% canopy cover of mixed tree and crop-based agroforestry.

In production forests with a closed tree canopy cover, soil erosion rates were low (Table 4 and Figure 8a.1,a.2,b.1). These production forests still had a protective understory vegetation that contributed to litter necromass and surface roughness (Table 3), controlling splash erosion. Upstream, with a reduction in tree cover, canopy soil erosion increased dramatically from 20 to 110 times the rates measured in forested plots (Table 4). Erosion rates in all plots increased with the amount of rainfall (Figure 8a.3,a.4). Midstream agroforestry systems had erosion rates ranging from 2.8 to 10.3 Mg ha^{-1} in the measurement period (Table 4). As annual rainfall is approximately three times what was recorded in the measurement period, with similar rainfall intensities, these erosion rates are to be multiplied by a factor of three, leading to 9–31 Mg ha^{-1} year^{-1}.

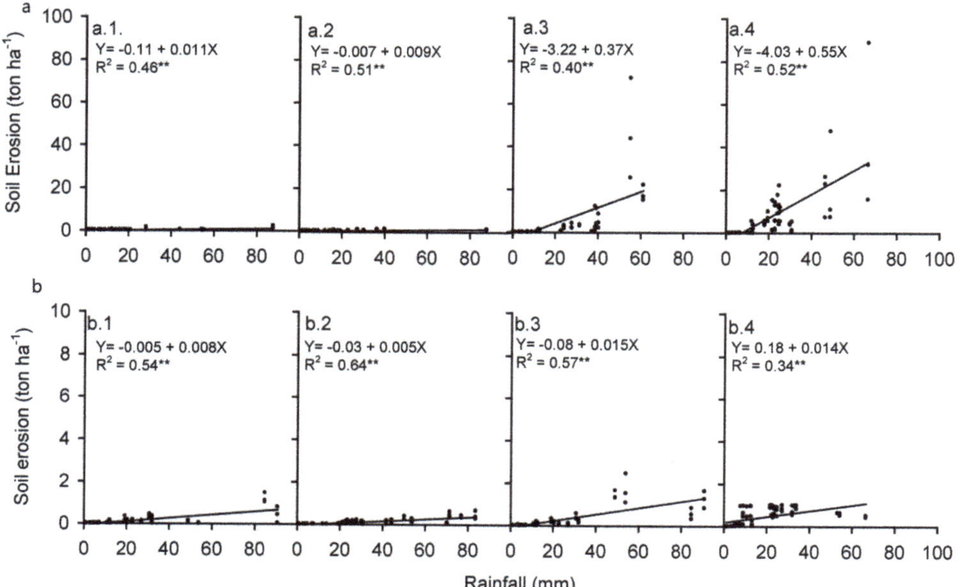

Figure 8. Soil erosion in relation to daily rainfall rates in production forest and agroforestry in (**a**) the upstream Rejoso Watershed, under (a.1) 55% canopy cover of Pine-based old production forest, (a.2) 40% canopy cover of pine-based young production forest, (a.3) 5% canopy cover of *Casuarina*-based agroforestry with cabbage crop, (a.4) 0% tree canopy cover of arable land (maize crop); (**b**) the midstream Rejoso Watershed under (b.1) 87% canopy cover of pine/mahogany-based old production forest, (b.2) 75% canopy cover of coffee-based agroforestry, (b.3) 52% canopy cover of clove-based agroforestry, (b.4) 26% canopy cover of mixed tree and crop-based agroforestry.

3.5. Thresholds for Infiltration-Friendly Land Use

Increasing tree canopy cover, while maintaining understory vegetation and litter necromass, is a strong indicator of watershed health and the main driver of low surface runoff (or high soil infiltration) and low soil erosion in production and agroforestry forest systems in the Rejoso Watershed (Figures 8a.1, 9a.1, 10b.1 and 11b.1).

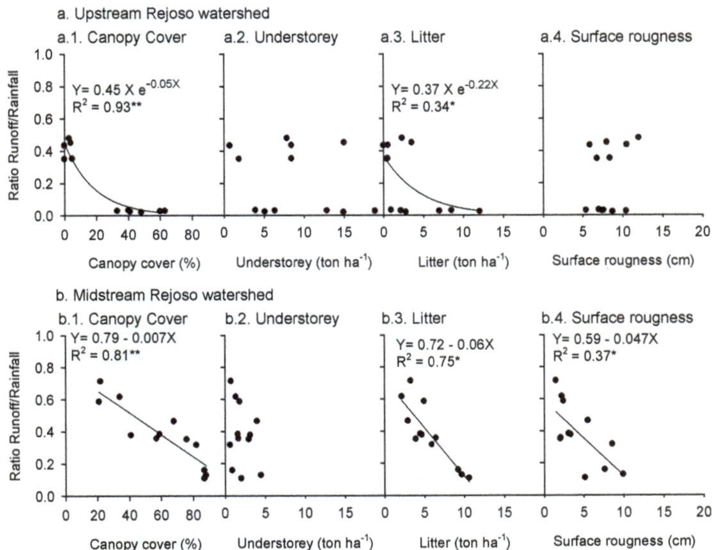

Figure 9. The runoff/rainfall ratio as a function of tree canopy cover, understory vegetation, litter necromass, and land surface roughness in the (**a**): upstream, (**b**): midstream.

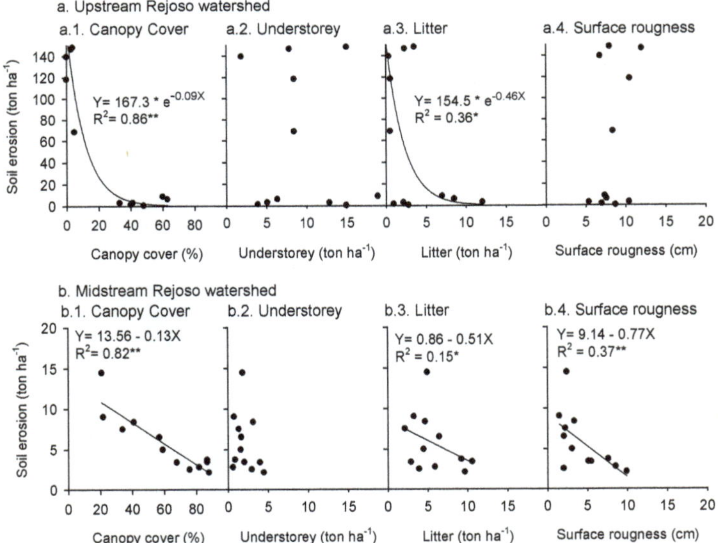

Figure 10. Soil erosion in relation to tree canopy cover, understory vegetation, litter necromass, and land surface roughness in the (**a**): upstream, (**b**): midstream.

Understory vegetation theoretically can reduce splash impacts on the soil and supports infiltration, as does the litter necromass present. However, the result of this study indicated that understory had no statistically significant relationships with runoff coefficient and soil erosion (Figures 9a.3,b.3 and 10a.3,b.3). Land surface roughness, in contrast to litter necromass, had no consistent relationship with runoff or erosion (Figures 9a.4 and 10a.4).

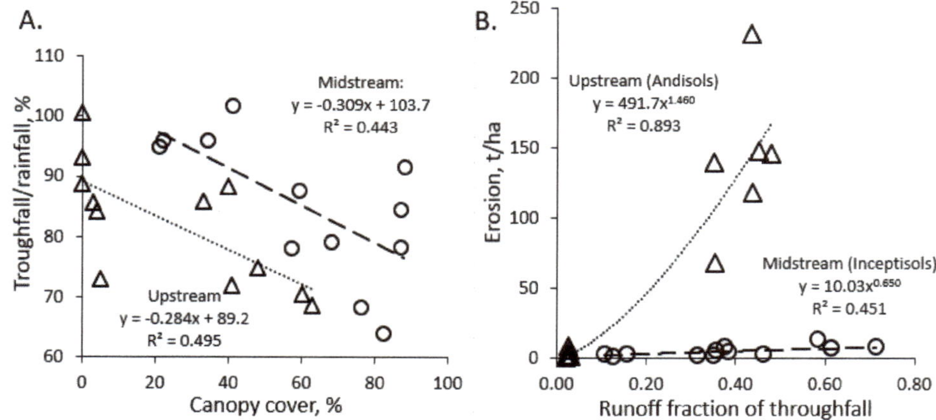

Figure 11. Comparison of canopy effects on throughfall in the two zones (**A**) and the relationship between erosion and surface runoff for the two soil types (**B**).

Summarizing (Figure 11), we found a similar slope (3% more canopy water retention per 10% canopy cover) but a 10% higher canopy retention overall in the upstream area (with lower rainfall intensity and more small events) and a strong difference between the two soil types in erosion per unit surface runoff, offsetting the higher infiltration in Andisols.

4. Discussion

Our study only covered two months' worth of data, rather than the recommended 3–5 years for such studies. Measurements included one-fourth of the mean annual rainfall, and the validity of the result may be primarily limited by the assumption that it represented a fair sample of the rainfall intensities that can be expected in the landscape. With a disproportional fraction of erosion normally associated with extreme events (compare the curvature of responses in Figure 11B), scaling up our comparison among land cover types to a multi-year basis may underestimate the relevance of controlling overland flows.

The first research question tested the hypothesis that, along the forest to open field agriculture continuum, there is a significant decrease in soil hydrological functions. The results of the present study confirmed that the conversion of high-density forest to land uses with a lower tree canopy significantly decreased soil infiltration rates (Table 2). The results of this study align with previous studies that showed that decreases in ground cover resulted in decreases in soil infiltration rates [51]. Forests and coffee agroforestry have been shown to reduce surface runoff and erosion compared with coffee monoculture [52]. Soil infiltration into Andisols both under deciduous and pine forest was higher than that in cropland in a study on the Canary Islands [53]. A study in China [54] found that the soil infiltration rate of forest was greater than that of agroforestry. A meta-analysis [55] concluded that converting any land use type with permanent vegetation cover (grassland, shrub, or forest) to seasonal cropland leads to a decline in the soil infiltration rate, harming soil and water conservation, while agroforestry improved the soil infiltration capacity compared to cropland and plantations.

The degradation of the soil hydrological functions of forest could be attributed to the decrease in soils' macroporosity, organic matter content, and increased soil bulk density (Table 2), which had relevance to the decreasing infiltration rates (Figure 8). Among various land use patterns, plant root activities are important factors affecting soil infiltration [56]. The reason why cropland has a lower infiltration rate than the land use types with a high density of trees compared with those with a low density of trees in forest may be verified by the fact that soils beneath the canopies of woody plants had a more extensive distribution of plant roots and a greater number of macropores, which are biologically produced pores [57,58], which created a positive feedback on infiltration [59,60]. The soil macroporosity, needed for effective infiltration, is the result of a continuous process of compaction and the filling in of macropores with fine soil particles, and the creation of biogenic channels (formed by old tree roots, earthworms, and other soil engineers) or abiotic processes (cracks). As no heavy machinery is used in any of these land use systems, compaction is restricted to human feet and motorbikes on specific tracks. The formation of old tree root channels can cause long time lags between land cover change and soil macroporosity [61,62], obscuring relations between current tree cover and soil hydrologic functions. "Fallows" were found to be intermediate between forests and grasslands in terms of infiltration in Madagascar [63]. Recovery of infiltration after the reforestation of grasslands in the Philippines was found to be a matter of decades rather than years [64]. In studies elsewhere in Indonesia, forest soils had more macropores and higher surface infiltration rates than monoculture coffee plantations [52]. Land use changes, especially from forest to cropland, have caused remarkable changes in soil properties, including the loss of organic matter and increases in bulk density [65], which lead to decreased infiltration rates [66]. Some researchers suggested a positive relationship between soil organic matter and infiltration rate [67,68].

Our study results can be compared to earlier studies in the volcanic uplands of West and East Java. Sediment delivery to streams increased after a clear-felling and replant operation in the Citarum Basin (W. Java), which involved the delayed flushing of material trapped during forest clearance and the incipient gullying of trails created by farmers involved in the replanting program, rather than by in-field erosion [69]. A study [70] of the Kali Konto catchment in East Java similarly concluded that "despite their relatively small areal extent (5% in the study area), rural roads, trails and settlements are significant producers of runoff and sediment at the catchment scale and should be included in watershed management programs designed to reduce catchment sediment yields and reservoir siltation." Soil conservation practices that transform slopes to relatively flat terrace beds and steep terrace risers are, in the absence of vegetation, still subject to erosion, with splash impacts on the terrace risers as a major cause [71,72].

The second research question came out with the hypothesis that the dominant factors that determine "infiltration friendliness" at the plot scale are tree canopy cover, understory vegetation, litter necromass, and land surface roughness. Our research shows that a number of land cover types had infiltration rates below the required rates at peak rainfall events. Among the four factors tested, tree cover and litter layer necromass could be used to define zone-specific thresholds for infiltration-friendly land use, but understory vegetation and surface roughness could not. Although slopes in the upper watershed are much steeper than in the midstream, the coarser texture and likely higher aggregate stability means that thresholds for canopy cover and litter necromass can be lower. A first "line of defense" of forests is the canopy retention of rainfall, prolonging the time for infiltration, as canopy dripping lasts beyond the rainfall event. Canopy retention of rainfall tends to be relatively high for small (but potentially frequent) rainfall events, and low for high rainfall intensities. Our throughfall results for the two zones corresponded with differences in observed intensity. A five-year study in the Amazon forests of Colombia [73] showed that throughfall ranged from 82 to 87% of gross rainfall in the forests studied (with a canopy cover of 83–91%) and varied with event-level gross rainfall, but also with forest structure, while stemflow contributed, on average, only 1.1% of gross rainfall in all forests. Throughfall is more spatially heterogeneous than rainfall, creating a challenge for its measurement. Roving, rather than fixed, location throughfall gauges led to narrower confidence intervals of throughfall fractions in longer-term studies [74] in lower montane rainforest in Puerto Rico, where throughfall was 75% and stemflow 4.1% of rainfall, with palms responsible for about 3% and other trees 1.1%. Spatial heterogeneity in throughfall can be expected to lead to uneven patterns of deep percolation and groundwater recharge in "patchy" forests [75]. Canopy interception can lead to direct evaporation, throughfall, or stemflow [76]. The ratio between throughfall and stemflow depends on the architecture of leaves (e.g., erect leaves favoring stemflow, pendulous leaves favoring throughfall) and stems. Storage along the stem pathway depends on bark properties [77]. Stemflow accounted for less than 3% of gross rainfall for tropical hardwoods in a study in Panama, while it was high for tall grasses [78]. High stemflow fractions have also been reported for bamboo, bananas, shaded coffee and cocoa, and understory shrubs [79–82]. Canopy interception and direct evaporation tend to be high in coastal areas with frequent light rainfall events, but low where tropical rainstorms are predominant and the canopy storage is rapidly saturated [83,84]. By creating throughfall drops that are larger than those of open-field rainfall, tree canopies may increase sub-canopy erosivity [13,84].

Many authors have emphasized that the key to hydrologic functions is in the soil rather than the aboveground parts of the forest [12]. Still, we found strong and direct relations with canopy cover. Positive effects of canopy cover on infiltration were related to raindrop interception in earlier studies [75]. Interception will (a) reduce the destructive power of rainwater splash on the ground surface (as long as the erosive canopy drips described earlier are avoided), (b) allow more time for infiltration as water reaches the surface more slowly, (c) keep a thin water film on the leaves that will (d) cool the surrounding air when it subsequently evaporates. It reduces the amount of water reaching the soil surface, but by increasing air humidity it also decreases transpiration demand when stomata are open. Coffee gardens close to forest had high macroporosity and infiltration rates relative to more compacted pasture and sugarcane land on volcanic slopes in Costa Rica [85]. Dye infiltration patterns in a comparison of natural forest and rubber plantations in Yunnan (China) showed [86] that the fine roots of understory vegetation promoted subsurface flow and reduced water erosion. The effects of trees on infiltration have been described as a "double-funneling" [87] with stemflow (dependent on the insertion angle of branches on the main stem), bringing water to the soil surface connection point for root-induced preferential flow [88,89].

A comparison of infiltration rates (median K_s values 16–98 mm h^{-1}) in broadleaf, pine-dominated, and mixed community-managed forest in Nepal [90] found the less intensively used pine-dominated site to be more conducive to vertical percolation than the other two forest types. These results were remarkable in relation to the negative local perceptions of the role of pine plantations on declining water resources.

Understory vegetation can theoretically reduce splash impacts on the soil and supports infiltration, as does the litter necromass present. However, the result of this study indicated that the understory shows no significant relationships with the runoff coefficient and soil erosion. This is possibly because surface runoff and erosion are largely controlled by land cover. The growth and development of the understory is determined by canopy cover. Likewise, the tree plantations in each plot are also diverse, so this also affects the diversity of the understory vegetation underneath. The result of this study indicates that the litter layer in the old production forest both upstream and midstream is significantly thicker than that other land uses (Table 3) and there is a significant correlation with the runoff coefficient and soil erosion (Figures 7 and 8, respectively). Litter is the parts of the body of the plant (in the form of leaves, branches, twigs, flowers, and fruit) that die (deciduous or pruned) and lie on the surface of the soil either intact or partially weathered. The role of litter in maintaining infiltration and soil erosion is through: (a) M=maintaining soil looseness by protecting the soil surface from rainwater, so that aggregates and soil macropores are maintained, (b) providing food sources for soil organisms, especially "soil engineers" (e.g., earthworms), so that the organism can live and develop in the soil, thus, the number of macro pores is maintained through the activity of these organisms, and (c) maintaining water quality in the river through the filtering of soil particles carried by surface runoff before entering the river. In a study in North China [91], the presence of the litter of *Quercus variabilis*, representing broadleaf litter, and *Pinus tabulaeformis*, representing needle leaf litter, reduced surface runoff rates by 29.5% and 31.3%, respectively. The overall effect of fast plus slow decomposing surface litter means the protection of the soil surface from splash erosion, surface roughness that reduces sediment entrainment, an energy source for soil biota, and a conducive microclimate [92,93].

Infiltration fractions depend on the scale of measurement and on variations in slope steepness, as overland flow can re-infiltrate on less steep foot-slopes in the case of the upper plots [94] or water infiltrates can re-emerge as surface flow depending on subsoil conductivity [95,96]. Such effects will need to be included if catchment level hydrology is to be predicted from plot-level measurements.

The land surface roughness also contributes to a high infiltration rate, reducing soil erosion. In the upstream, there is no significant different between land uses, but in the midstream, land surface roughness in agroforestry systems with tightly different canopies is significantly higher than rare canopies (Table 3). Without a high canopy cover (Table 3), this roughness was not able to control surface runoff and erosion in the upstream area. This is due to a steep slope in this plot. Both the production forest and agroforestry systems with high canopies maintained a relatively high land surface roughness compared with rare canopies in the midstream area. In the midstream, the land surface roughness was significantly correlated with the runoff coefficient and soil erosion. The role of surface roughness as a sediment filter may depend on frequent regeneration to counter homogenization [97]. Surface roughness in the landscape includes a cavity, the meandering of streams due to the presence of litter, necromass, tree trunks, and rocks, which provide opportunities for water flow to stop for longer periods and experience infiltration. This condition also functions as a sediment filter. This function needs to be managed through land management, so that surface roughness is maintained on the ground.

Shifts in local rainfall patterns between sub-watersheds make it difficult to disentangle the relative importance of land use and climate change through statistical pattern analysis without knowledge of the underlying processes [98,99]. The holy grail of scientific hydrology, connecting overall aggregated flow patterns to local extreme events and possible hysteresis, is still worth searching for even if a general solution might ultimately prove impossible to find [100]. For the deep seepage component of the hillslope and catchment water balance, we can expect that extreme events are less important than gradual changes that influence average flows, but empirical analysis of the uncertainties involved is still a challenge [101].

The third research question is, as an analysis, the answers to the previous two research questions with the hypothesis that it is not always that the upstream watershed area is more sensitive to hydrological disturbance due to changes in land use than the midstream, but the factor of soil properties also determines considerations in watershed hydrological management. From a land use policy perspective, our results suggest that maintaining high (~80%) canopy cover in the mid-slope farmer-controlled landscape under bench terracing, which does not match the slope criteria for designation as watershed protection forest, is important. In Indonesia, protection forest areas have the primary functions of the protection of life support systems to regulate water management, prevent flooding, control soil erosion, and maintain soil fertility [102].

Erosion rates of 9–31 Mg ha^{-1} year^{-1}, as estimated here, are a challenge, especially if the 400-year time frame of using up all soil, as used in Equation 7, is replaced by a tolerance equal to the rate of soil formation. A study in a high rainfall area with Inceptisols in Central Java [103] estimated that the rate of chemical weathering was around 0.85 Mg ha^{-1} yr^{-1} and used that as estimate of erosion rates that can be sustained indefinitely without affecting soil depth. Volcanic ash inputs add soil on top of the profile but may also be disproportionately included in what gets removed from the plots. Our measurements in Rejoso suggested that critical thresholds of the degree of canopy cover that is hydrologically desirable depend on soil and climatic conditions, which may vary over a relatively short distance. When the focus is on erosion and net sediment transport, the scale of consideration strongly influenced conclusions in the volcanic Way Besai Watershed in Sumatra as well [104]. With the higher rainfall intensities in midstream Rejoso and more erodible soils upstream, the risks for degradation from a downstream perspective are differentiated by zone. Combining our plot-level results with efforts of hydrologic modeling for the Rejoso catchment as a whole [105,106] can guide further advice to a local watershed forum on the measures and incentives needed to restore and protect the watershed as a whole.

The Indonesian legal requirement of 30% forest cover across all its local government entities [31] is a coarse translation of the hydrologic relations at risk. It clearly matters what the land cover in the "non-forest" parts of the landscape is and how vegetation interacts with soils and geomorphology in shaping rivers and groundwater flows [107,108]. Our findings for the Rejoso Watershed show that, within the agroforestry spectrum, hydrologic the thresholds of infiltration friendliness exist between the systems that are mostly "agro" and those that are mostly "forest", but higher tree cover systems are desirable.

5. Conclusions

Our results demonstrated that vegetation-based thresholds for adequate infiltration, given the existing rainfall intensities, differed between the middle and upper Rejoso Watershed. Despite steep slopes and low tree cover, the upper watershed, with its course soil texture (pseudo-sand/silt), low bulk density due to a high content of amorphic minerals, strong micro-aggregation and individual minerals, sponge-pores typical of Andosols, and land management practices that combine vegetable crops with a tree canopy cover of around 55%, can maintain infiltration and keep erosion at acceptable levels. In the midstream part of the catchment, despite gentle slopes under bench terracing, infiltration-friendly land use on the fine-textured Inceptisols required a canopy cover of 80%. Beyond tree canopy cover, litter layer necromass was found to be a good and easily observed indicator of infiltration rates, while understory vegetation and surface roughness may support infiltration, but are not sufficiently strong indicators.

Author Contributions: D.S., W.W., K.H., and M.v.N. designed the study. N.M. collected data in the midstream, A.L.R. collected data in the upstream, R.M.I. coordinated the data collection in the field, and were academically supervised by D.S., W.W., and K.H. D.S., K.H. and M.v.N. shaped the manuscript, which was approved by all co-authors.

Funding: Fieldwork for this research was funded by the Danone Ecosystem Fund via the World Agroforestry Centre, ICRAF, Indonesia office. This research was **also** partially funded by the Indonesian Ministry of Research, Technology and Higher Education under the WCU Program managed by **the** Institute of Research and Community Services, Universitas Brawijaya and Institut Teknologi Bandung.

Acknowledgments: Authors thank to the community of the Rejoso Watershed and the *"Rejoso Kita"* Forum. They also thank the Department of Soil Science, Faculty of Agriculture, University of Brawijaya and the Research Group of Tropical Agroforestry for their support of the research. The authors would like to thank the Social Investment Indonesia (SII) organization for connecting the local stakeholders during fieldwork. The manuscript benefitted from the comments of Sampurno Bruijnzeel and an anonymous reviewer.

Conflicts of Interest: The authors declare no conflict of interest.

Appendix A

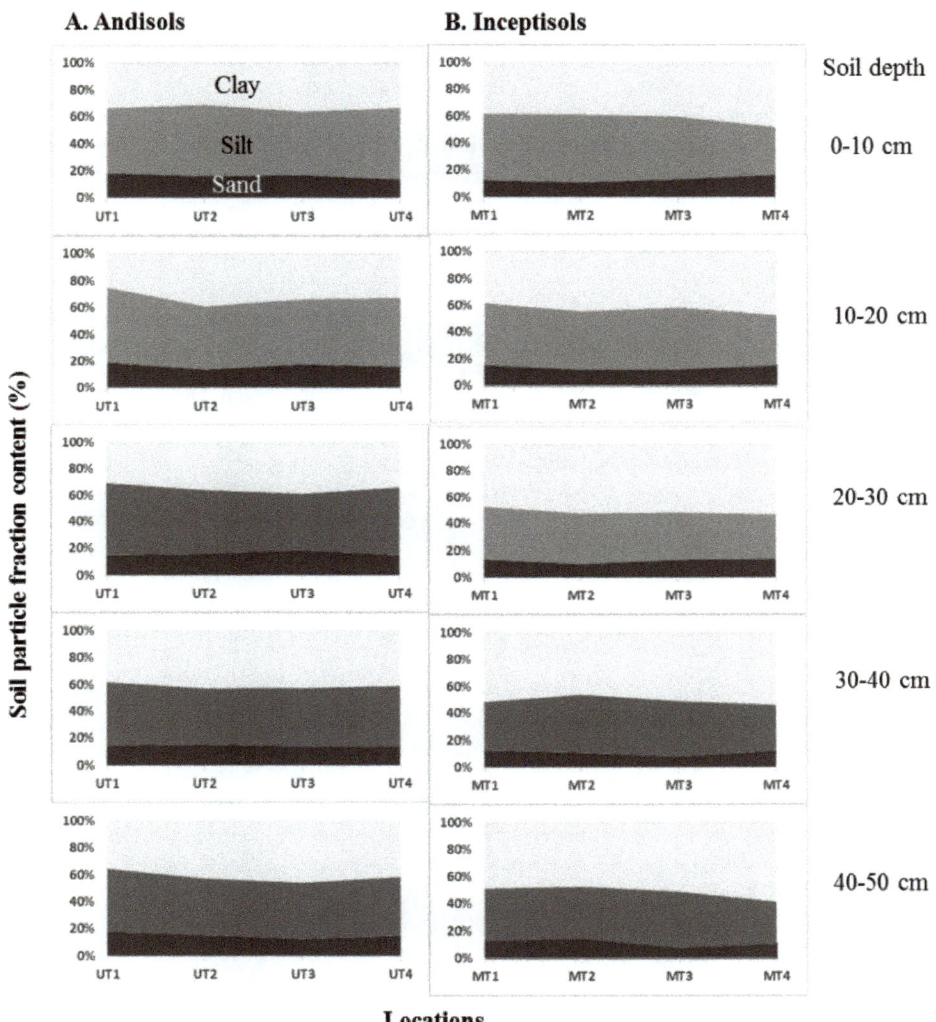

Figure A1. Soil texture in five different layers in runoff plot measurements.

References

1. UN-Water. *2018 UN World Water Development Report, Nature-Based Solutions for Water*; UN-Water: Nairobi, Kenya, 2018.
2. Ostovar, A.L. Investing upstream: Watershed protection in Piura, Peru. *Environ. Sci. Policy* **2019**, *96*, 9–17. [CrossRef]
3. Ali, M.; Hadi, S.; Sulistyantara, B. Study on land cover change of Ciliwung downstream watershed with spatial dynamic approach. CITIES 2015 International Conference, Intelligent Planning towards Smart Cities, CITIES 2015, 3–4 November 2015, Surabaya, Indonesia. *Procedia—Soc. Behav. Sci.* **2016**, *227*, 52–59. [CrossRef]
4. Jacobs, S.R.; Timbe, E.; Weeser, B.; Rufino, M.C.; Butterbach-Bahl, K.; Breuer, L. Assessment of hydrological pathways in East African montane catchments under different land use. *Hydrol. Earth Syst. Sci.* **2018**, *22*, 4981–5000. [CrossRef]

5. Van Noordwijk, M.; Bayala, J.; Hairiah, K.; Lusiana, B.; Muthuri, C.; Khasanah, N.; Mulia, R. Agroforestry solutions for buffering climate variability and adapting to change. In *Climate Change Impact and Adaptation in Agricultural Systems*; Fuhrer, J., Gregory, P.J., Eds.; CAB-International: Wallingford, UK, 2014; pp. 216–232.

6. Van Noordwijk, M.; Bargues-Tobella, A.; Muthuri, C.W.; Gebrekirstos, A.; Maimbo, M.; Leimona, B.; Bayala, J.; Ma, X.; Lasco, R.; Xu, J.; et al. Agroforestry as part of nature-based water management. In *Sustainable Development through Trees on Farms: Agroforestry in Its Fifth Decade*; van Noordwijk, M., Ed.; World Agroforestry (ICRAF): Bogor, Indonesia, 2019; pp. 261–287.

7. Van Noordwijk, M.; Kim, Y.S.; Leimona, B.; Hairiah, K.; Fisher, L.A. Metrics of water security, adaptive capacity, and agroforestry in Indonesia. *Curr. Opin. Environ. Sustain.* **2016**, *21*, 1–8. [CrossRef]

8. Filoso, S.; Bezerra, M.O.; Weiss, K.C.; Palmer, M.A. Impacts of forest restoration on water yield: A systematic review. *PLoS ONE* **2017**, *12*, e0183210. [CrossRef]

9. Bentley, L.; Coomes, D.A. Partial river flow recovery with forest age is rare in the decades following establishment. *Glob. Chang. Biol.* **2020**, *26*, 1458–1473. [CrossRef]

10. Zhang, J. Hydrological Response of a Fire-Climax Grassland and a Reforest Before and After Passage of Typhoon Haiyan (Leyte Island, the Philippines). Ph.D. Thesis, Free University, Amsterdam, The Netherlands, 2018.

11. Bruijnzeel, L.A. Hydrological functions of tropical forests: Not seeing the soil for the trees? *Agric. Ecosyst. Environ.* **2004**, *104*, 185–228. [CrossRef]

12. Peña-Arancibia, J.L.; Bruijnzeel, L.A.; Mulligan, M.; van Dijk, A.I. Forests as 'sponges' and 'pumps': Assessing the impact of deforestation on dry-season flows across the tropics. *J. Hydrol.* **2019**, *574*, 946–963. [CrossRef]

13. Wiersum, K.F. Effects of various vegetation layers of an Acacia auriculiformis forest plantation on surface erosion in Java, Indonesia. In *Soil Erosion and Conservation*; El-Swaify, S.A., Moldenhauer, W.C., Lo, A., Eds.; John Wiley & Sons: Hoboken, NJ, USA, 1985; pp. 79–89.

14. Hairiah, K.; Suprayogo, D.; Widianto, B.; Suhara, E.; Mardiastuning, A.; Widodo, R.H.; Prayogo, C.; Rahayu, S. Forests conversion into agricultural land: Litter thickness, earthworm populations and soil macro-porosity. *Agrivita* **2004**, *26*, 68–80.

15. Suprayogo, D.; Widianto, K.; Hairiah, I.N. *Watershed Management: Hydrological Consequences of Land Use Change during Development*; Universitas Brawijaya Press: Malang, Indonesia, 2017.

16. Creed, I.F.; van Noordwijk, M. *Forest and Water on a Changing Planet: Vulnerability, Adaptation and Governance Opportunities*; A Global Assessment Report. World Series Volume 38; IUFRO: Vienna, Austria, 2018.

17. Jones, J.A.; Wei, X.; Archer, E.; Bishop, K.; Blanco, J.A.; Ellison, D.; Gush, M.; McNulty, S.G.; van Noordwijk, M.; Creed, I.F. Forest-Water Interactions under Global Change. In *Forest-Water Interactions*; Ecological Studies 240; Levia, D.F., Carlyle-Moses, D.E., Michalzik, B., Nanko, K., Tischer, A., Eds.; Springer Nature: Picassopl, Switzerland, 2020; pp. 589–624. [CrossRef]

18. Liu, Y.; Cui, Z.; Huang, Z.; Miao, H.T.; Wu, G.L. The influence of litter crusts on soil properties and hydrological processes in a sandy ecosystem. *Hydrol. Earth Syst. Sci.* **2019**, *23*, 2481–2490. [CrossRef]

19. Zhang, J.; van Meerveld, H.I.; Tripoli, R.; Bruijnzeel, L.A. Runoff response and sediment yield of a landslide-affected fire-climax grassland micro-catchment (Leyte, the Philippines) before and after passage of typhoon Haiyan. *J. Hydrol.* **2018**, *565*, 524–537. [CrossRef]

20. Zhang, J.; Bruijnzeel, L.A.; Quiñones, C.M.; Tripoli, R.; Asio, V.B.; van Meerveld, H.J. Soil physical characteristics of a degraded tropical grassland and a 'reforest': Implications for runoff generation. *Geoderma* **2019**, *333*, 163–177. [CrossRef]

21. Van Noordwijk, M.; Lawson, G.; Hairiah, K.; Wilson, J. Root distribution of trees and crops: Competition and/or complementarity. In *Tree–Crop Interactions: Agroforestry in a Changing Climate*; CABI: Wallingford, UK, 2015; pp. 221–257.

22. Ellison, D.; Morris, C.E.; Locatelli, B.; Sheil, D.; Cohen, J.; Murdiyarso, D.; Gutierrez, V.; van Noordwijk, M.; Creed, I.F.; Pokorny, J. Trees, forests and water: Cool insights for a hot world. *Glob. Environ. Chang.* **2017**, *43*, 51–61. [CrossRef]

23. Ellison, D.; Wang-Erlandsson, L.; van der Ent, R.J.; van Noordwijk, M. Upwind forests: Managing moisture recycling for nature-based resilience. *Unasylva* **2019**, *251*, 13–26.

24. Van Noordwijk, M.; Tanika, L.; Lusiana, B. Flood risk reduction and flow buffering as ecosystem services—Part 1: Theory on flow persistence, flashiness and base flow. *Hydrol. Earth Syst. Sci.* **2017**, *21*, 2321–2340. [CrossRef]

25. Shannon, P.D.; Swanston, C.W.; Janowiak, M.K.; Handler, S.D.; Schmitt, K.M.; Brandt, L.A.; Butler-Leopold, P.R.; Ontl, T. Adaptation strategies and approaches for forested watersheds. *Rev. Artic. Clim. Serv.* **2019**, *13*, 51–64. [CrossRef]

26. Van Noordwijk, M.; Tanika, L.; Lusiana, B. Flood risk reduction and flow buffering as ecosystem services—Part 2: Land use and rainfall intensity effects in Southeast Asia. *Hydrol. Earth Syst. Sci.* **2017**, *21*, 2341–2360. [CrossRef]

27. Coster, C. Wortelstudien in de tropen. V. Gebergtehoutsoorten [Root studies in the Tropics. V. Tree species of the mountain region]. *Tectona* **1935**, *28*, 861–878.

28. Coster, C. De verdamping van verschillende vegetatievormen op Java [The transpiration of different types of vegetation in Java]. *Tectona* **1937**, *30*, 1–124.

29. Gonggrijp, L. De verdamping van gebergtebosch in West Java op 1750–200 m zeehoogte [The transpiration of mountain forest in West Java at an elevation of 1750–2000 m above sea level]. *Tectona* **1941**, *34*, 437–447.

30. Galudra, G.; Sirait, M. A discourse on Dutch colonial forest policy and science in Indonesia at the beginning of the 20th century. *Int. For. Rev.* **2009**, *11*, 524–533. [CrossRef]

31. Government of Indonesia. *Law (Undang-Undang) Number 26 Year 2007 about Spatial Planning*; Government of Indonesia: Pemerintah, Indonesia, 2007.

32. Van Noordwijk, M.; Creed, I.F.; Jones, J.A.; Wei, X.; Gush, M.; Blanco, J.A.; Sullivan, C.A.; Bishop, K.; Murdiyarso, D.; Xu, J.; et al. Climate-forest-water-people relations: Seven system delineations. In *Forest and Water on a Changing Planet: Vulnerability, Adaptation and Governance Opportunities: A Global Assessment Report (No. 38)*; International Union of Forest Research Organizations (IUFRO): Vienna, Austria, 2018; pp. 27–58.

33. Van Noordwijk, M.; Ekadinata, A.; Leimona, B.; Catacutan, D.; Martini, E.; Tata, H.L.; Öborn, I.; Hairiah, K.; Wangpakapattanawong, P.; Mulia, R.; et al. Agroforestry options for degraded landscapes in Southeast Asia. In *Agroforestry for Degraded Landscapes: Recent Advances and Emerging Challenges*; Dagar, J.C., Gupta, S.R., Teketay, D., Eds.; Springer: Singapore, 2020.

34. Sajikumar, N.; Remya, R.S. Impact of land cover and land use change on runoff characteristics. *J. Environ. Manag.* **2015**, *161*, 460–468. [CrossRef] [PubMed]

35. Anache, J.A.; Wendland, E.C.; Oliveira, P.T.; Flanagan, D.C.; Nearing, M.A. Runoff and soil erosion plot-scale studies under natural rainfall: A meta-analysis of the Brazilian experience. *Catena* **2017**, *152*, 29–39. [CrossRef]

36. USGS. Landsat 8 OLI (2019-10-18 (6, 53). U.S. Geological Survey. 2019. Available online: https://earthexplorer.usgs.gov/ (accessed on 9 June 2020).

37. Wischmeier, W.H. The USLE: Some reflections. *J. Soil Water Conserv.* **1984**, *39*, 105–107.

38. Hairsine, P.B.; Rosem, C.W. Modeling water erosion due to overland flow using physical principles: 1. Sheet flow. *Water Resour. Res.* **1992**, *28*, 237–243. [CrossRef]

39. Van Noordwijk, M.; van Roode, M.; McCallie, E.L.; Cadisch, G. Erosion and sedimentation as multiscale, fractal processes: Implications for models, experiments and the real world. In *Soil Erosion at Multiple Scales, Principles and Methods for Assessing Causes and Impacts*; Penning de Vries, F., Agus, F., Kerr, J., Eds.; CAB International: Wallingford, UK, 1998; pp. 223–253.

40. Knox County Tennessee. Stormwater Management Manual, Section on the Rational Method, Volume 2 Technical Guidance. Available online: https://dec.alaska.gov/water/wastewater/stormwater/swppp-dev-runoff-coefficient-values-rm (accessed on 26 April 2020).

41. Widianto, D.S.; Noveras, H.; Widodo, R.H.; Purnomosidhi, P.; van Noordwijk, M. Alih guna lahan hutan menjadi lahan pertanian: Apakah fungsi hidrologis hutan dapat digantikan sistem kopi monokultur? *Agrivita* **2004**, *26*, 47–52.

42. Gee, G.W.; Bauder, J.W. Particle Size Analysis. In *Methods of Soil Analysis Part 1, Physical and Mineralogical Methods*, 2nd ed.; Klute, A., Ed.; American Society of Agronomy: Madison, WI, USA, 1986; Volume 9, pp. 383–411.

43. Blake, G.R.; Hartge, K.H. Bulk density. In *Methods of Soil Analysis. Part I: Physical and Mineralogical Methods*; Klute, A.K., Ed.; American Society of Agronomy—Soil Science Society of America: Madison, WI, USA, 1986; pp. 363–375.

44. Nimmo, J.R. Porosity and pore size distribution. In *Encyclopedia of Soils in the Environment*; Hillel, D., Ed.; Elsevier: London, UK, 2004; pp. 295–303.

45. Walkley, A.; Black, I.A. An examination of the Degtjareff method for determining soil organic matter, and a proposed modification of the chromic acid titration method. *Soil Sci.* **1934**, *37*, 29–38. [CrossRef]

46. Bouwer, H. Intake rate: Cylinder infiltrometer. In *Methods of Soil Analysis*; Klute, A., Ed.; ASA Monograph 9; ASA: Madison, WI, USA, 1986; pp. 825–843.

47. Jennings, S.B.; Brown, N.D.; Sheil, D. Assessing forest canopies and understorey illumination: Canopy closure, canopy cover and other measures. *Forestry* **1999**, *72*, 59–73. [CrossRef]

48. Hairiah, K.; Ekadinata, A.; Sari, R.R.; Rahayu, S. Measurement of Carbon Stock: From land level to landscape. In *Practical Instructions*, 2nd ed.; Bogor, World Agroforestry Centre, ICRAF SEA Regional Office, University of Brawijaya (UB): Malang, Indonesia, 2011; p. 110. Available online: http://apps.worldagroforestry.org/sea/Publications/files/manual/MN0049-11.pdf (accessed on 10 April 2020).

49. Hoechsteetter, S.; Walz, U.; Dang, L.H.; Thinh, N.X. Effect of Topography and Surface Roughness in Analyses of Landscape Structure—A Proposal Modify the Existing Set of Landscape Metrics. *Landsc. Online* **2008**, *3*, 1–14. Available online: https://gfzpublic.gfz-potsdam.de/pubman/item/item_32854 (accessed on 10 April 2020). [CrossRef]

50. Landon, J.R. *Booker Tropical Soil Manual: A Handbook for Soil Survey and Agricultural Land Evaluation in the Tropics and Subtropics*; Routledge: Abingdon, UK, 2014.

51. Gifford, G.F.; Hawkins, R.H. Hydrologic impact of grazing on infiltration: A critical review. *Water Resour. Res.* **1978**, *14*, 305–313. [CrossRef]

52. Widianto, D.S.; Purnomosidhi, P.; Widodo, R.H.; Rusiana, F.; ZAini, Z.; Khasanah, N.; Kusuma, Z. Degradasi sifat fisik tanah sebagai akibat alih guna lahan hutan menjadi sistem kopi monokultur: Kajian perubahan makroporositas tanah. *Agrivita* **2004**, *26*, 60–67.

53. Neris, J.; Jiménez, C.; Fuentes, J.; Morillas, G.; Tejedor, M. Vegetation and land-use effects on soil properties and water infiltration of Andisols in Tenerife (Canary Islands, Spain). *Catena* **2012**, *98*, 55–62. [CrossRef]

54. Ma, Q.; Wang, J.; Zhu, S. Effects of precipitation, soil water content and soil crust on artificial Haloxylon Ammodendron forest. *Acta Ecol. Sin.* **2007**, *27*, 5057–5067. [CrossRef]

55. Sun, D.; Yang, H.; Guan, D.; Yang, M.; Wu, J.; Yuan, F.; Jin, C.; Wang, A.; Zhang, Y. The effects of land use change on soil infiltration capacity in China: A meta-analysis. *Sci. Total Environ.* **2018**, *626*, 1394–1401. [CrossRef] [PubMed]

56. Zimmermann, B.; Elsenbeer, H.; De Moraes, J.M. The influence of land-use changes on soil hydraulic properties: Implications for runoff generation. *For. Ecol. Manag.* **2006**, *222*, 29–38. [CrossRef]

57. Dunkerley, D. A Review of the Effects of Throughfall and Stemflow on Soil Properties and Soil Erosion. In *Precipitation Partitioning by Vegetation*; Springer: Cham, Germany, 2020; pp. 183–214.

58. Colloff, M.J.; Pullen, K.R.; Cunningham, S.A. Restoration of an ecosystem function to revegetation communities: The role of invertebrate macropores in enhancing soil water infiltration. *Restor. Ecol.* **2010**, *18*, 65–72. [CrossRef]

59. Reid, K.D.; Wilcox, B.P.; Breshears, D.D.; MacDonald, L. Runoff and erosion in a pinon–juniper woodland: Influence of vegetation patches. *Soil Sci. Soc. Am. J.* **1999**, *63*, 313–325. [CrossRef]

60. Bhark, E.W.; Small, E.E. Association between plant canopies and the spatial patterns of infiltration in shrubland and grassland of the Chihuahuan Desert, New Mexico. *Ecosystems* **2003**, *6*, 185–196. [CrossRef]

61. Van Noordwijk, M.; Heinen, M.; Hairiah, K. Old tree root channels in acid soils in the humid tropics: Important for crop root penetration, water infiltration and nitrogen management. In *Plant-Soil Interactions at Low pH*; Springer: Dordrecht, The Netherlands, 1991; pp. 423–430.

62. Wu, G.L.; Liu, Y.; Yang, Z.; Cui, Z.; Deng, L.; Chang, X.F.; Shi, Z.H. Root channels to indicate the increase in soil matrix water infiltration capacity of arid reclaimed mine soils. *J. Hydrol.* **2017**, *546*, 133–139. [CrossRef]

63. Zwartendijk, B.W.; Van Meerveld, H.J.; Ghimire, C.P.; Bruijnzeel, L.A.; Ravelona, M.; Jones, J.P.G. Rebuilding soil hydrological functioning after swidden agriculture in eastern Madagascar. *Agric. Ecosyst. Environ.* **2017**, *239*, 101–111. [CrossRef]

64. Van Meerveld, H.J.; Zhang, J.; Tripoli, R.; Bruijnzeel, L.A. Effects of reforestation of a degraded Imperata grassland on dominant flow pathways and streamflow responses in Leyte, the Philippines. *Water Resour. Res.* **2019**, *55*, 4128–4148. [CrossRef]

65. Sanchez, P.A. *Properties and Management of Soils in the Tropics*; Cambridge University Press: Cambridge, UK, 2019.

66. Mwendera, E.J.; Saleem, M.M. Infiltration rates, surface runoff, and soil loss as influenced by grazing pressure in the Ethiopian highlands. *Soil Use Manag.* **1997**, *13*, 29–35. [CrossRef]

67. Martens, D.A.; Frankenberger, W.T. Modification of Infiltration Rates in an Organic-Amended Irrigated. *Agron. J.* **1992**, *84*, 707–717. [CrossRef]

68. Osuji, G.E.; Okon, M.A.; Chukwuma, M.C.; Nwarie, I.I. Infiltration characteristics of soils under selected land use practices in Owerri, Southeastern Nigeria. *World J. Agric. Sci.* **2010**, *6*, 322–326.

69. Bons, C.A. Accelerated erosion due to clearcutting of plantation forest and subsequent Taungya cultivation in upland West Java, Indonesia. *Int. Assoc. Hydrol. Sci. Publ.* **1990**, *192*, 279–288.

70. Rijsdijk, A.; Bruijnzeel, L.S.; Sutoto, C.K. Runoff and sediment yield from rural roads, trails and settlements in the upper Konto catchment, East Java, Indonesia. *Geomorphology* **2007**, *87*, 28–37. [CrossRef]

71. Purwanto, E.; Bruijnzeel, L.A. Soil conservation on rainfed bench terraces in upland West Java, Indonesia: Towards a new paradigm. In *Advances in Geoecology*; Catena: Reiskirchen, Germany, 1998; pp. 1267–1274.

72. Van Dijk, A.I.J.M.; Bruijnzeel, L.A.; Wiegman, S.E. Measurements of rain splash on bench terraces in a humid tropical steepland environment. *Hydrol. Process.* **2003**, *17*, 513–535. [CrossRef]

73. Marin, C.T.; Bouten, W.; Sevink, J. Gross rainfall and its partitioning into throughfall, stemflow and evaporation of intercepted water in four forest ecosystems in western Amazonia. *J. Hydrol.* **2000**, *237*, 40–57. [CrossRef]

74. Holwerda, F.; Scatena, F.N.; Bruijnzeel, L.A. Throughfall in a Puerto Rican lower montane rain forest: A comparison of sampling strategies. *J. Hydrol.* **2006**, *327*, 592–602. [CrossRef]

75. Klos, P.Z.; Chain-Guadarrama, A.; Link, T.E.; Finegan, B.; Vierling, L.A.; Chazdon, R. Throughfall heterogeneity in tropical forested landscapes as a focal mechanism for deep percolation. *J. Hydrol.* **2014**, *519*, 2180–2188. [CrossRef]

76. Muzylo, A.; Llorens, P.; Valente, F.; Keizer, J.J.; Domingo, F.; Gash, J.H.C. A review of rainfall interception modelling. *J. Hydrol.* **2009**, *370*, 191–206. [CrossRef]

77. Van Stan, J.T.; Lewis, E.S.; Hildebrandt, A.; Rebmann, C.; Friesen, J. Impact of interacting bark structure and rainfall conditions on stemflow variability in a temperate beech-oak forest, central Germany. *Hydrol. Sci. J.* **2016**, *61*, 2071–2083. [CrossRef]

78. Friesen, P.; Park, A.; Sarmiento-Serrud, A.A. Comparing rainfall interception in plantation trials of six tropical hardwood trees and wild sugar cane *Saccharum spontaneum* L. *Ecohydrology* **2013**, *6*, 765–774.

79. Taniguchi, M.; Tsujimura, M.; Tanaka, T. Significance of stemflow in groundwater recharge. 1: Evaluation of the stemflow contribution to recharge using a mass balance approach. *Hydrol. Process.* **1996**, *10*, 71–80. [CrossRef]

80. Cattan, P.; Bussière, F.; Nouvellon, A. Evidence of large rainfall partitioning patterns by banana and impact on surface runoff generation. *Hydrol. Process.* **2007**, *21*, 2196–2205. [CrossRef]

81. Siles, P.; Vaast, P.; Dreyer, E.; Harmand, J.M. Rainfall partitioning into throughfall, stemflow and interception loss in a coffee (*Coffea arabica* L.) monoculture compared to an agroforestry system with Inga densiflora. *J. Hydrol.* **2010**, *395*, 39–48. [CrossRef]

82. Niether, W.; Armengot, L.; Andres, C.; Schneider, M.; Gerold, G. Shade trees and tree pruning alter throughfall and microclimate in cocoa (*Theobroma cacao* L.) production systems. *Ann. For. Sci.* **2018**, *75*, 38. [CrossRef]

83. Bruijnzeel, L.A.; Wiersum, K.F. Rainfall interception by a young Acacia auriculiformis (A. Cunn) plantation forest in west Java, Indonesia: Application of Gash's analytical model. *Hydrol. Process.* **1987**, *1*, 309–319. [CrossRef]

84. de Almeida, W.S.; Panachuki, E.; de Oliveira, P.T.S.; da Silva Menezes, R.; Sobrinho, T.A.; de Carvalho, D.F. Effect of soil tillage and vegetal cover on soil water infiltration. *Soil Tillage Res.* **2018**, *175*, 130–138. [CrossRef]

85. Toohey, R.C.; Boll, J.; Brooks, E.S.; Jones, J.R. Effects of land use on soil properties and hydrological processes at the point, plot, and catchment scale in volcanic soils near Turrialba, Costa Rica. *Geoderma* **2018**, *315*, 138–148. [CrossRef]

86. Nespoulous, J.; Merino-Martín, L.; Monnier, Y.; Bouchet, D.C.; Ramel, M.; Dombey, R.; Viennois, G.; Mao, Z.; Zhang, J.L.; Cao, K.F.; et al. Tropical forest structure and understorey determine subsurface flow through biopores formed by plant roots. *Catena* **2019**, *181*, 104061. [CrossRef]

87. Johnson, M.S.; Lehmann., J. Double-Funneling of trees: Stemflow and root-induced preferential flow. *Ecoscience* **2006**, *13*, 324–333. [CrossRef]

88. Gonzalez-Ollauri, A.; Stokes, A.; Mickovski, S.B. A novel framework to study the effect of tree architectural traits on stemflow yield and its consequences for soil-water dynamics. *J. Hydrol.* **2020**, *582*, 124448. [CrossRef]

89. Liang, W.L. Effects of Stemflow on Soil Water Dynamics in Forest Stands. In *Forest-Water Interactions*; Springer: Cham, Germany, 2020; pp. 349–370.

90. Badu, M.; Nuberg, I.; Ghimire, C.P.; Bajracharya, R.M.; Meyer, W.S. Negative Trade-Offs between Community Forest Use and Hydrological Benefits in the Forested Catchments of Nepal's Mid-hills. *Mt. Res. Dev.* **2020**, *39*, 822–832. [CrossRef]

91. Li, X.; Niu, J.; Xie, B. The Effect of Leaf Litter Cover on Surface Runoff and Soil Erosion in Northern China. *PLoS ONE* **2014**, *9*, e107789. [CrossRef]

92. Hairiah, K.; Sulistyani, H.; Suprayogo, D.; Purnomosidhi, P.; Widodo, R.H.; Van Noordwijk, M. Litter layer residence time in forest and coffee agroforestry systems in Sumberjaya, West Lampung. *For. Ecol. Manag.* **2006**, *224*, 45–57. [CrossRef]

93. Derpsch, R.; Franzluebbers, A.J.; Duiker, S.W.; Reicosky, D.C.; Koeller, K.; Friedrich, T.; Sturny, W.G.; Sá, J.C.M.; Weiss, K. Why do we need to standardize no-tillage research? *Soil Tillage Res.* **2014**, *137*, 16–22. [CrossRef]

94. Stomph, T.J.; De Ridder, N.; Steenhuis, T.S.; Van de Giesen, N.C. Scale effects of Hortonian overland flow and rainfall-runoff dynamics: Laboratory validation of a process-based model. *Earth Surf. Process. Landf.* **2002**, *27*, 847–855. [CrossRef]

95. Gomi, T.; Sidle, R.C.; Miyata, S.; Kosugi, K.I.; Onda, Y. Dynamic runoff connectivity of overland flow on steep forested hillslopes: Scale effects and runoff transfer. *Water Resour. Res.* **2008**, *44*, 1–16. [CrossRef]

96. Moreno-de las Heras, M.; Nicolau, J.M.; Merino-Martín, L.; Wilcox, B.P. Plot-Scale effects on runoff and erosion along a slope degradation gradient. *Water Resour. Res.* **2010**, *46*, 1–12. [CrossRef]

97. Rodenburg, J.; Stein, A.; van Noordwijk, M.; Ketterings, Q.M. Spatial variability of soil pH and phosphorus in relation to soil run-off following slash-and-burn land clearing in Sumatra, Indonesia. *Soil Tillage Res.* **2003**, *71*, 1–14. [CrossRef]

98. Pradiko, H.; Soewondo, P.; Suryadi, Y. The change of hydrological regime in upper Cikapundung Watershed, West Java Indonesia. *Procedia Eng.* **2015**, *125*, 229–235. [CrossRef]

99. Nurhayati, S.A.; Sabar, A.; Marselina, M. The effect of land use changes to discharge extremities in Cimahi Watershed–Upper Citarum Watershed, West Java. In *E3S Web of Conferences*; EDP Sciences: Ulis, France, 2020; Volume 148, p. 07002.

100. Beven, K. Searching for the Holy Grail of scientific hydrology: Q t=(S, R,? t) a as closure. *Hydrol. Earth Syst. Sci. Discuss.* **2006**, *10*, 609–618. [CrossRef]

101. Graham, C.B.; Van Verseveld, W.; Barnard, H.R.; McDonnell, J.J. Estimating the deep seepage component of the hillslope and catchment water balance within a measurement uncertainty framework. *Hydrol. Process.* **2010**, *24*, 3631–3647. [CrossRef]

102. Government of Indonesia. *Law (Undang-Undang) Number 41 Year 1999 about Forestry*; Government of Indonesia: Pemerintah, Indonesia, 1999.

103. Bruijnzeel, L.A. Evaluation of runoff sources in a forested basin in a wet monsoonal environment: A combined hydrological and hydrochemical approach. *Hydrol. Humid Trop. Reg. IAHS Publ.* **1983**, 140.

104. Verbist, B.; Poesen, J.; van Noordwijk, M.; Suprayogo, D.; Agus, F.; Deckers, J. Factors affecting soil loss at plot scale and sediment yield at catchment scale in a tropical volcanic agroforestry landscape. *Catena* **2010**, *80*, 34–46. [CrossRef]

105. Tanika, L.; Khasanah, N.; Leimona, B. *Simulasi Dampak Perubahan Tutupan Lahan Terhadap Neraca Air Di Das Dan Sub-Das Rejoso Menggunakan Model. Genriver*; Report; World Agroforestry Centre (ICRAF) Southeast Asia Regional Program: Bogor, Indonesia, 2018.

106. Ridwansyah, I.; Pawitan, H.; Sinukaban, N.; Hidayat, Y. Watershed Modeling with ArcSWAT and SUFI2 in Cisadane Catchment Area: Calibration and Validation of River Flow Prediction. *Int. J. Sci. Eng.* **2014**, *6*, 92–101. [CrossRef]

107. Liu, Z.; Ma, D.; Hu, W.; Li, X. Land use dependent variation of soil water infiltration characteristics and their scale-specific controls. *Soil Tillage Res.* **2018**, *178*, 139–149. [CrossRef]
108. Zhao, Y.; Wang, Y.; Wang, L.; Zhang, X.; Yu, Y.; Zhao, J.; Lin, H.; Chen, Y.; Zhou, W.; An, Z. Exploring the role of land restoration in the spatial patterns of deep soil water at watershed scales. *Catena* **2019**, *172*, 387–396. [CrossRef]

 *land*

Article

Groundwater-Extracting Rice Production in the Rejoso Watershed (Indonesia) Reducing Urban Water Availability: Characterisation and Intervention Priorities

Ni'matul Khasanah [1,*], Lisa Tanika [1,2], Lalu Deden Yuda Pratama [1], Beria Leimona [1], Endro Prasetiyo [1], Fitri Marulani [1], Adis Hendriatna [1], Mukhammad Thoha Zulkarnain [1], Alix Toulier [3,4,5] and Meine van Noordwijk [1,6,7]

[1] World Agroforestry (ICRAF), Bogor 16001, Indonesia; lisa.tanika@gmail.com (L.T.); lalu.deden.y@gmail.com (L.D.Y.P.); l.beria@cgiar.org (B.L.); e.prasetiyo@cgiar.org (E.P.); fitrimarulani@gmail.com (F.M.); a.hendriatna@cgiar.org (A.H.); m.zulkarnain@cgiar.org (M.T.Z.); M.vannoordwijk@cgiar.org (M.v.N.)
[2] Forest Ecology and Forest Management Group, Wageningen University & Research, 6708 PB Wageningen, The Netherlands
[3] Institut de physique du globe de Paris, Université de Paris, CNRS, 75005 Paris, France; alix.toulier@univ-reunion.fr
[4] Laboratoire GéoSciences Réunion, Université de La Réunion, 97744 Saint Denis, France
[5] HydroSciences Montpellier (HSM), Université Montpellier, IRD, CNR, 34090 Montpellier, France
[6] Plant Production Systems, Wageningen University & Research, 6700 AK Wageningen, The Netherlands
[7] Study Group Agroforestry, Faculty of Agriculture, Brawijaya University, Malang 65145, Indonesia
* Correspondence: n.khasanah@cgiar.org

 **check for updates**

Citation: Khasanah, N.; Tanika, L.; Pratama, L.D.Y.; Leimona, B.; Prasetiyo, E.; Marulani, F.; Hendriatna, A.; Zulkarnain, M.T.; Toulier, A.; van Noordwijk, M. Groundwater-Extracting Rice Production in the Rejoso Watershed (Indonesia) Reducing Urban Water Availability: Characterisation and Intervention Priorities. *Land* **2021**, *10*, 586. https://doi.org/10.3390/land10060586

Academic Editor: Krish Jayachandran

Received: 8 April 2021
Accepted: 26 May 2021
Published: 1 June 2021

Publisher's Note: MDPI stays neutral with regard to jurisdictional claims in published maps and institutional affiliations.

Abstract: Production landscapes depend on, but also affect, ecosystem services. In the Rejoso watershed (East Java, Indonesia), uncontrolled groundwater use for paddies reduces flow of lowland pressure-driven artesian springs that supply drinking water to urban stakeholders. Analysis of the water balance suggested that the decline by about 30% in spring discharge in the past decades is attributed for 47 and 53%, respectively, to upland degradation and lowland groundwater abstraction. Consequently, current spring restoration efforts support upland agroforestry development while aiming to reduce lowland groundwater wasting. To clarify spatial and social targeting of lowland interventions five clusters (replicable patterns) of lowland paddy farming were distinguished from spatial data on, among other factors, reliance on river versus artesian wells delivering groundwater, use of crop rotation, rice yield, fertiliser rates and intensity of rodent control. A survey of farming households (461 respondents), complemented and verified through in-depth interviews and group discussions, identified opportunities for interventions and associated risks. Changes in artesian well design, allowing outflow control, can support water-saving, sustainable paddy cultivation methods. With rodents as a major yield-reducing factor, solutions likely depend on more synchronized planting calendars and thus on collective action for effectiveness at scale. Interventions based on this design are currently tested.

Keywords: artesian wells; ecosystem services; landscape approach; *Oryza*; paddy cultivation; restoration; rodents; sustainable intensification; water balance; Mount Bromo-Tengger

1. Introduction

Ecosystem services are defined as benefits people obtain from ecosystems [1], directly as goods or indirectly as regulating, cultural, and supporting services dependent on well-functioning ecosystems [2]. There is growing evidence of significant adverse impacts from landscape degradation due to land use/cover changes, population growth, and anthropogenic pressures, aggravated by the impacts of global climate change, for example increasing variability of rainfall [3]. These issues deserve attention at the global, regional, and national levels; sustainable landscape management is needed, encompassing both upland and lowland issues, to combat landscape degradation and strengthen the resilience of communities to climate change.

Sustainable landscape management typically implies the application of a landscape-scale approach, which has increasingly been recognised as an opportunity to minimize negative trade-offs and reconcile conservation, agriculture, and rural-to-urban livelihoods [4,5]. A 'landscape(-scale) approach' emphasises stakeholder engagement, including smallholder perspectives on the achievement of multiple objectives: maintaining ecosystem services and goods while improving livelihoods and addressing 'development deficits'. It also implies integrated assessment of upland–lowland relations and flexible implementation [6]. Several landscape management schemes have been introduced to combat landscape degradation and to strengthen the resilience of communities, such as through ecosystem services co-investment schemes [7]. However, such schemes are still in the pilot stage and usually end when external support is withdrawn. Thus, 'upscaling' technologies and sustainability of interventions is indispensable. Internalization of externalities has to include the establishment of new norms of behaviour, beyond economic incentives in the initial phase. Building on the mixed success of 'scaling up' technologies that were successful in the locations where they were developed, but not as good elsewhere, Sinclair and Coe [8] identified the need for an 'option by context' approach to addressing the variability of social, economic, and ecological issues across geographies for research and development which involves smallholder farmers. Representing the context to characterize variability of farmer's practices in managing the land that is needed to operationalise the evaluation of options.

In the context of production landscapes, agriculture both depends on 'upstream' ecosystem services and influences (often negatively) services for stakeholders further 'downstream' [9]. For water-related services, the up- and downstream terminology can be taken literally [10] (i.e., as a spatial geographical location), in other services, it is used as a metaphor (i.e., upland as the supplier of ecosystem services, while lowland as the beneficiaries of such services). Land cover type and land use management, including the status of property rights [11] in the upland and lowland determine the quantity and quality of the ecosystem services generated and utilized in the landscape. Water availability that is naturally based on a flow from the uplands to adjacent lowlands, is influenced by the capacity of the watershed to filter and buffer the flows [12,13] in different parts of a landscape. Landscape managers have both legal and perceived rights to modify these flows, such as by abstracting water that may reduce extractable surface and groundwater flows. This activity may affect the water supply further down in the landscape, which at the end will raise complex issues of legal and perceived water rights of the lowland communities. Thus, the understanding of the hydrological relations, is fundamental to disentangle the social interactions and find solutions that manage conflicts and adverse trade-offs. The interactions between farmer practices and ecological subsystems need to be quantitatively understood to manage the overall resource in a fair and efficient way [14]. The scale of the overall resource availability and use needs to be connected to that of farmer decisions, i.e., access rights and appropriation, and that of collective action, essential for reliable solutions and interventions.

The Rejoso watershed in the Pasuruan District, East Java Province (Indonesia) has experienced progressive deforestation on the higher slopes of Mount Bromo-Tengger, land use/cover changes across all elevations, and unsustainable farming practices due to rapid population growth and anthropogenic pressures [15]. In combination, these changes have affected the watershed's function of maintaining ecosystem goods and services, including impact on the quality and quantity of water resources, i.e., depleting the water flows, increasing risks of droughts and floods, soil erosion, and landslides according to local stakeholders [15]. In addition, the government is implementing a national project to pipe the water from the Rejoso watershed, i.e., from the Umbulan artesian spring to supply the adjacent districts and cities, including the metropole of Surabaya, the 2nd largest city of Indonesia and East Java capital. Artesian conditions develop where the hydraulic head (pressure) from a confined aquifer is higher than the topographic surface, allowing the free flow of groundwater through artesian springs (and/or wells) [16]. There are both similarities and differences with the well-documented agricultural over-use of groundwater in India, where a reduction of the energy subsidy for pumps provides at least some incentives for farmers to only use pumps when needed [17]. The simplest forms or artesian wells flow 24 hours per day and 365 days per year.

Figure 1 illustrates the upper (>1000 m a.s.l), middle (100–1000 m a.s.l) and lower (<100 m a.s.l) zones of the Rejoso watershed supplying surface flows (rivers) and groundwater flows (aquifers). Artesian conditions develop in the lowland zone, where a mostly impermeable layer inherited from volcanic processes covers and confines the underlying aquifer (water-rock reservoir). The current data shows that the discharge of the Umbulan spring has been decreased from about 5000 L/s in 1980 to 3500 L/s in 2020, with a continuous trend towards further decline [18]. The attribution of this decrease across the upper, middle and lower zone has triggered debates (e.g., climate change affecting all zones vs. local anthropogenic impact) that led to the current research. In the decline of 1500 L/s, lowland flowing artesian wells and reduced recharge of aquifer by reduced upper and middle zone infiltration both may play a role. Sustainable landscape management in the Rejoso watershed will depend on appropriate incentives, rules and motivation across all zones, based on a detailed diagnosis and co-investment by stakeholders [15,19].

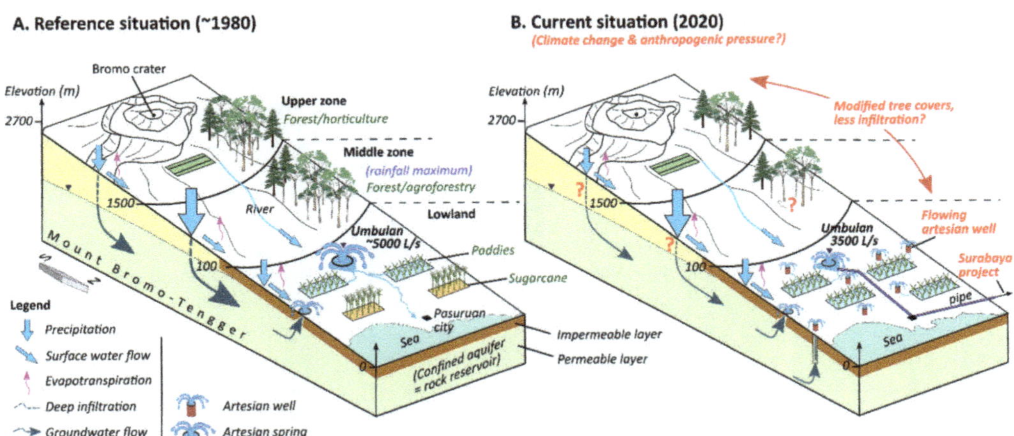

Figure 1. The simplified block diagram of the landscape and zone-specific water balance, illustrating (**A**): the historical reference scheme 1980, and (**B**): the current situation in 2020.

Through diagnostic studies, broad stakeholder participation and consultations with government agencies, proposals were formulated for performance based-payment schemes. These include managing tree and grass strips in horticultural farming systems in the upland part, increasing tree density and building of infiltration/sediment capture pits in the agroforestry farming systems of midstream smallholders [16]. Activities here target increased soil infiltration rates for groundwater recharge, control of soil erosion and increased on-site sedimentation. According to a study in the upper part of the Rejoso catchment [20], increasing tree canopy cover to values >55% in the upland and >80% in the midstream (highest rainfall elevation) qualified as 'infiltration-friendly' land use in the watershed, respectively, and can be expected to reduce runoff below 15% of rainfall. Groundwater recharge depends mainly on the balance between precipitation, evapotranspiration and runoff in each zone, but is also influenced by the seasons (wet and dry) [16].

In the lowland area, the Pasuruan district used to be a major sugarcane producer with good surface irrigation infrastructure, hosting since 1887 the national sugarcane research institute. However, since the last decades, most of the land has been converted to paddy fields, using additional groundwater resources mostly from flowing artesian wells (Figure 1B). In this zone, reported problems include diminishing areas of fertile soil for farming (rather than for urban expansion), high intensity of pest and diseases, and low paddy productivity indicate unsustainable agricultural practices. Furthermore, intensive use of groundwater to irrigate the paddy fields decreases the aquifer pressure and then the water productivity of artesian wells and springs as the sources of agriculture and domestic water for local communities [16]. Hence, better water management in irrigated areas is one of the targets for improved landscape-wide ecosystem services from a lowland, urban water user perspective. Five crucial root causes of unsustainable agricultural in Rejoso watershed have been identified [17] as unsynchronised planting calendars, inefficient use of groundwater, high chemical inputs, imbalanced fertiliser application, and conventional, suboptimal planting patterns. The average rice yield at the district and province level is about 5.8 ton per ha [21], which is lower than in other provinces, i.e., Bali and Central Java. East Java Province is the second largest (with about 19%) contributor to national paddy production, with 3% of national level produced in Pasuruan District [22]. Therefore, addressing the issues by introducing sustainable paddy cultivation (i.e., optimal use of chemical fertiliser, application of biopesticide, improved water management regimes and planting pattern) to increase productivity while reducing environmental impacts is essential. Current agricultural practices lead to high methane (CH_4) and nitrous oxide (N_2O) emissions, high intensity of pest and diseases and low agricultural yield. Water-saving techniques are expected to be financially and environmentally beneficial to smallholders by enhancing their resilience to shocks and improving the capacity of the production landscape to generate ecosystem services [23]. Nevertheless, introducing sustainable paddy cultivation beyond the current, conventional practices is a challenge, as behavioural changes and biophysical conditions vary.

Understanding the variability of farmer's practices in managing the land and cultivating paddy is, therefore, considered as an initial step towards pilot actions for the lowland zone with the potential to scale up sustainable paddy cultivation. To contextualize current practices and propose 'options by context' as restoration solutions, we thus needed a detailed characterisation of paddy farming and possible spatial patterns in cropping intensity and use of river versus groundwater for irrigation. By triangulation of quantitative spatial data analysis, qualitative insights from the participation of local farmers, communities, and government agencies, and a targeted, quantitative household survey, we hoped to understand the rationale(s) of farmers for considering and choosing specific practices. Scenarios for improved resource management at landscape scale require identification of the main sustainability risks and local perspectives, at the scale required for impacts to be noticeable. Our analysis of catchment-level water balance, patterns of land and water use, and specific practices used in paddy farming tried to answer questions at three levels:

1. Is there quantitative evidence that the lowland practices are co-responsible for the decrease of the Umbulan spring's discharge?
2. Is there relevant geographic variation between villages and hamlets in the farmers' practices in managing land and water in cultivating paddy?
3. Can a participatory survey of paddy cultivation and spatial data analysis for the development of characteristics identify options by context for upscaling sustainable paddy cultivation?

We expected that the combined use of quantitative and qualitative methods, together with the participatory approach used in this study, would enable a subsequent scaling-up phase that is salient, credible, and legitimate for all segments of the community at village and district levels.

2. Site Description and Methodology

2.1. The Rejoso Watershed

The Rejoso watershed has an area of 62,773 ha based on the boundaries set by the Watershed Management Agency (BPDAS). It covers 17 sub-districts: Bugul Kidul, Gading Rejo, Gondang Wetan, Grati, Kejayan, Kraton, Lekok, Lumbang, Nguling, Pasrepan, Pohjentrek, Purworejo, Puspo, Rejoso, Tosari, Tutur, and Winongan, on the lower, middle and upper slopes of Mount Bromo-Tengger, East Java, Indonesia. The artesian spring Umbulan is located in the lowland part of the watershed (Winongan sub-district).

Paddy fields and sugarcane plantations are dominant land covers in the lowland area, complex agroforest dominates in the mid-stream area, and horticulture and pine plantation are mostly found in the upland area of the watershed [15]. *Inceptisols* are the dominant soil type in the upland, midstream to the lowland area; a small area of *Entisols* is found in the lowland area.

Complementing studies in the middle and higher zones, our study developed a characteristic of paddy farming for the lowland area of the watershed, specifically, in the eleven villages of two sub-districts (Figure 2), Winongan (4341 ha) and Gondang Wetan (2692 ha) sub-districts (07°42′30–07°43′30″ NL and 112°54′30–112°57′0″ LE). The two sub-districts were selected based on parameters: the (high) number of artesian wells as one of the main sources to irrigate paddy fields, (high) area of paddy fields, and (low) yield. Artesian wells, flowing twenty-four hours per day are a specificity of the volcanic study area as the hydrogeology is represented by a shallow artesian basin.

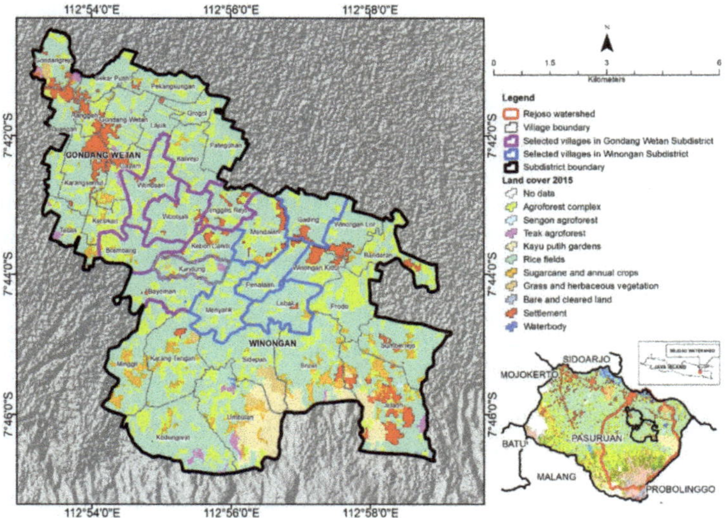

Figure 2. Delineation of the study area.

2.2. Annual Water Balance Model of Rejoso Watershed

A simple water balance has at the minimum to include precipitation (P), evapotranspiration (E), and runoff (river flow; Qs), with changes in soil moisture storage potentially negligible at annual time steps (Figure 3). Evapotranspiration can be expressed as a vegetation-dependent fraction ε of the climate-driven potential value Epot and be further constrained by the fraction of P that infiltrates into the soil. Runoff can be estimated as an infiltration-limited (or Hortonian) fraction ρ of P, plus a saturation-excess amount max $(0; P (1 - \rho) - \varepsilon$ Epot). This is an alternative (first used in [24]) to the commonly used Fuh–Budyko equation, which tends to underestimate discharge under low rainfall conditions.

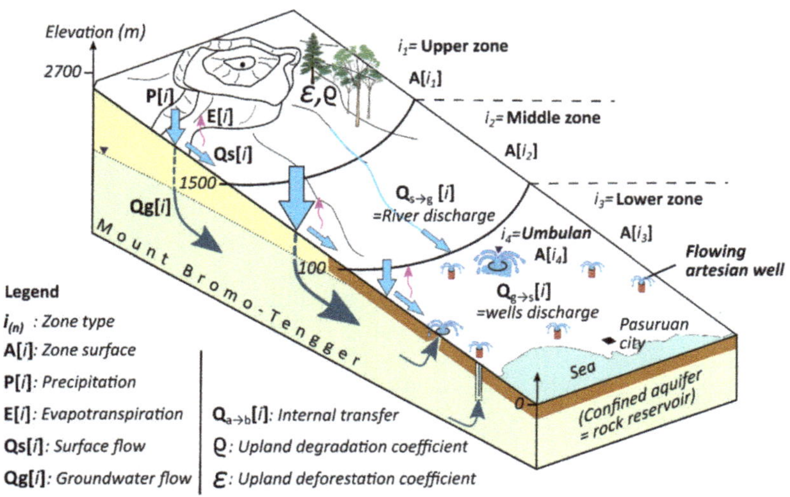

Figure 3. Simplified schema of the water balance components.

For the current analysis, a distinction is needed between surface flows (Q_S) and groundwater flows (Q_G) based on three possible transfers between surface and groundwater: deep infiltration (as fraction of water infiltrated, and reducing the saturation-excess river flow), seepage of water already on the river, and the resurfacing of groundwater in springs and wells. The annual water balance model (assuming no change in storage terms, expressed per unit area) linked four zones i ($n = 1$ to 4), respectively, labelled Upper, Middle, Lower, and Umbulan spring zone (with areas A[i] in (km^2)), and computed as follows:

$$Q_{Soutflow}[i] + Q_{Goutflow}[i] = P[i] - E[i] + Qs_{inflow}[i] + Q_{Ginflow}[i] \tag{1}$$

with as inputs for each zone i,

$$P[i] = \text{precipitation (mm/y)},$$

$Qs_{inflow}[i]$ = incoming surface flow (river) corrected for the relative areas of adjacent zones, defined as (A[$i-1$]/A[i]) × $Q_S[i-1]$) (mm/y),

$Q_{Ginflow}[i]$ = incoming groundwater flow corrected for the relative areas of adjacent zones, defined as (A[$i-1$]/A[i]) × $Q_G[i-1]$) (mm/y),

and as outputs:

E[i] = evapotranspiration defined as $\varepsilon[i] \times E_{pot}[i]$ with the E_{act}/E_{pot} ratio $\varepsilon[i]$ dependent on tree cover and crop type (mm/y),

$$Q_{Soutflow}[i] = \text{outgoing surface flow (river) (mm/y)},$$

$$Q_{Goutflow}[i] = \text{outgoing groundwater flow (mm/y)},$$

and as internal transfers from groundwater to surface water or vice versa:

$$Q_G[i] = A[i-1]/A[i]) \times Q_G[i-1] - Q_{G\to S}[i] + Q_{S\to G}[i]$$

In such a framework, we can represent "upland deforestation" as a decrease in ε and "upland degradation" as an increase in ρ. The shift from upland crops to paddy in the lowland as an increase in ε, plus wells that transfer ground to surface water. Estimates of the net transfers from surface to groundwater flows Qs→g[i] in the Upper and Middle zone were based on measured river discharge at the transition from Middle to Lower zone. Estimates of the net transfers from groundwater to surface flows Qg→s[i] in the Lower zone was derived from measured artesian well distribution and flow rates (Figure 4).

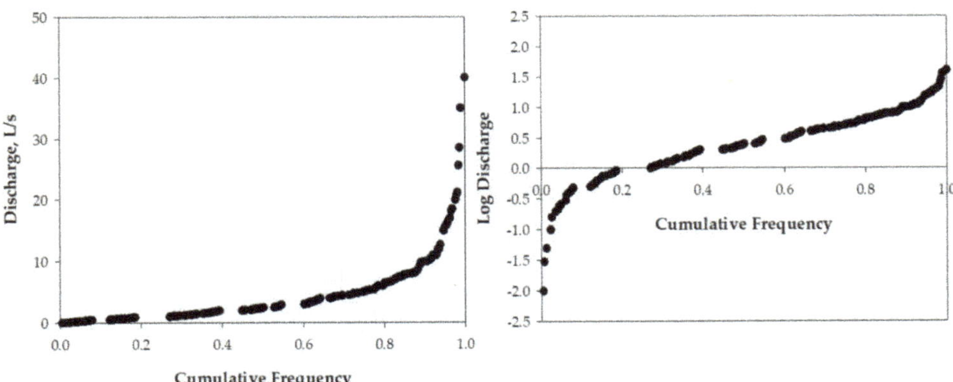

Figure 4. Measured flow of around 450 artesian wells in lowland Rejoso area; data source: [16].

After parameterizing this simple model (Appendix A), we compared five scenarios: (A) a historical reference scenario, (B) upland degradation, (C) lowland conversion to paddy with uncontrolled artesian wells, (D) combining the changes of (B + C, E) a restoration scenario with agroforestry in upper and middle zones and reduced groundwater use in the lowland paddy zone (Table 1).

Table 1. Scenarios of the Rejoso water balance model; t_w indicates the fraction of time the artesian wells are flowing; the evapotranspiration ratio ε was estimated from land cover composition and known temporal dynamics of Leaf Area Index of the vegetation; the ρ coefficient from existing runoff data.

Scenario, Description	Geographical Zoning	ε	ρ	#Wells (Number)	t_w
A. Baseline; using landcover data from 1990, before the expansion of paddy rice cultivation in the lowlands, and with a higher forest fraction in the middle and upper zone.	Upper Middle Lower	0.71 0.72 0.81	0.15 0.15 0.05	 10	 1
B. Upland degradation; keeping lowland conditions as in 1990, but reflecting the hydrological degradation in the upland and middle parts of the watershed that are caused by conversion of forest to horticulture and agroforestry.	Upper Middle Lower	0.71 0.76 0.81	0.23 0.23 0.05	 10	 1
C. Lowland dominated by paddy field and artesian wells; paddy field and unmanaged and unregulated artesian wells in the lowland, combined with upland conditions of 1990.	Upper Middle Lower	0.71 0.72 0.80	0.15 0.15 0.05	 600	 1
D. Upland degradation and intensive lowland for agriculture; using landcover data from 2015 for all zones, along with the artesian wells in the lowlands	Upper Middle Lower	0.71 0.76 0.80	0.23 0.23 0.05	 600	 1
E. Applied sustainable interventions in lowland; as negotiated interventions payment for ecosystem services for tree-based farms and soil-water conservation techniques are introduced in upland and middle parts. Water efficient and low emissions paddy cultivation, and good management of artesian wells are introduced and practiced.	Upper Middle Lower	0.71 0.76 0.80	0.19 0.19 0.05	 600	 0.2

2.3. Selected Villages for the Lowland Characteristic

The selection of the villages was based on the number of artesian wells, area of paddy fields and its yield, number of low-income families, and number of families with members of the family as farm labour. The eleven villages were Wonosari, Wonojati, Tenggilis Rejo, Kebon Candi, Brambang and Bayeman in Gondang Wetan Sub-district, and Gading, Mendalan, Penataan, Menyarik, and Lebak in Winongan Sub-district (Figure 2). Based on data of fourteen rainfall stations, the mean annual rainfall is approximately 1350 mm with relative humidity ranges from 68% to 83%. The rainfall is distributed with a peak in

January and a dry season in August, and the annual mean of maximum and minimum air temperatures are 20 °C and 34 °C, respectively.

2.4. Development of Paddy Farming Characteristic

Figure 5 presents the flow of development of paddy farming characteristic. The development of paddy farming characteristic used the cluster analysis approach. A cluster analysis is a process of grouping a set of parameters in such a way that parameters in the same group (a cluster) are more similar to each other than to those in other groups. Twelve parameters were collected through spatial data analysis and survey of paddy cultivation that both were verified through a participatory process reflecting the variability of the paddy field, farmers characteristics and their cultivation practices in the lowland area of the Rejoso watershed. These parameters were (1) area of paddy fields, (2) density of the channel network, (3) fraction of area with crop rotation, (4) intensity of pest (rodents), (5) rice yield, (6) dose of urea (46% N) fertiliser, (7) dose of compound fertiliser (15% of N, 15% of P, and 15% of K), (8) number of pesticide types applied, (9) existence of a water regulatory officer (*ulu-ulu*), (10) number of artesian wells, (11) river as the main water source, and (12) number of water sources. The development of paddy farming characteristic emphasises on the landscape-approach as the methods applied in this study engage the direct stakeholders, i.e., smallholders with multiple objectives of positive environmental impacts with substantial livelihood improvement reflecting by the selections of identified and analysed parameters, as part of the more comprehensive picture of the Rejoso watershed.

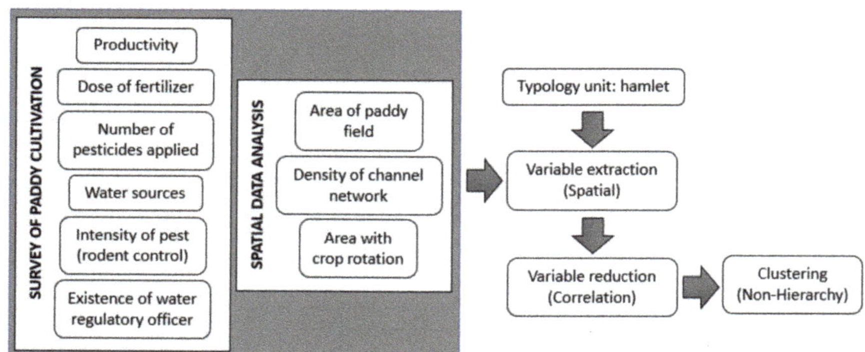

Figure 5. Flow diagram of cluster analysis to develop paddy farming characteristic.

2.4.1. Spatial Data Analysis

Figure 6 presents the workflow of paddy field and irrigation system mapping that consists of four main steps: (a) data gathering, (b) visual interpretation, (c) participatory mapping, (d) data analysis, and visualisation.

Primary data collected through a survey and participatory approach and secondary data were two main data used for the mapping. A drone survey was conducted to obtain aerial photographs of paddy field, while focus group discussions (FGD) and key informant interviews were the approaches to obtain locations of artesian wells and detailed information of paddy fields. The secondary data were topographic maps at 1:50.000 scale [25], irrigation systems data [26], and artesian well distribution [16].

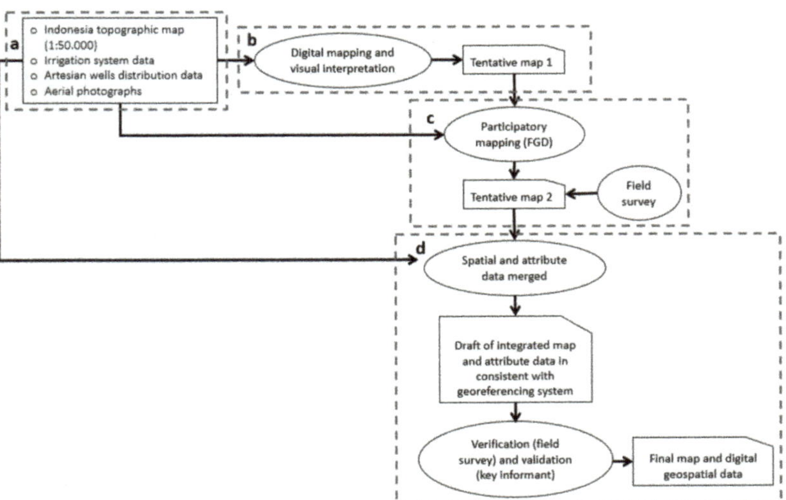

Figure 6. The workflow of participatory mapping scaled 2D mapping techniques.

Data on the area of paddy fields, the area with crop rotation, and irrigation system data were extracted from the aerial photographs through visual interpretation and convergence of the evidence approach. To correctly identify surface objects, several visual interpretation elements such as tone/colour, shape, size, pattern, texture, shadow, site, and associations were considered [27]. Artesian wells, irrigation systems, and paddy field data were overlaid with a topographic map and visualised at the village scale to optimise the information extraction process. The result of this process led to a first tentative map that was completed and validated through FGD. The FGD was attended by 128 participants (mostly male) from eleven discussions in eleven villages.

Participatory mapping is a map-making process that attempts to make visible the association between land and local communities by using the commonly understood and recognised language of cartography [28]. The participatory mapping method in this study used a 2D map to allow two-way dialogues between researchers and key informants to minimise distortions of mapped information [29]. This discussion was focused on information such as the location of artesian wells, irrigation system, hamlet boundaries, paddy fields and their owners, crop rotation, and the existence of farmer groups.

Detailed information gathered from the participatory mapping was used as an input for data compilation and first tentative map improvement in digital format using GIS (Figure 6). The process included (1) scanning the result of participatory mapping; (2) georeferencing; (3) reinterpreting data; (4) and inputting attribute data. The results of this process were tentative map 2, which was then validated by eleven key informants from eleven villages. The validated spatial data was compiled as geodatabase for further analysis.

The spatial data resulted from participatory mapping were (1) hamlets boundary, (2) percentage of paddy field area in each hamlet, (3) percentage of crop rotation area per paddy field area in each hamlet, (4) drainage density in each hamlet, (5) and the number of artesian wells in each hamlet. The length of the channel network was classified by channel width. Channels with less than 1 m widths were identified as trenches, while channels with more than 1 m widths were identified as irrigation channels and rivers. The rivers that cross the study areas are Kedung River, Palembon River, Sumbermade River, and Umbulan River. Drainage density is defined as the total length of channels (trench, irrigation channels, and river) per unit area of hamlet.

2.4.2. Survey of Paddy Cultivation

The survey of paddy cultivation aimed to gather information about paddy cultivation practices and related issues in the eleven villages of the two sub-districts, Gondang Wetan and Winongan. The survey was conducted from August to October 2019. The survey consisted of (1) survey preparation including the development of survey questionnaire and training on interview technique, (2) respondent selection and interview process, and (3) data cleaning and analysis.

The questionnaire was designed to survey five main parameters related to paddy cultivation practices and its issues: (1) yield, (2) dose of fertiliser, (3) number of pesticide types, (4) water sources, and (5) intensity of pest (rodents). For further analysis, the dose of fertiliser was divided into (a) urea fertiliser (46% of nitrogen) and (b) compound fertiliser (15% of nitrogen, 15% of phosphate and 15% of potassium), while water source parameter was divided into (a) river as the main water source, (b) number of water resources, and (c) existence of a water regulatory officer (*ulu-ulu*).

Table 2 presents the distribution and characteristics of respondents by village, age, and number and size of plots/fields owned/managed. In total, there were 461 respondents, the respondent's information gathered from chairs of farmer groups, and applying a snowball technique. The age of respondents varied from 22 to 83 years old and was 56 years old on average. The respondents of this study at least owned/managed one plot. The maximum number of plots owned/managed was 25. The average area owned/managed by the respondents was 0.25 ha. The smallest was 0.05 ha, and the largest was 2.25 ha.

Table 2. Respondent distribution and characteristic of paddy cultivation survey.

Sub-Districts	Villages	Number of Respondents	Age Distribution			Number of Plots Owned/Managed per Respondent			Owned/Managed Area per Plot (ha)		
			Max	Min	Avg	Max	Min	Avg	Max	Min	Avg
Gondang Wetan	Bayeman	34	80	31	59	6	1	3	0.80	0.10	0.26
	Brambang	32	70	26	51	8	1	4	2.25	0.10	0.27
	Tengglis Rejo	38	79	35	60	10	1	5	0.60	0.10	0.23
	Kebon candi	67	74	33	55	8	1	3	0.50	0.06	0.22
	Wonojati	93	83	22	54	17	1	18	2.00	0.05	0.27
	Wonosari	31	80	28	57	12	1	4	0.85	0.07	0.28
	Gading	31	70	33	55	25	1	8	0.50	0.05	0.18
Winongan	Lebak	33	73	25	51	6	1	3	1.10	0.09	0.35
	Mendalan	34	70	40	56	13	1	3	0.60	0.07	0.24
	Menyarik	32	70	29	57	6	1	2	1.33	0.08	0.20
	Penataan	36	83	32	56	5	1	2	0.60	0.10	0.26

2.5. Data Analysis

For paddy cultivation data, data analysis was performed after data cleaning is completed. The cleaning included filtering location of paddy field and domicile of farmers/respondents. We only considered respondents who stay and manage the farm in the eleven villages. In the data analysis process, the basic statistical analysis was used to analyse and explore the variation of the data at sub-district, village and hamlet scale. The basic statistical analysis included an average of yield, dose of fertiliser and number of types of pesticide; percent of respondents with perception on high intensity of rodent and use of artesian wells. Once the analysis completed, the result was interpreted by sub-district, village, and hamlet. Then, we used hamlet as the unit of cluster analysis considering the social characteristic of paddy cultivation management, the existence of farmer group and water regulatory officer in each hamlet. The information on hamlet was described on the spatial maps under the Results section.

As we worked with extensive data set, for the cluster analysis, we applied the K-means approach to cluster the data and used the elbow method to find the optimum number of clusters [30]. Before we clustered the data, we conducted a correlation analysis of the twelve parameters that were extracted at the hamlet level to verify the statistical independence of the twelve parameters used (Figure 5).

3. Results

3.1. Water Balance

Parameters for the water balance model were adjusted to account for the approximately 5000 L/s Umbulan spring flow in reference scenario A (see results in Table 3). Upland degradation alone (Scenario B) would likely lead to increased river flow (due to increased runoff triggered by lower evapotranspiration in the uplands) and some decrease of the Umbulan spring flow (due to lower infiltration, i.e., lower recharge on the mountain slope). Lowland conversion to paddy with current artesian wells (Scenario C) would decrease both river flow and Umbulan spring flow.

Table 3. Predicted discharge of the Umbulan spring, river and groundwater flow for five land use scenarios: A) a historical reference scenario, B) upland degradation, C) lowland conversion to paddy with uncontrolled artesian wells, D) combining the changes of B + C, and E) a restoration scenario with agroforestry in upper and middle zones and reduced groundwater use in the lowland paddy zone.

Scenario	Predicted Umbulan (L/s)	Predicted Mean River Flow, (m^3/s)			Predicted Groundwater Flow, (m^3/s)		
		Upper	Middle	Lower	Upper	Middle	Lower
A	5087	2.5	6.7	12.4	1.5	5.3	0.05
B	4310	2.7	7.3	12.1	1.3	4.4	0.04
C	4206	2.5	6.7	12.5	1.5	5.3	0.04
D	3496	2.7	7.3	12.1	1.3	4.4	0.03
E	4468	2.6	6.9	12.1	1.4	4.7	0.04

Combining the changes of scenario B and C (Scenario D), river flow would approximate that of scenario B, but the Umbulan flow would be reduced to approximately the level currently observed (31% reduction). Attribution of this reduction (A–D), with 1.3% interaction, would be for about 15% to the middle and upper zone, and 17% to the lowland. The restoration potential in Scenario E is estimated to nearly 4500 L/s, at which the planned offtake of 4000 L/s still leaves sufficient discharge for local use.

3.2. Area of Paddy Field and Crop Rotation

The paddy fields in eleven villages reached 980.2 ha in both targeted sub-districts, with 536.1 ha (54.7%) in Winongan Sub-district. The villages Gading, Mendalan, and Menyarik in Winongan Sub-district had the largest area of paddy fields, with 119.6 (12.2%), 118.7 (12.1%), and 118.6 (12.1%) ha, respectively. On the other hand, Kebon Candi Village in Gondang Wetan Sub-district had 56.4 ha of paddy fields, the smallest among all other villages (Table 4). At the hamlet level, Kemong Hamlet in Lebak Village (Winongan Sub-district) had the largest percentage of paddy fields reaching 91%, which meant that only 9% of the area was used for non-agricultural activities. Areas with smaller percentages of paddy fields were in Gondang Wetan Sub-districts, starting from 0 to 30.9% (Figure 7A).

Table 4. Distribution of area of paddy field (ha), the area with crop rotation (ha), number of artesian wells, and drainage density (km/km^2) in each village in two sub-districts.

Sub-Dis-Tricts	Villages	Area of Paddy (ha)	Artesian Wells per Ha Paddy	Artesian Well Water [#], mm/Day	Length of Chan-nels (km)	Drainage Density (km/km^2)	Main Water Sources (% Respondents)	
							River	Artesian Wells
	Bayeman	97.5	0.73	17.6	13.7	7.4	58	42
	Tenggilis Rejo	80.1	0.45	10.9	11.2	7.9	46	54
Gondang	Wonosari	74.3	0.35	8.5	8.4	5.9	99	1
Wetan	Brambang	60.5	0.41	10.0	8.3	5.6	38	62
	Kebon Candi	56.4	0.41	9.9	7.3	6.1	54	46
	Wonojati	75.3	0.25	6.1	9.6	5.9	70	30
	Lebak	91.4	0.33	7.9	11.2	6.5	90	10
	Penataan	87.8	0.33	8.0	10.1	7.9	91	9
Winongan	Gading	119.6	0.18	4.2	18.5	10.4	100	0
	Menyarik	118.6	0.18	4.3	16.6	8.1	94	6
	Mendalan	118.7	0.14	3.5	20.3	10.0	67	33

[#] Assuming a constant (24/365) median well flow rate of 2.8 L/s.

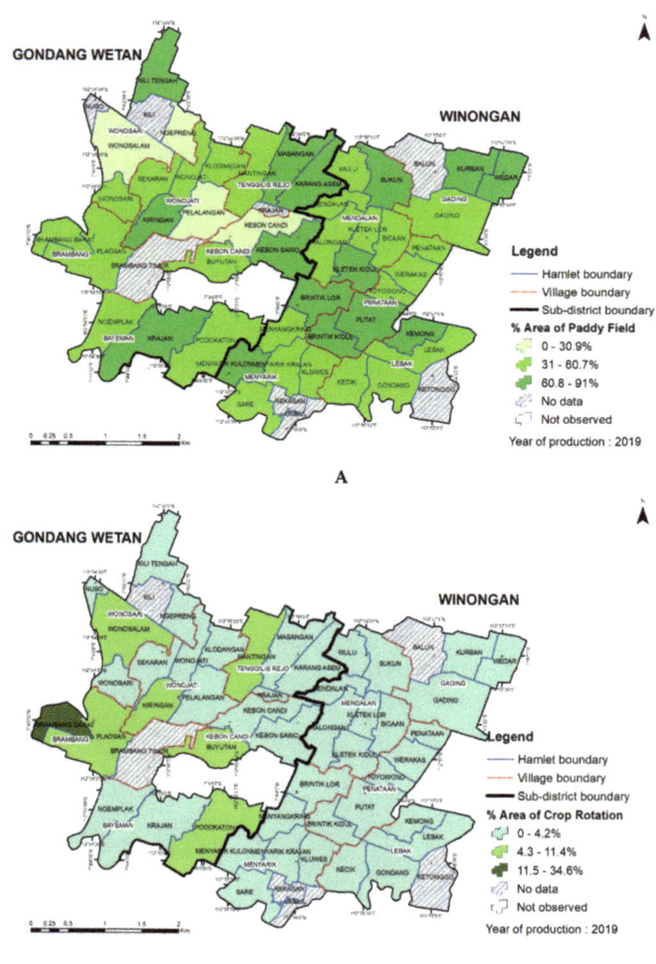

Figure 7. Spatial distribution of the percentage of paddy field area (**A**) and area with crop rotation (**B**) at hamlet level.

Crop rotation is a vital paddy farming practice to reduce the intensity of pests and diseases, and to improve soil fertility. Different crops rotated with paddy included corn, beans, taro, chilli, and other crops. Villages in Gondang Wetan Sub-district tended toward crop rotating more often compared to those in Winongan Sub-district. The area of crop rotation reached 30.1 ha, and 26.6 ha of it (88.5%) was found in Gondang Wetan Sub-district. Brambang Village in Gondang Wetan Sub-district was the village with the largest area of crop rotation with 10.9 ha, while Penataan Village in Winongan Sub-district had only 0.34 ha, the smallest area compared to other villages (Table 5). At the hamlet level, Brambang Barat Hamlet in the village of Brambang had the largest area of crop rotation, reaching 34.6% of the total paddy field area (Figure 7B). Hamlets in Gondang Wetan Sub-district had a percentage of area with crop rotation between 4.3–11%. Other hamlets were classified as having a low rotation area, which was less than or the same as 4.2% and it was relatively common in Winongan Sub-district.

Table 5. Distribution of rice yield (ton/ha), fertiliser application (kg/ha), number of type of pesticides applied, water sources (% respondent), and intensity of rodents (% respondent) in each village in two sub-districts.

Sub-Districts	Villages	Rice Yield (Ton/ha/Cropping Season)	Fertilizer Application (kg/ha)		Area with Crop Rotation (ha)	Number of Type of Pesticides Applied	Intensity of Rodents (% Respondents)
			Urea	Compound			
Gondang Wetan	Bayeman	6.7	354	308	4.7	4	61
	Tenggilis Rejo	4.1	601	370	2.5	5	96
	Wonosari	5.1	375	299	2.5	5	37
	Brambang	5.1	425	315	10.9	6	27
	Kebon Candi	4.2	461	348	2.3	6	55
	Wonojati	4.5	408	293	3.9	6	36
	Lebak	4.5	257	231	1.0	5	52
Winongan	Penataan	5.0	318	252	0.3	5	71
	Gading	6.2	411	268	0.9	4	92
	Menyarik	4.7	396	309	0.4	5	70
	Mendalan	4.3	332	268	0.7	4	94

3.3. Water Sources and Irrigation Systems

Rivers and artesian wells were two primary water sources to irrigate paddy fields in Gondang Wetan and Winongan Sub-districts. Water from those two sources flowed into the irrigation channel, but in some cases, the artesian wells were located inside the paddy fields itself. The total number of artesian wells reached 318 points spread across 11 villages, and 63% of it was found in Gondang Wetan Sub-district (Table 4). In line with this distribution, 89% of respondents in Winongan Sub-district mentioned that the river was more dominant than artesian wells as a water source to irrigate their paddy fields, while in Gondang Wetan Sub-district, they were only 61% of respondents (Table 4). Brambang Village was the village with the lowest percentage of respondents that mentioned river water as the primary source. Therefore, one paddy field could use river water and more than two artesian wells (shared with other farmers) (Figure 8D). At the hamlet level, most artesian wells were found in Podokaton Hamlet of Bayeman Village of Gondang Wetan Sub-district. Some hamlets were identified to have artesian wells between 8 to 17 wells. Other hamlets had less than eight wells. Nuso Hamlet at Wonosari Village was the only hamlet without artesian wells for irrigation, relying almost 100% on river water to irrigate paddy fields (Figure 8A,C).

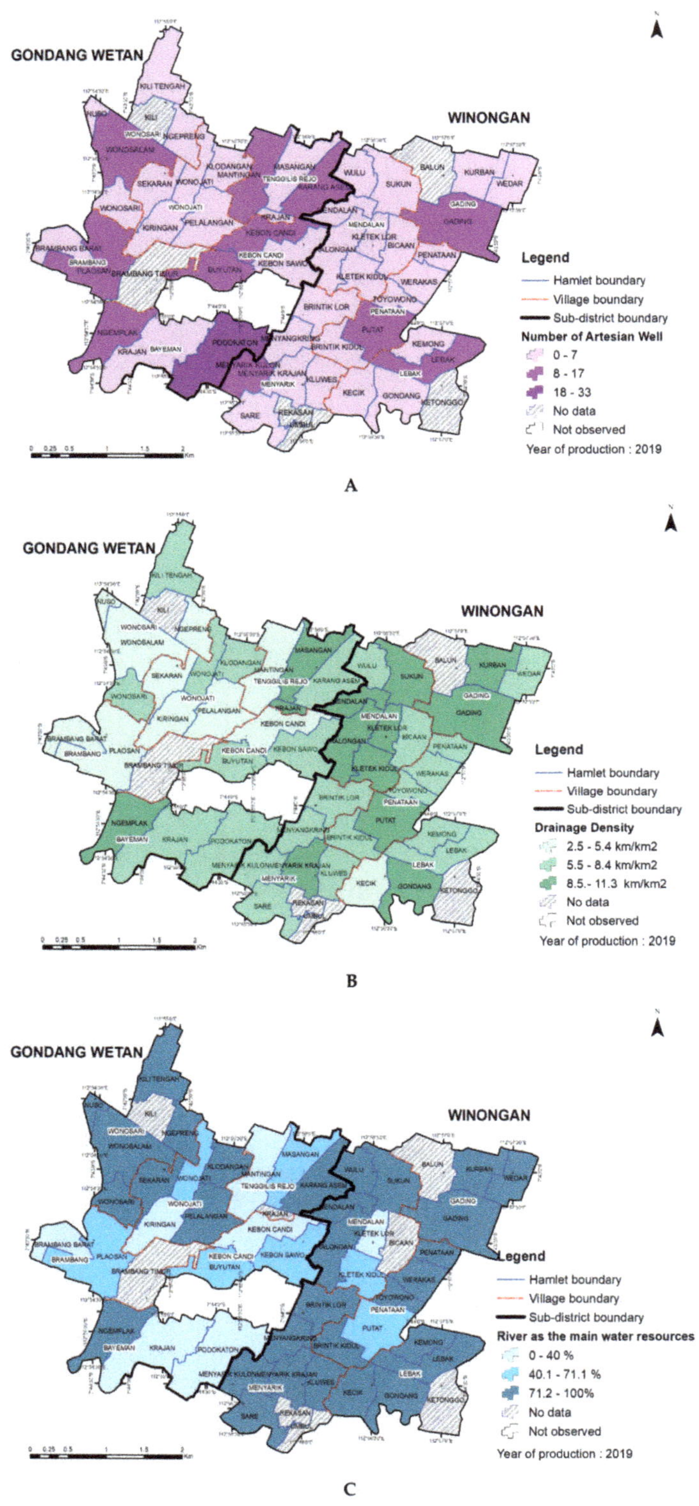

Figure 8. *Cont.*

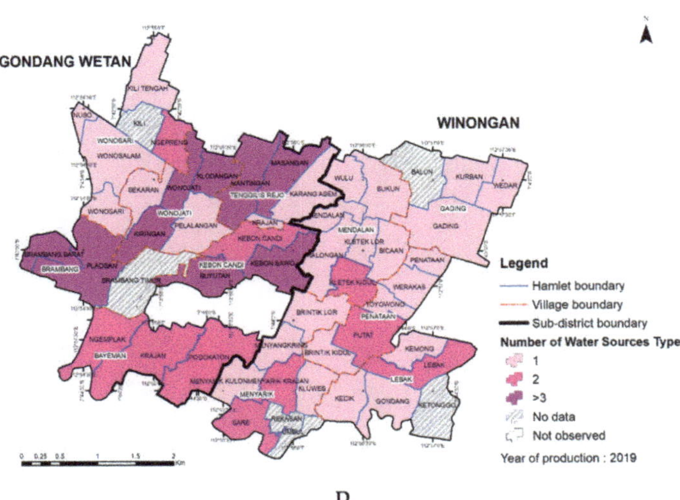

Figure 8. Spatial distribution of the number of artesian wells (**A**), drainage density (**B**), the hamlets that use the river as the main water resource (**C**), and the number of water resources (**D**) at hamlet level.

Winongan Sub-district had more extended irrigation channels compared to Gondang Wetan Sub-district, reaching 76.6 km in length (Table 4). Mendalan Village in Winongan Sub-district was the area with the most elongate irrigation channels (20.3 km). Kebon Candi Village in Gondang Wetan Sub-district had the shortest channels compared to other villages (7.27 km). At hamlet level, hamlets in Winongan Sub-districts such as Mendalan, Sukun, Kurban, Gading, Kalongan, Kletek Lor, Kletek Kidul, Putat, Krajan, Gondang, and two hamlets in Gondang Wetan Sub-district such as Karangasam and Ngemplak were classified as the highest drainage density, i.e., 8.5–11.3 km/km^2 (Figure 8B).

3.4. Rice Yield and Fertiliser Application

Farmers in Gondang Wetan Sub-district tended to apply a higher dose of fertiliser compared to farmers in the Winongan Sub-district (Table 5). The average dose of N fertiliser (Urea) applied by farmers in Gondang Wetan was 436 kg/ha/season and 322 kg/ha/season for compound fertiliser (NPK), while farmers in Winongan Sub-district applied N fertiliser about 343 kg/ha/season and compound fertiliser about 266 kg/ha/season. However, Gondang Wetan and Winongan Sub-districts had the same average of rice yield of 4.9 tons/ha/season, with distribution between 4.1-6.7 tons/ha/season (Table 5). Bayeman in Gondang Wetan and Gading in Winongan were villages that had the highest average of yield. Tenggilis Rejo in Gondang Wetan Sub-district was the village with the highest use of fertiliser, which was almost 600 kg/ha/season for N fertiliser and 370 kg/ha/seasons for compound fertiliser. It was indicated that high yield did not necessarily depend on high fertiliser rates (Table 5).

At the hamlet level, Krajan in Bayeman Village and Kurban in Gading Village were hamlets with the highest rice yield (almost 8 tons/ha/season) (Figure 9A). Karang Asem and Krajan in Tenggilis Rejo Village, and Kebon Sawo in Kebon Candi Village were some hamlets with the highest use of urea (above 540 kg/ha/season) and compound fertiliser (more than 348 kg/ha/season) (Figure 9B,C). Hamlets with low yield were Masangan, Karang Asem, Wulu, Bicaan, and Kebon Candi (Figure 9A).

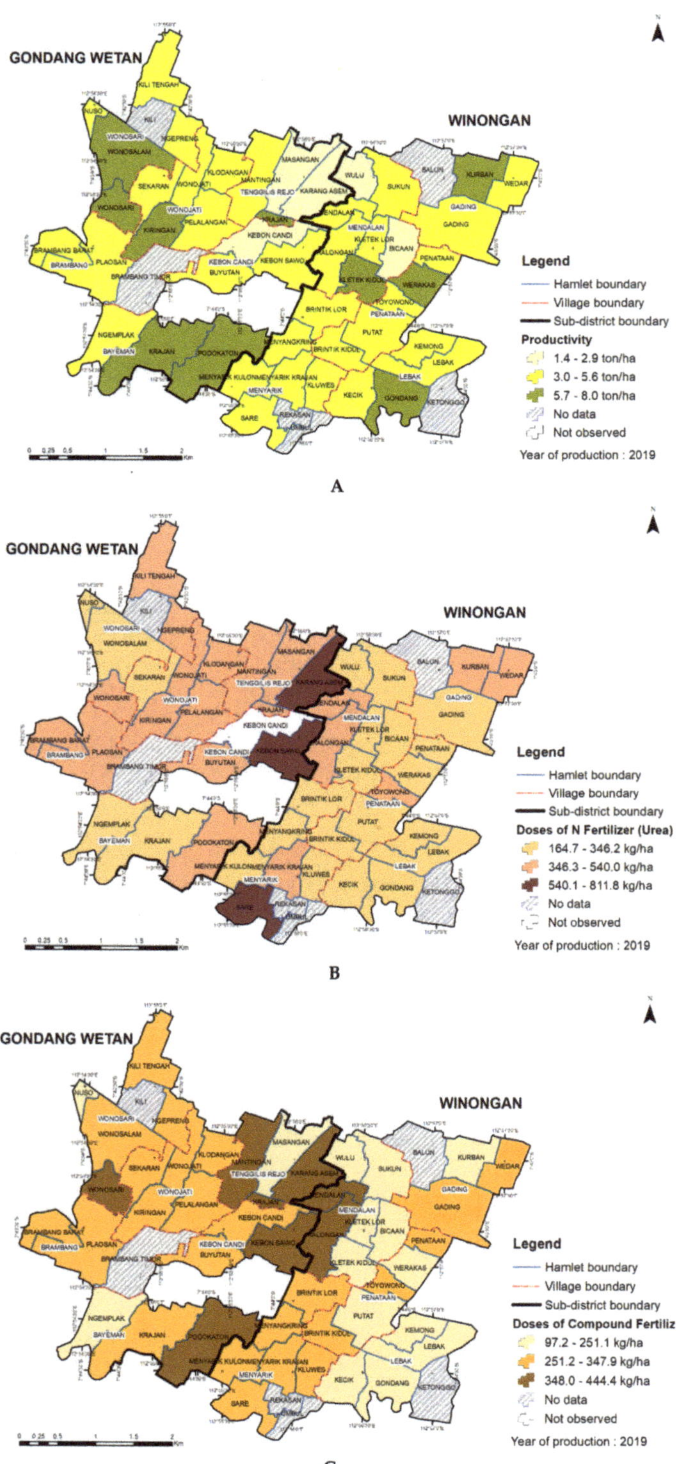

Figure 9. Spatial distribution of rice yield (**A**), urea application (**B**), and compound fertiliser (**C**) at hamlet level.

3.5. Intensity of Pest/Rodents and Number of Type of Pesticides Applied

Regarding the main problem of paddy cultivation, farmers in both Gondang Wetan and Winongan Sub-districts agreed that the high intensity of rodent attacks was the major problem in need of immediate attention. However, the respondents in Winongan Sub-district mentioned that rodents were the main problem, and their condition was worse than in Gondang Wetan (Table 5). Figure 10A shows that most paddy fields in Winongan Sub-district suffered from rodents. In Gondang Wetan Sub-district, only paddy fields in hamlets or villages next to Winongan Sub-district had a high intensity of rodent attacks, especially during rainy season. Until this study was undertaken, farmers were still struggling with how to deal with rodents. In terms of pests and disease control, there were more than 100 brands of pesticides used by farmers. Farmers in Gondang Wetan and Winongan Sub-districts used at least three brands, but some applied more than seven brands of pesticides in one season. In general, farmers in Gondang Wetan Sub-district applied more brands of pesticides compared to farmers in Winongan Sub-district (Figure 10B).

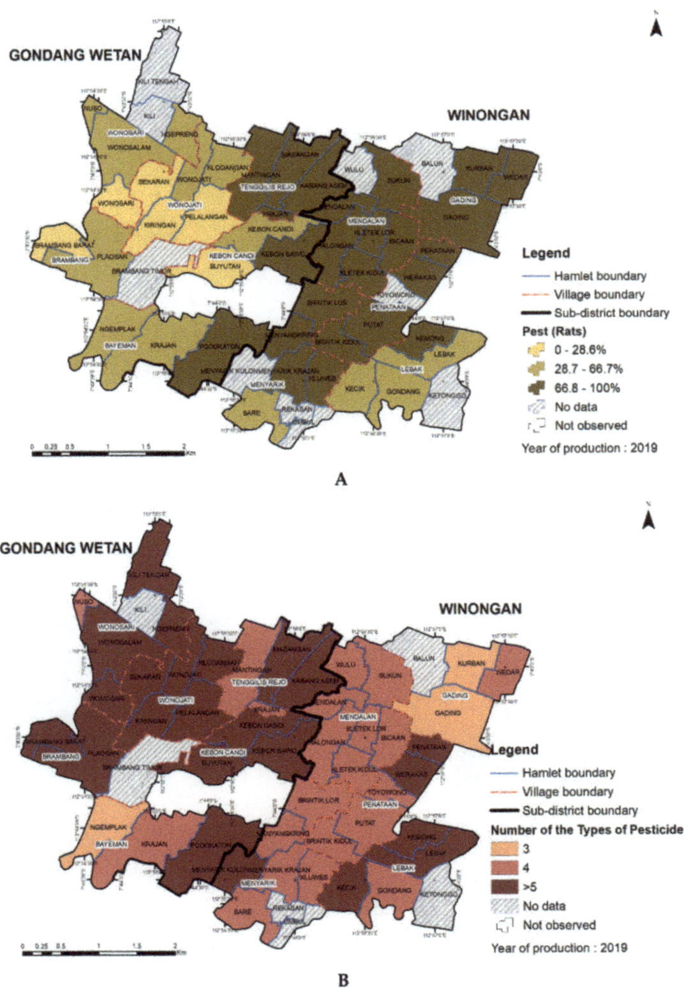

Figure 10. Spatial distribution of the intensity of pest/rodents (**A**) and number of types of pesticide applied (**B**) at hamlet level.

3.6. Characteristic of Paddy Farming

The twelve parameters (paddy field area, drainage density, number of artesian wells, the dose of N fertiliser, the dose of compound fertiliser, number of pesticide types, yield, the area with crop rotation, the intensity of rodents attack, number of water sources, river as the primary water source and presence of water regulatory officer) are independent to each other. The cluster analysis of the twelve parameters and the elbow method (Figure 11) allows us to have five clusters of paddy fields in eleven villages of Gondang Wetan and Winongan Sub-districts. Figure 12 presents the map of the resulting characteristic of paddy farms in eleven villages, and Tables 6 and 7 describe the characteristics of each cluster. Cluster 1 was in Gondang Wetan Sub-district, cluster 2 was mostly in Winongan Sub-district, and cluster 5 was spread evenly in both sub-districts (Figure 12). Meanwhile, clusters 3 and 4 consisted of only 1 hamlet, which was located in Gondang Wetan Sub-district.

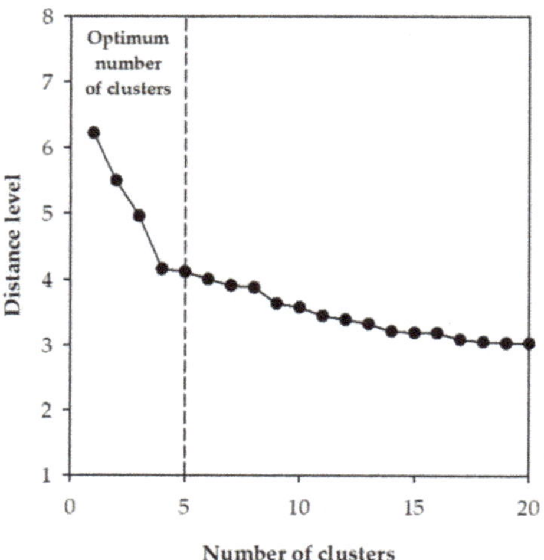

Figure 11. Dependence of distance level on the number of clusters as basis for elbow method of optimal cluster number to be used for the characteristic of paddy farming.

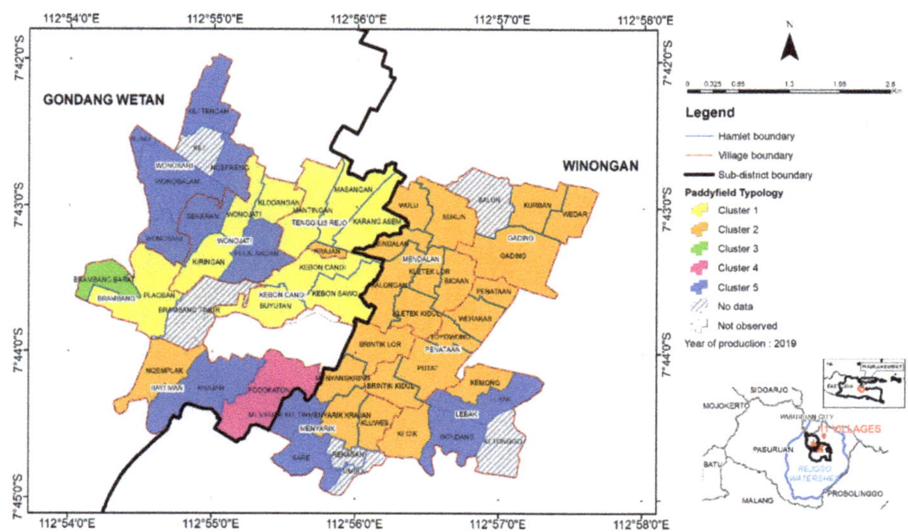

Figure 12. Characteristic of paddy farming in Gondang Wetan and Winongan Sub-districts.

Table 6. The result of cluster analysis and characteristic results for the 12 parameters.

No	Parameters	Unit	Clusters				
			1	2	3	4	5
1	Area of paddy field	%	High (218 ha)	High (407 ha)	Low (9 ha)	Low (29 ha)	Medium (230 ha)
2	Flow density (irrigation, river and trench)	km/km^2	Medium	High	Low	Medium	Low
3	Area with crop rotation	%	Medium	Low	High	High	Low
4	Intensity of pest (rodents)	% respondents	Medium	High	Low	High	Low
5	Rice yield	Ton/ha	Low	Low	Medium	High	Medium
6	Dose of urea fertiliser (46% of N)	kg/ha	High	Low	Medium	High	Low
7	Dose of compound fertiliser (15% of N, 15% of P, and 15% of K)	kg/ha	Medium	Low	High	High	Low
8	Types of applied pesticide	Number	Medium	Low	Medium	Medium	Medium
9	The presence of 'Ulu-ulu' as a water regulatory officer	Existing/not existing	Exist	Not exist	Exist	Not exist	Exist
10	Artesian wells	Number	High	Medium	Low	High	Medium
11	River as the main water source	% of respondents	Medium	High	Low	Low	High
12	Types of water sources	Number	High	Low	High	Medium	Low

Table 7. Description of each paddy field type.

Clusters	Description
1	High paddy field fraction, medium drainage density, and a high number of artesian wells. The dose of application of N fertiliser is high, and use of compound fertiliser is medium, but rice yield is low. The number of pesticide types is medium. The area with crop rotation and intensity of pests (rodents) is medium. The paddy field area with rivers as the main sources of water is medium, but still have a high number of water resources. There is an 'Ulu-ulu' as the water regulatory officer.
2	High paddy field fraction and drainage density, with a medium number of artesian wells. Doses of application N and compound fertiliser are low and rice yield is relatively low. The number of pesticide types is low. The area with crop rotation is low, hence the intensity of pests (rodents) is high. The rivers have a very important role as the main water source so that the number of other water sources is low. There is no 'Ulu-ulu' as the water regulatory officer.
3	Low paddy field fraction, low drainage density, and few artesian wells. Doses of application of N fertiliser are medium and those of compound fertiliser high while yield is at a medium level. The number of pesticide types is medium. The area with crop rotation area is high, and the intensity of pests (rodents) is low. The rivers have a small role as a source of water so that the number of other water sources is high. There is an 'Ulu-ulu' as the water regulatory officer.
4	Low paddy field fraction, with medium drainage density and a high number of artesian wells. The rate of application of N and compound fertiliser is high and followed by high yield. The number of pesticide types is medium. The area with crop rotation is high, but the intensity of pests (rodents) is still high. Rivers have a small role as a source of water, but the number of other water sources is medium. There is no 'Ulu-ulu' as the water regulatory officer.
5	Medium paddy field fraction, with low drainage density and a medium number of artesian wells. The dose of application of N and compound fertiliser is low, but yield is medium. The number of pesticide types is medium. The area with crop rotation is low, hence the intensity of pests (rodents) is low. The rivers have a critical role as the main water source so that the number of other water sources is low. There is an 'Ulu-ulu' as a water regulatory officer.

4. Discussion

4.1. Relevance of Reducing Groundwater Use in Lowland Zone

Our first question was seeking quantitative evidence that the lowland practices are co-responsible for the decrease of the Umbulan spring's discharge. The estimated total outflow of the artesian wells (over 2400 L/s in 450 measured wells; currently there may be 600 wells) does not fully account for the observed decline in the Umbulan spring flow record (from 5000 to 3500 L/s with a declining trend). The water balance results of Table 3 suggest that the lowland paddy production through its reliance on unconstrained artesian wells has been the major contributor to the observed decline of the Umbulan spring, but changes in the upper and middle zone also contribute. The water balance model includes some interactions between surface and groundwater flows, but models at a higher temporal resolution that include seasonal patterns of rainfall could refine the results in future. The attribution of effects as 55% lowlands, 49% upland and 4% interaction is, at the current level of detail in the analysis, only indicative; however, it is consistent with process understanding. Models that operate on a daily time step and at higher spatial resolution will be needed [31,32], but tend to require parametrization efforts that challenge current data availability.

4.2. Groundwater-Wasting Irrigation Methods: Understanding Farmer Decisions

The second question sought relevant geographic variation between villages and hamlets in the farmer's practices in managing land and water in cultivating paddy. The shift from rainfed sugarcane to irrigation-based paddies in Pasuruan district got a boost when relatively cheap groundwater drilling with bore holes of 10–100 m (or more) provided additional water, year-round. Without any control valve to manage the well's flow, excess groundwater was channelled back to the river and then lost to the sea. Agronomically, the water supply is considerably in excess of crop demand, with the average 10 mm/day of well supply per unit paddy area, two to three times the potential evapotranspiration rate, even without accounting for rainfall and river-based irrigation water. The high crop frequency, approaching three crops per year and currently reaching five crops per two years leaves short time for a break between crops. As the landscape as a whole is permanently saturated, the rice crop does not ripen off well and the harvested product is only of medium quality. Crop rotations with other crops are applied to only a small part of the area. Despite abundant water availability and intensive fertilization, however, rice yields of around 5 ton per ha in the survey were below the 5.8 ton per ha reported for the district [19] and representing a nearly 50% yield gap relative to the potential yield for irrigated rice in Indonesia of around 9.5 ton per ha [33]. Within our data, there was no indication that variation in yields were related to variation in fertilizer level (either Urea or compound fertilizer, the two were strongly correlated). Rodents were widely seen by farmers as the main yield-reducing factor, and despite ample use of pesticides, could not be controlled at farmer level. It appears that the current intensification pathway is reaching a dead end, where they produce large volumes of a medium quality product at considerable environmental costs and, even if these externalities are ignored, modest farmgate profitability.

Investment in closing the existing wells and replacing them by wells with improved design that can be turned off when not needed appears to be a cost-effective way for external stakeholders to recover the flow at the Umbulan spring (and the millions of households that can thus be supported). In addition, the irrigation from artesian wells only during the night can avoid water wasting from evaporation process during the day. A substantial reduction of groundwater wasting seems to be feasible without risking water shortages in critical periods for the crop. However, for a free flowing well the only decision is in its construction, whereas a controlled well requires agreements among farmers about when it will be opened.

4.3. Collective Action Aspects of Solutions

The third question was whether a participatory survey of paddy cultivation and spatial data analysis for the development of characteristic can identify options by context for upscaling sustainable paddy cultivation. Sustainable agriculture within a sustainable landscape context is beyond food production, it safeguards the increasing capacity of rural people to be self-reliant and resilient when facing changes and shocks and building strong rural institutions, including landscape governance, and their economies [34]. A well-known analysis of the Bali water temples in Indonesia, or 'subak' in the local language, highlights the importance of local institutions that secured synchronous rice planting [35]. Synchronous rice planting ensured landscape-wide breaks between cropping seasons in the traditional system, that effectively controlled rodents. When the water temples were abandoned and technical irrigation allowed for an increased cropping frequency, rodent problems came to the fore in Bali. It appears that with the unconstrained artesian wells in Rejoso, an even easier year-round availability of habitat and food supports rat populations beyond control.

Following this line of interpretation, we suggest that in the Rejoso watershed context, the lack of strong farmer institutions imply an inability to synchronise planting calendars and that this has become one of the principal causes of aggravating rodent pest attack. When farmers can collectively dry their paddy field, this cycle of fallow can reduce the rodent pressures. Thus, the management of rodent pest problems (which directly link to yield and income) can be considered as a collective driver to strengthen local institutions, which at the end leads to better water management and may allow win–win solutions of water-saving and yield increase to be feasible.

4.4. Sustainable Paddy Cultivation and Its Relevance to Global Agenda and Practices

Our efforts to understand existing constraints to a sustainable production landscape in the Rejoso watershed showed that among the eleven villages in the Gondang Wetan and Winongan Sub-districts, a substantial variation in farmer's practices in managing the land could already be found. Given this variation, a one-size-fits-all solution is not likely to work. Our current understanding of such variations of the paddy field enables the landscape managers and decision-makers to identify the potential area for upscaling specific solutions and to ensure that the adoptions of interventions run smoothly because the process has embraced potential constraints and solutions according to the perspectives of smallholders and local communities.

Our approach aligns with a global agenda. The innovations of sustainable production landscape management consider the dual goals of reducing environmental impacts while increasing productivity. Discussion of the interconnected dimensions of sustainable production landscape is not new, while understandings and solutions towards actions to transform the environmentally sustainable food production systems are still unresolved [36].

The research by Pretty et al. [37] on the adoption of practices and technologies for environmentally sustainable with substantial benefits for the rural poor hold promising advances. The 208 projects were derived from 52 countries of the South resulted in approximately 8.98 million household farming 28.92 million ha representing 3.0% of the 960 million ha of arable and permanent crops in Africa, Asia and Latin America, adopted and practised sustainable agriculture. Knowledge from the literature confirms that increasing paddy productivity while reducing environmental impacts is doable. Likewise, it is attainable that farmers practice more efficient water use [38] and emit less CH_4 and N_2O [39,40].

4.5. Potential of Development of Paddy Farming Characteristic for Intervention Scenario and Upscaling

The five clusters identified here represent variation of farmer's practices in managing the land and cultivating paddy. Each cluster provides unique information with different degrees of the constituent parameters, yet still presenting the whole targeted landscape. Intervention scenarios might include incentives for sustainable cultivation, such as insurance of stable agricultural inputs, microcredit, agricultural insurance, market transparency, and capacity strengthening in farmer group management. Table 8 presents the analysis and the risks for intervention and upscaling in each paddy field cluster. Clusters with high risk, when targeted for conducting innovative interventions, will provide 'gold standards' of success, compared to the ones with low risks.

The proposed characteristic of paddy farming that considers the variation of farmer's practices in managing the land and cultivating paddy corresponds to the requirement to implement an 'options by context' approach [8]. Rather than selections based on proximity to the central power, or 'low-hanging fruit' ones, we expect that through the 'options by context', a more robust selection of locations and intervention scenarios allow for higher adoption rates with expected results and calculated risks of the innovations at each cluster. However, ongoing implementation efforts will have to provide the test of effectiveness. We identified some limitations in the methods applied.

Table 8. Analysis and risk for intervention and upscaling of each paddy field cluster, based on feedback in local focus group discussions.

Clusters	Analysis and Risk
1	**Analysis:** Cluster 1 has the potential for upscaling as there are large areas of paddy fields and high numbers of artesian wells. Another interesting fact is to understand the high level of fertiliser use, but the yield is low. The risk of technology failure due to pest is medium, and there is a potential for crop rotation to increase soil fertility. Institutionally, the potential for better water management can be explored by the presence of an 'Ulu-ulu'. **The risk for intervention and upscaling:** Medium to low.
2	**Analysis:** Similar to Cluster 1, Cluster 2 has the potential for upscaling. Contrary to Cluster 1, Cluster 2 is lower in yield as most of the agricultural inputs are low, but the pest prevalence is high. There is no 'Ulu-ulu' in this area. **The risk for interventions and upscaling:** High, but when the intervention is successful, this will provide a high standard for successful upscaling as well.
3	**Analysis:** With less paddy field area, low drainage density, and a low number of artesian wells, Cluster 3 might not be promising for intervention and upscaling. No obvious challenges regarding paddy cultivation. **The risk for interventions and upscaling:** Medium to low, but with a limited area of paddy field, interventions might not be attractive.
4	**Analysis:** Similar to Cluster 3, the size of the paddy field area is the limiting factor for upscaling. Factors contributing to high yield interesting to analyse: is it about the fertiliser? Or crop rotation? Cluster 4 may function as a learning site, especially for the application of crop rotation. **The risk for interventions and upscaling:** High to medium due to pest intensity.
5	**Analysis:** Cluster 5 can provide another option for intervention and for upscaling with average, mild conditions on several aspects of paddy cultivation. The role of 'Ulu-ulu' might be interesting to be observed. **The risk for interventions and upscaling:** Low.

4.6. Implications for Methodology

From the survey of paddy cultivation, identification of respondents by combining available information of farmer group and its member and a snowball technique is an ideal approach. However, the unavailability of up-to-date farmer group data was a challenge, and the impacts were on the length of time in finding a respondent and level of respondent representation in each hamlet. During the data analysis, the domicile of the farmer and the location of the paddy field can be unmatched. The domicile of some respondents can be outside of the eleven villages. Considering that the interventions would be delineated according to the sub-district jurisdictional boundary, the data of seven percent of the respondents who stayed outside the eleven villages were eliminated. Hence, we suggested that the filtering process in selecting respondents should be better from the beginning, and the number of respondents in each hamlet should be adequately represented if resources allow. Apart from the above limitation, we ensured that information gained from the farmer representatives were well obtained, and a structured questionnaire complemented the interviews.

Analysis of spatial data based on aerial photograph/drones produced a high-resolution image. However, the direct georeferencing method that we applied referred that orthorectification was processed without ground control point (GCP) and independent checkpoints (ICP), resulting in a mild shift of location and take effect of Root Mean Square Error (RMSE) value. Although the shifting is about 2–2.5 m compared to orthorectification using GCP and ICP [41], in the future, we suggested applying GCP and ICP for similar analysis to increase geometric accuracy when resources allow. Other analyses to increase the degree of spatial data accuracy focused on verification of the area of paddy fields with annual crop rotation. Multi-temporal images should be used, or a detailed survey of farmers should be conducted to obtain more accurate area estimates of paddy fields with annual crop rotation.

5. Conclusions

The production landscape of the Rejoso watershed has problems of unsustainable agricultural development, particularly on its lowland part, where paddy is the primary land use. The unrestricted use of artesian wells to irrigate rice paddies is reducing the pressure on and water yield of artesian wells for urban water users, while the actual rice yields achieved are below potentials achieved elsewhere. The introduction of water-saving technology, with modification of conventional paddy cultivation, and better design and management of artesian wells more control over the wells, can target yield improvement as well as positive environmental impacts. Considerable variation was found to exist within this paddy-dominant production landscape. Our analysis of the variation of farmer's practices in managing the land and cultivating paddy was based on a survey of paddy cultivation and spatial data analysis complemented and verified by the participatory approach indicating the application of landscape-approach in its development. The characteristics of paddy farming as the results reflect the requirement to implement an 'option by context' approach within a landscape for targeting effective, yet efficient, interventions and upscaling technological improvement. The characteristic of paddy farming encompassed five clusters of paddy farms. Clusters were characterised based on the relative area of paddy fields, the density of irrigation networks, area with crop rotation, rice yield, the dose of fertiliser, number of pesticide types, water sources, and the intensity of pests/rodents. Hamlets within a cluster are similar in characteristics of farmer's practices and have unique, contextual conditions. We discussed the potentials and risks of such characteristics for further implementation and upscaling. Clusters with high risk, when targeted for conducting innovative interventions, will provide 'gold standards' of success, compared to the ones with low risks. The information is expected to be useful for the landscape managers and decision-makers in targeting, considering, and budgeting the interventions that are relevant for sustainable landscape management. Future applications of a spatially differentiated intervention approach to innovate in the direction of sustainable paddy cultivation based on water saving is expected to reduce environmental impacts while increasing productivity. On-the-ground empirical action research activities engaging research organisations, private sectors and financing institutions are ongoing.

Author Contributions: N.K., L.T. and B.L. designed the study. L.D.Y.P. collected and analysed spatial data and was supervised by A.H. and M.T.Z. F.M. and E.P. collected paddy cultivation data and were supervised by L.T. L.T. analysed paddy cultivation data. N.K., L.T., L.D.Y.P., A.H., M.T.Z. writing the original draft. N.K., B.L., L.T. and M.v.N. shaped the original draft into the manuscript, which was approved by all co-authors, A.T. provided the ecohydrological backgrounds of the Rejoso watershed to improve the manuscript. All authors have read and agreed to the published version of the manuscript.

Funding: The research was funded by the Danone Ecosystem Fund through "Sustainable, Low Carbon Emissions and Water-Efficient Agriculture of Rejoso Watershed" project.

Conflicts of Interest: The authors declare no conflict of interest.

Appendix A. Calibrating the Annual Water Balance Model

Data for mean of annual precipitation (P) were obtained from 13 rainfall stations in lower zone (7 stations), middle zone (5 stations) and upper zone (1 station) (Table A1). Precipitation for each zone (lower zone < 100 m a.s.l., middle zone 100–1000 m a.s.l. and upper zone m a.s.l.) was then generated based on correlation of elevation and annual mean of rainfall (Figure A1). Potential evapotranspiration E_{pot} were generated using Thornthwaite equation using temperature data of Accu weather (Table A2).

Table A1. Mean of annual rainfall for 13 rainfall stations.

Stations	Zone	Elevation (m a.s.l.)	Mean of Annual Rainfall (mm/year) *)
P3GI	Lower	5	1143.49
Kedawung	Lower	7	1041.40
Gondang wetan	Lower	8	1000.30
Kawis rejo-Rejoso	Lower	8	1138.33
Winongan	Lower	10	1259.06
Gading-Winongan	Lower	12	1307.53
Ranu Grati	Lower	14	1242.07
Sidepan-Umbulan-Winongan	Lower	25	652.33
Wonorejo-Lumbang	Middle	100	1590.71
Lumbang	Middle	137	1485.71
Panditan-Rejoso	Middle	600	2255.14
Puspo	Middle	640	2243.86
Tosari	Upper	1045	1539.27

*) 1990–2015.

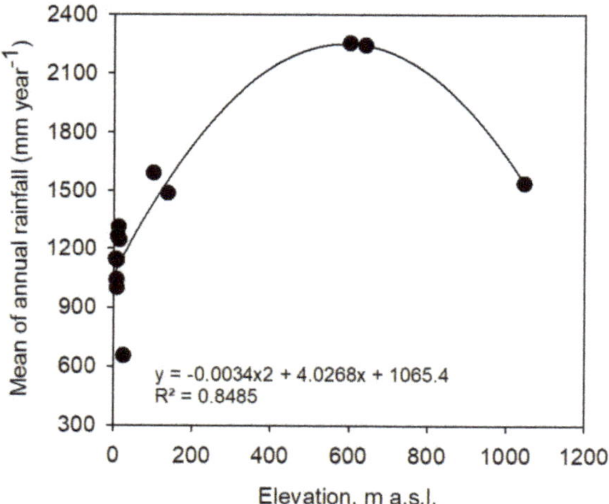

Figure A1. Correlation between mean of annual rainfall and elevation of 13 rainfall stations.

Table A2. Potential evapotranspiration of three zones.

Zones	Mean of Annual Temperature (°C)			Total Potential Evapotranspiration (mm year^{-1})		
	1990	2015	2020	1990	2015	2020
Upper	17	17	17	770.4	770.4	770.4
Middle	25	25	25	1367.8	1367.8	1367.8
Lower	27	27	27	1726.8	1726.8	1726.8

Parameter estimates for actual/potential evapotranspiration (ε) and runoff coefficient (ρ) were derived as area-weighted average from estimate for the specific land use types in the GenRiver model, calibrated for a number of Indonesian watersheds [29].

References

1. Millennium Ecosystem Assessment. *A Report of the Millennium Ecosystem Assessment. Ecosystems and Human Well-Being*; Island Press: Washington, DC, USA, 2005.
2. Costanza, R.; d'Arge, R.; deGroot, R.; Farber, S.; Grasso, M.; Hannon, B.; Limburg, K.; Naeem, S.; Oneill, R.V.; Paruelo, J.; et al. The value of the world's ecosystem services and natural capital. *Nature* **1997**, *387*, 253–260. [CrossRef]

3. Olsson, L.; Barbosa, H.; Bhadwal, S.; Cowie, A.; Delusca, K.; Flores-Renteria, D.; Hermans, K.; Jobbagy, E.; Kurz, W.; Li, D.; et al. Land Degradation. In *Climate Change and Land an IPCC Special Report on Climate Change, Desertification, Land Degradation, Sustainable Land Management, Food Security, and Greenhouse Gas Fluxes in Terrestrial Ecosystems*; Shukla, P.R., Skea, J., Buendia, E.C., Masson-Delmotte, V., Pörtner, H.-O., Roberts, D.C., Zhai, P., Slade, R., Connors, S., van Diemen, R., et al., Eds.; IPCC: Geneva, Switzerland, 2019; in press.

4. Minang, P.A.; van Noordwijk, M.; Freeman, O.E.; Mbow, C.; de Leeuw, J.; Catacutan, D. (Eds.) *Climate-Smart Landscapes: Multifunctionality in Practice*; World Agroforestry Centre (ICRAF): Nairobi, Kenya, 2015; p. 405. Available online: http://www.worldagroforestry.org/publication/climate-smart-landscapes-multifunctionality-practice (accessed on 12 November 2020).

5. Estrada-Carmona, N.; Hart, A.K.; Declerck, F.A.J.; Harvey, C.A.; Milder, J.C. Integrated landscape management for agriculture, rural livelihoods, and ecosystem conservation: An assessment of experience from Latin America and the Caribbean. *Landsc. Urban Plan.* **2014**, *129*, 1–11. [CrossRef]

6. Wu, J. Landscape sustainability science: Ecosystem services and human well-being in changing landscapes. *Landsc. Ecol.* **2013**, *28*, 999–1023. [CrossRef]

7. Van Noordwijk, M.; Leimona, B.; Jindal, R.; Villamor, G.B.; Vardhan, M.; Namirembe, S.; Catacutan, D.; Kerr, J.; Minang, P.A.; Tomich, T.P. Payments for environmental services: Evolution toward efficient and fair incentives for multifunctional landscapes. *Annu. Rev. Environ. Resour.* **2012**, *37*, 389–420. [CrossRef]

8. Sinclair, F.; Coe, R. The options by context approach: A paradigm shift in agronomy. *Exp. Agric.* **2019**, *55*, 1–13. [CrossRef]

9. Díaz, S.; Settele, J.; Brondízio, E.; Ngo, H.; Guèze, M.; Agard, J.; Arneth, A.; Balvanera, P.; Brauman, K.; Butchart, S.; et al. *Summary for Policy Makers of the Global Assessment Report on Biodiversity and Ecosystem Services of the Intergovernmental Science-Policy Platform on Biodiversity and Ecosystem Services*; IPBES: Bonn, Germany, 2020.

10. Van Noordwijk, M.; Speelman, E.; Hofstede, G.J.; Farida, A.; Abdurrahim, A.Y.; Miccolis, A.; Hakim, A.L.; Wamucii, C.N.; Lagneaux, E.; Andreotti, F.; et al. Sustainable agroforestry landscape management: Changing the game. *Land* **2020**, *9*, 243. [CrossRef]

11. Swallow, B.M.; Garrity, D.P.; van Noordwijk, M. The effects of scales, flows and filters on property rights and collective action in watershed management. *Water Policy* **2002**, *3*, 457–474. [CrossRef]

12. Van Noordwijk, M.; Van Roode, M.; McCallie, E.L.; Lusiana, B. Erosion and sedimentation as multiscale, fractal processes: Implications for models, experiments and the real world. In *Soil Erosion at Multiple Scales*; CAB International: Wallingford, UK, 1998; pp. 223–253.

13. van Noordwijk, M.; Poulsen, J.G.; Ericksen, P.J. Quantifying off-site effects of land use change: Filters, flows and fallacies. *Agric. Ecosyst. Environ.* **2004**, *104*, 19–34. [CrossRef]

14. van Noordwijk, M. Integrated natural resource management as pathway to poverty reduction: Innovating practices, institutions and policies. *Agric. Syst.* **2019**, *172*, 60–71. [CrossRef]

15. Amaruzaman, S.; Khasanah, N.; Tanika, L.; Dwiyanti, E.; Lusiana, B.; Leimona, B.; Janudianto, N. *Landscape Characteristics of Rejoso Watershed: Assessment of Land Use—Land Cover Dynamic, Farming System and Community Resilience*; World Agroforestry Centre (ICRAF) Southeast Asia Regional Program: Bogor, Indonesia, 2018.

16. Toulier, A.; Baud, B.; de Montety, V.; Lachassagne, P.; Leonardi, V.; Pistre, S.; Dautria, J.; Hendrayana, H.; Fajar, M.; Muhammad, A.; et al. Multidisciplinary study with quantitative analysis of isotopic data for the assessment of recharge and functioning of volcanic aquifers: Case of Bromo-Tengger volcano, Indonesia. *J. Hydrolol. Reg. Stud.* **2019**, *26*, 100634. [CrossRef]

17. Scott, C.A.; Shah, T. Groundwater overdraft reduction through agricultural energy policy: Insights from India and Mexico. *Int. J. Water Resour. Dev.* **2004**, *20*, 149–164. [CrossRef]

18. Toulier, A. Multidisciplinary Study for the Characterization of Volcanic Aquifers Hydrogeological Functioning: Case of Bromo-Tengger Volcano (East Java, Indonesia). Ph.D. Thesis, Montpellier University, Montpellier, France, 2019.

19. Leimona, B.; Khasanah, N.; Lusiana, B.; Amaruzaman, S.; Tanika, L.; Hairiah, K.; Suprayogo, D.; Pambudi, S.; Negoro, F.S. *A Business Case: Co-Investing for Ecosystem Service Provisions and Local Livelihoods in Rejoso Watershed*; World Agroforestry Centre: Bogor, Indonesia, 2018.

20. Suprayogo, D.; van Noordwijk, M.; Hairiah, K.; Meilasari, N.; Rabbani, A.L.; Ishaq, R.M.; Widianto, W. Infiltration-friendly agroforestry land uses on volcanic slopes in the Rejoso Watershed, East Java, Indonesia. *Land* **2020**, *9*, 240. [CrossRef]

21. The Central Statistics Agency. *Provinsi Jawa Timur Dalam Angka*; BPS Jawa Timur: Surabaya, Indonesia, 2020.

22. The Central Statistics Agency. *Luas Panen dan Produksi Padi di Indonesia 2019 No. 16/02/Th. XXIII*; BPS—Statistics Indonesia: Jakarta Pusat, Indonesia, 2020.

23. Khumairoh, U. On Complex Rice Systems. Ph.D. Thesis, Wageningen University, Wageningen, The Netherlands, 2019.

24. Creed, I.F.; van Noordwijk, M. *Forest and Water on a Changing Planet: Vulnerability, Adaptation and Governance Opportunities. A Global Assessment Report*; IUFRO World Series 38; IUFRO: Vienna, Austria, 2018.

25. Geospatial Information Agency. *Peta Digital Rupa Bumi Indonesia Skala 1:50,000*; Geospatial Information Agency: Bogor, Indonesia, 2014.

26. Pasuruan District Water-Related Public-Works Agency. *Irrigation network of Pasuruan District*; Pasuruan District Water-Related Public-Works Agency: Pasuruan, Indonesia, 2019.

27. Lillesand, T.M.; Kiefer, R.W. *Remote Sensing and Image Interpretation*, 6th ed.; Wiley India Pvt: New Delhi, India, 2011.

28. IFAD. *Good Practices in Participatory Mapping: A Review Prepared for the International Fund for Agricultural Development (IFAD)*; IFAD: Rome, Italy, 2009.

29. Cadag, J.R.; Gaillard, J.C. Integrating knowledge and actions in disaster risk reduction: The contributiin of participatory mapping. *R. Geogr. Soc. Area* **2011**, *44*, 100–109. [CrossRef]

30. Kassambara, A. Practical Guide to Cluster Analysis in R: Unsupervised Machine Learning. Available online: http://www.sthda.com/english/2017 (accessed on 30 December 2020).

31. Van Noordwijk, M.; Tanika, L.; Lusiana, B. Flood risk reduction and flow buffering as ecosystem services-Part 2: Land use and rainfall intensity effects in Southeast Asia. *Hydrol. Earth Syst. Sci.* **2017**, *21*, 2341–2360. [CrossRef]

32. Arnold, J.G.; Moriasi, D.N.; Gassman, P.W.; Abbaspour, K.C.; White, M.J.; Srinivasan, R.; Santhi, C.; Harmel, R.D.; van Griensven, A.; Van Liew, M.W.; et al. SWAT: Model use, calibration, and validation. *Am. Soc. Agric. Biol. Eng.* **2012**, *55*, 1491–1508.

33. Agus, F.; Andrade, J.F.; Edreira, J.I.R.; Deng, N.; Purwantomo, D.K.; Agustiani, N.; Aristya, V.E.; Batubara, S.F.; Hosang, E.Y.; Krisnadi, L.Y.; et al. Yield gaps in intensive rice-maize cropping sequences in the humid tropics of Indonesia. *Field Crops Res.* **2019**, *237*, 12–22. [CrossRef]

34. Van Mansvelt, J.D.; van der Lubbe, M.J. *Checklist for Sustainable Landscape Management: Final Report of the EU Concerted Action AIR3-CT93-1210: The Landscape and Nature Production Capacity of Organic/Sustainable Types of Agriculture*; Elsevier: Amsterdam, The Netherlands, 1999. [CrossRef]

35. Lansing, J.S.; Thérèse, A. The functional role of Balinese water temples: A response to critics. *Hum. Ecol.* **2012**, *40*, 453–467. [CrossRef]

36. Beddington, J.; Asaduzzaman, M.; Clark, M.; Fernández, A.; Guillou, M.; Jahn, M.; Erda, L.; Mamo, T.; Van Bo, N.; Nobre, C.A.; et al. *Achieving Food Security in the Face of Climate Change: Final Report from the Commission on Sustainable Agriculture and Climate Change*; CGIAR Research Program on Climate Change, Agriculture and Food Security (CCAFS): Copenhagen, Denmark, 2012.

37. Pretty, J.N.; Morison, J.I.L.; Hine, R.E. Reducing food poverty by increasing agricultural sustainability in developing countries. *Agric. Ecosyst. Environ.* **2003**, *95*, 217–223. [CrossRef]

38. Subari, S.; Joubert, M.D.; Sofiyuddin, H.A.; Triyono, J. Pengaruh Perlakuan Pemberian Air Irigasi pada Budidaya SRI, PTT dan Konvensional terhadap Produktivitas Air. *J. Irig.* **2012**, *7*, 28–42. [CrossRef]

39. Cai, Z.; Xing, G.; Yan, X.; Xu, H.; Tsuruta, H.; Yagi, K.; Minami, K. Methane and nitrous oxide emissions from rice paddy fields as affected by nitrogen fertilizers and water management. *Plant Soil* **1997**, *196*, 7–14. [CrossRef]

40. Wang, C.; Lai, D.Y.F.; Sardans, J.; Wang, W.; Zeng, C.; Peñuelas, J. Factors Related with CH4 and N2O Emissions from a Paddy Field: Clues for Management implications. *PLoS ONE* **2017**, *12*, e0169254. [CrossRef]

41. Widyaningrum, E.; Fajari, M.; Octariady, J. Accuracy Comparison of VHR Systematic ortho Satellite Imageries Against VHR Orthorectified Imageries Using GCP. *Int. Arch. Photogramm. Remote Sens. Spat. Inf. Sci. ISPRS Arch.* **2016**, *41*, 305–309. [CrossRef]

Article

Fruit Tree-Based Agroforestry Systems for Smallholder Farmers in Northwest Vietnam—A Quantitative and Qualitative Assessment

Van Hung Do [1,2,*], Nguyen La [1], Rachmat Mulia [1], Göran Bergkvist [2], A. Sigrun Dahlin [3], Van Thach Nguyen [1], Huu Thuong Pham [1] and Ingrid Öborn [2,4]

[1] World Agroforestry (ICRAF) Vietnam, 249A Thuy Khue Street, Thuy Khue Ward, Tay Ho District, Hanoi, Vietnam; l.nguyen@cgiar.org (N.L.); r.mulia@cgiar.org (R.M.); n.thach@cgiar.org (V.T.N.); p.thuong@cgiar.org (H.T.P.)

[2] Department of Crop Production Ecology, Swedish University of Agricultural Sciences (SLU), P.O. Box 7043, 750 07 Uppsala, Sweden; goran.bergkvist@slu.se (G.B.); ingrid.oborn@slu.se (I.Ö.)

[3] Department of Soil and Environment, Swedish University of Agricultural Sciences (SLU), P.O. Box 7014, 750 07 Uppsala, Sweden; sigrun.dahlin@slu.se

[4] World Agroforestry Centre (ICRAF) Headquarters, UN Avenue, P.O. Box 30677, Nairobi 00100, Kenya

* Correspondence: d.hung@cgiar.org

Received: 30 October 2020; Accepted: 13 November 2020; Published: 17 November 2020

Abstract: Rapid expansion of unsustainable farming practices in upland areas of Southeast Asia threatens food security and the environment. This study assessed alternative agroforestry systems for sustainable land management and livelihood improvement in northwest Vietnam. The performance of fruit tree-based agroforestry was compared with that of sole cropping, and farmers' perspectives on agroforestry were documented. After seven years, longan (*Dimocarpus longan* Lour.)-maize-forage grass and son tra (*Docynia indica* (Wall.) Decne)-forage grass systems had generated 2.4- and 3.5-fold higher average annual income than sole maize and sole son tra, respectively. Sole longan gave no net profit, due to high investment costs. After some years, competition developed between the crop, grass, and tree components, e.g., for nitrogen, and the farmers interviewed reported a need to adapt management practices to optimise spacing and pruning. They also reported that agroforestry enhanced ecosystem services by controlling surface runoff and erosion, increasing soil fertility and improving resilience to extreme weather. Thus, agroforestry practices with fruit trees can be more profitable than sole-crop cultivation within a few years. Integration of seasonal and fast-growing perennial plants (e.g., grass) is essential to ensure quick returns. Wider adoption needs initial incentives or loans, knowledge exchange, and market links.

Keywords: fruit tree-based agroforestry; economic benefits; ecosystem services; farmer perspectives; resource competition; systems improvement; uptake and expansion

1. Introduction

The United Nations sustainable development goals and Agenda 2030 include poverty eradication, ending hunger, and environmental restoration, among other objectives [1]. Related targets are to implement resilient agricultural practices that increase productivity and production, and to maintain ecosystems that strengthen the capacity for adaptation to climate change and risks and improve land health [2]. Agroforestry, a planned combination of trees and crops with or without livestock on the same land, is increasingly being recognised as a sustainable system to reconcile agricultural production and environmental protection [3,4]. When combined with contour planting on sloping uplands, agroforestry is an effective land-use system to reduce soil erosion and maintain soil fertility [5,6].

In addition, as an integrated and more permanent farming system, agroforestry can generate diverse economic, ecological, and social benefits [3,7] beyond those provided by sole-crop farming systems.

Mountainous areas in the lower Mekong region are experiencing severe forest and land degradation, driven by expansion of unsustainable farming practices [8]. For example, in northwest Vietnam, sole-maize cultivation is widespread over hills and fragile sloping land [9,10]. The northwest region is home to ethnic minorities with a poverty rate of about 14% in 2016, or 8% higher than the average poverty rate at the national level, according to the 2017 statistic book of Vietnam. Around 60% of land in the region has a slope of ≥ 30% [11]. Soil degradation in the region is acute, resulting in low crop productivity [12–16].

Driven by high economic benefits, smallholder fruit-tree cultivation has recently expanded in several provinces in northwest Vietnam [17]. For example, the total area of fruit-tree plantations in Dien Bien, Yen Bai, and Son La provinces reached 58,464 ha in 2018, a 51.4% increase compared with 2015. The main fruit commodities are longan (*Dimocarpus longan* Lour.), mango (*Mangifera indica* L.) and plum (*Prunus domestica* L.). There is also some production of son tra (*Docynia indica* (Wall.) Decne), also called H'Mong apple, which is native to the region and one of 50 special fruits of Vietnam [18]. Son tra is a multipurpose tree, restoring natural forest cover and producing fruit [19].

Despite recent developments, farmers in the northwest region generally lack technical knowledge of agroforestry [9,20], including fruit tree-based agroforestry, in terms of adequate species composition, optimal plant arrangement and spacing, and management practices to optimise delivery of products and ecosystem services over time. Good management could better utilise potential economic, social, and environmental benefits of diversified tree-based farming systems. Farmers in the region usually develop "temporary" agroforestry by combining fruit trees and annual crops such as maize or cassava, and vegetables, in the early years of planting before tree canopy closure, most often in the first to third year after tree planting [21]. Reliable scientific-based information on permanent combinations of fruit trees and annual crops is necessary to promote agroforestry systems that can offer long-term and diverse income sources through product diversification to farmers in the region.

This study assessed the performance of two fruit-tree agroforestry systems in order to obtain knowledge on sustainable farming systems for the region. Quantitative and qualitative approaches were used to assess the agroforestry systems: longan–maize–forage grass and son tra–forage grass. Specific objectives were (i) to evaluate the productivity and profitability of agroforestry systems compared with sole-tree and annual crop systems over the seven years after establishment and (ii) to survey farmers on the performance of fruit tree-based systems to identify possibilities for improvement and wider-scale development.

2. Materials and Methods

2.1. Site Description

On-farm experiments with two agroforestry systems, longan (*Dimocarpus longan* Lour.)–maize (*Zea mays* L.)–forage grass and son tra (*Docynia indica* (Wall.) Decne.) –forage grass, were carried out on three farms each, using farms as replicates. The farms were situated in Van Chan district (21.56° N, 104.56° E; 374 m a.s.l.) in Yen Bai province and Tuan Giao district (21.56° N, 103.50° E; 1267 m a.s.l.) in Dien Bien province, northwest Vietnam (Figure 1). The climate at both sites is sub-humid tropical, with a rainy season from April to October and a dry season from November to March. Mean annual temperature is 18.6 °C and 21 °C; and annual rainfall is 1200–1600 mm and 1700–2000 mm in Tuan Giao and Van Chan, respectively. Mean slope of the experimental plots was 27% at both sites.

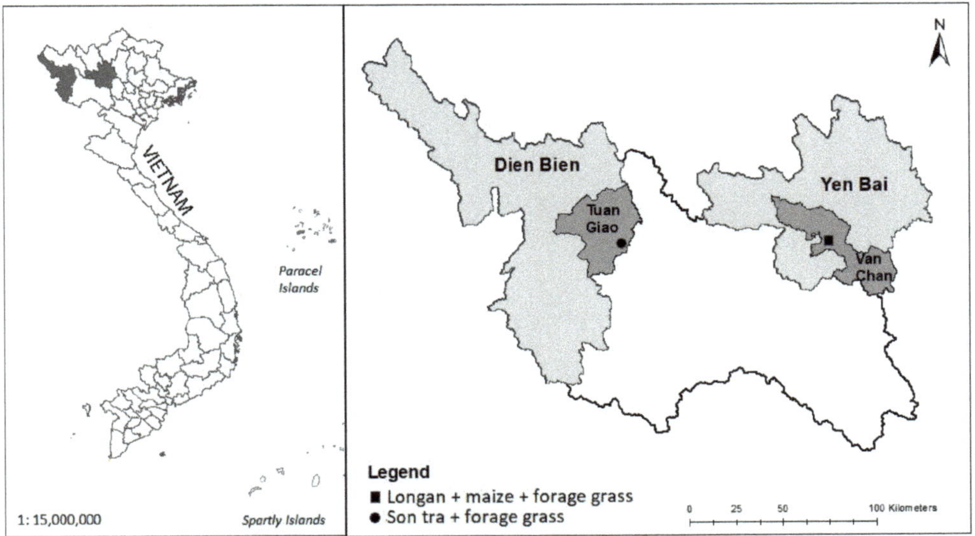

Figure 1. Location of the agroforestry experiments with longan-maize-forage grass in Van Chan District, Yen Bai province, and son tra-forage grass in Tuan Giao District, Dien Bien province, north-west Vietnam. Replicate trials were established on three farms in each district.

The soil profile at each site was characterised at the start of the experiments. The soils at Van Chan were silty clay loams, with, on average, pH 4.7, 1.7% soil organic matter (SOM), 0.12% total nitrogen (N), 0.02% total phosphorus (P), and 0.50% total potassium (K). The soil at Tuan Giao was silty clay, with, on average, pH 4.6, 3.8% SOM, and total N, P, and K of 0.24%, 0.02%, and 0.85%, respectively. SOM and total N, and P, and K were determined by the Walkley–Black method [22], Kjeldahl method [23], and digestion with mixed strong acids [24,25], respectively. Available soil P (Bray II) [26] was 5 mg kg^{-1} at Van Chan and 9.2 mg kg^{-1} at Tuan Giao.

2.2. Field Experiment Design

Both experiments were designed as randomised complete blocks with three replicates on three different farms. At Van Chan, the experiment lasted seven years (2012–2018). The agroforestry system consisted of longan, maize and guinea grass (*Panicum maximum* Jacq.) (LMG) and was compared with sole-crop maize (SM) and sole-crop longan (SL) (Figure 2a). The sole-crop longan was planted with 5 m row spacing and 5 m spacing between trees within rows (400 trees ha^{-1}). In the LMG system, longan was planted at 5 m spacing in double rows along contour lines, with 15 m between two double rows (240 trees ha^{-1}). Guinea grass was planted in double rows 0.5 m from the trees, and the distance between two rows was 0.5 m. The seed rate, row spacing, and distance between plants for sole-crop maize was 15 kg ha^{-1}, 0.65 m, and 0.3 m, respectively. The seed rate was 10–20% lower in the LMG system, since maize was not sown in the grass strips or within 0.5 m from the canopy of longan, so maize plants were sown with the same row spacing and plant distance in both systems. The longan variety used in the experiment was late maturing. The maize variety used in all cropping systems was the hybrid PAC 999.

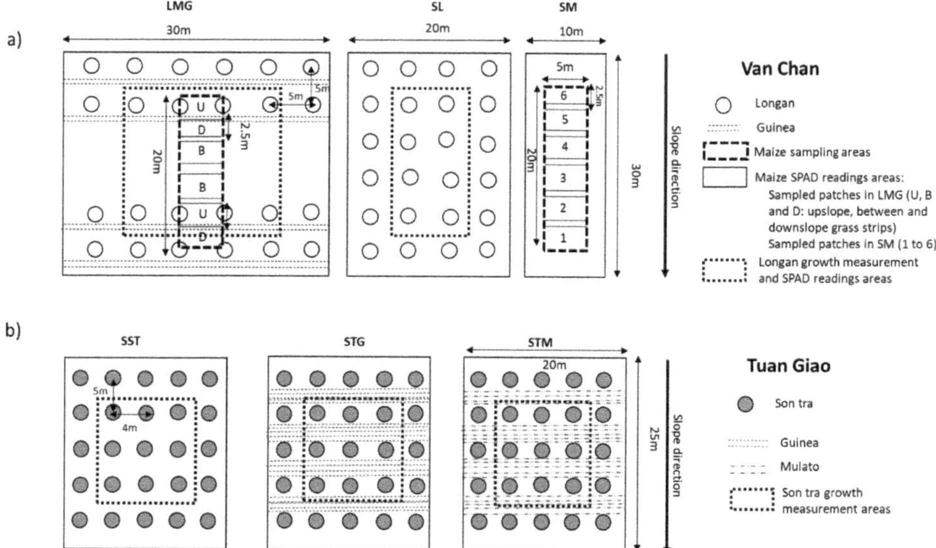

Figure 2. Design of field experiments: (**a**) Van Chan: sole-crop maize (SM), sole-crop longan (SL), and longan–maize–forage grass (LMG), U: upslope grass strips, D: downslope grass strips, B: between grass strips. The plot area was 300 m² for sole-crop maize, 600 m² for sole-crop longan, and 900 m² for the LMG agroforestry system; (**b**) Tuan Giao: sole-crop son tra (SST), son tra–guinea grass (STG), and son tra–mulato grass (STM). Plot area was 500 m².

The experiment at Tuan Giao lasted six years (2013–2018) and comprised three treatments: sole-crop son tra (SST), son tra–guinea grass (STG), and son tra–mulato grass (*Brachiaria* sp.) (STM). In all treatments, son tra was planted with 5 m row spacing and with 4 m spacing between trees within rows (500 trees ha^{-1}). Seven rows of guinea grass or mulato grass were planted between two rows of son tra in the STG and STM system, respectively (Figure 2b). The distance between the grass rows was 0.5 m and the strips were 1 m from the son tra rows. Grafted son tra seedlings were used, while guinea grass and mulato grass cuttings were obtained from a nursery.

Mineral NPK fertiliser was applied annually to maize in SM and LMG (NPK 5–10–3) as a basal application, with a topdressing with urea (46% N) and potassium chloride (48.6% K) at maize stage 6–7 fully expanded leaves (50%) and before silking (50%). In the SL and LMG treatments, 15 kg of composted animal manure and 1 kg of mineral fertiliser (NPK 5–10–3) were applied per longan tree in year 1. In years 2–7, 1 kg mineral fertiliser (NPK 5–10–3) was applied per tree, while in years 5–7, 20 kg of animal manure were applied per tree. In SST, STG, and STM, son tra received 15 kg composted animal manure and 1 kg mineral fertiliser (NPK 5–10–3) per tree in year 1 and an annual topdressing of 0.9 kg mineral fertiliser (NPK 5–10–3) per tree in years 2–6. In both experiments, the purpose of planting grass strips was to utilise nutrients in runoff, and therefore no nutrients were applied to the forage grasses. For more information about the experiments in Van Chan and Tuan Giao, see Table S1 in Supplementary Materials (SM).

2.3. Data Collection in the Field Trials

2.3.1. Tree Growth and Tree/Maize/Grass Yield Determination

Eight longan and nine son tra trees in each plot were measured every three months for the whole experimental period to determine base diameter (consistently measured at a height of 10 cm from soil surface because the trees were still small in the early years of experiments), canopy diameter, and plant height. Fresh weight biomass production of forage grasses was measured monthly by harvesting a

5 m forage grass strip per plot and weighing the biomass. Maize grain production was measured by harvesting a 5 × 20 m sub-area within each plot, air-drying the cobs outdoors before shelling and weighing. Fruit yield per plot was determined by collecting and weighing the fruit of all trees at harvest.

2.3.2. Competition for Resources in the Longan–Maize–Forage Grass System

An in-depth study of the variation in plant N concentration, growth, and productivity was carried out in year 7 of the experiment at Van Chan. Maize stover (stems, leaves, cobs, and covers) and grain were harvested at physiological maturity and weighed to determine their fresh weight. Fresh sub-samples of these materials were weighed and dried to constant weight. The ratio between fresh and dry weight was calculated and used to calculate the total harvested dry weight of each material. Within the LMG plots, measurements and sampling were performed in duplicate patches at three positions on the plots; 2.5 m upslope of the grass strips, between grass strips (4 m distance), and 2.5 m downslope of the grass strips (marked U, B, and D, respectively, in Figure 2a). The sampled area of each patch was 2.5 × 5 m. Similar sampling of patches was carried out in SM.

Plant N status was monitored in LMG and SM using a soil plant analysis development (SPAD) 502 Plus chlorophyll meter to determine the amount of chlorophyll present in plant leaves [27], as a proxy for N concentration [28]. The SPAD readings and maize plant height measurements were carried out at four vegetative stages of the maize crop (3–4, 6–7, and 10–11 fully expanded leaves, and silking). In each sampled patch, five maize plants along a diagonal were used for measurements on each occasion. The third, sixth, ninth, and index leaves were used as standard leaves for the stages 3–4, 6–7, and 10–11 fully expanded leaves and silking, respectively. The SPAD readings were taken at two-thirds of the distance from the leaf tip towards the stem [29]. In grass, the SPAD readings were carried out on 10 new fully expanded leaves [30] and height measurements were made on 10 grass plants every month in a 5 m section of each grass strip before cutting during the maize season. For longan, the SPAD readings were taken on eight longan trees within LMG and SL (Figure 2a) at the beginning and end of the maize season. One fully expanded mature leaf on the east, west, south, and north side of each tree was selected. The third leaflet position from the terminal leaf of each fully expanded mature leaf was used as the standard leaf for SPAD readings [31].

2.3.3. Land Equivalent Ratio

A land equivalent ratio (LER) was used to compare yields in the different treatments, with LER greater than 1.0 indicating that the mixed system (intercrop) was more advantageous than the sole crop. LER was calculated as [32]:

$$\text{LER} = \text{Intercrop1/Sole crop1} + \text{Intercrop2/Sole crop2} + \ldots\ldots \tag{1}$$

The fresh yield of sole-crop guinea grass and sole-crop mulato grass was calculated from their average reported dry biomass yield, i.e., 30 ton ha^{-1} year^{-1} [33] and 18.5 ha^{-1} year^{-1} [34], respectively, assuming a dry matter content of 23% [35] and 21% [36], respectively. The LER of the LMG, STG, and STM systems was calculated annually.

2.3.4. Profitability

Cost-benefit analysis was performed for each agroforestry and sole-crop treatment, taking into account details of investment costs, maintenance costs, and revenue from products sold across monitoring years. Net profit was calculated by subtracting all input costs from gross income. Annual inputs included fertiliser, pesticide, labour, planting materials, etc. Total annual income was calculated based on yield and the price obtained for the different products at harvest. Data on the cost of inputs and market prices for products were obtained from the provincial extension departments covering the study sites (see Table S2 in Supplementary Materials). Net profits of each system were calculated as:

$$N = T - I \qquad (2)$$

where N is net profit, T is total income, and I is total cost of all inputs, all in USD ha^{-1} year^{-1}.

2.4. Selection of Participants for Farmer Group Discussions

Farmers' perceptions and aspirations for the agroforestry systems involving longan–maize–forage grass (in Yen Bai) and son tra–forage grasses (in Dien Bien) were documented in group discussions carried out in January 2020. For each agroforestry system, two villages were selected: one village that hosted an experiment (experiment-hosting village) and a nearby village (non-hosting village) (Table S3 in Supplementary Materials). In each village, farmers who were familiar with or had observed the agroforestry system in the field experiment were selected and divided into three groups based on resources and gender (poor female, poor male, non-poor mixed female and male). Farmers hosting the experiments were interviewed individually, using the same open-ended questions as in the group discussions. In total, there were six different farmer groups at each study site, three in the experiment-hosting village and three in the non-hosting village, plus the three farmers hosting the experiments at each site (experiment-hosting famers). The Vietnamese government's poverty scale [37] was used to capture responses from farmers experiencing different levels of poverty. The questions (see Table S4 in Supplementary Materials) were posed by an interview team, including three researchers from World Agroforestry (ICRAF) in Vietnam who served as facilitators. All interviews were recorded and the responses were transcribed and translated into English by the researchers after each group discussion. The responses from farmers belonging to the different groups were then analysed to identify the consensus or most common responses to each question within each group. Thus, responses from individual farmers are not presented. The main ideas expressed in responses were identified and grouped into themes/categories reflecting farmers' perceptions of the two agroforestry systems tested in terms of tree, maize, and grass performance related to competition for resources, economic and ecological benefits, markets, constraints to adoption, and potential of agroforestry as a future option for the region.

2.5. Statistical Analysis

The software R (version 3.6.1) was used for all statistical analyses. Repeated measures ANOVA with the mixed model was used to assess the effects of the different treatments on maize, grass, and tree performance; yield; and profitability over the years. Log-transformation was used to normalise the data where necessary. When a significant difference was indicated in F-tests, lsmeans was used to identify significant ($p < 0.05$) differences between means. Repeated measures ANOVA was also applied to compare SPAD values and growth of maize in LMG and SM plots in year 7 of the experiment at Van Chan. ANOVA was used to compare the yield of maize at different positions relative to the grass strips within LMG in the last year, and then Tukey's HSD test was used to identify positions that were significantly different from other positions.

3. Results

3.1. Tree Growth

There was a significant effect by cropping systems on growth of longan trees. Base diameter, canopy diameter, and height in the sole-crop (SL) system were significantly greater ($p < 0.05$) than in the LGM system (Figure 3a). By the end of year 7, the base diameter of longan in SL and LMG had increased by 9 and 7 cm, respectively, since planting, and the height of longan trees was about 148 cm in SL and 121 cm in LGM, i.e., a height increase of 36 and 32 cm year^{-1} in SL and LGM, respectively.

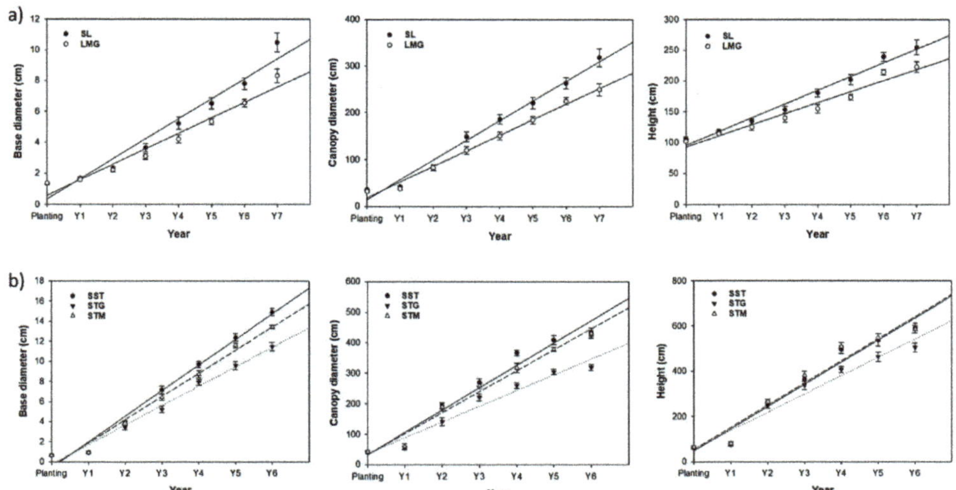

Figure 3. Regression lines describing tree growth (mean and standard error): (**a**) Growth of longan in the sole-tree system (SL) and longan–maize–forage grass (LMG) system; (**b**) growth of son tra in the sole-tree system (SST), son tra–guinea grass (STG) system, and son tra–mulato grass (STM) system.

The base diameter of son tra trees was significantly greater ($p < 0.05$) in the sole-tree system than in the systems with forage grass (STM and STG) (Figure 3b). Both tree height and canopy diameter were affected by the cropping system, with an interaction between cropping system and year ($p < 0.05$). Three years after planting, the canopy diameter and height of son tra trees were similar in the agroforestry and sole-tree systems. However, from year 4 to 6, canopy diameter and tree height were significantly higher ($p < 0.05$) in the sole-tree and STM systems than in the STG system (Figure 3b).

3.2. Yield and Land Equivalent Ratio

During the first three years, the products in LMG were primarily maize cobs and forage-grass biomass (Table 1). The grass started yielding from year 2. The products became more diversified from year 4, when longan started to bear fruit, and yield increased during subsequent years. There was no significant effect from the cropping system, or interaction between treatments and year, on maize yield. However, the yield of longan was significantly higher in the sole-tree system than in LMG, and there was a significant interaction between treatment and year ($p < 0.05$). From year 2 to 7, LER of the LMG system ranged from 1.1 to 1.9 (Figure 4a).

In the STG and STM agroforestry systems, the guinea grass and mulato grass were harvested from year 2 (2014), with high yield (Table 1). The agroforestry practices had more products from year 3, when son tra started to bear fruit. However, there was a significant effect from the cropping system on the productivity of son tra ($p < 0.05$), with fruit yield being significantly lower in agroforestry than in the sole-crop system. LER of the agroforestry practices from year 2 to 6 ranged from 0.5 to 1.1 for STG and 0.6 to 1.8 for STM (Figure 4b).

Table 1. Yield of maize (dry grain), longan, and son tra (fresh fruit) in sole-crop/tree systems (SM, SL, SST) and of these crops and forage grasses (fresh matter) in agroforestry systems (LMG, STG, STM) in the seven years of the field experiments.

Crop/Trees	Cropping System	2012	2013	2014	2015	2016	2017	2018	Mean
					Yield (Ton ha^{-1})				
Maize	SM	5.9 (±0.1)	4.7 (±0.4)	4.1 (±0.3)	4.2 (±0.3)	4.3 (±0.1)	4.2 (±0.2)	4.6 (±0.5)	4.6 (±0.2)
	LMG	5.5 (±0.3)	5.3 (±0.1)	3.9 (±0.2)	4.1 (±0.2)	4.1 (±0.1)	4.0 (±0.2)	4.2 (±0.4)	4.5 (±0.3)
By cropping system					*p*-value = 0.33				
Cropping system x year					*p*-value = 0.35				
Longan	SL				0.35 (±0.2)	0.32 (±0.3)	0.47 (±0.1)	3.04 (±1.1)a	1.04 (±0.7)a
	LMG				0.06 (±0.03)	0.18 (±0.2)	0.38 (±0.2)	0.90 (±0.3)b	0.30 (±0.1)b
By cropping system					*p*-value = 0.02				
Cropping system x year					*p*-value = 0.03				
Guinea grass	LMG		4.4 (±4.4)	19.5 (±2.3)	15.9 (±1.2)	18.2 (±1.1)	18 (±1.5)	14.6 (±4.1)	15 (±2.3)
Son tra	SST	na			0.6 (±0.4)	5.6 (±4.1)	2.1 (±1.6)	8.7 (±7.6)	4.2 (±1.8)a
	STG	na			0.2 (±0.1)	1.8 (±0.8)	0.2 (±0.2)	1.8 (±1.3)	1.0 (±0.5)b
	STM	na			0.2 (±0.1)	0.9 (±0.4)	0.5 (±0.4)	4.7 (±3.4)	1.6 (±1)ab
By cropping system					*p*-value = 0.03				
Cropping system x year					*p*-value = 0.75				
Guinea grass	STG	na		67 (±26.3)	61 (±11)	55 (±6.3)	56 (±2.9)	67 (±5.9)	61 (±2.6)
Mulato grass	STM	na		58 (±29)	65 (±4)	74 (±7.6)	64 (±4.7)	66 (±4)	65 (±2.5)
By cropping system					*p*-value = 0.62				
Cropping system x year					*p*-value = 0.85				

SM: sole-crop maize, SL: sole-crop longan, LMG: longan–maize–forage grass, SST: sole-crop son tra, STG: son tra–guinea grass, STM: son tra–mulato grass; na: not applicable since the experiment was established in 2013. Values are mean ± standard error; different letters indicate significant differences (*p* < 0.05).

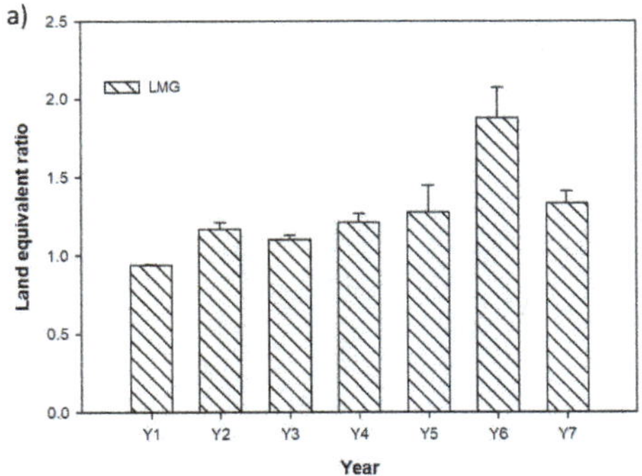 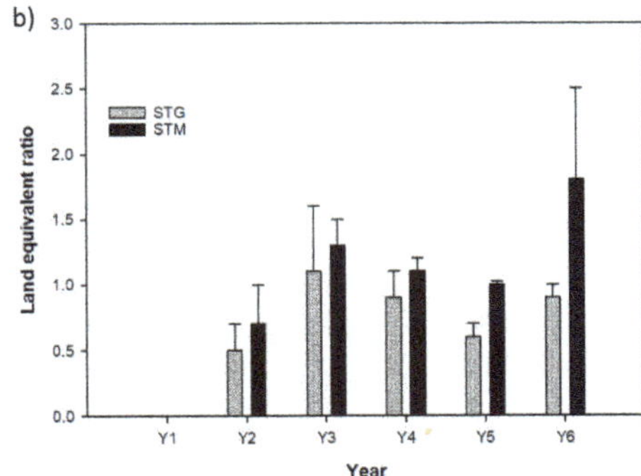

Figure 4. Land equivalent ratio (LER) of the agroforestry practices in each year of the experiment, expressed as mean and standard error (bars): (**a**) Longan–maize–forage grass (LGM); (**b**) son tra–guinea grass (STG) and son tra–mulato grass (STM).

3.3. Leaf Nitrogen Content and Competition in LMG

The SPAD value was significantly higher in sole-crop maize than in the LMG system ($p < 0.05$) from maize development stages 6–7 to silking, while maize plant height was significantly higher from 10–11 fully expanded leaves to silking (Table 2). However, the biomass of maize, including grain and stover, was not significantly different between the sole-crop and agroforestry systems.

Table 2. Dry yield, height, and SPAD readings of maize in the longan–maize–forage grass system (LMG) and the sole-crop system (SM) in year 7 of the experiment.

			Maize Growth Stage				At Maturity	
	Cropping System	3–4 Leaves	6–7 Leaves	10–11 Leaves	Silking		Cropping System	Dry Yield (Ton Ha^{-1})
SPAD	SM	38.0	44.6a	52.1a	57.7a	Grain	SM	4.6
	LMG	38.5	41.3b	47.8b	54.3b		LMG	4.2
p-value			<0.001			*p*-value		0.25
Height (cm)	SM	28.4	65.2	112.1a	230.4a	Stover	SM	5.6
	LMG	26.9	61.4	96.3b	218b		LMG	4.9
p-value			<0.001			*p*-value		0.09

Different letters indicate significant differences ($p < 0.05$).

The height of maize upslope, downslope, and between grass strips in LMG during year 7 was not significantly different from the height of maize in SM at the stages of 3–4 and 6–7 fully expanded leaves (Figure 5a). However, in later development stages, maize growth was significantly higher ($p < 0.05$) between two grass strips than immediately upslope or downslope of the grass. In stages 6–7 and 10–11 fully expanded leaves and silking, the SPAD readings of maize between grass strips were also significantly ($p < 0.05$) higher than those upslope and downslope of grass strips. The average SPAD readings for longan trees were not significantly different between LMG and SL (Figure 5a). Meanwhile, the average SPAD readings of guinea grass recorded 43.4 within LMG. This indicates that competition for N took place at positions where trees, crops, and grass were close to each other within the LMG system.

In LMG, the yield of maize grain between grass strips was 24% higher ($p < 0.05$) than in SM and about 62% higher than in upslope and downslope maize in LMG (Figure 5b). Yield of stover was also significantly higher (53–59%) between grass strips than for maize upslope and downslope of grass strips. Overall, the results clearly showed competition between grass, longan, and maize upslope and downslope of the grass strips within the LMG system in year 7.

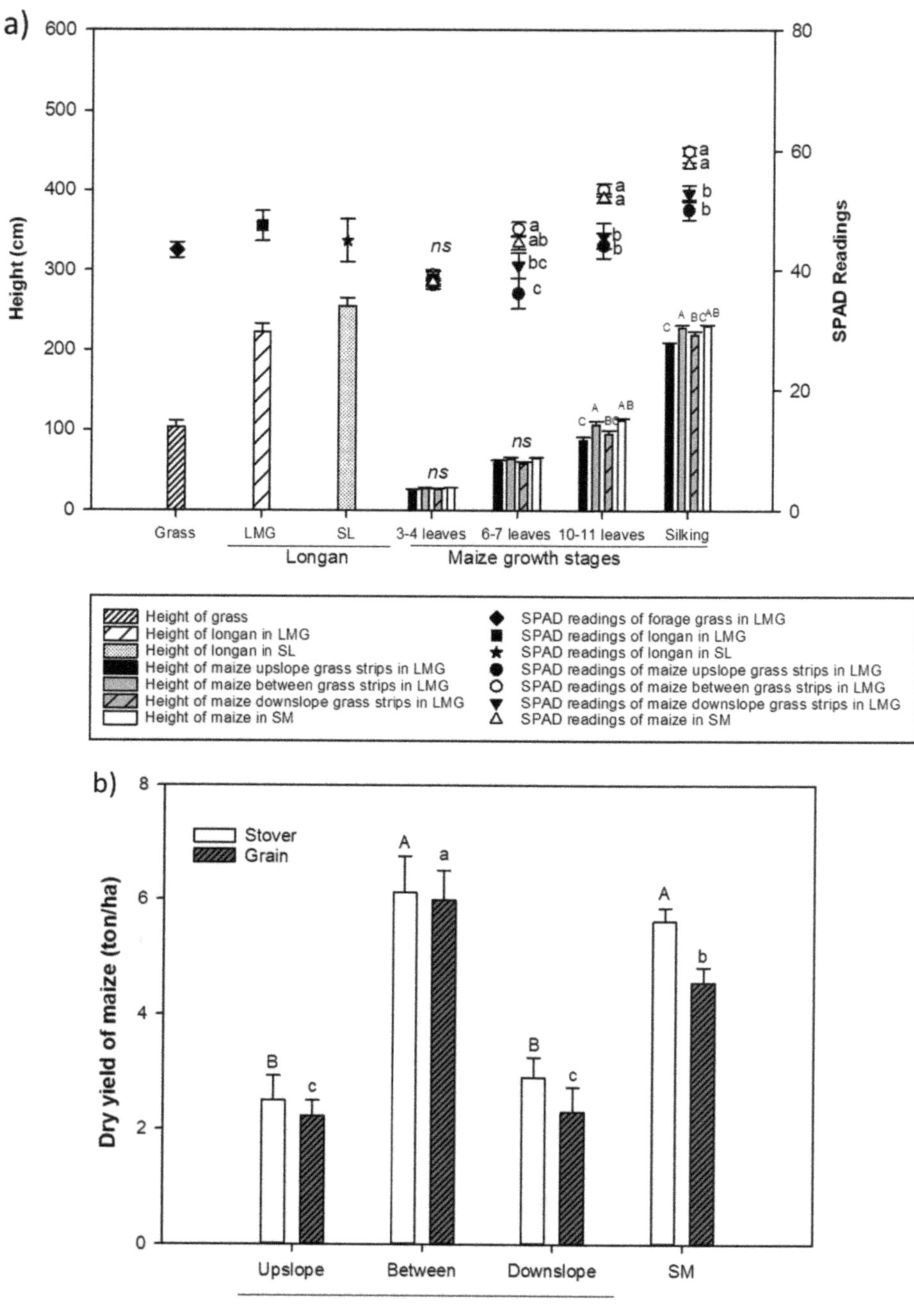

Figure 5. (**a**) Height of the tree and crop components and SPAD (soil plant analysis development) readings in the longan–maize–forage grass (LMG), sole-longan (SL) and sole-maize (SM); (**b**) dry yield of maize growing in different positions (upslope, between, downslope) relative to the grass strips within LMG in year 7. Values are means and standard errors. Bars with different upper case (stover) and lower-case (grain) letters indicate significant differences ($p < 0.05$).

3.4. Profitability

Sole maize had a mean annual investment cost of 670 USD ha^{-1}, while that of the sole-longan and the LMG system was 3.7-fold and 3.2-fold higher, respectively. The average maintenance cost of SL and LMG was 300 and 863 USD ha^{-1} year^{-1}, respectively (Figure 6a). The net profit was related to the cropping system, with an interaction between cropping system and year ($p < 0.05$). The mean net profit of LMG (1018 USD ha^{-1}) was 2.4-fold higher than for SM, while the SL system only achieved a positive profit from year 6 (Table 3). The trend of decreasing net profit of SM across year was partially due to the decreasing selling price of maize over time (presented in the Supplementary Materials Table S2) and lower maize yield in the subsequent compared to the initial years of experiment. From year 2, the net profit from LGM was equal to that from SM, while from year 4 the net profit from LGM was significantly ($p < 0.05$) higher than for SM and SL. In addition, the cumulative profit from LMG was positive from year 2 and higher than that from SM from year 4 (Figure 6a). In contrast, the cumulative profit from SL was still negative in year 7.

Table 3. Net profit from the agroforestry systems and the corresponding sole crop/tree.

Cropping System	Net Profit (USD ha^{-1})							
	2012	2013	2014	2015	2016	2017	2018	Mean (±SE)
SM	1118a	611a	388a	246b	233b	196b	190b	425.9 (±118.9)b
SL	−2463c	−355b	−229b	40b	−41b	112b	947ab	−284.4 (±336.8)c
LMG	−391b	839a	1550a	1179a	1380a	1404a	1168a	1018.2 (±231.6)a
By cropping system					*p*-value < 0.001			
Cropping system x year					*p*-value < 0.001			
SST	na	−1422	−290	42	2120	538	3238	704.5 (±632.4)b
STG	na	−1772	3297	2853	3069	2381	3570	2232.9 (±746.6)a
STM	na	−1772	2661	3018	4067	3147	4773	2648.7 (±857.1)a
By cropping system					*p*-value = 0.005			
Cropping system x year					*p*-value = 0.72			

SM: sole-crop maize, SL: sole-crop longan, LMG: longan–maize–forage grass, SST: sole-crop son tra, STG: son tra–guinea grass, STM: son tra–mulato grass; na: not applicable since the experiment was established in 2013. Values are means; different letters indicate significant differences ($p < 0.05$).

In the year of establishment, the total input costs were approximately 1772 USD ha^{-1} for both STG and STM, but lower (1422 USD ha^{-1}) for SST. In the following years, STG and STM required higher investment than the sole-tree system, mainly deriving from labour costs for forage-grass harvesting (Figure 6b). There was a significant effect ($p < 0.05$) of cropping system on net profit, with the mean net profit in STG (2233 USD ha^{-1}) and STM (2649 USD ha^{-1}) being around 3.2- and 3.7-fold higher, respectively, than in SST (Table 3). The SST system gave a positive net profit from year 3, but the cumulative profit from STG and STM was positive and higher than from SST from year 2 (Figure 6b).

3.5. Farmers' Perceptions and Aspirations for Fruit Tree-Based Agroforestry

3.5.1. Tree and Crop Performance in Agroforestry

Most farmers were fully aware of possible effects of competition for resources (light, water, nutrients) on the performance of tree and crop components within the agroforestry systems (Figure 7). All interviewees in Van Chan noted that growth and productivity of maize in the longan–maize–forage grass system were lower than in sole-maize cultivation. They attributed this to close distance between trees, crops, and grass leading to competition in the agroforestry system. However, non-hosting village groups claimed that longan trees performed better in agroforestry than in sole cultivation since they believed that longan trees utilised the nutrients applied to the maize. However, the experiment-hosting village, the experiment-hosting farmers in Van Chan, and all interviewees in Tuan Giao reported that growth and productivity of trees in both agroforestry systems were lower than when trees were grown separately.

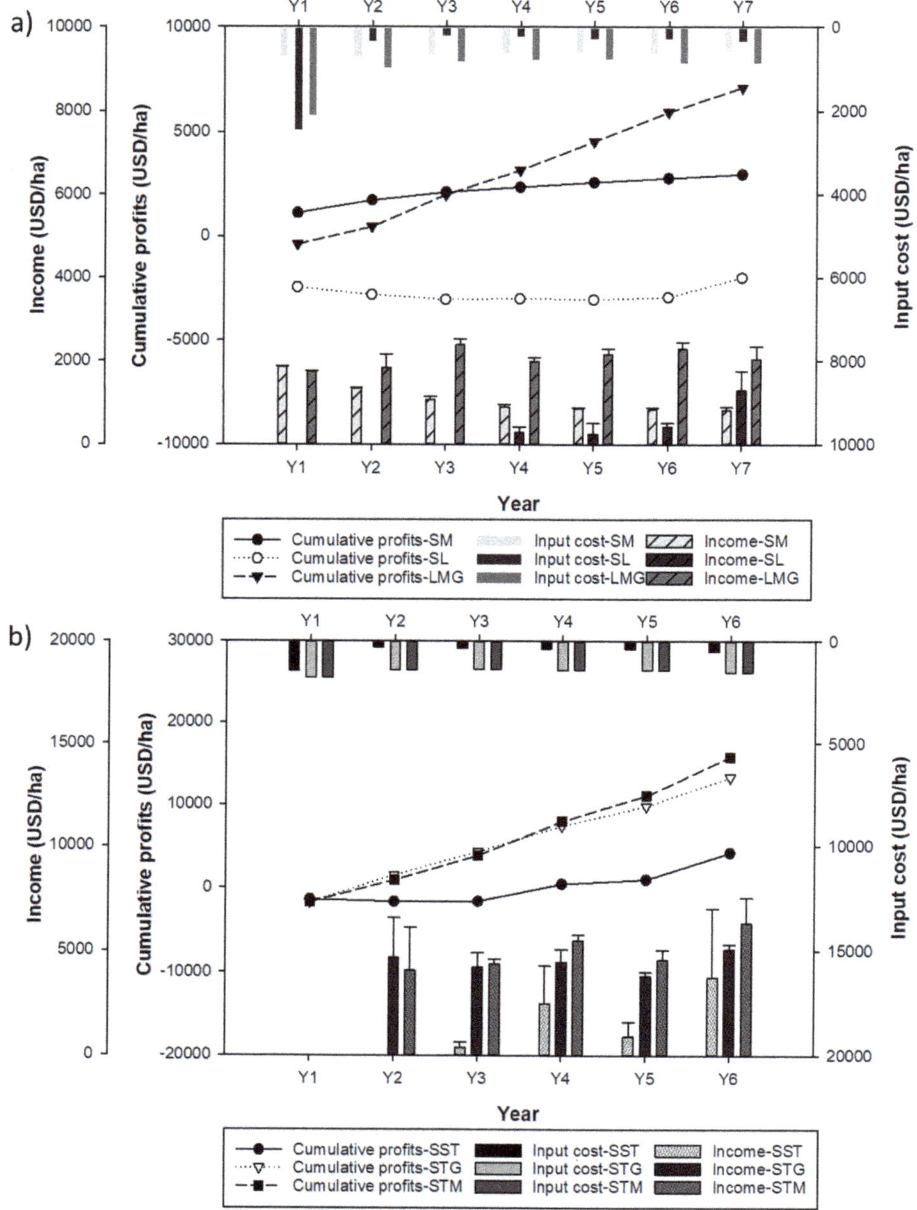

Figure 6. Input costs, income, and cumulative profit from: (**a**) the longan–maize–forage grass (LGM) compared with sole maize (SM) and sole longan (SL); (**b**) the son tra–guinea grass (STG) and son tra–mulato grass (STM) compared with sole son tra (SST).

The interviewees also suggested that the agroforestry systems could be optimised through better management of trees and crops (Figure 7). The groups proposed different solutions to improve the efficiency, such as adding more fertilisers to plants suffering from nutrient deficiency in areas where trees, crops, and grass affected each other's nutrient availability, reducing tree density and pruning to reduce shading. In addition, modifying the planting distance between trees and grass was suggested by groups from both sites. The farmers interviewed also suggested less-competitive crops for the agroforestry systems, e.g., legume species with biological N-fixation such as soybean and groundnut in LMG in Van Chan (3 of 7 groups), and upland rice or cucumber in STG and STM in Tuan Giao (2 of 7 groups).

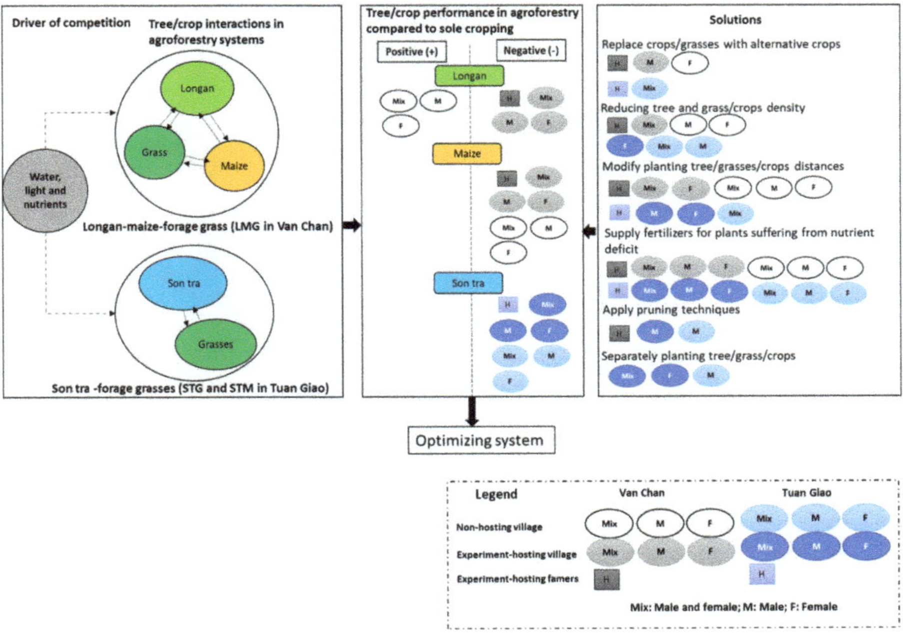

Figure 7. Farmers' perception of the performance of trees and crops in the agroforestry experiments in Van Chan and Tuan Giao compared with that of sole crops/trees. Open-ended questions were used in group interviews with non-hosting and experiment-hosting villages; experiment-hosting farmers were interviewed individually.

3.5.2. Benefits of the Agroforestry Systems

All famer groups in Van Chan and Tuan Giao shared the opinion that the experimental agroforestry systems produced earlier and more diverse products and gave higher economic benefit than the sole crop/tree (Figure 8a). Most interviewees reported that after 3–4 years, when the trees began to bear fruit, the income from agroforestry was much higher than from sole-crop cultivation. They also reported ecological benefits of the agroforestry systems in terms of reduced erosion, weed control, enhanced soil moisture and fertility, and greater resilience to extreme weather conditions (drought, snow, and frost) compared with sole-crop cultivation (Figure 8a). However, no group mentioned any benefits regarding pests and diseases, while only one group (the host farmers in Van Chan) mentioned terrace formation as an advantage (Figure 8a,b). Female and mixed groups in Tuan Giao claimed that the soil was less fertile and soil moisture lower in agroforestry than sole-tree cultivation, because the very dense forage grass used much water and nutrients within the agroforestry system (Figure 8b). Only the groups in Van Chan and the host farmer group in Tuan Giao expressed appreciation of the reduced labour requirement for harvesting forage from the grass strips in the agroforestry system (Figure 8b). These groups mentioned the possibility of using the forage to feed livestock, produce green manure, and provide earlier income when sold on the local market.

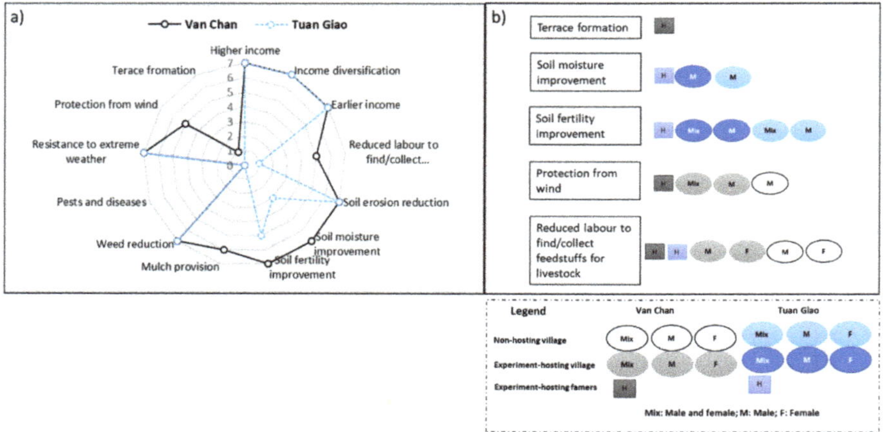

Figure 8. (**a**) Farmers' perceptions about benefits of agroforestry systems and number of farmer groups mentioning each of the identified benefits; (**b**) perceived benefits and the farmer groups in Van Chan and Tuan Giao that mentioned each. Open-ended questions were used in group interviews with non-hosting and experiment-hosting villages; experiment-hosting farmers were interviewed individually.

3.5.3. Constraints to Uptake of Agroforestry

Most of the farmer groups recognised and listed constraints to the uptake of agroforestry and proposed possible solutions to improve uptake in the local region (Figure 9). At both Tuan Giao and Van Chan, all groups indicated that the investment costs were higher than for sole-crop cultivation, making it difficult for poor households to adopt agroforestry. Management of pests and diseases in agroforestry was also more complicated, with more tree and crop components. An unstable market and low prices for products were other constraints to the uptake of agroforestry in the region.

All groups in Van Chan indicated that harsh weather events such as drought and lack of awareness among farmers of the benefits of agroforestry (4 of 7 groups) were the main drawbacks to the uptake of agroforestry. In Tuan Giao, all farmer groups considered that it would be difficult to combine traditional free grazing of livestock on crop residues with agroforestry. The forage grass was not considered valuable, since in this area with only free-grazing livestock farmers are not accustomed to collecting fodder. Extreme weather such as snow and frost and lack of techniques for implementing agroforestry were reported as other constraints to the adoption of agroforestry.

The farmers interviewed proposed solutions to address these issues (Figure 9). At Van Chan and Tuan Giao, all farmer groups mentioned training in agroforestry techniques, support in obtaining seedlings and fertilisers, and financial support or access to low-interest loans/credits as important incentives for implementing agroforestry. Development of market links for agroforestry products and a stable market were also considered key factors for agroforestry adoption by all farmer groups, but the suggested schemes differed. In Van Chan, the interviewees envisaged creating a stable market by building a farmers' cooperative to improve product quality to meet market demand and a processing factory to produce secondary products from longan fruit. The interviewees wanted maize replaced with other, higher-value annual crops. In Tuan Giao, the interviewees wanted a market link to a processing factory that would buy and add value to son tra and create a stable market.

All farmer groups in Van Chan and Tuan Giao saw a need for plant protection interventions to control pests and weeds as a way to reduce the labour costs of implementing agroforestry. According to farmers in Tuan Giao, shifting from free grazing to captive grazing and promoting livestock production to utilise the forage grass would increase the feasibility of agroforestry in the region. Although drought is a major concern in Van Chan, only the experiment-hosting farmers and the female farmer groups mentioned construction of water storage facilities as a solution. They saw a need for an electric pump and water tanks on the top of hills to supply water for tree/crops during drought periods.

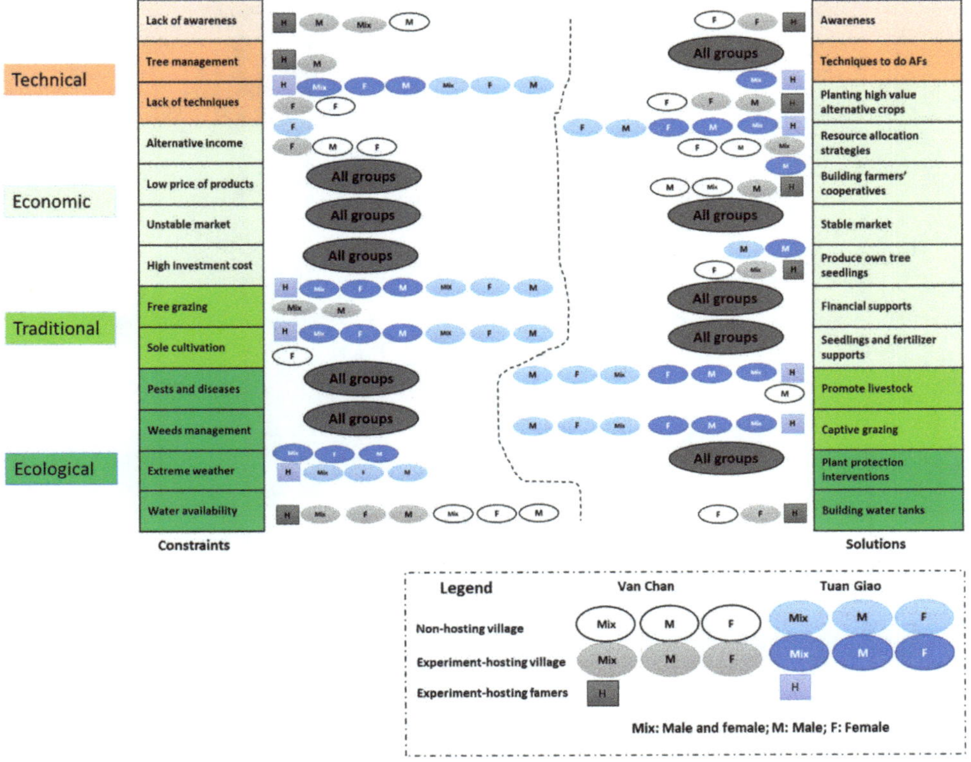

Figure 9. Farmers' perceptions of constraints (left) and solutions (right) to the uptake of agroforestry (AF) in Van Chan and Tuan Giao, and (centre) the farmer groups that mentioned the respective constraint/solution. Open-ended questions were used in group interviews with non-hosting and experiment-hosting villages; experiment-hosting farmers were interviewed individually.

All farmer groups interviewed mentioned a need to reduce the investment costs of agroforestry (Figure 9), e.g., by producing their own fruit-tree seedlings (3 of 7 groups in Van Chan and male groups in Tuan Giao). Some groups suggested offsetting the investment costs by planting higher-value crops to replace maize in Van Chan (4 of 7 groups) and forage grasses in Tuan Giao (2 of 7 groups). In addition, all farmers in Tuan Giao and 3 of 7 groups in Van Chan (Figure 9) indicated that resource allocation strategies could help reduce the maintenance cost of implementing agroforestry. They believed that during the first three years of the experiments, when the trees had not yet produced fruit, the farmers prioritised the annual crops and grasses to generate annual income. Later, when the trees were maturing and bearing fruit, farmers prioritised the trees.

3.5.4. Factors Enabling Expansion

The farmers at both Van Chan and Tuan Giao indicated that large-scale annual crop production on sloping land is an unstable system (land degradation, low yield). However, the ownership of land by local farmers is suited to implementing agroforestry. In addition, agroforestry has potential in both areas because it can bring economic and ecological benefits for local farmers. The local climate conditions are suitable for longan trees in Van Chan and for son tra trees in Tuan Giao, so both species can produce high yield. Recently, many farmers in Van Chan have shifted from sole-maize production to fruit trees and intercropping of fruit trees with annual crops, while farmers in Tuan Giao expressed interest in grafted son tra seedlings because they start to produce fruit rapidly. Local farmers saw potential for intercropping high-value trees (e.g., longan, mango, plum) and high-value crops (e.g., medicinal plants, soybean, green bean) in Van Chan, or amomum (*Amomun xanthioides* Wall.) in Tuan Giao (Figure 10).

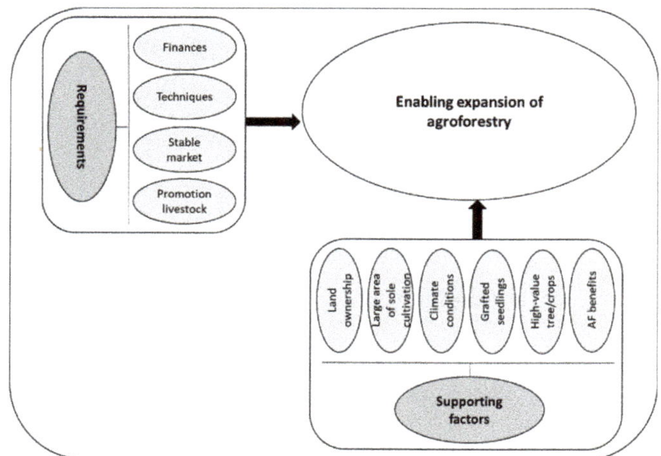

Figure 10. Farmers' perspectives about factors enabling expansion of agroforestry in Van Chan and Tuan Giao.

However, based on the interview responses, techniques to implement agroforestry, a stable market for products, and financial support for farmers in the establishment year(s), in combination with expansion of livestock production, would be required to expand agroforestry in northern upland areas of Vietnam (Figure 10).

4. Discussion

4.1. Effects of Competition for Resources on Tree and Crop Performance in Agroforestry and Ways to Improve the Systems

Total income was higher in the agroforestry systems than in the sole-cropping systems studied, but individual crop components generally grew more slowly in agroforestry systems than in sole-crop/tree systems, most likely due to competition for light, water, and nutrients [38]. The tree species in maize agroforestry systems may contribute differently to tree–crop interactions, e.g., leguminous tree species have been shown to compete less with maize for N than non-leguminous species [39–41]. The presence of tree roots, especially in the maize-cropping zone, also affects the competition with maize, and is determined by e.g., inherent rooting patterns, management, and soil conditions [41,42]. Conversely, maize restricts root development of trees in the cropping zone of agroforestry systems. A study on maize-based agroforestry systems in the sub-humid highlands of western Kenya indicated that the length of fine roots of intercropped trees (*Grevillea robusta* and *Senna spectabilis*) decreased in the maize root zone because of competition and damage to tree roots during weed hoeing [43]. In addition, maize uses the C4 photosynthetic pathway and is sensitive to shading [44] and may therefore be more negatively affected by tree shading in agroforestry systems than C3 species. Such competition was evident in the LMG system in our study, with slower growth and lower yield of longan and maize in areas where trees and crops were close to each other. This was particularly evident in year 7, when SPAD measurements showed competition for N between trees, crops, and grass growing close to each other (Table 2 and Figure 5).

In our experiments, the grass component of the agroforestry systems was competitive and negatively affected N uptake and growth of trees and maize. A previous study of maize intercropped with guinea grass in northwest Vietnam [45] found that aboveground biomass of maize at positions downslope and upslope of grass strips was around 60% and 40% lower, respectively, than that of maize 3 m from grass strips and sole maize, as we found for the LMG system (year 7). The farmer groups interviewed confirmed that maize downslope and upslope of grass strips showed lower growth and yield compared with maize farther from grass strips and sole maize, and that longan also had lower growth and yield as an intercrop than as a sole crop.

In our experiment, the yield from sole-longan planting was 2–4-ton ha^{-1} at the seventh year after tree planting. However, higher yield can be expected with e.g., improved irrigation. For example, in Hung Yen province of the Red River Delta region of Vietnam, the longan yield could reach 20 ton per ha^{-1} in the eighth year after tree planting [46]. Thanks to better market access including for export, partially due to the proximity to Hanoi as the country's capital and urban centre, the farmers in the province could derive high income from selling longan, and they partially allocate the income to improving irrigation systems [46]. The farmers in the province have been cultivating longan for decades.

However, the degree of competition may differ between grass species. A study in Costa Rica showed that when guinea grass and mulato grass were planted 0.9 m from *Eucalyptus deglupta* they produced similar grass biomass, but root length density (RLD) at 0–0.4 m depth was up to three-fold higher under guinea grass than under mulato grass [47]. At 0–0.4 m depth but 0.45 m from *E. deglupta* trees, RLD of guinea grass was up to four-fold higher than that of mulato grass. Thus *E. deglupta* growth was significantly reduced by the presence of guinea grass, and to a lesser extent by mulato grass, compared with sole-crop *E. deglupta* [47]. The STG and STM systems in our study confirmed the competition from guinea grass and mulato grass strips with the trees. In these systems, the forage grasses were planted 1 m from son tra rows, resulting in lower growth and yield of son tra trees with guinea grass than with mulato grass or sole-tree cultivation, while the two grasses produced similar grass biomasses.

It is possible to reduce competition between trees and crops by pruning the trees [41], as proposed by farmer groups in our study. Another option may be to intercrop C3 crops instead of C4 crops, as previous studies have indicated that yields of C3 crops are less reduced in agroforestry systems [48,49]. In our study, farmer groups suggested improving the efficiency of the agroforestry systems by planting legume species such as soybean and groundnut instead of maize in LMG, and by planting upland rice or cucumber to replace forage grasses in STG and STM. Greater planting distance between trees, crops, and grass strips would reduce competition. Supplying more fertiliser to plants suffering from nutrient deficiency in competition zones was also suggested in the group interviews.

4.2. Productivity Benefits and Ecosystem Services of Agroforestry Systems

Evaluation of the agroforestry systems tested in this study indicated that they provided earlier products than sole-tree systems and more diverse products than sole-maize systems. They also gave higher total productivity for farmers than the sole-crop systems from the second year onwards. During the first three years, total productivity was mainly from forage grasses and maize, with the LMG, STG, and STM systems giving forage-grass biomass for farmers from the second year. The products became more diverse from year 4, when the trees started to bear fruit, with yield increasing in subsequent years.

We found that the LMG agroforestry system was more productive than sole maize and longan from year 2 onwards, as indicated by LER ranging from 1.1 to 1.9 (Figure 4a). In a previous study on agroforestry systems based on apple (*Malus domestica*), e.g., apple/maize, apple/peanut, and apple/millet, LER was found to be 1.2–1.3 after the apple trees started bearing fruit from year 6 [50]. In our study, LER of the STM system was >1.0 from year 3, when the son tra started bearing fruit. However, in the STG system LER was <1.0, which can probably be explained by competition, as previously shown [47]. Other studies on forage grasses have reported that guinea grass [33] produces more biomass than mulato grass [34] in sole-grass cultivation. It may therefore affect the LER of the STG agroforestry system. Management of tree and crop components of a fruit tree-based agroforestry system thus needs to change from the year of establishment to when trees are maturing and high-producing, so that farmers can overcome competition effects and optimise the efficiency of land use [50]. In this study, the farmer groups interviewed suggested that a resource allocation strategy could improve the productivity of different components of the agroforestry systems. In the first three years, when the trees had not yet produced fruit, their main priority was the annual crop and grasses, whereas they paid more attention to the trees when they started bearing fruit. The farmers needed the short-term income from annual crops to support the long-term benefits from the fruit trees.

Growing forage grasses can be an incentive to improve smallholder livestock production by improved the daily weight gain of cattle and reducing labour in finding feedstuffs [51]. In this study, farmer groups confirmed that growing forage grasses reduces the labour requirement for finding/collecting feedstuffs for livestock in areas where captive grazing is common practice. This may be particularly beneficial for rural women in the study region, as 60% of the workload in farming is carried out by women [11]. In areas where free grazing is common practice like in Tuan Giao district, farmers will be less motivated/perceive less benefit from growing forage grass. This can be a "temporary" constraint for agroforestry adoption in the areas because along with population growth and higher demand for agricultural lands, the area of free-gazing lands will become more limited in the future. Therefore, we strongly considered fodder grass as one of main components of the tested agroforestry systems. Moreover, agroforestry systems with grass have been identified as the most suitable practice for northwest Vietnam to reconcile livelihood and erosion control [9].

Sole-maize cultivation on steep slopes in the northwest region of Vietnam produced annual soil loss that reached up to 174 ton ha^{-1} [15]. However, growing forage grass along the contour lines can play a significant role in reducing soil loss, especially on the steep slopes of the study region [15]. All experiments in our study were conducted in lands with about 27% slope, and measurement of soil erosion was not part of our study. However, a study in the northwest region that measured and compared soil erosion rate in agroforestry and sole-crop plantations clearly showed that soil erosion was substantially reduced in agroforestry [52]. The study found that the erosion rate in longan–mango–maize–forage grass agroforestry was 43% lower than that measured in sole-maize cultivations, and the rate in son tra–coffee–forage grass was 34% lower than that measured in sole-coffee plantations. All agroforestry systems and sole-coffee plantations observed in the study were three years old. A higher reduction in the soil erosion rate can be expected in more mature agroforestry such as in our experiments that have larger tree-canopy cover.

Ecological benefits or ecosystem services noted by farmers in this study were the effect of grass strips in reducing soil erosion and maintaining soil moisture and fertility, but also in forming terraces on the steep slopes [52]. In steeply sloping areas, the terraces formed could significantly increase agricultural productivity and enhance water-use efficiency when combined with other agricultural techniques [53].

4.3. Economic Benefits of Agroforestry Systems and Possibilities for Improvement

The agroforestry systems evaluated here showed higher profitability than the sole-crop systems from year 2 onwards. However, the initial investment cost for agroforestry was high: 2122 USD ha^{-1} for LMG and 1772 USD ha^{-1} for STG and STM. Farmers in the region lack the financial resources to shift to new practices [10]. New practices thus need to be shown to be safe and ensure food security before smallholders risk changing from their current system. The main incentive for farmers to adopt agroforestry is increasing yield and stable prices for their products. When comparing production and profitability, a cycle of some years must be considered, because it takes longer to establish perennial trees than annual crops and the financial input is higher in agroforestry systems. Therefore, initial investment funding (possibly organised by farmers themselves), subsidies, or loans will be necessary to compensate for the high investment and maintenance costs in the first few years of agroforestry [16].

The farmer groups interviewed proposed some ways to make implementation of agroforestry more profitable. First, the establishment of agroforestry will require financial support or access to low-interest loans/credits. In addition, implementing agroforestry with fodder-grass strips would become more beneficial for local people if changing from free to captive grazing were promoted. To achieve both in the study region, local farmers can seek support from the Vietnamese government through e.g., the National Target Programme (NTP) on New Rural Development [54] or the NTP-Sustainable Poverty Reduction and 135 Programme [55]. In addition, they can seek loans (low interest rate) from formal actors such as the Vietnam Bank for Agriculture and Rural Development, the Vietnam Bank for Social Policy, and People's Credit Funds [56].

Second, the farmers interviewed suggested producing their own low-cost tree seedlings to reduce the investment cost. These could be grown in community nurseries, where all members share costs and provide inputs [57]. The project Agroforestry for Livelihoods of Smallholder Farmers in Northwest Vietnam (2012–2016), together with relevant stakeholders, has provided training for farmers on the establishment and management of smallholder and group nurseries, producing tree seedlings by seedling propagation, grafting, and marcotting techniques. The project has published various technical extension materials on producing different tree-species seedlings suitable for local conditions. These technical sources could be useful for local farmers producing their own tree seedlings [58].

Third, the interviewees believed that they could achieve stable production by forming growers' cooperatives and could improve product quality to meet market demand. The cooperatives could provide production services, including inputs for farm households, fertilisers, feed ingredients, plant protection chemicals, and vaccines for livestock. They could also mediate between entrepreneurs and farmers, representing and protecting the rights of farmer members in contracting to supply raw materials to processing enterprises and export agricultural products [11]. In rural development work, agricultural service cooperatives can make a significant contribution [11]. Recently, the Vietnamese government introduced a programme to develop 15,000 cooperatives and effective agricultural cooperative unions in rural areas, with the government providing institutions, mechanisms, and policies to support the programme [59]. This offers an opportunity for farmers in the region to develop cooperatives to ensure stable production of agricultural products.

5. Conclusions

Agroforestry systems based on fruit trees, grass, and crops had higher productivity, higher profitability, and earlier returns on investment than sole-crop fruit systems, but also higher initial investment costs. The agroforestry systems produced a diversity of products and provided ecosystem services such as erosion control and soil fertility improvement. However, challenges such as higher investment cost and an unstable market for agroforestry products make it uncertain whether agroforestry can be easily promoted in the area.

During development of the agroforestry systems, there were negative effects on growth and productivity of the different components, most likely due to competition. There was evidence of competition for nitrogen between tree, grass, and crop components at positions upslope and downslope of the grass strips. These competition effects need to be considered when designing agroforestry systems and formulating management regimes.

Future fruit tree-based agroforestry systems should apply adaptive management while the agroforestry system is maturing and consider measures such as widening the planting distance between trees, crops, and grass; supplying fertiliser to plant components suffering from nutrient deficiency; and pruning trees in competition zones. Introducing high-value crops or biological N-fixing species to reduce competition and support the growth of trees can also be considered in order to optimise the systems.

To enable uptake and expansion of agroforestry in northwest Vietnam, financial support to meet the higher investment costs for agroforestry and for better value chains with market stability are prerequisites for farmers. Local farmers can produce their own tree seedlings to reduce the investment cost for agroforestry in the region.

Supplementary Materials: The following are available online at http://www.mdpi.com/2073-445X/9/11/451/s1, Table S1: Fertilisation regime applied in the sole-crop and agroforestry systems in Van Chan and Tuan Giao; Table S2: Cost of cropping inputs and prices paid for products at the study sites, 2012–2018 (data provided by the provincial extension department); Table S3: Groups selected for farmer group discussions (FGD); Table S4: List of questions used in farmer group discussions.

Author Contributions: Conceptualization, V.H.D., I.Ö., and N.L.; methodology, V.H.D., I.Ö., and N.L.; formal analysis, V.H.D.; investigation, V.H.D., H.T.P., and V.T.N; writing—original draft preparation, V.H.D.; writing—review and editing, I.Ö., G.B., A.S.D., R.M., N.L., H.T.P., and V.T.N. All authors have read and agreed to the published version of the manuscript.

Funding: This study was part of the project "Developing and promoting market-based agroforestry and forest rehabilitation options for northwest Viet Nam" (short name AFLI) led by World Agroforestry (ICRAF) and funded by the Australian Centre for International Agricultural Research and the CGIAR Research Programme on Forests, Trees and Agroforestry, in cooperation with the Swedish University of Agricultural Sciences (SLU) within the Swedish Research Council Formas funded project 2019-00376.

Acknowledgments: We highly appreciate the support of researchers and technical staff from the AFLI project at World Agroforestry (ICRAF) and national partners in Vietnam. Our thanks to researchers at the Northern Mountainous Agriculture and Forestry Science Institute and Forest Science Centre of Northwest Vietnam for their assistance. We are grateful to Chu Van Tien, Tran Quang Khanh, and Vu Dinh Chi, who hosted experiments in Van Chan, to Giang Dung Vu, Mua A Do, and Vang A Sua, who hosted experiments in Tuan Giao, and to the local authorities and farmers in the study sites for their support during field research.

Conflicts of Interest: The authors declare no conflict of interest. The funders had no role in the design of the study; in the collection, analyses, or interpretation of data; in the writing of the manuscript, or in the decision to publish the results.

References

1. United Nations General Assembly. Transforming Our World: The 2030 Agenda for Sustainable Development, A/RES/70/1, New York. 2015. Available online: https://www.un.org/ga/search/view_doc.asp?symbol=A/RES/70/1=E (accessed on 10 June 2020).

2. United Nations. Sustainable Development Goals: 17 Goals to Transform Our World. 2015. Available online: https://www.un.org/sustainabledevelopment/sustainable-development-goals/ (accessed on 27 October 2020).

3. Catacutan, D.C.; Van Noordwijk, M.; Nguyen, T.H.; Öborn, I.; Mercado, A.R. *Agroforestry: Contribution to Food Security and Climate-Change Adaptation and Mitigation in Southeast Asia. White Paper*; World Agroforestry Centre (ICRAF): Bogor, Indonesia, 2017.

4. Van Noordwijk, M.; Bayala, J.; Hairiah, K.; Lusiana, B.; Muthuri, C.; Khasanah, N.; Mulia, R. Agroforestry Solutions for Buffering Climate Variability and Adapting to Change. In *Climate Change Impact and Adaptation in Agricultural Systems*; Fuhrer, J., Gregory, P.J., Eds.; CABI Publishing: Wallingford, UK, 2014; pp. 216–232.

5. Roshetko, J.M.; Mercado, A.R.; Martini, E.; Prameswari, D. *Agroforestry in the Uplands of Southeast Asia. Policy Brief no. 77. Agroforestry Options for ASEAN Series no. 5*; World Agroforestry (ICRAF) Southeast Asia Regional Program: Bogor, Indonesia, 2017; Available online: http://www.awg-sf.org/www2/wp-content/uploads/2017/12/ANNEX-10.-PB5-Upland-AF-in-SEA.pdf/ (accessed on 27 October 2020).

6. Tacio, H.D. Sloping Agricultural Land Technology (SALT): A Sustainable Agroforestry Scheme for the Uplands. *Agrofor. Syst.* **1993**, *22*, 145–152. [CrossRef]

7. Nair, P.R.; Nair, V.D.; Kumar, B.M.; Showalter, J.M. Carbon Sequestration in Agroforestry Systems. *Adv. Agron.* **2010**, *108*, 237–307. [CrossRef]

8. Schreinemachers, P.; Holger, L.; Fröhlich, H.L.; Clemens, G.; Stahr, K. From Challenges to Sustainable Solutions for Upland Agriculture in Southeast Asia. In *Sustainable Land Use and Rural Development in Southeast Asia: Innovations and Policies for Mountainous Areas*; Fröhlich, H.L., Schreinemachers, P., Stahr, K., Clemens, G., Eds.; Springer: New York, NY, USA, 2013; pp. 3–23.

9. Hoang, L.T.; Roshetko, J.M.; Huu, T.P.; Pagella, T.; Mai, P.N. Agroforestry—The Most Resilient Farming System for the Hilly Northwest of Vietnam. *Int. J. Agric. Syst.* **2017**, *5*, 1–23. [CrossRef]

10. Zimmer, H.; Le Thi, H.; Lo, D.; Baynes, J.; Nichols, J.D. Why Do Farmers Still Grow Corn on Steep Slopes in Northwest Vietnam? *Agrofor. Syst.* **2017**, *92*, 1721–1735. [CrossRef]

11. Staal, S.; Tran, D.T.; Nguyen, D.P.; Vu, D.H.; Nguyen, S.; Nguyen, T.T.L.; Nguyen, T.S.; Le, N.T.; Ngo, T.H.; Truong, Q.C.; et al. *A Situational Analysis of Agricultural Production and Marketing, and Natural Resources Management Systems in Northwest Vietnam*; International Livestock Research Institute for CGIAR Research Program: Kenya, Nairobi, 2014; Available online: http://humidtropics.cgiar.org/wp-content/uploads/downloads/2014/09/Situational-Analysis-NW-Vietnam-Report.pdf/ (accessed on 10 September 2020).

12. Wezel, A.; Luibrand, A.; Thanh, L.Q. Temporal Changes of Resource Use, Soil Fertility and Economic Situation in Upland Northwest Vietnam. *Land Degrad. Dev.* **2002**, *13*, 33–44. [CrossRef]

13. Clemens, G.; Fiedler, S.; Cong, N.D.; Van Dung, N.; Schuler, U.; Stahr, K. Soil Fertility Affected by Land Use History, Relief Position, and Parent Material Under a Tropical Climate in NW-Vietnam. *Catena* **2010**, *81*, 87–96. [CrossRef]

14. Schmitter, P.; Dercon, G.; Hilger, T.; Le Ha, T.T.; Thanh, N.H.; Lam, N.; Vien, T.D.; Cadisch, G. Sediment Induced Soil Spatial Variation in Paddy Fields of Northwest Vietnam. *Geoderma* **2010**, *155*, 298–307. [CrossRef]

15. Tuan, V.D.; Hilger, T.; Macdonald, L.; Clemens, G.; Shiraishi, E.; Vien, T.D.; Stahr, K.; Cadisch, G. Mitigation Potential of Soil Conservation in Maize Cropping on Steep Slopes. *Field Crop. Res.* **2014**, *156*, 91–102. [CrossRef]

16. Do, H.; Luedeling, E.; Whitney, C. Decision Analysis of Agroforestry Options Reveals Adoption Risks for Resource-Poor Farmers. *Agron. Sustain. Dev.* **2020**, *40*, 1–12. [CrossRef]

17. General Statistics Office of Vietnam. *Statistical Yearbook of Dien Bien, Yen Bai and Son La province*; General Statistics Office of Vietnam: Hanoi, Vietnam, 2018.

18. Hoang, T.L.; Mamo, A.E. *Son tra, the H'mong Apple*; World Agroforestry (ICRAF): Ha Noi, Vietnam, 2015; Available online: http://blog.worldagroforestry.org/index.php/2015/07/01/son-tra-the-hmong-apple/ (accessed on 27 October 2020).

19. Ministry of Agriculture and Rural Development. *Decision No. 4961/QD-BNN-TCLN. Promulgating a List of Main Tree Species for Planting Production Forests and a List of Main Tree Species for Afforestation According to Forest Ecological Regions*; Ministry of Agriculture and Rural Development: Hanoi, Vietnam, 2014; Available online: https://thuvienphapluat.vn/van-ban/tai-nguyen-moi-truong/Quyet-dinh-4961-QD-BNN-TCLN-2014-Danh-muc-cay-chu-luc-trong-rung-san-xuat-vung-sinh-thai-lam-nghiep-259936.aspx/ (accessed on 30 September 2020).

20. Do, T.; Mulia, R. Constraints to Smallholder Tree Planting in the Northern Mountainous Regions of Vietam: A Need to Extend Technical Knowledge and Skills. *Int. For. Rev.* **2018**, *20*, 43–57. [CrossRef]

21. Nguyen, M.P.; Mulia, R.; Nguyen, Q.T. *Fruit Tree Agroforestry in Northwest Vietnam: Sample Cases of Current Practices and Benefits (A Report in Vietnamese Language)*; World Agroforestry (ICRAF): Ha Noi, Vietnam, 2020.

22. TCVN 8941:2011. Soil Quality—Determination of Total Organic Carbon–Walkley Black Method. 2011. Available online: https://vanbanphapluat.co/tcvn-8941-2011-chat-luong-dat-cac-bon-huu-co-tong-so-phuong-phap-walkley-black#van-ban-goc/ (accessed on 11 September 2020).

23. TCVN 6498:1999. Soil Quality—Determination of Total Nitrogen–Modified Kjeldahl Method. 1999. Available online: https://vanbanphapluat.co/tcvn-6498-1999-chat-luong-dat-xac-dinh-nito-tong-phuong-phap-kendan#van-ban-goc/ (accessed on 11 September 2020).

24. TCVN 8940:2011. Soil Quality—Determination of Total Phosphorus–Colorimetry Method. 2011. Available online: https://vanbanphapluat.co/tcvn-8940-2011-chat-luong-dat-xac-dinh-phospho-tong-so-phuong-phap-so-mau#van-ban-goc/ (accessed on 11 September 2020).

25. TCVN 8660:2011. Soil Quality—Method for Determination of Total Potassium. 2011. Available online: https://vanbanphapluat.co/tcvn-8660-2011-chat-luong-dat-phuong-phap-xac-dinh-kali-tong-so#van-ban-goc/ (accessed on 11 September 2020).

26. TCVN 8942:2011. Soil Quality—Determination of Available Phosphorus—Bray and Kurtz (Bray II) Method. 2011. Available online: https://vanbanphapluat.co/tcvn-8942-2011-chat-luong-dat-xac-dinh-phospho-de-tieu-phuong-phap-bray-kurtz#van-ban-goc/ (accessed on 7 July 2020).

27. Minolta. *SPAD-502 Owner's Manual*; Industrial Meter Division, Minolta Corporation: Ramsey, NJ, USA, 1989.

28. Wolfe, D.W.; Henderson, D.W.; Hsiao, T.C.; Alvino, A. Interactive Water and Nitrogen Effects on Senescence of Maize. II. Photosynthetic Decline and Longevity of Individual Leaves. *Agron. J.* **1998**, *80*, 865–870. [CrossRef]

29. Argenta, G.; Da Silva, P.R.F.; Sangoi, L. Leaf Relative Chlorophyll Content as an Indicator Parameter to Predict Nitrogen Fertilization in Maize. *Ciência Rural* **2004**, *34*, 1379–1387. [CrossRef]

30. Viana, M.C.M.; Da Silva, I.P.; Freire, F.M.; Ferreira, M.M.; Da Costa, É.L.; Mascarenhas, M.H.T.; Teixeira, M.F.F. Production and Nutrition of Irrigated Tanzania Guinea Grass in Response to Nitrogen Fertilization. *Rev. Bras. Zootec.* **2014**, *43*, 238–243. [CrossRef]

31. Sritontip, C.; Khaosumain, Y.; Changjeraja, S. Different Nitrogen Concentrations Affecting Chlorophyll and Dry Matter Distribution in Sand-Cultured Longan Trees. *Acta Hortic.* **2014**, 227–233. [CrossRef]

32. Mead, R.; Willey, R.W. The Concept of a 'Land Equivalent Ratio' and Advantages in Yields from Intercropping. *Exp. Agric.* **1980**, *16*, 217–228. [CrossRef]

33. Cook, B.G.; Pengelly, B.C.; Brown, S.D.; Donnelly, J.L.; Eagles, D.A.; Franco, M.A.; Hanson, J.; Mullen, B.F.; Partridge, I.J.; Peters, M.; et al. *Tropical Forages: An Interactive Selection Tool*; [CD-ROM]; CSIRO, DPI&F (Qld), CIAT, ILRI: Brisbane, Australia, 2005.

34. Argel, P.J.; Miles, J.W.; Guiot García, J.D.; Cuadrado Capella, H.; Lascano, C.E. *Cultivar Mulato II (Brachiaria hybrid CIAT 36087): A High-Quality Forage Grass, Resistant to Spittlebugs and Adapted to Well-Drained, Acid Tropical Soils*; International Center for Tropical Agriculture (CIAT): Cali, Colombia, 2008; pp. 1–3. Available online: http://ciat-library.ciat.cgiar.org/Articulos_CIAT/mulato_ii_ingles.pdf/ (accessed on 25 March 2020).

35. Aganga, A.A.; Tshwenyane, S. Potentials of Guinea Grass (*Panicum maximum*) as Forage Crop in Livestock Production. *Pak. J. Nutr.* **2004**, *3*, 1–4.

36. Bacorro, T.J.; Reyes, P.M.; Loresco, M.M.; Angeles, A.A. Herbage Dry Matter Yield, Nutrient Composition and In Vitro Gas Production of Mulato II Mombasa Grasses at 30-and 45-Day Cutting Intervals. *Philipp. J. Vet. Anim. Sci.* **2018**, *44*, 86–89.

37. Vietnamese Government. *Decision No. 59/2015/QD-TTg. Promulgating Multiple Dimensional Poverty Levels Applicable during 2016-2020*; Vietnamese Government: Hanoi, Vietnam, 2015. Available online: https://thuvienphapluat.vn/van-ban/Van-hoa-Xa-hoi/Decision-No-59-2015-QD-TTg-promulgating-multiple-dimensional-poverty-levels-applicable-301414.aspx/ (accessed on 9 September 2020).

38. Malezieux, E.; Crozat, Y.; Dupraz, C.; Laurans, M.; Makowski, D.; Lafontaine, H.O.; Rapidel, B.; De Tourdonnet, S.; Valantin-Morison, M. Mixing Plant Species in Cropping Systems: Concepts, Tools and Models. A Review. *Agron. Sustain. Dev.* **2009**, *29*, 43–62. [CrossRef]

39. Okorio, J.; Byenkya, S.; Wajja, N.; Peden, D. Comparative Performance of Seventeen Upperstorey Tree Species Associated With Crops in the Highlands of Uganda. *Agrofor. Syst.* **1994**, *26*, 185–203. [CrossRef]

40. Bertomeu, M. Growth and Yield of Maize and Timber Trees in Smallholder Agroforestry Systems in Claveria, Northern Mindanao, Philippines. *Agrofor. Syst.* **2011**, *84*, 73–87. [CrossRef]

41. Nyaga, J.; Muthuri, C.W.; Barrios, E.; Öborn, I.; Sinclair, F.L. Enhancing Maize Productivity in Agroforestry Systems Through Managing Competition: Lessons From Smallholders' Farms, Rift Valley, Kenya. *Agrofor. Syst.* **2017**, *93*, 715–730. [CrossRef]

42. Van Noordwijk, M.; Lawson, G.; Hairiah, K.; Wilson, J. Root Distribution of Trees and Crops: Competition and/or Complementarity. In *Tree-Crop Interactions: Agroforestry in a Changing Climate*, 2nd ed.; Ong, C.K., Black, C.R., Wilson., J., Eds.; CAB International: Wellington, UK, 2015; pp. 221–257.

43. Livesley, S.J.; Gregory, P.; Buresh, R. Competition in Tree Row Agroforestry Systems. 1. Distribution and Dynamics of Fine Root Length and Biomass. *Plant Soil* **2000**, *227*, 149–161. [CrossRef]

44. Chirko, C.P.; Gold, M.A.; Nguyen, P.; Jiang, J. Influence of Direction and Distance From Trees on Wheat Yield and Photosynthetic Photon Flux Density (Qp) in a Paulownia and Wheat Intercropping System. *For. Ecol. Manag.* **1996**, *83*, 171–180. [CrossRef]

45. Tuan, V.D.; Hilger, T.; Vien, T.D.; Cadisch, G. Nitrogen Recovery and Downslope Translocation in Maize Hillside Cropping as Affected by Soil Conservation. *Nutr. Cycl. Agroecosystems* **2014**, *101*, 17–36. [CrossRef]

46. Hung, V.; Hung, N. Effect of Pruning and Flower and Fruit Thinning on Growth, Yield and Quality of 'PHM 99.1.1' Late Longan in Hung Yen Province. In *VI International Symposium on Lychee, Longan and Other Sapindaceae Fruits*; Van Hau, T., Mitra, S.K., Drang Khanh, T., Eds.; Acta Hortic: Brussels, Belgium, 2019; pp. 143–148.

47. Schaller, M.; Schroth, G.; Beer, J.; Jiménez, F.; Jiménez, F. Root Interactions Between Young Eucalyptus Deglupta Trees and Competitive Grass Species in Contour Strips. *For. Ecol. Manag.* **2003**, *179*, 429–440. [CrossRef]

48. Rao, M.R.; Nair, P.K.R.; Ong, C.K. Biophysical Interactions in Tropical Agroforestry Systems. *Agrofor. Syst.* **1997**, *38*, 3–50. [CrossRef]

49. Thevathasan, N.; Gordon, A. Ecology of Tree Intercropping Systems in the North Temperate Region: Experiences from Southern Ontario, Canada. *Agrofor. Syst.* **2004**, *61*, 257–268. [CrossRef]
50. Xu, H.; Bi, H.; Gao, L.; Yun, L. Alley Cropping Increases Land Use Efficiency and Economic Profitability Across the Combination Cultivation Period. *Agronomy* **2019**, *9*, 34. [CrossRef]
51. Bush, R.D.; Young, J.R.; Suon, S.; Ngim, M.S.; Windsor, P.A. Forage Growing as an Incentive to Improve Smallholder Beef Production in Cambodia. *Anim. Prod. Sci.* **2014**, *54*, 1620–1624. [CrossRef]
52. Birgitta, S.; Hanna, T. Impact of Agroforestry on Soil Loss Mitigation in the Sloping Land of Northwest Vietnam. Master's Thesis, Swedish University of Agricultural Sciences, Uppsala, Sweeden, May 2020.
53. Chai, Q.; Gan, Y.; Turner, N.C.; Zhang, R.-Z.; Yang, C.; Niu, Y.; Siddique, K.H. Water-Saving Innovations in Chinese Agriculture. *Adv. Agron.* **2014**, *126*, 149–201. [CrossRef]
54. Vietnamese Government. *Decision No. 800/2010/QD-TTg. Approving the National Target. Program. on Building a New Countryside During 2010-2020*; Vietnamese Government: Hanoi, Vietnam, 2010. Available online: https://thukyluat.vn/vb/decision-no-800-qd-ttg-approving-the-national-target-program-on-building-1ad9a.html?hl=en#VanBanTA/ (accessed on 9 September 2020).
55. Vietnamese Government. *Decision No. 551/2013/QD-TTg. Approved the 135 Program. on Supporting Investment in Infrastructure, Production Development for Extremely Difficult Communes, Border Communes, Safety Zone Communes*; Vietnamese Government: Hanoi, Vietnam, 2013. Available online: http://vanban.chinhphu.vn/portal/page/portal/chinhphu/hethongvanban?class_id=2&_page=1&mode=detail&document_id=166543/ (accessed on 9 September 2020).
56. JICA. *Data Collection Survey on the Needs for Agriculture Productivity Improvement in Northern Six Provinces*; Japan International Cooperation Agency, Vietnam Ministry of Agriculture and Rural Development: Hanoi, Vietnam, 2016. Available online: https://openjicareport.jica.go.jp/pdf/12261343.pdf/ (accessed on 9 September 2020).
57. FAO. *Sustainable Forest Management (SFM) Toolbox*; The Food and Agriculture Organisation of the United Nations: Rome, Italy, 2018; Available online: http://www.fao.org/sustainable-forest-management/toolbox/modules/agroforestry/basic-knowledge/en/ (accessed on 9 September 2020).
58. La, N.; Catacutan, D.C.; Nguyen, M.P.; Do, V.H. *Agroforestry for Livelihoods of Smallholder Farmers in Northwest. Vietnam—Final Report, Canberra ACT 2601*; Australian Government, Australian Center for International Agricultural Research: Canberra, Australia, 2019; pp. 10–34. Available online: https://aciar.gov.au/publication/technical-publications/agroforestry-livelihoods-smallholder-farmers-northwest-Vietnam-final-report/ (accessed on 10 September 2020).
59. Vietnamese Government. *Decision No. 461/2018/QD-TTg. Approving the Scheme on the Development of 15,000 Cooperatives and Unions of Agricultural Cooperatives With Effective Operation up to 2020*; Vietnamese Government: Hanoi, Vietnam, 2018. Available online: https://thuvienphapluat.vn/van-ban/Doanh-nghiep/Quyet-dinh-461-QD-TTg-2018-De-an-phat-trien-15000-hop-tac-xa-lien-hiep-hop-tac-xa-nong-nghiep-380877.aspx/ (accessed on 9 September 2020).

 *land*

Article

Local Knowledge about Ecosystem Services Provided by Trees in Coffee Agroforestry Practices in Northwest Vietnam

Mai Phuong Nguyen [1,*], Philippe Vaast [1,2], Tim Pagella [3] and Fergus Sinclair [3,4]

[1]   World Agroforestry, Vietnam Office, Hanoi 100000, Vietnam; Philippe.vaast@cirad.fr
[2]   UMR Eco&Sols, CIRAD, Université Montpellier, 34090 Montpellier, France
[3]   School of Natural Sciences, Bangor University, Bangor, Gwynedd LL57 2DG, UK;
      t.pagella@bangor.ac.uk (T.P.); F.Sinclair@cgiar.org (F.S.)
[4]   World Agroforestry, Headquarter Office, P.O. Box 30677, Nairobi 00100, Kenya
*   Correspondence: n.maiphuong@cgiar.org

Received: 30 October 2020; Accepted: 30 November 2020; Published: 2 December 2020

Abstract: In recent decades in northwest Vietnam, Arabica coffee has been grown on sloping land in intensive, full sun monocultures that are not sustainable in the long term and have negative environmental impacts. There is an urgent need to reverse this negative trend by promoting good agricultural practices, including agroforestry, to prevent further deforestation and soil erosion on slopes. A survey of 124 farmers from three indigenous groups was conducted in northwest Vietnam to document coffee agroforestry practices and the ecosystem services associated with different tree species used in them. Trees were ranked according to the main ecosystem services and disservices considered to be locally relevant by rural communities. Our results show that tree species richness in agroforestry plots was much higher for coffee compared to non-coffee plots, including those with annual crops and tree plantations. Most farmers were aware of the benefits of trees for soil improvement, shelter (from wind and frost), and the provision of shade and mulch. In contrast, farmers had limited knowledge of the impact of trees on coffee quality and other interactions amongst trees and coffee. Farmers ranked the leguminous tree species *Leucaena leucocephala* as the best for incorporating in coffee plots because of the services it provides to coffee. Nonetheless, the farmers' selection of tree species to combine with coffee was highly influenced by economic benefits provided, especially by intercropped fruit trees, which was influenced by market access, determined by the proximity of farms to a main road. The findings from this research will help local extension institutions and farmers select appropriate tree species that suit the local context and that match household needs and constraints, thereby facilitating the transition to a more sustainable and climate-smart coffee production practice.

Keywords: agroforestry coffee; ecosystem services; shade tree species; pairwise ranking; Vietnam

---

1. Introduction

Coffee is the one of the most important commodities globally and contributes to the income of millions of smallholder farmers. Approximately twenty-five million farmers grow coffee in sixty developing countries [1]. There are approximately 10.5 million ha of coffee around the world consisting of both Arabica (*Coffea arabica*) and Robusta (*Coffea canephora*) varieties [2]. Vietnam is the second largest world producer after Brazil [3]. It is also the largest producer of Robusta coffee, accounting for 14.5% of global production and a share of about 40% of global trade [4].

Coffee was first introduced to Vietnam by the French in 1857. Between 1975 and 2010, the area of coffee planted in Vietnam increased from 134,000 ha to 513,000 ha. By 2016, the total area of coffee had reached 650,000 ha [5]. Vietnam exports over 90% of its total production but the value remains low mainly because low-quality beans rather than processed coffee are exported at a low price. The annual export volume was approximately 1.8 million tons with a value of USD 3.5 billion in 2018 [6].

Whilst Robusta is the most popular coffee variety grown in Vietnam, it is mainly Arabica that is cultivated in mountainous areas, generally at elevations above 600 m above sea level (masl). Arabica production accounts for approximately 7% of the total coffee area of Vietnam [7], including approximately 3995 ha in Dien Bien [8] and 17,128 ha in Son La [9] provinces in 2018.

Smallholder farmers have converted large areas of annual crops to coffee in northwest Vietnam, which has transformed their livelihoods from subsistence to commercial commodity dependent [10]. Most of the Arabica plantations in the northwest are full sun monoculture on sloping land, which is not sustainable in the long term because of soil degradation. The long-term sustainability of unshaded coffee is likely to be limited given climate change, resulting in higher temperatures and more variable weather patterns. Increasing temperatures, uneven precipitation and increased frequency and severity of extreme weather events such as storms, floods or frost may have negative impacts on Arabica coffee production.

Coffee was originally cultivated under moderate to heavy shade in Ethiopia, which was gradually changed over much of the area, grown under light or no shade to increase coffee yield, at least in the short term. However, reducing shade in coffee plots makes them more vulnerable to water run-off [11] and soil erosion [12] biodiversity loss [13] and climate change [14]. Shade coffee, which is a form of agroforestry where coffee is cultivated under a tree canopy [15], is generally considered to be more environmentally friendly. Studies have shown increased biodiversity, higher levels of natural pollination [16], greater erosion control, and higher carbon sequestration [13,17–19]. Boundary tree planting in coffee farms has potential for climate mitigation and adaptation [14,20].

Expanding the area of shaded coffee in Northwest Vietnam offers a potential mechanism to provide more sustainable cultivation, both economically and environmentally, when compared with either unshaded coffee or other monoculture crops such as maize. There has been an increased interest, globally, in shaded coffee given the sensitivity of coffee, particularly Arabica, to climate change [21,22]. Coffee cultivation, with between 20 to 40% shade and appropriate species and management, has been found to have a regulated microclimate which, in turn, leads to high coffee yield and quality [23]. However, poorly designed tree cover in coffee can lead to low coffee productivity [24]. Shade systems provide additional benefits beyond the direct impacts on coffee production in the shape of a more balanced supply of ecosystem services [25,26]. Shade trees in coffee can increase the sustainability of coffee production as well as the economic resilience of households to agricultural price volatility through production and revenue diversification. Furthermore, they facilitate the adaptation of rural communities to climate change via the adoption of more 'climate-smart' farming practices.

There have been a number of studies looking at farmers' knowledge about coffee agroforestry [27–29] however, to date, there has been no study assessing farmers' attitudes and knowledge associated with coffee agroforestry in northwest Vietnam. Understanding farmers' knowledge of the benefits that trees potentially provide alongside any trade-offs or disservices is critical for developing interventions that enable more resilient landscapes [30], whilst reconciling production and conservation objectives. This is particularly important in the context of northwest Vietnam given the broad spectrum of ethnicities at play and the way in which factors such as gender might influence tree species selection.

To address this knowledge gap, the principal aim of this study was to explore farmers' perceptions of the range of benefits associated with shade trees in coffee systems and then to use this information to customize a decision-support tool (the 'Shade Tree Advice tool' developed by researchers from World Agroforestry, French Agricultural Research Centre for International Development (CIRAD) and International Institute of Tropical Agriculture (IITA)) to local conditions in Vietnam. The original tool was developed using data from Uganda, Ghana and China; http://www.shadetreeadvice.org [31,32]. The 'Shade Tree Advice' tool is aimed primarily at extension services, members of farmers' cooperatives and non-governmental organizations working in northwest Vietnam. The tool supports farmers in selecting the most appropriate tree species that are both adapted to local ecological conditions and match household needs and constraints.

2. Materials and Methods

2.1. Study Sites

Son La and Dien Bien are the only two provinces in northwest Vietnam where coffee is planted, and Son La is the second largest Arabica coffee producer in Vietnam. This study was conducted in seven communes of Son La and Dien Bien (Figure 1). The criteria for commune selection were coffee planting area, ethnicity, proximity to main road to compare the different characteristics of coffee agroforestry practices, and willingness to share perceptions about tree services and tree species ranking. Information on these criteria were provided by district extension workers. The communes were selected from the most well known coffee planting areas in two provinces. Coffee agroforestry households were suggested by commune extension workers. As Thai, H'mong and Kinh are the major ethnic groups in the region, households from these three ethnicities that were actively involved in coffee agroforestry were selected for the surveys with specific care taken to ensure gender balance amongst interviewees. The Kinh are the largest ethnic group in Vietnam but account for less than 30% of the population in the northwest. Kinh people generally live at lower altitudes (below 600 m) while Thai people are generally found at altitudes of 500 to 800 m and H'mong people generally live in areas above 800 m. The proximity to main roads were categorized into three levels: 0–2, 2–5 km or further than 5 km. As farmers often sell their fruit products, such as mango or plum, along the main road, this represents an indicator of their proximity to markets.

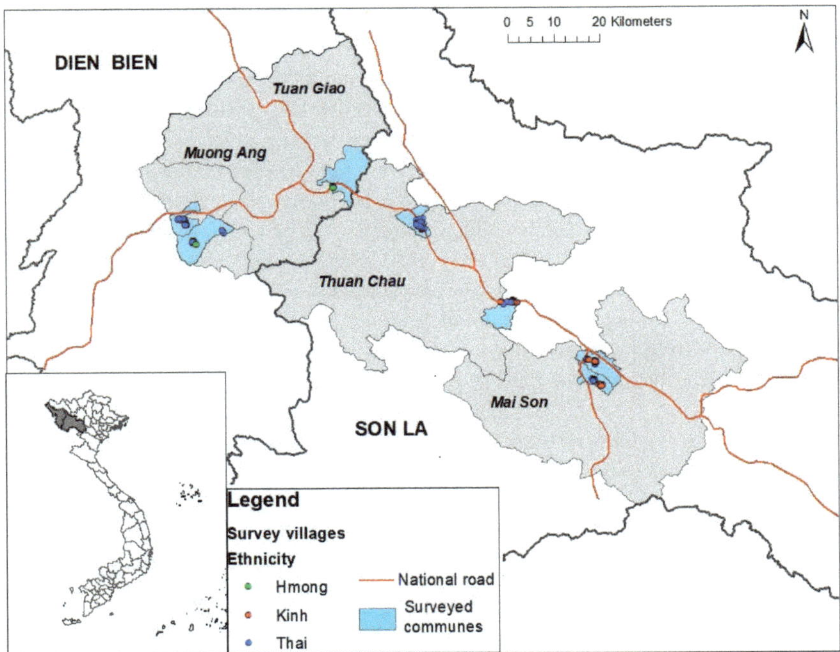

Figure 1. Map of the study sites and the communes surveyed in the Son La and Dien Bien provinces of northwest Vietnam.

The elevation of the study sites ranges from 300 to 2000 masl on 5 to 50% slopes. Annual temperature ranges from 21 to 24 °C, and annual precipitation ranges from 1500 to 2500 mm. Rainy season is from

April to September. The main soil type is Ferrasols, and average soil layer thickness ranges from 50 cm to 1 m. These natural conditions are favorable for Arabica coffee. The upper limit frost in winter is the most serious constraint for Arabica coffee cultivation at high elevations. In addition to coffee, the main agricultural land uses are annual crops (upland rice (*Oryza sativa*), maize (*Zea mays*), fruit tree plantations such as longan (*Dimocarpus longan*), plum (*Prunus salicina*), and mango (*Mangifera indica*). There are also limited areas of planted forest and secondary natural forest.

2.2. Data Collection

Data were collected through two rounds of surveys between March to May 2018, following a well established tree attribute ranking methodology [31,33]. The first survey consisted of a household interview and tree inventory in coffee agroforestry farms in March 2018. A total of 124 households were surveyed; consisting of 16 H'mong farmers, 25 Kinh farmers and 83 Thai farmers (with those proportions largely representative of the ethnic distributions at the study sites). Sicty-eight men and 56 women participated in the interview process. A sub-section of 50 farms were inventoried to provide a baseline of potential shade tree species found in existing coffee agroforestry plots. The survey questions captured social and economic characteristics, coffee plot descriptions, including information about trees in agroforestry plots, including those with coffee as the major crop and non-coffee plots (where annual crops or tree plantations were the main productive component), the benefits that trees provide to coffee, and coffee management. Additional information on farmer perceptions of ecosystem services and disservices was captured via focus group discussions. These were also used to explore how perceptions might vary in relation to farmers' age, ethnicity and gender.

From the first survey, all the tree species on coffee farms were documented and the most dominant 25 species were selected for the second survey in May 2018. Based on the initial discussions with farmers, the perceived benefits of shade trees were grouped into 11 broad topic areas (which effectively acted as proxies for ecosystem services). These topics were tree effects on: coffee production; soil moisture, soil fertility, soil erosion, biodiversity, micro-climate regulation, wind control, frost control, effects of and on shade provision, the value of mulch provision and effects on the use of fertilizer (which relates to soil fertility and greenhouse gas emission). The same farmers involved in the previous survey were invited for the second set of interviews. There were 118 farmers from three ethnic groups Kinh, Thai and H'mong, of which were 64 men and 54 women, that were involved in the second survey. The second survey interview focused on ranking exercises [33]. Interviewees were asked to select up to 10 preferred species to grow with coffee. They must have been managing, or at least been familiar with these selected species. Pictures of both tree species and tree services were printed on cards with which farmers then ranked the performance of their selected trees for each topic area (stating whether their effect was positive or negative). Results of from individual farmer ranking of trees were recorded in the ranking sheet for later analysis. During this exercise, farmers also provided the explanation for their rankings.

A final workshop was conducted in July 2018 in Mai Son district (Son La) and Muong Ang district (Dien Bien) with 25 farmers to feedback the analysis results.

2.3. Data Analysis

The results of tree species inventory were analyzed using the principles of BiodiversityR packages in R studio and combined with Excel for visualization. Species accumulation was presented with the trendline of a data plot using logarithm function and first order Jackknife asymptote [34]. Several species accumulation analyses were made for three ethnic groups, coffee farms and non-coffee farms (for example, maize, rice, orchard plantation) to compare the biodiversity among Thai, H'mong, Kinh farms, coffee and non-coffee agroforestry plots. Coffee monoculture plots were excluded from analyses, as there were no tree species in those plots.

Following the ranking in the second survey, an analysis was undertaken using the Bradley–Terry model in R studio [35]. Ranking was converted into pairwise comparison as input data for the model.

The model was run for eleven tree services for three ethnic groups, three distances to road (near, medium and far), and gender (male and female). Species, ranked less than 10 times, were excluded from the results. The scores reflect the comparison of performance rather than the absolute values. Scores were normalized to values between 0 and 1, where 1 is the maximum value for the best tree species. Ranking data was uploaded to the online database at www.shadetreeadvice.org. This study followed the same methods as those used for other databases from Ghana [31] and China [32] to ensure that the results were comparable.

3. Results

3.1. Farm Characteristics of Coffee Agroforestry Systems

All the interviewed farmers were between 25 and 50 years old with a mean age of 39 years. Among the three ethnic groups, Kinh people had the highest mean annual income while Thai farmers had the next highest and H'mong the lowest at approximately USD 1762 per year. Mean farm size was about 1.5 ha; Kinh farmers have smaller cultivation areas compared to those of the H'mong group. Coffee was cultivated on about two thirds of their lands (approximately 1 ha on average) and the remaining areas were dedicated to annual crops for family and livestock consumption. Almost all their coffee areas were 'shaded' coffee, i.e., coffee intercropped with trees such as fruit trees, timber trees, nut trees or *L. leucocephala* (Table 1). These were considered coffee agroforestry plots. Fruits were often sold at home or on the side of the main road. Timber and coffee were collected by middlemen at home or sometimes at the local market. All the Kinh villages were near the main road, at a distance of 0–2 km. Thai farmers typically lived at distances of 0–5 km from a main road. H'mong farmers mostly lived far from main roads (5–7 km) except for the H'mong group in Toa Tinh commune, Dien Bien province, because of the newly built highway near their village.

Table 1. Social and plot characteristics of the coffee farms; data are the mean values with standard errors.

Social and Farm Characteristics	Total Population	Kinh	Thai	H'mong
Total number of respondents	124	25	83	16
Mean agricultural income (USD/year/household)	4571 ± 390	8762 ± 1286	3810 ± 324	1762 ± 276
Mean farm area (ha)	1.50 ± 0.09	1.60 ± 0.26	1.30 ± 0.10	1.90 ± 0.30
Mean number of coffee plots	2-3	2-3	2-3	2-3
Mean area of coffee plots (ha)	1.00 ± 0.09	1.10 ± 0.16	1.00 ± 0.08	1.20 ± 0.20
Mean area of agroforestry coffee plots (ha)	0.80 ± 0.06	1.00 ± 0.12	0.80 ± 0.07	1.00 ± 0.21
Mean distance to main road		0–2 km	0–5 km	0–3 km (Tuan Giao district) or 5–7 km (Muong Ang district)

3.2. Tree Species Inventory in Coffee Agroforestry Systems

Forty-seven tree species were identified during the farm inventories. These included fruit trees, timber trees, nut trees, and shade trees. All inventoried farms were located from 580 to 1000 masl. Two native timber tree species could not be botanically identified and are only referred to by their local name. Nearly half of the tree species were native (Table 2). Frequencies of tree species were calculated by dividing the number of farms a tree species was observed on by the total number of farms inventoried and expressed as a percentage. Many species were quite rare (37% were encountered on only 1% of the coffee farms while a further 37% were found on 2–10% of coffee farms). The most abundant species, found on 13–46% of farms, constituted 26% of the total species encountered. The majority of the most abundant species were commercial fruit and nut trees except for the leguminous *L. leucocephala* and *Melia azedarach* (a valuable timber tree). There were 25 exotic species out of the 48 species recorded. Longan and mango are native species, however, only grafted hybrid varieties are being planted on farms.

Table 2. Characterization of the tree species mentioned by farmers as growing on their coffee farms and observed during the farm inventory in the provinces of Son La and Dien Bien, northwest Vietnam.

No.	English Name	Scientific Name	Dominant Function	Exotic/Native	Frequencies of Species (%)
1	Longan	*Dimocarpus longan*	Commercial fruit	Exotic	46
2	Plum	*Prunus salicina*	Commercial fruit	Native	43
3	Mango	*Mangifera indica*	Commercial fruit	Exotic	42
4	Leucaena	*Leucaena leucocephala*	Shade	Exotic	23
5	Jackfruit	*Artocarpus heterophyllus*	Family consumption/ Commercial fruit	Native	22
6	Pomelo	*Citrus grandis*	Commercial fruit	Exotic	22
7	Melia	*Melia azedarach*	Commercial timber	Exotic	20
8	Peach	*Prunus persica*	Commercial fruit	Native	15
9	Macadamia	*Macadamia* spp.	Commercial nut	Exotic	15
10	Avocado	*Persea americana*	Commercial fruit	Exotic	13
11	Docynia indica	*Docynia indica*	Commercial fruit	Native	9
12	Guava	*Psidium guajava*	Commercial fruit	Native	8
13	Orange	*Citrus sinensis*	Commercial fruit	Exotic	6
14	Lime	*Citrus aurantifolia*	Family consumption	Native	6
15	Litchi	*Litchi chinensis*	Commercial fruit	Exotic	6
16	Eucalyptus	*Eucalyptus* spp.	Commercial timber	Exotic	6
17	Vernicia montana	*Vernicia montana*	Commercial timber	Native	6
18	Apricot	*Prunus mume*	Commercial fruit	Native	5
19	Tamarind	*Tamarindus indica*	Family consumption	Native	4
20	Chukrasia	*Chukrasia tabularis*	Commercial timber	Exotic	4
21	Dalbergia	*Dalbergia tonkinensis*	Commercial timber	Exotic	4
23	Manglietia	*Manglietia conifera*	Commercial timber	Exotic	3
22	Michelia	*Michelia mediocris*	Commercial timber	Exotic	3
24	Local pear	*Pyrus granulosa*	Family consumption	Native	2
25	Oroxylum indicum	*Oroxylum indicum*	Timber/flowers	Native	2
26	Lucuma	*Pouteria lucuma*	Family consumption	Native	1
27	Fig	*Ficus auriculata*	Timber/fruit	Native	1
28	Baccaurea sapida	*Baccaurea sapida*	Timber/fruit	Native	1
29	Bischofia javanica	*Bischofia javanica*	Timber	Native	1
30	Papaya	*Carica papaya*	Family consumption	Native	1
31	Star apple	*Chrysophyllum cainito*	Family consumption	Exotic	1
32	Star fruit	*Averrhoa carambola L.*	Family consumption	Exotic	1
33	Pomegranate	*Punica granatum*	Family consumption	Exotic	1
34	Indian Jujube	*Indian Jujube*	Family consumption	Native	1
35	Pine	*Pinus latteri*	Commercial timber/resin	Exotic	1
36	Styphnolobium japonicum	*Styphnolobium japonicum*	Timber	Native	1
37	Teak	*Tectona grandis*	Timber	Exotic	1
38	Alstonia scholaris	*Alstonia scholaris*	Timber	Exotic	1
39	Syzygium nervosum	*Syzygium nervosum*	Timber/leaf/flower	Exotic	1
40	Khaya senegalensis	*Khaya senegalensis*	Timber	Exotic	1
42	Dillenia Indica	*Dillenia Indica*	Timber	Exotic	1
43	Tea	*Camellia sinensis*	Leaf	Exotic	1

Table 2. Cont.

No.	English Name	Scientific Name	Dominant Function	Exotic/Native	Frequencies of Species (%)
44	Zanthoxylum rhetsa	*Zanthoxylum rhetsa*	Seed/timber	Native	1
45	Agarwood	*Aquilaria malaccensis*	Timber/resin	Exotic	1
46	Schima wallichii	*Schima wallichii*	Timber	Native	1
47	Local timber tree (mý) *		Timber	Native	1
48	Local timber tree (thro) *		Timber	Native	1

* Scientific names of those species that have not been identified.

Trees were planted in different settings such as in rows or scattered within the coffee plots or along plot boundaries. Coffee–tree intercropping practices were characterized as four common types including: coffee–fruit trees, coffee–timber trees, coffee–nut trees, and coffee–*L. leucocephala* that could be combined on one farm. In all instances, trees were planted at regular spacing in rows if there were one to two species being intercropped with coffee. Trees were planted at 5 m × 5 m for plum or mango, macadamia, *L. leucocephala*, and up to 10 m × 10 m for avocado. If more than three species were intercropped with coffee, farmers often planted them scattered between coffee rows or along the boundaries of the coffee plots. Most of the coffee-multiple fruit tree systems were found in home gardens and coffee—native timber trees were more frequently planted on plots far from the homestead. Tree products in both cases were for family consumption. Coffee with commercial trees were planted at medium to close distance to home for the ease of management, harvest and transport.

3.3. Tree Species Richness in Coffee Farms

There were significant differences in the species and area accumulations between coffee agroforestry plots and non-coffee agroforestry plots (*n* = 124, Figure 2). Non-coffee plots were fields with annual food crops or tree plantations as the main productive component. First order Jackknife asymptote showed that the extrapolated value of the total tree species richness of agroforestry coffee plots was 48. All tree species encountered during the inventories also appeared in coffee agroforestry plots. Species richness and the accumulated area of coffee agroforestry plots were almost double those of non-coffee plots as was the species richness per ha.

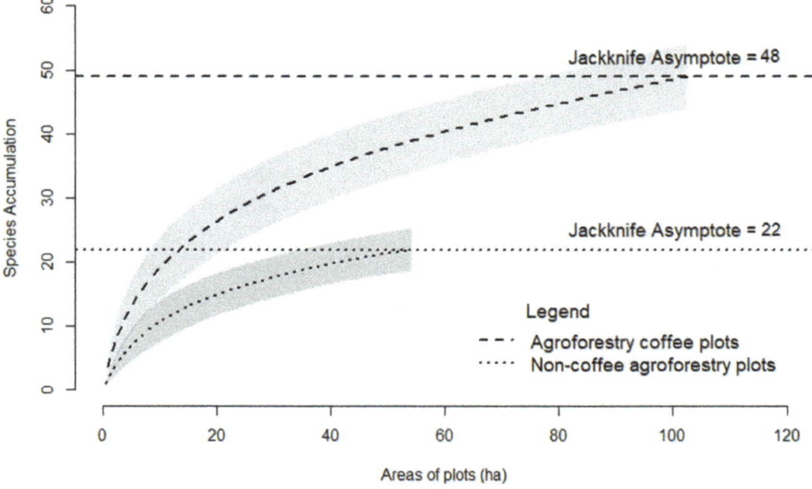

Figure 2. Species accumulations and first order Jackknife asymptotes by area from coffee agroforestry plots and non-coffee plots (orchards, annual crops, timbers).

Higher species richness was found on Thai farmers' coffee agroforestry plots compared to those of the Kinh and H'mong ethnic groups. First order Jackknife asymptotes from Kinh coffee plots was the same as that of H'mong coffee plots (Figure 3). This reflects the higher number of tree species in coffee plots of the Thai group in comparison to other groups as the Thai people live at mid-elevation which is suitable for most of the species encountered at all elevations (Table 2).

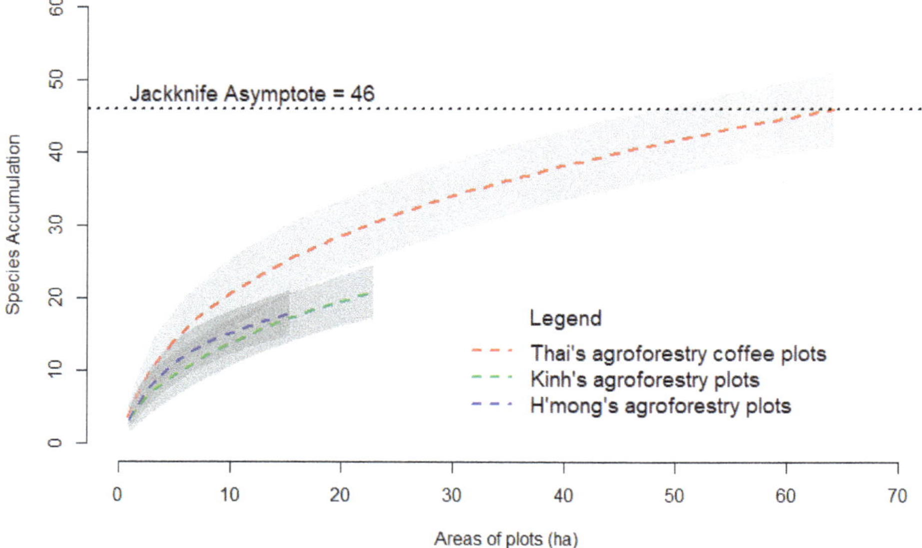

Figure 3. Species accumulations and first order Jackknife asymptotes by area from all coffee agroforestry plots of Thai, H'mong and Kinh ethnic groups.

3.4. Variation in Shade Tree Species Composition of Coffee Farms in Relation to Proximity to Main Roads

When farmers selected tree species to be intercropped with coffee in agroforestry plots, the primary factor that influenced their selection was their commercial value, i.e., their value for the provision of fruits or timbers that could be sold. There was a number of variables related to this factor including market price, proximity to market (indicated in the present study, by proximity to a main road because farmers usually sold fruits on the side of the main road). Among the three ethnic groups, the H'mong were generally at the furthest distance from the road compared with the other two groups. The Kinh people had good road networks because they live in the lower altitudes. The proximity to the road was identified as a critical factor for determining tree species selection by farmers. Fifty-two of the sampled farms were in communes near the main road (0–2 km), 54 farms were at medium distance to the main road (2–5 km) and 18 farms were were far from the main road (5–7 km) (Figure 4). High-value commercial species (fruits, timbers and nuts) were commonly integrated with coffee in the close and medium distance farms (accounting for 81 and 66% of the total trees species, respectively). Commercial species were less commonly (54%) far from the main road. In these farms, more native timbers were intercropped in the coffee plots. The leguminous species, *L. leucocephala*, was more common in the farms far from the main road.

3.5. Farmer Perspectives on Ecosystem Services Associated with Trees in Coffee Systems

Farmers had in-depth knowledge of the benefits trees provided to coffee. Most of them were aware of observable ecosystem services such as reducing soil erosion, improving soil fertility, enhancing biodiversity, preventing damage from wind and frost, and providing shade and mulch.

Some ecosystem services were linked to one another, such as shade provision, mulch provision and soil moisture or frost control and wind control. Trees with big leaves and wide crowns were most generally associated with these benefits. Interestingly, farmers had limited experience or knowledge on the effects of tree species on coffee quality. In addition, their responses suggest that they were not highly knowledgeable about the light and nutrient interactions compared to other services with more than half of their answers consisting of 'don't know', indicating knowledge gaps (Figure 5).

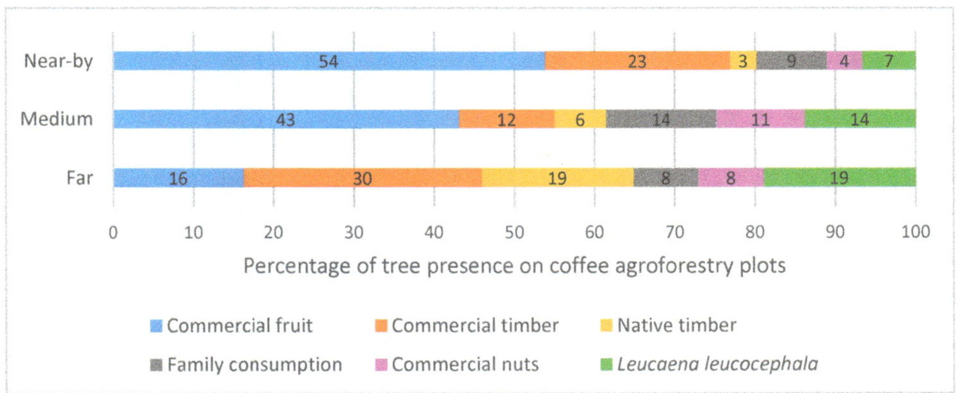

Figure 4. Percentage of tree species present in coffee agroforestry plots with respect to proximity to the road (Note: number of farmers near the road was 52, medium to road was 54, far from road is 18).

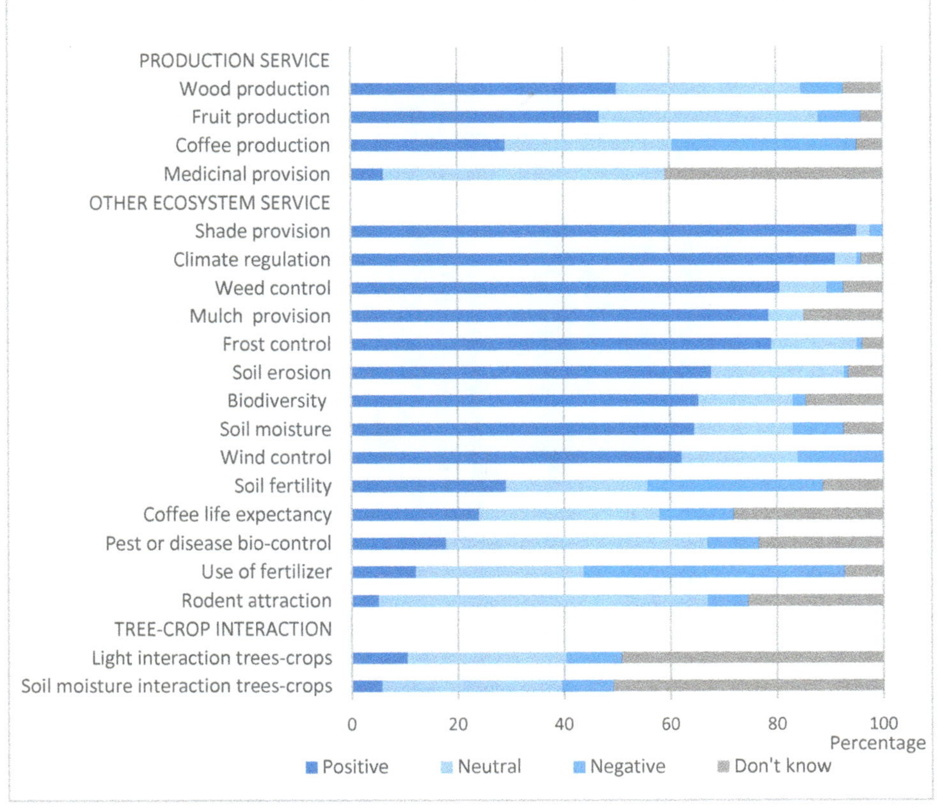

Figure 5. Farmers' perspectives on tree services to coffee in agroforestry practices in the two provinces of Son La and Dien Bien, northwest Vietnam.

When taking ethnicity into account in the analyses, the proportion of four answers were quite similar among the three ethnic groups, except for coffee production, soil fertility, use of fertilizers (Kinh farmers were more negative about these), wood production (H'mong farmers were more positive about this), biodiversity (H'mong farmers were more negative about this), and climate regulation (Thai and Kinh farmers were more positive about this) (Appendix A).

3.6. Tree Species Pairwise Ranking

Among 48 tree species encountered in coffee plots, only 25 species were intercropped in more than 2% of coffee farms. These species were selected for tree pairwise ranking exercises. Accumulated values for each species are shown in Figure 6. The leguminous shade tree species (*L. leucocephala*) was the highest valued species based on its services provided to coffee. *Dimocarpus longan* was the second highest valued species but was the most common species intercropped across coffee farms, followed by mango due to the high commercial value of their fruit. Timber trees were not preferred as farmers stated that timber trees compete for nutrients and water with coffee. Details on the ranking for each tree service are presented in Appendix B.

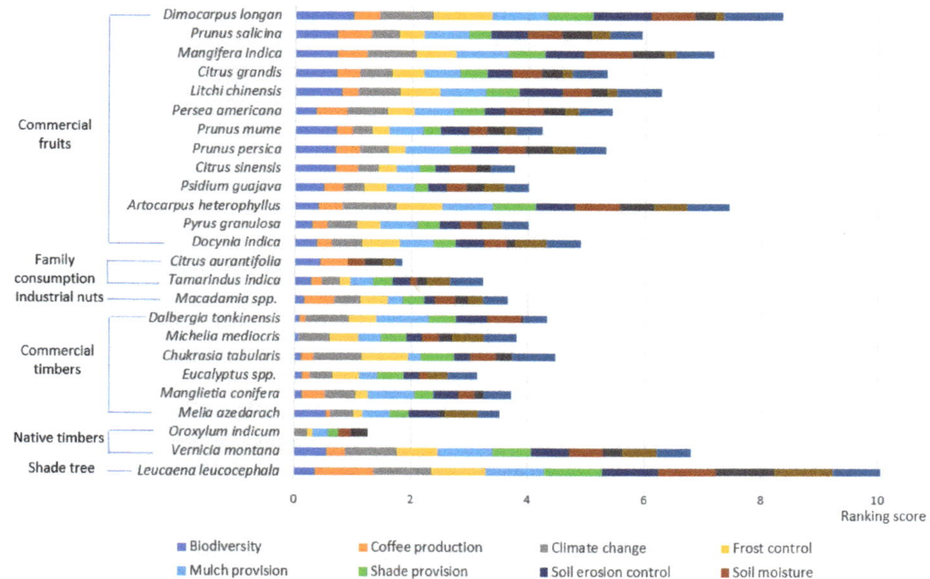

Figure 6. Accumulated tree species ranking scores for various ecosystem services based on the whole group of interviewed male and female farmers in the provinces of Son La and Dien Bien, northwest Vietnam.

Variation in how ranking changed in relation to distance to road was also analyzed and showed differences in tree ranking for coffee production (Table 3). The farmer group living near the main road ranked fruit trees such as plum and longan highly because the fruit selling prices were high, and it was easy to sell them by the roadside. They also liked the shade provided by these trees. They explained that the economic value of *L. leucocephala* was much lower than commercial fruit trees, although this tree species was good for coffee. Therefore, when their coffee farms were far from the main road, where market forces held less sway, they generally grew more *L. leucocephala* with coffee.

Table 3. Best tree species ranked by farmers with respect to the tree services provision to coffee in agroforestry practices (grouped by ethnicity and proximity to main road).

Tree Services	By Ethnic Group			By Proximity to Road			Overall Ranking
	Kinh	Thai	H'mong	Nearby	Medium	Far	
Coffee production	Leucaena *	Leucaena	Leucaena	Plum	Leucaena	Leucaena	Leucaena
Soil fertility	Leucaena *	Leucaena	Leucaena	Leucaena	Leucaena	Leucaena	Leucaena
Shade provision	Longan	Leucaena	Leucaena	Longan	Leucaena	Leucaena	Leucaena
Climate regulation	Longan	Leucaena	Mango	Longan	Leucaena	Leucaena	Leucaena
Soil moisture	Longan	Leucaena	Leucaena	Jackfruit	Leucaena	Leucaena	Leucaena
Soil erosion	Longan	Longan	Longan	Longan	Longan	Longan	Longan
Wind control	Longan	Longan	Longan	Longan	Longan	Longan	Longan
Frost control	Longan	Leucaena	Longan	Longan	Leucaena	Longan	Longan
Mulch provision	Plum	Leucaena	Longan	Longan	Jackfruit	Leucaena	Longan
Biodiversity	Longan	Longan	Mango	Longan	Longan	Longan	Longan
Use of fertilizer	Jackfruit	Leucaena	Leucaena	Leucaena	Leucaena	Peach	Leucaena

(Note: Leucaena: *L. leucocephala*, Longan: *D. longan*, Jackfruit: *A. heterophyllus*, Plum: *P. salicina*, Peach: *P. percia*; Mango: *M. indica*) (*) indicates low level of confidence in ranking results.

The leguminous tree, *L. leucocephala*, was introduced in coffee plantations first by the French since they introduced coffee cultivation in Vietnam in the 19th century. In the 1980s, *L. leucocephala* was promoted to be intercropped with coffee by Son La coffee company in Son La and Dien Bien provinces. After the company went bankrupt, farmers continued planting shade trees by themselves. More recently, some of them still keep *L. leucocephala* in their coffee plantations while an increasing number of them replaced it with commercial fruit trees. Farmers in Son La replaced *L. leucocephala* with fruit trees such as plum and longan which provide more immediate and constant income.

Ninety percent of *L. leucocephala* was found in Dien Bien province and grown by Thai farmers ($n = 29$). A few H'mong farmers also planted *L. leucocephala* intercropped with coffee and other trees. Some H'mong farmers stated that they learned the technique and took the seedlings from fellow farmers in Thai villages. Son La province is famous for fruit production while fruit tree cultivation has not been well developed in some areas of Dien Bien province despite some fruit tree support programs from the local government. Farmers in Dien Bien said that it was hard to manage fruit trees as young seedlings and fruits during harvesting season were often stolen. Harsh weather like drought and frost were also mentioned as major constraints in planting fruit trees. Most fruits for domestic consumption in Dien Bien were imported from Son La and other provinces.

Results of pairwise comparisons between men and women did not show any significant difference with the overall ranking. Men and women had quite similar average ranking for the best tree species with respect to all tree services. The main differences in their ranking were related to timber trees with respect to soil erosion, mulch provision, use of fertilizer and coffee production (Appendix C). Rankings from the three ethnic groups showed remarkable differences, mainly for timber trees and the local fruit tree son tra (*D. indica*) (Appendix D).

Longan (*D. longan*) was a highly popular fruit tree species in the northwest. From the tree inventory, *D. longan* was the most abundant species, appearing in 46% of surveyed farms. Most of the *D. longan* trees were planted by Kinh and Thai farmers. *D. longan* was highly ranked for coffee agroforestry practices because of its good services to coffee such as providing shade, litterfall to keep the soil humidity, reducing wind and frost impact.

All ranking data were uploaded to the online tool supporting decision-making process for selecting shade tree species on coffee farms together with data from other countries. The tool is available at https://www.shadetreeadvice.org/. Vietnamese data can be found under the Vietnam country category, northwest region, Arabica variety and medium zone. After the users select the ecosystem services that they wish to prioritize, the tool produces outputs suggesting appropriate shade tree species to be considered to be integrated into their coffee practices.

4. Discussion

4.1. Tree Diversity in Coffee Agroforestry Systems

Shaded coffee systems were well recognized for biodiversity enhancement according to the farmers' perception. Tree species richness from agroforestry coffee plots was almost double that of non-coffee plots, supporting the notion that shaded coffee systems have higher tree diversity compared to other agricultural land uses [23]. The highest diversity of trees found on coffee plots were associated with Thai farmers (Figure 3), which can be explained by the favorable conditions of their mid-elevation location for various tree species compared to other two groups. Thai people generally live at a medium elevation from 600 to 800 m and are near or at medium distance to main roads.

Among 48 recorded species from the tree inventory, 10 species are the most abundant as being planted in more than 10% of total inventoried coffee plots. This is relatively low compared to the 162 tree species found in the Yunnan, China [32], 165 tree species in Mount Kenya [36] and 100 species associated with coffee in Muranga, Kenya, 29 of which were sufficiently abundant to be ranked [28]. However, this finding was similar to the 36 species that farmers had knowledge of in coffee farms in Costa Rica [27] and the number of species found by Nyaga et al. [37] in agroforestry in the Rift Valley in Kenya with 44 species (55% native). However, the primary functions of trees in Vietnam were commercial fruits with a ready market while the most abundant species in the Rift Valley in Kenya were firewood and fast growing fodder or fertilizer species [37].

4.2. Tree Ranking and Farmer's Perception on Tree Services or Disservices

In northwest Vietnam, farmers did not like planting timber trees with coffee. They only planted timber trees when they had no other choices for fruits (either because of unsuitable conditions or distance from the market). Farmers' ranked timber trees lowest indicating competitive rather than complementary relationships with coffee. This was consistent with the tree frequencies (Table 2) where timber species appeared in less than 6% of surveyed farms. Farmers were knowledgeable about the benefits that trees can provide to coffee as documented in other continents [33,38]. In this study, farmers could confidently describe the effects of trees on climate regulation, soil erosion, shade provision, mulch provision, frost and wind control.

Unlike in the Central Highlands of Vietnam where coffee was blamed for deforestation [39], in the northwest, most coffee farms were established on annual crop land or fallow land [10]. Therefore, farmers could clearly explain the benefits, based on direct observation, of integrating trees on their coffee farms on steep slopes, such as soil becoming soft and moist, and having darker brown color compared to a more yellow color in the past. Farmers were much less clear about tree shade interactions with the coffee plants, particularly with regard to the competition between coffee and trees for light and water (Figure 5), although they were aware that 'too much' shade made coffee beans take too long to ripen, and that the coffee produced fewer cherries but the size of the cherries was bigger. Coffee quality remains a big knowledge gap because farmers do not drink the coffee they produce. These are areas where better information about the effects of different shade species on the quality of the coffee may lead to the greater diversification of shade trees—provided that quality influences the price farmers receive for their coffee.

4.3. The Relevance of Tree Knowledge and Tree Biophysical Suitability

Nitrogen-fixing species are common in many coffee regions of the world such as *Erythrina* spp. and *Inga* spp. in Central America, although they have no timber values [40]; and *L. leucocephala* was ranked as best species for most ecosystem services, but was only planted in 23% of coffee fields, mainly in Dien Bien. As observed by farmers, *L. leucocephala* grew well with coffee and provided good shade as its leaf size was not too small or too large. Furthermore, *L. leucocephala* leaves were soft, which was good for soil fertility because they decompose quickly. Although legume species are good for coffee, this also depends on the management of legume trees on farms [12]. Farmers in Dien Bien expressed their concern about too much shade from *L. leucocephala* trees. Coffee density under shade was not discussed with farmers while it appeared as a factor significantly influencing performance. Medium intensity coffee management under shade is recommended in Central America [38].

Fruit tree cultivation in coffee farms is also popular in other countries thanks to their high profit. Farmers in Yunnan ranked fruit trees including *D. longan* quite high [30]. *Dimocarpus longan* is restricted to elevations lower than 600 m whereas Arabica coffee is suitable at elevations of 500 masl and above. Thai and H'mong farmers ranked *D. longan* as a good species for some ecosystem services and they expressed the wish to expand coffee–*D. longan* systems if they had available land. This raises a concern about the suitability of *D. longan* at higher elevations, as it may lead to the low productivity of the trees and integrated coffee–*D. longan* cultivation—although farmers stated that *D. longan* has a large crown and wide root system which negatively interact with coffee. The farmers' selection of tree species in coffee agroforestry systems was highly influenced by the economic benefits of intercropped trees and market access, particularly the proximity of farms to a main road. The oversupply of *D. longan* could lead to market price reduction in the near future. Moreover, the farmers' choice of trees were just a few species (*L. leucocephala*, longan, peach, plum, jackfruit), and hence the low biodiversity of the coffee landscape can be foreseen.

Farmers ranked timber species as the worst species to integrate in coffee plots. They explained that timber trees often compete with coffee for water and nutrition, and provide heavy shade and hence were not good for coffee. Poor quality road networks make farmers reluctant to grow timber trees as the transportation costs are high. This is quite similar to Central America where timber trees are also less common in coffee farms although they can provide additional values for farmers [41]. Compared to fruit trees, timber trees provide long-term benefit and could be considered as analogous to a savings bank account for the next generation. This suggests a need for further studies to focus on the economic benefits and trade-offs associated with integrating various types of trees in coffee farms, such as legume trees for shade and fertility versus timber or fruit trees for their commercial value.

4.4. Gender, Ethnicity and Tree Selection

Both men and women farmers from the three ethnic groups: Kinh, Thai, H'mong, ranked *L. leucocephala* as the best tree species providing multiple services to coffee. Overall ranking shows that the most preferred options were *D. longan* or plum and jackfruit if the farms were close to the main road. *Leucaena leucocephala* was preferred if the farms were further away from the road. Thus, market access appears to be a main driver of tree species selection. Moreover, coffee was produced only for trade, as the farmers did not consume it themselves and hence had no knowledge about tree species or the management effects on the quality of the final product; that is, the cup of coffee is mostly consumed in consuming countries. In this commercial system, market factors seemed to be the most important consideration rather than ethnicity, gender or the local knowledge about different tree species.

5. Conclusions

This study suggests that integrating trees into monoculture coffee systems can bring multiple benefits to the environment and rural livelihoods in northwest Vietnam. From the field surveys, it was observed that most of the smallholder coffee area was under agroforestry, but highly dependent on plot location, ecological suitability and market access. The tree diversity of coffee agroforestry plots were double those of agroforestry with annual crops, orchards or timber plantations. There were observable differences between the three ethnic groups. Thai farms had higher tree species richness in comparison with Kinh and H'mong groups—but it was not clear the degree to which this was driven by cultural rather than other contextual factors (such as proximity to roads and elevation). Commercial fruit trees were more common when farmers lived near or at a medium distance to main roads. Timber trees and a legume tree, *L. leucocephala*, were more common on coffee farms which were far from the road.

Farmers had detailed knowledge about observable tree services such as soil erosion control and mulch provision but they appeared to lack knowledge about the effects of shade trees on soil fertility, pest and disease control, and interestingly, on the competition or complementarity between coffee and trees in respect of light and water capture. This is where integrating local knowledge and scientific knowledge can provide richer advice to farmers on selecting appropriate tree species for their coffee farms.

By uploading the data from this research to the online tool http://shadetreeadvice.org, the information is available to help farmers, researchers and policy makers to choose suitable tree species according to the services that they can provide. Further improvement of the tool could include a focus on management aspects with farmers ranking additional information, such as the labor requirement for species associated with the timing and severity of tree pruning. This could be used to create a more comprehensive knowledge management system that goes beyond tree species selection.

Author Contributions: Conceptualization, P.V., T.P., M.P.N. and F.S.; methodology, M.P.N., P.V. and T.P., data collection and analysis, M.P.N.; original draft preparation, M.P.N.; review and editing, T.P., P.V., F.S.; funding acquisition, P.V. and F.S. All authors have read and agreed to the published version of the manuscript.

Funding: The study was funded by CANSEA—a Research and Development Network on Agroecology Transition in South East Asia project financed by Agence Française de Dévleoppement (AfD) and FTA (CGIAR Research Program on Forest, Trees and Agroforestry).

Acknowledgments: This research was supported by AFLi II project (Developing and Promoting market-based agroforestry and forest rehabilitation options for northwest Vietnam) and CCAFS (CGIAR Research Program on Climate Change, Agriculture and Food Security). We thank our colleagues Nguyen Thi Thanh Hai, Nguyen Van Huan from Northern Mountainous Agriculture and Forestry Science Institute (NOMAFSI) CANSEA, Pham Thanh Van, Nguyen Van Cuong, Duong Minh Tuan (World Agroforestry—Vietnam office), Master student Mathilde Lépine for their contribution in farmer interview, farmer ranking, focus group discussion and Clement Rigal for guidance in data analysis.

Conflicts of Interest: The author declares no conflict of interest.

Appendix A

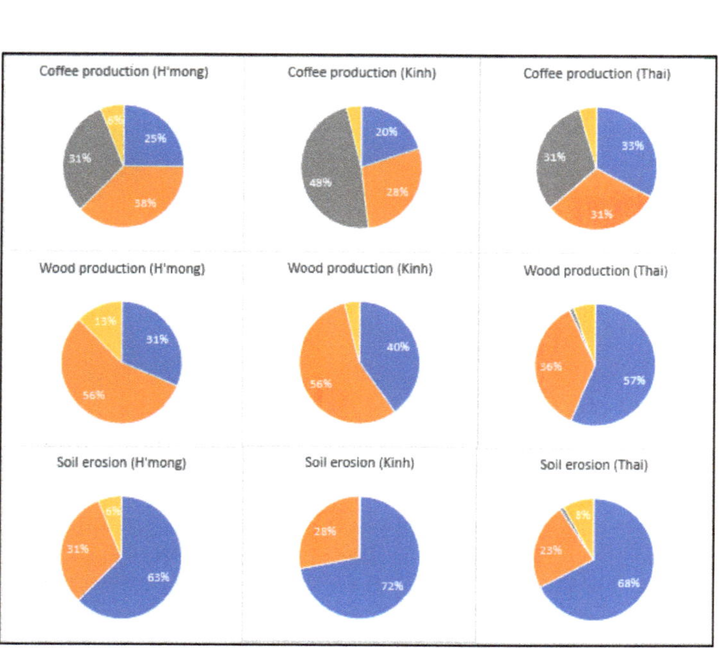

Figure A1. *Cont.*

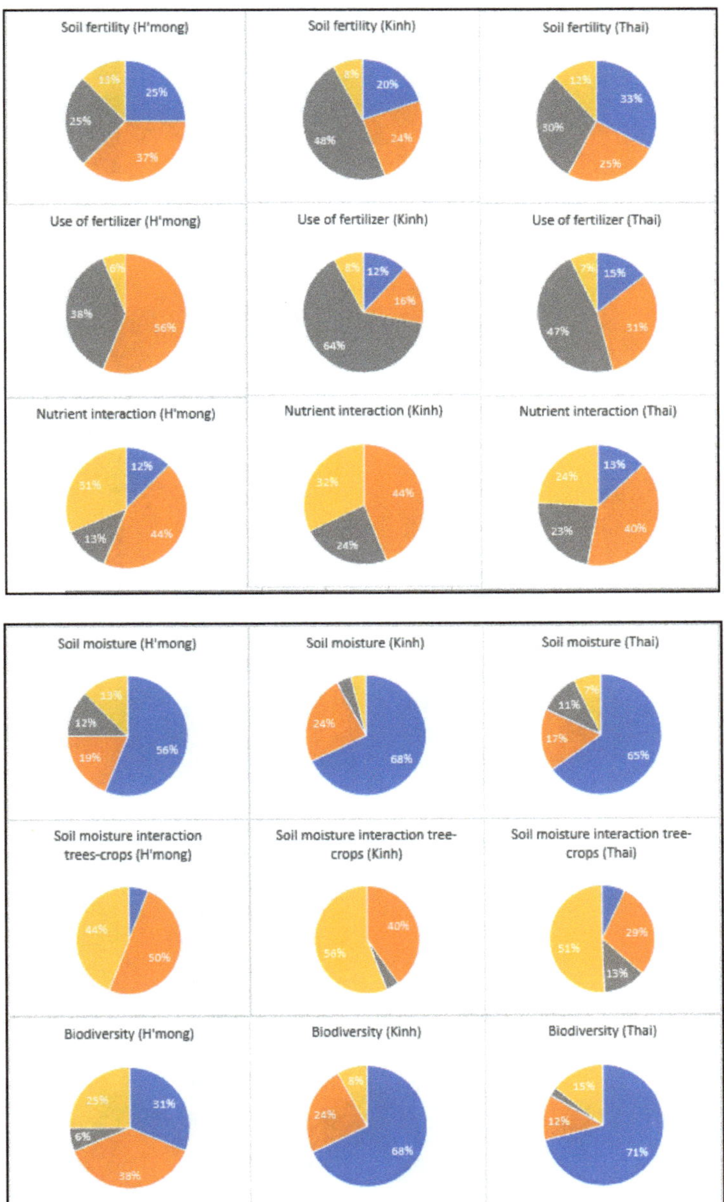

Figure A1. *Cont.*

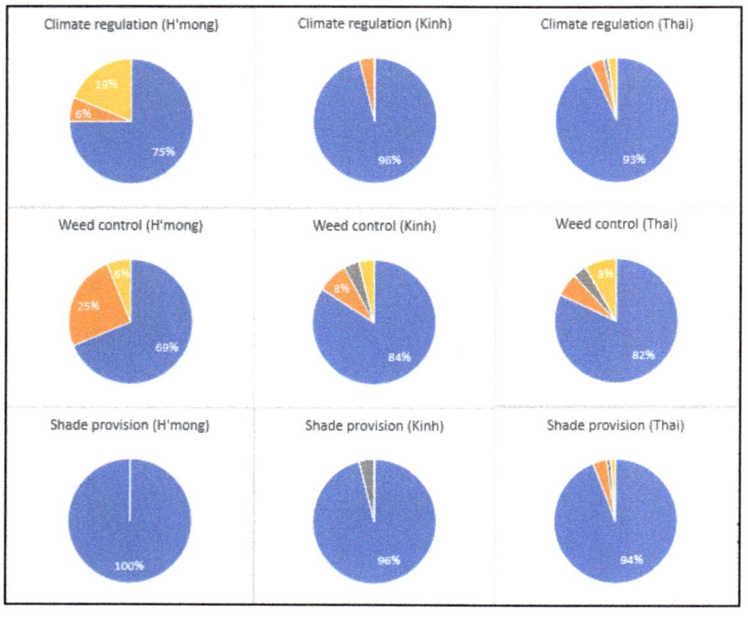

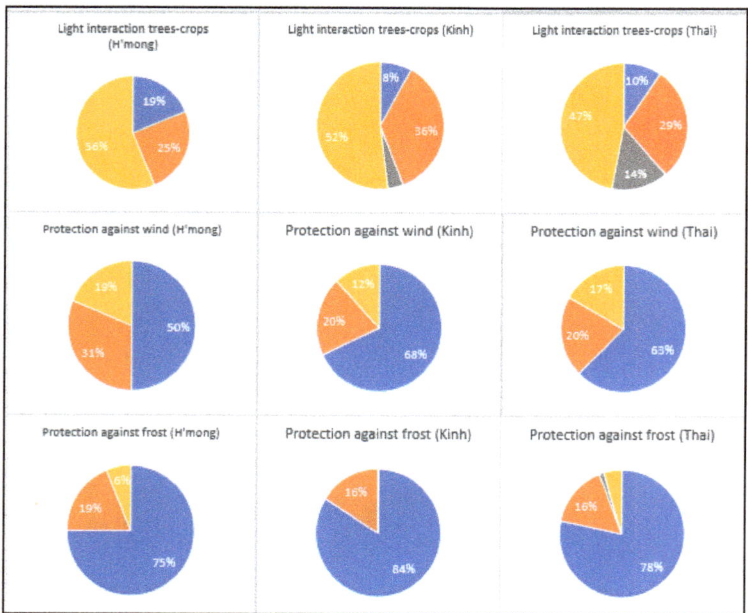

Figure A1. *Cont.*

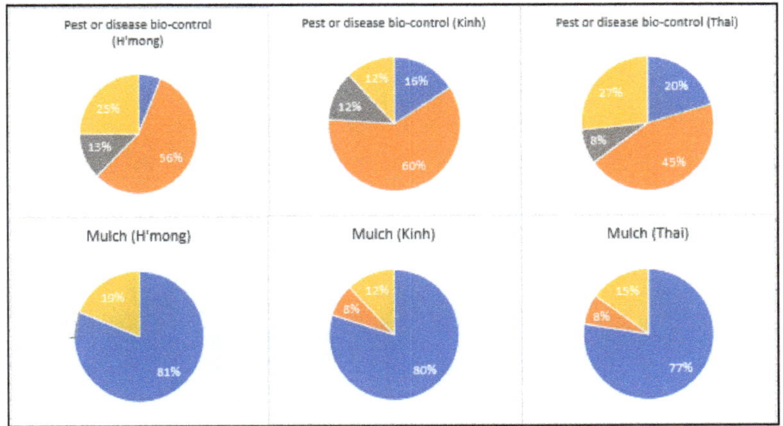

Figure A1. Farmers' perception of tree services for coffee in coffee agroforestry systems by ethnicity.

Appendix B

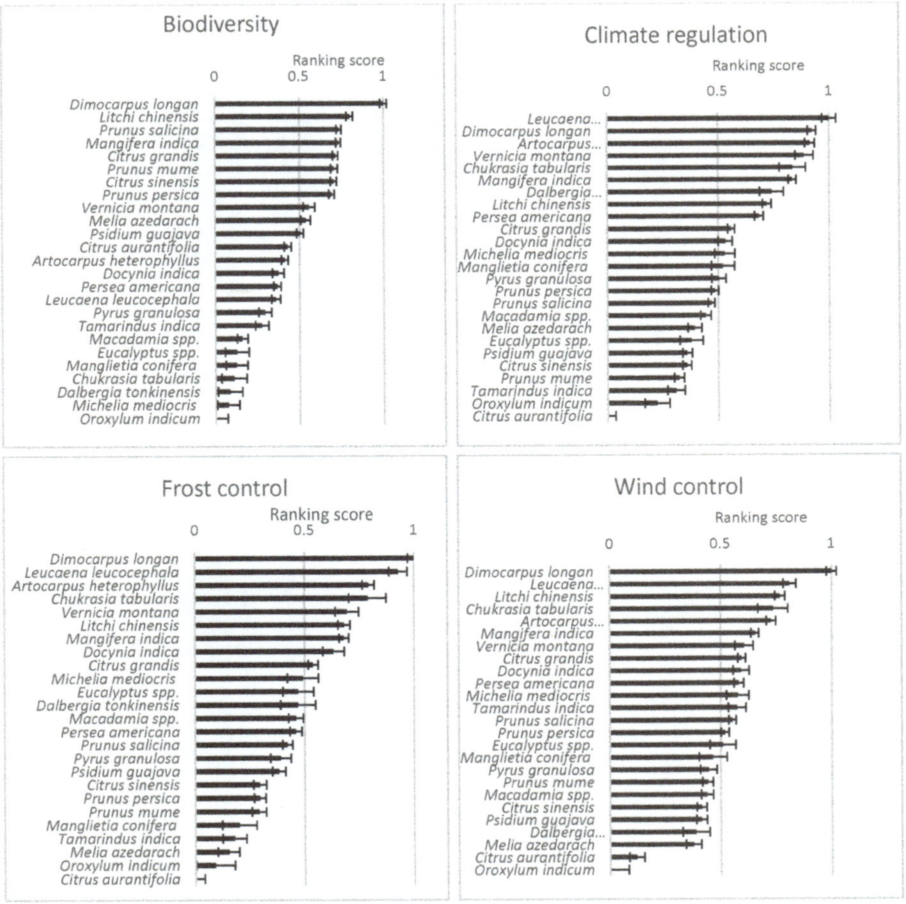

Figure A2. *Cont.*

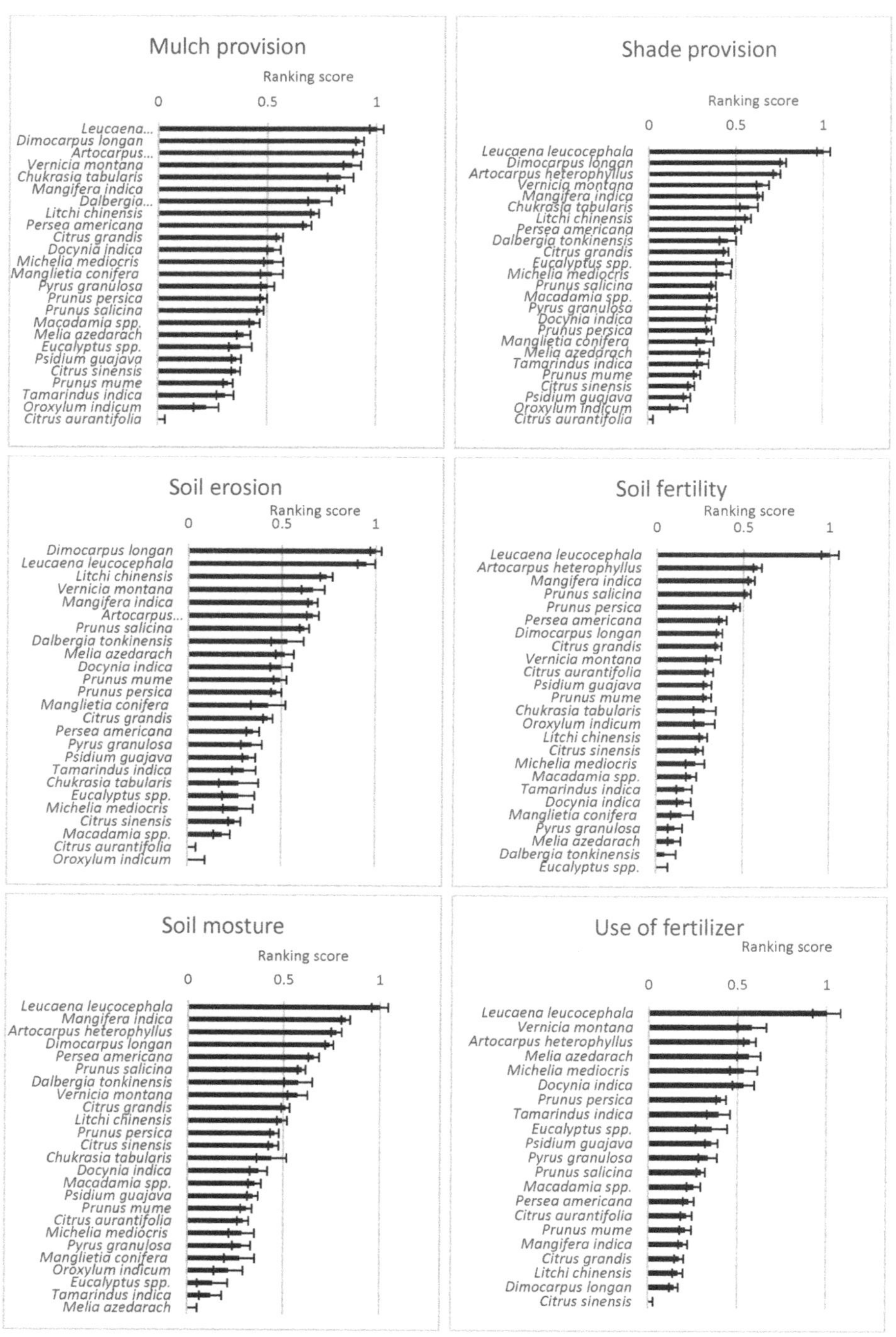

Figure A2. *Cont.*

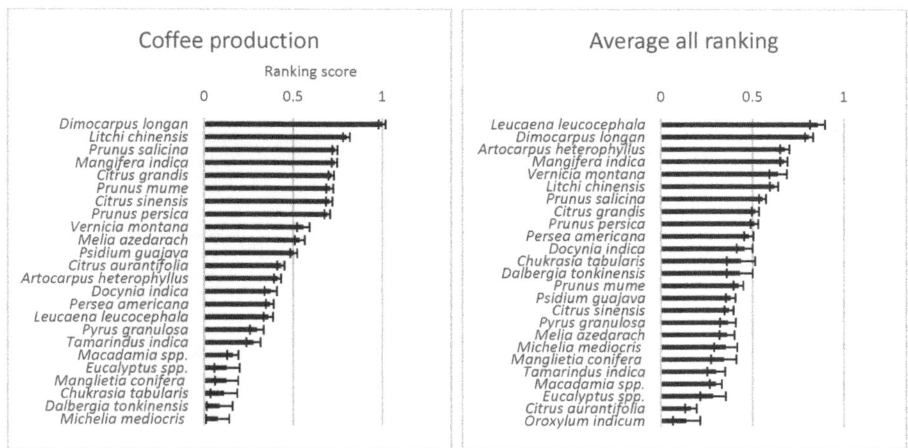

Figure A2. Pairwise ranking of tree species contributing to various tree services in coffee agroforestry systems with all groups combined.

Appendix C

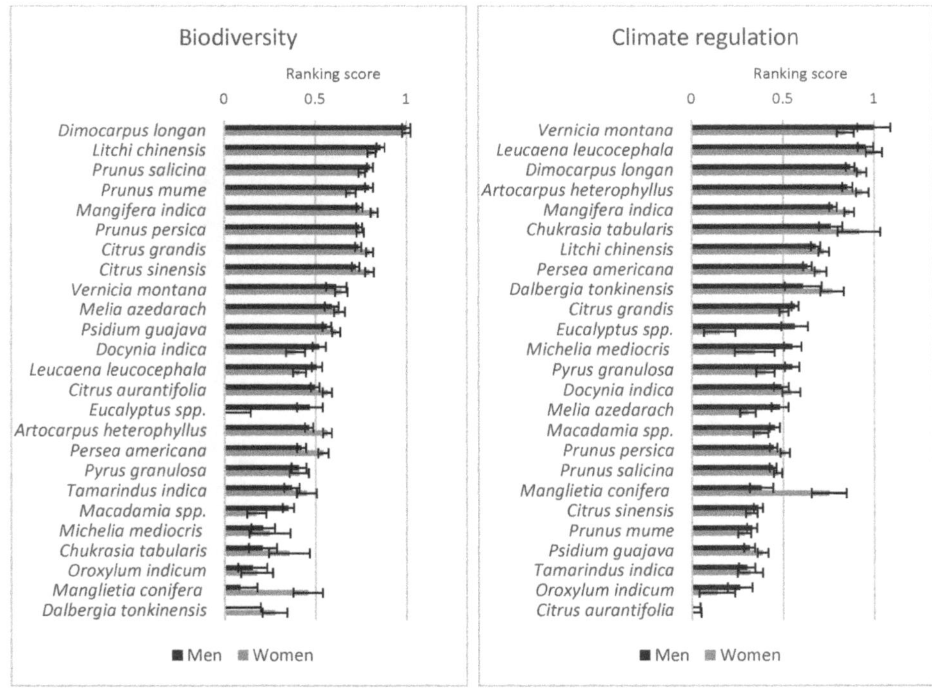

Figure A3. *Cont.*

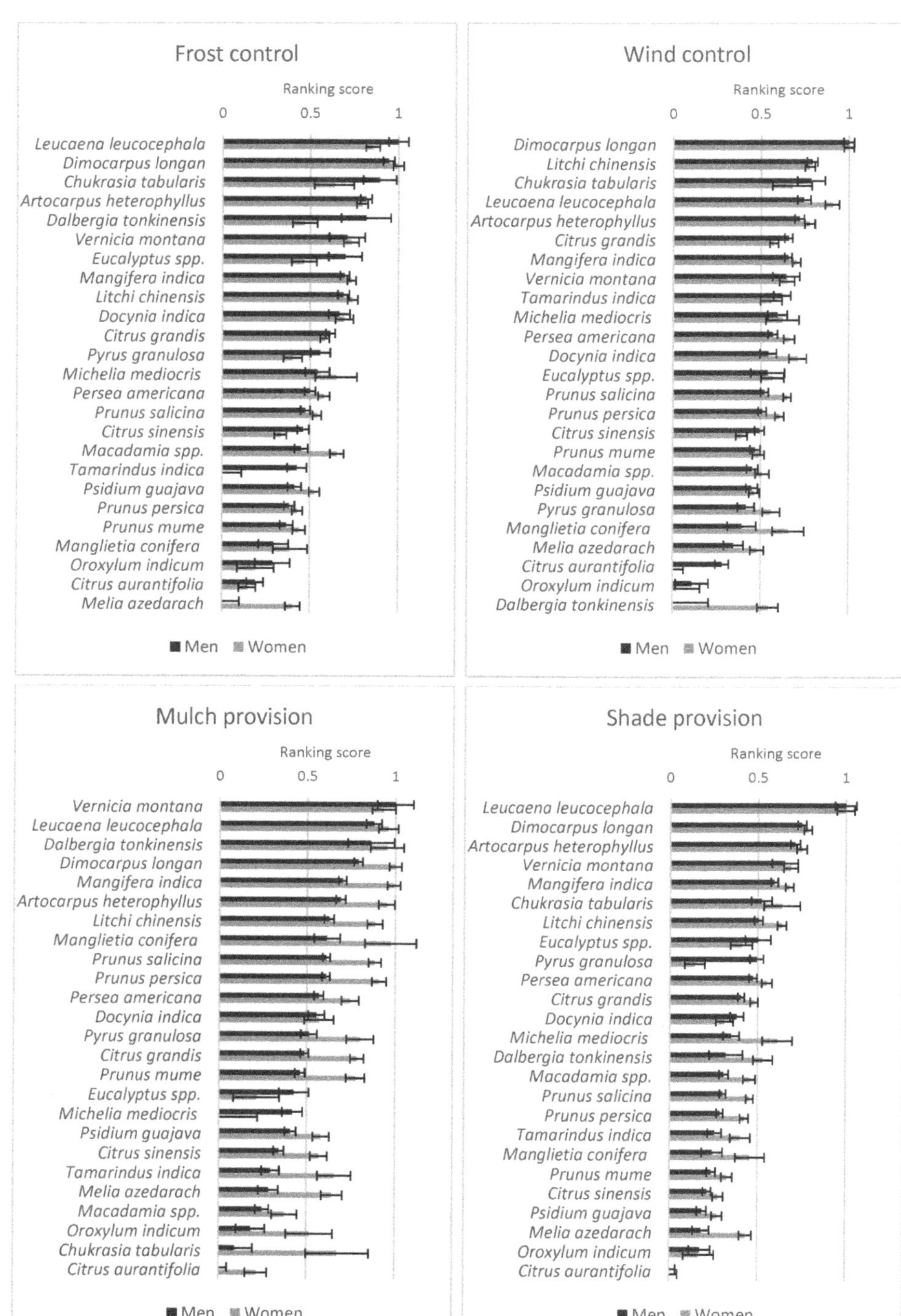

Figure A3. *Cont.*

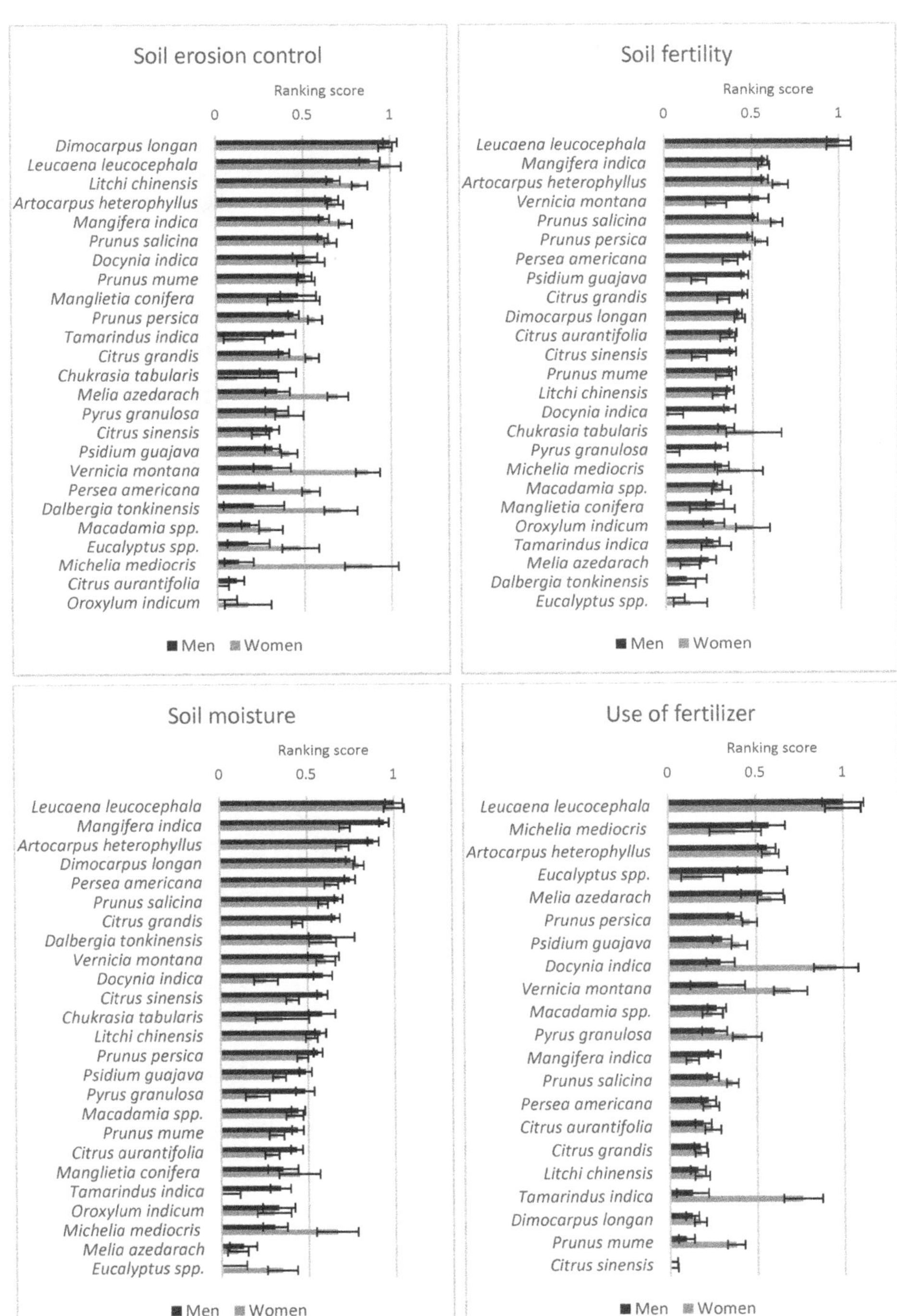

Figure A3. *Cont.*

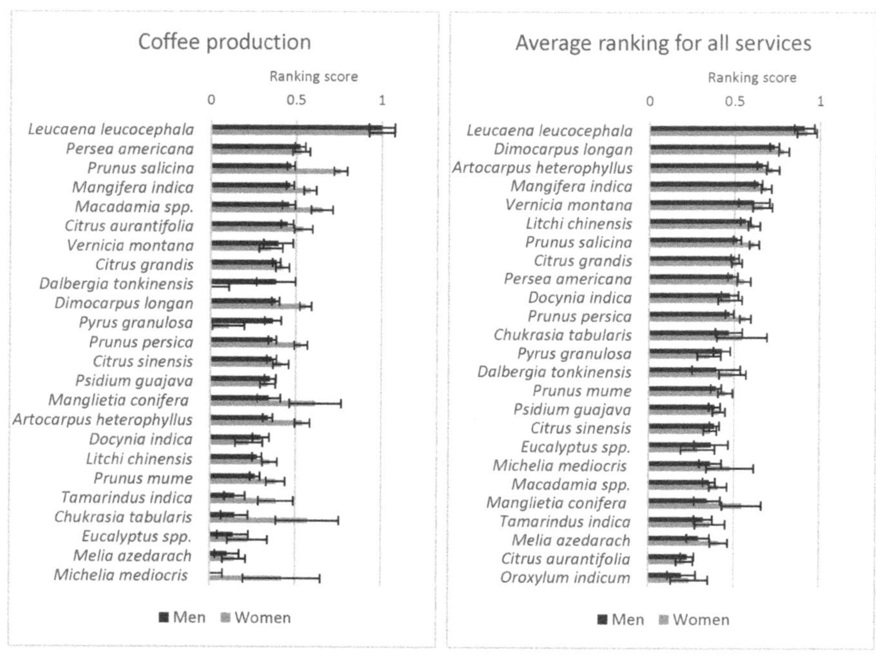

Figure A3. Pairwise ranking of tree species contributing to different tree services in coffee agroforestry systems by gender.

Appendix D

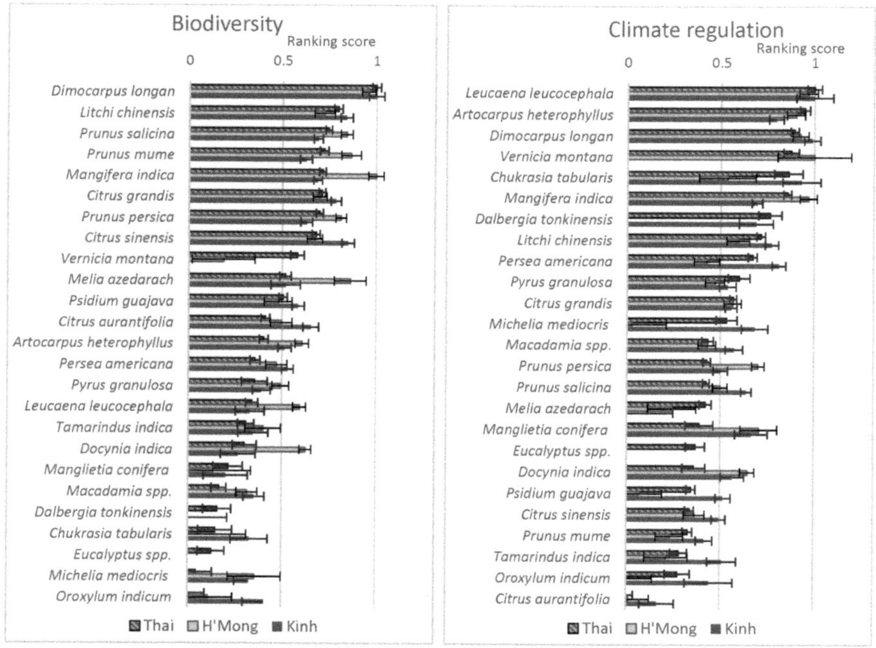

Figure A4. *Cont.*

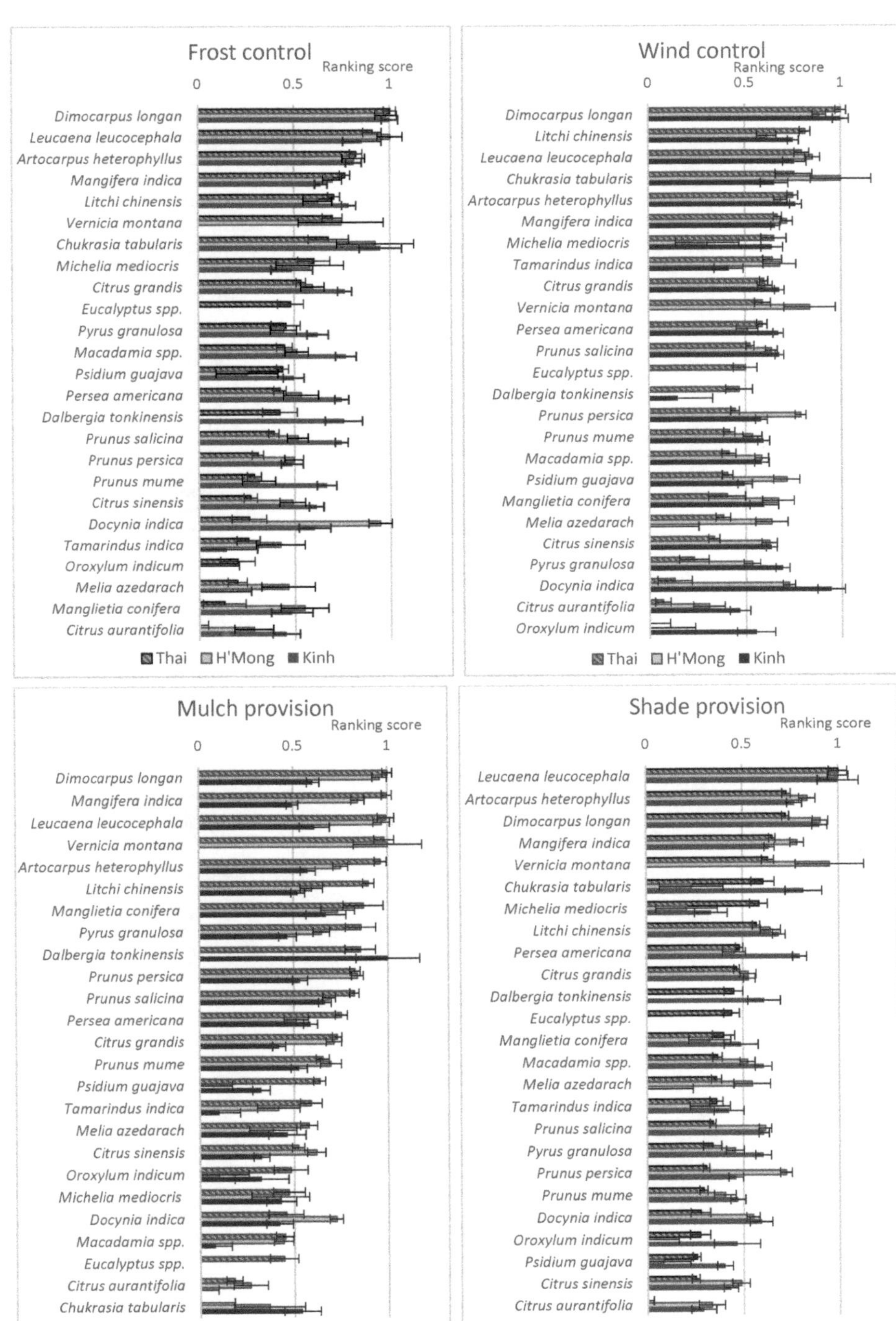

Figure A4. *Cont.*

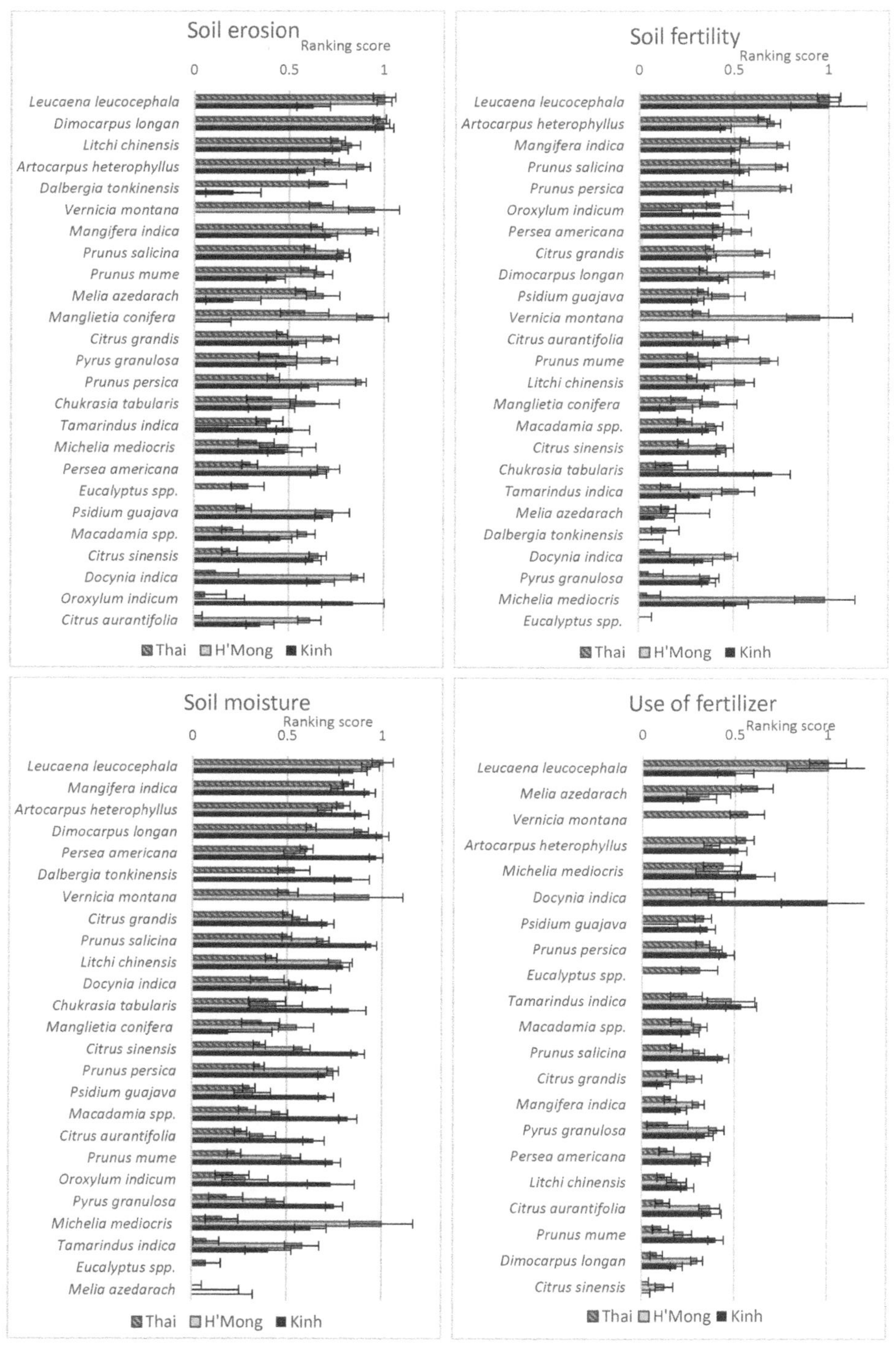

Figure A4. *Cont.*

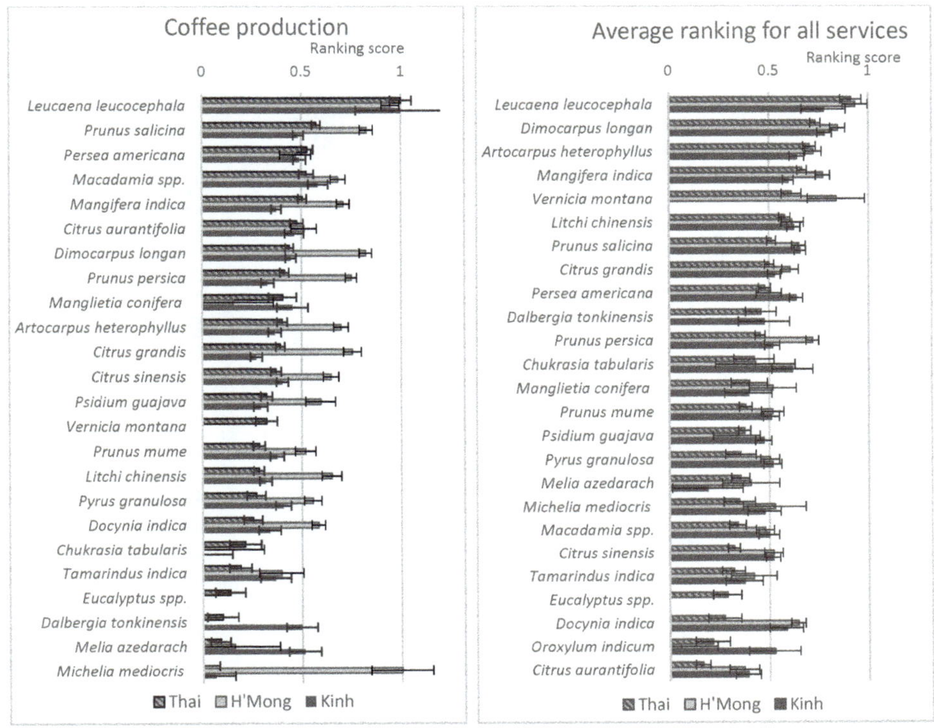

Figure A4. Pairwise ranking of tree species in coffee agroforestry systems by tree service and by ethnic group (Thai, H'Mong and Kinh).

References

1. Tucker, C.M. *Coffee Culture: Local Experiences, Global Connections*, 2nd ed.; Taylor & Francis: Abingdon, UK, 2017; ISBN 9781317392255.
2. Food and Agriculture Organization of the United Nations. *FAOSTAT Statistical Database 2018*; Food and Agriculture Organization of the United Nations: Rome, Italy, 2018.
3. International Coffee Organization. *ICO Annual Review 2010/11 2011*; International Coffee Organization: London, UK, 2011.
4. Amarasinghe, U.A.; Hoanh, C.T.; D'haeze, D.; Hung, T.Q. Toward sustainable coffee production in Vietnam: More coffee with less water. *Agric. Syst.* **2015**, *136*, 96–105. [CrossRef]
5. Vietnam Ministry of Agriculture and Rural Development. *Report on Crop Production in Vietnam*; Crop Production Department: Ha Noi, Vietnam, 2017.
6. Vietnam Customs. *Vietnam International Merchandise Trade Statistics 2018*; Vietnam Customs: Ha Noi, Vietnam, 2018.
7. International Coffee Organization. *Country Coffee Profile: Vietnam*; International Coffee Organization: London, UK, 2019.
8. General Statistics Office of Dien Bien Province. *Agriculture, Fishery and Forestry Census 2018*; General Statistics Office of Dien Bien Province: Vietnam, 2018.
9. General Statistics Office of Son La Province. *Agriculture, Fishery and Forestry Census 2018*; General Statistics Office of Son La Province: Vietnam, 2018.
10. Nghiem, T.; Kono, Y.; Leisz, S.J. The Case of Coffee Production in the Northern Mountain Region of Vietnam. *Land* **2020**, *9*, 56. [CrossRef]
11. Perfecto, I.; Rice, R.A.; Greenberg, R.; Van Der Voort, M.E. Shade coffee: A disappearing refuge for biodiversity: Shade coffee plantations can contain as much biodiversity as forest habitats. *Bioscience* **1996**, *46*, 598–608. [CrossRef]

12. Haggar, J.; Barrios, M.; Bolaños, M.; Merlo, M.; Moraga, P.; Munguia, R.; Ponce, A.; Romero, S.; Soto, G.; Staver, C.; et al. Coffee agroecosystem performance under full sun, shade, conventional and organic management regimes in Central America. *Agrofor. Syst.* **2011**, *82*, 285–301. [CrossRef]

13. Philpott, S.M.; Arendt, W.J.; Armbrecht, I.; Bichier, P.; Diestch, T.V.; Gordon, C.; Greenberg, R.; Perfecto, I.; Reynoso-Santos, R.; Soto-Pinto, L.; et al. Biodiversity loss in Latin American coffee landscapes: Review of the evidence on ants, birds, and trees. *Conserv. Biol.* **2008**, *22*, 1093–1105. [CrossRef]

14. Vaast, P.; Harmand, J.-M.; Rapidel, B.; Jagoret, P.; Deheuvels, O. Coffee and Cocoa Production in Agroforestry—A Climate-Smart Agriculture Model. In *Climate Change and Agriculture Worldwide*; Torquebiau, E., Ed.; Springer: Dordrecht, The Netherlands, 2016; pp. 209–224.

15. Sinclair, F.L. A general classification of agroforestry practice. *Agrofor. Syst.* **1999**, *46*, 161–180. [CrossRef]

16. Boreux, V.; Kushalappa, C.G.; Vaast, P.; Ghazoul, J. Interactive effects among ecosystem services and management practices on crop production: Pollination in coffee agroforestry systems. *Proc. Natl. Acad. Sci. USA* **2013**, *110*, 8387–8392. [CrossRef]

17. Somarriba, E.; Harvey, C.A.; Samper, M.; Anthony, F.; González, J.; Staver, C.; Rice, R.A. Biodiversity conservation in neotropical coffee (Coffea arabica) plantations. In *Agroforestry and Biodiversity Conservation in Tropical Landscapes*; Island Press: Washington, DC, USA, 2004; pp. 198–226.

18. Jha, S.; Bacon, C.M.; Philpott, S.M.; Rice, R.A.; Méndez, V.E.; Läderach, P. A Review of Ecosystem Services, Farmer Livelihoods, and Value Chains in Shade Coffee Agroecosystems. In *Integrating Agriculture, Conservation and Ecotourism: Examples from the Field*; Campell, W.B., Ortiz, S.L., Eds.; Springer: Dordrecht, The Netherlands, 2011; pp. 141–208.

19. Guillemot, J.; le Maire, G.; Munishamappa, M.; Charbonnier, F.; Vaast, P. Native coffee agroforestry in the Western Ghats of India maintains higher carbon storage and tree diversity compared to exotic agroforestry. *Agric. Ecosyst. Environ.* **2018**, *265*, 461–469. [CrossRef]

20. Rahn, E.; Läderach, P.; Baca, M.; Cressy, C.; Schroth, G.; Malin, D.; van Rikxoort, H.; Shriver, J. Climate change adaptation, mitigation and livelihood benefits in coffee production: Where are the synergies? *Mitig. Adapt. Strateg. Glob. Chang.* **2014**, *19*, 1119–1137. [CrossRef]

21. Davis, A.P.; Gole, T.W.; Baena, S.; Moat, J. The Impact of Climate Change on Indigenous Arabica Coffee (Coffea arabica): Predicting Future Trends and Identifying Priorities. *PLoS ONE* **2012**, *7*. [CrossRef]

22. Schroth, G.; Laderach, P.; Dempewolf, J.; Philpott, S.; Haggar, J.; Eakin, H.; Castillejos, T.; Moreno, J.G.; Pinto, L.S.; Hernandez, R.; et al. Towards a climate change adaptation strategy for coffee communities and ecosystems in the Sierra Madre de Chiapas, Mexico. *Mitig. Adapt. Strateg. Glob. Chang.* **2009**, *14*, 605–625. [CrossRef]

23. Vaast, P.; Van Kanten, R.; Siles, P.; Dzib, B.; Franck, N.; Harmand, J.M.; Genard, M. Shade: A Key Factor for Coffee Sustainability and Quality. In Proceedings of the 20th International Conference on Coffee Science, Bangalore, India, 11–15 October 2004.

24. Mancuso, M.A.C.; Soratto, R.P.; Perdoná, M.J. Shaded coffee production. *Colloq. Agrar.* **2013**, *9*, 31–44. [CrossRef]

25. Meylan, L.; Gary, C.; Allinne, C.; Ortiz, J.; Jackson, L.; Rapidel, B. Evaluating the effect of shade trees on provision of ecosystem services in intensively managed coffee plantations. *Agric. Ecosyst. Environ.* **2017**, *245*, 32–42. [CrossRef]

26. Roupsard, O.; Van Den Meersche, K.; Allinne, C.; Vaast, P.; Rapidel, B.; Avelino, J.; Jourdan, C.; Le Maire, G.; Bonnefond, J.-M.; Harmand, J.-M.; et al. Eight years studying ecosystem services in a coffee agroforestry observatory. Practical applications for the stakeholders. In Proceedings of the World Coffee Summit, San Salvador, Salvador, 31 May–3 June 2017.

27. Cerdán, C.R.; Rebolledo, M.C.; Soto, G.; Rapidel, B.; Sinclair, F.L. Local knowledge of impacts of tree cover on ecosystem services in smallholder coffee production systems. *Agric. Syst.* **2012**, *110*, 119–130. [CrossRef]

28. Lamond, G.; Sandbrook, L.; Gassner, A.; Sinclair, F.L. Local knowldge of tree attreibutes underpins sepcies selection on coffee farms. *Exp. Agric.* **2019**, *55*, 35–49. [CrossRef]

29. Valencia, V.; West, P.; Sterling, E.J.; García-Barrios, L.; Naeem, S. The use of farmers' knowledge in coffee agroforestry management: Implications for the conservation of tree biodiversity. *Ecosphere* **2015**, *6*, 1–17. [CrossRef]

30. Dumont, E.S.; Bonhomme, S.; Pagella, T.F.; Sinclair, F.L. Structured stakeholder engagement leads to development of more diverse and inclusive agroforestry options. *Exp. Agric.* **2019**, *55*, 252–274. [CrossRef]

31. Van Der Wolf, J.; Jassogne, L.; Gram, G.I.L.; Vaast, P. Turning local knowledge on agroforestry into an online decision-support tool for tree selection smallholger's farms. *Exp. Agric.* **2019**, *55*, 50–66. [CrossRef]

32. Rigal, C.; Vaast, P.; Xu, J. Using farmers' local knowledge of tree provision of ecosystem services to strengthen the emergence of coffee-agroforestry landscapes in southwest China. *PLoS ONE* **2018**, *13*. [CrossRef]

33. Smith Dumont, E.; Gassner, A.; Agaba, G.; Nansamba, R.; Sinclair, F. The utility of farmer ranking of tree attributes for selecting companion trees in coffee production systems. *Agrofor. Syst.* **2019**, *93*, 1469–1483. [CrossRef]

34. Kindt, R.; Coe, R. *Tree Diversity Analysis: A Manual and Software for Common Statistical Methods for Ecological and Biodiversity Studies*; World Agroforestry Centre: Nairobi, Kenya, 2005.

35. Turner, H.; Firth, D. Bradley-terry models in R: The bradleyterry2 package. *J. Stat. Softw.* **2012**, *48*. [CrossRef]

36. Carsan, S.; Stroebel, A.; Dawson, I.; Kindt, R.; Swanepoel, F.; Jamnadass, R. Implications of shifts in coffee production on tree species richness, composition and structure on small farms around Mount Kenya. *Biodivers. Conserv.* **2013**, *22*, 2919–2936. [CrossRef]

37. Nyaga, J.; Barrios, E.; Muthuri, C.W.; Öborn, I.; Matiru, V.; Sinclair, F.L. Evaluating factors influencing heterogeneity in agroforestry adoption and practices within smallholder farms in Rift Valley, Kenya. *Agric. Ecosyst. Environ.* **2015**, *212*, 106–118. [CrossRef]

38. Soto-Pinto, L.; Villalvazo-López, V.; Jiménez-Ferrer, G.; Ramírez-Marcial, N.; Montoya, G.; Sinclair, F.L. The role of local knowledge in determining shade composition of multistrata coffee systems in Chiapas, Mexico. *Biodivers. Conserv.* **2007**, *16*, 419–436. [CrossRef]

39. Meyfroidt, P.; Lambin, E.F.; Erb, K.H.; Hertel, T.W. Globalization of land use: Distant drivers of land change and geographic displacement of land use. *Curr. Opin. Environ. Sustain.* **2013**, *5*, 438–444. [CrossRef]

40. Beer, J. Litter production and nutrient cycling in coffee (Coffea arabica) or cacao (Theobroma cacao) plantations with shade trees. *Agrofor. Syst.* **1988**, *7*, 103–114. [CrossRef]

41. Vaast, P.; van Kanten, R.; Siles, P.; Angrand, J.; Aguilar, A. Biophysical Interactions Between Timber Trees and Arabica Coffee in Suboptimal Conditions of Central America. In *Toward Agroforestry Design*; Jose, S., Gordon, A.M., Eds.; Springer: Dordrecht, The Netherlands, 2008; pp. 133–146.

Publisher's Note: MDPI stays neutral with regard to jurisdictional claims in published maps and institutional affiliations.

 *land*

Article

Gendered Species Preferences Link Tree Diversity and Carbon Stocks in Cacao Agroforest in Southeast Sulawesi, Indonesia

Rika Ratna Sari [1,2,*] , **Danny Dwi Saputra** [1,2] , **Kurniatun Hairiah** [1],
Danaë M. A. Rozendaal [2,3] , **James M. Roshetko** [4] and **Meine van Noordwijk** [2,4]

[1]   Soil Science Department, Faculty of Agriculture, Universitas Brawijaya, Malang 65145, Indonesia;
      danny_saputra@ub.ac.id (D.D.S.); kurniatun_h@ub.ac.id (K.H.)
[2]   Plant Production Systems Group, Wageningen University and Research, 6708 PB Wageningen,
      The Netherlands; danae.rozendaal@wur.nl (D.M.A.R.); m.vannoordwijk@cgiar.org (M.v.N.)
[3]   Centre for Crop Systems Analysis, Wageningen University and Research, 6708 PB Wageningen,
      The Netherlands
[4]   World Agroforestry Centre, ICRAF Southeast Asia, Bogor 16001, Indonesia; j.roshetko@cgiar.org
*   Correspondence: rr.sari@ub.ac.id; Tel.: +62-341-55-3623

Received: 4 March 2020; Accepted: 2 April 2020; Published: 3 April 2020

Abstract: The degree to which the maintenance of carbon (C) stocks and tree diversity can be jointly achieved in production landscapes is debated. C stocks in forests are decreased by logging before tree diversity is affected, while C stocks in monoculture tree plantations increase, but diversity does not. Agroforestry can break this hysteresis pattern, relevant for policies in search of synergy. We compared total C stocks and tree diversity among degraded forest, complex cacao/fruit tree agroforests, simple shade-tree cacao agroforestry, monoculture cacao, and annual crops in the Konawe District, Southeast Sulawesi, Indonesia. We evaluated farmer tree preferences and the utility value of the system for 40 farmers (male and female). The highest tree diversity (Shannon–Wiener H index 2.36) and C stocks (282 Mg C ha^{-1}) were found in degraded forest, followed by cacao-based agroforestry systems (H index ranged from 0.58–0.93 with C stocks of 75–89 Mg ha^{-1}). Male farmers selected timber and fruit tree species with economic benefits as shade trees, while female farmers preferred production for household needs (fruit trees and vegetables). Carbon stocks and tree diversity were positively related ($R^2 = 0.72$). Adding data from across Indonesia ($n = 102$), agroforestry systems had an intermediate position between forest decline and reforestation responses. Maintaining agroforestry in the landscape allows aboveground C stocks up to 50 Mg ha^{-1} and reduces biodiversity loss. Agroforestry facilitates climate change mitigation and biodiversity goals to be addressed simultaneously in sustainable production landscapes.

Keywords: carbon storage; cacao agroforestry; farmer tree preference; utility value

1. Introduction

The global relevance of managing landscapes simultaneously for resilient production (climate change adaptation), carbon (C) storage (climate change mitigation), biodiversity, and watershed functions has raised interest in the degree to which such functions tend to correlate [1,2], with some authors claiming causal links beyond correlations [3]. Synergy between combatting climate change (mitigation) and conserving biodiversity in production landscapes is desirable, and may be attained in agroforestry systems [4]. Theoretically, there is a positive feedback between biodiversity and ecosystem functioning [5,6], as trait diversity reduces vulnerability to external shocks, enhances niche differentiation and productivity [7], and thus carbon stocks. However, there are plateaus for each function considered at relatively low diversity levels which can be maintained by a few species in a particular functional group [8].

Biological diversity promotes various ecosystem functions at multiple spatial and temporal scales [9]. Within forest ecosystems, it allows species to access more available resources, and facilitate other species, resulting in enhanced resistance to disturbance and generally enhanced stability [10] at the system level. Various tree species from multiple aboveground strata provide different types of litter input, which contribute to soil surface protection and supply organic matter to support the nutrient cycle. Litter input maintains soil fertility, promotes biomass growth and enhances carbon dioxide sequestration from the atmosphere. The patterns are more complex within intact tropical forests [4,6,11,12]. Several studies concluded that there was a positive relationship between biodiversity and ecosystem services (including C stock) in intact tropical forest [6,12,13], while other studies found weak [14–16] or even a negative [17] relationships. These differences may be caused by differences in forest types and the biodiversity metrics that were used [9]. However, the degree of causality and the relevance of such relationships beyond (modified) natural forests is still uncertain and its relevance in the face of global climate change is debated [18].

In managed tropical forest landscapes, relationships between biodiversity and C storage may be less clear. Intact tropical forests are known for both their high biodiversity (including tree diversity) and high C stocks [12,19]. Converting forests to agricultural use may start with logging and/or land clearing through slash-and-burn agriculture that reduces plot-level aboveground biodiversity as well as C stock to zero, but in subsequent recovery the two characteristics are not necessarily aligned (Figure 1). Logging tropical forests reduces the terrestrial C stock, before it negatively affects biodiversity [20,21] (the "degradation leg"; Figure 1). Murdiyarso et al. [22] assessed the relationship between relative C stocks and relative plot-level biodiversity for a range of forest-derived land uses in the humid tropics, and found that the loss of C stocks was higher than that of plant species richness (data from 3 continents). Restoration can take place through replanting in plantations (low diversity), but high biodiversity as a result of natural seed dispersal will usually be achieved after C stocks increased (the "restoration leg"; Figure 1). Natural regeneration (secondary succession) can simultaneously increase C stocks and diversity over time, but C stocks recover faster than diversity [23]. Agroforestry systems are expected to occupy an intermediate position between the degradation and restoration leg, because they support a higher diversity.

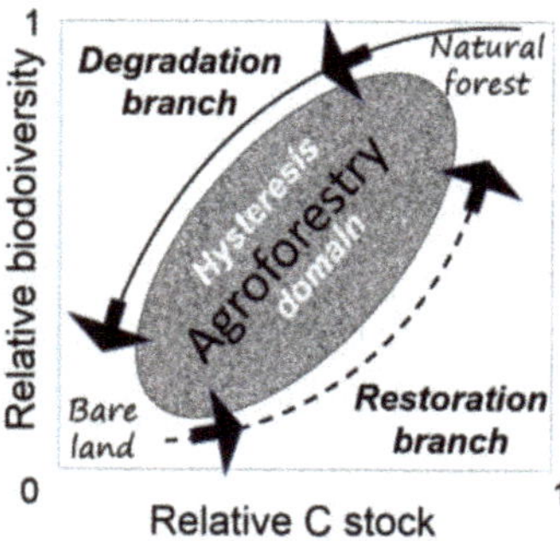

Figure 1. Relationship between carbon stocks and biodiversity across forest-derived land uses with agroforestry intermediate to degradation and restoration legs (modified from [22]).

The agroforestry literature for various climatic and biogeographical zones covers three orders of magnitude of tree diversity: 1–10, 10–100 and 100–1000 tree species [24]. Agroforests in Indonesia can harbour hundreds of tree species where a high influx of seeds from surrounding forests provides opportunities for farmers to selectively retain what appears to be of value. Forest conversion to other land-use systems at the forest/agriculture interface in Sulawesi [25] has followed a dual economy track of providing for local needs as well as income [26], with limited outsourcing of local food production when market-based income became reliable and the terms of trade favourable [27]. While becoming the largest cacao producing region of Indonesia, Sulawesi boomed in cacao-based land-use systems after 1980 [28], with further increased from 55,000 ha in 1990 to 230,000 ha in 2010, particularly in South and Southeast Sulawesi [29]. Due to its suitability to a broad range of climatic conditions, cacao may be as great a threat to tropical rainforests [18,30] as the more widely debated oil palm. Subsequent intensification in cacao agroforestry systems leads to shade tree removal, shifting the systems to become monoculture. The loss of diversity influences litter production, which is a major input to soil organic matter and influences the soil quality of the systems, plant growth and biomass production (including C stocks) [31,32]. Therefore, their potential to provide goods and services for local livelihoods, as well as global environmental services, may be reduced [33]. Expected cacao yield benefits may be short-lived. By 2013, cacao production in Indonesia decreased to 420,000 tons per year (around 40% of total production in 1990s) due to pests, diseases, and poor land management [34], and failed cacao intensification programs.

Cacao is traditionally grown in agroforestry systems with limited fertilizer inputs. Most of the cacao in Sulawesi is produced by smallholders in agroforestry systems with various shade trees, including fruit trees, timber and multipurpose tree species (MPTS) such as *Gliricidia sepium* (*'mother of cacao'*). The diverse mixture of trees in agroforestry systems has great potential for climate change mitigation through increased carbon storage, biodiversity conservation, and economic benefits from a range of products from shade trees [35]. Including shade trees may also maintain other ecosystem services that forests provide, such as temperature and humidity regulation, input of organic matter [36], and water and nutrient cycling [37].

However, management intensity directly affects stand composition and structural complexity [33], which influences the ecosystem services provided by agroforestry systems. The way farmers cultivate and integrate trees and cacao is likely tightly linked to their knowledge [30]. Farmers also determine how long agroforestry systems persist in the landscape. In addition, different knowledge, experience, and strategies of male and female farmers may lead to different shade tree management choices [38].

In this study, we explore the potential of agroforestry systems in Indonesia for supporting biodiversity maintenance and carbon storage, accounting for farmer preference in tree selection. We address the following research questions: (1) To what extent do cacao-based agroforestry, remaining forest and cacao monoculture systems differ in C stocks and species diversity? (2) What are the shade tree preferences based of male and female farmers? (3) Do C stocks in agroforestry systems correlate with tree diversity? We hypothesize that tree diversity and C stocks in agroforestry systems are positively related and that they are distinct from forest degradation and restoration curves due to farmer management actions (Figure 1).

2. Materials and Methods

2.1. Study Site

Field work was conducted in the Konawe District, Southeast Sulawesi, Indonesia, in the Konaweha watershed (3°15′0″–5°13′0″ S and 121°22′30″–122°31′0″ E), where cacao (*Theobroma cacao*) is commonly cultivated. In the greater study area, forests have been converted to intensive agricultural systems. After a few years of annual crop cultivation (maize, paddy and patchouli), land is often converted to cacao monoculture systems or simple cacao agroforestry systems. In the older plots (complex cacao agroforestry), fruit trees were inter-planted with cacao to increase income and address daily needs. Cacao is spaced at 3m × 3m, and the age of cacao ranges from 9–14 years. The average annual rainfall recorded is generally 1500–1900 mm [39] and the average temperature varies from 24 to 31 °C.

We selected plots using stratified sampling methods. We conducted initial rapid surveys in three villages (Lawonua, Wonuahoa, and Asinua Jaya) to characterize land-use system domination by analysing 45 land use system points. Based on this, we selected five land-use systems which represented the whole watershed: degraded forest (DF), complex cacao agroforestry (CAF) combined with fruit trees, simple shade cacao agroforestry (SAF), monoculture cacao (CM), and annual crops (CR). Per land-use system, three plots of 20 × 100 m were included in the study, based on common criteria: a minimum cacao age of 9 years, slope ranging from 0% to 15%, and similar soil texture (silt loam to silty clay loam). Subplots of 20m × 20m (sub-plot) were used for C stocks measurements for trees with 5–30 cm DBH. If a big tree (with DBH > 30 cm) was found in a subplot, then we extended the plot size to 100 m × 20 m (plot) for measuring the large tree component only (tree or necromass for trees > 30 cm DBH). Tree diversity was measured in the plot of 100 m × 20 m.

2.2. Plot Measurements

In each plot, the diameter at breast height (1.3 m above soil surface; DBH) of all trees with DBH > 5 cm was measured, and trees were identified by species. All trees < 5 cm DBH were included as understorey. Tree density was calculated based on the number of trees per ha. Plot basal area, the proportion of the area occupied by trees per ha, was calculated (m^2 ha^{-1}) as $\frac{(\frac{1}{4}\pi\Sigma_i DBH^2)}{land\ area}$, where DBH$_i$ is the DBH of tree i. Cacao agroforestry systems were selected in which cacao contributed less than 80% to the total basal area of the system [40]. We included both cacao and shade trees when calculating tree density, basal area, tree diversity and C stocks in the plots. Biodiversity was estimated using the Shannon–Wiener diversity index (H) = $-\sum_{i=1}^{s}(p_i)(\ln p_i)$, where p_i is the proportion of the number of individuals of species i divided by the total number of individuals [41].

2.3. C stock Estimation

We assessed C stocks using the RaCSA (Rapid Carbon Stock Appraisal) protocol [42], by quantifying five carbon pools including tree biomass, understorey biomass, necromass, litter and soil organic C [43]. Aboveground tree biomass was estimated using the allometric equation for humid tropical forest [44] and species-specific allometric equations for some plant species found in agroforestry systems, such as cacao [45], banana [46], and palms (Table 1). Wood density (WD) values were obtained from the wood density database of the World Agroforestry Centre (http://db.worldagroforestry.org/wd). Ten samples of understory biomass and litter (leaves and branches) were collected within each plot using a 0.5 × 0.5 m frame (Figure 2). Litter samples were oven-dried for 48 hours at 60 °C. Root biomass was estimated using the default shoot:root ratio of 4:1 in the tropics [47]. Biomass was converted into carbon by multiplying by 0.46 [42]. Ten soil samples per plot were collected at the same point beneath the understorey and litter frame at 3 depths (0–10 cm, 10–20 cm, and 20–30 cm). A composite soil sample for each plot was used to determine soil organic C. Soil C content was determined based on the Walkley and Black method [48]. The soil C stock was calculated by multiplying soil C content (%) with the bulk density (g cm^{-3}). Total C stocks (Mg ha^{-1}) were calculated as the sum of aboveground C (tree biomass, understorey biomass, necromass) and belowground C (roots and soil C stock (Mg ha^{-1})).

Table 1. Allometric equations for several plant species (AGB$_{est}$ = estimated aboveground biomass (kg); D = diameter at breast height (cm); WD = wood density (g cm^{-3}).

Plant species	Equation	Source
Tropical trees (moist forest)	AG$_{Best}$ = WD*exp($-1.499 + 2.148\ln(D) + 0.148\ln(D) + 0.207(\ln(D))^2 - 0.0281(\ln(D))^3$)	Chave et al., 2005
Cacao	AG$_{Best}$ = 0.1208 D$^{1.98}$	Yuliasmara, 2008
Banana	AG$_{Best}$ = 0.030 D$^{2.13}$	Arifin, 2001
Palms	AG$_{Best}$ = 0.118 D$^{2.53}$	Brown, 1997

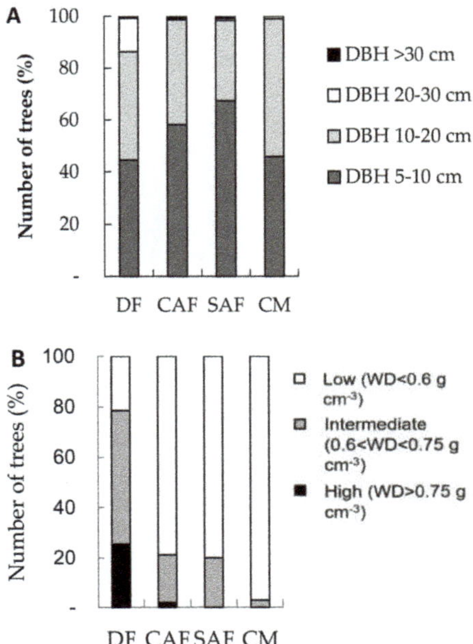

Figure 2. (**A**) The distribution of tree diameter for degraded forest (DF), cacao complex agroforestry (CAF), cacao simple agroforestry (SAF), and cacao monoculture (CM); (**B**) Number of trees based on wood density (WD) classes.

2.4. Farmer Preferences for Shade Tree Species

Informal in-depth (semi-structured) interviews of approximately 90 minutes were conducted with each farmer to assess whether farmers have preferences for certain tree species. We selected 20 agroforestry farm families from three villages using purposive sampling technique, which included farmers who managed the included agroforestry systems. We separately interviewed male and female farmers from each household. The preferences of both men and women determine what plant species are planted, as both male and female are actively involved in land management. In total, 40 farmers were interviewed, some of the farmers were Tolaki people (local people) and the others were Bugis people (immigrants from another province). Each interview consisted of two parts, with approximately 50 minutes focused on the question which trees farmers selected to maintain (remnant forest trees), plant, or tolerate in their agroforestry systems. The second part of approximately 40 minutes focused on the utility value of their shade trees, which ranged from 1 (not important) to 4 (very important).

2.5. Compilation of C Stocks and biodiversity Data from Previous Research

We evaluated additional published [15,49] and unpublished [50–54] data on tree diversity and C stocks that were collected using the same methodology, in order to compare the relationship between C stocks and tree diversity across a broader range of land-use systems. Data for 7 plots in primary forest, 23 plots in degraded forest, 42 plots in agroforestry systems, and 18 plots in plantations from Indonesia were included.

2.6. Data Analysis

Tree diversity, basal area and carbon stocks were compared among cacao agroforestry systems, degraded forest, and cacao monocultures (land-use systems) using a one-way ANOVA. Post hoc multiple comparisons between land-use systems were performed using Fisher's protected LSD. Prior to analysis, basal area, tree biomass, understorey biomass, necromass and litter biomass were log (base10) transformed to meet the normality assumption for analysis of variance. Differences were regarded as significant at $\alpha = 0.05$. We used linear and power regressions to evaluate the relationship between C stocks and tree diversity in agroforestry systems (at the local and regional scale). All statistical analyses were performed using Genstat (18th edition).

3. Results

3.1. Tree Diversity

Tree diversity was significantly different ($P < 0.05$) among land-use systems. The highest tree diversity (number of species and species diversity) was found in DF followed by cacao-based agroforestry systems. The Shannon–Wiener index (H) in DF was low (2.36), ranging from 2.26 to 2.52, but was substantially higher than values in cacao agroforestry and monoculture (Table 2). A total of 28 species were recorded in DF, dominated by *Myrtaceae, Fagaceae, Salicaceae, Clusiaceae, Elaeocarpaceae*, and *Primulaceae*. The five dominant tree species were *Sloanea sp., Castanopsis buruana, Fagraea fragrans, Homalium foetidum*, and *Metrosideros petiolata*. In contrast, only 18 species were found in CAF and SAF, where most species were fruit trees planted by farmers such as *Durio zibethinus, Lansium domesticum, Nephelium lappaceum*, and *Mangifera indica*. The shade tree *Gliricidia sepium* was also common. Other species, such as *Albizia procera* and *Fagraea fragrans*, were remnant forest trees.

Table 2. Structure and diversity of cacao systems compared to degrded forest (H = Shannon index; DF = Degraded forest; CAF = Cacao complex agroforestry; SAF = Cacao simple agroforestry; CM = Cacao monoculture).

Land Use Systems	Tree Density, Trees ha^{-1}	Total BA, m^2 ha^{-1}	Number of Species	H	Dominant/Codominant Species	Benefit
DF	1275[a]	18.42[a]	28	2.36[a]	*Metrosideros petiolata, Homalium foetidum*	Timber
CAF	1317[a]	9.14[b]	18	0.93[b]	*Theobroma cacao, Durio zibethinus, Lansium domesticum*	Fruits
SAF	1267[a]	7.63[b]	4	0.58[b]	*Theobroma cacao, Gliricidia sepium*	Fruits, Fodder
CM	900[b]	8.32[b]	2	0.24[c]	*Theobroma cacao*	Fruits

Note: Values not followed by the same letter were statistically different at P < 0.05.

3.2. Vegetation Structure

The highest tree density (including both cacao and shade trees) was found in CAF (1317 trees ha^{-1}) followed by DF and SAF (Table 2). Most of the shade trees in CAF were fruit trees, such as *D. zibethinus, L. domesticum, N. lappaceum*, and *M. indica*. Shade trees were planted less frequently in SAF than in CAF. Most of the shade trees planted in SAF were *G. sepium*, which is an important fodder resource for livestock. Total basal area significantly ($P < 0.01$) differed among land-use systems. The highest tree basal area was recorded in DF followed by the cacao agroforestry systems and monoculture (Table 2). Trees from 5 to 20 cm DBH (approximately 85% of all trees) dominated the DF plots (Figure 2A), but some trees were larger (> 20 cm DBH; 14% of all trees). Larger trees belonged to species medium-to-very-heavy WD, such as *H. foetidum* and *M, petiolata*. Cacao-based land-use systems were dominated by trees from 5 to 10 cm DBH (57% of all trees) and 10 to 20 cm DBH (42% of all trees). A few large trees (DBH > 30 cm) were found in CAF and SAF, but these accounted for just 1% of the total number of trees. However, almost 20% of the trees in CAF had high WD, which were *N. lappaceum* (WD > 0.9 g cm^{-3}) and *Citrus sinensis* (WD 0.78 g cm^{-3}) trees of 10–20 cm DBH (Figure 2B).

3.3. Tree Biomass, Necromass and C Stocks

The highest total tree biomass found was in DF (Table 3), whereas tree biomass in CM had the lowest value because of its low shade tree density. The total necromass found in CAF was relatively high compared to the other two cacao systems (41.5 Mg ha^{-1}). Land management was less intensive, because farmers did not transport the necromass. Ethnicity influenced the way farmers managed their cacao systems. Most of the CAF was managed by Tolaki people who tend to minimize the intensification of their land because of the lack of labour. In contrast, the Bugis people had more intensely managed cacao systems. Average litter biomass in cacao agroforestry systems was 5.6 Mg ha^{-1} and ranged from 4.3–7.0 Mg ha^{-1}. Thus, biomass stored in the litter layer contributed little to aboveground biomass across all land-use systems.

Table 3. Tree biomass, understorey biomass, necromass, and litter in cacao-based agroforestry systems (DF = Degraded forest; CAF = Cacao complex agroforestry; SAF = Cacao simple agroforestry; CM = Cacao monoculture; CR = Annual crops).

Land Use Systems	Tree, Mg ha^{-1}	Understorey, Mg ha^{-1}	Woody necromass, Mg ha^{-1}	Litter, Mg ha^{-1}
DF	288.41[a]	0.64[b]	137.54[a]	6.98[a]
CAF	52.64[b]	0.27[b]	41.53[ab]	7.02[a]
SAF	54.00[b]	0.11[b]	5.68[b]	5.51[a]
CM	35.25[bc]	0.19[b]	3.31[b]	4.26[a]
CR	0[c]	0.91[a]	0[b]	0[b]

Note: Values not followed by the same letter were statistically different at P < 0.05

The total C stock differed significantly ($P < 0.01$) among land-use systems. Total C stock was highest in DF (282 Mg ha^{-1}), followed by CAF (89 Mg ha^{-1}), SAF (75 Mg ha^{-1}), CM (56 Mg ha^{-1}) and CR (40 Mg ha^{-1}) (Figure 3). Aboveground C stocks (tree biomass, understorey biomass, necromass, and litter) contributed 80% to the total in DF, more than half of the total C stock (62%) in CAF, 48% in SAF and 43% in CM. Shade trees (fruit trees) in CAF contributed almost 30% of the aboveground C stock, while in SAF, the contribution was higher (almost 40%). Larger trees were found in SAF, but they had lower WD compared to CAF.

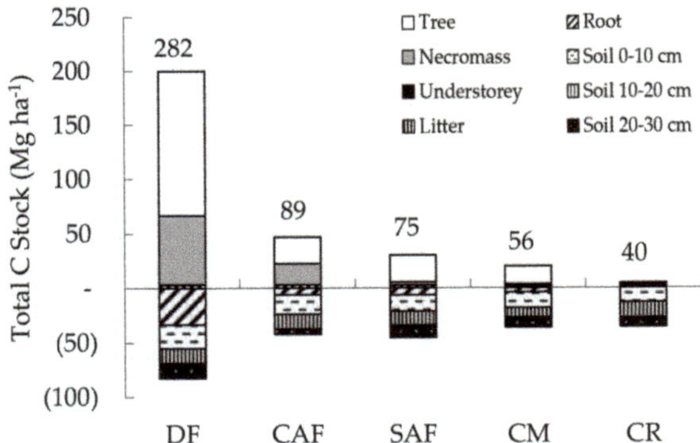

Figure 3. Total carbon stock in cacao-based agroforestry systems in 5 C pools compared to degraded forest and annual crop systems (DF = Degraded forest; CAF = Cacao complex agroforestry; SAF = Cacao simple agroforestry; CM = Cacao monoculture; CR=Annual crops).

Belowground C stocks (roots and soil organic matter) accounted for 18% of the total in DF, 41% in CAF, 52% in SAF and 57% in CM. Soil organic carbon was on average 38.4 Mg ha^{-1} and did not significantly differ ($P > 0.05$) among land-use systems.

3.4. Relationships between C Stocks and Tree Diversity in Agroforestry Systems

Within the cacao-based agroforestry systems in this study, we found a strong and positive relationship between the total C stock and tree diversity (Figure 4A). However, there was a weak relationship between tree diversity and soil C stock ($R^2 = 0.33$). When comparing the relationship between C stocks and tree diversity across a larger range of agroforestry systems in Indonesia (Figure 4B), variation was larger, but C stocks and tree diversity tended to be positively related.

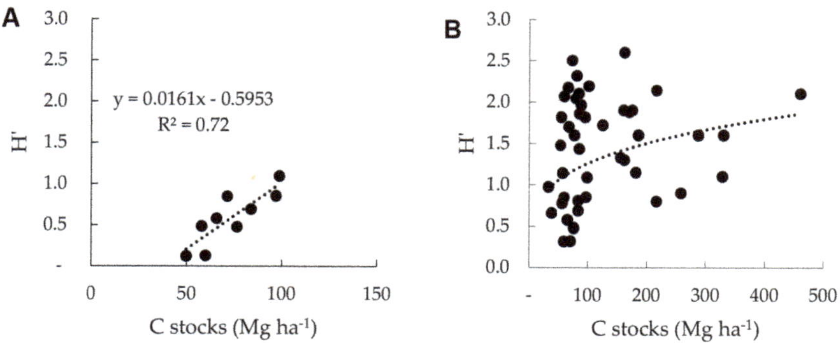

Figure 4. (**A**) The relation between total C stock and tree diversity (H index) in the cacao-based agroforestry systems in this study; (**B**) In agroforestry systems across Indonesia (primary data plus [15,49,52–54]).

We assessed whether C stocks and diversity were related across a wider range of land-use systems in Indonesia: plantations, agroforestry systems (complex and simple), and degraded and primary forest (Figure 5). The wide envelope of points appears to be bounded by an upper degradation leg and a lower restoration leg. We found a positive and relatively strong relationship between C stocks and diversity in primary and degraded forest ($R^2 = 0.50$) but weak relationships for agroforestry systems ($R^2 = 0.034$) and plantations ($R^2 = 0.032$). However, most of the agroforestry systems were in the area between the degradation and restoration leg, as agroforestry systems had higher diversity than monocultures (plantations).

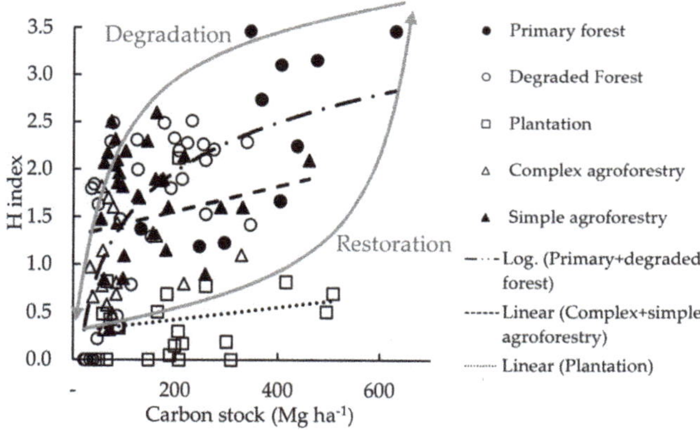

Figure 5. Total C stock and tree diversity (Shannon–Wiener index) in different land-use systems, n = 102 (primary data plus [15,49,52–54]).

3.5. Farmer Tree Preferences and Utility Value of Cacao-based Agroforestry Systems

The major reasons for selecting tree species in the systems were economic benefits and production for subsistence, but preference for certain species differed between male and female farmers (Table 4). Female farmers preferred shade tree species that provide food, such as fruit trees and vegetables, and/or social and cultural services. Banana and coconut, for example, are not only food crops, but are also used as traditional ceremonial ornaments. Male farmers preferred species with long-term economic benefits such as timber (*Tectona grandis* and *Anthocephalus cadamba*) and fruit trees of high economic value such as *D. zibethinus*, *L. domesticum*, and *N. lappaceum*. Most of the selected tree species preferred by male and female farmers had medium to high WD (WD > 0.6 g cm^{-3}).

Table 4. Top-ranked plant species in cacao agroforestry systems that are preferred by male and female farmers in the Konawe District, Southeast Sulawesi.

No	Species	Main Benefits	Frequency (% of Plots)	Farmer's Preference (Rank)	
				Male	Female
1	*Theobroma cacao* (cacao)	Bean	100	1	1
2	*Pogostemon cablin* (patchouli)	Oil	78	2	2
3	*Gliricidia sepium* (mother of cacao)	Fodder (leave), climbed tree for pepper	78	11	9
4	*Musa sp* (banana)	Fruits, vegetable (flower), cultural services (leaf), toys (trunk)	67	-	4
5	*Cocos nucifera* (coconut)	Fruits, cultural services (leaf), roof (leaf), toys (trunk)	56	-	5
6	*Capsicum annum* (chili pepper)	Vegetable/spice	56	-	7
7	*Piper nigrum* (pepper)	Vegetable/spice	44	10	6
8	*Fagraea fragrans* (tembesu)	Timber	44	3	10
9	*Tectona grandis* (teak)	Timber	44	4	11
10	*Anthocephalus cadamba* (jabon)	Timber	44	5	-
11	*Durio zibethinus* (durian)	Fruits	44	6	3
12	*Lansium domesticum* (langsat)	Fruits	22	7	-
13	*Mangifera indica* (mango)	Fruits	22	8	-
14	*Nephelium lappaceum* (rambutan)	Fruits	22	9	-

There were, in total, nine species of remnant forest trees in cacao-based agroforestry systems, 23 species of planted trees and four species of crops, according to the farmers. Almost 90% of the remnant trees that are maintained by the farmers are timber species such as *F. fragrans*, *Calophyllum* sp., *A. procera*, *Agathis* sp., and *Acacia* sp. *Arenga pinnata*, a palm species, was maintained for palm sugar production.

Farmer perspectives on the value of cacao agroforestry systems varied. The utility value of cacao-based agroforestry systems was considered to increase with time (next 30 years), especially in terms of providing food, medicine, handicraft, and recreation with an average score of 3 (Figure 6). Most farmers argued that in the future, they can no longer depend on forests because of continuous degradation. Farmers begin to recognize the importance of agroforestry systems in the provision of daily necessities. On the other hand, supply of firewood, handicraft, and decorations/ rituals were thought to become less important. Most of the respondents claimed that within the next three decades, firewood will not be required anymore because of the planned distribution of gas stoves for cooking. Likewise, some farmers predict a decrease in the production of handicrafts (woven bamboo hats and bags) due to competition from inexpensive imported products, as well as a decrease in the production of plant material for rituals and decoration because of less demand.

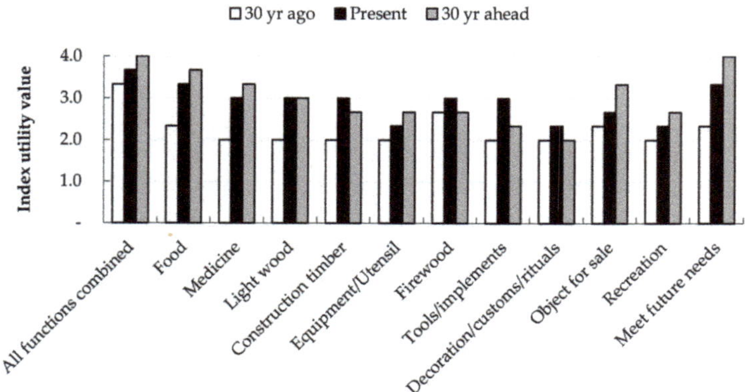

Figure 6. The utility value of cacao-based agroforestry systems based on its products 30 years ago, currently, and 30 years ahead based on farmer interviews.

4. Discussion

4.1. Land Use Affects Tree Diversity and C Stocks

Forest conversion to intensive agricultural systems led to a significant decrease in biodiversity and C stocks due to the loss of trees. However, planting shade trees, thus the establishment of (simple and complex) agroforestry systems, can gradually increase tree diversity. Compared to another study of cacao-based agroforestry systems in mid-west Ghana, the diversity in Konawe was still low [55], but systems had higher tree density than cacao plantations in Talamanca, Costa Rica (149 shade trees ha^{-1} and 475 cacao trees ha^{-1}) [35].

C stocks in annual crop systems were nearly 85% lower than in DF due to tree biomass removal (Figure 4). The higher tree density and diversity in cacao-based systems resulted in higher carbon stocks. Cacao land-use systems can increase from a total carbon stock of 56 Mg ha^{-1} in a monoculture to 75–89 Mg ha^{-1} if cacao is planted in a more complex agroforestry system. Shade trees played an important role in contributing to C stocks in agroforestry systems [56]. These results are similar to the C stocks of cacao-*Gliricidia* plots (56 Mg C ha^{-1}) in Central Sulawesi [31]. Nevertheless, we found lower C stocks than in a cacao agroforest in Mekoe, southern Cameroon, which reached a C stock of 250 Mg ha^{-1} [57]. Cacao-based agroforestry systems in Konawe had also lower carbon stocks than much older agroforestry systems in Central America (117 ± 47 Mg ha^{-1}) [58].

Soil C stocks in the Konawe district were relatively low (38 Mg ha^{-1}) compared to DF and another cacao study in southern Bahia, Brazil which had more soil carbon (57 Mg ha^{-1}) [36]. Land-use change from forest to cropping systems can decrease belowground C through soil surface exposure, a decrease in organic matter input and soil organic C degradation. Changes due to conversion of degraded forest into agricultural systems (CR) caused a soil C loss of approximately 8.5 Mg ha^{-1} in the top 10 cm. The relatively large soil C debt in Southeast Sulawesi, as we showed here, agrees with results of a recent global study by Sanderman et al. [59].

Above- and belowground C stock are linked in a virtuous cycle where organic matter accumulation (from litter and roots) increases the buffering to climate variability through a relatively stable microclimate and additional soil water storage, further enhancing plant growth. The type of tree species, and their abundance, will influence litter quality and quantity, which in turn affects soil organic matter, soil fertility and tree growth [32]. However, the estimation of the belowground part of biomass remains uncertain, with other authors estimating lower aboveground stocks for cocoa assuming an above-:below- ground biomass ratio of 7:1 [60]. We found no significant relationship between tree diversity and soil C stocks. A recent review that included our case study data suggested that increases in soil bulk density tended to conceal changes in C concentration per unit soil dry weight, as C stocks are the product of these two parameters [61]. This result aligns with Wartenberg et al. [62] who also found that there was no relation among tree diversity, soil C-content and total soil N level.

A recent summary of soil C data for Indonesian forest transitions that included the current data points concluded that increased bulk density partly conceals changes in soil organic C concentrations (per unit dry soil) within the globally agreed accounting of changes in soil C stock in the top 30 cm of the soil [61].

4.2. Gendered Tree Preferences Increased Diversity, C Stocks and the Utility Value of Agroforestry Systems

Farmer preferences on shade tree species relate to the enhancement of their livelihood and are represented in their land management [8]. Gendered species choices for shade trees resulted in species-rich agroforestry systems because male and female farmers have different knowledge on, and experience with, land management and utilization [38]. Similar results were found in another study in the southern Philippines, where male smallholder farmers favoured fruit trees, while female farmers preferred food crops [63]. Overall, tree selection by farmers was influenced by how much benefit the trees provide [64,65], ease of maintenance, and their drought tolerance to avoid income loss as a result of climatic fluctuations.

Differing WD, as represented by farmers' choices, not only enriches the diversity but also contributes to C sequestration. Compared to other agricultural systems, the diversity of shade trees will remain in agroforestry systems for a long time since farmers usually plant a new tree before the old tree is harvested, which maintains C stocks. The longer trees persist in the system, the more C they will store as they increase in size [55]. Management practices geared towards high shade tree density in cacao agroforestry systems can maintain ecosystem functions that forests provide, such as aboveground C sequestration [56], enhancing the belowground C stock through organic matter addition from litter and roots, and increase resilience against changes to climate change [66].

4.3. Relationships between C Stocks and Tree Diversity within and Across Land-use Systems

The relationship between C stocks and tree diversity within agroforestry systems in our research site was positive but not as strong as that within forests. A positive relationship between C stocks and biodiversity in forests has been demonstrated at different sub-climate [9] and global scales [67]. The relation between C stocks and diversity in natural forests is driven by ecological processes: high diversity allows more niches to be occupied, supporting a higher level of plant productivity [8]. In agroforestry systems, in contrast, farmers select the trees in the system, and therefore determine the diversity of agroforestry systems.

Forest degradation through logging reduces carbon stocks before affecting diversity [20,21], while a sustained loss of tree biomass leads to species loss and gradually decreases biodiversity. In terms of restoration, increases in biodiversity will usually be achieved after increases in C stocks have been achieved [23]. People tend to choose the most productive species in plantation systems to rapidly increase C stocks over time, resulting in the restoration leg with relatively high C stocks and low diversity (Figure 5). Our results suggest an intermediate position for agroforestry systems, between forest decline and restoration responses (Figures 1 and 5), with some plots demonstrating intermediate diversity and low C stocks, and some intermediate C stocks but low tree diversity, breaking the hysteresis pattern.

5. Conclusions

The diverse mixture of shade trees with medium to high wood density contributes to C stock maintenance in cacao-based agroforestry systems. We found a positive relationship between C stocks and tree diversity, which suggests that management practices in cacao agroforestry systems increase both C stocks and tree diversity. Gendered species preferences for shade tree species in cacao-based agroforestry systems contributed to increasing the tree diversity of the system. Farmer preferences for tree species depended on the benefit that shade tree species provide, where male and female farmers prioritized economic benefits and subsistence production, respectively.

Agroforestry systems with a mixture of diverse tree species provide more functions, both for the environment and the local people in fulfilling daily needs, including income generation. Less forest area in the future will increase the importance of trees in agroforestry systems [68]. Looking 30 years ahead, in or beyond the lifetime of the trees currently planted, farmers saw a greater need for agroforests to provide products that so far were derived from forests, and a necessary need for an increase in shade tree functional diversity. The maintenance of C stocks, biodiversity conservation and locally relevant ecosystem services (protection of the soil surface through a litter layer) will benefit from such choices.

Author Contributions: Conceptualization, R.R.S., M.V.N., and K.H.; methodology, R.R.S., D.D.S., K.H., and M.V.N.; writing-original draft preparation, R.R.S.; writing-review and editing, M.V.N., D.M.A.R., J.M.R., and K.H.; data analysis, R.R.S., D.M.A.R., and D.D.S.; field supervision, K.H., J.M.R.; writing supervision, M.V.N., and D.M.A.R.; funding acquisition, J.M.R. All authors have read and agreed to the published version of the manuscript.

Funding: Most of this research was supported through the Agroforestry and Forestry in Sulawesi: Linking Knowledge to Action (AgFor) project (Contribution Arrangement No. 7056890) funded by the Government of Canada, represented by the Minister of International Development acting through Global Affairs Canada under the leadership of the World Agroforestry Center (ICRAF)-Southeast Asia.

Acknowledgments: We acknowledge the World Agroforestry Centre for facilitating this research. Many thanks to WCU (World Class University) Program -SAMA (Scenarios Analysis for Management of Agroforests) year 2019, Universitas Brawijaya, for supporting the writing process and Directorate General of Resources for Science, Technology and Higher Education of the Republic of Indonesia for supporting this study. We are grateful to the local community in the Wonuahoa, Lawonua, and Asinua Jaya villages for the kind help, support and warm family welcome.

Conflicts of Interest: The authors declare no conflict of interest.

References

1. Van Noordwijk, M. Integrated Natural Resource Management as Pathway to Poverty Reduction: Innovating Practices, Institutions and Policies. *Agric. Syst.* **2019**, *172*, 60–71. [CrossRef]

2. Barrios, E.; Valencia, V.; Jonsson, M.; Brauman, A.; Hairiah, K.; Mortimer, P.E.; Okubo, S. Contribution of Trees to the Conservation of Biodiversity and Ecosystem Services in Agricultural Landscapes. *Int. J. Biodivers. Sci. Ecosyst. Serv. Manag.* **2018**, *14*, 1–16. [CrossRef]

3. Cavanaugh, K.C.; Gosnell, J.S.; Davis, S.L.; Ahumada, J.; Boundja, P.; Clark, D.B.; Mugerwa, B.; Jansen, P.A.; O'Brien, T.G.; Rovero, F.; et al. Carbon storage in tropical forests correlates with taxonomic diversity and functional dominance on a global scale. *Glob. Ecol. Biogeogr.* **2014**, *23*, 563–573. [CrossRef]

4. Van de Perre, F.; Willig, M.R.; Presley, S.J.; Andemwana, F.B.; Beeckman, H.; Boeckx, P.; Cooleman, S.; de Haan, M.; De Kesel, A.; Dessein, S.; et al. Reconciling biodiversity and carbon stock conservation in an Afrotropical forest landscape. *Sci. Adv.* **2018**, *4*, eaar6603. [CrossRef] [PubMed]

5. Loreau, M.; Hector, A. Partitioning selection and complementarity in biodiversity experiments. *Nature* **2001**, *412*, 72–76. [CrossRef] [PubMed]

6. Hicks, C.; Woroniecki, S.; Fancourt, M.; Bieri, M.; Garcia Robles, H.; Trumper, K.; Mant, R. *The Relationship between Biodiversity, Carbon Storage and the Provision of other Ecosystem Services: Critical Review for the Forestry Component of the International Climate Fund*; Department for International Development: Cambridge, UK, 2014; p. 119.

7. Cardinale, B.J.; Matulich, K.L.; Hooper, D.U.; Byrnes, J.E.; Duffy, E.; Gamfeldt, L.; Balvanera, P.; O'Connor, M.I.; Gonzalez, A. The functional role of producer diversity in ecosystems. *Am. J. Bot.* **2011**, *98*, 572–592. [CrossRef] [PubMed]

8. Swift, M.J.; Izac, A.M.N.; van Noordwijk, M. Biodiversity and ecosystem services in agricultural landscapes—Are we asking the right questions? *Agric. Ecosyst. Environ.* **2004**, *104*, 113–134. [CrossRef]

9. Lecina-Diaz, J.; Alvarez, A.; Regos, A.; Drapeau, P.; Paquette, A.; Messier, C.; Retana, J. The positive carbon stocks-biodiversity relationship in forests: Co-occurrence and drivers across five subclimates. *Ecol. Appl.* **2018**, *28*, 1481–1493. [CrossRef]

10. Healy, C.; Gotelli, N.J.; Potvin, C. Partitioning the effects of biodiversity and environmental heterogeneity for productivity and mortality in a tropical tree plantation. *J. Ecol.* **2008**, *96*, 903–913. [CrossRef]

11. Sullivan, M.J.; Talbot, J.; Lewis, S.L.; Phillips, O.L.; Qie, L.; Begne, S.K.; Chave, J.; Cuni-Sanchez, A.; Hubau, W.; Lopez-Gonzalez, G.; et al. Diversity and carbon storage across the tropical forest biome. *Sci. Rep.* **2017**, *7*, 39102. [CrossRef]

12. Poorter, L.; van der Sande, M.T.; Thompson, J.; Arets, E.J.M.M.; Alarcón, A.; Álvarez-Sánchez, J.; Ascarrunz, N.; Balvanera, P.; Barajas-Guzmán, G.; Boit, A.; et al. Diversity enhances carbon storage in tropical forests. *Glob. Ecol. Biogeogr.* **2015**, *24*, 1314–1328. [CrossRef]

13. Conti, G.; Díaz, S. Plant functional diversity and carbon storage—An empirical test in semi-arid forest ecosystems. *J. Ecol.* **2013**, *101*, 18–28. [CrossRef]

14. Mandal, R.A.; Dutta, I.C.; Jha, P.K.; Karmacharya, S. Relationship between carbon stock and plant biodiversity in collaborative forests in Terai, Nepal. *ISRN Bot.* **2013**, *2013*, 625767. [CrossRef]

15. Natalia, D.; Arisoesilaningsih, E.; Hairiah, K. Are high carbon stocks in agroforests and forest associated with high plant species diversity? *Agrivita* **2017**, *39*, 74–82. [CrossRef]

16. Kirby, K.R.; Potvin, C. Variation in carbon storage among tree species: Implications for the management of a small-scale carbon sink project. *For. Ecol. Manag.* **2007**, *246*, 208–221. [CrossRef]

17. Murray, J.P.; Grenyer, R.; Wunder, S.; Raes, N.; Jones, J.P. Spatial patterns of carbon, biodiversity, deforestation threat, and REDD+ projects in Indonesia. *Conserv. Biol.* **2015**, *29*, 1434–1445. [CrossRef]

18. Dewi, S.; van Noordwijk, M.; Zulkarnain, M.T.; Dwiputra, A.; Hyman, G.; Prabhu, R.; Gitz, V.; Nasi, R. Tropical forest-transition landscapes: A portfolio for studying people, tree crops and agro-ecological change in context. *Int. J. Biodivers. Sci. Ecosyst. Serv. Manag.* **2017**, *13*, 312–329. [CrossRef]

19. Sanchez, J.L.M.; Cabrales, L.C.; Académica, D.; Biológicas, D.C.; Juárez, U.; Tabasco, A.D. Is there a relationship between floristic diversity and carbon stocks in tropical vegetation in Mexico? *Afr. J. Agric. Res.* **2012**, *7*, 2584–2591. [CrossRef]

20. Kalaba, F.K.; Quinn, C.H.; Dougill, A.J.; Vinya, R. Floristic composition, species diversity and carbon storage in charcoal and agriculture fallows and management implications in Miombo woodlands of Zambia. *For. Ecol. Manag.* **2013**, *304*, 99–109. [CrossRef]

21. Rutishauser, E.; Hérault, B.; Baraloto, C.; Blanc, L.; Descroix, L.; Sotta, E.D.; Ferreira, J.; Kanashiro, M.; Mazzei, L.; D'Oliveira, M.V.N.; et al. Rapid tree carbon stock recovery in managed Amazonian forests. *Curr. Biol.* **2015**, *25*, R787–R788. [CrossRef]

22. Murdiyarso, D.; van Noordwijk, M.; Wasrin, U.R.; Tomich, T.P.; Gillison, A.N. Environmental benefits and sustainable land-use options in the Jambi transect, Sumatra. *J. Veg. Sci.* **2002**, *13*, 429–438. [CrossRef]

23. Martin, P.A.; Newton, A.C.; Bullock, J.M. Carbon pools recover more quickly than plant biodiversity in tropical secondary forests. *Proc. Biol. Sci.* **2013**, *280*, 20132236. [CrossRef] [PubMed]

24. Van Noordwijk, M.; Rahayu, S.; Gebrekirstos, A.; Kindt, R.; Tata, H.L.; Muchugi, A.; Ordonnez, J.C.; Xu, J. Tree diversity as basis of agroforestry. In *Sustainable Development through Trees on Farms: Agroforestry in Its Fifth Decade*; Van Noordwijk, M., Ed.; World Agroforestry Centre (ICRAF) Southeast Asia Regional Program: Bogor, Indonesia, 2019; pp. 17–44.

25. Kehlenbeck, K.; Maass, B.L. Crop diversity and classi cation of homegardens in Central Sulawesi, Indonesia. *Differences* **2004**, *63*, 53–62.

26. Dove, M.R. *The Banana Tree at the Gate: A History of Marginal Peoples and Global Markets in Borneo*; Yale University Press: New Haven, CT, USA, 2011; Volume 3, p. 332.

27. Van Noordwijk, M.; Agus, F.; Dewi, S.; Purnomo, H. Reducing emissions from land use in Indonesia: Motivation, policy instruments and expected funding streams. *Mitig. Adapt. Strateg. Glob. Chang.* **2014**, *19*, 677–692. [CrossRef]

28. Schroth, G.; Harvey, C.A. Biodiversity conservation in cocoa production landscapes: An overview. *Biodivers. Conserv.* **2007**, *16*, 2237–2244. [CrossRef]

29. Rahmanulloh, A.; Sofiyuddin, M. *Profitability of Land-Use Systems in South Sulawesi and Southeast Sulawesi*; World Agroforestry Centre: Bogor, Indonesia, 2012; p. 16. [CrossRef]

30. Dumont, E.S.; Gnahoua, G.M.; Ohouo, L.; Sinclair, F.L.; Vaast, P. Farmers in Cote d'Ivoire value integrating tree diversity in cocoa for the provision of ecosystem services. *Agrofor. Syst.* **2014**, *88*, 1047–1066. [CrossRef]

31. Smiley, G.L.; Kroschel, J. Temporal change in carbon stocks of cocoa-gliricidia agroforests in Central Sulawesi, Indonesia. *Agrofor. Syst.* **2008**, *73*, 219–231. [CrossRef]

32. Triadiati, T.S.; Guhardja, E.; Sudarsono, Q.I.; Leuschner, C. Litterfall production and leaf-litter decomposition at natural forest and cacao agroforestry in Central Sulawesi, Indonesia. *Asian J. Biol. Sci.* **2011**, *4*, 2011.

33. Deheuvels, O.; Rousseau, G.X.; Quiroga, G.S.; Franco, M.D.; Cerda, R.; Mendoza, S.J.V.; Somarriba, E. Biodiversity is affected by changes in management intensity of cocoa-based agroforests. *Agrofor. Syst.* **2014**, *88*, 1081–1099. [CrossRef]

34. Benton, T. *Sustainable Intensification*; International Plant Nutrition Institute (IPNI): Indonesia, 2016; pp. 95–109. [CrossRef]

35. Cerda, R.; Deheuvels, O.; Calvache, D.; Niehaus, L.; Saenz, Y.; Kent, J.; Vilchez, S.; Villota, A.; Martinez, C.; Somarriba, E. Contribution of cocoa agroforestry systems to family income and domestic consumption: Looking toward intensification. *Agrofor. Syst.* **2014**, *88*, 957–981. [CrossRef]

36. Monroe, P.H.M.; Gama-Rodrigues, E.F.; Gama-Rodrigues, A.C.; Marques, J.R.B. Soil carbon stocks and origin under different cacao agroforestry systems in southern Bahia, Brazil. *Agric. Ecosyst. Environ.* **2016**, *221*, 99–108. [CrossRef]

37. Jose, S. Agroforestry for ecosystem services and environmental benefits: An overview. *Agrofor. Syst.* **2009**, *76*, 1–10. [CrossRef]

38. Mulyoutami, E.; Roshetko, J.M.; Martini, E.; Awalina, D. Gender roles and knowledge in plant species selection and domestication: A case study in South and Southeast Sulawesi. *Int. For. Rev.* **2015**, *17*, 99–111. [CrossRef]

39. BPS. *Konawe in Fugures*; Konawe, B.K., Ed.; Statistics of Konawe Regency: Regency, Indonesia, 2015; p. 286.

40. Hairiah, K.; Sulistyani, H.; Suprayogo, D.; Purnomosidhi, P.; Widodo, R.H.; van Noordwijk, M. Litter layer residence time in forest and coffee agroforestry systems in Sumberjaya, West Lampung. *For. Ecol. Manag.* **2006**, *224*, 45–57. [CrossRef]

41. Shannon, C.E. A mathematical theory of communication. *Bell Syst. Tech. J.* **1948**, *27*, 379–423. [CrossRef]

42. Hairiah, K.; Dewi, S.; Agus, F.; Velarde, S.; Andree, E.; Rahayu, S.; van Noordwijk, M. *Measuring Carbon Stocks*; World Agroforestry Centre: Bogor, Indonesia, 2011; p. 95.

43. Eggleston, S.; Buendia, L.; Miwa, K.; Ngara, T.; Tanabe, K. *2006 IPCC Guidelines for National Greenhouse Gas Inventories*; Eggleston, S., Buendia, L., Miwa, K., Ngara, T., Tanabe, K., Eds.; IPCC: Geneva, Switzerland, 2006; p. 6.

44. Chave, J.; Andalo, C.; Brown, S.; Cairns, M.A.; Chambers, J.Q.; Eamus, D.; Fölster, H.; Fromard, F.; Higuchi, N.; Kira, T.; et al. Tree allometry and improved estimation of carbon stocks and balance in tropical forests. *Oecologia* **2005**, *145*, 87–99. [CrossRef]

45. Yuliasmara, F.; Wibawa, A.; Prawoto, A.A. Carbon stock in different ages and plantation system of cocoa: Allometric approach. *Pelita Perkeb.* **2009**, *25*, 86–100. [CrossRef]

46. Arifin, J. *Estimasi Cadangan Karbon pada Berbagai Sistem Penggunaan Lahan di Kecamatan Ngantang, Malang*; Brawijaya University: Malang, Indonesia, 2001.

47. Mokany, K.; Raison, R.J.; Prokushkin, A.S. Critical analysis of root: Shoot ratios in terrestrial biomes. *Glob. Chang. Biol.* **2006**, *12*, 84–96. [CrossRef]

48. Walkley, A.; Black, I.A. An examination of the degtjareff method for determining soil organic matter, and a proposed modification of the chromic acid tritation method. *Soil Sci.* **1934**, *37*, 29–38. [CrossRef]

49. Markum, M.; Soesilaningsih, E.A.; Suprayogo, D.; Hairiah, K. Plant species diversity in relation to carbon stocks at Jangkok watershed, Lombok Island. *AGRIVITA* **2013**, *35*, 45–63. [CrossRef]

50. Faranisa, R. Keanekaragaman Flora Pada Berbagai Tingkat Kepadatan Tanah di Hutan Pegunungan Taman Nasional Bromo Tengger Semeru (TNBTS). Bachelor's Thesis, Brawijaya University, Malang, Indonesia, 2017.

51. Kendom, M. Estimasi Emisi Karbon di DAS Casteel Timur Berdasarkan Perubahan Tutupan Hutan Alami di Wilayah Kabupaten Asmat, Papua. Master's Thesis, Brawijaya University, Malang, Indonesia, 2013.

52. Hairiah, K.; Suprayogo, D.; Saputra, D.D. *Reklamasi Lahan Pertanian Pascaerupsi Gunung Kelud: Efisiensi Serapan N Dalam System Agroforestry Kopi-Kakao*; Brawijaya University: Malang, Indonesia, 2017; p. 53.

53. Kurniawan, S. Bio-Physical Characteristics, Vegetation, Biomass, Carbon Stock, and Carbon Sequestration in the Bangsri Micro Watershed-East Java. In *Final Report CCCD Project*; Brawijaya University and Research group of Tropical Agroforestry: Malang, Indonesia, 2018; p. 53.

54. Prayogo, C. *Biodiversity and Ecosystem Management in the Bangsri Micro Watershed-East Java*. *Final Report CCCD Project*; Brawijaya University and Research Group of Tropical Agroforestry: Malang, Indonesia, 2018; p. 65.

55. Dawoe, E.; Asante, W.; Acheampong, E.; Bosu, P. Shade tree diversity and aboveground carbon stocks in Theobroma cacao agroforestry systems: Implications for REDD + implementation in a West African cacao landscape. *Carbon Balance Manag.* **2016**, *11*, 17. [CrossRef] [PubMed]

56. Middendorp, R.S.; Vanacker, V.; Lambin, E.F. Impacts of shaded agroforestry management on carbon sequestration, biodiversity and farmers income in cocoa production landscapes. *Landsc. Ecol.* **2018**, *33*, 1953–1974. [CrossRef]

57. Duguma, B.; Gockowski, J.; Bakala, J. Smallholder cacao (*Theobroma cacao* Linn.) cultivation in agroforestry systems of West and Central Africa: Challenges and opportunities. *Agrofor. Syst.* **2001**, *51*, 177–188. [CrossRef]

58. Somarriba, E.; Cerda, R.; Orozco, L.; Cifuentes, M.; Dávila, H.; Espin, T.; Mavisoy, H.; Ávila, G.; Alvarado, E.; Poveda, V.; et al. Carbon stocks and cocoa yields in agroforestry systems of Central America. *Agric. Ecosyst. Environ.* **2013**, *173*, 46–57. [CrossRef]

59. Sanderman, J.; Hengl, T.; Fiske, G.J. Soil carbon debt of 12,000 years of human land use. *Proc. Natl. Acad. Sci. USA* **2017**, *114*, 9575–9580. [CrossRef]

60. Norgrove, L.; Hauser, S. Carbon stocks in shaded Theobroma cacao farms and adjacent secondary forests of similar age in Cameroon. *Trop. Ecol.* **2013**, *54*, 15–22.

61. Hairiah, K.; van Noordwijk, M.; Sari, R.R.; Saputra, D.D.; Suprayogo, D.; Kurniawan, S.; Prayogo, C.; Gusli, S. Soil carbon stocks in Indonesian (agro) forest transitions: Compaction conceals lower carbon concentrations in standard accounting. *Agric. Ecosyst. Environ.* **2020**, *294*, 106879. [CrossRef]

62. Wartenberg, A.C.; Blaser, W.J.; Gattinger, A.; Roshetko, J.M.; Van Noordwijk, M.; Six, J. Does shade tree diversity increase soil fertility in cocoa plantations? *Agric. Ecosyst. Environ.* **2017**, *248*, 190–199. [CrossRef]

63. Ureta, J.U.; Evangelista, K.P.A.; Habito, C.M.D.; Lasco, R.D. Exploring gender preferences in farming system and tree species selection: Perspectives of JESAM smallholder farmers in southern Philippines. *J. Environ. Sci. Manag.* **2016**, *I*, 56–73.

64. Lasco, R.D.; Espaldon, M.L.O.; Habito, C.M.D. Smallholder farmers' perceptions of climate change and the roles of trees and agroforestry in climate risk adaptation: Evidence from Bohol, Philippines. *Agrofor. Syst.* **2015**, *90*, 521–540. [CrossRef]

65. Mustapha, R.I.; Jimosh, S.O. Farmer's preferences for tree species on agroforestry system in Ijebu North Local Government Area, Ogun State, Nigeria. *J. Agric. For. Soc. Sci.* **2012**, *10*, 176–187.

66. Nair, P.K.R.; Garrity, D. *Agroforestry—The Future of Global Land Use*; Springer International Publishing: New York, NY, USA, 2012.

67. Liang, J.; Crowther, T.W.; Picard, N.; Wiser, S.; Zhou, M.; Alberti, G.; Schulze, E.D.; McGuire, A.D.; Bozzato, F.; Pretzsch, H.; et al. Positive biodiversity-productivity relationship predominant in global forests. *Science* **2016**, *354*, aaf8957. [CrossRef] [PubMed]

68. Roshetko, J.M.; Snelder, D.J.; Lasco, R.D.; Van Noordwijk, M. Future challenge: A Paradigm shift in the forestry sector. In *Smallholder Tree Growing for Rural Development and Environmental Services*; Snelder, D.J., Lasco, R.D., Eds.; Springer: Berlin/Heidelberg, Germany, 2008; pp. 453–485.

 *land*

Article

Agroforestry Innovation through Planned Farmer Behavior: Trimming in Pine–Coffee Systems

Edi Dwi Cahyono [1,*], Salsabila Fairuzzana [1], Deltanti Willianto [1], Eka Pradesti [1], Niall P. McNamara [2], Rebecca L. Rowe [2]  and Meine van Noordwijk [3]

[1] Faculty of Agriculture, Universitas Brawijaya, Malang 65145, Indonesia; salsabilazatira@gmail.com (S.F.); deltantiwww@gmail.com (D.W.); eka_pradesti@student.ub.ac.id (E.P.)

[2] UK Centre of Ecology & Hydrology, Lancaster LA14AP, UK; nmcn@ceh.ac.uk (N.P.M.); rebrow@ceh.ac.uk (R.L.R.)

[3] World Agroforestry (ICRAF), Bogor 16155, Indonesia; m.vannoordwijk@cgiar.org

* Correspondence: edidwi.fp@ub.ac.id

Received: 26 August 2020; Accepted: 22 September 2020; Published: 30 September 2020

Abstract: Knowledge transfer depends on the motivations of the target users. A case study of the intention of Indonesian coffee farmers to use a tree canopy trimming technique in pine–based agroforestry highlights path-dependency and complexity of social-ecological relationships. Farmers have contracts permitting coffee cultivation under pine trees owned by the state forestry company but have no right to fell trees. A multidisciplinary international team of scientists supported farmers at the University of Brawijaya Forest in East Java to trial canopy trimming to improve light for coffee production while maintaining tree density. Data were collected using surveys through interviews, case study analysis using in-depth interviews, focus group discussions and nonparticipant observations. Using the Theory of Planned Behavior, we found that though farmer attitudes toward trimming techniques were positive, several factors needed to be scrutinized: perceived limited socio-policy support and resources. While there is hope that canopy trimming can improve coffee production and local ecosystem services, a participatory and integrative extension and communication strategy will be needed. In the relationship between farmers as agents and forest authorities as principals, any agroforestry innovation needs to incorporate knowledge and concerns in the triangle of farmers, policymakers and empirical science.

Keywords: agroforestry; innovation transfer; trimming; intention; participatory and integrative research-extension; stakeholders

---

1. Introduction

Innovators—and especially developers of innovations that others are supposed to use—may assume that once innovations are transferred they are inherently beneficial and accessible to the potential adopters [1–4]. In other words, they are subject to a "pro-innovation bias". By starting from the needs of the targeted adopters a different design of an innovation process may lead to alternate outcomes [5–8]. How benefits and costs of any innovation are distributed is key to the acceptance of innovations, but groups, rituals, affiliation, status and power are major drivers of human decision-making in a social context [9]. In a context of agroforestry, trees and farmers interact, at the plot, landscape and policy scales [10]. Ecological interactions between trees and other agroforestry components, both complementary and antagonistic, can be a mirror for the social interactions between forest institutions and farmers. This is especially true in a specific form of agroforestry where forestry institutions own the land and are interested in permanent tree cover and timber harvests. While farmers may benefit indirectly or directly from timber production, they are primarily benefiting from and interested in understory crops leading to potentially antagonistic relationships. The principal–agent theory, as developed in institutional economics, analyses the strategic interactions between principals who delegate, often limited in time or scope, a conditional grant of authority to an agent [11,12]. An agent (e.g., farmers or farmer groups) is an individual or entity that has the authority to act on behalf of the principal (party or parties, e.g., forestry institution) who has power over the agent. A dispute—namely the principal–agent problem—arises when the actions of agents pursue different interests or even are in conflict with those of the principal [12]. Forms of "social forestry" (locally known as "tumpangsari" and internationally as "taungya") as implemented on densely populated Java (Indonesia) have such "principal–agent" interactions between the forestry principals, who under the social forestry program have granted increased authority to farmer to act as "agents" for the "principal" [13]. This principle–agent interaction faces the inherent challenges of this type of relationship, with little harmony in management objectives expected and strict enforcement of forestry rules essential, but not always feasible leading to the risk of agent shirking and slippage. Analysis [14] suggests that other types of benefit-sharing arrangements are needed before a much wider range of locally adapted solutions can be found, which more fully exploit the opportunities of tree–crop synergy. Social forestry programs in Indonesia have tried to bring more balance to the stakes of the forestry principals and farmer agents, including timber sharing arrangements [15]. However, technical innovation may still be constrained by the institutional context and its history (path dependency), as our case study explores.

For many decades, the interaction between Indonesian forest authorities and forest-edge farmers who have tried to expand their coffee into existing forests has been strongly antagonistic in Indonesia, as also evident elsewhere [16–18]. Only after the 1997 "reformasi" period and the 1998 forestry law, the "tumpangsari"—contracts that allow farmers to grow annual food crops among newly planted trees for a few years—were extended to allow longer-term shade-tolerant perennials such as coffee or fodder for a cut-and-carry system to be grown [19–21]. For the state forest company authorized to manage these lands, the survival and growth of the tree plantation has been the priority and farmers risked being left out of future contracts if trees were found to have been damaged. The intercrops were contractually treated as a "share-crop", with around one-third of the yield belonging to the "landlord," (in this case, the state forest company). These rules gave the state forest company a stake in the productivity of intercrops (different from when a fixed "land rent" would have been charged) but reduced the profitability for the farmer. Any intervention, such as social innovations from forest authorities or technical innovations from development agencies can destabilize the agent–farmer economic structure and cause disputes with the existing authority-principal policies, which cause resistance towards agroforestry innovation [22,23]. A balance needs to be found between agents (or principals) who may be profiteers, prove-it accountants or visionary prophets [24,25]. Profiteers are profit agents (entrepreneurs) as part of the national economy; prove-it accountants are intermediary agents that encourage transparent processes for mutual benefit. Prophets see themselves as representative of the local community to voice their interests, needs and perspectives in the complexity of the external environment. The roles of NGOs (prove-it accountants) as an intermediary for marginal groups to negotiate with the authorities (profiteers) have been identified [26,27]. On the other hand, efforts to put forward local forest community interest ("the prophet") form a challenge [28] that open spaces for more a deeper study.

Our case study focuses on an area of forest on the slopes of the Arjuna volcano in Malang regency, Java where principal–agent interactions had been ongoing for two decades. The forest management was handed over to Brawijaya University (UB) to be operated as an "education forest" and the new university management team proposed a change in forest-farmer relations. Rather than trying to collect the coffee share crop revenue directly, they asked all farmers to sell their coffee (purportedly for going market prices) to the new university UB-Forest Management company. It was anticipated that through better coffee processing and targeting other market segments enough profit would be made to offset the foregone sharecropping payments. UB-Forest management was also looking to return profit or benefit to the farmers in other ways, e.g., community electricity provision which may not be fully recognized or thought valuable by the farmers. In practice, farmers quickly reverted to selling most of their coffee to outside traders, as they could achieve a better price than selling to the UB-Forest Management company.

As part of the wider interaction between UB and the forest community, UB researchers independent of the UB-Forest management team proposed to address technical constraints around farmer yield and income (Figure 1). One such constraint was the low levels of light for the promotion of healthy understory coffee growth due to the overlying dense pine forest canopy. Our overarching hypothesis was that while farmers perceived the introduced innovation positively, some individual and external factors may hamper their intention to practice it.

Figure 1. Coffee plants under the pine forest canopy and the interdisciplinary science experts.

Our specific research questions were (1) what farmers thought about innovative efforts to provide more light to understory crops, (2) how this relates to their knowledge and expectations and (3) under what conditions it could be sufficiently profitable for them to undertake this innovation. We explored these questions with reference the theory of planned behavior, as we will describe here, before providing details of the site and specific research methods.

2. Methods

2.1. Theory of Planned Behavior

The theory of planned behavior (TPB) (Figure 2) can reveal the cognitive, affective and motivational dynamics of farmers facing new techniques. TPB is commonly employed in the fields of health and economics [29,30], but rarely in the field of agroforestry, agriculture or environmental management. A recent study using TPB in such a field was carried out in Serbia [31]. Some researchers argue that self-efficacy is interrelated with the three antecedent variables to measure intentions [32,33]. For this reason, we included self-efficacy as part of the variable of perceived behavioral control.

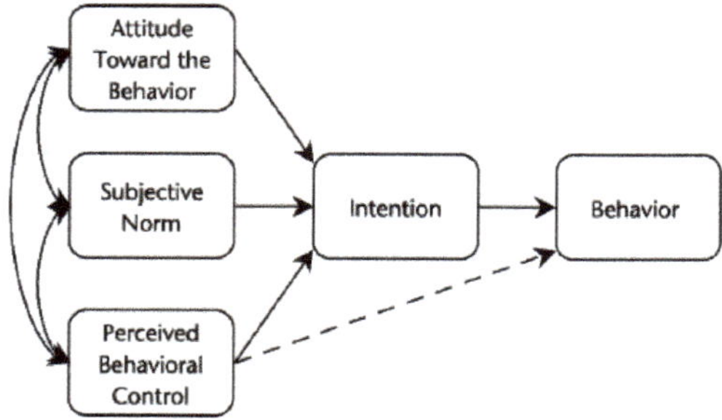

Figure 2. Theory of planned behavior, adapted from [32], showing the three antecedent interacting aspects: (1) attitude towards behavior/behavioral beliefs; (2) subjective norms/normative beliefs; and (3) perceived behavioral control, intention and behavior/action.

The single construct of intention may be compared with the three antecedent variables of intention. In addition, as can be seen in the model, Perceived Behavioral Control (PBC) is a unique variable that acts as an antecedent for intention and actual behavior as well. Intention is interpreted as volitional control or willingness to take certain actions [34–36]. In this study, intention variables were treated as the consequence or impact of attitudes, subjective norms and perceived behavioral control. We also measure intention solely (as a single variable), to be exemplified farmer intentions in applying trimming techniques; the score can be compared with the three antecedent variables of intention.

Typically, previous studies used "perception" and/or "attitude "as a determinant of adoption [37–40]. The TBP allows the variables of attitudes, subjective norms, and, perceived behavioral control to be combined as determinants of intention [41,42]. Furthermore, the self-efficacy variable (self-confidence) may be added to TBP to predict intentions. Hence, TBP can describe the behavior that is influenced by external factors (nonvolitional behavior) and self-efficacy as the individual's internal factor [43].

In the context of agroforestry, we expect that an intention to use pine trimming is associated with a positive attitude (expectations of benefit), strong social support and perceived low constraints of farmers to implement such techniques. In practice, the possible low scores of intention variables can lead to the rearrangement of social and communicative interventions to increase the adoption of trimming techniques. Given the current challenges and opportunities in the UB-Forest, there are simultaneous efforts to explore: (1) techno-ecological options for pine–coffee agroforestry revitalization (such as increasing light intensity, improving crop and soil management (trimming leaves used as litter/organic fertilizer); (2) socioeconomic options (the integration of stakeholders, such as to the financing of production factors); and (3) renewing the existing Indonesia's forestry policy framework to overcome past conflicts (sharing systems, alternative marketing channels and land management regulations) and achieving effective joint management (involving farmers, local governments and forest authority).

2.2. Place, Location and Research Participants

Currently, a variety of collaborative schemes on community forest management exist in Java. The collaboration is between forest management and members of the surrounding communities. UB-Forest management inherited contracts made by the state forest company in response to massive forest land annexation by the community at the end of the 20th century due to political reforms. Cultivation of coffee under pine (or mahogany) trees is common in UB-Forest. At the end of 2015, The UB-Forest management through the University of Brawijaya (UB) received the mandate to administer the forest on behalf of the Indonesian Minister of Forestry and Environment. This production forest—which is located on the slopes of Mount Arjuno (altitude 1200–1900 m in the Karangploso District, Malang regency) has three specific objectives: education, research and ecotourism. Historically, the pioneer farmers settled around the UB Forest in the 1920s and the present farmers are the descendants from the early farmers. Their livelihoods are highly dependent on the agroforestry system, in which they also cultivate food crops and vegetables along with the coffee under the pine trees.

The research location was the Sumbersari Hamlet in the enclave area of UB-Forest (Figure 3).

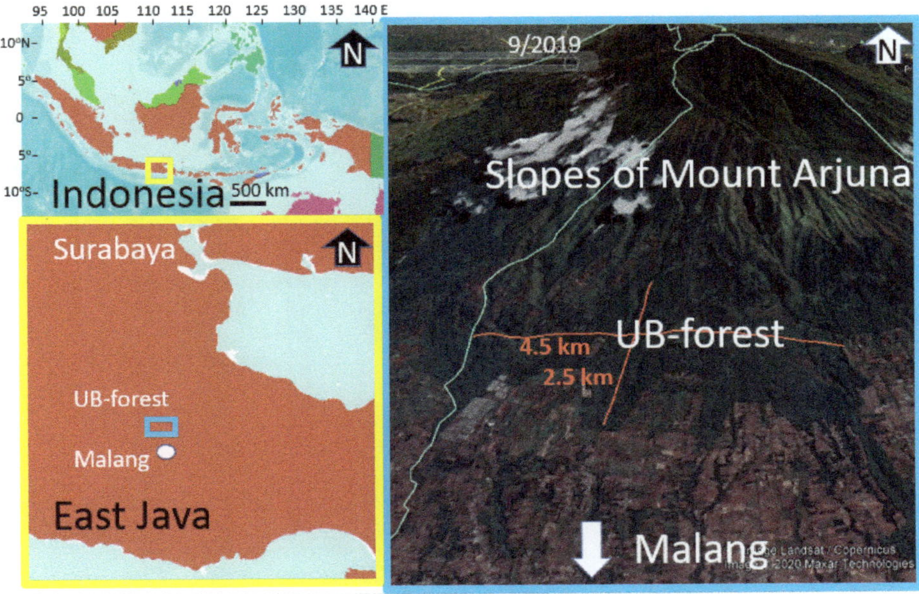

Figure 3. Research location UB Forest in East Java, Indonesia.

The hamlet is located in the Kalisari Watershed of Malang Regency, East Java. Data were collected in the period of March–May 2019. The senior researchers were those from the fields of agricultural extension/communication, soil science and agroforestry. Communication was established with a local key person to get wider access to the sites and respondents. Approval for this study was obtained from the management of UB-Forest and local regional authorities. Anonymity and voluntary participation of the participants were ensured. The research participants for the survey comprised 22 farmers out of an overall 30 within the hamlet, who were cultivating coffee under the pine. The research participants were verbally informed about the purpose, use and possible impacts of the research; in the local context written consent was not deemed to be required. Several members of the hamlet, including its leader, were involved in testing the innovation in the forest.

2.3. Research Design and Methods

The research design used the four stages of the action research approach (Figure 4): plan, act, observe and reflect [44]. It used mixed methods, during the observation stage combining quantitative (through a survey) and qualitative data collection (case study).

Plan—Preparations, site visit. Researchers from UB-UKCEH-ICRAF conducted site visits and a series of discussions under the umbrella of the agroforestry research group at UB to identify issues to be solved through action research. UB and UKCEH accessed funding through different research funding schemes to support the project.

Act—Intervention on-site and communicate with farmers. To overcome low coffee productivity under the pine trees, canopy pine trimming techniques were disseminated. This technique was considered new, as it had never been applied by local farmers. Strategies to communicate this innovation was by (1) "advocacy" through personal communication to a key farmer as the gatekeeper to gain trust; (2) "behavior development" through group meetings with farmers followed by a demonstration of pine trimming techniques on the site. The communication for the advocacy strategy was designed to ensure the key farmer understood the intended benefits of using the technique. Then, he acted as the "innovation endorser"—influencing other farmers to adopt the technique. For this reason, he was also asked to demonstrate the techniques to his peers.

Figure 4. Stages of action research for transferring the trimming techniques.

Observe—Collection data on sites/evaluate evidence. Farmer responses were measured using a mixed-methods sequential explanatory design [45]. This design covers two phases: (1) quantitative data collection through a "survey" of the coffee growers (executed March—April 2019); (2) qualitative data collection using a single instrument case study (May 2019) regarding the grower perceptions of the trimming technique in the context of UB Forest management. The qualitative data collection helped to get a deeper understanding of the quantitative database. This was especially important when dealing with extreme cases or demographic factors that may be associated with farmer intentional behaviors.

Reflect—Reflection was based on the research findings/evidence, which was applied through (1) individual/researcher evaluation; (2) discussions between researchers; and (3) stakeholder views through research finding seminar (involving resource person; a representative from university, governmental agents, NGOs, etc.); this is also a way to increase external validity. Subsequently, assertions/affirmation was conducted by clarifying the findings using the theory of planned behavior and other references and followed by giving options for further study and intervention approach.

2.4. Research Instruments

For the survey, a closed questionnaire (quantitative database) was administered to all participants (22 farmers), covering variables of intention that were (1) attitude toward trimming (6 indicators, e.g., covering statements such as "I believe that pine trimming . . . is necessary because the litters can become mulch"; (2) subjective norms (8 indicators, e.g., "In my opinion . . . local community supports trimming"; and (3) perceived behavioral control (9 indicators, e.g., "I am confident to trim, because . . . it is easy to trim at 10 m high". The five-point Likert scale technique was employed: "strongly disagree" (score 1), "disagree" (2), "neutral" (3), "agree" (4) and "strongly agree" (5). The statements were based on theoretical concepts and adjusted with pine–coffee agroforestry issues as exposed by a key farmer (Appendix A); Additional questions were about farmers" demographic status (age, education, farming systems and experience, source of coffee farming information). The related questions for in-depth interviews can be seen in Appendix B. To reduce nonresponse error, respondents were interviewed face-to-face to overcome their unfamiliarity to fill out the questionnaire.

2.5. Data Validity

For the quantitative data, content validity and reliability tests were managed by experts, discussions with farmers and statistical tests. The steps to develop indicators for intentional aspects: (1) the variables were derived from published theories [46–48] and indicators arranged based on discussions with peers and leading researchers and (2) from extracting key concepts from key farmers during initial interview sessions. The validity of the statement items in the questionnaire was justified by members of the research team. Questions were then tested for clarity in a pilot study with 10 farmers following which sentences within the questionnaire were revised to increase participants" understanding. Statistically, item validity is measured by the correlation coefficient between an item's score and the total score of all items at a certain significant level (10%). The validity of the intention related items was tested with the SPSS statistical package resulting in a Pearson's correlation value (r) of > 0.36, indicating the items have high validity.

As a measure of the reliability of items, internal consistency was also tested. This technique examines the correlation between multiple items to measure a construct. The calculation of Cronbach's alpha Coefficient was rendered for item sections in the TBP questionnaire. The high instrument reliability was indicated by a coefficient of more than 0.80; any item coefficient of less than 0.30 need to be eliminated to avoid an unreliable indicator [49]. The Cronbach's alpha value was 0.85, indicating high reliability of the items.

2.6. Data Collection Methods

As a follow-up, after analyzing quantitative data, the researchers interviewed key farmers again to cross-check noteworthy quantitative data (such as the mode or extreme data) as a case for presentation (embedded analysis) [50]. Other techniques were used to triangulate: (1) nonparticipant observation (by observing the agroforestry setting/location); (2) member-checks" that is discussions with the key farmer and other farmers regarding the findings; (3) focus group discussions (FGD). The first FGD was conducted with 12 coffee farmers at the local meeting house to discuss the technical and social aspects of pine–based agroforestry cultivation. The second FGD was carried out in the local village meeting hall (*Balai desa*), by inviting key farmers (those who were the first to grow coffee or "the innovators") and coffee farmer representatives (those who planted coffee afterward or "the followers") and the stakeholders (representatives of the neighborhood and citizen associations, village officer, rural forest community member, village and sub-district officers, UB Forest employee, Perhutani/state-own forestry company agent and UB agroforestry research group members). In this activity, farmers were welcome to express their views on various issues, such as the relationship between the local community and the forest management authority, the history and quality of partnership, farmer economy and policy. These steps were useful to explore and describe sensitive issues and complexity of farmer livelihoods as the context which may influence their intentional behaviors. Field notes, camera imagery, and audio recordings were used to record the data. The data were recorded on a CD for storage only while useful for further intended research. Data are presented in an anonymized way to protect respondents.

2.7. Data Analysis

For the data analysis, descriptive statistics (frequencies, percentages, mean scores and standard deviations) were calculated to summarize the survey data. Separate tabulation was applied to capture qualitative data. The qualitative findings (statements, observation, photos) being used to validate and provide context to the qualitative results. Research findings were also presented in a series of seminars with members of the research team and stakeholders as part of the reflection and interpretation process.

3. Results

3.1. Social Context and the Demographic Factors

The agroforestry land in the UB-Forest area managed by the respondents of this study totaled 33.04 ha and the total area of UB-Forest is 544 ha. All respondent-farmers managed crops both in UB Forest 13.5 ha (pine-coffee agroforestry), but also in the state-owned forest (Perhutani) 18 ha (pine-vegetable agroforestry) and other forested areas 1.5 ha (pine-vegetables). Although, the survey data shows that the average of land areas managed by the respondent was relatively limited, at 1.2 ha (in Perhutani land); 0.7 ha (in UB Forest) and 0.5 ha (in farmer home gardens). The demographic data (Figure 5) show that 15 of 22 farmers (68%) have been farming within the agroforestry scheme for more than 20 years. Most respondents (41%) were in the >50-year age class and 77% of the farmers have had limited access to education being educated only to the elementary school level.

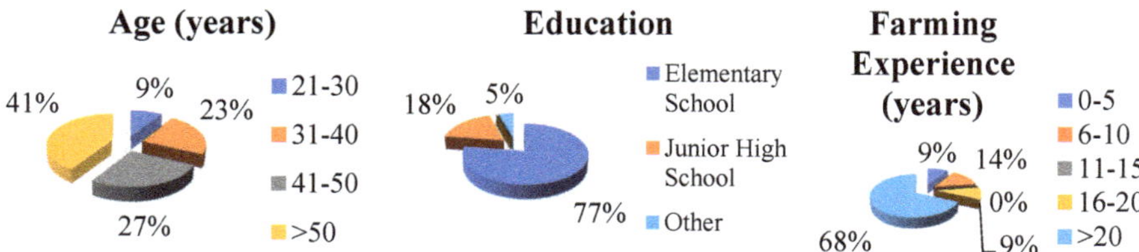

Figure 5. Demographic characteristics of respondents (n = 22).

Moreover, living in relatively remote areas (enclave) had become a constraint for their economic development. The results of an interview with a farmer revealed that in the past (especially before the reform era) the forest authorities did not allow farmers to grow coffee under the pine canopies, arguing that "forests cannot be turned into gardens." The coffee planted by farmers as an agroforestry system was the result of farmers" struggles to survive in these enclaved forest areas, despite causing conflicts with the forest authority at that time. In a lower level of preference, farmers also cultivate other types of crops, such as horticulture and food staples in sufficient quantities for their daily needs. The current use of chemical fertilizers from outside is still difficult as before, though they need it. One farmer stated, "apart from being far away from obtaining fertilizer, it is also expensive," indicating their high dependence on local efforts to obtain adequate coffee yields.

We asked the respondents about their experience in accessing the information on coffee cultivation. The data on Table 1 shows that 19 of 22 respondents (86%) had participated in extension activities discussing coffee cultivation. The sources of information were from public agricultural workers and/or from Perhutani personnel or experts from UB. Only 4 of 22 farmers (18%) received information on coffee cultivation from farmer groups and 7 of 22 (32%) received the information from other sources.

Table 1. Transfer of information on coffee cultivation (n = 22).

No	Source of Information	Number of Respondents Receiving the Information
1	Office of agriculture	19
2	State forest Company/Perhutani	19
3	University staff	19
4	Forum/farmer Group	4
5	Others	7

3.2. Attitudes toward the Trimming

As has been noted, the primary aim of this study was to measure the intention levels of farmers to adopt pine trimming techniques. We calculated the supporting variables: attitudes toward the trimming, subjective norms (social supports) and perceived behavioral control (obstacles in doing trimming) using a Likert scale from one (strongly disagree) to five (strongly agree). The study showed that 79% of respondents agree or strongly agree that pine trimming "provides benefits" for farmers (item #5; Table 2). The perceived benefits primarily related to the pine litters being transformed into mulch (item #1); and to a lesser extent benefits were expected to arise from fertilizing the soil, increasing light at the level of the coffee canopy or use of branches as firewood (items #1, #3 and #6). Moreover 83% of respondents agreed or strongly agreed that trimming can "increase coffee production" (#2).

Table 2. Attitudes towards pine trimming (n = 22), value based on a Likert scale from one (strongly disagree) to five (strongly agree).

No	Statements: I Believe that Pine Trimming …	% "Agree" or "Strongly Agree"	Means (M) of Attitudes	Standard Deviation (Sd)
1	■ … is necessary because the leaf litters can be made into mulch.	84.55	4.23	0.53
2	■ … increases coffee production.	82.73	4.14	0.64
3	■ … fertilizes the soil.	81.82	4.09	0.68
4	■ … increases the incoming light.	81.82	4.09	0.87
5	■ … provides benefits.	79.09	3.95	0.72
6	■ … is useful because the branches can be used as firewood	78.18	3.91	0.68

It is worth noting that the mean score of "provides benefits" (#5) is slightly lower than most of the other indicators (except benefits of getting firewood), indicating the possibility that the farmers are not just considering the direct "benefits" to coffee yields, as is the objective of the question, but are instead considering the balance between benefits and risks (Table 2).

As seen in Table 2, there is a considerable variation in the perception scores. An in-depth interview with a farmer revealed some uncertainty as to the level of benefit of pine trimming, "If it is only trimmed, then it does not have much effect … only the number of leaves on the coffee plants increases." This indicates that although considered generally useful, there is an uncertain view of the trimming technique.

3.3. Subjective Norms of Trimming

Farmers were asked about the level of social support related to pine trimming, which reflects the subjective norm variable (Table 3). When confronted with the statement "experts help farmers know how to trim through demonstration" (item #2), 81% of respondents said agree or strongly agree. This is consistent with the response to the statement "I understand the discussion about trimming (#1): 83% expressed agree or strongly agree. To a lesser extent, there were only 75% of respondents who agreed or strongly agreed that "hamlet personnel support trimming" (indicator #8). It is interesting to note that a relatively moderate response was given regarding the perceived support by UB-Forest management, both in terms of support for trimming and especially for land maintenance (indicators #5 and #7).

The findings show that respondents felt that they received support not only from UB-UKCEH-ICRAF personnel, but also from the local community and their families. On the other hand, the support from the UB-Forest management and local officers (hamlet personnel) were moderate or even considered limited. This is likely because of perceived problems of the newly revised coffee sharing mechanism and also possible latent conflicts in the community. Moreover, a complicated relationship was experienced by farmers with *Perhutani*, the former state forest management entity.

Table 3. Subjective norms/social supports of pine trimming (n = 22).

No	Statements: In My Opinion …	%"Agree" and "Strongly Agree"	Means/M of Subjective Norms	Standard Deviation/Sd
1	■ … I understand the discussion about trimming.	82.73	4.14	0.35
2	■ … experts help farmers know how to trim through demonstration.	81.82	4.09	0.53
3	■ … the local community supports trimming.	80.91	4.05	0.38
4	■ … my family supports trimming.	80.91	4.05	0.58
5	■ … UB Forest supports trimming.	79.09	3.95	0.58
6	■ … discussions with friends encourage me to look for additional information about trimming.	78.18	3.91	0.81
7	■ … UB Forest takes care farmers to take care of the agroforestry lands.	76.36	3.82	0.40
8	■ … hamlet personnel support trimming	74.55	3.73	0.70

A key farmer expressed the historical context plainly:

"These coffee plants … in [the year of] 96 was very difficult because at that time the Perhutani did not allow us to plant coffee under the shade … under the stands of pine."

"The reason was that the forest cannot be used as a garden. However, I think because we live in a forest area … the impression is that it destroys the forest, but we must earn a livelihood for our survival."

"For this reason, I tried to plant whatever crops I planted at that time. The plant that was able to grow and produce just coffee. It means that the people here are still looking for income in the forest area."

Hence, this study shows that farmers receive various levels of support from different parties. Relatively high support from the UB-UKCEH-ICRAF is probably because of the action of disseminating the trimming techniques through site demonstrations and discussions with farmers on some occasions.

3.4. Perceived Behavioral Control of Trimming

We noted that perceived behavioral control influences farmer intentions to undertake to trim. We use two primary variables to measure this self-efficacy (individual factors) and non-self-efficacy (other factors). The indicators of self-efficacy covering aspects of farmer confidence to trim because of expected improvements in their livelihoods, easiness of trimming techniques or suitability with local conditions, etc. Meanwhile, the indicators of non-self-efficacy include the availability of finance, time, information or other resources. The results are presented in Table 4.

When confronted with the statement "if funds are available" (item #1) and "the demonstration by experts encourages me to trim" (#2), 83% and 80% of respondents stated "agree"/A and "strongly agree"/SA, respectively. Furthermore, the number of respondents who strongly agreed or agreed however is reduced when responding to the statement "it helps improve my life" (#3); and "it is easy to understand" (#4).

Table 4. Perceived behavioral control/PBC of trimming (n = 22).

No	Statements "I Am Confident to Trim … "	%"Agree" and "Strongly Agree"	Means/M of PBC	Standard Deviation/ Sd
1	■ … if funds are available.	82.73	4.14	0.35
2	■ … because the demonstration by experts encourage me to trim.	80.00	4.00	0.76
3	■ … because it helps improve my life.	77.27	3.86	0.64
4	■ … because it is easy to understand.	76.36	3.82	0.85
5	■ … because I will do trimming in the farming land.	76.36	3.82	0.91
6	■ … because trimming is suitable with local condition.	75.45	3.77	0.69
7	■ … because it is easy to do.	70.00	3.50	1.10
8	■ … because it is cheap.	67.27	3.36	1.05
9	■ … it is easy to trim at 10 m high.	56.36	2.82	1.05

The more statements cover aspects of follow-up or practicality, the less the farmer's intention to do the trimming ("I will do trimming in the farming land": agree or strongly agree = 76). Less than 77% of respondents agree or strongly agree to practice the trimming. To exemplify suitability to local conditions (#6); availability of funds (#8) and time (#7). These responses are also consistent with a finding in our study, in which 75.45% of respondents stated "my time is limited to trim. "The lowest agreement is for the ease of trimming at a height of 10 m (#9), indicating the potential safety risk of this work (Figure 6). The key farmer highlighted this case:

"As you get older … you only dare [to go up] in part … If you go up maybe up to three to five trees you can still be brave, but not for the whole day … If more than that [five trees] I start to tremble … often slides. It means that the conditions are not possible to carry out all day … even though one hectare has more than 600 trees."

Figure 6. Demonstration of pine trimming by a key farmer.

Only a moderate number of farmers (76%) "will do trimming in the farming land" (#5, M = 3.82; Sd = 0.91). However, further analysis shows that proper extension methods may increase the farmer's intention to seek additional information—or apply trimming. Among these methods were interpersonal communication, such as discussions with friends on pine trimming (#6); especially discussions with the experts; and provision of demonstration plots (#2).

Last, but not least, there are strong indications that external factors were also key points to raise the level of farmer intention. The statement "if funds are available", then they will be confident to trim (A/SA = 83%). Similarly, when the trimming is associated with other external resources, such as local condition (#6), easiness to do (#7), money (#8) and potential risk (#9) (agree or strongly agree < 76%). These indicators suggest the conditions that may hamper the application of trimming, as expressed by a farmer: "To be honest, if I trim it, I'm not happy … the pine remains small, but high. Aside from work safety, the risk is falling … "*suloyo*" [local term for "sluggish, weak or helpless"]. Don't let it happen."

4. Discussion

The process to change individual behavior, commonly seen in an innovation/technological transfer, tends to pro-innovation bias [3]. The assumption is that new technology is better than the previous one. One of the problems is that for targeted individuals the nature of technology is perceived as complicated, as is evident in this study. Nevertheless, our findings have compiled evidence that shows trimming is complex from a farmer's perspective. An individual's intentions are also influenced by considerations of individual capacity, including technical complexity and resource availability to operate. This is particularly evident when such innovation is interfaced with marginal farmers, as in the case of pine–based agroforestry community in Java. We have used the theory of planned behavior as the lens to understand farmer intention to undertake the innovation of pine canopy trimming. We found that the value of the three factors of intention varied.

The attitude/behavioral beliefs—Measurements of the theme or aspect of attitude indicate that, to some extent, the economic benefit is an important element in forming farmers" attitudes towards pine trimming. However, income stability and production costs are also significant factors, regardless of plant productivity. The additional cost for trimming could be a sensitive factor as limitations of family labor and the age of farmers would necessitate the employment of outside labor to complete the trimming. The sensitivity of the extra costs for a small enterprise has long been identified as a sensitive factor for marginal farmers [50].

From the ecological perspective, trimming is perceived useful for (1) increasing light intensity for coffee plants; and (2) fertilizing and mulching the soil to retaining moisture. The literature shows the significant roles of mulch for improving microclimates for plant growth while suppressing weeds [51,52]. Retaining soil moisture is particularly valuable for dry land agroforestry such as in the UB Forest. Therefore, from the techno-ecological views, farmers are responsive to the innovation; their motivation is to increase productivity under the pine trees—vegetables, food crops, and especially coffee plants. The current pine canopy has suppressed coffee growth; the optimum light intensity should be 60–80% [53] but is lower in UB due to over-shading by the pine. This can result in the mortality of coffee tree branches and problems of shortening the harvesting period [54,55].

The subjective norms—We have noted that subjective norms are a belief to fulfill the social expectation. The support from the UB-UKCEH-ICRAF is shown in the trimming demonstration site to help farmers' awareness of and interest in the trimming. However, limited assistance from the UB-Forest management, possibly because of limited involvement at the early phase of the project, is a potential challenge although engagement with this UB-UKCEH-ICRAF project is now ongoing. Moreover, farmers are unhappy as their previous proposals to the forest management to get agricultural inputs were neglected, as expressed during an interview session. Farmers have never practiced trimming, perhaps because of the perceived strict rules on the treatment of the pine tree by the former and current forest management. Hence, the trimming technique itself may be simple in practice, but there are socio-psychological barriers due to the considerations of past rules.

Perceived behavioral control—As mentioned, self-efficacy is a belief to do something. We use this concept to measure farmer confidence to apply the trimming technique to predict the adoption in the future [56]. Generally, farmers are confident to trim because it may improve their livelihoods and due to the availability of information on trimming [57]. In our case, the information brought by the UB-UKCEH-ICRAF personnel is understood to dictate that trimming is only for the lower dry branches on the trees to a height of 10–12 m. However, in practice, farmers are uncomfortable trimming to 12 m. The data show that most farmers were seniors, which may have physical and health constraints. The old farmers were only able to trim three to five trees per day roughly, as revealed in the interview sessions; and only one to three trees for those above 7 m. Furthermore, farmers were unsure of their ability to perform trimming over a large area, with an area of one hectare containing 600 trees, regardless of their long experience of farming (over 20 years). In summary, it is easy for farmers to understand the trimming procedures, but not to practice it. Furthermore, the non-self-efficacy factor—that is the combination of externality aspects [33,58]—is also a significant consideration for intentions. We found that trimming is perceived to be expensive (when done by somebody else) and time-consuming; though having ecological benefits.

Hence, the trimming technique is perceived as complicated because of sound technical problems and limited supports from the stakeholders, regardless of the plant productivity and ecological benefits. Hence, the question is with the farmers" resource scarcity and social supports, will they be willing to apply the technique in the future? Discussions with the farmers revealed suggestions for the option of loans (from the forest management) to pay workers to trim. The other alternative is to using machinery to fasten the trimming process, which is probably cheaper, faster and safer. The use of machinery is something that researchers will investigate as part of any future forest management strategy, for example, telescopic pruners can reach 8 m in height. In-depth discussions reveal an interesting thing; farmers had proposed alternative techniques beyond trimming—such as pine thinning by removal of selected trees. The advantages proposed by farmers were (a) increasing light intensity for coffee plants (just as the trimming); (b) no need to add fertilizers; (c) no need to climb pine trees at extended heights; and (d) cheap and easy to do. A leading farmer experienced that without thinning, the soil fertility and strength of pine trees would decrease with time, as exemplified at a certain forest location (identified as plot #84); plant growth as stunted and at the risk of falling during heavy rains. He argued that it would be dangerous for individuals under the trees and considering that the forest is an also educational and recreational destination. He recommends that thinning needs to be accomplished every 10–15 years. The research team also heard of other farmers" innovative techniques on agroforestry management that are beyond our study at this stage.

5. Implications

Based on our experience at UB Forest, the trimming innovation has a prospect to be adopted, but with caution. On one side there is a sound theoretical basis that trimming can increase the productivity of pine–coffee agroforestry and improve the microclimate, suggesting the innovation has productivity-ecological benefits. However, it may be hampered by limited support and scarcity of farmer resources and conditions (such as old age), indicating issues on psycho-social burdens and practicality. The theory of planned behavior is a useful theoretical lens to identify the opportunity and obstacles of applying trimming techniques from the farmer's point of view. Furthermore, we have shared these issues with stakeholders (researchers, local government officers, UB forest management, key farmers, etc.) through a series of seminars. As a reflection, the attendance and engagement of the stakeholders including the government agents were high at every seminar and showing support to the research team and the findings of this study.

For the implications, the trimming program, which has been employed over two years, we have highlighted options: keep the innovation by (1) providing subsidy or loans for the farmers to hire workers to trim or buy trimming machinery, and/or (2) integrate with the farmer techniques. The first option has conditional acceptance because of perceived financial and technical burdens. The second option has a higher chance of acceptance and implementation by farmers, first, for one reason, it would better align with the farmer's needs and capabilities. For other reasons, it would also enhance collaboration between researchers and farmers. Both options, however, have a challenge of how to cooperate with the stakeholders. For this reason, the plan and knowledge learned from action research need to be shared with the stakeholder from the beginning of any agroforestry innovation development. To reflect, this experience has improved the understanding of researchers and the collaborative options that would enhance the capacities for all the social actors in the agroforestry system. Based on this research we would encourage the application of the second option in the future.

We learn that technical constraints (the majority were senior farmers and low income) and lack of optimal external support may be important limitations in the future if there are no specific interventions. The new intervention should be not only on technical assistance, but also the sharing of information or knowledge and collaborations between stakeholders. To sustain, the communicative intervention processes require a participatory and integrative approach. Co-management of agroforestry innovations may be developed throughout the stages, from the creation, pilot/local implementation and scaling up. We need to explore any aspirations and local knowledge, contemplating the specific demographic conditions and needs of the marginal farmers at the UB Forest and the demand for stakeholder integration. Our experience may be useful as a lesson learned for policymakers and executives to comprehend the social-ecological dynamic of agroforestry management, let alone the Javanese farmers who reside close to forest lands.

As a reflection, this action research is useful, but needs improvement. We notice that the process of transferring agroforestry innovation cannot only consider innovation and farmers, but also the forest management authority. The key to accelerating the innovation transfer process is the use of an integrated communication and extension strategy. For this reason, it is necessary to establish connections between external scientists as initiators of innovation, farmers as practitioners and relevant authorities [59,60], in the case of our study is the UB Forest management. Moreover, a study shows that farmers" knowledge is often overlooked in the transfer of innovation [61,62]. In our case, farmers can observe biophysical problems in the context of their socioeconomic complexity. A farmer revealed:

"Regarding the soil … the farmers should fertilize it every six months … the nutrients have been reduced because [it has been absorbed by] pine roots. I have tried like that [fertilizing], and then I calculated the maintenance costs. I calculated and the results were even. Finally, I let it go, I did not care—'simalakama'!" (emphasis added).

Simalakama is the name of a bitter fruit for traditional medicinal ingredients. People usually say, "as if eating simalakama fruits," a local parable to express a dilemma when someone wants to do something or not.

More than this, farmers—through various considerations—have their rationality and solution preference. The results of in-depth interviews and FGDs show that outside of trimming, farmers have crop management preferences through thinning techniques. After 10 to 15 years, it is believed that pine plants need to be sparsely spaced, with a spacing of twice the current one (currently in 2×3 m). Farmers recommend a 2×6 m spacing for pine trees because with more incoming sunlight, then—according to farmer experience—there are simultaneous benefits. The advantages for productivity and ecology are: "soil fertility is increasing"; "reduced acidity"; "better coffee plant growth"; "bigger plant"; "sturdy, when exposed to the wind the plants will be stronger" and "groundwater storage is better". For social benefits, farmers framed it as: "both [parties] can get [benefits]; forestry gets, the community gets too." Other techniques, vice versa, jargonized as *"Sana untung, sini buntung"* [the party over there gets profits, the community is in disadvantaged]" (results of in-depth interviews and an FGD with farmers).

Consequently, the transfer of innovation does not only come from scientists focused on productivity-ecological aspects [63], but also those embracing the aspirations of farmers and the interest of powerholders. Therefore, knowledge of practical and political interests is needed simultaneously. This means that various modes of innovation transfer can be combined: from scientists (linear, top-down), farmers" needs (bottom-up) and interests from politics/UB Forest (lateral). Since its inception, bidirectional extension and communication strategies [64] or participatory strategies need to be developed although previous studies have shown that it is not easy for change agents to do so [6]. Included in the method action research is the need to apply transdisciplinary research, where farmers" local knowledge needs to be examined, confirmed and incorporated with those from scientific knowledge [65]—as an agroforestry knowledge pool. To sum up, efforts to balance the distinct roles of the tripartite agents for profit—prove-it—prophets may be more easily achieved through the participatory-collaborative approach [56,66].

Specific types of land management contracts between forest authorities and farmers looking for land to grow their food crops became known in 19th Century southeast Asian teak growing traditions under the name "taungya"—originally referring to indigenous swidden-fallow practices. The current pine–coffee management contracts represent considerable development since the earlier schemes [14,67], but the unequal power distribution of the past is still loading the dice against schemes that are both fair and efficient.

6. Conclusions

Based on the study, the adoption of the trimming innovation is prospective, though it needs adaptations. On one hand, on some levels trimming is believed to increase the productivity of pine–coffee agroforestry and maintain the microclimate, suggesting the productivity-ecological benefits of the innovation. However, this may be hampered by limited support and scarcity of farmer resources and conditions, indicating issues on psychosocial burdens and practicality. We have shared these issues with stakeholders (researchers, local government officers, UB forest management, key farmers, etc.) through a series of seminars. Practically, the trimming program that has been managed for two years, has two main options: keep the innovation by providing subsidy or loans for the farmers, and/or integrate with farmer techniques. Further options for innovation are necessary, including direct pruning of coffee and crop fertilization strategies, some of which are being tested now.

We learned that technical constraints (most senior farmers and low income) and lack of optimal external support may be an important limitation in the future if there are no specific interventions. a new intervention should be not only on technical assistance, but also the sharing of information or knowledge and collaborations between stakeholders. Sustaining the communicative intervention processes requires a participatory and integrative approach. Co-management of agroforestry innovations may be developed throughout the stages, from the creation, local implementation and scaling-up. This approach is crucial because stakeholders have roles and specific rights or interest in making decisions. Moreover, we had limited initiatives from farmers to be integrated with agroforestry innovation development. For these reasons, we need to explore more any local knowledge and practices to meet with the specific conditions and needs of the marginal farmers and their unique relations with forest management. Considering that this study had only a relatively small number of participants, a larger number of participants is needed to increase the generalizability of the results. However, our experience may be useful as a lesson learned for policymakers and executives to comprehend the social-ecological dynamic of agroforestry management, as well as the Javanese farmers who reside adjacent to forest lands.

Author Contributions: E.D.C.; M.v.N.; N.M. and R.R. conceptualized the study. E.D.C. designed the study, while S.F. and D.W. collected quantitative data; E.D.C. gathered qualitative data as well as acted as their undergraduate supervisor. Data curation and graph was prepared by E.P., D.W. and S.F. Then, E.D.C.; and M.v.N. assisted by E.P.; S.F.; and D.W. wrote the original manuscript preparation. M.v.N.; N.P.M.; R.L.R. and E.D.C. reviewed, added and edited the final manuscript. All authors have read and agreed to the published version of the manuscript.

Funding: This research received funding from Faculty of Agriculture Brawijaya University, Grant number DIPA-042.01.2.400919/2019 and the the UK Natural Environment Research Council grant Sustainable Use of

Natural Resources to Improve Human Health and Support Economic Development (SUNRISE, grant number NE/R000131/1).

Acknowledgments: Acknowledgment is given to the agroforestry group members from the Sumbersari community at UB Forest for their generosity in participating in this study and to the stakeholders of various institutions in their involvement during research result seminars.

Conflicts of Interest: The authors declare no conflicting interests. The funders had no role in the design of the study; in the collection, analyses or interpretation of data; in the writing of the manuscript or in the decision to publish the results.

Appendix A. Survey Instruments Used for Measuring Farmer Intention

Answer options

1 = strongly disagree
2 = disagree
3 = do not know (neutral)
4 = agree
5 = strongly agree

NO.	Could you please respond (V) to the following statements with one of the answer options?	1	2	3	4	5
	I believe that pine trimming …					
1.	… provides benefits.					
2.	… increases the incoming light.					
3.	… fertilizes the soil.					
4.	… is necessary because the leaf litters can be made into mulch.					
5.	… is useful because the branches can be used as firewood.					
6.	… increases coffee production.					
	In my opinion …					
7.	… I understand the discussion about trimming.					
8.	… experts help farmers know how to trim through demonstration.					
9.	… my family supports trimming.					
10.	… discussions with friends encourage me to look for additional information about trimming.					
11.	… the local community supports trimming.					
12.	… hamlet personnel support trimming.					
13.	UB Forest takes care farmers to take care of the agroforestry lands.					
14.	UB Forest supports trimming.					
	I am confident to trim, because …					
15.	… it is easy to understand.					
16.	… it is easy to do.					
17.	… it is easy to trim at 10 m high.					
18.	… it is cheap.					
19.	…					
20.	…					
21.	… it helps improve my life.					
22.	… the demonstration by experts encourage me to trim.					
23.	I will do trimming in the farming land.					

Appendix B. Guideline Questions for In-Depth Interviews

1. What types of plants or crops do you grow under the pine shade?
2. What are the responses from the forest management when you grow coffee under this shade?
3. Give your thoughts on the pine trimming technique recommended by the experts.
4. Please describe your experience when trimming pine.
5. How do you feel when trimming?
6. What obstacles do you face when you want to do trimming?
7. Do you have any experiences other than trimming to increase your coffee production?
8. What are the advantages of these alternative methods?

References

1. Padel, S. Conversion to organic farming: A typical example of the diffusion of an innovation? *Sociologiaruralis* **2001**, *41*, 40–61. [CrossRef]
2. Sveiby, K.E.; Gripenberg, P.; Segercrantz, B.; Eriksson, A.; Aminoff, A. Unintended and undesirable consequences of innovation. In Proceedings of the XX ISPIM Conference, The Future of Innovation, Vienna, Austria, 21–24 June 2009.
3. Rogers, E.M. *Diffusion of Innovations*; Simon and Schuster: New York, NY, USA, 2010.
4. Costantini, V.; Liberati, P. Technology transfer, institutions and development. *Technol. Forecast. Soc. Chang.* **2014**, *88*, 26–48. [CrossRef]
5. Pannell, D.J.; Marshall, G.R.; Barr, N.; Curtis, A.; Vanclay, F.; Wilkinson, R. Understanding and promoting adoption of conservation practices by rural landholders. *Aust. J. Exp. Agric.* **2006**, *46*, 1407–1424. [CrossRef]
6. Cahyono, E.D.; Agunga, R. Policy and practice of participatory extension in Indonesia: A case study of extension agents in Malang District, East Java Province. *J. Int. Agric. Ext. Educ. (JIAEE)* **2016**, *23*, 38–57. [CrossRef]
7. Agunga, A.; Cahyono, E.D.; Buck, E.; Scheer, S. Challenges of Implementing Participatory Extension in Indonesia. *J. Commun. Media Res.* **2016**, *8*, 20–45.
8. Cahyono, E.D. Persepsi Dan Tipe Keputusan Petani Madura Terhadap Inovasi Agens Hayati-PGPR. In *Proceeding Konferensi Nasional Penyuluhan&Komunikasi Pembangunan*; Cahyono, E.D., Ed.; Fakultas Pertanian Universitas Brawijaya: Malang, Indonesia, 2016; pp. 284–292.
9. Hofstede, G.J. GRASP agents: Social first, intelligent later. *AI Soc.* **2019**, *34*, 535–543. [CrossRef]
10. van Noordwijk, M. *Sustainable Development through Trees on Farms: Agroforestry in Its Fifth Decade*; World Agroforestry (ICRAF): Bogor, Indonesia, 2019.
11. Hawkins, D.G.; Lake, D.A.; Nielson, D.L.; Tierney, M.J. Delegation Under Anarchy: States, International Organizations, and Principal-Agent Theory. In *Delegation and Agency in International Organizations*; Cambridge University Press: New York, NY, USA, 2006; p. 21.
12. Eisenhardt, K.M. Agency theory: An assessment and review. *Acad. Manag. Rev.* **1989**, *14*, 57–74. [CrossRef]
13. Peluso, N.L. 'Traditions' of forest control in Java: Implications for social forestry and sustainability. *Nat. Resour. J.* **1992**, *32*, 883–918. [CrossRef]
14. Van Noordwijk, M.; Tomich, T.P. Agroforestry Technologies for Social Forestry: Tree-Crop Interactions and Forestry-Farmer Conflicts. In *Social Forestry and Sustainable Forest Management*; Hasanu, S.H., Sabarnurdin, S., Sumardi, HeruIswantoro, Eds.; PerumPerhutani: Jakarta, Indonesia, 1995; pp. 168–193.
15. Djamhuri, T.L. Community participation in a social forestry program in Central Java, Indonesia: The effect of incentive structure and social capital. *Agrofor. Syst.* **2008**, *74*, 83–96. [CrossRef]
16. Sunderland, T.C.H.; Deakin, E.L.; Kshatriya, M. Conclusions: Agrarian Change—A Change for the Better? In *Agrarian Change in Tropical Landscapes*; Center for International Forestry Research (CIFOR): Bogor, Indonesia, 2016.
17. McCarthy, J.F.; Cramb, R.A. Policy narratives, landholder engagement, and oil palm expansion on the Malaysian and Indonesian frontiers. *Geogr. J.* **2009**, *175*, 112–123. [CrossRef]
18. Sanders, A.J.; Ford, R.M.; Mulyani, L.; Larson, A.M.; Jagau, Y.; Keenan, R.J. Unrelenting games: Multiple negotiations and landscape transformations in the tropical peatlands of Central Kalimantan, Indonesia. *World Dev.* **2019**, *117*, 196–210. [CrossRef]

19. Potter, L.; Lee, J. *Tree planting in Indonesia: Trends, Impacts and Directions*; CIFOR Occasional Paper No. 18; CIFOR: Bogor, Indonesia, 1998; 85p.

20. Bachriadi, D.; Sardjono, M.A. Conversion or Occupation?: The Possibility of Returning Local Communities' Control Over Forest Lands in Indonesia. In *Luce Project on Green Governance Fellowship Paper*; University of California: Berkeley, CA, USA, 2005.

21. Siscawati, M.; Banjade, M.R.; Liswanti, N.; Herawati, T.; Mwangi, E.; Wulandari, C.; Tjoa, M.; Silaya, T. *Overview of Forest Tenure Reforms in Indonesia*; Working Paper 223; CIFOR: Bogor, Indonesia, 2017. [CrossRef]

22. Jerneck, A.; Olsson, L. More than trees! Understanding the agroforestry adoption gap in subsistence agriculture: Insights from narrative walks in Kenya. *J. Rural Stud.* **2013**, *32*, 114–125. [CrossRef]

23. Place, F.; Ajayi, O.C.; Torquebiau, E.; Detlefsen Rivera, G.; Gauthier, M.; Buttoud, G. Improved Policies for Facilitating the Adoption of Agroforestry. In *Agroforestry for Biodiversity and Ecosystem Services—Science and Practice*; Kaonga, M.L., Ed.; Intech Open: Rijeka, Croatia, 2013; pp. 113–128. [CrossRef]

24. Sahide, M.A.K.; Fisher, M.R.; Supratman, S.; Yusran, Y.; Pratama, A.A.; Maryudi, A.; Kim, Y.S. Prophets and profits in Indonesia's social forestry partnership schemes: Introducing a sequential power analysis. *For. Policy Econ.* **2020**, *115*, 102160. [CrossRef]

25. van Noordwijk, M. Prophets, Profits, Prove It: Social Forestry under Pressure. *One Earth* **2020**, *2*, 394–397. [CrossRef]

26. Rival, L. From Global Forest Governance to Privatised Social Forestry: Company–Community Partnerships in the Ecuadorian Choco. In *Privatising Development*; Martinus Nijhoff Publishers: Leiden, The Netherlands, 2005; pp. 253–270.

27. Pham, T.T.; Campbell, B.M.; Garnett, S.; Aslin, H.; Hoang, M.H. Importance and impacts of intermediary boundary organizations in facilitating payment for environmental services in Vietnam. *Environ. Conserv.* **2010**, *37*, 64–72. [CrossRef]

28. Taylor, P.L. Conservation, community, and culture? New organizational challenges of community forest concessions in the Maya Biosphere Reserve of Guatemala. *J. Rural Stud.* **2010**, *26*, 173–184. [CrossRef]

29. Wang, J.; Chu, M.; Deng, Y.; Lam, H.; Tang, J. Determinants of pesticide application: An empirical analysis with theory of planned behavior. *China Agric. Econ. Rev.* **2017**, *10*, 608–625. [CrossRef]

30. Satsios, N.; Hadjidakis, S. Applying the Theory of Planned Behaviour (TPB) in Saving Behaviour of Pomak Households. *Int. J. Financ. Res.* **2018**, *9*, 122–133. [CrossRef]

31. Despotović, J.; Rodić, V.; Caracciolo, F. Factors affecting farmers' adoption of integrated pest management in Serbia: An application of the theory of planned behavior. *J. Clean. Prod.* **2019**, *228*, 1196–1205. [CrossRef]

32. Ajzen, I. *Attitudes, Personality, and Behavior*; Open University Press: Maidenhead, UK, 2005.

33. Ramdhani, N.; Carver, C.S.; Scheier, M.F.; Segerstrom, S.C.; Solberg Nes, L.; Evans, D.R.; Khatimah, H. Penyusunanalatpengukurberbasis theory of planned behavior. *BuletinPsikologi* **2010**, *4*, 55–69. [CrossRef]

34. Shin, Y.H.; Hancer, M. The role of attitude, subjective norm, perceived behavioral control, and moral norm in the intention to purchase local food products. *J. Foodserv. Bus. Res.* **2016**, *19*, 338–351. [CrossRef]

35. Zeweld, W.; Van Huylenbroeck, G.; Tesfay, G.; Speelman, S. Smallholder farmers' behavioural intentions towards sustainable agricultural practices. *J. Environ. Manag.* **2017**, *187*, 71–81. [CrossRef] [PubMed]

36. Popa, B.; Niţă, M.D.; Hălălişan, A.F. Intentions to engage in forest law enforcement in Romania: An application of the theory of planned behavior. *Forest Policy Econ.* **2019**, *100*, 33–43. [CrossRef]

37. Kahmir, H.R.; Muharam, A. Sifat Inovasi dan Aplikasi Teknologi Pengelolaan Terpadu Kebun Jeruk Sehat dalam Pengembangan Agribisnis Jeruk di Kabupaten Sambas, Kalimantan Barat. *J. Hortik.* **2008**, *18*, 477–490. Available online: http://ejurnal.litbang.pertanian.go.id/index.php/jhort/article/viewFile/899/743 (accessed on 22 September 2020).

38. Meijer, S.S.; Catacutan, D.; Ajayi, O.C.; Sileshi, G.W.; Nieuwenhuis, M. The role of knowledge, attitudes and perceptions in the uptake of agricultural and agroforestry innovations among smallholder farmers in sub-Saharan Africa. *Int. J. Agric. Sustain.* **2015**, *13*, 40–54. [CrossRef]

39. Zolkepli, I.A.; Kamarulzaman, Y. Social media adoption: The role of media needs and innovation characteristics. *Comput. Human Behav.* **2015**, *43*, 189–209. [CrossRef]

40. Malahayatin, D.M.; Cahyono, E.D. Faktor kesesuaian dengan kebutuhan petani dalam keputusan adopsi inovasi pola tanam jajar legowo (Studi kasus petani padi di Kecamatan Widang, Kabupaten Tuban). *J. Ekon. Pertan. Agribisnis* **2017**, *1*, 56–61. [CrossRef]

41. Fishbein, M.; Ajzen, I. *Intention and Behavior: An Introduction to Theory and Research*; Addison-Wesley: Reading, MA, USA, 1975.

42. Rezaei, R.; Safa, L.; Damalas, C.A.; Ganjkhanloo, M.M. Drivers of farmers' intention to use integrated pest management: Integrating theory of planned behavior and norm activation model. *J. Environ. Manag.* **2019**, *236*, 328–339. [CrossRef]

43. Sheppard, B.H.; Hartwick, J.; Warshaw, P.R. The theory of reasoned action: A meta-analysis of past research with recommendations for modifications and future research. *J. Consum. Res.* **1988**, *15*, 325–343. [CrossRef]

44. Burns, A. Action Research. In *Qualitative Research in Applied Linguistics*; Palgrave Macmillan: London, UK, 2009; pp. 112–134.

45. Creswell, J.W.; Creswell, J.D. *Research Design: Qualitative, Quantitative, and Mixed Methods Approaches*; Sage Publications: Newbury Park, CA, USA, 2017.

46. Ajzen, I. From Intentions to Actions: A Theory of Planned Behavior. In *Action Control*; Springer: Berlin/Heidelberg, Germany, 1982; pp. 11–39.

47. Notani, A.S. Moderators of perceived behavioral control's predictiveness in the theory of planned behavior: A meta-analysis. *J. Consum. Psychol.* **1998**, *7*, 247–271. [CrossRef]

48. Sussman, R.; Gifford, R. Causality in the theory of planned behavior. *Personal. Soc. Psychol. Bull.* **2019**, *45*, 920–933. [CrossRef] [PubMed]

49. De Vaus, D. *Surveys in Social Research*; Routledge: London, UK, 2013.

50. Taneo, J.M.; Cahyono, E.D. Analisis factor social ekonomis adopsi teknologi pupuk urea tablet. *Wawasan* **1999**, *7*, 8–16.

51. Mayun, I.A. Efek mulsa jerami padi dan pupuk kendang sapi terhadap pertumbuhan dan hasil bawang merah di daerah pesisir. *Agritrop* **2007**, *26*, 33–40.

52. Kader, M.A.; Senge, M.; Mojid, M.A.; Ito, K. Recent advances in mulching materials and methods for modifying soil environment. *Soil Tillage Res.* **2017**, *168*, 155–166. [CrossRef]

53. Utomo, S.B. *Dinamika Suhu Udara Siang Malam Terhadap Foto Respirasi Fase Generative Kopi Robusta di Bawah Naungan Yang Berbeda Pada Sistem Agroforestry*; Tesis: Universitas Jember. 2011. Available online: http://digilib.unej.ac.id (accessed on 3 November 2019).

54. Bote, A.D.; Struik, P.C. Effects of shade on growth, production and quality of coffee (*Coffea arabica*) in Ethiopia. *J. Hortic. For.* **2011**, *3*, 336–341.

55. Daras, U.; Sobari, I. *Pemangkasan Tanaman Kopi Dan Pemeliharaan Pohon Penaung*; Balai Penelitian Tanaman Industri dan Penyegar: Sukabumi, Indonesia, 2018.

56. McGinty, M.M.; Swisher, M.E.; Alavalapati, J. Agroforestry adoption and maintenance: Self-efficacy, attitudes and socio-economic factors. *Agroforest. Syst.* **2008**, *73*, 99–108. [CrossRef]

57. Binam, J.N.; Place, F.; Djalal, A.A.; Kalinganire, A. Effects of local institutions on the adoption of agroforestry innovations: Evidence of farmer managed natural regeneration and its implications for rural livelihoods in the Sahel. *Agric. Food Econ.* **2017**, *5*, 2. [CrossRef]

58. Sparks, P.; Guthrie, C.A.; Shepherd, R. The dimensional structure of the perceived behavioral control construct. *J. Appl. Soc. Psychol.* **1997**, *27*, 418–438. [CrossRef]

59. Sokolovska, N.; Fecher, B.; Wagner, G.G. Communication on the Science-Policy Interface: An Overview of Conceptual Models. *Publications* **2019**, *7*, 64. [CrossRef]

60. Böcher, M. Advanced approaches for a better understanding of scientific knowledge transfer in forest and forest-related policy. *For. Policy Econ.* **2020**, *114*, 102165. [CrossRef]

61. Cahyono, E.D. Participatory Communication and Extension for Indigenous Farmers: Empowering Local Paddy Rice Growers in East Java. In *Communicating for Social Change*; Palgrave Macmillan: Singapore, 2019; pp. 213–233.

62. Kankeu, R.S.; Demaze, M.T.; Krott, M.; Sonwa, D.J.; Ongolo, S. Governing knowledge transfer for deforestation monitoring: Insights from REDD+ projects in the Congo Basin region. *For. Policy Econ.* **2020**, *111*, 102081. [CrossRef]

63. Do, T.H.; Krott, M.; Böcher, M. Multiple traps of scientific knowledge transfer: Comparative case studies based on the RIU model from Vietnam, Germany, Indonesia, Japan, and Sweden. *For. Policy Econ.* **2020**, *114*, 102134. [CrossRef]

64. Roux, D.J.; Rogers, K.H.; Biggs, H.C.; Ashton, P.J.; Sergeant, A. Bridging the science–management divide: Moving from unidirectional knowledge transfer to knowledge interfacing and sharing. *Ecol. Soc.* **2006**, *11*, 4. [CrossRef]

65. Cahyono, E.D. *Inovasi Budidaya Agroforestri Kopi-Pinus di UB Forest: Intensi Petani Untuk Mengadopsi Teknik Trimming (Pemangkasan) Cabang Pinus Yang Diintroduksikan Oleh Kolaborasi UB-CEH*; Fakultas Pertanian Universitas Brawijaya: Malang, Indonesia, 2019.

66. Haggar, J.; Ayala, A.; Díaz, B.; Reyes, C.U. Participatory design of agroforestry systems: Developing farmer participatory research methods in Mexico. *Dev. Pract.* **2001**, *11*, 417–424. [CrossRef]

67. Smiet, A.C. Forest ecology on Java: Conversion and usage in a historical perspective. *J. Trop. For. Sci.* **1990**, *2*, 286–302.

 *land*

Article

Gendered Migration and Agroforestry in Indonesia: Livelihoods, Labor, Know-How, Networks

Elok Mulyoutami [1,*] **, Betha Lusiana** [1] **and Meine van Noordwijk** [1,2]

[1] World Agroforestry Centre (ICRAF), Bogor 16001, Indonesia; b.lusiana@cgiar.org (B.L.); m.vannoordwijk@cgiar.org (M.v.N.)

[2] Plant Production Systems, Wageningen University, 6708 PB Wageningen, The Netherlands

* Correspondence: eloknco@gmail.com

Received: 31 October 2020; Accepted: 16 December 2020; Published: 18 December 2020

Abstract: Migration connects land use in areas of origin with areas of new residence, impacting both through individual, gendered choices on the use of land, labor, and knowledge. Synthesizing across two case studies in Indonesia, we focus on five aspects: (i) conditions within the community of origin linked to the reason for people to venture elsewhere, temporarily or permanently; (ii) the changes in the receiving community and its environment, generally in rural areas with lower human population density; (iii) the effect of migration on land use and livelihoods in the areas of origin; (iv) the dynamics of migrants returning with different levels of success; and (v) interactions of migrants in all four aspects with government and other stakeholders of development policies. In-depth interviews and focus group discussions in the study areas showed how decisions vary with gender and age, between individuals, households, and groups of households joining after signs of success. Most of the decision making is linked to perceived poverty, natural resource and land competition, and emergencies, such as natural disasters or increased human conflicts. People returning successfully may help to rebuild the village and its agricultural and agroforestry systems and can invest in social capital (mosques, healthcare, schools).

Keywords: coffee; cocoa; Java; livelihoods; rural–urban; remittances; returning migrants; Sumatra; Sulawesi

1. Introduction

Contrary to the long-term attachment to place that prevails in "myths of origin" and cultural constructs of place-based identity [1,2], humans have a history of dispersal and migration [3], as reflected in our complex DNA and linguistic signatures [4]. Migration has been the demographic basis of the expansion of our species spreading to all parts of the world, adapting to a wide range of circumstances and learning how to cope with variability and diversity. Both cultural and genetic evidence suggests that human dispersal and migration were not a one-way process and that links to areas of origin were maintained through any means of communication and transport that was accessible in given periods of human history and development. Cultures have absorbed newcomers while migrants kept a cultural attachment to areas of origin, creating the rich "unity in diversity" fabric that characterizes many Asian countries [5]. Human dispersal and migration have had ethnobotanical consequences, with a large number of semi-domesticated trees and crops spreading along with humans as well as the knowledge of how to use them [6,7]. Working across Indonesia, anecdotes of how agroforestry and local forest management practices were inspired by experiences elsewhere during "circular migration" or were traced back to migrants from other parts of the country are commonly heard [8] but appear not to have been systematically analyzed. In the current "lockdown" response to the COVID-19 pandemic [9], both positive and negative aspects of such human movement call for a more nuanced analysis as part of the Sustainable Development Goals [10]. Migration can involve a radical decision to uproot and try one's luck elsewhere for a variety of social, political, economic, and environmental reasons [11,12] or be a more gradual process wherein temporary and circular migration precedes "permanent" migration at the individual or household level, with family members potentially following suit [13]. The decision to migrate circularly or permanently is usually taken step by step in response to perceived success or failure and interacting with external circumstances, such as policies in both the area of origin and area of the temporary new residence. While many studies have zoomed in on specific parts of the decisions to move and/or return, a holistic perspective requires a human life-cycle approach that is not easy to obtain experimentally but can be constructed by combining separately studied pieces of a larger puzzle.

Each year, around half a million Indonesians travel abroad to work, half of those to the Middle East. They are typically women from small cities or villages with primary education and limited work experience, hired to perform domestic work [14]. Econometric analysis of data on emigration rates of countries at different stages of economic development has revealed inverse U-shaped responses. With a GDP per capita at purchasing power parity of USD 3893 in 2018 [15], Indonesia as an emerging lower-middle-income country is approaching the income level at which international migration is expected to peak (USD 5–10,000). Although merely based on cross-sectional evidence, the "migration hump" is widely interpreted as a causal relationship. However, this interpretation is contested [16,17]. The fact that income growth increases the opportunity cost of migration and also eases liquidity constraints—two opposite forces at play—may explain the "hump". Further analysis of a migration matrix for all of Indonesia [18] found roles for ethnic networks in groups such as the Sundanese from West Java and the Buginese from South Sulawesi that reduced the fixed costs of international migration from rural areas of Indonesia and contributed to the recorded way income elasticity of migration varies depending on the exposure of the given type of landholdings to variability of rainfall and rice prices. Although these ethnic communities may be isolated from native ethnic populations when residing outside their historical homelands, they still have strong ties to the broader Sundanese and Buginese networks with the potential for connections to international labor markets. The choices involved in internal migration within large countries like Indonesia may, however, differ from those for international migration. A study in East Java found that individuals with access to water, health insurance, or markets, or those living in villages that have a large proportion of non-irrigated land being used for non-agricultural activities, were less likely to seek employment elsewhere [19].

Internal migration in Indonesia has been linked to productivity growth [20]. Between 1979 and 1988, the Transmigration Program relocated 2 million voluntary migrants (hereafter, transmigrants) from the Inner Islands of Java and Bali to newly created agricultural settlements in the Outer Islands (with Sumatra, Kalimantan, Sulawesi, and Papua being the largest). The success rate depended on the agro-ecological similarity between the areas of origin and migration targets [21], while diverse communities, rather than migrants from a single origin, could be linked to stronger integration and success in the new environment [22].

Migration decisions relate to gender-specific expectations and have gender-specific consequences for those who move, those who stay behind, and those who return. The latter can enrich local livelihoods with new knowledge, norms, and expectations, apart from the financial resources they may have. However, returning migrants will need to invest in social capital to earn their place back in the local community. In cases where males are primarily involved in migration to Africa and South Asia, effectson the source areas may be a "feminization" of agriculture [23,24], however, the opposite case—where females preferentially engage—is common in Asia as well. Distinct feminization of labor migration in Southeast Asia, participating in gender-segmented global labor markets, has significantly altered care arrangements, gender roles and practices, as well as family relationships within households [25]. Elsewhere, youth migration has become an issue [26]. What these all have in common is that selective migration of either gender or age segments leads to a reshuffling of roles in local livelihoods and households, with a gradual change in associated "identities". Burgers [27] described the consequences of female migration for the matrilineally inherited rice fields in Kerinci, Sumatra with a complex adaptive change to customary rules.

Anecdotal evidence, as reflected in journalists' accounts and literary reflections, suggests that the full spectrum of success to failure and exploitation exists. Policy responses that try to minimize risks of exploitation and trafficking need to be constructed without reducing opportunities for livelihood benefits, which is no easy task. Our analysis of case studies was aimed at obtaining a more complete understanding of the drivers and consequences of migration from densely populated rural parts of Java in Indonesia to overseas urban and rural target areas, using a gender lens. Gender-specific aspects can be expected to apply to the decision to move (for young males, young females, families); their roles and opportunities in the new areas of temporary residence interacting with local communities and businesses; their relations back home, including remittances, invitations to join, sharing of knowledge and experience; the consequences of their migration on those who stay behind; and the consequences in case they decide to return home with any assets they may have acquired, their new skills, and norms of behavior, as shown in Figure 1. Each of these aspects deserves a fully fledged study but, even with the limited evidence available, a system approach to the whole picture is needed to guide policy development and to inform public discourse, where strongly polarized opinions pro and con tend to dominate.

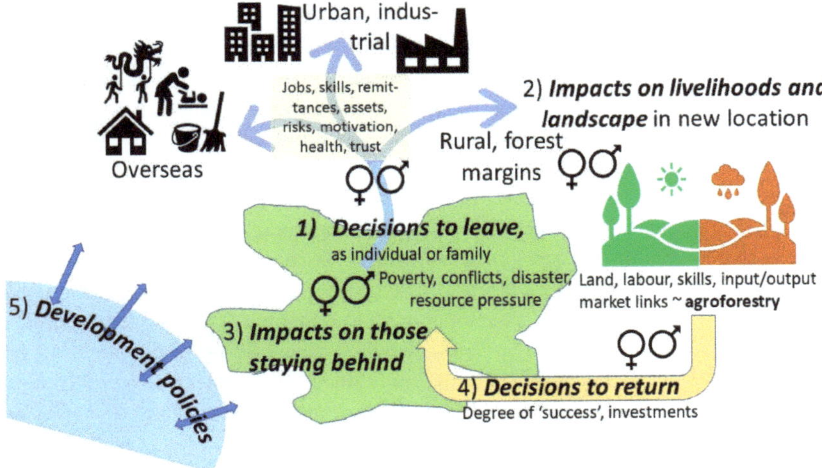

Figure 1. Gender-specific aspects before, during, and after (circular) migration decisions concerning land use in the area of origin.

While migration decisions—leaving and returning—are part of a large body of ethnographic studies, few studies have analyzed both the source and target areas of specific rural-to-rural migration patterns in Indonesia. We here describe two such cases, which allowed us partial answers to the following questions.

(1) What are the gendered patterns of movement concerning age and life histories in both source and target areas? Do gender norms of behavior influence land-use patterns differentially in source and target areas?

(2) How do (temporary) migrants compare the positive and negative aspects of home and temporary abodes?

(3) How do returnees reintegrate and modify land use, gender norms, and culture of the areas from which they originated?

We will first describe the two cases as such and then draw comparisons between them. The discussion will also touch on interactions with the government and other stakeholders concerning development policies.

2. Methods and Locations

2.1. Methods

This paper is based on anthropological fieldwork by the first author analyzing gender aspects of social development across Indonesia and, in particular, in two cases where migration source areas (West Java and South Sulawesi) could be linked to specific migration target areas (Lampung and Southeast Sulawesi, respectively), as shown in Table 1. Details of some of the field studies synthesized here have been published elsewhere [28–30]. Their *post hoc* combination into a single study, based on emerging opportunities rather than prior design, is new here.

Table 1. Characteristics of migration in the study area.

Study Area	Source	Area of Destination	Ethnicity	Type of Migration	Type of Work in the Destination Area
West Java	Ciamis	Sumberjaya, West Lampung	Sundanese	Temporary	Land-based and off farm
Southeast Sulawesi	Bone, Bulukumba, Soppeng, Sinjai, Wajo, Jeneponto, Maros in South Sulawesi	Kolaka and Konawe in Southeast Sulawesi	Bugis	Temporary	Land-based and off farm
			Bugis	Permanent	Land-based

The two case studies were each explored in two phases. Phase 1 aimed at understanding the migration context in the origin areas; Phase 2 on understanding the in-migration pattern and challenge in receiving communities. The West Java study was conducted in 2016 while the Southeast Sulawesi study was conducted in 2013.

In-depth individual interviews, individual structured interviews through a household survey, participatory observation, and structured group discussions were the methods used to obtain primary data used for this paper. The primary data could explore migration phenomenon, networks, patterns, and challenges at micro- and meso-levels. Specific to Southeast Sulawesi, the migration network was deeply analyzed at these levels. Stratified purposive sampling was applied for each method, for the individual structured interviews and structured group discussions.

From the characteristics of migration in each area, as shown in Table 1, we identified each community household according to some typologies considering the status of migration (migrant or stayer), the reason for migration and its destination areas and, most importantly, the gender aspect. In the case study of South to Southeast Sulawesi, we interviewed 65 respondents—in the Southeast Sulawesi we only interviewed migrants and in South Sulawesi we interviewed both migrant and non-migrant families. In the case study of West Java to Lampung, we interviewed in West Java alone 120 respondents, categorized as migrant and non-migrant.

Snowball sampling was used in particular for in-depth individual interviews and social network analysis. In-depth interviews were conducted to understand the historical and social realities that described the established migration chain and network. Secondary data, such as the results of a population census by the Central Statistical Bureau (Biro Pusat Statistik: BPS) [31,32], related literature and documentation, were used to support an explanation of the migration phenomenon, mainly at meso- and macro-levels.

The results of the household surveys were analyzed using descriptive statistics and interpreted qualitatively combined with information from focus group discussions and in-depth interviews to explain migrants' and non-migrants' characteristics, social typology, and actors involved in the migration chains. T-test statistics were employed to determine the significant differences between males and females and differing situations of migration. The analysis of migration or social networks was conducted using NodeXL software, which can measure and visualize the relationships of actors.

2.2. Study Sites: Background and Context

2.2.1. Indonesia

Indonesia is an archipelagic nation with the world's fourth-largest population (around 260 million people). The nation has recently reached lower-middle-income status. The average annual population growth was 1.27% for 2005–2010 and is expected to decline to 0.82% for 2020–2025. In 2015, 52.6% of the population lived in urban areas. Urban growth is, at 4.1% y^{-1}, the highest in Southeast Asia. The sex ratio of recent migration has been 110.3 (males per 100 females).

In terms of human population density, the largest contrast is between (i) densely populated, volcanic Java and Bali, with fertile soils and wet-rice agricultural traditions; (ii) Sumatra with population densities around the national average and strong tree-crop traditions, with coffee in the mountains and rubber and oil palm in the lowlands; and (iii) the rest of the archipelago with lower human population densities, with the exception of South Sulawesi, which approximates Java. Gender roles in agriculture tend to vary with regional contexts [28,29]. In upland areas, farming systems have mostly shifted from rice swiddens to a reliance on tree crops, such as cocoa, coffee, and rubber [33] and oil palm in areas with suitable rainfall.

2.2.2. Case Study One: The South to Southeast Sulawesi Connection

Our first case study analyzed migration histories, patterns, and networks from South to Southeast Sulawesi and other areas in rural areas in Indonesia and abroad. Cocoa began to boom and experience a "golden age" in Lawonua around 1997–1998 following the rise of prices worldwide, owing to a decrease in production in the Ivory Coast that led to global shortages. Li [34] described the same condition in Central Sulawesi, which also experienced a large arrival wave from South Sulawesi during the same period. Indonesia became a promising candidate for "major cocoa producer" at the time.

Working in the context of a regional development program underway at the time, we found that Lawonua Village in Besulutu Sub-district, Kolaka District, Southeast Sulawesi, with ongoing cocoa expansion, was a destination area for migrants from South Sulawesi. In the area, we had contacts that allowed surveys to be undertaken. The migrant percentage and composition in Lawonua were similar to that of the sub-district as a whole. The flow of migration into the village had been continuous over the preceding few years, which assisted our study in tracing the identity of the migrant community at origin. The tracing was conducted in the context of creating a community profile, which included physical, social, and economic conditions. The tracing was not only conducted at the location of migration (their current place of living) but also included the conditions (physical, social, and economic) in their origin village.

From the tracing process in Lawonua, 60% of migrants (which consisted of 40% of the total village population) came from Kalobba Village in Tellu Limpoe Sub-district, Sinjai District, South Sulawesi, as shown in Figure 2. Kalobba, characterized by limited resources and medium agricultural technology, is classified as a "suburban" area with limited land resources owing to pressure from outsiders. Competition for land causes a fairly high number of outgoing migrations from this village.

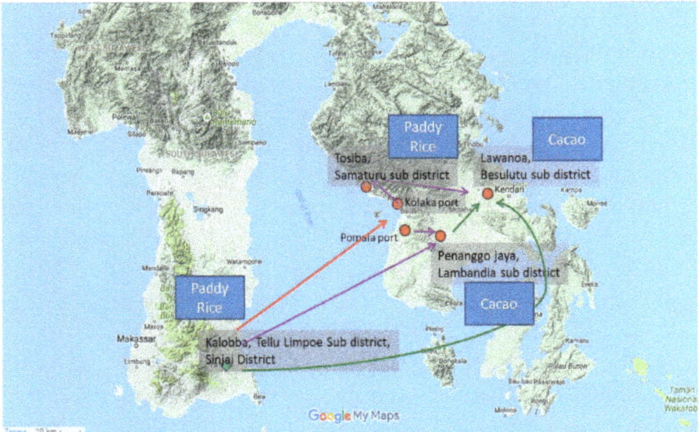

Figure 2. The source and destination of migrants in our first case study, in South (left) and Southeast Sulawesi (right), respectively.

2.2.3. Case Study Two: The West Java to Lampung Connection

Our second case study analyzed migration patterns, dynamics of migrant and stayer communities, migration decision making, and gender relations, with a focus on Ciamis District, West Java, where there is a close migration connection to coffee-growing landscapes in West Lampung District, Lampung, as shown in Figure 3. In Ciamis, the study was conducted in two villages in Panjalu Sub-district and two villages in Rajadesa Sub-district. Panjalu is in the northwest of the district and is the capital of Ciamis. Rajadesa is in the eastern part. Both villages are categorized as agricultural communities that rely heavily on farming as their source of livelihoods.

Figure 3. The source (right) and destination (left) of migrants in our second case study, West Java and Lampung.

3. Results

3.1. Migration Decision Making

Within the community of origin, decisions to migrate or stay in the village were mainly due to economic opportunity. Many migrants chose to migrate based on the capital they had, the support provided by their extended family and neighbors who could reduce the need for money for taking care of family members who stay in the village, or even the cost of living in the migration destination. There were four types of migrants in the community of origin.

(1) Off-farm out-migrants without capital. This group of out-migrants had low–middle economic status and were landless or had low levels of land ownership. They usually tried to find work to meet their daily needs as well as accumulate capital to establish farms. Their migration destinations were mostly in urban areas nearby, both in Sulawesi and West Java. This out-migrant group generally consisted of some family members (either women or men); however, most were male. Most women who out-migrated were typically unmarried. When they married, they usually chose to stop working for money, preferring to take care of their children and the household instead. Other family members who did not migrate and chose to remain in their village usually maintained businesses and/or cultivated farms.

(2) Off-farm out-migrants with capital. This group of middle–up out-migrants owned medium-sized areas of land. They usually had capital from previous work, from the sale of crops, or from inheritance. This out-migrant group could feature entire families migrating to cities or other prospective rural areas, who would only return to their village of origin for holidays. Most of these out-migrants were male. If there were women who out-migrated, they usually went with other family members.

They were mostly interested in off-farm livelihood sources, e.g., selling secondhand iron in West Java, while in Lampung and Southeast Sulawesi, selling clothes and other household items. The success of previous out-migrants in this line of work, as well as the networks created, attracted others from their village to do the same.

(3) and (4) Land-based out-migrants with and without capital. Land-based migrants without capital consisted of low–middle economic status who were mostly landless or had limited land in the origin community. They moved to the destination as on-farm labor, mostly working on their relatives' land, with some developing patron–client relationships. Rajadesa Village members in West Java migrated to Sumberjaya, Lampung to support their relatives in growing coffee. In Sulawesi, the relationship between a landowner and their followers became one of the doors through which a large number of migrants arrived. Landowners (land-based out migrants with capital) had some funds to open cocoa plantations and recruit followers from their home village to clear the land, plant, and care for the cocoa. When the plantation began to produce—after five years—a sharing system was implemented. The landowners maintained relationships with their clients to support their economic activities as well as maintain the power of their networks. The patrons' clients, who generally came from lower socio-economic groups, had better income sources through the patron compared to those in their home village, which no longer attracted their attention. Furthermore, through the patron's support, their migration to the new area became lower in risk and the cost required to migrate was reduced. The reciprocal relationship (reciprocity), although it was not entirely symmetrical (it was often highly asymmetric), was still able to improve their income. A summarized excerpt from an interview that illustrates this pattern follows.

> *Around 1997, HS, a landlord from South Sulawesi, purchased a large amount of land in Southeast Sulawesi. HS [a patron] later recruited his men [clients] to manage his land. His men were given approximately four hectares each for cocoa plantations. During the first six months, HS's men were given a living allowance of approximately 20 kg of rice per month, salted fish, and a few other staples. After six months, HS's men subsisted on seasonal crops grown in their plantations. After producing cocoa—after about 5 years—the harvest was divided: one part for the landowner and another part for the workers. Nowadays, the land here still belongs to the landlord, the clients may get a small part of the land as theirs, though the other land was still owned by HS. [30]*

3.2. Social Network, The Instrument of Migration Decision Making

As detailed elsewhere [30], three main network models reflected strategies used by migrants in their decision to migrate. The first model was a kinship-based network of either close relatives or immediate family, as well as distant relatives or extended family (53.84%). This strategy was commonly deployed by migrant communities who tended to be more mature, had sufficient capital to start migrating, as well as knowledge of cocoa or coffee cultivation as a requirement for planting. However, some landless people also decided to migrate as on-farm labor for other migrant communities who had established their farming practices in destination areas. Those migrants could be temporary, permanent, or seasonal. An excerpt from an interview that illustrates this pattern follows.

> Mr T, a resident of T, Lawonua Village, stated, *"I was visited by my uncle ... and he said, come on, move ... What are you doing staying in this village? You can plant a cocoa plantation there and the price of the land is cheap, not as expensive as here. Sell your plantation or cows here, it's enough to buy land there". [30]*

The second was a network set up to gain profit (44.12% of the total network). This network was either run in balance or not and built through a patron–client mechanism. A capital owner who later acts as a patron needs workers who are his inferiors as clients. The patron provides jobs and financial support, including the cost of migrating and supporting the clients' living needs in the early days of migration. These clients need the patron to improve the economic conditions of their families through managing their land as well as minimizing their migration risks. In this type of relationship, often the client's decision to migrate was not voluntary but forced owing to economic pressure and the vertical relationship with the patron.

The third network is a pattern of relationships that emerge owing to similarity of purpose. Generally, this pattern is characterized by identity, location of origin, and current residence similarities and was generally found in migrant communities who had been pioneer settlers in Southeast Sulawesi and Lampung. These groups built a network of neighborhood or identity similarity among community origin who had the same goal of increasing the number of plantations by expanding to villages that still had available land.

In the case of West Java, the first settlers in Lampung joined the transmigrant program in the early 1960s [35]. With their success in managing coffee systems apparent by the late 1980s, migrants joined from West Java originally as laborers and opened new coffee plots when they had access to capital [36].

These various relationship patterns often overlapped. A vertical relationship pattern, such as the patron–client relationship, could be reinforced by the patterns of kinship and neighborhoods, which were horizontal. For example, in the patron–client relationship, the kinship between the two often enlarged the client's decision to migrate not only because of economic need but also reluctance to reject an offer from a relative. Moreover, the overlapping relationship was also enforced by brokers, intermediaries, or ones who bridge the various groups of migrants from different regions to select land and encourage them to move. An intermediary or broker is an actor who can bridge and build the trust of the individual or group of individuals who initially were not interconnected. They facilitate social interaction, increase a community's economic activity, and minimize the risks of migration. On the other hand, the broker or intermediary is often associated with exploitation, transfer of risk into profits for intermediaries, and the accumulation of profit.

3.3. Type of In-Migrant

Migrant characteristics in receiving communities vary, however, among (1) new migrants (first-time movers), (2) recurrent migrants (multiple movers), and (3) follow-up migrants (family movers), as shown in Table 2. New migrants (pioneer migrants) are categorized as migrants who come directly from the area of origin to the receiving community and who have not migrated to other areas before. They may have a connection to the receiving community, through their family or neighbor who has lived in those areas, or they just decided to move, driven by the motivation to obtain land and increase their incomes.

Recurrent migrants or multiple movers are those who have already moved into, and out of, the region in surrounding receiving communities more than once. They may have some experience abroad in Malaysia or in Kalimantan, Indonesia. Recurring migrants in Southeast Sulawesi were generally migrants moving from an area near to their current receiving communities but decided to move to other areas in the same province to find better economic opportunities. A summarized excerpt from an interview that illustrates this pattern follows.

> *P, 68 years-old, a cocoa farmer from Kampala Village in Sinjai District, South Sulawesi sailed to Konaweha, Kolaka District, Southeast Sulawesi in the 1970s. He crossed the Gulf of Bone with his youngest child, aged 6 years, bringing five sacks of rice, two cows, and money amounting to IDR 35,000 (≈USD 3.50). After sailing for three days and two nights, he arrived at Kolaka and immediately visited his uncle who had already moved to Konaweha Village. He was helped by his uncle to look for flat land to be used to grow rice. After two days, he found land owned by a native resident who had received the land from the government but was unable to cultivate it. The land was sold cheaply to P. Over time, P's desire for land increased, especially for providing land for his children to equip them for the future. In 1995, P sought land in Lawonua (their current location) assisted by SF. Once the land was obtained, P did not move to Lawonua but still lived in Kolaka. However, in 2000, P finally decided to move to Lawonua with his wife and child. [30]*

Table 2. Migrant types and characteristics in Southeast Sulawesi [30].

Migrant Types	%	Origin Areas	Educational Background (% of Population)				Migrant Age Group (%)				Gender	
			Primary	Lower Secondary	Upper Secondary	Tertiary	16–24	35–29	40–54	>54	M	F
New migrants	45.83	Bone, Bulukumba, Sinjai, Soppeng, Wajo in South Sulawesi	44.4	44.4	11.1	0	1.39	22.22	18.06	5.56	14	20
Recurrent migrants	38.89	Bone, Bulukumba, Soppeng, Sinjai, Wajo, Jeneponto, Maros in South Sulawesi	67.9	14.3	10.7	7	-	11.11	18.06	5.56	13	14
Follow-up migrants	12.5	Bone, Bulukumba, Soppeng, Sinjai, Wajo in South Sulawesi	69.7	15.2	12.1	3	2.78	6.94	2.78	-	6	3

These recurring migrants migrated with the motivation to increase the amount of land they owned. Some of them had financial difficulties and intended to sell land in Pinanggo or Konaweha to overcome their problems. Land in the area was more expensive than land in Lawonua. Selling the land in their area and looking for land in a cheaper place could solve their problems. Some other migrants who were able to accumulate capital deliberately looked for land in new areas to increase the number of their plantations. Recurring migrants usually owned land. The acquired land was later shared as inheritance or dowry for the marriages of their offspring.

Follow-up migrants are descendants of migrants who have lived for a long time in receiving communities. Most of them were born in their origin areas and later moved to the receiving communities as toddlers or teens after their parents had already moved there. They are follow-up migrants who have not received a share of the family land nor been able to obtain their own.

3.4. Gender Relations in Communities of Origin

In describing the gender and age specificity of migration patterns, we used distinctions based on the status of who had migrated, as shown in Table 3. From the two case studies, we could see that for land-based livelihood options, youth or single men and adult or married men migrated by themselves and left their families in the origin area, who might join later. However, the discussion has shown that no married women left their villages for land-based options but perhaps to work in urban areas.

Table 3. Livelihood options based on the type of migration, age, and gender.

Gender of the Migrant	Age Classification	Livelihood Option		Type of Migration	Condition in Origin Areas
		Off-Farm Based	Land-Based		
Only men migrated	Youth (un-married)	Industrial work in urban and rural areas abroad	Plantation labor	SeasonalPermanent	- Abandoned land
	Adults (married)	Business (sales)	Sharecropping Land-rent	Seasonal	- Spouse or an extended family member manages the land
Only women migrated	Majority youth and unmarried	Industrial and domestic workers (urban and abroad)	No	Seasonal	- Spouse or an extended family member manages the land
Men migrated first, women following	Adults (married)	Starting a new business–urban areas	Sharecropping Land-rent Land-purchase	Permanent (combination with seasonal)	- Spouse or an extended family manage the land - Abandoned land
Whole family migrated	All ages (family)	Industrial Starting new business	Sharecropping Land-rent Land-purchase Plantation labor	Permanent	- Abandoned land

Source: Household survey in West Java, Southeast and South Sulawesi, and focus group discussions in West Java and South Sulawesi.

From the case study in West Java, of the family members left behind in the origin communities, some were maintaining their agricultural systems and there were some changes in the roles of family members—when men migrate, the women or other family members maintain their agricultural plots. Statistical analysis of the difference in women's roles showed that women's decision making was significantly increased in paddy rice cultivation and regular maintenance; in particular, for fertilizer and pesticide purchasing, as shown in Table 4. Decision making on other agricultural practices was slightly increased although not significantly. Although women's roles increased in maintaining plots, decision making regarding timber cultivation was still considered to be the man's domain. An excerpt from an interview illustrates this pattern.

> *"When father migrated, I had the responsibility to cultivate our timber garden. Usually, I weed the plot. To me, cultivating a timber garden is not so difficult, just weeding. If we need to harvest it, we need to wait for my husband, who usually knows how to calculate the timber prices and who can negotiate the price. I don't know anything about selling timber and how to calculate the price of timber"* (in-depth interview, Rajadesa Village, West Java).

When we compare the situation of land-based out-migrant and off-farm out-migrant situations, the role of women in decision making is greater in the off-farm, in which the men move to urban areas and have less involvement in agriculture. For land-based out-migrant communities, men and women still worked together on agricultural practices and most decisions were made jointly, or by the man alone, as shown in Table 5.

Table 4. Women's workload and power related to decision making based on different periods, West Java.

Category	Activity and Decision Making	When Male Stayed	When Male Out-Migrated	*p*-Value	Significance
All	All decisions	4.48	5.01	0.00	√
Farming	Paddy rice	4.11	6.79	0.00	√
	Annual crops	2.78	3.11	0.44	
	Trees	2.78	3.11	0.44	
	Planting	3.64	4.05	0.25	
	Farm labor	4.24	4.78	0.18	
	Fertilizer/pesticide application	3.20	3.44	0.54	
	Purchasing fertilizer/pesticide	3.01	3.93	0.05	√
	Marketing products	4.03	4.49	0.27	
Household	Daily consumption	8.52	9.12	0.04	√
	Schooling	5.09	5.48	0.08	
	Purchase house	4.41	3.64	0.02	√
	Purchase land	4.53	3.94	0.03	√
	Purchase vehicle	4.33	3.63	0.02	√
	Sell land	4.67	4.00	0.08	
	Financial management	6.17	6.53	0.37	
	Home and childcare	7.76	8.71	0.00	√

Source: Household survey, West Java (2016).

Table 5. Women's workload and power related to decision making based on the type of migration, West Java.

Category	Activity and Decision Making	Off-Farm Based Out-Migration	Land-Based Out-Migration	*p*-Value	Significance
All	All decisions	4.95	4.43	0.00	√
Farming	Paddy rice	5.08	5.87	0.19	
	Market products	4.88	3.60	0.02	√
	Farm labor	4.79	3.90	0.08	
	Annual crops	4.27	2.91	0.00	√
	Purchasing fertilizer/pesticide	4.19	2.43	0.00	√
	Planting time	4.02	3.21	0.07	
	Fertilizer/pesticide application	4.02	2.50	0.00	√
	Trees	3.59	2.04	0.00	√
Household	Daily consumption	8.76	8.93	0.62	
	Schooling	5.28	5.23	0.87	
	Purchase house	4.00	4.05	0.91	
	Purchase land	4.21	4.17	0.89	
	Purchase vehicle	4.12	4.04	0.82	
	Sell land	4.83	4.07	0.08	
	Financial management	6.58	6.04	0.28	
	House and childcare	8.55	7.83	0.07	

Source: Household survey, West Java (2016).

We might see that this migration owing to land-based activities does not improve women's role in decision making. However, we could see that this means less burden for women. When the male migrates and women are responsible for the land, the extended family may be involved, or the woman hires labor. However, for a poor family, women's work in agriculture will be increased. We see that in land-based migration, women can still make decisions in certain cases but for more strategic decisions they needed to discuss it with the man, as the land manager. Additionally, in land-based migration, at certain times, i.e., in seasons when source areas need more labor, the male might return to the village to help while for off-farm migrants, opportunities to return to the village were less.

3.5. Returnee Migrants

Some of the migrants who had accumulated considerable capital moved back to their place of origin, cleared land, and started to cultivate. The remittances and investment of the returnee migrant could be in various forms, such as building a mosque or road for the villages of origin. In the West Java study, some of the migrants once back in their village, in addition to managing their agricultural land there, kept their land in Lampung or other destination areas, employing landless farmers from their village of origin to cultivate it. Hence, inducing follow-up migrants.

In West Java, returnee migrants from Lampung used their collected capital and skills they built during the migration experience to manage the coffee system in the Gunung Sawal protected area under a community-based management scheme. De Royer [37] stated that while there seems to be no clear correlation between the beginning of the community forestry program in 2008 and the mass return of migrants, the program has been perceived by migrants as a great opportunity to make money from the skills they obtained in Lampung while re-establishing themselves in West Java. The Gunung Sawal region has since developed into the coffee hub of West Java with a very well-developed market. Here we can see a clear link between the return of migrants and the rapid expansion of land under coffee cultivation during the last two decades. People converted their mixed tree gardens into coffee gardens; those who were not migrants learned from those who were.

In Sulawesi, returnee migrants did not move back to cultivate their land. Some of them tried to find better luck elsewhere with cheaper, or any available, land, thereby creating multiple-time migrants. A few of them used their network and their knowledge of land information to connect people from the origin areas to available land in destination areas, acting as intermediaries or brokers. However, the legality of such land in the market is often "grey". Some land "ownership" does not have a strong legal basis supported by a letter or certificate from the National Land Agency.

4. Discussion

These various observations can be interpreted as pieces of a larger puzzle in which migration decisions become linked at human life-cycle scales, as shown in Figure 1, based on the five questions raised in the introduction. We split the first question ("What are the gendered patterns of movement concerning age and life histories in both source and target areas? Do gender norms of behavior influence land-use patterns differentially in source and target areas?") into two parts.

4.1. Q1A. Which Conditions within the Community of Origin Have Been the Main Trigger for People to Try Their Luck Elsewhere (Overseas, Urban, or Rural), at Least Temporarily?

In the two study cases, economic needs were the main trigger for migration. The lack of arable land and lack of capital to intensify farming systems pushed the pioneer farmers in West Java and South Sulawesi to search for land elsewhere. The choice of target areas was usually decided through existing kinship or neighborhood networks. The broker or intermediary mechanism was formed through informal relationships and occurred at almost all stages of migration. In South Sulawesi, during the era of crop development implemented by government programs (the "Green Revolution", agricultural intensification, and agricultural extension), migration occurred spontaneously into those communities that wanted to improve their livelihoods by attracting more labor [38]. Kinship networks were so strongly binding that the functions of intermediaries were not visible. After the boom in cocoa commodity development in Sulawesi, the rate of migration rapidly increased, the primary purpose of which was to improve incomes through the expansion of cocoa plantations. Migration increased massively and spread to other areas; migrants from different villages began to arrive. The roles of intermediaries became stronger during this phase, using networks of kinship, neighborhood, and friendship.

What occurred in the two study areas was similar to that described in a study of motivations to migrate conducted by Amacher and Hyde [39] in remote areas in the Philippines. They indicated that the direction of migration to remote areas was largely determined by the availability of land that was accessible for migrants, resulting in a preference for areas with low population density. Accessible in this context meant that the land had no clear ownership status so that they could easily start cultivating it.

4.2. Q1B. What Effects Did Migration Have on Gender-Specific Land Use and Livelihoods in the Areas of Origin?

Feminization of agriculture as described elsewhere [40] under dominantly male migration patterns, appeared to be less common in the two study areas. What occurred was often a pattern where elderly people or other extended family members taking care of grandchildren stayed behind and struggled to maintain their agricultural practices. In the case of the West Java study, male farmers still had the decision-making power on tree-based farming in the origin area. Male farmers would return during harvesting or for any other activity that required their presence. Vice versa, females could undertake temporary migration to help harvesting or other laborious activities. However, in the day-to-day decision making in managing households and annual crops, there were significant differences in decision making compared to when the male was not on migration. Despite the changing roles in the households of a migrant community, balancing the needs of agricultural labor, work seasons, and family needs might be achieved through mutual adjustments of marriage partners [41]. This could be the key to the success of livelihood actions in the areas of origin adapting to change.

Hecht (2015) [42] indicated that off-farm labor migration could result in changes in land uses and forest dependence, following shifts in gender and generational relations. Our findings indicate, though not significant, that women's burden was heavier in the off-farm-based migration situation and abandonment of land could occur more readily. Further exploration is needed to see patterns of work and the success of working relationships in non-land-based migration.

4.3. Q2. How Do (Temporary) Migrants Compare the Positive and Negative Aspects of Home and Temporary Abodes?

The influx of migrants into new areas can change the type of farming systems. For example, the Bugis migrants entering Southeast Sulawesi in the era of the rice-focused Green Revolution began to switch their interest to cocoa cultivation. It became easier for them to find new areas that were suitable for planting cocoa because of their migration experience, under the schemes inspired by the Green Revolution, which turned out to be compatible with cocoa cultivation. New technology (such as herbicides and hand tractors) introduced in the Green Revolution for rice fields reduced the length of time on the farm, and therefore, the opportunities to manage more land became greater. Their mobility was increased as well, owing to less time being consumed on-farm and were motivated even more to look for land and manage cocoa plantations in other places. In some of the cases, experience with more intensified land use in the source area (e.g., West Java) enriched agriculture and agroforestry in the new environment (e.g., South Sumatra).

The influx of migrants also triggered a new type of job. The farmers in Southeast Sulawesi who were no longer interested in farming and planting began to turn to other sources of income (off-farm). Those who understood land matters later became intermediaries in land markets, drawing on the pool of people they knew, be they relatives or neighbors, to buy land. Absorption of new labor can start as paid labor and a patron–client relationships or share-cropping but also involve land renting and buying within customary ownership rules (rarely involving formal land certification).

In both Southeast Sulawesi and Lampung, migrants from more densely populated South Sulawesi and Java, respectively, did not arrive on an empty slate but in landscapes where local farming communities of Semendo (a southward expansion of a group in the neighboring province of South Sumatra) and Tolaki ethnic identity had converted forests for extensive land use. The migrants brought traditions of more intensive agroforestry management of coffee and cocoa, leading to a synergy where the group with longer presence opened forest and the newcomers bought existing gardens and intensified them before they returned to fallow vegetation. In the surveys of local ecological knowledge in Sumberjaya in Lampung, the two groups were distinguishable in knowledge and concepts [43].

Similarly, experience in farming cultivation can have an impact on the farming systems in the origin areas. Such is the case with the West Java migrants who, upon returning from Lampung and gaining access to forest land through a social forestry scheme, were able to manage coffee systems that were considered to be more environmentally friendly than timber systems, which were the traditional farming systems in the area.

4.4. Q3. How Do Returnees Reintegrate and Modify Land Use, Gender Norms, and Culture of the Areas from Which They Originated?

There is a tendency that migrants who have failed may find other land uses or urban labor elsewhere, rather than returning home. For those who succeed, these land-based migrants may return home if arable land still exists, as in the case of the West Java study. The migrants, returning or not, will continue supporting the home villages by building a mosque or supporting other social infrastructure development, such as building schools or daycare centers.

In the case of South Sulawesi, the migrant farmers continued exploring other regions to gain more land. In current circumstances, there is very little forest area left that is accessible to migrants. Most of the accessible land is already owned by local communities or is in areas in which conflict between companies, governments, and communities has already emerged. Acquiring land requires contracts with (and payments to) the native or migrant communities who have access (de facto and/or de jure). The actions of land brokers and the migration process grew rapidly, spreading to areas where ample land was available with less population density and less strict local institutions. Simbune Village in Kolaka District firmly rejected migrants. The village head stated that a ban on selling land to new settlers was strictly implemented. It was proven to be the case by the lead researcher of this study, who visited the village in 2014. All land was still controlled by residents.

Ruf [44] analyzed the effects of cocoa development on migration. The model of migration for the development of cocoa plantations was mostly adopted by Bugis migrants and several other groups. Various methods were used to facilitate migrants in obtaining land for cocoa plantations, especially for those who had limited capital. The systems included informal "forest rent" payments to forest authorities, share-cropping, and patron–client relationships. This situation strengthened migrant desire to gain more land in other areas, when they felt to have failed in migrating a previous time or when they needed to expand their livelihood.

4.5. Interactions with Development Policies?

Overall, our analysis suggests that positive aspects in both areas of origin and receiving areas may prevail, with the exchange of knowledge between areas of different land-use intensities spreading agroforestry practices. The latter may well be more effective than the routes through formal knowledge and extension and in some cases is combined with tree germplasm exchange. Feminization of agriculture through preferentially male-based migration is not common in Indonesia but age-based consequences are common, in both urban (or overseas) migration and dispersal to areas of lower population density.

Costs and benefits of social capital in the context of rural–urban migration are organized by gender [45]. Opportunities for urban jobs tend to be gender-specific, with higher perceived success for unmarried women to work in the export factories of West Java so as to remit money to their natal households and, at the same time, gather experience of modernity, has had consequences for the young men that stayed behind [46].

Spontaneous and organized population movements have long been used as a means of promoting a country's goals of development and national integration. At the local level, on the other hand, these movements have frequently done the opposite, fueling local grievances, sharpening group distinctions, and at times creating "sons-of-the-soil" conflicts [47]. In different policy phases associated with environmental governance in Lampung in Indonesia, migrants were defined initially as pioneer entrepreneurs, bringing progress to Indonesia's hinterland but, subsequently, as forest squatters, threatening the cultural and ecological integrity of the province [41]. Rural migrants attempted to resolve their problematic positioning through multi-local livelihoods, which combine access to non-local income through temporary migration with the maintenance of a foothold that signals belonging and legitimate entitlement to state resources.

Migrants coming from more densely populated areas with more intensive agriculture bring know-how and customs that can be new to the areas where they settle but also differ from government regulations. In an irrigation-based Balinese migrant society in Sulawesi, the traditional ("subak") institutions for water management linked to synchronized cropping cycles clashed with the state-regulated water users' association that separated technical water management from the wider scope of subak. Adjustment of perceived property rights to land, water, and irrigation infrastructure was feasible within a transmigration setting but conflicts persist where farmers of different ethnic and religious backgrounds farm close together [48].

Bilocality—in which an individual will spend part of the year in a rural area and the other in an urban area—is increasingly common in Central Java, addressing lack of income in rural areas. It generates remittances but also contributes in limited ways to the commonly anticipated rural development outcomes [49].

A review of the literature on remittances [50] concluded that land use as a driver of migration, livelihood strategies, and the use of remittances in investment in land use need to be studied more coherently than has commonly been the case. Financial remittances are only part of the way relationships are maintained, as may be clear from our case studies.

5. Conclusions

Synthesizing across two case studies in Indonesia, we focused on five aspects of (circular) migration.

(1) The conditions within the community of origin that encourage people to migrate are diverse. Most of the decision making to migrate is linked to natural resource and land competition and emergencies, such as natural disasters or increased human conflict. However, decisions are facilitated by networks that take over existing social obligations within the community.

(2) The effect of migration on land use and livelihoods in the areas of origin mean extensification or lowering of the labor/land ratio. Feminization of agriculture, as reported elsewhere, appears to be less common in the study areas. There are differences in decision making between women with migrating and non-migrating spouses but not in managing tree-based systems.

(3) The changes in the receiving community and its environment generally imply intensification. A new way of managing more efficient farming systems was commonly found in both study areas. Absorption of new labor as follow-on migration through expansion of the agricultural area encourages the existence of new jobs as a land broker owing to increases in land renting and buying.

(4) Migrants return with different levels of external success. People coming back with success may help to rebuild the village and its agricultural system and could invest in social capital (mosques, healthcare, schools). Some who have failed may find other land uses or urban labor options elsewhere.

(5) The interaction of migrants with the government and other stakeholders concerning development policies is largely implicit. Government programs on rural and agricultural development often support migrants in the initial phases of migration with the technical know-how of new commodities or farming systems that migrants practice in the origin areas. The development of good

transportation infrastructure allows successful farmers to easily manage agricultural land in both the origin and target areas. Overall, circular migration facilitates the exchange of know-how within the broader Indonesian agroforestry traditions of rural land use on the forest margins, with substantial reliance on economically important tree crops.

Author Contributions: Conceptualization, E.M., B.L., M.v.N.; methodology, E.M., B.L.; formal analysis, E.M., B.L.; investigation, E.M.; writing—original draft preparation, E.M., B.L., M.v.N.; writing—review and editing, E.M., B.L., M.v.N. All authors have read and agreed to the published version of the manuscript.

Funding: This research was funded by World Agroforestry (ICRAF) as part of its Indonesia program.

Acknowledgments: We acknowledge the willingness of all interviewed respondents in sharing their experience, context, and considerations.

Conflicts of Interest: The authors declare no conflict of interest. The funders and authorities involved had no role in the design of the study; in the collection, analyses, or interpretation of data; in the writing of the manuscript, or in the decision to publish the results.

References

1. De Royer, S.; Visser, L.E.; Galudra, G.; Pradhan, U.; Van Noordwijk, M. Self-identification of indigenous people in post-independence Indonesia: A historical analysis in the context of REDD+. *Int. For. Rev.* **2015**, *17*, 282–297. [CrossRef]
2. Rahmati, S.H.; Tularam, G.A. A critical review of human migration models. *Clim. Chang.* **2017**, *3*, 924–952.
3. Manning, P.; Trimmer, T. *Migration in World History*; Routledge: New York, NY, USA, 2020.
4. Baker, J.L.; Rotimi, C.N.; Shriner, D. Human ancestry correlates with language and reveals that race is not an objective genomic classifier. *Sci. Rep.* **2017**, *7*, 1572. [CrossRef] [PubMed]
5. McColl, H.; Racimo, F.; Vinner, L.; Demeter, F.; Gakuhari, T.; Moreno-Mayar, J.V.; Van Driem, G.; Wilken, U.G.; Seguin-Orlando, A.; De la Fuente Castro, C.; et al. The prehistoric peopling of Southeast Asia. *Science* **2018**, *361*, 88–92. [CrossRef]
6. Iskandar, J.; Iskandar, B.S.; Partasasmita, R. The impact of social and economic change on domesticated plant diversity with special reference to wet rice field and home-garden farming of West Java, Indonesia. *Biodivers. J. Biol. Divers.* **2018**, *19*, 515–527. [CrossRef]
7. Mulyoutami, E.; Rismawan, R.; Joshi, L. Local knowledge and management of simpukng (forest gardens) among the Dayak people in East Kalimantan, Indonesia. *For. Ecol. Manag.* **2009**, *257*, 2054–2061. [CrossRef]
8. De Foresta, H.; Kusworo, A.; Michon, G.; Djatmiko, W.A. *Ketika Kebun Berupa Hutan: Agroforest khas Indonesia Sebuah Sumbangan Masyarakat*; ICRAF: Bogor, Indonesia, 2000.
9. Duguma, L.A.; van Noordwijk, M.; Minang, P.A. COVID-19 and social-ecological system resilience at the forest-agriculture interface. *LAND* **2020**. in review.
10. van Noordwijk, M.; Duguma, L.A.; Dewi, S.; Leimona, B.; Catacutan, D.C.; Lusiana, B.; Öborn, I.; Hairiah, K.; Minang, P.A. SDG synergy between agriculture and forestry in the food, energy, water and income nexus: Reinventing agroforestry? *Curr. Opin. Environ. Sustain.* **2018**, *34*, 33–42. [CrossRef]
11. Black, R.; Adger, W.N.; Arnell, N.W.; Dercon, S.; Geddes, A.; Thomas, D. The effect of environmental change on human migration. *Glob. Environ. Chang.* **2011**, *21*, S3–S11. [CrossRef]
12. McLeman, R.; Smit, B. Migration as an adaptation to climate change. *Clim. Chang.* **2006**, *76*, 31–53. [CrossRef]
13. Fan, C.C. *China on the Move: Migration, the State, and the Household*; Routledge: New York, NY, USA, 2007.
14. Farbenblum, B.; Taylor-Nicholson, E.; Paoletti, S. *Migrant Workers' Access to Justice at Home: Indonesia*. Public Faculty Scholarship at Penn Law. 495. Available online: https://scholarship.law.upenn.edu/faculty_scholarship/495 (accessed on 20 September 2020).
15. Available online: https://www.worldbank.org/en/country/indonesia/overview (accessed on 20 September 2020).
16. Benček, D.; Schneiderheinze, C. *Higher Economic Growth in Poor Countries, Lower Migration Flows to the OECD: Revisiting the Migration Hump with Panel Data*; Working Paper 2145; Kiel Institute for the World Economy: Kiel, Germany, 2020; ISSN 2195-7525.
17. Clemens, M. *The Emigration Life Cycle: How Development Shapes Emigration from Poor Countries*; Center for Global Development: Bonn, Germany, 2020; Available online: https://www.cgdev.org/sites/default/files/emigration-life-cycle-how-development-shapes-emigration-poor-countries.pdf (accessed on 20 September 2020).

18. Bazzi, S. Wealth heterogeneity and the income elasticity of migration. *Am. Econ. J. Appl. Econ.* **2017**, *9*, 219–255. [CrossRef]

19. Syafitri, W. Determinants of labour migration decisions: The case of East Java, Indonesia. *Bull. Indones. Econ. Stud.* **2013**, *49*, 385–386. [CrossRef]

20. Bryan, G.; Morten, M. The aggregate productivity effects of internal migration: Evidence from Indonesia. *J. Polit. Econ.* **2019**, *127*, 2229–2268. [CrossRef]

21. Bazzi, S.; Gaduh, A.; Rothenberg, A.D.; Wong, M. Skill transferability, migration, and development: Evidence from population resettlement in Indonesia. *Am. Econ. Rev.* **2016**, *106*, 2658–2698. [CrossRef]

22. Bazzi, S.; Gaduh, A.; Rothenberg, A.D.; Wong, M. *Unity in Diversity?: Ethnicity, Migration, and Nation Building in Indonesia*; Centre for Economic Policy Research: Jakarta, Indonesia, 2017.

23. Deshingkar, P. Environmental risk, resilience and migration: Implications for natural resource management and agriculture. *Environ. Res. Lett.* **2012**, *7*, 015603. [CrossRef]

24. Pattnaik, I.; Lahiri-Dutt, K.; Lockie, S.; Pritchard, B. The feminization of agriculture or the feminization of agrarian distress? Tracking the trajectory of women in agriculture in India. *J. Asia Pac. Econ.* **2018**, *23*, 138–155. [CrossRef]

25. Lam, T.; Yeoh, B.S. Migrant mothers, left-behind fathers: The negotiation of gender subjectivities in Indonesia and the Philippines. *Gend. Place Cult.* **2018**, *25*, 104–117. [CrossRef] [PubMed]

26. Mueller, V.; Doss, C.; Quisumbing, A. Youth migration and labour constraints in African agrarian households. *J. Dev. Stud.* **2018**, *54*, 875–894. [CrossRef]

27. Burgers, P.P. Livelihood dynamics, the economic crisis, and coping mechanism in Kerinci district, Sumatera. In *Rural Livelihoods, Resources and Coping with Crisis in Indonesia. A Comparative Study*; University of Utrecht: Utrecht, The Netherlands, 2008; pp. 71–90.

28. Colfer, C.J.P.; Achdiawan, R.; Roshetko, J.M.; Mulyoutami, E.; Yuliani, E.L.; Mulyana, A.; Moeliono, M.; Adnan, H. The balance of power in household decision-making: Encouraging news on gender in southern Sulawesi. *World Dev.* **2015**, *76*, 147–164. [CrossRef]

29. Mulyoutami, E.; Roshetko, J.M.; Martini, E.; Awalina, D. Gender roles and knowledge in plant species selection and domestication: A case study in South and Southeast Sulawesi. *Int. For. Rev.* **2015**, *17*, 99–111.

30. Mulyoutami, E.; Wahyuni, E.S.; Kolopaking, L.M. *Agroforestry and Forestry in Sulawesi Series: Unravelling Rural Migration Networks: Land-Tenure Arrangements among Bugis Migrant Communities in Southeast Sulawesi*; Working Paper 225; World Agroforestry Centre (ICRAF): Bogor, Indonesia, 2016; p. 58. [CrossRef]

31. Sub Direktorat Statistik Ketahanan Wilayah. *Statistik Potensi Desa Indonesia 2014*; Badan Pusat Statistik: Jakarta, Indonesia, 2014.

32. BPS. Diseminasi Hasil Sensus 2010. Available online: https://sp2010.bps.go.id/index.php/metadata/index (accessed on 20 September 2020).

33. van Noordwijk, M.; Mulyoutami, E.; Sakuntaladewi, N.; Agus, F. *Swiddens in Transition: Shifted Perceptions on Shifting Cultivators in Indonesia*; ICRAF: Bogor, Indonesia, 2008.

34. Li, T.M. Local histories, global markets: Cocoa and class in upland Sulawesi. *Dev. Chang.* **2002**, *33*, 415–437.

35. Pasya, G.; Mudikdjo, K.; Tjondronegoro, S.M.P.; Kusmana, C.; Nurbaya, S. Analysis of Causes dan Styles of Environmental Conflict in Protection Forest Zone Management (A Case Study in Register 45B Bukit Rigis Protection Forest Zone, Lampung Province). *Forum Pascasarj.* **2011**, *34*, 215–229.

36. Verbist, B.; Pasya, G. Perspektif sejarah status kawasan hutan, konflik dan negosiasi di Sumberjaya, Lampung Barat, Propinsi Lampung. *AGRIVITA* **2004**, *26*, 20–28.

37. De Royer, S.; Pradhan, U.; Fauziyah, E.; Sulistyati, T. *Community-Based Forest Management: Who Is Benefiting? Community-Based Forest Management, Coffee and Migration Patterns in Ciamis Regency, West Java, Indonesia*; Brief 44; World Agroforestry Centre (ICRAF): Bogor, Indonesia, 2014.

38. Ruf, F. Cacao migrants from boom to bust. In *Agriculture in Crisis: People, Commodities, and Natural Resources in Indonesia, 1996–2000*; Gerard, F., Ruf, F., Eds.; Routledge: London, UK, 2001.

39. Amacher, G.S.; Cruz, W.; Grebner, D.; Hyde, W.F. Environmental motivations for migration: Population pressure, poverty, and deforestation in the Philippines. *Land Econ.* **1998**, 92–101. Available online: https://www.jstor.org/stable/3147215 (accessed on 20 September 2020). [CrossRef]

40. Tamang, S.; Paudel, K.P.; Shrestha, K.K. Feminization of Agriculture and its Implications for Food Security in Rural Nepal. *J. For. Livelihood.* **2014**, *12*, 20–32.

41. Elmhirst, R. Displacement, resettlement, and multi-local livelihoods: Positioning migrant legitimacy in Lampung, Indonesia. *Crit. Asian Stud.* **2012**, *44*, 131–152. [CrossRef]

42. Hecht, S.; Yang, A.L.; Basnett, B.S.; Padoch, C.; Peluso, N.L. *People in Motion, Forests in traNsition: Trends in Migration, Urbanization, and Remittances and Their Effects on Tropical Forests*; CIFOR: Bogor, Indonesia, 2015; Volume 142.

43. Joshi, L.; Schalenbourg, W.; Johansson, L.; Khasanah, N.M.; Stefanus, E.; Fagerström, M.H.; van Noordwijk, M. Soil and water movement: Combining local ecological knowledge with that of modellers when scaling up from plot to landscape level. In *Belowground Interactions in Tropical Agroecosystems*; CAB International: Wallingford, UK, 2004; pp. 349–364.

44. Ruf, F. Tree crops as deforestation and reforestation agents: The case of cacao in Cote d'Ivoire and Sulawesi. In *Agricultural Technologies and Tropical Deforestation*; Angelsen, A., Kaimowitz, D., Eds.; CABI Publishing: Wallingford, UK; Center for International Forestry Research: Bogor, Indonesia, 2001; pp. 291–316.

45. Silvey, R.; Elmhirst, R. Engendering social capital: Women workers and rural–urban networks in Indonesia's crisis. *World Dev.* **2003**, *31*, 865–879. [CrossRef]

46. Elmhirst, R. Tigers and gangsters: Masculinities and feminised migration in Indonesia. *Popul. Space Place* **2007**, *13*, 225–238. [CrossRef]

47. Côté, I. Internal migration and the politics of place: A comparative analysis of China and Indonesia. *Asian Ethn.* **2014**, *15*, 111–129. [CrossRef]

48. Roth, D. Property and authority in a migrant society: Balinese irrigators in Sulawesi, Indonesia. *Dev. Chang.* **2009**, *40*, 195–217. [CrossRef]

49. Rudiarto, I.; Hidayani, R.; Fisher, M. The bilocal migrant: Economic drivers of mobility across the rural-urban interface in Central Java, Indonesia. *J. Rural Stud.* **2020**, *74*, 96–110. [CrossRef]

50. Cole, R.; Wong, G.; Brockhaus, M. *Reworking the Land: A Review of Literature on the Role of Migration and Remittances in the Rural Livelihoods of Southeast Asia*; CIFOR: Bogor, Indonesia, 2015.

Publisher's Note: MDPI stays neutral with regard to jurisdictional claims in published maps and institutional affiliations.

 *land*

Article

Effects of Agroforestry and Other Sustainable Practices in the Kenya Agricultural Carbon Project (KACP)

Ylva Nyberg [1,*] , Caroline Musee [2], Emmanuel Wachiye [2], Mattias Jonsson [3], Johanna Wetterlind [4] and Ingrid Öborn [1,5]

[1]   Department of Crop Production Ecology, Swedish University of Agricultural Sciences (SLU), P.O. Box 7043, SE-75007 Uppsala, Sweden
[2]   Vi Agroforestry, P.O. Box 2006, Kitale 30200, Kenya; caroline.musee@viagroforestry.org (C.M.); emmanuel_wachiye@wvi.org (E.W.)
[3]   Department of Ecology, SLU, P.O. Box 7044, SE-75007 Uppsala, Sweden; mattias.jonsson@slu.se
[4]   Department of Soil and Environment, SLU, P.O. Box 234, SE-53223 Skara, Sweden; johanna.wetterlind@slu.se
[5]   World Agroforestry (ICRAF), UN Avenue, P.O. Box 30677-00100 Nairobi, Kenya; ingrid.oborn@slu.se
*   Correspondence: ylva.nyberg@slu.se; Tel.: +46-702-580-488

Received: 22 September 2020; Accepted: 9 October 2020; Published: 13 October 2020

Abstract: With growing global demand for food, unsustainable farming practices and large greenhouse gas emissions, farming systems need to sequester more carbon than they emit, while also increasing productivity and food production. The Kenya Agricultural Carbon Project (KACP) recruited farmer groups committed to more Sustainable Agricultural Land Management (SALM) practices and provided these groups with initial advisory services on SALM, farm enterprise development and village savings and loan associations. Recommended SALM practices included agroforestry, cover crops, mulching, composting manure, terracing, reduced tillage and water harvesting. The effects of the KACP on the uptake of SALM practices, maize yield, perceived food self-sufficiency and savings during the initial four years were assessed comparing control and project farmers using interviews, field visits and measurements. Farmers participating in the KACP seemed to have increased uptake of most SALM practices and decreased the use of practices to be avoided under the KACP recommendations. Agroforestry and terraces showed positive effects on maize yield. During all four years, the KACP farms had higher maize yield than control farms, but yield differences were similar in 2009 and 2012 and there was no overall significant effect of the KACP. In 2012, the KACP farms had higher food self-sufficiency and tended to have higher monetary savings than control farms.

Keywords: adaptation; carbon sequestration; Kisumu; Bungoma; payment for ecosystem services; village savings and loan associations

---

1. Introduction

Managing the trade-off between short-term provision of human needs in terms of water, food, shelter and fibre for a growing population and maintaining the capacity of the planet to provide these services in the future is a growing global challenge [1–3]. World agriculture is currently both contributing to global environmental change [3,4] and being severely affected by it [5]. It is, therefore, necessary and urgent to transform farming systems from primarily focusing on productivity to instead having social, economic and environmental sustainability at the core of their development [6]. Most food production world-wide is currently unsustainable and needs modification [7–10]. For example, nutrients need to be recycled, external inputs reduced and farming systems diversified [7,10–12]. Simultaneously, anthropogenic CO_2 emissions are changing the global climate and will continue to do so throughout the 21st century [13]. Crop production and carbon sequestration are ecosystem services with great potential for generating synergies in agricultural landscapes worldwide. Crop production can maintain higher levels of soil carbon through, for example, more perennial crops or returning more crop residues and organic material to fields, and this higher soil carbon level can increase yield [14,15].

The increasing concentration of atmospheric CO_2 is largely attributable to emissions from fossil fuel combustion or land-use change, in the latter case mainly through soil organic carbon (SOC) depletion [13]. Conversion of natural land to agriculture and subsequent soil degradation from, for example, erosion have resulted in average losses of 50–70% of the original SOC pool in agricultural soils [16]. In some areas, this has led to declining yield. There is thus a need for sustainable intensification, where production can increase while the environmental footprint of agriculture decreases [17,18]. This is particularly relevant in developing countries, where most emissions are land-based and where carbon sources can be turned directly into sinks [19]. Sub-Saharan Africa is an area with considerable opportunities for sustainable intensification [17]. Kenya, as a fast-developing country in East Africa, has the potential to take the lead in reaching the Paris Agreement goals on climate change mitigation [20]. In Kenya, 70% of the rural population relies on agriculture for their employment and more than a quarter of the gross domestic product is derived from agriculture [21]. The Kenyan National Climate Change Action Plan 2018–2022 sets goals for agriculture where the main mitigation actions proposed for climate change are limited burning in croplands and more use of conservation tillage and agroforestry [21]. Kenya has the goal of converting 281,000 ha of existing arable and grazing land into agroforestry by 2030 as a climate change mitigation action, and of making climate change-related information and advice an integral part of the Kenyan agricultural advice system [22]. Kenyan smallholders are aware of agroforestry and other sustainable management practices [23] that could increase the soil carbon pool [15,24] and help increase production and income [22,25–28]. The Clean Development Mechanism, which allows countries with carbon emission reduction commitments to implement these reductions in developing countries, can create incentives for smallholders in developing countries to sell carbon sequestered in, for example, agroforestry systems to industrialised countries [29]. Similar incentives can be created within developing countries, through the nationally determined contributions [30]. Engel and Muller [31] identified climate-smart agriculture (e.g., agroforestry) as the most promising practice to be promoted by Payments for Ecosystem Services (PES) among smallholders with limited income. However, the use of carbon finance to incentivise this type of bio-carbon storage is still very low, due to the absence of institutional frameworks, reliable sources of carbon finance and involvement of public and private sector actors [19,32,33]. The number of smallholders needed to achieve an area of land large enough to compensate for project transaction costs also makes carbon finance projects practically unattainable at the current market price for carbon [34,35]. Low carbon prices mean that the incentive for farmers is not the carbon payment, but the benefits arising from emission-reducing farm management. Mbow et al. [36] question whether smallholder farmers can benefit from carbon payments at all and how advisory services can succeed in effectively promoting climate-smart agriculture. More research is needed to identify suitable types of climate-smart agrosystems for different user groups and their results and impacts [36].

The Kenya Agricultural Carbon Project (KACP), a soil and tree carbon project implemented by the non-government organisation (NGO) Vi Agroforestry within the Lake Victoria basin, is targeting small-holder farmers (with <2.5 ha). A majority of the farmers are aiming to be self-subsistent from their farming, largely depending on maize, beans and dairy production, while some are combining the agriculture with off-farm employment or casual jobs. The aim of the KACP is carbon sequestration through the uptake of Sustainable Agricultural Land Management (SALM) practices, enabling smallholder farmers to access the carbon market, as well as increasing yield and productivity and enhancing resilience to climate variability and change [37]. SALM practices include, for example, cover crops and agroforestry to increase biomass production, use of biomass for mulching and composting instead of burning, and avoiding soil erosion through, for example, terracing, reduced tillage and water harvesting [38] (Table A1). The project provided a dedicated advisory package to promote the use of SALM practices. It also included training in farm enterprise development and village savings and loan associations (VSLA), a microfinance intervention for the accumulation of regular savings within a group and rotating credit opportunities, following definitions by, for example, Bouman [39]. The VSLA system is based on trust among people who know each other and is run by the people themselves in one-year cycles with no outside support except for the initial training. A number of studies have reported on the KACP as a novel type of intervention [40–46]. However, no previous study has examined the effects of different SALM practices on maize yield or the actual maize yield response on farms taking part in the project.

The overall aim of this study was to assess the effects of the KACP on farm productivity and livelihoods during the initial four years (2009–2012) using the uptake of SALM practices, maize yield, food self-sufficiency and savings as indicators. Specific objectives were to (i) assess the level and type of SALM implementation among farmers participating in the KACP over time and compare it with that of neighbouring non-participating farms (control farms); (ii) determine the relationship between SALM practices and maize productivity, and compare the productivity over time and between the KACP farms and control farms; and (iii) quantify the level of food self-sufficiency and savings of the KACP farmers and control farmers. Mechanisms for the spread of knowledge and practices between neighbouring farmers were also discussed.

2. Materials and Methods

2.1. Characterisation of the Study Areas and Background to the KACP

The Kenya Agricultural Carbon Project started in 2009 and was implemented in two areas of Kenya, both with around 22,500 ha of potential project area (Figure 1). These were Kisumu (southern part including Kisumu and Siaya counties) and Bungoma (northern part including Bungoma county), with 28 administrative locations (Table A2). The region has a bimodal rainfall pattern with two cropping seasons, normally March–July and August–December (Figure A1). The KACP methodology was developed together with the BioCarbon Fund of the World Bank [37]. In the KACP, smallholder farmers receive carbon credits for soil carbon, which makes it unique since most other community and smallholder carbon projects only support carbon sequestration through tree planting [46]. However, the participating farmers were not given any financial support from the KACP during the study period (2009–2012) and the benefits of using SALM practices were instead emphasised, while the carbon revenue was small and presented to farmers as a co-benefit. Carbon credits were generated and claimed using the approved Verified Carbon Standard methodology, based on background data on soils and climate together with data on the management practices used by the farmers, instead of analysing carbon content in soil samples [47,48]. Carbon revenues were post-paid to farmers in 2014 for the years 2010–2014. Validation and verification were conducted periodically by external teams on emission reductions reported within the KACP [47].

From 2009 to 2012, the advisory system had a fixed number of 28 Vi Agroforestry field advisors, with one advisor in every administrative location, covering approximately 70 km^2. These field advisors offered advisory services to farmers and facilitated monitoring. They identified registered farmer groups or facilitated group formation, and recruited, informed, trained and contracted smallholder farmers to implement a free choice of SALM practices on their farms. Regular interaction and monitoring of project activities by advisors proved important to avoid misunderstandings and identify risks and challenges in agricultural production as early as possible. No new farmer groups were recruited after 2013, so the number of advisors was reduced to four by 2017, to maintain the information flow. The monitoring data were consolidated in maps and tables at the farmer group level for internal monitoring by farmers. They were also collated at the end of each cropping season and entered in databases and ArcGIS for monitoring and evaluation by Vi Agroforestry.

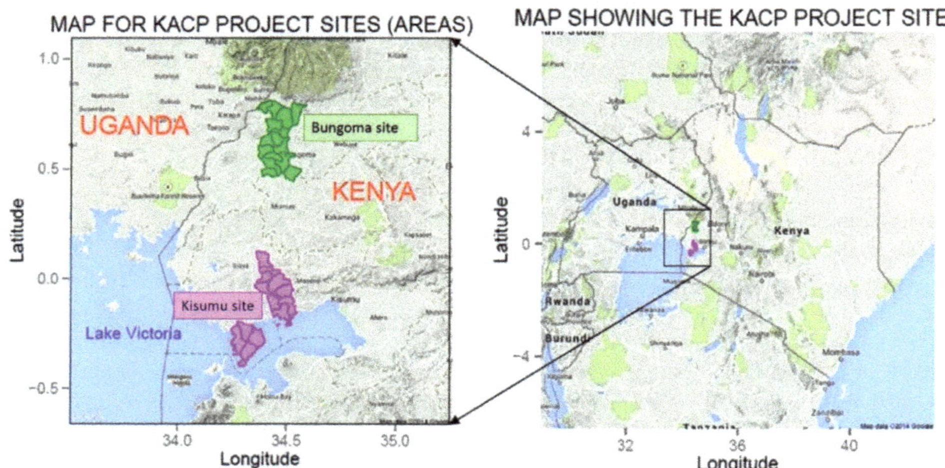

Figure 1. Left: Kenya Agricultural Carbon Project (KACP) areas; the Bungoma sites in Bungoma County (green) and the Kisumu sites in Kisumu and Siaya Counties (purple). Right: The KACP areas within East Africa.

2.2. Farm Sampling and Data Collection

The project started with 660 farmer groups involving 10,873 voluntary farmers [49]. By the end of 2012, the KACP had recruited 1555 verified farmer groups with 26,535 farmers, who implemented SALM on 16,490 ha of eligible cropland or grazing land. Among the initial 10,873 project farms, 200 were selected and monitored more closely by the KACP field advisors, in permanent farm monitoring (PFM). To select the 200 PFM farms (100 in each of the project areas), the areas were stratified into agro-ecological zones (AEZ) and the number of farms in each zone was decided in relation to the zone size. A systematic grid of 1.5×1.5 km^2 for unaligned systematic sampling was applied and the area was divided into clusters formed by the areas between four intersection points of the grid. PFM farms were randomly selected within the clusters according to the agro-ecological stratification [48].

In this study, the PFM (hereafter called "project") data collected by Vi Agroforestry between 2009 and 2012 were used to determine the uptake rate and correlations of SALM practices to maize yield (Figure 2). In addition, in 2012 a set of 160 control farms (80 in each project area) was selected from the 28 locations of the KACP for the purposes of the present study. When selecting control farms, the project farms were used as reference points and the second farm to the north of the project farm was selected if the owners consented. If that farmer belonged to any group that had worked with Vi Agroforestry, it was skipped and the next farm to the north was asked instead, and so on. Thus, only farms that had not previously worked with Vi Agroforestry were selected as control farms. Data collected from all 360 farms (200 project and 160 control farms) included field size, yield and SALM methods used for all maize fields on the farms during two seasons and four years (2009–2012). However, for the control farms, SALM data were only collected for 2012 and maize yield data for 2009–2011 were obtained retrospectively through farmer recall interviews in 2012 (Figure 2). Data on species, numbers and sizes of farm trees were collected, and months of food self-sufficiency and amount and frequency of savings were ranked by the farmers themselves for 2012 only. All parameters were compared between project and control farms except the effects of individual SALM practices on maize yield, which was done across all farms. The project farmers were monitored for two more years (2013–2014) after this study and the project was then converted to a solely self-monitoring system by all farmers.

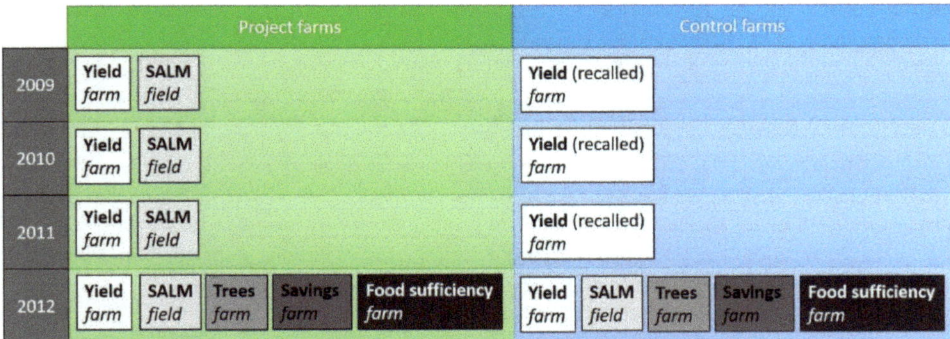

Figure 2. Available data for project and control farms over the years analysed and the scale of the analysis. All data were compared between project and control farms when possible, except for the effects of SALM and trees on maize yield.

2.3. Data Analysis

All statistical analyses were conducted in R 3.4.2 [50]. Correlations between SALM practices and yield were analysed on the field level and for all available years (2009–2012 for project farms, 2012 for control farms). The effects of each SALM practice on maize yield were examined at the field level. For all nine management practices, a linear mixed effect model (the lme function in the nlme package in R) was created, with maize yield as the response variable, usage (or not) of the nine practices, area, year and season as fixed factors, and farm and pair of farms as random factors. The model included recommended SALM practices, such as (1) no tillage, (2) crop residues for direct mulch, (3) raw manure composting, (4) cover crops, (5) terracing and (6) water harvesting structures, and practices to be avoided due to higher carbon emissions, such as (7) removal of crop residues from fields, (8) applying raw manure to fields and (9) burning of residues. Interactions between different practices were not considered, due to large variations in the degree of implementation of each practice. A model simplification procedure was then used to compare and select the model that best explained the variation in the data. The model comparison was carried out using a step-wise Akaike information criterion (AIC) (allowing both forward and backward comparisons). The best-fit model was run using the lmer function in the lme4 package in R to identify potential significant effects of SALM practices.

The effects of agroforestry SALM practices, represented by the number and function/s of farm trees, on maize yield were analysed in a similar way as described above for the SALM practices. The only difference was that the maize yield data used were farm averages from 2012 for each of the two rainy seasons and study areas. For trees, the fixed factors included were area, season and the total number of trees, fodder trees, timber trees and fruit trees. Farm was included as a random factor.

For comparisons of mean seasonal maize yield per farm over time, a linear mixed effect model (lmer function in lme4 package) was used, with four fixed factors (treatment, area, year, season), which were tested as direct effects and with all two-way interactions included in the model. Only project and control farms forming pairs were included in that analysis, which resulted in 78 pairs of farms in Kisumu and 79 in Bungoma. Both individual farms and pairs of farms (project and control) were set as random factors in the model. The paired farms were analysed for four years (2009–2012), and two seasons per year. Contrasts were used to compare the differences in 2009 and 2012 between project and control farms.

The data on food self-sufficiency were categorised into four levels: <6 months food self-sufficiency, 6–7 months, 8–9 months and 10–12 months. A Chi-2 test was used to identify dependencies between food self-sufficiency and KACP participation. The significance level for all analyses was set to $p < 0.05$.

3. Results

3.1. Uptake of SALM and Other Practices

Uptake of SALM was studied for all four years for project farms, but only for 2012 for control farms (Figure 2). Project farmers responded well to the advisory services within the KACP and started to implement several of the SALM practices (Figure 3). In 2012, about 60% of the fields were under mulch and terracing, compared with 25% and 40%, respectively, in 2009. Water harvesting increased from around 10% in 2009 to 40% in 2012 and composting of raw manure increased from 50% to 65% (Figure 3). The three most popular SALM practices among project farmers were using crop residues as mulch, composting raw manure before application and terracing fields (Figure 3).

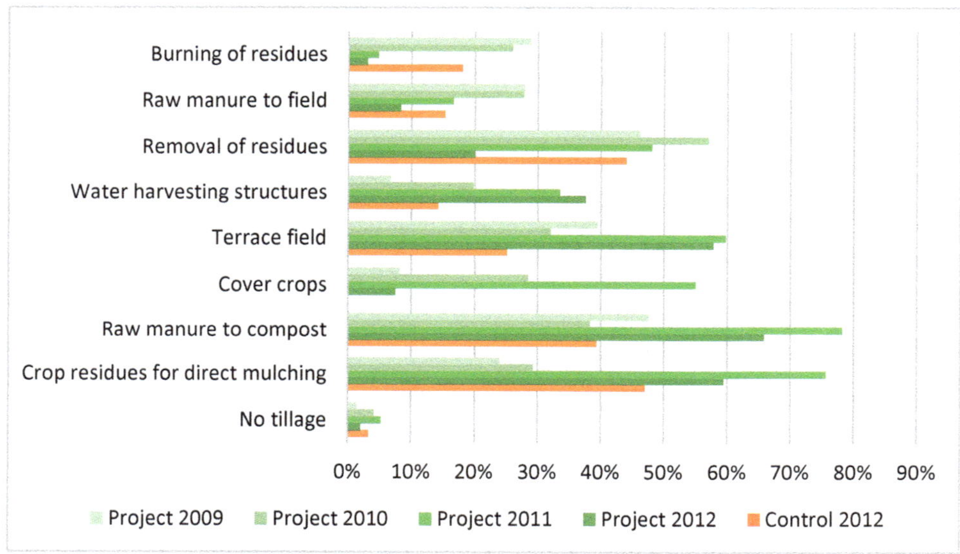

Figure 3. Diagram showing SALM uptake and implementation in fields in the two areas of Kisumu and Bungoma by project farmers and control farmers between 2009 (project start) and 2012. Data for control groups only available for 2012. Data for project farms in Kisumu in 2009 were lost due to a computer malfunction. N = 183 for control 2012, n = 208 for project 2009, n = 481 for project 2010, n = 591 for project 2011 and n = 495 for project 2012.

Some measures never became commonly used, such as no tillage and cover crops. Cover crops increased from 10% in 2009 to 55% in 2011 but decreased again to 10% in 2012. Information on the use of SALM practices on control farms was only available for 2012, at which time crop residues were used for direct mulching on around 50% of control farmers' fields, compared with 25% on project fields in 2009 and 60% in 2012. Raw manure composting was also quite common on control farms, while terracing, use of cover crops and creation of water harvesting structures were rare. The measures to be avoided according to the KACP recommendations were reduced on project farms. Removal of residues decreased from 50% to 20%, raw manure application from 30% to 10% and burning of residues from 30% to close to zero. On the control farms, removing residues from fields and burning of residues were still used to a relatively high degree. In general, by 2012, the project farms had on average more of the promoted practices and fewer practices to be avoided according to the KACP recommendations compared with both the start of the project in 2009 and with the control farms.

Tree planting, especially agroforestry, was another promoted practice taken up by farmers. The majority of trees (counted only in 2012) on the farms were young and planted within the project period. On average, farms in Bungoma had a higher number of trees (98 for project farms, 57 for control farms) than farms in Kisumu (74 for project farms, 40 for control farms) (Figure 4). Most trees on the farms were timber trees (64% on project farms, 75% on control farms). Apart from having on average more trees (86) per farm compared with control farms (48), project farms also had a larger average proportion (19%) of fodder trees than control farms (4%) (Figure 4). In terms of species, on average control farms had more Eucalyptus spp. and project farms had more N-fixing species like Sesbania sesban, Acacia spp. and Calliandra spp. The most common species overall were Grevillea robusta, Markhamia spp. and Albizia spp.

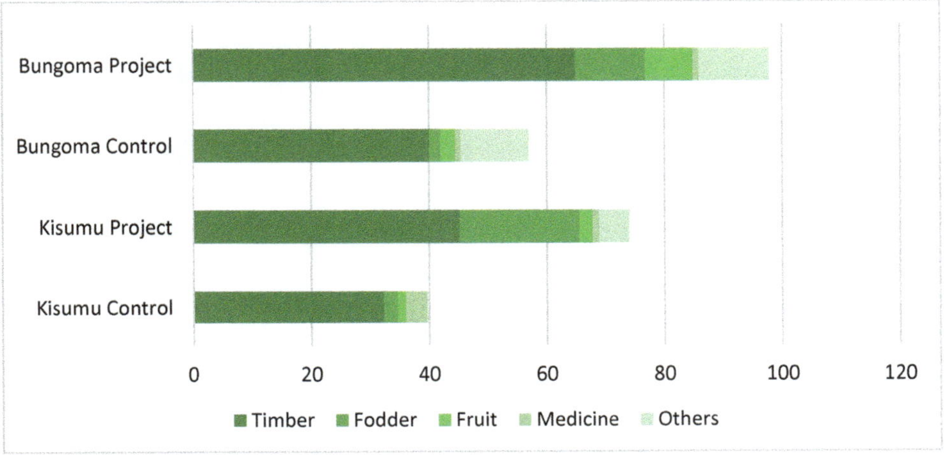

Figure 4. The average number of trees per farm on project and control farms in 2012, divided between fodder, fruit, medicine, timber and other tree types. In total, 65% of all trees were 10 cm or less in diameter at breast height, meaning that they were probably planted within the project period (since 2009). Eucalyptus comprised 4.6% of trees on control farms and 2.7% on project farms. Data from a tree survey in Kisumu (n = 37 control farms, n = 90 project farms) and Bungoma (n = 38 control farms, n = 60 project farms).

3.2. Effects of SALM Practices on Maize Yield

Among the recommended SALM practices, only terracing had a significant ($p = 0.0004$) positive effect on maize yield. However, the management effects were small compared with the highly significant differences ($p < 0.0001$) between years, seasons and regions. Terracing was the only practice that was part of the best explanatory model, together with area, season and year (Yield ~ Area + Year + Season + Terracing + (1|Place/Farm)). When analysing effects from trees, first-season maize yield was not affected by any factor other than region, with Bungoma having a higher yield. Second-season maize yield had the best-fit model that included region and the total number of trees per farm (Yield 2 ~ Region + Total trees + (1|Farm)). Second-season maize yield increased with the increasing total number of trees ($p = 0.02$).

3.3. Maize Productivity

Maize yield varied widely between the areas and was on average 1572–2675 kg ha^{-1} in Bungoma, compared with 725–1661 kg ha^{-1} in Kisumu, among project farms in the first season of the four years (Table 1). There was also a wide variation in second-season yield, which was 518–1054 kg ha^{-1} and 152–678 kg ha^{-1} for Bungoma and Kisumu control farms, respectively. Maize productivity analyses showed higher yield ($p < 0.0001$) for project farms than control farms, higher first-season than second-season yield ($p < 0.0001$) and higher yield ($p < 0.0001$) in Bungoma than Kisumu (Table 1, Figure 5). The four study years also differed ($p < 0.0001$) in terms of yield, with mostly increasing trends (Figure 5). Apart from the main effects, there were three significant interactions (Figure 6): (1) yield differences between project and control farms were larger ($p < 0.0001$) in 2010–2011 than in 2009 or 2012 (Figure 6a); (2) first-season yield increased more ($p = 0.004$) than second-season yield in 2012 (Figure 6b); and (3) there was a larger difference ($p = 0.005$) in the first-season yield compared with the second-season yield between the regions (Figure 6c). Project farms increased their yield mostly in 2010 when the first-season yield declined in control farms. The second-largest increase for project farms was in 2011, while control farms had a one-year lag with their largest increase in 2011 and second-largest in 2012 (although the yield was still lower than on project farms). However, in total, the difference in yield between project and control farms was similar in 2009 and 2012 ($p = 0.15$), and therefore the yield gap between project and control farms did not change significantly during the four initial years of the KACP.

Table 1. Maize yield (kg ha^{-1}) for all project and control farms (measured on field level) in Bungoma and Kisumu. Values are mean ± standard deviation for all four years (2009–2012) and both seasons (1 and 2); n = number of farms included in the analysis.

Year	Season	PROJECT Bungoma N = 96	CONTROL Bungoma N = 55	PROJECT Kisumu N = 85	CONTROL Kisumu N = 44	ALL FARMS N = 280
2009	1	1572 ± 1211	1169 ± 741	725 ± 572	494 ± 397	1069 ± 981
2010	1	1925 ± 1356	1141 ± 934	1055 ± 1037	266 ± 166	1239 ± 1203
2011	1	2647 ± 1756	1222 ± 994	1371 ± 907	702 ± 390	1724 ± 1479
2012	1	2675 ± 1384	2015 ± 1279	1661 ± 1406	1075 ± 840	1939 ± 1397
2009	2	1094 ± 1206	518 ± 411	565 ± 532	321 ± 249	638 ± 781
2010	2	1416 ± 1431	706 ± 469	902 ± 1023	152 ± 101	900 ± 1105
2011	2	1646 ± 1052	783 ± 526	957 ± 757	354 ± 297	945 ± 862
2012	2	1563 ± 941	1054 ± 714	937 ± 744	678 ± 567	982 ± 784

3.4. KACP Effects on Savings and Food Sufficiency

Project farmers added to their savings more often on average than control farmers (Figure 7a) and also added larger amounts per occasion (Figure 7b). More than 45% of project farmers added to savings two or more times per month, while the corresponding figure for control farmers was 21%. Project farmers saved to a larger extent (72%) than control farmers (52%) and Kisumu farmers saved on average more than Bungoma farmers. In the VSLAs within the KACP, farmers were able to borrow up to three times their savings to use for investments.

Among Bungoma farmers, 39% had farm inputs as their main expenditure, compared with just 14% for Kisumu farmers, 60% of whom had food as their main cost. In Bungoma, there was a difference between project and control farmers in that 51% of project farmers spent most on education for their children, while 25% of control farmers still had to spend most on food and thereby only 27% had education as their main expenditure. The majority of farmers in both Kisumu and Bungoma had their main source of income from agricultural products.

Only 4% of control farmers in Kisumu (31% in Bungoma) had enough food for 10–12 months, compared with 16% of project farmers in Kisumu (51% in Bungoma). Moreover, 46% of control farmers in Kisumu and 25% in Bungoma had enough food for less than six months, while the corresponding values for project farmers were 19% and 7%, respectively. In general, farmers in Bungoma had more months of food sufficiency than farmers in Kisumu. Chi-2 tests revealed that food sufficiency was significantly higher ($p < 0.001$) overall for project farmers than control farmers.

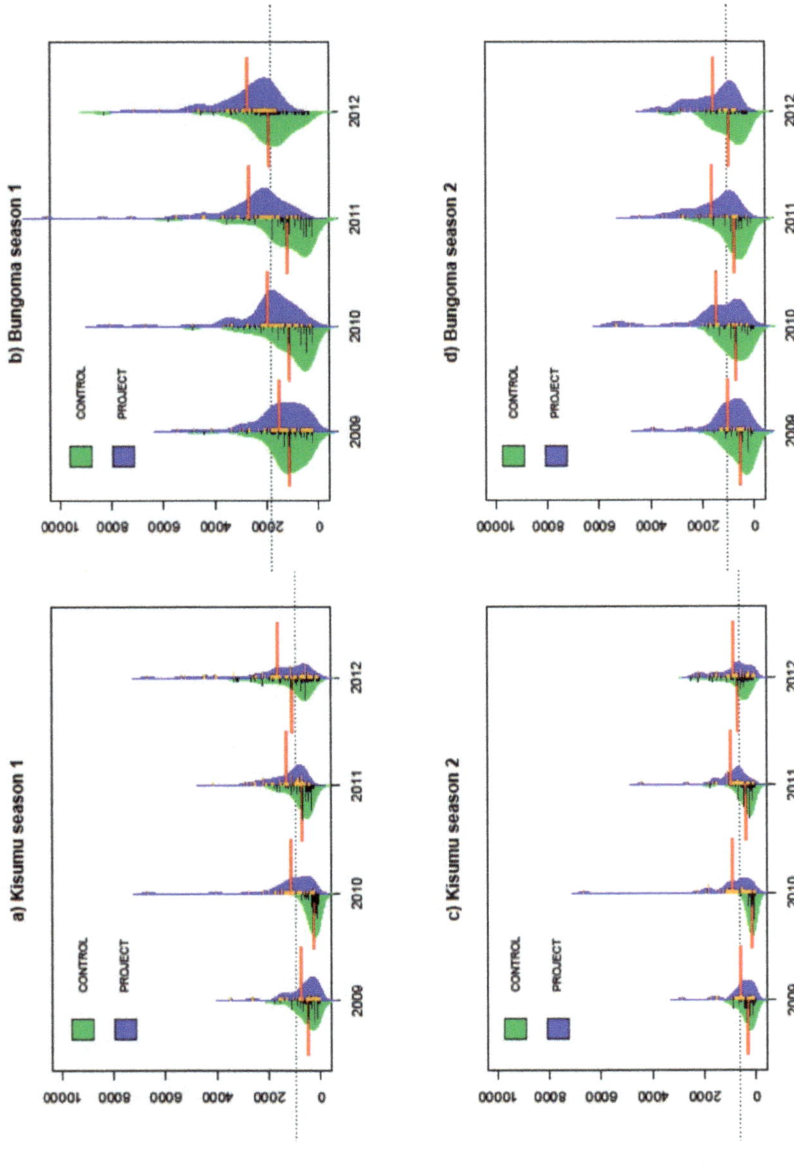

Figure 5. First-season (**a**,**b**) and second-season (**c**,**d**) maize yield kg ha^{-1} on project (blue) and control (green) farms in Kisumu (**a**,**c**) and Bungoma (**b**,**d**), 2009–2012. The lines indicate mean values for each distribution and the dotted line shows the mean for each sub-plot.

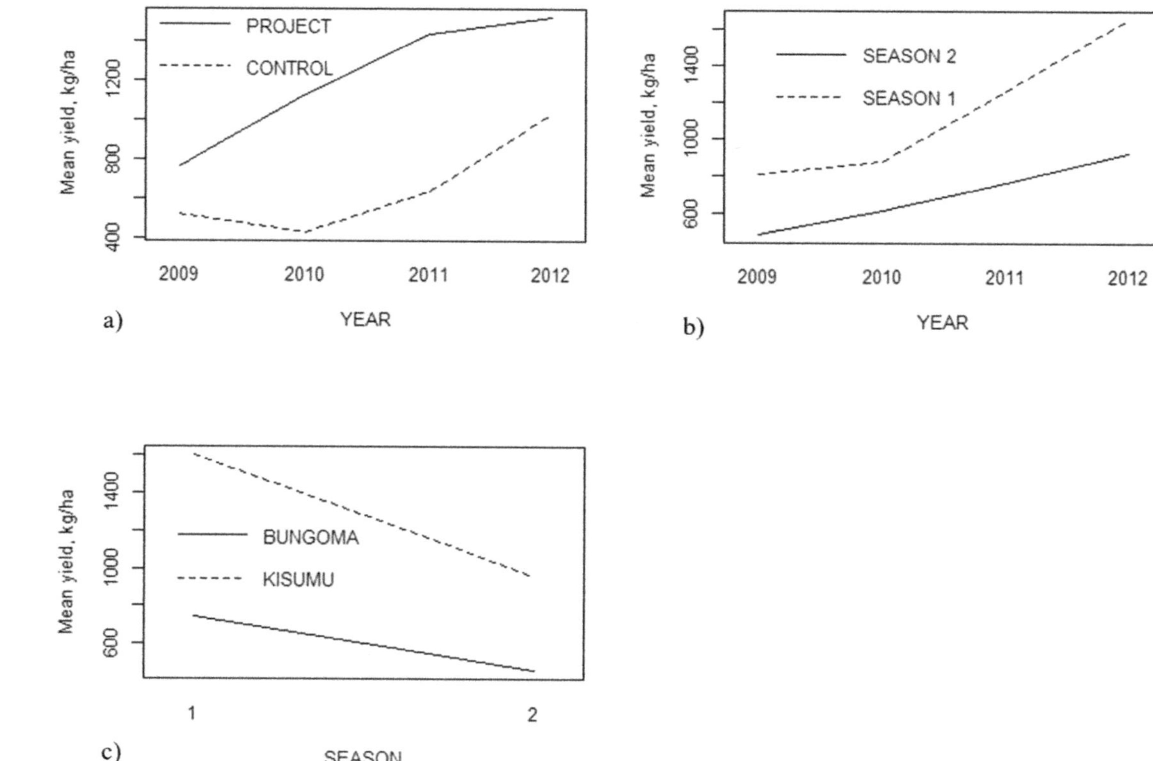

Figure 6. Diagrams showing the three significant interactions identified, between: (**a**) year and treatment ($p < 0.0001$), (**b**) year and season ($p = 0.004$) and (**c**) region and season ($p = 0.005$) on maize yield in kg ha^{-1} for project and control farms.

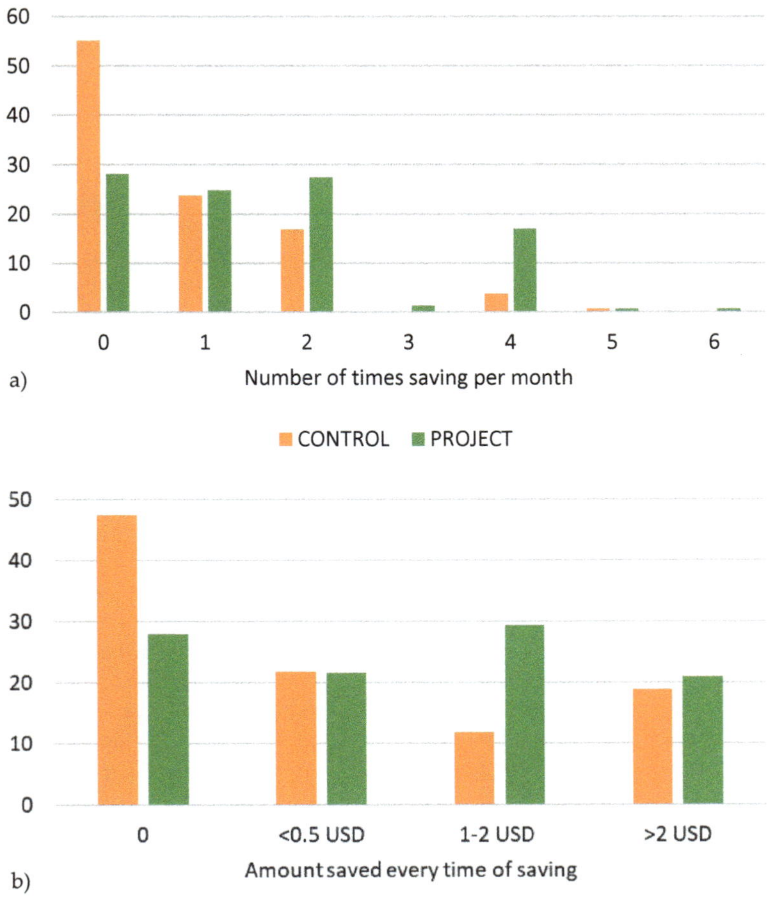

Figure 7. (**a**) Percentage (*y*-axis) of project and control farm households saving money on different numbers of occasions per month in 2012. (**b**) Percentage of project and control farm households saving different amounts on every saving occasion in 2012.

4. Discussion

4.1. Mulching, Terracing and Trees More Popular Than No Tillage

By 2012, there was a large difference in the use of terraces between project (60%) and control fields (25%) (Figure 3). Terraces can greatly enhance soil moisture recharge [51] but require major initial labour inputs [52]. Cover crop uptake among project farmers increased until 2011, but declined to the initial level in 2012, possibly due to crop rotation patterns or to dissatisfaction with the practice itself (Figure 3). Cover crops/green manure grown for six weeks (planted two weeks earlier than maize), and then incorporated into soil can significantly improve soil fertility by adding nitrogen, increasing maize yield in the same season [53]. However, convincing farmers to grow cover crops mainly for soil fertility reasons was apparently difficult in the KACP (Figure 3), as in other projects [54]. Different SALM practices had different degrees of uptake, for example, mulching and terracing became popular, while no tillage did not. Minimum tillage practices have varying and sometimes negative or contradictory outcomes for smallholders in the region [55,56] and the willingness to use minimum tillage is lower than, for example, mulching [57]. It is also debatable whether minimum tillage options can act as a carbon sink [58,59].

Advice on SALM practices to be used, and practices to be avoided, seemed to be adhered to more by project farmers in 2012 compared with the initial use in 2009, and also compared with control farmers and with the uptake of practices in a similar study [60]. Use of soil fertility management practices is generally low among smallholders in Western Kenya [60] but is influenced positively by, for example, plot size, market and labour access, off-farm earnings and knowledge [60,61]. This implies that the poorest farmers may have less possibility to use such practices.

Trees in the agricultural landscape, that is agroforestry, can have both positive and negative effects on production and other ecosystem services, but most studies show a net positive effect [62]. Fodder trees were promoted by the KACP advisors, which can explain why project farms had a larger share of leguminous fodder trees than control farms (Figure 4). A lack of fodder trees on control farms could also explain the lower use of crop residues for mulch, as more crop residues might be needed for livestock feed. The larger number of trees per farm in Bungoma than Kisumu may be because the farms were larger in Bungoma. It is also easier for tree seedlings to survive in Bungoma since most farmers control their grazing animals and prevent them from browsing on tree seedlings.

4.2. Terraces and Fodder Trees Increased Maize Productivity

Of the six SALM practices and three non-SALM practices analysed, only terraces had a significant positive effect on maize yield. The differences in SALM practices between 2009 and 2010 (when the yield increase was largest for project farms) were mainly the increased use of cover crops and water harvesting structures. However, neither practice showed any significant correlation with yield. The four-year study period can be expected to be sufficient to reveal the effects of most management practices [63]. While the main aim of SALM practices in the KACP is to reduce greenhouse gas emissions and sequester more carbon, they can also improve water availability for crops and soil fertility, both of which improve yield [17,64]. However, few SALM practices showed any significant effects on maize yield in this study. One drawback related to terraces is the relatively high labour demand needed upon establishment, which may restrict farmers from using the measure.

The intensity of the different SALM practices was not studied (since data was not available), but differences in intensity could have masked relationships between management and yield. Most recommended practices are dependent on the type, quantity and quality of, for example, mulch, compost, water harvesting or cover crops. Project farmers, for example, reduced burning and removal of residues (for fodder or fuel) in order to use more as mulch, but 40% of residues were still used for purposes other than mulching in 2012 (Figure 3), suggesting that much is used as fodder and could be returned as manure. Mulching is known to increase and stabilise maize yield [55,57]. However, practices related to crop residues (mulching, composting, green manure, etc.) require trade-offs between several uses, and the end-use often reflects more immediate concerns such as feeding livestock rather than improving soil fertility in the longer term [65,66]. With more fodder trees on the farm, feed and fuel could come from sources other than crop residues.

The use of agroforestry SALM practices was popular in project farms, which had on average around double the number of trees as control farms. A possible drawback with tree–maize intercropping is competition for light, water and nutrients unless the trees are adequately pruned above and below ground [67]. However, the number and type of trees did not have any effect on the first-season maize yield. Timber trees, which were the most common type used, are often planted in woodlots and therefore rarely interfere with crops, while fodder trees are more often intercropped within fields. In this study, second-season maize yield was positively affected by the total number of trees, possibly because farms with many trees had a larger share of fodder trees that are leguminous and can fix the nitrogen from the atmosphere, which has been found earlier [67,68]. Apart from increasing second-season maize yield, trees could provide other services (animal feed, fruits, timber and firewood) which were not considered in this study. Further, a combination of trees and grass is effective in holding and maintaining the soil on hillslopes and terraces [69]. Agroforestry is known to be both knowledge and labour intensive which can limit the uptake of the practice [70].

4.3. Maize Productivity Increased but SALM Practices Only Part of the Explanation

Average maize yield on small, subsistence-focused farms in western Kenya was 0.9 tonnes ha^{-1} during the study period [71]. Nation-wide, mean annual maize yield in Kenya in the four years 2009–2012 was 1294, 1725, 1584 and 1737 kg ha^{-1} [72], compared with 1707, 2139, 2669 and 2921 kg ha^{-1} for both seasons and all farmers in this study. The first cropping season usually has more rain days and larger amounts of rainfall ("long rains") than the second season ("short rains") in both study areas, which largely explains the higher productivity in first-season maize (Figures 5 and A1). Farmers who composted their manure applied it to fields once per year at the beginning of the first season, which may be another factor in the yield differences between seasons. Overall, Bungoma had higher amounts of rainfall and a cooler climate (Table A2 and Figure A1), leaving more water available for crops instead of being lost through evapotranspiration. Therefore, productivity was significantly higher in Bungoma and the pattern was the same for all years and seasons (Figure 5).

Water is not the only limiting factor for maize growth in Western Kenya, as nitrogen and phosphorus are also main limiting factors [64]. Small-scale farms with subsistence-focused production tend to have resource flows showing net nutrient export and low values of carbon and nitrogen especially [71]. Smallholders in the KACP were encouraged to practise integrated soil fertility management, with careful use of inorganic fertiliser and more focus on organic fertilisers. This advice is debatable since many researchers view inorganic fertilisers as important in increasing yield and boosting biomass production for higher carbon cycling [73,74]. The use of inorganic fertilisers may partly explain the different yield patterns for project and control farms (Figure 6a). However, the use of inorganic fertilisers was not included in this study and therefore needs further studies for a better understanding of its effects with or without SALM practices.

The reason why the first-season yield increased more than second-season yield in 2012 (Figure 6b) was likely that yield actually decreased in 2010 for control farms, and also that SALM use was higher for first-season crops and high for both project and control farmers in 2012. The greater yield increase in Bungoma than Kisumu for first-season crops (Figure 6c) could be because Bungoma farmers concentrate on agriculture, while Kisumu farmers were traditionally fishermen and also have more off-farm job opportunities close by, making them less dependent on farming [66]. Farmers who are more focused on farming are likely to be more eager to invest in their farm and use new practices. The Bungoma area (known as "Kenya's breadbasket") also has more responsive soils and a more favourable climate than the Kisumu area.

4.4. Role of Control Farms in Interpreting Results

Immediate effects from SALM practices related to water-holding capacity [64,75] could explain the initial higher yield increases for project farms. Possible reasons why the control farms increased their yield more than project farms in 2011–2012 (Figure 6a) could be reductions in inorganic fertiliser use in project farms, control farms copying practices from project farms or project farmers paying more attention to their crops during the first years of the project, but then relaxing. Learning from neighbours and friends (farmer-to-farmer) is often the most common [76] and most effective [77] way of learning new practices in this region, which could explain both the relatively high SALM uptake in 2012 and delayed yield increases on control farms compared with project farms. Farmer-to-farmer uptake within the KACP areas was reported by Hughes et al. [42], who found significant uptake of certain practices within the project area (not only among the targeted farmer groups) than in neighbouring areas outside the project. However, this does not mean that advisory services were not worth the effort. Farmers generally want better access to advisory services [78], especially in the early stages of adoption [77].

The project and control farmers were all within the KACP areas and lived in the same villages. For control farms, the use of the different SALM practices was only studied in 2012 and by then their use of practices recommended by the KACP was on average higher than the initial (2009) use on project farms. Maize productivity at the start of the project (2009) was higher on project farms than control farms and, while maize yield increased over the years, the difference between project and control farms was similar in 2012. Thus, comparisons of maize yield initially and at the end of the monitoring period showed no overall difference between project and control farms. The value of having control farms in addition to the baseline production in 2009 was to ensure that the yield differences were not normal annual fluctuations due to weather patterns. Control farms were two or three farms away from their paired project farm, so soils and weather conditions were probably similar. However, there might be other differences between project and control farms.

Participation in the KACP was voluntary and participants were generally both asset-poor and income-poor. However, earlier research has found higher adoption of SALM practices among the less poor in the project [41]. This may indicate that it is not the poorest of the poor that risk venturing into such projects, as has been found elsewhere [79]. However, they can perhaps not be expected to participate unless they are specifically targeted with up-front financial support to overcome their risk aversion. In earlier studies, the main factors limiting uptake of SALM practices within the KACP were found to be lack of labour to implement SALM practices, knowledge on how to implement them and land availability [41,44]. In addition, women had problems finding the time to attend training [43,44].

4.5. KACP Farms Had Higher Savings and Food Sufficiency

Kisumu farmers were able to save on average more than Bungoma farmers. This was most likely partly due to more off-farm income opportunities in Kisumu and partly because the yield was lower in Kisumu, so farmers needed cash savings to buy food. On average, project farmers saved to a larger extent, more often and in larger amounts than control farmers, probably because the VSLA concept in the KACP provided opportunities to save (Figure 7a,b). The VSLA methodology has earlier shown positive results for several household indicators, such as the development of income-generating enterprises and access to education and food [80]. Preliminary data from a study in 2016 also indicated higher income from tree products, better fuelwood access and higher milk yield among the KACP farmers compared with neighbouring control areas [42]. All of these can increase the scope of cash savings by households.

Project farmers at both sites had significantly higher food sufficiency than control farmers. This can be directly related to differences in maize yield, but maize is just one of many food crops. The improved food sufficiency may also be attributable to other farm commodities such as sorghum, beans and cassava, yields of which probably also benefit from terracing and other SALM practices. Improved yields and better food sufficiency among the KACP farmers have been reported in an earlier study of women and men farmers [44]. The higher level of food sufficiency in Bungoma most likely reflects the greater farm size and higher average yields in this area than in Kisumu. Agroforestry practices could have added to food sufficiency through, for example, fruits. Earlier studies report maintained yields in combination with such additional benefits from agroforestry [81].

4.6. Limitations of the Study

It is a strength that control farms were included in the study, rather than just monitoring changes over time from the baseline on the project farms. However, the study has some limitations partly related to the control farms. A possibility of farmers participating in the KACP being different, perhaps better-off, than those who did not participate emerged early in the project, so an attempt was made to avoid potential self-selection bias by adding a control group of farmers in 2012. However, this meant that no control farms were monitored from the beginning of the project and there is uncertainty in the data on maize yield based on control farmer recall for the first three years (2009–2011). Moreover, uptake of SALM practices on control farms was only assessed in 2012 and thus data were lacking for 2009–2011. The identified control farms were also perhaps too close to project farms, which may have given a non-representative control group. Another limitation of this study was the "weak definitions" of uptake of practices. The uptake data were binary (practiced or not practiced) and the quantities of mulch, manure, etc. applied were not recorded, which created large variation in the effects of farmers' uptake of practices.

4.7. Implications and Recommendations

Overall, the KACP had a tendency for positive effects regarding uptake and effect of certain SALM practices, development of maize yield, food sufficiency and savings. However, to obtain stronger evidence that the results are due to the intervention, large-scale projects like this would need to include control farms in the design from the beginning, together with clearly defined quantitative indicators of uptake of practices. This could be done jointly by development agencies and research institutions, in order to achieve an optimal design and reduce the costs of monitoring and evaluation. Some of the practices promoted in the KACP were also questionable (e.g., no tillage and careful use of inorganic fertiliser), especially considering the relatively poor target group. A primary aim of the advisory services in such projects should be yield and productivity increases and building resilience, while climate change mitigation objectives should be viewed as a secondary goal [41]. The target group should not be expected to take any risks. Of the SALM practices promoted, only terraces and trees showed positive (but relatively small) effects on yield. Thus, other parts of the advisory services, for example, micro-credit access [82], might have played a greater role in the increased maize yield and savings and need to be explored in future research.

5. Conclusions

This is the first research study on the yield effects of promoted practices in an agricultural soil carbon project for smallholders in sub-Saharan Africa. Advisory services provided to farmers participating in the KACP appeared to increase the uptake of some of the soil management and tree planting (especially fodder trees) practices and to lower the use of non-sustainable practices. Indications of some uptake through farmer-to-farmer horizontal learning were observed for control farmers.

Higher inclusion of trees (agroforestry) and the use of terraces were the only practices that correlated positively with higher maize yield. The lack of effect of practices was likely due to vague definitions of the practices within the project. Maize yield increased from the start of the KACP and continued over the four years studied and was higher among project farmers than control farmers during all years. The yield increase was similar for project and control farmers but the timing differed, since project farms achieved their main increase during the first years, while control farms obtained their main increase during later years.

Participants in the KACP had higher food self-sufficiency and the results indicated greater ability to save money, both more frequently and in greater amounts. Thus, apart from carbon revenues, the KACP farmers seemed to use more sustainable management practices, had higher maize yield, had better food self-sufficiency and tended to save more money than control farmers. This improved the preparedness of the smallholder farmers participating in the KACP and increased their resilience for future challenges. However, it was not possible to determine whether those farmers were already on this path to improvement before joining the KACP.

Author Contributions: Conceptualization, Y.N. and I.Ö.; methodology, I.Ö., Y.N., C.M. and E.W.; software, Y.N., C.M. and E.W.; validation, Y.N., C.M., E.W. and I.Ö.; formal analysis, Y.N.; investigation, C.M. and E.W.; resources, I.Ö.; data curation, Y.N.; writing—original draft preparation, Y.N.; writing—review and editing, Y.N., C.M., E.W., J.W., M.J. and I.Ö.; visualization, Y.N., E.W.; supervision, J.W., M.J. and I.Ö.; project administration, Y.N.; funding acquisition, I.Ö. All authors have read and agreed to the published version of the manuscript.

Funding: This research was funded by the Swedish Ministry for Foreign Affairs, as part of its special allocation on global food security; the Formas-Sida programme "Sustainable development in developing countries" (220-2009-2073); and the Swedish University of Agricultural Sciences (SLU).

Acknowledgments: Farmers and advisors around Kisumu and Bungoma are gratefully acknowledged for their comprehensive efforts. Special thanks to Bo Lager, Fred Marani, Wangu Mutua, Amos Wekesa, Peter Wachira and the rest of the Vi Agroforestry team. Thanks also to Johannes Forkman, who assisted in statistical analyses.

Conflicts of Interest: The authors declare no conflict of interest. All procedures performed in the study involving human participants were in accordance with the ethical standards of the institutional and/or national research committee and with the 1964 Helsinki Declaration and its later amendments, or comparable ethical standards. Informed consent was obtained from all individual participants in the study. The funders had no role in the design of the study; in the collection, analyses, or interpretation of data; in the writing of the manuscript; or in the decision to publish the results.

Appendix A

Table A1. Sustainable Agricultural Land Management (SALM) practices promoted and implemented in the Kenya Agricultural Carbon Project (KACP) (after [31]).

SALM Categories	SALM Practices
Nutrient management	Mulching, composting, cover crops, nitrogen-fixing crops, manure, restricted chemical fertilisers and chemical management
Soil and water conservation	Terraces, contour bunds, broad beds and furrows, semi-circular bunds, trash lines, diversion ditches and cut-off drains, retention ditches, pitting, trenches, tied ridges, grass strips, irrigation, roof catchment, ground surfaces and rocks, irregular surfaces, tanks, birkas, pans, ponds, dams, wells and boreholes, ecological sanitation, kitchen water
Agronomic practices	Crop rotation, intercropping, green manure, contour strip cropping, relay cropping, use of improved crop varieties
Agroforestry	Plant trees amongst crops, trees and livestock, trees, crops and livestock, trees and insects, trees and water animals, woodlots, boundary planting, dispersed interplanting, fruit orchards
Tillage and residue management	No-tillage/zero-tillage, reduced tillage, pitting systems, stubble and residue mulch tillage, dibble stick planting, strip and spot tillage, ripping, ridge and furrow tillage, residue management
Land restoration and rehabilitation	Natural regeneration, assisted natural regeneration, enrichment planting, fire management, agroforestry
Integrated livestock management	Improved feeding and watering, housing, stall management systems, improved waste management, pest and disease control, improved breeding practices
Sustainable energy	Biomass, biogas, farm residues, energy-efficient stoves, sustainable charcoal production
Integrated pest management	Biological pest control, use of crop-resistant varieties, alternative agricultural practices (spraying, use of fertilisers, pruning), mechanical pest control, pesticides, cultural methods, pest management plan

Table A2. Characteristics of the Kenya Agricultural Carbon Project (KACP) sites at Kisumu and Bungoma.

Parameter	Kisumu Site	Bungoma Site	Data Source
Counties	Siaya, Kisumu	Bungoma	[83]
Divisions	Wagai, Kombewa, Madiany	Bumula, Malakisi, Sirisia	[83]
Location	South Gem, West Gem, North West Gem, North East Gem, South West Gem, South Central Seme, North Central Seme, South West Seme, West Seme, West Uyoma, Central Uyoma, East Uyoma, South Uyoma	Bumula, Kabula, Mabusi, South Bukusu, Khasoko, Siboti, Mukwa Kimaeti, Napara, Kibuke, Malakisi, Namubila, Lwandanyi, Sirisia, Namwela	[83]
Agro-ecological zones [1]	LM1, LM2, LM3, LM4	UM1, UM2, LM1, LM2, LM3	[84]
Soils	Clay content about 39%	Clay content about 20%	[85]
Altitude (m above sea level)	1200–1500 m	1200–1850 m	[86]
Major crops	Maize & sorghum	Maize & sugarcane	Data from project farms n = 200
Average farm size	0.7 ha	1.1 ha	Data from project farms n = 200
Average household size	3.6 adults and 3.2 children	3.7 adults and 4.4 children	Data from project farms n = 200
Mean temperature range and mean annual precipitation	17.4 °C–29.8 °C; 1326 mm	14 °C–27.6 °C; 1884 mm	[83]
Project locations	0°7′45.53″ N, 34°23′38.56″ E; 0°23′34.29″ S; 34°17′58.55″ E	0°27′0.12″ N, 34°31′14.87″ E; 0°48′18.13″ N, 34°24′54.61″ E	[83]
Population density (persons km^{-2})	333 and 465 in Siaya and Kisumu County, respectively	454 in Bungoma County	[87]

[1] LM1 = Lower Midland 1; LM2 = Lower Midland 2; LM3 = Lower Midland 3; LM4 = Lower Midland 4; UM1 = Upper midland 1; UM2 = Upper midland 2.

a) Bungoma site rainfall data

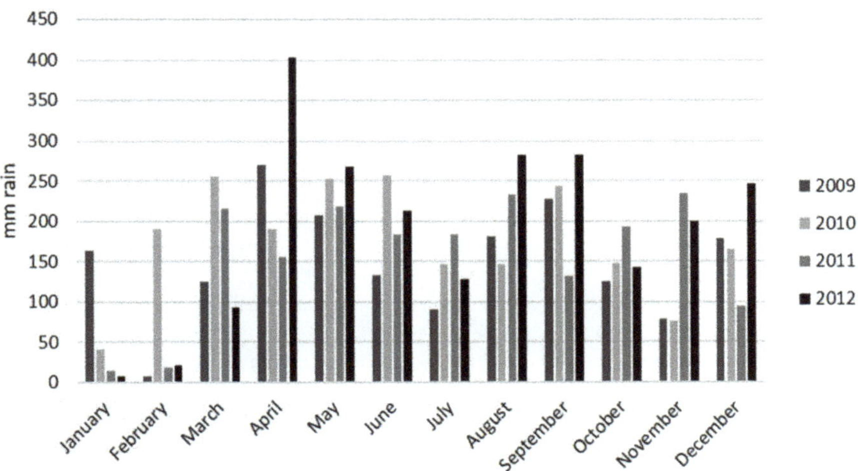

b) Kisumu site rainfall data

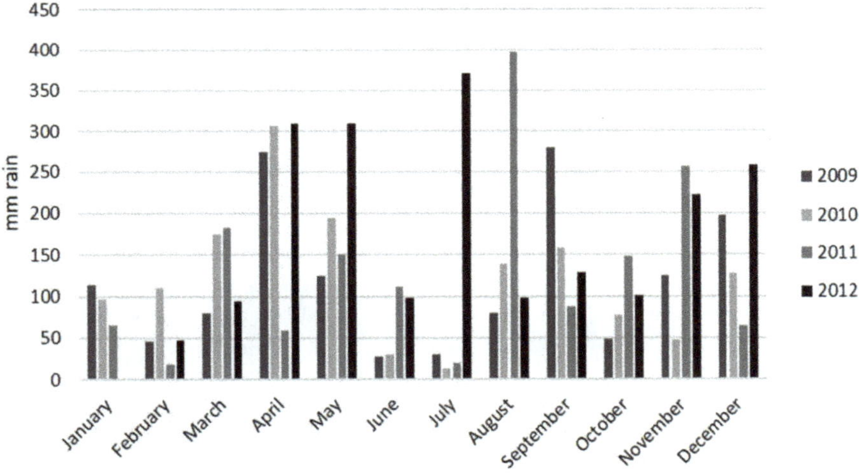

Figure A1. Monthly rainfall data in the study areas, 2009–2012; (**a**) Bungoma site (Kakamega meteorological station; total annual rainfall 2009–2012 = 1789, 2113, 1876 and 2287 mm, respectively). (**b**) Kisumu site (Kisumu meteorological station; total annual rainfall 2009–2012 = 1426, 1477, 1563, and 2036 mm, respectively).

References

1. Elliott, J.; Deryng, D.; Müller, C.; Frieler, K.; Konzmann, M.; Gerten, D.; Glotter, M.; Flörke, M.; Wada, Y.; Best, N.; et al. MConstraints and Potentials of Future Irrigation Water Availability on Agricultural Production under Climate Change. *Proc. Natl. Acad. Sci. USA* **2014**, *111*, 3239–3244. [CrossRef] [PubMed]
2. Foley, B.J.; Melton, F.; Middleton, E.; Hain, C.; Anderson, M.; Allen, R.; McCabe, M.F.; Hook, S.; Baldocchi, D.; Townsend, P.A.; et al. The Future of Evapotranspiration: Global Requirements for Ecosystem Functioning, Carbon and Climate Feedbacks, Agricultural Management, and Water Resources. *Water Resour. Res.* **2017**, *53*, 2618–2626.

3. Foley, J.A.; Ruth, D.; Gregory, P.; Asner, C.B.; Gordon, B.; Stephen, R.; Carpenter, F.; Stuart, C.; Michael, T.; Coe, G.; et al. Global Consequences of Land Use. *Science* **2005**, *309*, 570–574. [CrossRef]

4. Godfray, H.; Charles, J.; Tara, G. Food Security and Sustainable Intensification. *Philos. Trans. R. Soc. B* **2014**, *369*, 20120273. [CrossRef]

5. IPCC. Summary for Policymakers. In *Climate Change 2014: Impacts, Adaptation, and Vulnerability. Part A: Global and Sectoral Aspects. Contribution of Working Group II to the Fifth Assessment Report of the Intergovernmental Panel on Climate Change*; Field, C.B., Barros, V.R., Dokken, D.J., Mach, K.J., Mastrandrea, M.D., Bilir, T.E., Chatterjee, M., Ebi, K.L., Estrada, Y.O., Genova, R.C., et al., Eds.; Cambridge University Press: Cambridge, UK; New York, NY, USA, 2014; pp. 1–32.

6. Johan, R.; Williams, J.; Daily, G.; Noble, A.; Matthews, N.; Gordon, L.; Wetterstrand, H.; DeClerck, F.; Shah, M.; Steduto, P. Sustainable Intensification of Agriculture for Human Prosperity and Global Sustainability. *Ambio* **2017**, *46*, 4–17.

7. Bergström, L.; Bowman, B.T.; Sims, J.T. Definition of Sustainable and Unsustainable Issues in Nutrient Management of Modern Agriculture. *Soil Use Manag.* **2005**, *21*, 76–81. [CrossRef]

8. Horrigan, L.; Robert, S.; Walker, P. How Sustainable Agriculture Can Address the Environmental and Human Health Harms of Industrial Agriculture. *Environ. Health Persp.* **2002**, *110*, 445–456. [CrossRef]

9. Will, S.; Richardson, K.; Rockström, J.; Sarah, E.; Fetzer, C.I.; Bennett, M.E.; Biggs, R.; Carpenter, S.R.; de Vries, W.; de Cynthia, W. Planetary Boundaries: Guiding Human Development on a Changing Planet. *Science* **2015**, *347*. [CrossRef]

10. Wivstad, M.; Dahlin, A.S.; Grant, C. Perspectives on Nutrient Management in Arable Farming Systems. *Soil Use Manag.* **2005**, *21*, 113–121. [CrossRef]

11. Altieri, M.; Clara, A.; Nicholls, I.; Rene, M. Technological Approaches to Sustainable Agriculture at a Crossroads: An Agroecological Perspective. *Sustainability* **2017**, *9*, 349. [CrossRef]

12. Buttel, F.H. Sustaining the Unsustainable. In *Handbook of Rural Studies*; Paul, C., Ed.; Terry Marsden and Patrick Mooney; SAGE Publications Ltd.: London, UK, 2006.

13. IPCC. Summary for Policymakers. In *Global Warming of 1.5 °C. An IPCC Special Report on the Impacts of Global Warming of 1.5 °C above Pre-Industrial Levels and Related Global Greenhouse Gas Emission Pathways, in the Context of Strengthening the Global Response to the Threat of Climate Change, Sustainable Development, and Efforts to Eradicate Poverty*; Masson-Delmotte, P., Zhai, H.O., Pörtner, D., Roberts, J., Skea, P.R., Shukla, A., Pirani, W., Moufouma-Okia, C., Péan, R., Pidcock, S., et al., Eds.; World Meteorological Organization: Geneva, Switzerland, 2018.

14. Abraha, M.; Hamilton, S.K.; Chen, J.; Robertson, G.P. Ecosystem Carbon Exchange on Conversion of Conservation Reserve Program Grasslands to Annual and Perennial Cropping Systems. *Agric. For. Meteorol.* **2018**, *253*, 151–160. [CrossRef]

15. Lal, R. Soil Carbon Sequestration Impacts on Global Climate Change and Food Security. *Science* **2004**, *304*, 1623–1627. [CrossRef]

16. Lal, R. Global Potential of Soil Carbon Sequestration to Mitigate the Greenhouse Effect. *Crit. Rev. Plant. Sci.* **2003**, *22*, 151–184. [CrossRef]

17. Mueller, N.; James, D.; Gerber, S.; Matt, J.; Deepak, K.R.; Navin, R.; Jonathan, A.F. Closing Yield Gaps through Nutrient and Water Management. *Nature* **2012**, *490*, 254–257. [CrossRef] [PubMed]

18. Öborn, I.; Vanlauwe, B.; Phillips, M.; Thomas, R.; Brooijmans, W.; Atta-Krah, K. *Sustainable Intensification in Smallholder Agriculture: An. Integrated Systems Research Approach*; Öborn, I., Vanlauwe, B., Phillips, M., Thomas, R., Brooijmans, W., Atta-Krah, K., Eds.; Earthscan Food and Agriculture Series: London, UK; Routledge Taylor and Francis Group: New York, NY, USA, 2017.

19. Swallow, M.B.; Goddard, T.W. Value Chains for Bio-Carbon Sequestration Services: Lessons from Contrasting Cases in Canada, Kenya and Mozambique. *Land Use Policy* **2013**, *31*, 81–89. [CrossRef]

20. Paris Agreement. *Paper Presented at the Conference of the Parties to the United Nations Framework Convention on Climate Change*; 21st Session; FCCC: Paris, France, 2015.

21. Government, K. *National Climate Change Action Plan 2018-2022*; Ministry of Environment and Forestry: Nairobi, Kenya, 2018.

22. Government, Kenya. *National Climate Change Action Plan 2013–2017. 258*; Ministry of Environment and Mineral Resources: Nairobi, Kenya, 2013.

23. Ylva, N.; Jonsson, M.; Ambjörnsson, E.L.; Wetterlind, J.; Öborn, I. Smallholders' Awareness of Adaptation and Coping Measures to Deal with Rainfall Variability in Western Kenya. *Agroecol. Sustain. Food* **2020**, *44*, 1–29.

24. Sikstus, G.; Sumeni, S.; Sabodin, R.; Muqfi, I.H.; Nur, M.; Hairiah, K.; Useng, D.; van Noordwijk, M. Soil Organic Matter, Mitigation of and Adaptation to Climate Change in Cocoa–Based Agroforestry Systems. *Land* **2020**, *9*, 323.

25. Elizabeth, B.; Ringler, C.; Okoba, B.; Roncoli, C.; Silvestri, S.; Herrero, M. Adapting Agriculture to Climate Change in Kenya: Household Strategies and Determinants. *J. Environ. Manag.* **2013**, *114*, 26–35.

26. FAO. *Adapting to Climate Change through Land and Water Management in Eastern Africa—Results of Pilot Projects in Ethiopia, Kenya and Tanzania*; FAO, Ed.; FAO: Rome, Italy, 2014.

27. Amy, Q.; Neufeldt, H.; McCabe, J.T. Building Livelihood Resilience: What Role Does Agroforestry Play? *Clim. Dev.* **2019**, *11*, 485–500.

28. Vermeulen, S.J.; Aggarwal, P.K.; Ainslie, A.; Angelone, C.; Campbell, B.M.; Challinor, A.J.; Hansen, J.W.; Ingram, J.S.I.; Jarvis, A.; Kristjanson, P. Options for Support to Agriculture and Food Security under Climate Change. *Environ. Sci. Policy* **2012**, *15*, 136–144. [CrossRef]

29. Nair, P.K.R.; Vimala, D.; Nair, M.; Kumar, B.; Solomon, G.H. Soil Carbon Sequestration in Tropical Agroforestry Systems: A Feasible Appraisal. *Environ. Sci. Policy* **2009**, *12*, 1099–1111. [CrossRef]

30. UNFCCC. Nationally Determined Contributions (Ndcs). United Nations Framework Convention on Climate Change. Available online: https://unfccc.int/process-and-meetings/the-paris-agreement/nationally-determined-contributions-ndcs#eq-2 (accessed on 30 August 2019).

31. Stefanie, E.; Muller, A. Payments for Environmental Services to Promote Climate-Smart Agriculture? Potential and Challenges. *Agric. Econ.* **2016**, *47*, 173–184.

32. Kelley, H.; Goldstein, A. *Ahead of the Curve—State of the Voluntary Carbon Markets 2015*; Peters-Stanley, M., Gonzalez, G., Eds.; Ecosystem Marketplace—A Forest trends initiative: Washington, DC, USA, 2015.

33. Stringer, L.C.; Dougill, A.J.; Thomas, A.D.; Spracklen, D.V.; Chesterman, S.; Speranza, C.I.; Rueff, H.; Riddell, M.; Williams, M.; Beedy, T.; et al. Challenges and Opportunities in Linking Carbon Sequestration, Livelihoods and Ecosystem Service Provision in Drylands. *Environ. Sci. Policy* **2012**, *19–20*, 121–135. [CrossRef]

34. Kelley, H.; Goldstein, A. *Raising Ambition: State of the Voluntary Carbon Markets 2016*; Ecosystem Marketplace: Washington, DC, USA, 2016.

35. Henry, M.H.; Tittonell, P.; Manlay, R.J.; Bernoux, M.; Albrecht, A.; Vanlauwe, B. Biodiversity, Carbon Stocks and Sequestration Potential in Aboveground Biomass in Smallholder Farming Systems of Western Kenya. *Agric. Ecosyst. Environ.* **2009**, *129*, 238–252. [CrossRef]

36. Cheikh, M.; van Noordwijk, M.; Luedeling, E.; Neufeldt, H.; Minang, P.A.; Kowero, G. Agroforestry Solutions to Address Food Security and Climate Change Challenges in Africa. *Cur. Opin. Environ. Sustain.* **2014**, *6*, 61–67.

37. Johannes, W. *Project Information Document (Pid) Kenya Agricultural Carbon Project (Kacp)*; World Bank: Washington, DC, USA, 2010.

38. Amos, W.; Jönsson, M. Sustainable Agriculture Land Management. A Training Material. Vi Agroforestry. Available online: http://www.viagroforestry.org/who-we-are/resources/publications/ (accessed on 8 January 2015).

39. Bouman, F.J.A. Rotating and Accumulating Savings and Credit Associations: A Development Perspective. *World Dev.* **1995**, *23*, 371–384. [CrossRef]

40. Atela, J.O. *The Politics of Agricultural Carbon Finance: The Case of the Kenya Agricultural Carbon Project, STEPS Working Paper 49*; STEPS Centre: Brighton, UK, 2012; 44p. Available online: www.steps-centre.org (accessed on 5 June 2017).

41. Cavanagh, C.J.; Chemarum, A.K.; Vedeld, P.O.; Petursson, J.G. Old Wine, New Bottles? Investigating the Differential Adoption of 'Climate-Smart' Agricultural Practices in Western Kenya. *J. Rural Stud.* **2017**, *56*, 114–123. [CrossRef]

42. Karl, H.; Morgan, S.; Baylis, K.; Oduol, J.; Smith-Dumont, E.; Vagen, T.; Mutemi, M.; LePage, C.; Kegode, H. Assessing the Downstream Socioeconomic Impacts of Agroforestry in Kenya. In *ICRAF Working Paper No 291*; World Agroforestry: Nairobi, Kenya, 2018.

43. Lee, J. Farmer Participation in a Climate-Smart Future: Evidence from the Kenya Agricultural Carbon Market Project. *Land Use Policy* **2017**, *68*, 72–79. [CrossRef]

44. Lee, J.L.; Martin, A.; Kristjanson, P.; Wollenberg, E. Implications on Equity in Agricultural Carbon Market Projects: A Gendered Analysis of Access, Decision Making, and Outcomes. *Environ. Plan. A* **2015**, *47*, 2080–2096. [CrossRef]

45. Seth, S.; Heiner, K.; Kapukha, M.; Kiguli, L.; Masiga, M.; Kalunda, P.N.; Ssempala, A.; Recha, J.; Wekesa, A. Building Local Institutional Capacity to Implement Agricultural Carbon Projects: Participatory Action Research with Vi Agroforestry in Kenya and Ecotrust in Uganda. *Agric. Food Sec.* **2016**, *5*, 13.

46. Öborn, I.; Wekesa, A.; Natongo, P.; Kiguli, L.; Wachiye, E.; Musee, C.; Kuyah, S.; Neves, B. Who Enjoys Smallholder Generated Carbon Benefits? In *Co-Investment in Ecosystem Services: Global Lessons from Payment and Incentive Schemes*; Namirembe, S., Leimona, B., van Noordwijk, M., Minang, P., Eds.; World Agroforestry Centre: Nairobi, Kenya, 2017; pp. 1–10.

47. Mutua, W. Kenya Agricultural Carbon Project Monitoring Report. Vi Agroforestry. Available online: https://www.vcsprojectdatabase.org/#/project_details/12252012 (accessed on 12 December 2018).

48. Matthias, S.; Tennigkeit, T.; Zanchi, G.; Bird, N. *Technical Guidelines—Activity Baseline and Monitoring Survey Guideline for Sustainable Agricultural Land Management Practices (Salm)*; UNIQUE forestry consultants: Freiburg, Germany, 2010.

49. Lager, B. *Vcs Project Description Kenya Agricultural Carbon Project. Kisumu*; Vi Agroforestry: Kisumu, Kenya, 2012.

50. R: A Language and Environment for Statistical Computing. R Foundation for Statistical Computing, Vienna, Austria. Available online: http://www.r-project.org/index.html (accessed on 5 May 2019).

51. Wei, W.; Feng, X.; Yang, L.; Chen, L.; Feng, T.; Chen, D. The Effects of Terracing and Vegetation on Soil Moisture Retention in a Dry Hilly Catchment in China. *Sci. Total Environ.* **2019**, *647*, 1323–1332. [CrossRef]

52. Tejendra, C.; Raizada, M.N. Agronomic Challenges and Opportunities for Smallholder Terrace Agriculture in Developing Countries. *Front. Plant. Sci.* **2017**, *8*. [CrossRef]

53. Ngome, A.F.E.; Becker, M.; Mtei, K.M. Leguminous Cover Crops Differentially Affect Maize Yields in Three Contrasting Soil Types of Kakamega, Western Kenya. *J. Agric. Rural Dev. Trop.* **2011**, *112*, 1–10.

54. Sommer, R.; Mukalama, J.; Kihara, J.; Koala, S.; Winowiecki, L.; Bossio, D. Nitrogen Dynamics and Nitrous Oxide Emissions in a Long-Term Trial on Integrated Soil Fertility Management in Western Kenya. *Nutr. Cycl. Agroecosyst.* **2016**, *105*, 229–248. [CrossRef]

55. Biamah, E.K.; Gichuki, F.N.; Kaumbutho, P.G. Tillage Method and Soil and Water Conservation in East Africa. *Soil Till Res.* **1993**, *27*, 105–123. [CrossRef]

56. Ndoli, A.; Baudron, F.; Sida, T.S.; Schut, A.G.T.; van Heerwaarden, J.; Giller, K.E. Conservation Agriculture with Trees Amplifies Negative Effects of Reduced Tillage on Maize Performance in East Africa. *Field Crop. Res.* **2018**, *221*, 238–244. [CrossRef]

57. Kiboi, M.N.; Ngetich, K.F.; Diels, J.; Mucheru-Muna, M.; Mugwe, J.; Mugendi, D.N. Minimum Tillage, Tied Ridging and Mulching for Better Maize Yield and Yield Stability in the Central Highlands of Kenya. *Soil Till. Res.* **2017**, *170*, 157–166. [CrossRef]

58. Potma, G.; Carlos, D.R.J.; Mishra, U.; Furlan, F.J.F.; Ferreira, L.A.; Inagaki, T.M.; Romaniw, J.; Ferreira, A.; Briedis, C. Soil Carbon Inventory to Quantify the Impact of Land Use Change to Mitigate Greenhouse Gas Emissions and Ecosystem Services. *Environ. Pollut.* **2018**, *243*, 940–952. [CrossRef]

59. Sommer, R.; Paul, B.K.; Mukalama, J.; Kihara, J. Reducing Losses but Failing to Sequester Carbon in Soils—the Case of Conservation Agriculture and Integrated Soil Fertility Management in the Humid Tropical Agro-Ecosystem of Western Kenya. *Agric. Ecosyst. Environ.* **2018**, *254*, 82–91. [CrossRef]

60. Kamau, M.; Smale, M.; Mutua, M. Farmer Demand for Soil Fertility Management Practices in Kenya's Grain Basket. *Food Sec.* **2014**, *6*, 793–806. [CrossRef]

61. Phuong, L.T.H.; Biesbroek, G.R.; Sen, L.T.H.; Wals, A.E.J. Understanding Smallholder Farmers' Capacity to Respond to Climate Change in a Coastal Community in Central Vietnam. *Clim. Dev.* **2018**, *10*, 701–716. [CrossRef]

62. Shem, K.; Öborn, I.; Jonsson, M.; Dahlin, A.S.; Barrios, E.; Muthuri, C.; Malmer, A.; Nyaga, J.; Magaju, C.; Namirembe, S.; et al. Trees in Agricultural Landscapes Enhance Provision of Ecosystem Services in Sub-Saharan Africa. *IJBESM* **2016**, *12*, 255–273.

63. Esteban, L.M.; Zema, D.A.; Carrà, B.G.; Cerdà, A.; Plaza-Alvarez, P.A.; Cózar, J.S.; Gonzalez-Romero, J.; Moya, D.; de las Heras, J. Short-Term Changes in Infiltration between Straw Mulched and Non-Mulched Soils after Wildfire in Mediterranean Forest Ecosystems. *Ecol. Eng.* **2018**, *122*, 27–31.

64. Kihara, J.; Bationo, A.; Waswa, B.; Kimetu, J.M.; Vanlauwe, B.; Okeyo, J.; Mukalama, J.; Martius, C. Effect of Reduced Tillage and Mineral Fertilizer Application on Maize and Soybean Productivity. *Exp. Agric.* **2011**, *48*, 159–175. [CrossRef]

65. Castellanos-Navarrete, A.; Tittonell, P.; Rufino, M.C.; Giller, K.E. Feeding, Crop Residue and Manure Management for Integrated Soil Fertility Management—A Case Study from Kenya. *Agric. Syst.* **2015**, *134*, 24–35. [CrossRef]

66. Lagerkvist, C.J.; Shikuku, K.; Okello, J.; Karanja, N.; Ackello-Ogutu, C. A Conceptual Approach for Measuring Farmers' Attitudes to Integrated Soil Fertility Management in Kenya. *NJAS Wagen. J. Life Sci.* **2015**, *74–75*, 17–26. [CrossRef]

67. Ndoli, A.; Baudron, F.; Antonius, G.; Schut, T.; Athanase, M.; Giller, K.E. Disentangling the Positive and Negative Effects of Trees on Maize Performance in Smallholdings of Northern Rwanda. *Field Crop. Res.* **2017**, *213*, 1–11. [CrossRef]

68. Abubeker, H.; Talore, D.G.; Tesfamariam, E.H.; Friend, M.A.; Mpanza, T.D.E. Potential Use of Forage-Legume Intercropping Technologies to Adapt to Climate-Change Impacts on Mixed Crop-Livestock Systems in Africa: A Review. *Reg. Environ. Chang.* **2017**, *17*, 1713–1724.

69. Kurniatun, H.; Widianto, W.; Suprayogo, D.; van Noordwijk, M. Tree Roots Anchoring and Binding Soil: Reducing Landslide Risk in Indonesian Agroforestry. *Land* **2020**, *9*, 256.

70. Ylva, N.; Wetterlind, J.; Jonsson, M.; Öborn, I. The Role of Trees and Livestock in Ecosystem Service Provision and Farm Priorities on Smallholder Farms in the Rift Valley, Kenya. *Agric. Syst.* **2020**, *181*, 102815.

71. Diwani, T.N.; Asch, F.; Becker, M.; Mussgnug, F. Characterizing Farming Systems around Kakamega Forest, Western Kenya, for Targeting Soil Fertility-Enhancing Technologies. *J. Plant. Nutr. Soil Sci.* **2013**, *176*, 585–594. [CrossRef]

72. FAO. Faostat—Maize Yields Kenya 2009–2012. In *Food and Agriculture Data*; FAOSTAT: Food and Agriculture Organization of the United Nations: Rome, Italy, 2013. Available online: http://www.fao.org/faostat/en/#data/QC (accessed on 17 May 2019).

73. Yongshan, C.; Camps-Arbestain, M.; Shen, Q.; Singh, B.; Cayuela, M.L. The Long-Term Role of Organic Amendments in Building Soil Nutrient Fertility: A Meta-Analysis and Review. *Nutr. Cycl. Agroecosyst.* **2018**, *111*, 103–125.

74. Vanlauwe, B.; Barrios, E.; Robinson, T.; van Asten, P.; Zingore, S.; Gérard, B. System Productivity and Natural Resource Integrity in Smallholder Farming: Friends or Foes? In *Sustainable Intensification in Smallholder Agriculture: An. Integrated Systems Research Approach*; Öborn, I., Vanlauwe, B., Phillips, M., Thomas, R., Brooijmans, W., Atta-Krah, K., Eds.; Routledge: London, UK, 2017; pp. 159–176.

75. Erenstein, O. Smallholder Conservation Farming in the Tropics and Sub-Tropics: A Guide to the Development and Dissemination of Mulching with Crop Residues and Cover Crops. *Agric. Ecosys. Environ.* **2003**, *100*, 17–37. [CrossRef]

76. Kalungu, J.W.; Filho, W.L. Adoption of Appropriate Technologies among Smallholder Farmers in Kenya. *Clim. Dev.* **2018**, *10*, 84–96. [CrossRef]

77. Pramila, K.; Patnam, M. Neighbors and Extension Agents in Ethiopia: Who Matters More for Technology Adoption? *Am. J. Agric. Econ.* **2014**, *96*, 308–327.

78. Stefanovic, J.O.; Yang, H.; Zhou, Y.; Kamali, B.; Ogalleh, S.A. Adaption to Climate Change: A Case Study of Two Agricultural Systems from Kenya. *Clim. Dev.* **2017**, *11*, 319–337. [CrossRef]

79. Helth, L.J.; Rasmussen, O.D. Can Microfinance Reach the Poorest: Evidence from a Community-Managed Microfinance Intervention. *World Dev.* **2014**, *64*, 460–472.

80. Brannen, C.F. An Impact Study of the Village Savings and Loan Association (Vsla) Program in Zanzibar, Tanzania. Bachelor's Degree, Wesleyan University, Middletown, CT, USA, 2010.

81. Kearney, S.P.; Fonte, S.J.; García, E.; Siles, P.; Chan, K.M.A.; Smukler, S.M. Evaluating Ecosystem Service Trade-Offs and Synergies from Slash-and-Mulch Agroforestry Systems in El Salvador. *Ecol. Indic.* **2017**, *105*, 264–278. [CrossRef]

82. Moushumi, C.; Ajayi, O.C.; Hellin, J.; Neufeldt, H. Climate Change Adaptation and Social Protection in Agroforestry Systems: Enhancing Adaptive Capacity and Minimizing Risk of Drought in Zambia and Honduras. In *ICRAF Working Papers*; World Agroforestry Centre: Nairobi, Kenya, 2011.

83. Andersson, A. Kenya Agricultural Carbon Project Monitoring Report. Vi Agroforestry. 2016. Available online: https://www.vcsprojectdatabase.org/#/project_details/1225 (accessed on 1 September 2018).

84. FAO. *Agro-Ecological Zoning Guidelines*; Food and Agriculture Organisation of the United Nations: Rome, Italy, 1996. Available online: http://www.fao.org/docrep/W2962E/W2962E00.htm (accessed on 13 September 2014).

85. FAO/IIASA/ISRIC/ISSCAS/JRC. Harmonized World Soil Database (Version 1.1). FAO, Rome, Italy and IIASA, Luxenburg, Austria. Available online: http://www.iiasa.ac.at/Research/LUC/External-World-soil-database/HTML/ (accessed on 15 March 2018).

86. USGS. Shuttle Radar Topography Mission (Srtm) 1 Arc-Second Global. Elevation Data. US Geological Survey. Available online: https://lta.cr.usgs.gov/SRTM1Arc (accessed on 3 February 2018).

87. KNBS. Population Distribution by Sex, Number of Households, Area and Density by County and District. In *2009 Population and Housing Census*; Kenya National Bureau of Statistics 5, Ed.; KNBS: Nairobi, Kenya. Available online: https://www.knbs.or.ke/download/population-distribution-by-sex-number-of-households-area-and-density-by-county-and-district/ (accessed on 14 August 2017).

 *land*

Article

A Discounted Cash Flow and Capital Budgeting Analysis of Silvopastoral Systems in the Amazonas Region of Peru

Stephanie Chizmar [1],*, Miguel Castillo [1], Dante Pizarro [2], Hector Vasquez [3], Wilmer Bernal [3], Raul Rivera [1], Erin Sills [1], Robert Abt [1], Rajan Parajuli [1] and Frederick Cubbage [1]

[1] Department of Forestry and Environmental Resources, North Carolina State University, 2800 Faucette Dr., Raleigh, NC 27695, USA; mscastil@ncsu.edu (M.C.); rrivera4@ncsu.edu (R.R.); sills@ncsu.edu (E.S.); bobabt@ncsu.edu (R.A.); rparaju@ncsu.edu (R.P.); cubbage@ncsu.edu (F.C.)

[2] Livestock and Climate Change Center, Universidad Nacional Agraria La Molina, Lima 15000, Peru; dpizarro@lamolina.edu.pe

[3] Departamento de Zootecnia, Universidad Nacional Toribio Rodriguez de Mendoza de Amazonas, Chachapoyas 477002, Peru; hvasquez@untrm.edu.pe (H.V.); wilmer.bernal@untrm.edu.pe (W.B.)

* Correspondence: sjchizma@ncsu.edu

Received: 28 August 2020; Accepted: 24 September 2020; Published: 25 September 2020

Abstract: Silvopasture is a type of agroforestry that could deliver ecosystem services and support local livelihoods by integrating trees into pasture-based livestock systems. This study modeled the financial returns from silvopastures, planted forests, and conventional cattle-pasture systems in Amazonas, Peru using capital budgeting techniques. Forests had a lower land expectation value (USD 845 per hectare) than conventional cattle systems (USD 1275 per hectare) at a 4% discount rate. "Typical" model silvopastures, based on prior landowner surveys in the Amazonas region, were most competitive at low discount rates. The four actual silvopastoral systems we visited and examined had higher returns (4%: USD 1588 to USD 9524 per hectare) than either alternative pure crop or tree system, more than likely through strategies for generating value-added such as on-site retail stands. Silvopasture also offers animal health and environmental benefits, and could receive governmental or market payments to encourage these practices.

Keywords: silvopasture; economics; financial analysis; carbon payment; Peru

1. Introduction

Expanding forest cover remains a global priority to combating trends in deforestation and global climate change. Countries in Latin America, where the livestock sector generates 58–70% of overall agricultural emissions, are particularly interested in mitigating climate change through integrating forest cover into agricultural production systems [1]. Agroforestry operations combine forest or horticultural species and pasture or cropland to make mixed land-use systems that produce commercial benefits to landowners [2]. Silvopasture, a branch of agroforestry, is a strategic and managed agroecosystem in which livestock, forage, and trees or shrubs are integrated to help improve individual components [3,4]. Silvopastoral systems (SPS) diversify earnings to landowners by generating products on various harvest schedules, from daily as in the case of milk and cheese to multi-year for forest products such as fuelwood, posts, and boards [5].

As such, silvopasture has been identified as a key integrated landscape approach that could potentially increase returns to landowners by generating timber and non-timber products, while also improving conditions for forage production and cattle [6]. This most likely occurs when the components of the system have complementary relationships such that tree and cattle–forage production are mutually beneficial, which has been found true at low tree densities in the USA [7]. However, this may not be the case in the absence of established markets and technical knowledge networks. In other words, without access to knowledge and markets, silvopastoral practices may not achieve the social and environmental benefits they have been documented to produce [8].

Recently, the Peruvian government defined its Nationally Determined Contributions, which contemplate reducing 30% of the greenhouse gas emissions projected for the year 2030. The government recommended considering strategies such as the recovery of degraded soils with SPS in the Peruvian Amazon. Incorporating SPS in the Peruvian Amazon has the potential to mitigate 1344 metric tons of CO_2e through intervention on 102,000 hectares [9]. Peru also committed to restore 3.2 million hectares as part of the Bonn Challenge's mission to restore 150 million hectares of the world's degraded and deforested lands by 2020 [10]. Silvopasture is one strategy aimed at increasing forest cover in agricultural systems, augmenting a region's carbon sequestration potential, especially in deforested and degraded ecosystems [1].

Amazonas, a jurisdiction of Northeastern Peru, has an economically disadvantaged rural population that depends on agriculture and animal husbandry as primary income sources. Poverty rates exceed 50% and malnutrition levels are 30% or more in parts of Amazonas [11], suggesting a need for governmental intervention aimed at improving the livelihoods of the lower income populace. The rural population of Amazonas, encompassing over half of the total population [12], has opportunities to improve the welfare of landowners, farmers, and associated laborers. Rural landowners, both onsite and absentee, may generate additional income flows from timber and non-timber products by incorporating trees into their agricultural systems. Rural households may also become eligible for payments for ecosystem services through helping to expand tree cover and thereby provide habitat, sequester carbon, or provide other benefits.

The main goal of this paper is to evaluate the financial returns of Peruvian production systems, including silvopasture, typical cattle-pasture operations, and planted tree monocultures, through the application of discounted cash flow and capital budgeting analyses. The main hypothesis of SPS states that mixed-use systems can diversify farm income, reduce biophysical and financial risks, and perhaps, increase total farm returns by providing timber and non-timber products in addition to forage for livestock production. We postulated that the degree of tree and forage competition determines the profitability of SPS. In other words, at a modest tree density, trees may complement forage systems; however, at high tree densities, the production of forage will decline due to competition for resources, especially light. The collaborative team determined productivity data as well as prices of inputs and outputs as reported by local silvopasture practitioners and industry experts to model whole-system cash flows. Finally, we assessed the potential impact of incentive payments on a range of actual and typical landowner income for both absentee and onsite landowners. The rest of this paper is structured in the following five sections: (i) a review of the literature on silvopasture, particularly in tropical and subtropical regions; (ii) the methods we utilized to determine cash flows for multiple land-use systems; (iii) the results of the discounted cash flow and capital budgeting analyses; (iv) a discussion on the profitability of different land-uses; and (v) concluding remarks.

2. Literature Review

Loker [13] modeled low-input agroforests to determine their suitability as alternative, sustainable production systems for small-to-medium farmers in the Amazon Basin, as landowners in the region are attracted to cattle management for its multiple economic benefits. Landowners in Uruguay recognized that shelter from silvopasture benefits cattle, fostering increased calving rates [14]. In Argentina, small-scale farmers perceived cash flow diversification through the sale of livestock and forest products as the main advantage of silvopasture [15]. Silvopastures also likely provide better microenvironments, reducing climatic-induced stress, such that grazing animals are "happier" and gain weight at a faster rate than when grazing in traditional open pastures [3]. For many species of livestock, deviations in core body temperature more than 2 to 3 °C negatively impact performance, productivity, and fertility, potentially leading to decreased successful pregnancies [16]. Faster and increased weight gain from improved microenvironments as well as augmented crop growth may lead to higher profits for landowners, as well as better animal welfare in Amazonas.

Agroforests potentially sequester more carbon than pastures or field crops growing under similar ecological conditions [17]. Oliva et al. [18] measured 337.2 tons of carbon stored in *Pinus patula* silvopastures with 8- to 10-year-old trees in Amazonas. Dube et al. [19] reported that silvopastures in Patagonia, Chile, as well as in Minas Gerais, Brazil [20] required less time than a plantation monoculture to reach similar carbon gains above and below ground. The researchers attributed gains in carbon to the positive interactions between cattle, tree, and pasture components, including increased tree growth. Cubbage et al. [21] reviewed global SPS in eight regions, including areas in Argentina, Uruguay, Chile, and Brazil, and reported that silvopasture is financially competitive with alternative land-use systems and offers biophysical and financial diversity and resilience, attributes critical with the increased occurrences of extreme weather events due to climate change.

Payments for ecosystem (environmental) services (PES) compensate landowners for managing their property to sequester carbon, protect biodiversity, or provide watershed services. Some scientists have argued that in tropical systems the greatest potential for carbon sequestration is through the establishment of tree-based systems on degraded pastures [22]. Programs that provide incentives for forest management on degraded productive lands and supply additional sources of revenue may aid profitability as well as conservation of land-use systems in Amazonas. Orefice et al. [23] found that silvopastures in the U.S. North were financially superior to open pasture or thinned forests. Bruck et al. [24] observed that SPS are less financially profitable than pasture in the southern and western U.S. The study, however, did not account for any animal welfare or joint output effects.

Silvopastures can also shift forage production for livestock and soil quality, resulting in an expanded grazing season and livestock diets higher in protein [3,23]. Livestock provides essential elements such as nitrogen, phosphorus, and potassium through nutrient cycling to fertilize forages and trees, which may reduce dependence on additional external inputs [25,26]. Pent and Fike [25] suggested there is a complementary biophysical relationship between forage production for livestock and trees, assuming a modest tree density. Silvopastoral systems with relatively low tree densities required reduced weeding, provided increased available nitrogen, improved the microclimate, reduced erosion control costs, and fostered better animal health such as increased pregnancy success rates [26].

3. Methods

The Molinopampa district, located in the northeastern section of the Chachapoyas Province, in the southern portion of the Amazonas Region, has an oceanic climate (dry forest lower montane tropical according to Holdridge's Life Zone) [27]. Approximately 75% of the population in Molinopampa make a living in agriculture, ranching, hunting, and silviculture [28]. The Huayabamba Valley, in the Rodríguez de Mendoza Province, is characterized as having a tropical savanna and warm-humid climate (moisture forest premontane tropical according to Holdridge's Life Zone) [27]. Past surveys of the study area [28–31] found that 61% and 43% of the silvopasture producers in the Molinopampa district and the Huayabamba Valley, respectively, own farms that are on less than 10 hectares (ha). Conversely, 10-ha to 30-ha (30-ha+) farms represented 28% and 42% (11% and 16%) of the surveyed systems in the Molinopampa district and the Huayabamba Valley, respectively. The field sites for this study, which participated in the past surveys, are located on privately-owned farms, are managed as part of cooperatives with multiple institutions, and range from 10-ha to 65-ha in size [28–31]. Therefore, our sample farms are among the larger size class ownerships for these similar regions, but still within a representative range.

Universidad Nacional Agraria La Molina (UNALM), El Porvenir Research Station (INIA), the Universidad Nacional Toribio Rodríguez de Mendoza de Amazonas (UNTRM), and North Carolina State University (NCSU) have been collaborating with private landowners over the past five years to better understand the dynamics of applied agroforestry systems [28]. In July 2017, the collaborative team interviewed three landowners and one land manager well-known locally for practicing silvopasture and who have established relationships with the local university UNTRM. We selected these farms since they have established connections to the researchers and are most likely local leaders in their farm practices. Analyzing examples of successful programs may help to illuminate cost-effective characteristics and strategies that could be adapted throughout Amazonas. Cash flow estimations based on the four landowner responses, as well as average costs and benefits from discussions among co-authors, were predicted for each site in addition to "typical" model systems representing average farms in the region documented in the previously mentioned surveys [28–31].

The study collaborators gathered production function data for the case studies and typical model systems, organized in a database, to estimate values from soils, crops, and grazing. We estimated the associated economic impacts of inputs and outputs over a 25-year time horizon, both to calculate Net Present Value and Land Expectation Value. We then performed analyses in Microsoft Excel for Microsoft 365 using discounted cash flows associated with SPS, conventional grazing, and planted forests. The spreadsheets included necessary measures of productivity, product prices and costs, and management schedules, among other site-specific information, to calculate the net returns of annual activities. The operations were modeled as whole land-use regimes due to their complexity and the available information.

Tools in capital budgeting analysis such as Net Present Value (NPV), Land Expectation Value (LEV), Annual Equivalent Income (AEI), and Benefit-Cost Ratio (BCR) allow for comparison of land-use systems. Real discount rates of 4%, 8%, and 12% were used in all financial formulas to represent a range of the opportunity cost of the next best investment. Discount rates account for the future value of income that would equate to income earned in the present through representing preferences for incurring costs and benefits now or in the future, i.e., they capture the opportunity cost of the investment. Since discount rates are often unknown to landowners, most analyses use the internal rate of return of the next best option for comparisons, such as a bank investment [32]. The Internal Rate of Return (IRR) was not relevant and inapplicable for this analysis for ongoing farms since annual benefits always exceeded costs. We excluded land values from NPV and LEV calculations. As such, each metric represents use-values, or income-based values, of the land. We then compared these metrics to the alternative values, market prices for land in the region, to facilitate a discussion on the financial competitiveness of different land-use systems.

NPV measures the amount of capital that an investment returns at a given discount rate through summing the total expenditures and subtracting them from total income [33]. Formula (1) demonstrates how NPVs are calculated in terms of farm and forest costs and revenues.

$$NPV = \sum_{i=0}^{t} \frac{(B-C)}{(1+i)^t} \tag{1}$$

where *B* and *C* represent the annual total benefits and costs, respectively, of the land-use system, *i* signifies the interest or discount rate, and *t* is the year of the cash flow.

Similar to NPV, LEV utilizes expenditures, income, and a discount rate to measure the expected cash flow of a land-use in perpetuity. LEV has four assumptions to be viable: (1) identical costs and revenues in all rotations; (2) land-use will be maintained in perpetuity; (3) land requires identical regeneration costs at the beginning of each rotation; (4) the land value does not enter the calculation. In other words, landowners must replicate the following activities, including identical costs of inputs, prices of outputs, and cash flow schedules, into the future: purchase inputs, produce goods and services, extract products, and use or sell products. LEV allows researchers to compare systems of different rotation ages as well as how much an investor is willing to pay for the land at a given discount rate by assuming land-use will be continued indefinitely [26]. Formula (2) describes the method of estimating a production system's periodic and perpetual net returns in the present value [34].

$$LEV = NPV + \frac{NPV}{(1+i)^{T}-1}$$
$$\text{Or } LEV = \frac{V_n}{[(1+i)^{n}-1]}$$
(2)

where NPV equates to the net present value of the system, T is the final year of the system's rotation, V_n equates to the net future value of the system at n number of years per period, and i again signifies the interest or discount rate.

AEI expresses NPV or LEV in annual payments equally distributed over the life of the investment (Formula (3)). AEI allows comparison of long-term timber investments with seasonal returns from agriculture by expressing the income of each alternative in annual payments [32].

$$AEI = LEV * i$$
(3)

where LEV represents the system's land expectation value and i signifies the interest or discount rate.

BCR relates the total discounted benefits to the total discounted costs. The relation describes present value benefits and costs as a unitless proportion rather than as a difference such as in the case of NPV. The proportion (Formula (4)) reveals the return landowners receive per dollar invested [26].

$$BCR = \frac{\sum_{i=0}^{t} B_p}{\sum_{i=0}^{t} C_p}$$
(4)

where B_p and C_p represent the benefits or revenues and costs or expenditures, respectively, in present value terms.

For price transformations of Peruvian wood products, the co-authors converted from Nuevo Sol per product to USD per cubic meter of wood product. Derived residual stumpage prices, equating to 50% of the roadside value, were calculated to account for logging and transportation when not given labor amounts for transport and/or sale. As a result, the income generated by the primary landowner as well as additional laborers is captured. For this reason, primary landowner labor was not included in the analysis. In other words, the determined income includes earned income from self-employment of the primary landowner. However, for farms with more than 15 heads of cattle, average annual income data were used to proportionately increase income to outside laborers, additional landowners, and/or land managers, bore as a cost in the modeled scenarios. The average annual income of a farm laborer in Amazonas at the time of the study was USD 11,111.11 [28–31]. The most common land-uses in Amazonas, Peru are dual-production systems of cattle and dairy and cultivation of horticultural and forest products such as coffee, guava, citrus fruits, eucalyptus, and cedar [29,30]. Agroforestry systems in Amazonas are increasingly incorporating multi-strata systems and improved fallows [29,35,36]. The study collaborators represented this trend through the selection of field sites of various production complexity.

4. Results

Table 1 summarizes the operations based on actual farms and survey observations. The first three rows review typical model systems defined by our project cooperators and previously completed surveys [28–31]. The last four rows describe systems observed in field visits. Both typical and actual farms included a variety of predominantly exotic cattle breeds including Brown Swiss and Holstein and planted non-native Eucalypts and pines.

Table 1. Description of evaluated scenarios in Amazonas, Peru.

Scenario	Management Intensity *	Tree Species	Tree Growth (m^3ha^{-1} year^{-1})	Cattle Breeds	Number of Cattle (Lactating Cows)
Typical Model Systems (based on prior surveys)					
10-ha Planted Forest Only	Low	*Pinus patula*	5	-	0 (0)
10-ha Cattle-Pasture Only	Medium	-	0	Brown Swiss, Simmental, Holstein	16 (5)
10-ha SPS	Medium	*Alnus acuminata, Eucalyptus globulus, P. patula*	5	Brown Swiss, Simmental, Holstein	16 (5)
Actual Systems (field sites)					
10-ha Cattle + Trees + Fruit	Medium	*E. globulus*	5	Holstein	21 (6)
25-ha Cattle + Trees + Fruit + Store	High	*P. patula, Cupressus macrocarpa, E. globulus, A. acuminate*	5	Holstein, Simmental, Brown Swiss	38 (8)
30-ha Cattle + Trees	High	*E. globulus, A. acuminate, P. patula*	5	Brown Swiss	49 (24)
65-ha Cattle + Trees + Restaurant	Medium	*P. patula, C. macrocarpa*	5	Brown Swiss	38 (11)

* In terms of inputs including labor.

The income from the typical and actual SPS predominately comes from the production of agricultural and horticultural products. Table 2 separately identifies the present value income (at a 4% discount rate) from wood products as well as agricultural and horticultural products for the one SPS scenario based on the past surveys and the four actual systems. Table 3 displays the capital budgeting results for the typical and actual regimes. Returns for the modeled systems varied from positive to negative net profits depending on management choices, local markets, and the discount rate.

Table 2. Total present value income by product type (discounted at 4%).

SPS Scenario	Total Income at 4%			
	Wood Products		**Ag. and Hort. Products**	
Typical SPS	USD	3369	USD	148,060
10-ha C + T + F	USD	2616	USD	322,077
25-ha C + T + F + S	USD	4657	USD	715,614
30-ha C + T	USD	2616	USD	217,352
65-ha C + T + S	USD	2322	USD	67,951

C—Cattle; T—Trees; F—Fruit; S—Store (25-ha farm) or Restaurant (65-ha farm).

Table 3. Net present values (NPV) and land expectation values (LEV) of the evaluated scenarios.

Scenario	NPV (USD/ha)			LEV (USD/ha)		
	4%	**8%**	**12%**	**4%**	**8%**	**12%**
Typical Forest	527.87	−210.61	−583.94	844.74	−246.62	−620.44
Typical Cattle-Pasture	796.88	318.64	69.53	1275.25	373.12	73.87
Typical SPS	992.52	321.91	2.52	1588.33	376.96	2.68
10-ha C + T + F	5794.18	4045.62	3040.01	9272.43	4737.36	3230.01
25-ha C + T + F + S	3626.78	1433.91	276.45	5803.93	1679.08	293.73
30-ha C + T	5951.49	1840.67	−261.03	9524.17	2155.39	−277.35
65-ha C + T + S	4603.27	2615.55	1555.65	7366.60	3062.77	1652.88

For the three regimes modeled based on prior farm owner surveys [28–31], pure timber investments provided the lowest returns to landowners (LEV: 4% = USD 844.74; 8% = −USD 246.62; 12% = −USD 620.44 per ha). Dual-purpose cattle grazing systems earned about 50% more than the returns of plantation forests at the lowest discount rate (LEV: 4% = USD 1275.25; 8% = USD 373.12; 12% = USD 73.87 per ha). The "typical" silvopasture scenario based on survey findings generated the highest landowner profitability of the hypothetical regimes (LEV: 4% = USD 1588.33; 8% = USD 376.96; 12% = USD 2.68 per ha). At a 12% discount rate, however, the wait for timber income was penalized more. Consequently, the ranking of systems in respect to landowner profitability changed, signaling conventional cattle–forage regimes to earn more net income than conventional SPS.

The four actual agroforestry systems that we surveyed in Amazonas provided landowners high returns. Again, it is worth noting that these farms were selected by the project co-investigators, and probably are local leaders in their farm practices. The 10-ha farm including planted trees for fruit and timber production and cattle grazing earned relatively high net returns (LEV: 4% = USD 9272.43; 8% = USD 4737.36; 12% = USD 3230.01 per ha). The second smallest farm visited, a 25-ha system with planted timber and fruit trees, cattle grazing, and a store on-site for sale of final products, provided acceptable, positive returns at low discount rates (LEV: 4% = USD 5803.93; 8% = USD 1679.08; 12% = USD 293.73 per ha). The second largest agroforest, a management-intensive 30-ha cattle grazing regime with a "living-fence" of planted trees, generated positive returns at low discount rates but failed to return positive earnings at a 12% discount rate (LEV 4% = USD 9524.17; 8% = USD 2155.39; 12% = −USD 277.35 per ha). The largest property analyzed, a 65-ha farm with multiple cattle grazing paddocks and tree plantations, produced acceptable incomes at all discount rates (LEV: 4% = USD 7366.60; 8% = USD 3062.77; 12% = USD 1652.88 per ha).

Figure 1 demonstrates the AEI of the conventional and monoculture regimes. At 4%, the values of all typical systems in Peru ranged from USD 33.79 to USD 51.01/ha/year. Forest monocultures experienced negative returns at higher discount rates. Figure 2 shows the AEI values for mixed land-use regimes. Typical silvopastures experienced positive annual returns at all discount rates. The calculated annual net revenue of actual systems ranged between USD 134.33 and USD 378.99/ha/year at 8%. The AEI may be multiplied by the number of acres to estimate the system's annual income, including earned income for the self-employed primary landowner. Only the two largest actual operations, the 30-ha (4% discount rate) and 65-ha systems (all discount rates), earned more than the average annual income for farmers in Amazonas when accounting for total net system returns. Thus, the two larger systems were the only scenarios that support sustainable livelihoods for the landowner(s).

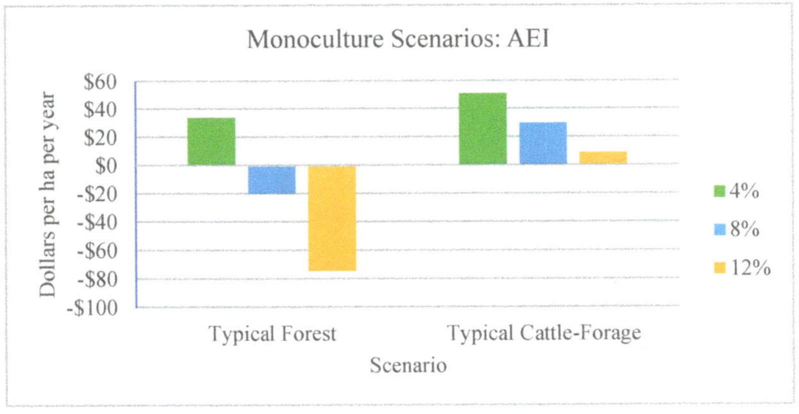

Figure 1. Annual equivalent income (AEI) of traditional forest and cattle systems.

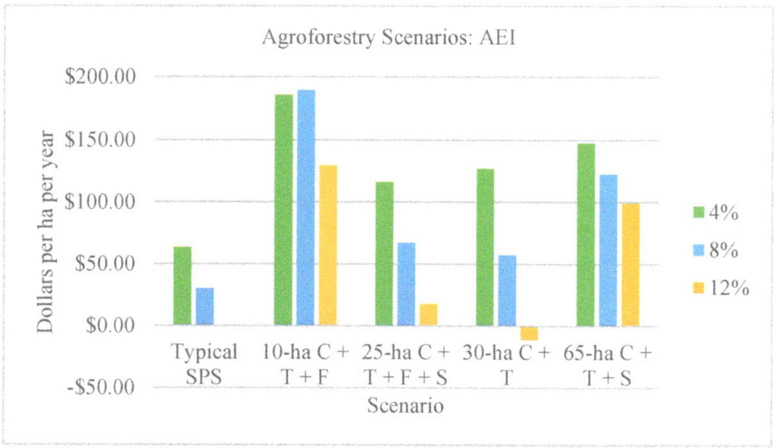

Figure 2. Annual equivalent income (AEI) of agroforestry systems.

Figure 3 displays the BCRs, a measure of the financial returns received per dollar spent in an investment, of all systems analyzed in the study. Tree plantations offered profitable returns per dollar invested at the lowest discount rate, as observed in its BCR, 1.27. At 8% and 12%, the BCR of forest monocultures was less than 1, signifying a loss per dollar invested (0.87, 0.61), due to the rapid diminution of future values. Dual-purpose cattle–forage regimes were cost-intensive and generated approximately USD 1.11 per dollar invested at the lowest discount rate (BCR: 4%= 1.11, 8% = 1.07, 12% = 1.02). The returns per dollar invested in silvopastures included a combination of costs and benefits from each industry (BCR: 4% = 1.16, 8% = 1.08, 12% = 1.00). The returns per dollar invested in the 10-ha actual farm increased as the discount rate increased (BCR: 4% = 1.62, 8% = 1.66, 12% = 1.71). The 25-ha mixed-use farm required high management intensity and generated just over a dollar per dollar spent at 12% (BCR: 4% = 1.27, 8% = 1.15, 12% = 1.04). The 30-ha silvopasture with an absentee landowner lost USD 0.02 per dollar invested at the highest discount rate (BCR: 4% = 1.24, 8% = 1.11, 12% = 0.98). The 65-ha silvopastoral system with a restaurant and timber sale on site generated over USD 2 per dollar invested at all discount rates (BCR: 4% = 2.54, 8% = 2.33, 12% = 2.12).

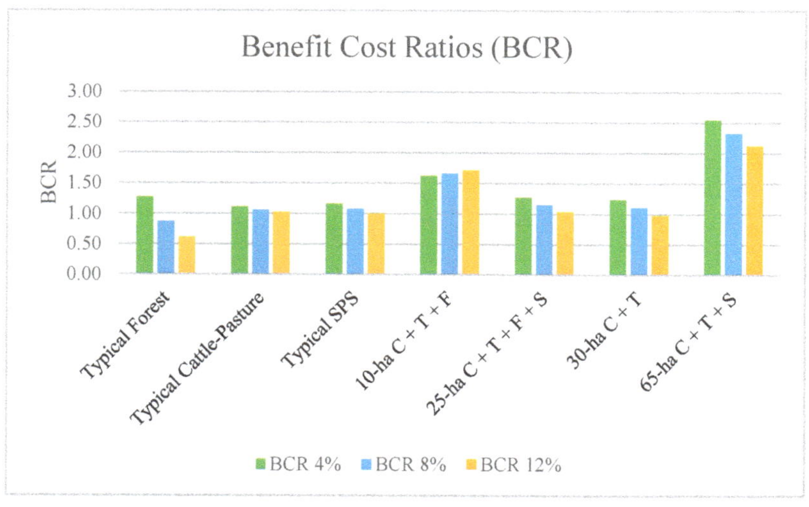

Figure 3. Benefit-cost ratios (BCR) for the evaluated scenarios.

The production systems analyzed required varying degrees of establishment intensity (Table 4). Of the three model scenarios, pure forest investments demanded the least in establishment inputs (USD 1130.50/ha). However, due to the longer turn-around time before generating a positive net income, forest monocultures also were associated with the longest payback period, in non-present value terms (8 years). While traditional cattle-pasture regimes required higher cost inputs for establishment (USD 1197.53/ha), they produced positive net returns sooner than planted forests (3 years). Establishing the typical cattle-pasture operation and the typical SPS required similar costs, differentiated only by the inputs needed to plant trees in the SPS. Cattle purchases were not included in this metric as they occur after pasture and trees are established. Establishment costs of hypothetical and actual silvopastures ranged from USD 623.99 to USD 1183.10 per ha and required 1 to 5 years to pay back, in non-present value terms.

Table 4. Establishment costs (not discounted) and payback periods for all scenarios.

Scenario	Establishment Cost (USD/ha)	Payback Period (Years)
Typical Forest	USD 1130.50	8
Typical Cattle + Pasture	USD 1197.53	3
Typical SPS	USD 1203.35	4
10-Ha C + T + F	USD 915.56	1
25-Ha C + T + F + S	USD 1183.10	1
30-Ha C + T	USD 1082.35	5
65-Ha C + T + S	USD 623.99	5

Cost-share payments at various rates pay landowners part of the establishment costs necessary to develop a tree plantation or other conservation practices. These are common in the United States now and were in Latin America in the past, and increase the net present value exactly by the amount paid in year 0, as well shorten the payback period needed to generate positive net earnings. An assumed one-time 50% cost-share payment during the first years of USD 87.61 per hectare increased the LEV of the typical silvopasture scenario to USD 90.29 per hectare at a 12% discount rate. The cost-share payments in the initial years increased the NPV or LEV by the amount of the payment, *ceteris paribus*. Payments for environmental services (PES), such as carbon sequestration and water quality protection, can help increase the returns of agroforestry systems and provide landowners an equal choice between land uses. Mixed systems, both hypothetical and actual, experienced higher returns than the monoculture alternatives at discount rates of 4% and 8%. At 12%, typical silvopastures earned USD 71.19 per ha less in present value terms than conventional cattle–forage systems and required an annuity of USD 12.60 per ha for 10 years to breakeven.

Using a growth rate of 5 m^3/ha/year and an approximate biomass volume of 42.7% dry wood volume [18], we estimated the net carbon flow of each tree-based land-use system. The baseline comparison to show additional carbon storage was traditional cattle–pasture systems. The derived aboveground carbon accumulation rate for pure planted forests in Amazonas was 2.56 tons/ha/year (9.4 metric tons CO_2e/ha/year). Typical silvopastures with 15–20% trees store 0.38–0.51 tons of carbon/ha/year (1.39–1.88 metric tons CO_2e/ha/year). In a 25-year rotation, landowners may sequester 34.75–47 metric tons CO_2e/ha.

5. Discussion

In Amazonas, Peru, we modeled three representative crop, forest, and silvopasture systems (SPS) based on previous farm owner surveys, interviewed four landowners or land managers as empirical case studies, and estimated their financial returns using capital budgeting techniques. The case study systems included dual-purpose dairy cattle with some forests, and often with on-farm retail stands. The span of systems analyzed in this study provided a mix of inputs and returns. The increased profitability of complex agroforestry scenarios to private landowners may have resulted from the reduced management costs, compared to monoculture systems, and the resulting increased system productivity through complementary biophysical characteristics. Some owners also made and sold cheese, yogurt, and jams at their farms—adding value and a handy outlet for their raw products. The "typical" model systems identified by the co-authors and based on past surveys [28–31] were moderately profitable. The actual operations visited demonstrated that with increased system complexity, and some home stores with added value for the products, returns could increase as well, at least by the best landowners. With a limited sample of farms, more replications are required to better determine impacts from returns to scale and added value products.

Planted forests appeared to have the most opportunity to be cost-effective for landowners with limited capital, generating the highest ratio of returns in proportion to total expenditures, as seen through the system's BCR. Low-input land-use methods may be appealing to those interested in systems with moderate profitability and limited costs. However, tree plantations earned relatively low returns at all discount rates, compared to cattle or mixed-use systems, required longer payback periods to cover establishment costs, and had less certain and established markets. PES incentives such as cost-share payments encourage landowners to invest in land-use systems by reducing initial costs and increasing income sooner. The calculated cost-share payment necessary to breakeven with conventional cattle systems was less than the U.S. national average rental payment, USD 133.86 per ha, offered in the Conservation Reserve Program (CRP) [37]. With PES, the payment amount needs to be at least the difference in returns between incorporating forest cover through silvopasture and the conventional more profitable use to provide an equal choice to landowners.

The returns for the 10-ha dairy–fruit SPS were the greatest in our small sample in Amazonas. However, we did not include on-farm home labor of the primary landowner as a cost, whether active onsite or absentee. The profits essentially represented the landowner income, so their returns per hectare were high, but the total area was relatively small. Nevertheless, only the 30-ha and the 65-ha farms earned a greater annual income than the average annual income of farmers in the region. Generally, systems utilized family labor when possible. As system size increases, the need for additional labor increases. The 25-ha and 65-ha systems, including a store and restaurant, respectively, did not earn as high of returns as the 10-ha and 30-ha systems at the lowest discount rate. The 10-ha and 65-ha systems used mostly family labor, while the other two systems employed more outside laborers. Nonetheless, salaries were deducted from the returns to represent the opportunity cost of all necessary laborers, both familial and hired, outside of the primary landowner in the analysis.

The varying results suggested that different mixes of production systems along with management intensities play a role in determining landowner profitability. Both the 10-ha and 65-ha regimes were associated with relatively low-intensity management, in terms of inputs, including but not limited to their self-sufficiency in labor. However, the 10-ha farm, both smaller in size and less profitable per dollar invested, generated higher net returns in present value terms than the 65-ha system at all discount rates. This smaller farm along with the larger 25-ha farm sold fruit and home-made cheese, which increased returns. This corroborated that larger investments with more value—added can increase discounted present values, but they required more technical and managerial time and skill. The relationship between the resource variables and the capital budgeting results requires more analysis to determine their marginal impacts on returns per unit of land.

Small landowners often have limited capital and need frequent streams of income, which limits long-term investments. In addition, all systems other than the two largest actual farms earned less than the average annual income for farm laborers. Likewise, it took more time to cover establishment costs in forest systems when compared to cattle and agroforestry systems. Tremblay et al. [38] found fruit-bearing agroforests generated positive profits after 7 years, reinforcing this study's estimates. The difference in returns as well as the turn-around time for landowners to generate positive profits represent challenges that governmental policies can alleviate in order to encourage investment in expanding forest cover in agricultural systems [39].

For example, PES such as offsets for forest carbon storage have the potential to increase and diversify returns as well as make less profitable land uses more attractive. The average price of offsets for carbon stored in forests in 2016 was USD 5.10 per ton CO_2e [40]. This means that with favorable carbon markets, silvopasture practitioners in Amazonas may be able to earn USD 177–USD 240/ha per 25-year rotation. Governmental intervention may be necessary through PES programs such as payments for carbon sequestration to make less profitable land-use systems competitive with conventional cattle-forage systems [39]. This would be consistent with the ambitions announced by the Peruvian authorities, in their plans to reduce net carbon emissions in the country. PES may also supplement annual income to landowners to make their profits comparable to income earned by workers in the farming industry, thus, ensuring an adequate income to landowners of rural lands.

Moreover, since the cost of land purchase was not considered in the analysis, we were able to compare LEV values with the average sale price of land to determine if the expected returns from the systems correlated with market prices for agricultural land. The average price of productive land for sale in Northern Peru, based on three lots on the market at the time of the study, was quite expensive, at USD 9618 per hectare [41]. At a 4% discount rate, our typical systems based on prior farm surveys and expert opinion earned less, but the four actual systems met or exceeded that level. None of the LEVs were greater than USD 9618 at the 12% discount rate.

Accordingly, using discount rates of 4%, 8%, and 12% provided a range of potential returns from land-use systems with various opportunity costs. Without the established literature and known landowner discount rates, using multiple discount rates allowed us to estimate returns with multiple preferences in mind. For instance, lower discount rates give more weight to long-term returns and less weight to short-term costs and benefits. Meanwhile, higher discount rates represent higher opportunity costs of alternative investments and/or involve riskier investments. For the scope of the study, we were able to see which land-use practices were more sensitive to high discount rates. Forests, which returned periodic, long-term revenues and required high establishment costs, experienced negative returns at 8% and 12% discount rates. Conventional cattle–forage systems in Amazonas appeared more resilient to high discount rates with relatively high returns at each discount rate. The lower-input case studies, the 10-ha and 65-ha regimes, maintained high returns at all discount rates, unlike the input—intensive 25-ha and 30-ha systems.

To our knowledge, no other projects have analyzed the economic returns of SPS compared to planted trees monocultures and traditional cattle grazing regimes in Amazonas, Peru. Our findings from Amazonas compare favorably with those of other global SPS research. In the Amazon lowlands of Ecuador, SPS featuring traditional and improved species of forage earned negative NPVs at 8%, −USD 356.17 per hectare and −USD 206.58 per hectare, respectively, but higher returns than traditional cattle ranching [42]. Other studies also highlighted the potential of added value from fruit–tree-based systems. For example, Tremblay et al. [38] recognized the economic advantages of silvopasture as a sustainable alternative land-use for small-scale agriculture in the Tapajos region of the Brazilian Amazon. At a 10% discount rate, the net present values of the agroforests returned USD 17,800 for the medium-sized orchard and USD 21,844 for the larger orchard. In addition, Hoch et al. [43] estimated positive NPVs at 3% and 12% discount rates for successful intercropping studies in Brazil, Bolivia, Peru, and Ecuador as part of a non-governmental agency's initiative.

Silvopasture systems are posited to provide significant risk reduction benefits because they produce at least two different commodities, which would have different, uncorrelated market price cycles given their different product life-cycle time spans [21]. In financial investment portfolio theory, any two or more uncorrelated investments will lead to larger overall investment returns. There is not much empirical literature on such financial risk benefits of silvopasture systems, and the complexity of finding historical returns and distributions and using them to estimate risk was beyond the scope of this study. We could have varied input costs and output prices for sensitivity or risk analyses, but given the many different factors of production and the lack of empirical cost distributions for any of our inputs, varying some arbitrary combination of these would not be supportable for this research.

Frey et al. [15] surveyed small-scale farmers in northeast Argentina, and found they believed that diversified cash flows, and thus, less risk of single commodity price fluctuations, were significant advantages of silvopasture. The mixed-use systems analyzed by Dube et al. [19,20] in Chile and Brazil showed positive potential for benefits by risk reduction from multiple products and carbon storage benefits. Similar approaches could be facilitated in Peru with financial assistance. Comparable studies acknowledged establishment costs to adopt silvopasture as well as increased labor as barriers for landowners with limited income [38,43,44]. A trade-off exists between the benefits of diverse mutually compatible species and its increased demand of specialized labor [13]. However, they also recognized the potential of timber sales to cover upfront and additional management costs [44], in addition to increased social and conservation benefits that are not always financially captured from silvopastures [38,45].

6. Conclusions

We estimated the financial returns for silvopasture systems in Amazonas, Peru based on previous farm surveys and representative models and four selected case studies of successful SPS farms based on local contacts with the co-authors. Pure cattle-pasture systems had larger returns in the synthetic calculations, but for the four integrated case studies that had pasture, forests, and some processing or retail parts of the value chain, silvopasture systems all had the greatest returns. However, the markets for wood are local and informal, so prices and quantity purchased depend on farm labor and individual negotiations, not established timber markets.

We made the analyses using the best available market prices or the estimated market equivalent of factor costs and output prices. We did not estimate social welfare or shadow prices for factors of production or output. Per strict economics jargon, we performed a financial analysis using discounted cash flows and capital budgeting criteria. We have referred to the research as an economic analysis as well, using the terms interchangeably per common usage [32]. Purists suggest that comprehensive economic analyses should include all the social (e.g., shadow prices, social welfare, non-market values) estimates of goods and services, but we used the most common convention and term of an economic or financial analysis for market prices, as is applied almost universally.

The sampled landowners captured excellent economic returns through SPS systems and diverse farm product sales, but they were exceptional opinion and business leaders. Achieving similar success for silvopasture with smaller typical landowners will require more research into the best practices for limited income farmers, extension advice, and implementation of good farm management practices as well. The opportunity cost of alternative investments will determine whether diversifying land-use through agroforestry will increase landowner returns.

Government support through direct payments could increase the profitability of silvopastures in order to gain ecosystem service benefits such as carbon storage or improved water quality. Cost-share payments may make land uses that expand forest cover more desirable to landowners by reducing establishment costs. Cost-share payments also have the potential of shortening payback periods so limited income landowners can reach a net positive cash flow sooner. When cost-share payments are not enough to make silvopasture more profitable, other PES, such as carbon storage, may be employed to provide a stream of annual benefits over some time period, e.g., the first ten years when revenue is at its lowest. Payments to landowners for carbon sequestration represent a promising solution to the low profits of silvopasture regimes at high discount rates. However, with low carbon prices, the absence of mature markets, and competitive resource relationships, carbon payments may not be sufficient to make silvopasture competitive with conventional cattle-pasture systems. In these cases, multiple forms of aid may be necessary to encourage landowners to adopt land-uses such as agroforestry. Overall, mixed-use systems in Amazonas represent essential models of land-use regimes which can be altered and/or replicated to produce supportable livelihoods for landowners.

Author Contributions: Conceptualization, M.C. and F.C.; methodology, F.C. and R.R.; formal analysis, S.C.; interviews and investigation, S.C., F.C., D.P., H.V., W.B.; resources, F.C.; data curation, S.C., F.C., D.P., W.B. and H.V.; writing—original draft preparation, S.C.; writing—review and editing, S.C., F.C., M.C., D.P., W.B., R.R., H.V., and R.P.; supervision, F.C., R.A., R.P. and E.S.; funding acquisition, S.C., R.A. and F.C. All authors have read and agreed to the published version of the manuscript.

Funding: The Southern Forest Resource Assessment Consortium (SOFAC) and the Bruce and Barbara Zobel Endowment for International Forestry Studies provided the funds for this study.

Acknowledgments: We would like to thank the landowners that graciously allowed us to visit and use their farms in this analysis and the Peruvian collaborators that provided their guidance and expertise. We appreciate the suggestions and comments from both our own and *Land*'s reviewers and editors.

Conflicts of Interest: The authors declare no conflict of interest. The funders did not have any influence on the methods or results of the study.

References

1. Aynekulu, E.; Suber, M.; van Noordwijk, M.; Arango, J.; Roshetko, J.M.; Rosenstock, T.S. Carbon Storage Potential of Silvopastoral Systems of Colombia. *Land* **2020**, *9*, 309. [CrossRef]
2. Zomer, R.J.; Neufeldt, H.; Xu, J.; Ahrends, A.; Bossio, D.; Trabucco, A.; Wang, M. Global Tree Cover and Biomass Carbon on Agricultural Land: The Contribution of Agroforestry to Global and National Carbon Budgets. Scientific Reports, 6. 2016. Available online: http://search.proquest.com.prox.lib.ncsu.edu/docview/1817890421?pq-origsite=summon (accessed on 9 January 2018).
3. Orefice, J.N.; Carroll, J. Silvopasture—It's not a Load of Manure: Differentiating Between Silvopasture and Wooded Livestock Paddocks in the Northeastern United States. *J. For.* **2017**, *115*, 71. Available online: http://search.proquest.com.prox.lib.ncsu.edu/docview/1863558982?pq-origsite=summon&accountid=12725 (accessed on 7 January 2018). [CrossRef]
4. USDA National Agroforestry Center. Silvopasture: An Agroforestry Practice [PowerPoint Slides]. 2012. Available online: https://nac.unl.edu/practices/silvopasture.htm (accessed on 1 April 2017).
5. Cotta, J.N. Revisiting Bora fallow agroforestry in the Peruvian Amazon: Enriching ethnobotanical appraisals of non-timber products through household income quantification. *Agroforest. Syst.* **2017**, *91*, 17–36. Available online: https://link-springer-com.prox.lib.ncsu.edu/article/10.1007%2Fs10457-016-9892-4 (accessed on 30 March 2017). [CrossRef]
6. Lacerda, A.E.B.; Hanisch, A.L.; Nimmo, E.R. Leveraging Traditional Agroforestry Practices to Support Sustainable and Agrobiodiverse Landscapes in Southern Brazil. *Land* **2020**, *9*, 176. [CrossRef]
7. Chizmar, S.; Sills, E.; Cubbage, F.; Castillo, M.; Abt, R. A Comparative Economic Assessment of Silvopasture Systems in the Amazonas Region of Peru and in North Carolina. Master's Thesis, North Carolina State University, Raleigh, NC, USA, 2018. Available online: http://www.lib.ncsu.edu/resolver/1840.20/35059 (accessed on 5 May 2018).
8. Van Noordwijk, M.; Gitz, V.; Minang, P.A.; Dewi, S.; Leimona, B.; Duguma, L.; Pingault, N.; Meybeck, A. People-Centric Nature-Based Land Restoration through Agroforestry: A Typology. *Land* **2020**, *9*, 251. [CrossRef]
9. Ministry of Environment [MINAM]. Intended Nationally Determined Contribution from the Republic of Peru. 2015. Available online: http://www4.unfccc.int/ndcregistry/PublishedDocuments/Peru%20First/iNDC%20Per%C3%BA%20english.pdf (accessed on 18 May 2018).
10. Bonn Challenge. n.d. Commitments: Peru. Available online: http://www.bonnchallenge.org/content/peru (accessed on 1 May 2017).
11. Sistema Nacional de Evaluation Acreditacion y Certificacion de la Calidad Educativa [SINEACE]. Caracterización de la Región Amazonas. 2017. Available online: https://www.sineace.gob.pe/wp-content/uploads/2017/08/PERFIL-AMAZONAS.pdf (accessed on 18 May 2018).
12. City Population. Amazonas. 2 October 2018. Available online: https://www.citypopulation.de/php/peru-admin.php?adm1id=01 (accessed on 1 February 2018).
13. Loker, W.M. Where's the beef? Incorporating Cattle into Sustainable Agroforestry Systems in the Amazon Basin. *Agroforest. Syst.* **1994**, *25*, 227–241. Available online: https://link.springer.com/article/10.1007/BF00707462 (accessed on 21 May 2018). [CrossRef]
14. Bussoni, A.; Juan, C.; Fernandez, E.; Boscana, M.; Cubbage, F.; Bentancur, O. Integrated beef and wood production in Uruguay: Potential and limitations. *Agroforest. Syst.* **2015**, *89*, 1107–1118. Available online: https://link-springer-com.prox.lib.ncsu.edu/article/10.1007%2Fs10457-015-9839-1 (accessed on 4 May 2017). [CrossRef]

15. Frey, G.E.; Fassola, H.E.; Pachas, A.N.A.; Colcombet, L.; Lacorte, S.M.; Pérez, O.; Renkow, M.; Warren, S.T.; Cubbage, F.W. Perceptions of silvopasture systems among adopters in northeast Argentina. *Agric. Syst.* **2012**, *105*, 21–32. Available online: https://www-sciencedirect-com.prox.lib.ncsu.edu/science/article/pii/S0308521S\times$11001272 (accessed on 1 February 2018). [CrossRef]

16. Walthall, C.L.; Hatfield, J.; Backlund, P.; Lengnick, L.; Marshall, E.; Walsh, M.; Adkins, S.; Aillery, M.; Ainsworth, E.A.; Ammann, C.; et al. Climate Change and Agriculture in the United States: Effects and Adaptation. In *USDA Technical Bulletin 1935*; USDA: Washington, DC, USA, 2012; pp. 1–193. Available online: https://www.usda.gov/oce/climate_change/effects_2012/CC%20and%20Agriculture%20Report%20(02-04-2013)b.pdf (accessed on 1 May 2017).

17. Nair, P.K.R. Methodological challenges in estimating carbon sequestration potential of agroforestry systems. In *Carbon Sequestration Potential of Agroforestry Systems: Opportunities and Challenges*; Kumar, B.M., Nair, P.K.R., Eds.; Springer: Dordrecht, The Netherlands, 2011; pp. 3–16. Available online: https://link-springer-com.prox.lib.ncsu.edu/chapter/10.1007%2F978-94-007-1630-8_1 (accessed on 8 April 2017).

18. Oliva, M.; Mirano, L.C.; Leiva, S.; Collazos, R.; Salas, R.; Vásquez, H.V.; Quintana, J.L.M. Reserva de carbono en un sistema silvopastoril compuesto de Pinus patula y herbáceas nativas. *Sci. Agropecu.* **2017**, *8*, 149–157. [CrossRef]

19. Dube, F.; Thevathasan, N.V.; Zagal, E.; Gordon, A.M.; Stolpe, N.B.; Espinosa, M. Carbon sequestration potential of silvopastoral and other land use systems in the Chilean Patagonia. In *Carbon Sequestration Potential of Agroforestry Systems: Opportunities and Challenges*; Kumar, B.M., Nair, P.K.R., Eds.; Springer: Dordrecht, The Netherlands, 2011; pp. 101–127. Available online: https://link-springer-com.prox.lib.ncsu.edu/chapter/10.1007%2F978-94-007-1630-8_6 (accessed on 15 June 2017).

20. Dube, F.; Couto, L.; Silva, M.L.; Leite, H.G.; Garcia, R.; Araujo, G.A. A simulation model for evaluating technical and economic aspects of an industrial eucalyptus-based agroforestry system in Minas Gerais, Brazil. *Agroforest. Syst.* **2002**, *55*, 73–80. Available online: https://link-springer-com.prox.lib.ncsu.edu/article/10.1023%2FA%3A1020240107370 (accessed on 15 June 2017). [CrossRef]

21. Cubbage, F.W.; Balmelli, G.; Bussoni, A.; Noellemeyer, E.; Pachas, A.N.; Fassola, H.; Colcombet, L.; Rossner, B.; Frey, G.; Dube, F.; et al. Comparing silvopastoral systems and prospects in eight regions of the world. *Agroforest. Syst.* **2012**, *86*, 303–314. [CrossRef]

22. Montagnini, F.; Nair, P.K.R. Carbon sequestration: An underexploited environmental benefit of agroforestry systems. *Agroforest. Syst.* **2004**, *61*, 281–295. Available online: https://link-springer-com.prox.lib.ncsu.edu/content/pdf/10.1023%2FB%3AAGFO.0000029005.92691.79.pdf (accessed on 1 February 2018).

23. Orefice, J.N.; Smith, R.G.; Carroll, J.; Asbjornsen, H.; Howard, T. Forage productivity and profitability in newly-established open pasture, silvopasture, and thinned forest production systems. *Agroforest. Syst.* **2019**, *93*, 51–65. [CrossRef]

24. Bruck, S.R.; Bishaw, B.; Cushing, T.L.; Cubbage, F.W. Modeling the financial potential of silvopasture agroforestry in eastern North Carolina and northeastern Oregon. *J. For.* **2019**, *177*, 13–20. [CrossRef]

25. Pent, G.J.; Fike, J.H. Winter Stockpiled Forages, Honeylocust Pods, and Lamb Performance in Hardwood Silvopastures & Sheep Performance and Behavior in Silvopasture Systems. In Proceedings of the Agroforestry for a Vibrant Future: Connecting People, Creating Livelihoods, Sustaining Places 15th NAAC on the Campus of Virginia Tech in Blacksburg, Blackburg, VA, USA, 27–29 June 2017.

26. Mercer, D.E.; Frey, G.E.; Cubbage, F.W. Economics of agroforestry. In *Handbook of Forest Economics*; Kant, S., Alavalapati, J.R.R., Eds.; Routledge: New York, NY, USA, 2014; pp. 188–209. Available online: https://www.srs.fs.usda.gov/pubs/ja/2014/ja_2014_mercer_001.pdf (accessed on 9 October 2017).

27. Holdridge, L.R. *Life Zone Ecology*; Tropical Science Center: San Jose, CA, USA, 1967.

28. Gómez, C.A. Sistemas Silvopastoriles: Investigación e Innovación en el caso de Amazonas y San Martín. In *I Curso Internacional Ganadería y Agroecosistemas Tropicales Sostenibles Para Enfrentar el Cambio Climático: Sistemas Silvopastoriles*; Print: Pucallpa, Peru, 2017.

29. Pizarro, D.; Vasquez, H.; Bernal, W.; Fuentes, E.; Alegre, J.; Castillo, M.S.; Gomez, C. Assessment of silvopasture systems in the northern Peruvian Amazon. *Agroforest. Syst.* **2020**, *94*, 173–183. [CrossRef]

30. Alegre, J.C.; Sánchez, Y.; Pizarro, D.M.; Gómez, C.A. *Manejo de los Suelos con Sistemas Silvopastoriles en las Regiones de Amazonas y San Martín*; Print: Lima, Peru, 2019.

31. Echevarría, M.G.; Pizarro, D.M.; Gómez, C.A. *Alimentación de Ganadería en Sistemas Silvopastoriles de la Amazonia Peruana*; Print: Lima, Peru, 2019.

32. Cubbage, F.W.; Davis, R.R.; Frey, G.E.; Behr, D.C.; Sills, E.O. Financial and economic evaluation guidelines for international forestry projects. In *Tropical Forestry Handbook*; Pancel, L., Kohl, M., Eds.; Springer Publishing: Berlin, Germany, 2016. [CrossRef]

33. Sharrow, H. Natural Resource Economics: Considering the Time Element of Investments. 2008. Available online: http://www.doctorrange.com/PDF/TimeinNRInvest.pdf (accessed on 1 April 2017).

34. Bullard, S.H.; Straka, T.J. *Basic Concepts in Forest Valuation and Investment Analysis*, 3rd ed.; Faculty Publications: Nacogdoches, TX, USA, 2011; p. 460. Available online: http://scholarworks.sfasu.edu/forestry/460 (accessed on 19 September 2020).

35. Porro, R.; Miller, R.P.; Tito, M.R.; Donovan, J.A.; Vivan, J.L.; Trancoso, R.; Van Kanten, R.F.; Grijalva, J.E.; Ramirez, B.L.; Gonçalves, A.L. Agroforestry in the Amazon Region: A pathway for balancing conservation and development. In *Agroforestry—The Future of Global Land Use*; Nair, P.K.R., Garrity, D., Eds.; Springer: Dordrecht, The Netherlands, 2012; pp. 391–427. Available online: https://link-springer-com.prox.lib.ncsu.edu/book/10.1007%2F978-94-007-4676-3 (accessed on 1 April 2017).

36. Alegre, J.C. Sistema agroforestal multiestrato. Recuperación de suelos degradados en la amazonía. *Rev. Leisa* **2015**, *31*, 28–30. Available online: http://www.leisa-al.org/web/index.php/volumen-31-numero-1/1091-sistema-agroforestal-multiestrato-recuperacion-de-suelos-degradados-en-la-amazonia (accessed on 1 April 2017).

37. USDA Farm Service Agency. The Conservation Reserve Program: 49th Signup Results. USDA. 2016. Available online: https://www.fsa.usda.gov/Assets/USDA-FSA-Public/usdafiles/Conservation/PDF/SU49Book_State_final1.pdf (accessed on 1 April 2017).

38. Tremblay, S.; Lucotte, M.; Reveret, J.; Davidson, R.; Mertens, F.; Sousas Passos, C.J.; Romaña, C.A. Agroforestry systems as a profitable alternative to slash and burn practices in small-scale agriculture of the Brazilian Amazon. *Agroforest. Syst.* **2015**, *89*, 193–204. [CrossRef]

39. Vosti, S.A.; Braz, E.M.; Carpentier, C.L.; d'Oliveira, M.V.; Witcover, J. Rights to forest products, deforestation and smallholder income: Evidence from the western Brazilian Amazon. *World Dev.* **2003**, *31*, 1889–1901. [CrossRef]

40. Hamrick, K.; Gallant, M. Unlocking Potential State of the Voluntary Carbon Markets 2017. Forest Trends' Ecosystem Marketplace. 2017. Available online: https://www.cbd.int/financial/2017docs/carbonmarket2017.pdf (accessed on 4 January 2018).

41. Realigro Real Estate. Peru. 2015. Available online: https://peru.realigro.com/for-sale/farmland/#annunci (accessed on 9 October 2017).

42. Ramírez, A.; Seré, C.; Uquillas, J. An economic analysis of improved agroforestry practices in the Amazon lowlands of Ecuador. *Agroforest. Syst.* **1992**, *17*, 65–86. Available online: https://link.springer.com/article/10.1007/BF00122928 (accessed on 1 April 2017).

43. Hoch, L.; Pokorny, B.; de Jong, W. Financial attractiveness of smallholder tree plantations in the Amazon: Bridging external expectations and local realities. *Agroforest. Syst.* **2012**, *84*, 361–375. [CrossRef]

44. Adams, M.L. Pasture extensification in the Southern Ecuadorian Andes: Appraisal and recommendations. *J. Sustain. For.* **2009**, *28*. [CrossRef]

45. McDermott, M.E.; Rodewald, A.D. Conservation value of silvopastures to Neotropical migrants in Andean forest flocks. *Biol. Conserv.* **2014**, *175*. [CrossRef]

 *land*

Article

Carbon Storage Potential of Silvopastoral Systems of Colombia

Ermias Aynekulu [1],*, Marta Suber [1], Meine van Noordwijk [1], Jacobo Arango [2], James M. Roshetko [1] and Todd S. Rosenstock [1]

[1] World Agroforestry (ICRAF), United Nations Avenue, P.O. Box 30677, Nairobi 00100, Kenya; m.suber@cgiar.org (M.S.); m.vannoordwijk@cgiar.org (M.v.N.); j.roshetko@cgiar.org (J.M.R.); t.rosenstock@cgiar.org (T.S.R.)

[2] International Centre for Tropical Agriculture (CIAT), Km 17, Recta Cali–Palmira CP, Apartado Aéreo 6713, Cali 763537, Colombia; j.arango@cgiar.org

* Correspondence: e.betemariam@cgiar.org; Tel.: +254-20-722-4356; Fax: +254-20-722-4001

Received: 8 July 2020; Accepted: 11 August 2020; Published: 2 September 2020

Abstract: Nine Latin American countries plan to use silvopastoral practices—incorporating trees into grazing lands—to mitigate climate change. However, the cumulative potential of scaling up silvopastoral systems at national levels is not well quantified. Here, we combined previously published tree cover data based on 250 m resolution MODIS satellite remote sensing imagery for 2000–2017 with ecofloristic zone carbon stock estimates to calculate historical and potential future tree biomass carbon storage in Colombian grasslands. Between 2000 and 2017, tree cover across all Colombian grasslands increased from 15% to 18%, with total biomass carbon (TBC) stocks increasing from 0.41 to 0.48 Pg. The range in 2017 carbon stock values in grasslands based on ecofloristic zones (5 to 122 Mg ha^{-1}) suggests a potential for further increase. Increasing all carbon stocks to the current median and 75th percentile levels for the respective eco-floristic zone would increase TBC stocks by about 0.06 and 0.15 Pg, respectively. Incorporated into national C accounting, such Tier 2 estimates can set realistic targets for silvopastoral systems in nationally determined contributions (NDCs) and nationally appropriate mitigation actions (NAMAs) implementation plans in Colombia and other Latin American countries with similar contexts.

Keywords: agroforestry; carbon sequestration; climate change mitigation; grazing management; land restoration; nationally determined contribution; silvopastoral; tree cover

1. Introduction

The ability to limit global warming below 2 °C, the target set by the Paris Climate Agreement, is predicated on significant greenhouse gas (GHG) emission reductions in the agriculture and land use sector (AFOLU) [1,2]. While land use accounts for nearly 25% of net annual anthropogenic GHG emissions [3], it also offers many options to mitigate climate change. Estimates suggest that land use interventions could generate up to 30% of the emission reductions and carbon sequestration needed to meet the Paris Agreement's ambition [4,5]. Most agriculture and land use mitigation interventions target the reduction of emissions from livestock systems, rice systems, deforestation, and nitrogen fertilizer. However, the integration of trees in crop and pasture lands, also known as agroforestry, is another potentially significant intervention option [6].

Nine Latin American countries have identified silvopastoral systems (SPS), agroforestry systems where trees are managed or planted in pasturelands, as a priority action to mitigate climate change within their nationally determined contributions (NDCs) to the United Nations Framework Convention on Climate Change (UNFCCC) [7,8] and the Nationally Appropriate Mitigation Actions (NAMAs), two documents that prescribe national climate response priorities. The focus on livestock in Latin American countries is due to the economic importance of the sector and its leverage on national emissions. Indeed, the livestock sector accounts for 58–70% of the overall agricultural emissions in Latin America [9], making greening milk and meat production a critical aspect of meeting national goals.

Silvopastoral systems sequester carbon in biomass increasing the amount of carbon in the landscape. Silvopastoral systems accumulate more than 2 Mg C per ha per year above and below ground biomass and additional 0.5 Mg C per ha per year in the soil [10]. Grazing grasslands occupy more than 550 million hectares of land in Latin America. Therefore, scaling up silvopastoral systems would seemingly have the potential to increase carbon in the landscape despite the modest change per unit of land. Estimates of the climate change mitigation benefits of scaling up silvopastoral systems at the national level, however, are largely unavailable. Zomer et al. [11,12] provided a first estimate of t C stocks in tree biomass in crop lands at global scale but did not specifically consider trees in grasslands (i.e., silvopastoral systems). Griscom et al. [5], on the other hand, estimate silvopastoral C but mix these estimates together with other land management options to estimate a total mitigation potential of land restoration. However, countries need estimates of silvopastoral potential to set targets for climate action [13].

This study estimated the climate change mitigation benefits of expanding silvopastoral systems in Colombia to inform realistic inclusion of silvopastoral systems in NDC and NAMA planning options for Colombia and other Latin American countries with similar contexts. Expanding on the earlier analysis for agricultural lands [11,12], we explored historical C stocks for Columbian grassing lands in 2000 and 2017, before considering options to increase C stock density with the silvopastoral systems in the local context. Specific questions were: (i) What changes in tree cover in Columbian pastureland are evident from satellite imagery? (ii) What changes in C stock do such changes in tree cover imply, using ecofloristic-zone-specific C density estimates? (iii) Evaluating the existing statistical distribution, which further changes may be feasible?

Colombia was selected because of its active promotion of silvopastoral systems. For example, the Colombian government and National Livestock Federation (FEDEGAN) have a strategy to reduce GHG emissions from the agricultural sector by 13.46 Mt CO_2e yr^{-1} by 2030. The conversion of current grasslands to SPS is among the priority mitigation activities [14]. Importantly, Tapasco et al. [15] also identified SPS as the most promising policy option for achieving this goal, and Lerner et al. [16] notes that the country is in an ideal position to create integrated plans for sustainable cattle intensification, including SPS, conservation, and restoration initiatives.

2. Methods

2.1. Conceptual Approach

For this analysis, a methodology earlier applied to agricultural lands [8] was adapted to spatially quantify carbon sequestration from tree cover in grasslands and estimate potential carbon stocks under widespread implementation of SPS. The steps were to define the extent of grasslands in the target area; derive the current level of tree cover; estimate the carbon stocks in above and belowground biomass; estimate the potential increase with shifts in tree cover from the current to the 25th percentile, median, and 75th percentile of 2017 biomass carbon stocks; and quantify the difference between current and potential tree cover and carbon sequestration ("silvopastoral carbon gap"). A detailed description of the methodology used for this analysis is found in Zomer et al. [11,12].

2.2. Study Area

Given that both the Colombian government and independent researchers have identified SPS as a priority in achieving Colombia's agricultural emission reduction goals [15], we focused on Colombia as a case study. As of 2015, Colombia had more than 12 million ha of grassland (13% of the national land area) scattered throughout tropical moist deciduous forests (77% of grassland), tropical rainforest

(12%), and other ecofloristic zones (Figure 1). The largest grazing areas are found in the sparsely populated Llanos Orientales and Orinoquia regions of the Orinoco River basin. In other regions of the country, smaller grassland areas are interspersed with various land uses, including croplands and settlements.

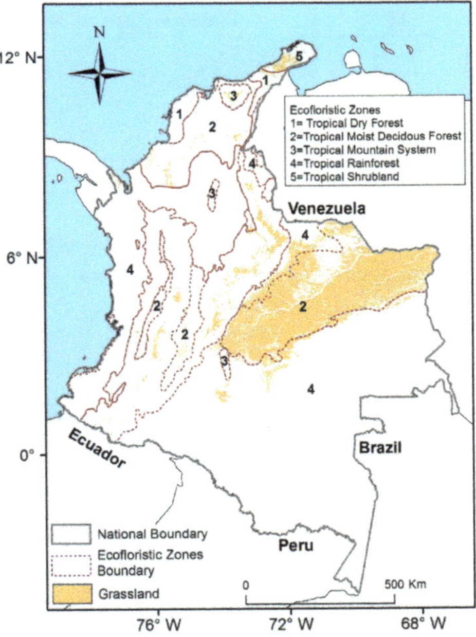

Figure 1. Grassland cover in 2015 (source: European Space Agency, 2017a) and ecofloristic zones of Colombia.

2.3. Data Collection and Analysis

Grassland area was derived at 300 m spatial resolution from the European Space Agency's (ESA) annual global land cover data. Per ESA recommendations, the ESA land use classes grass land, mosaic herbaceous cover > 50% and tree and shrub < 50% were combined to produce the grassland designation used in this analysis [17]. Percent tree cover was derived from previously published tree cover data based on the 250 m resolution Moderate-Resolution Imaging Spectroradiometer (MODIS) 44B Version 6 Vegetation Continuous Field remote sensing product [18]. This product is a continuous, quantitative representation of global land surface cover at 250 m spatial resolution. Ground cover gradations are based on percent tree cover, percent non-tree cover, and percent non-vegetated (bare) [18]. A detailed description of the methodology used for this analysis is found in Zomer et al. [11,12].

We used the default Intergovernmental Panel on Climate Change (IPCC) Tier 1 total carbon stock values for each ecofloristic zone as the minimum potential carbon stock values for 0% tree cover [19] (Table 1). We then estimated the total carbon sequestration potential of SPS by ecofloristic zone using the methodology recommended by IPCC Tier 1 and detailed in Zomer et al. [12]. The biomass carbon value of the equivalent (or most similar) Global Land Cover (GLC) 2000 Mixed Forest class [19] was used as a surrogate aboveground biomass carbon value for each ecofloristic zone to simulate full tree cover (100%). Belowground carbon sequestration potential was calculated using aboveground sequestration potential values and the root/shoot ratio for each of the ecofloristic zones as per IPCC Tier 1 values (Table 1) [19]. Below and aboveground C sequestration potential values were then summed to determine total potential carbon stock values for 100% tree cover. A carbon fraction value of 0.47 was uniformly applied to determine the carbon content of dry woody biomass. We then assumed a linear increase in biomass carbon from 0% to 100% tree cover for each ecofloristic zone. We used the Tier 1 value when tree cover was 0% and the maximum value for Mixed Forest when tree cover was 100%.

Table 1. IPCC Tier 1 values of above ground carbon stocks in grassland under different ecofloristic zones.

Ecofloristic Zone	Area		Above Ground Carbon Stocks Mg ha^{-1}		Root to Shoot Ratio
	ha	%	Minimum	Maximum	
Tropical dry forest	67,758	0.6	4	126	0.28
Tropical Moist deciduous forest	9,487,070	77	8	128	0.24
Tropical mountain system	1,076,967	8.8	6	87	0.27
Tropical rainforest	1,469,597	12	8	193	0.37
Tropical shrubland	206,495	1.7	4	126	0.28
Total	12,307,887	100			

We used the 2015 grassland area as a basis for 2000, 2008, and 2017 tree cover and biomass assessments to control for any change in grassland area over time. We used the 2017 25th, median, and 75th percentile carbon stock values in each ecofloristic zone to calculate carbon gaps and the current carbon sequestration potential. We identified areas with carbon stock values below the 25th percentile in each ecofloristic zone as those with the greatest potential to increase carbon stock. Similarly, we considered areas with carbon stock values between the 25th percentiles and medium values as having medium potential and areas between the median and 75th percentile values as having low potential to increase carbon stocks. Areas with greater than the 75th percentile tree cover values were considered as not suitable for increasing carbon stocks. This approach is particularly useful for spatially targeting SPS-based mitigation actions [20].

3. Results

3.1. Estimates of Tree Cover on Grassland in 2000, 2008, and 2017

Tree cover increased between 2000 and 2017. For presentation purposes, we classified these results into tree cover classes (Table 2). Land area with tree cover greater than 10% increased from 55% in 2000 to 74% in 2008 and 73% in 2017. The percentage of grasslands with tree cover > 30% increased with time, from 10% in 2000 to 12% in 2008 and 13% in 2017. Total tree cover across all grasslands increased by about 20% from 2000 to 2017.

Table 2. Grassland (ha) tree cover (%) in 2000, 2008, and 2017.

Tree Cover (%)	2000		2008		2017	
	ha	(%)	ha	%	ha	%
≤10	5,507,090	45	3,210,585	26	3,319,775	27
11–20	3,883,571	32	4,861,344	39	4,700,027	38
21–30	1,715,737	14	2,708,461	22	2,709,611	22
>30	1,201,488	10	1,527,496	12	1,578,474	13
Total	12,307,887	100	12,307,887	100	12,307,887	100

Tree cover increased on grassing lands between 2000 and 2017, but tree cover dynamics and the relative increase or decrease depended on the location. Colombian grassing lands show between 0% and 84% tree cover (Figure 2). The relative change on any given parcel increased up to 76% over this period while others decreased by 80%.

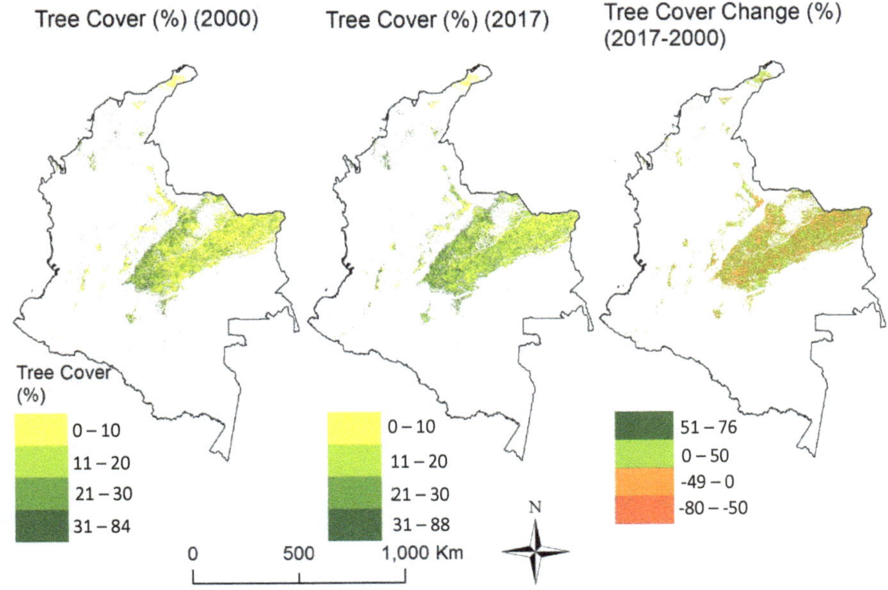

Figure 2. Tree cover change from 2000 to 2017 in grasslands.

3.2. Biomass Carbon Stocks

The TBC stock increased by 17% from 2000 to 2017 (Table 3). TBC stocks in Colombian grassland were 0.41 and 0.48 petagrams (Pg) C in 2000 and 2017, respectively (Table 3).

Table 3. Average and total biomass carbon stocks on Colombian grassland in 2000 and 2017.

Biomass Carbon	Average (SD) (Mg ha^{-1})			Total (Pg C)		
	2000	2017	Change	2000	2017	Change
Above ground	27 (14)	31 (14)	4	0.33	0.38	0.05
Total biomass	34 (18)	39 (18)	5	0.41	0.48	0.07

Land area with a C stock < 10 Mg ha^{-1} decreased from 2000 to 2017, and land area with C stocks of 26–100 Mg ha^{-1} increased (Table 4).

Table 4. Land area carbon stocks in 2000 and 2017.

Total Carbon Stocks (Mg ha^{-1})	2000 ha	%	2008 ha	%	2017 ha	%
≤10	489,333	4	610,508	5	355,760	3
11–25	4,967,851	40	2,715,698	22	3,063,775	25
26–50	4,937,931	40	6,199,358	50	6,207,735	50
51–75	1,185,878	10	1,968,009	16	1,796,833	15
76–100	514,348	4	575,421	5	643,622	5
>100	212,545	2	238,892	2	240,161	2
Total	12,307,887	100	12,307,887	100	12,307,887	100

The grassland biomass carbon stock maps are indicated in Figure 3.

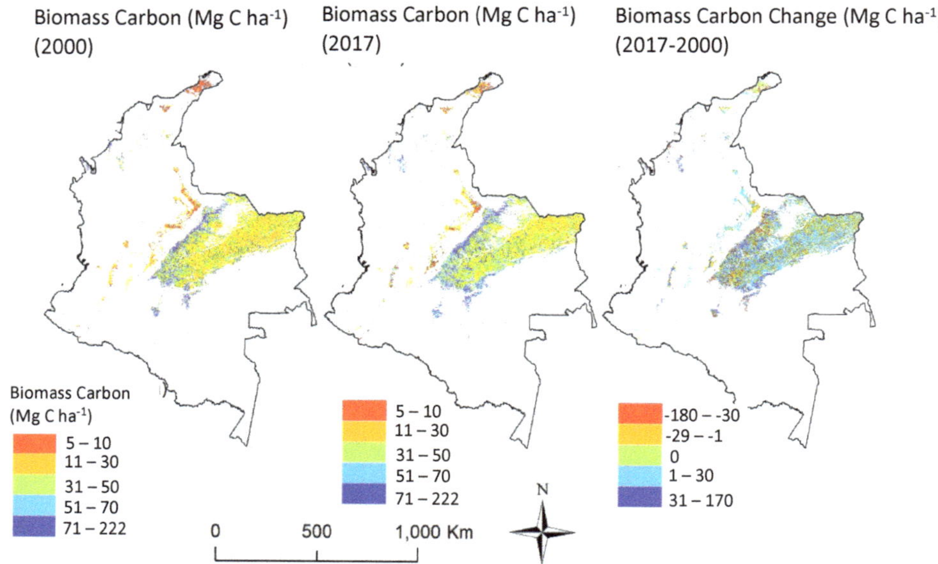

Figure 3. Biomass carbon stocks change (Mg ha^{-1}) from 2000 to 2017.

3.3. Carbon Stock Gaps by Ecofloristic Zones

Average 2017 carbon stock values ranged from 5 and 122 Mg ha^{-1}. The highest average stock per hectare was recorded in the tropical rainforest ecofloristic zone, which accounts for 12% of total grassland. The highest carbon stock density and variation occurred in tropical rainforest, while the lowest carbon stock per hectare and the lowest variation were found in shrubland ecofloristic zones (Figure 4). About 73% of the total grassland had carbon stocks less than the < 75th percentile. Increasing these stocks to the current 75th percentile level would increase total stocks by approximately 0.15 Pg C (0.57 Pg CO$_2$e). By the same token, about 46% of grasslands have carbon stocks lower than the median (Figure 4). Increasing these carbon stocks to the median value would increase total stocks by 0.06 Pg C (0.2 Pg CO$_2$e).

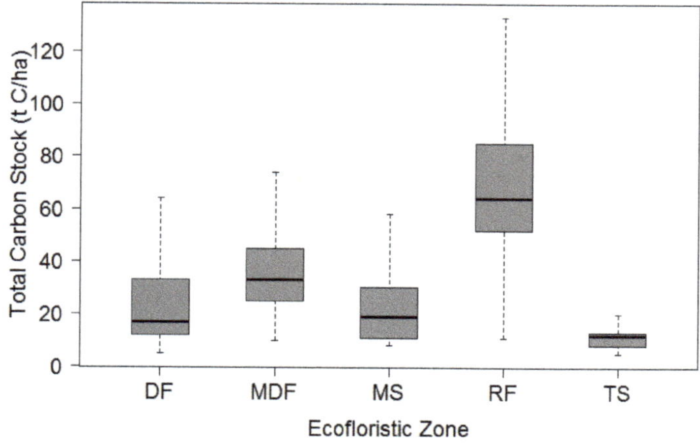

Figure 4. Variation in percentile carbon stocks in tropical dry forest (DF), moist deciduous forest (MDF), mountain system (MS), rain forest (RF), and tropical shrubland (TS).

Figure 5 indicates the spatial distribution of grazing lands with different levels of biomass carbon gaps. This helps identify priority areas for silvopastoral system interventions in Colombia.

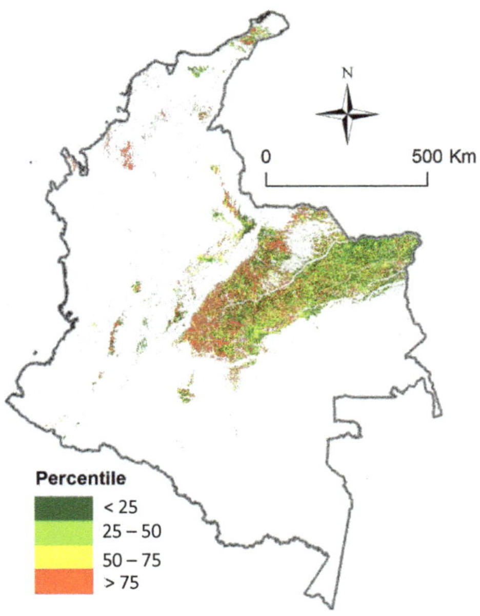

Figure 5. Priority areas for increasing Colombian grassland carbon stocks.

4. Discussion

Tree cover across all grasslands showed an increasing trend between 2000 and 2017. Trends in tree cover in grasslands found here support the same conclusions of more general studies on tree cover change, e.g., Sánchez-Cuervo et al. [21], who found that woody vegetation showed an increasing trend from 2001 to 2010 at the national scale in Colombia. This was mainly attributed to woody regrowth of trees following land abandonment resulting from armed conflicts, economic development, and increase in rainfall [21,22]. However, Sánchez-Cuervo et al. [21] reported that woody cover decreased by about 5% between 2000 and 2010, which was mainly due to intensive management of grasslands. The difference between these earlier results and ours may simply be a result of our study analyzing a longer time period, and the inconsistency in trends may be a function of changing management dynamics.

Zomer et al. [11,12] reported that the contribution of trees on global arable land was over 4 times higher than when estimated with IPCC default values [19]. Chapman et al. [23] also reported that trees in the global crop and pasture lands store about 3.07 and 3.86 Pg carbon in their aboveground biomass, respectively. Similarly, this study found that carbon stocks in the grassland of Colombia has likely also been underestimated. We found mean values of 34 Mg C ha^{-1} in 2000 and 39 Mg C ha^{-1} in 2017, which is more than four times larger than the IPCC Tier 1 global estimate of 8 Mg C ha^{-1} [19]. On the contrary, the global study by Liu et al. [24] reported that the TBC in grasslands and croplands did not show significant change during 1993–2010. Liu et al. [24] used harmonized Vegetation Optical Depth (VOD) data for 1993 onwards derived from a series of passive microwave satellite sensors to estimate above ground canopy in forest and non-forest land use types [25].

Zomer et al. [11,12] reported an average biomass carbon stock of 53 Mg ha^{-1} in Colombian arable land in 2000, demonstrating that trees in productive systems can significantly increase carbon stocks in Colombia. In a meta-analysis, Feliciano et al. [26] found that SPS that use controlled grazing practices and appropriate pasture species can increase average aboveground carbon sequestration by 2.29–6.54 Mg ha^{-1} yr^{-1}. López-Santiago et al. [27] reported that *Leucaena leucocephala* (shrub legume) and *Panicum maximum* grass based SPS contain higher aboveground (19.6 ± 1.6 Mg ha $^{-1}$) and belowground (7.7 ± 0.90 Mg ha^{-1}) biomass compared with deciduous tropical forest and grass monoculture systems in Mexico. This led to a rapid expansion of SPS in Mexico. Our study showed that average Colombian grasslands contained 34 Mg ha^{-1} in the same year. While 36% less than that of arable land, these significant levels of carbon in silvopastoral systems suggest a large opportunity for climate change mitigation, especially when considering the existing and planned areal extent of silvopastoral systems in Colombia. About 13% and 12% of the total area of Colombia was grazing and cultivated lands, respectively, in 2015 [28].

The potential carbon stock of SPS is notable for its ability to accumulate carbon over relatively short time frames. Increasing soil carbon in Colombian grasslands could take decades [29]. In contrast, carbon stocks in tree biomass can be accumulated in less than a decade. Trees have the additional benefit in that they help maintain and increase soil carbon stocks. Given the swift action necessary to limit global warming below the <2° climate change goal (by 2030 according to the IPCC), integrating trees into multi-use landscapes like SPS is an important pathway to meeting mitigation goals in a timely manner. Importantly, trees also regulate soil conditions, including organic matter content, fertility, structure, erosion resistance, and moisture content, during extreme events such as heavy precipitation and drought. Maintaining a hospitable soil environment has a dramatic effect on the presence of N-fixing bacteria, which are crucial in cropping systems with little or no N fertilizer inputs [12]. In turn, all of these mechanisms extend and improve productivity, both in terms of quality and quantity.

Silvopastoral system grasslands go beyond carbon stock potential to offer myriad co-benefits. SPS have direct positive impacts on the livelihoods of producers and environmental quality [30]. Trees create micro-climates that help protect crops and livestock from sun, wind, and extreme temperatures [31,32]. SPS have also been shown to reduce plant and animal production seasonality, reduce ruminal methane production, and increase biodiversity and natural pest control mechanisms as compared to conventional systems [7]. It is important to note that there may be significant trade-offs associated with integrating trees into some production systems. Smallholders rely on these systems as their source of livelihood. As such, increasing tree cover in SPS must be tested locally as one of the suites of potential climate-smart agricultural solutions with the aim of developing a project portfolio that minimizes tradeoffs and maximizes co-benefits.

The management of silvopastoral system trees can minimize trade-offs through complementary uses that augment climate resiliency and diversify household income. Complementary uses may include, timber plantations, living fences, tree alleys, windbreaks, fuelwood, perennial crops, and fodder banks [30,33]. For example, *Albizia saman* (syn. *Samanea saman*) trees improved forage production under their canopy [34], and their palatable pods are suitable as a dry season feed supplement [35]. Smallholder agroforestry systems can continue to increase their carbon stocks while also producing timber through select harvesting of high economic value tree species [36]. Colombian pastoral systems face serious complications because of climate change. The country is well poised to create national initiatives around SPS and other mitigation initiatives [16]. Indeed, the Colombian livestock sector will need to increase carbon efficiency in order to remain competitive on the international market. Our study reveals significant potential for addressing climate change mitigation and an array of co-benefits through implementation of SPS in Colombian grasslands. This suggests that SPS can significantly contribute to the NDCs and NAMAs of Colombia and other Latin American countries, as well as addressing other national commitments such as the United Nations 2030 Sustainable Development Goals and national regulatory targets.

However, scaling up SPS will not be straightforward. Calle et al. [37] demonstrated the importance of adequate technical assistance to farmer adoption of SPS in Colombia. Similarly, it is equally important to consider the quality and source of genetic material when promoting tree planting programs [38]. Market incentives including factors related to lower costs and/or higher benefits are important for silvopastoral system technology adoption [39]. In spite of the growing international demand for carbon-efficient and environmentally friendly animal products [16], very few producers will plant trees in grasslands just for the sake of climate change mitigation [40]. However, awareness of the implications of land use decisions and the increased productivity and resiliency of SPS over conventional systems, in conjunction with adequate technical and policy support [15], has been shown to incentivize adoption [41].

The MODIS data are low-resolution and may overestimate carbon stock estimates in grasslands by including different land uses like wooded areas and forest patches into grazing land pixels [23]. However, our study showed the potential of SPS to mitigate climate by sequestering more carbon in woody biomass. We assumed a linear increase in biomass carbon stocks with increasing tree cover, which may not be accurate in all cases. For example, species functional traits play a role in carbon dynamics by influencing species productivity [42].

We considered grasslands as proxy for silvopastoral systems as maps detailing silvopastoral systems were not identified. It is important to considered trade-offs across expected ecosystem services from increasing tree cover in grassland biomes. For example, a study on the effects of increased woody vegetation following fire suppression in Brazilian Cerrado reported a decline in plant and ant species by 27% and 35%, respectively [43]. Bond et al. [44] and Parr et al. [45] also reported that a large-scale increase in tree cover in grasslands could impact African grassland biomes. Several similar comments [46,47] were generated by the optimistic analysis of the global tree restoration potential by Bastin et al. [48].

As discussed in Lusiana et al. [49], however, common measures of pixel-level uncertainty in both classification and assigned C stock do not stand in the way of fairly narrow confidence intervals for aggregated data as used in national GHG accounting. The specific reduction of uncertainty with aggregation depends on the spatial structure of land cover change. For an Indonesian forest margin landscape, the study concluded that for the 1 km^2 scale of aggregation error was reduced below a defined tolerance. Further studies of this type on SPS in Columbia would be relevant.

The study is not without limitations. Data used in this study were generated on a global scale and thus may have some inaccuracies when down-scaled to a national analysis. Tree cover was measured via remote sensing and interpreted from MODIS Vegetation Continuous Fields (VCF) product; as such, it is an estimate of percentage crown cover, not tree density or tree biomass per se, and is likely to underestimate or overestimate tree cover [11]. Remotely sensed data should be validated with ground-level measurements [10], but this was not possible for logistical reasons during this study. We assumed a linear increase in biomass carbon stocks with increasing tree cover, which provides an approximation of biomass. We considered grasslands as proxy for silvopastoral systems as silvopastoral systems are a type of grazing system. Our estimates can inform the potential for further increase in C storage in SPS, but do not indicate yet what types of policy change and farmer innovation will be needed to achieve this.

5. Conclusions

We estimated carbon stocks in Colombian grasslands using remotely sensed tree cover data from MODIS and IPCC Tier 1 values. The results help clarify current and potential grassland carbon stocks at a national scale. Tree cover and carbon stocks increased significantly from 2000 to 2017. There remains high spatial variability, and regions with low tree cover have significant potential for increasing carbon stocks. This approach, along with ground-level data validation, could be useful in creating targets for planning NDCs and NAMAs in Latin American countries. Co-benefits include improved productivity and socioeconomic outcomes, climate resilience, and environmental conservation. Moving forward, it will be essential to create an enabling environment for the implementation of SPS in terms of policy, land governance, investment, capacity development, and international cooperation.

Author Contributions: Conceptualization, E.A., M.S., J.A. and T.S.R.; Formal analysis, E.A.; Funding acquisition, T.S.R.; Investigation, E.A. and T.S.R.; Methodology, E.A., M.S. and M.v.N.; Project administration, T.S.R.; Resources, M.S., J.A., T.S.R.; Writing—original draft, E.A., M.v.N. and T.S.R.; Writing—review and editing, E.A., M.S., M.v.N., J.A., J.M.R. and T.S.R. All authors have read and agreed to the published version of the manuscript.

Funding: This research was funded by the United States Agency for International Development (USAID) to the CGIAR Research Program on Climate Change, Agriculture and Food Security (CCAFS)'s Flagship on Low Emissions Development. The APC was funded separately by CCAFS. CCAFS is supported by the CGIAR Trust Fund and through bilateral funding agreements. For details please visit https://ccafs.cgiar.org/donors. The views expressed in this document cannot be taken to reflect the official opinions of these organizations. We acknowledge the support from Land Water and Ecosystems (WLE) during drafting of the manuscript.

Conflicts of Interest: The authors declare no conflict of interest. The funders had no role in the design of the study; in the collection, analyses, or interpretation of data; in the writing of the manuscript; or in the decision to publish the results.

References

1. Wollenberg, E.; Richards, M.; Smith, P.; Havlík, P.; Obersteiner, M.; Tubiello, F.N.; Herold, M.; Gerber, P.; Carter, S.; Reisinger, A.; et al. Reducing emissions from agriculture to meet the 2 °C target. *Glob. Chang. Biol.* **2016**, *22*, 3859–3864. [CrossRef] [PubMed]

2. Smith, P. Soil carbon sequestration and biochar as negative emission technologies. *Glob. Chang. Biol.* **2016**, *22*, 1315–1324. [CrossRef]

3. Roe, S.; Streck, C.; Obersteiner, M.; Frank, S.; Griscom, B.; Drouet, L.; Fricko, O.; Gusti, M.; Harris, N.; Hasegawa, T.; et al. Contribution of the land sector to a 1.5 °C world. *Nat. Clim. Chang.* **2019**, *9*, 817–828. [CrossRef]

4. Witkowski, K.; Medina, D. *Agriculture in the New Climate Action Plans of Latin America (Intended Nationally Determined Contributions)*; Inter-American Institute for Cooperation on Agriculture: San Jose, Costa Rica, 2016.

5. Griscom, B.W.; Adams, J.; Ellis, P.W.; Houghton, R.A.; Lomax, G.; Miteva, D.A.; Schlesinger, W.H.; Shoch, D.; Siikamäki, J.V.; Smith, P.; et al. Natural climate solutions. *Proc. Natl. Acad. Sci. USA* **2017**, *114*, 11645–11650. [CrossRef] [PubMed]

6. IPCC. *Climate Change and Land: An IPCC Special Report on Climate Change, Desertification, Land Degradation, Sustainable Land Management, Food Security, and Greenhouse Gas Fluxes in Terrestrial Ecosystems*; Shukla, P.R., Skea, J., Buendia, E.C., Masson-Delmotte, V., Pörtner, H.-O., Roberts, D.C., Zhai, P., Slade, R., Connors, S., van Diemen, R., et al., Eds.; IPCC: Geneva, Switzerland, 2019.

7. Cardona, C.A.C.; Naranjo Ramíirez, J.F.; Tarazona Morales, A.M.; Murgueitio Restrepo, E.; Charaá Orozco, J.D.; Ku Vera, J.; Solorio Saánchez, F.J.; Flores Estrada, M.X.; Solorio Saánchez, B.; Barahona Rosales, R. Contribution of intensive silvopastoral systems to animal performance and to adaptation and mitigation of climate change. *Rev. Colomb. Cienc. Pecu.* **2014**, *27*, 76–94.

8. Calle, Z.; Murgueitio, E.; Chará, J.; Molina, C.H.; Zuluaga, A.F.; Calle, A. A Strategy for Scaling-Up Intensive Silvopastoral Systems in Colombia. *J. Sustain. For.* **2013**, *32*, 677–693.

9. Van Dijk, S.; Tennigkeit, T.; Wilkes, A. *Climate-Smart Livestock Sector Development: The State of Play in NAMA Development*; CGIAR Research Program on Climate Change, Agriculture and Food Security (CCAFS): Copenhagen, Denmark, 2015.

10. Cardinael, R.; Umulisa, V.; Toudert, A.; Olivier, A.; Bockel, L.; Bernoux, M. Revisiting IPCC Tier 1 coefficients for soil organic and biomass carbon storage in agroforestry systems. *Environ. Res. Lett.* **2018**, *13*, 12. [CrossRef]

11. Zomer, R.; Trabucco, A.; Coe, R.; Place, F.; van Noordwijk, M.; Xu, J. *Trees on Farms: An Update and Reanalysis of Agroforestry's Global Extent and Socio-Ecological Characteristics*; World Agroforestry Centre (ICRAF): Bogor, Indonesia, 2014.

12. Zomer, R.J.; Neufeldt, H.; Xu, J.; Ahrends, A.; Bossio, D.; Trabucco, A.; Van Noordwijk, M.; Wang, M. Global Tree Cover and Biomass Carbon on Agricultural Land: The contribution of agroforestry to global and national carbon budgets. *Sci. Rep.* **2016**, *6*, 29987. [CrossRef]

13. Rosenstock, T.S.; Wilkes, A.; Jallo, C.; Namoi, N.; Bulusu, M.; Suber, M.; Mboi, D.; Mulia, R.; Simelton, E.; Richards, M.; et al. Making trees count: Measurement and reporting of agroforestry in UNFCCC national communications of non-Annex I countries. *Agric. Ecosyst. Environ.* **2019**, *284*, 106569. [CrossRef]

14. FEDEGAN. *Cifras de Referencia del Sector Ganadero Colombiano*; FEDEGAN Federacioún Colombiana de Ganaderos: Bogota, Colombia, 2018.

15. Tapasco, J.; LeCoq, J.F.; Ruden, A.; Rivas, J.S.; Ortiz, J. The Livestock Sector in Colombia: Toward a Program to Facilitate Large-Scale Adoption of Mitigation and Adaptation Practices. *Front. Sustain. Food Syst.* **2019**, *3*, 61. [CrossRef]

16. Lerner, A.M.; Zuluaga, A.F.; Chará, J.; Etter, A.; Searchinger, T. Sustainable Cattle Ranching in Practice: Moving from Theory to Planning in Colombia's Livestock Sector. *Environ. Manag.* **2017**, *60*, 176–184. [CrossRef] [PubMed]

17. ESA. *Land Cover CCI: Product User Guide Version 2.0*; UCL-Geomatics: London, UK, 2017.

18. Dimiceli, C.; Carroll, M.; Sohlberg, R.; Kim, D.H.; Kelly, M.; Townshend, J.R.G. *MOD44B MODIS/Terra Vegetation Continuous Fields Yearly L3 Global 250m SIN Grid V006*; LP DAAC: Sioux Falls, SD, USA, 2020.

19. Ruesch, A.; Gibbs, H.K. *New IPCC Tier-1 Global Biomass Carbon Map For the Year 2000*; Oak Ridge National Laboratory: Oak Ridge, YN, USA, 2008.

20. Heiskanen, J.; Liu, J.; Valbuena, R.; Aynekulu, E.; Packalen, P.; Pellikka, P. Remote sensing approach for spatial planning of land management interventions in West African savannas. *J. Arid Environ.* **2017**, *140*, 29–41. [CrossRef]

21. Sánchez-Cuervo, A.M.; Aide, T.M.; Clark, M.L.; Etter, A. Land Cover Change in Colombia: Surprising Forest Recovery Trends between 2001 and 2010. *PLoS ONE* **2012**, *7*, e43943. [CrossRef]

22. Fagua, J.C.; Baggio, J.A.; Ramsey, R.D. Drivers of forest cover changes in the Chocó-Darien Global Ecoregion of South America. *Ecosphere* **2019**, *10*, e02648. [CrossRef]

23. Chapman, M.; Walker, W.S.; Cook-Patton, S.C.; Ellis, P.W.; Farina, M.; Griscom, B.W.; Baccini, A. Large climate mitigation potential from adding trees to agricultural lands. *Glob. Chang. Biol.* **2020**, *26*, 4357–4365. [CrossRef]

24. Liu, Y.Y.; Van Dijk, A.I.J.M.; De Jeu, R.A.M.; Canadell, J.G.; McCabe, M.F.; Evans, J.P.; Wang, G. Recent reversal in loss of global terrestrial biomass. *Nat. Clim. Chang.* **2015**, *5*, 470–474. [CrossRef]

25. Liu, Y.Y.; De Jeu, R.A.M.; McCabe, M.F.; Evans, J.P.; Van Dijk, A.I.J.M. Global long-term passive microwave satellite-based retrievals of vegetation optical depth. *Geophys. Res. Lett.* **2011**, *38*. [CrossRef]

26. Feliciano, D.; Ledo, A.; Hillier, J.; Nayak, D.R. Which agroforestry options give the greatest soil and above ground carbon benefits in different world regions? *Agric. Ecosyst. Environ.* **2018**, *254*, 117–129. [CrossRef]

27. López-Santiago, J.G.; Casanova-Lugo, F.; Villanueva-López, G.; Díaz-Echeverría, V.F.; Solorio-Sánchez, F.J.; Martínez-Zurimendi, P.; Aryal, D.R.; Chay-Canul, A.J. Carbon storage in a silvopastoral system compared to that in a deciduous dry forest in Michoacán, Mexico. *Agrofor. Syst.* **2019**, *93*, 199–211. [CrossRef]

28. ESA. *300 m Annual Global Land Cover Time Series from 1992 to 2015*; ESA Climate Change Initiative: Oxford, UK, 2017.

29. Steiner, J.L.; Franzluebbers, A.J.; Neely, C.; Ellis, T.; Aynekulu, E. Enhancing Soil and Landscape Quality in Smallholder Grazing Systems. In *Enhancing Soil and Landscape Quality in Smallholder Grazing Systems*; Lal, R., Stewart, B.A., Eds.; CRC Press: Boca Raton, FL, USA, 2014; pp. 63–112.

30. Montagnini, F.; Ibrahim, M.; Murgueitio Restrepo, E. Silvopastoral Systems and Climate Change Mitigation in Latin America. *Bois For. Trop.* **2013**, *316*, 3–16. [CrossRef]

31. Dinesh, D.; Campbell, B.; Bonilla-Findji, O.; Richards, M. (Eds.) *10 Best Bet Innovations for Adaptation in Agriculture: A Supplement to the UNFCCC NAP Technical Guidelines*; CGIAR Research Program on Climate Change, Agriculture and Food Security (CCAFS): Wageningen, The Netherlands, 2017.

32. Eekhout, J.; de Vente, J. Assessing the effectiveness of Sustainable Land Management for large-scale climate change adaptation. *Sci. Total Environ.* **2018**, *654*, 85–93. [CrossRef] [PubMed]

33. Chara, J.; Reyes, E.; Peri, P.; Otte, J.; Arce, E.; Schneider, F. *Silvopastoral Systems and Their Contribution to Improved Resource Use and Sustainable Development Goals: Evidence from Latin America*; The Food and Agriculture Organization of the United Nations: Roma, Italy, 2018.

34. Durr, P.; Gangel, J. Enhanced forage production under Samanea saman in a subhumid tropical grassland. *Agrofor. Syst.* **2002**, *54*, 99–102. [CrossRef]

35. Durr, P. The biology, ecology and agroforestry potential of the raintree, Samanea saman (Jacq.) Merr. *Agrofor. Syst.* **2001**, *51*, 223–237. [CrossRef]

36. Roshetko, J.; Lasco, R.; Delos Angeles, M. Smallholder agroforestry systems for carbon storage. *Mitig. Adapt. Strateg. Glob. Chang.* **2007**, *12*, 219–242. [CrossRef]

37. Calle, A.; Montagnini, F.; Zuluaga, A.F. Farmer's perceptions of silvopastoral system promotion in Quindío, Colombia. *Bois For. Trop.* **2009**, *300*, 79–94. [CrossRef]
38. Roshetko, J.; Dawson, I.; Urquiola, J.; Lasco, R.; Leimona, B.; Weber, J.; Bozzano, M.; Lillesø, J.; Graudal, L.; Jamnadass, R. To what extent are genetic resources considered in environmental service provision? A case study based on trees and carbon sequestration. *Clim. Dev.* **2018**, *10*, 755–768. [CrossRef]
39. Iiyama, M.; Derero, A.; Kelemu, K.; Muthuri, C.; Kinuthia, R.; Ayenkulu, E.; Kiptot, E.; Hadgu, K.; Mowo, J.; Sinclair, F.L. Understanding patterns of tree adoption on farms in semi-arid and sub-humid Ethiopia. *Agrofor. Syst.* **2017**, *91*, 271–293. [CrossRef]
40. Mbow, C.; Van Noordwijk, M.; Luedeling, E.; Neufeldt, H.; Minang, P.A.; Kowero, G. Agroforestry solutions to address food security and climate change challenges in Africa. *Curr. Opin. Environ. Sustain.* **2014**, *6*, 61–67. [CrossRef]
41. Murgueitio, E.; Calle, Z.; Uribe, F.; Calle, A.; Solorio, B. Native trees and shrubs for the productive rehabilitation of tropical cattle ranching lands. *For. Ecol. Manag.* **2011**, *261*, 1654–1663. [CrossRef]
42. Lohbeck, M.; Winowiecki, L.; Aynekulu, E.; Okia, C.; Vågen, T.G. Trait-based approaches for guiding the restoration of degraded agricultural landscapes in East Africa. *J. Appl. Ecol.* **2018**, *55*, 59–68. [CrossRef]
43. Abreu, R.C.R.; Hoffmann, W.A.; Vasconcelos, H.L.; Pilon, N.A.; Rossatto, D.R.; Durigan, G. The biodiversity cost of carbon sequestration in tropical savanna. *Sci. Adv.* **2017**, *3*, e1701284. [CrossRef] [PubMed]
44. Bond, W.J.; Stevens, N.; Midgley, G.F.; Lehmann, C.E.R. The Trouble with Trees: Afforestation Plans for Africa. *Trends Ecol. Evol.* **2019**, *34*, 963–965. [CrossRef] [PubMed]
45. Parr, C.L.; Gray, E.F.; Bond, W.J. Cascading biodiversity and functional consequences of a global change-induced biome switch. *Divers. Distrib.* **2012**, *18*, 493–503. [CrossRef]
46. Friedlingstein, P.; Allen, M.; Canadell, J.G.; Peters, G.P.; Seneviratne, S.I. Comment on "The global tree restoration potential". *Science* **2019**, *366*, eaay8060. [CrossRef]
47. Skidmore, A.K.; Wang, W.; de Bie, K.; Pilesjo, P. Comment on "The global tree restoration potential". *Science* **2019**, *366*, eaaz0111. [CrossRef]
48. Bastin, J.F.; Finegold, Y.; Garcia, C.; Mollicone, D.; Rezende, M.; Routh, D.; Zohner, C.M.; Crowther, T.W. The global tree restoration potential. *Science* **2019**, *364*, 76–79. [CrossRef]
49. Lusiana, B.; van Noordwijk, M.; Johana, F.; Galudra, G.; Suyanto, S.; Cadisch, G. Implications of uncertainty and scale in carbon emission estimates on locally appropriate designs to reduce emissions from deforestation and degradation (REDD+). *Mitig. Adapt. Strateg. Glob. Chang.* **2014**, *19*, 757–772. [CrossRef]

 *land*

Article

Enhancing Vietnam's Nationally Determined Contribution with Mitigation Targets for Agroforestry: A Technical and Economic Estimate

Rachmat Mulia [1,*], Duong Dinh Nguyen [2], Mai Phuong Nguyen [1], Peter Steward [3], Van Thanh Pham [1], Hoang Anh Le [4], Todd Rosenstock [3] and Elisabeth Simelton [1]

[1] World Agroforestry (ICRAF), Hanoi 100000, Vietnam; N.Maiphuong@cgiar.org (M.P.N.); P.ThanhVan@cgiar.org (V.T.P.); E.Simelton@cgiar.org (E.S.)

[2] Institute of Geography, Vietnam Academy of Science and Technology, Hanoi 100000, Vietnam; ndduong@ig.vast.vn

[3] World Agroforestry (ICRAF), P.O. Box 30677, Nairobi 00100, Kenya; P.Steward@cgiar.org (P.S.); T.Rosenstock@cgiar.org (T.R.)

[4] Department of Science, Technology and Environment, Ministry of Agriculture and Rural Development, Hanoi 100000, Vietnam; anhlh.khcn@mard.gov.vn

* Correspondence: r.mulia@cgiar.org

Received: 30 October 2020; Accepted: 15 December 2020; Published: 17 December 2020

Abstract: The Nationally Determined Contributions (NDCs) of several non-Annex I countries mention agroforestry but mostly without associated mitigation target. The absence of reliable data, including on existing agroforestry practices and their carbon storage, partially constrains the target setting. In this paper, we estimate the mitigation potential of agroforestry carbon sequestration in Vietnam using a nationwide agroforestry database and carbon data from the literature. Sequestered carbon was estimated for existing agroforestry systems and for areas into which these systems can be expanded. Existing agroforestry systems in Vietnam cover over 0.83 million hectares storing a 1346 ± 92 million ton CO_2 equivalent including above-, belowground, and soil carbon. These systems could be expanded to an area of 0.93–2.4 million hectares. Of this expansion area, about 10% is considered highly suitable for production, with a carbon sequestration potential of 2.3–44 million ton CO_2 equivalent over the period 2021–2030. If neglecting agroforestry's potential for modifying micro-climates, climate change can reduce the highly suitable area of agroforestry and associated carbon by 34–48% in 2050. Agroforestry can greatly contribute to Vietnam's 2021–2030 NDC, for example, to offset the greenhouse gas emissions of the agriculture sector.

Keywords: agriculture sector; carbon sequestration; cost efficiency; land suitability; potential expansion areas; representative concentration pathway

---

1. Introduction

In 2015, all signatory countries of the landmark Paris Agreement pledged to reduce their national greenhouse gas (GHG) emissions and enhance resilience to climate change. The Nationally Determined Contributions (NDCs) are the blueprints outlining mitigation and adaptation efforts, and the Paris Agreement requires each party to prepare, communicate, and pursue their NDCs with domestic or international support. NDCs describe post-2020 climate programs, and parties can prioritize sectors which substantially contribute to their national emissions. Most signatory countries submitted their first NDCs to the United Nations Framework Convention on Climate Change (UNFCCC) by 2016 [1], with the option to amend before a final submission in 2020.

Agriculture features prominently in the NDCs of non-Annex I countries, which are mostly developing countries [2]. About 40% of 148 non-Annex I countries include mitigation measures for the agriculture sector in their NDCs, with half of these mentioning integrated systems such as agroforestry [3]. However, the NDCs of many countries do not elaborate these mitigation measures into concrete actions and associated targets.

Agroforestry, an integrated agricultural system with crops and trees, can substantially reduce GHG emissions through carbon sequestration [4,5]. Agroforestry also increases farmer adaptation to climate change through, e.g., diversified products and sources of income, resource use efficiency, and improved micro-climates [6–9]. Yet the absence of reliable data, including on types and distribution of existing practices and their carbon storage, partially constrain estimation of mitigation and adaptation potential from agroforestry at national scale [10].

Agriculture is a significant source of emissions in Vietnam. Total GHG emissions from Vietnam's agricultural sector reached an 88.3 million ton CO_2 equivalent (mil tCO_2e) in 2010, accounting for 35.8% of total national emissions [11], and are projected to increase to 109 mil tCO_2e by 2030 under baseline conditions. Vietnam's first NDC submitted to the UNFCCC in 2016 identified 15 mitigation measures for the agriculture sector but excluded agroforestry. The measures focused on improving the efficiency of inputs, plot management practices, such as transforming conventional water management in rice to alternate wetting drying for reducing methane production, and waste treatment, such as converting livestock waste into biogas, and are expected to reduce the projected 2030 baseline emissions by 6–42% [12]. Vietnam has recently included agroforestry in its revised NDC submitted to the UNFCCC in 2020 as part of the Land Use, Land Use Change, and Forestry (LULUCF) sector. However, the revised NDC only specifies the purpose of agroforestry measure, namely for "enhancing carbon stocks and conserving lands", without elaboration on activities and associated mitigation or adaptation targets.

A methodological framework for estimating mitigation and adaptation potential of agroforestry and associated targets at a national or sub-national scale for climate programs such as NDCs is necessary. In this paper, focusing on the mitigation potential, we describe such a framework that we used to provide a technical and economic estimate of mitigation potential from agroforestry in Vietnam by 2030 by assessing its capacity for offsetting GHG emissions through carbon sequestration. We took into account above-ground carbon (AGC), below-ground carbon (BGC), and soil carbon (SOC) sequestered in existing areas and potential expansion areas of agroforestry across the country. We considered the impact of future climate change scenarios on the expansion area and carbon sequestration potential from agroforestry. The economic estimate accounts for the investment cost required to fulfill the mitigation potential over 2021–2030.

2. Materials and Methods

2.1. Methodological Framework

Agroforestry mitigation potential was estimated at a national scale and can be disaggregated into a sub-national scale using eight ecological regions (Figure 1a), differentiated based on geographical characteristics, topographical features, ecosystem types, and climate [13]. These eight regions are North West, North East, Red River Delta, North Central Coast, South Central Coast, Central Highlands, South East, and Mekong River Delta. The detailed characteristics of each region, including soil condition and dominant land uses, are given in [13].

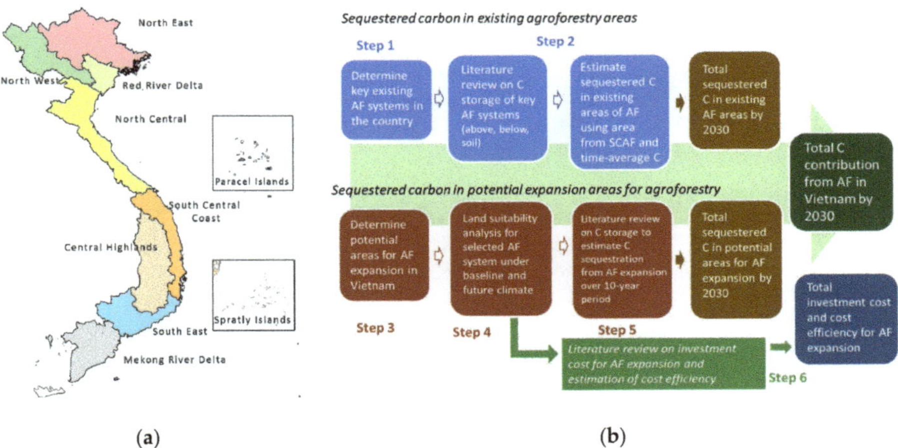

(a)   (b)

Figure 1. (**a**) Eight regions of Vietnam. (**b**) Methodological framework for estimating mitigation potential from agroforestry (AF) in Vietnam.

The assessment of agroforestry mitigation potential consists of six steps (Figure 1b). The first two steps pertain to existing agroforestry areas and the following four pertain to potential expansion areas for agroforestry. The sequestered carbon in the potential expansion areas was assessed over a ten-year period (2021–2030), and the impact of future climate scenarios on extent areas for agroforestry system expansion was investigated using representative concentration pathway (RCP) 4.5 and RCP 8.5 scenarios. The aggregate of sequestered carbon across existing and potential agroforestry areas constitutes the total carbon contribution from agroforestry by 2030.

We used the information on types and distribution of existing agroforestry practices in Vietnam from the Spatially Characterized Agroforestry (SCAF) database (http://scafs.worldagroforestry.org/), as a basis for estimating sequestration contributions from existing areas of agroforestry and for selecting agroforestry systems for expansion. For the carbon estimation, we used input carbon storage data reported in the literature or estimated carbon storage per land area using relevant allometric equations, stem diameter, and crop density data from the literature. The input AGC data are species-specific and mostly specific to Vietnam. On the other hand, the input BGC and SOC are system- but not species-specific. For example, the BGC of most of the assessed agroforestry systems was estimated using the BGC/AGC partitioning factor for agroforestry from [14], and the SOC using the SOC sequestration rate from [10].

2.1.1. Step 1: Determine Key Existing Agroforestry Systems in Vietnam

The SCAF database provides information on 48 agroforestry practices observed in 2013–2014 across 42 out of the 63 provinces in Vietnam. Most of the 48 practices can be classified into eight key systems based on their main perennial crop component, excluding the "other systems" (Table 1). The total area of the eight key systems is 820,000 hectares (ha) or about 91% of the total agroforestry system area. Examples of the eight key systems are illustrated in Figure 2. We cannot find a more recent database than SCAF for existing agroforestry systems in Vietnam.

Table 1. Existing agroforestry systems in Vietnam.

Agroforestry System *	Total Area (10³ ha) **	Common System Components	Main Regions
Melaleuca (*Melaleuca cajuputi*)-based	245.5	Fresh-water inland forest with paddy rice, sugarcane, bananas, and fish	Mekong River Delta
Robusta coffee (*Coffea canephora*)-based	245.3	*Cassia siamea*, black pepper, fruit trees such as durian and avocado, and nuts such as macadamia	Central Highlands, South East
Rhizophora (*Rhizophora* spp.)-based	149	Mangrove system with shrimp farming	Mekong River Delta
Acacia-based	129.5	*Acacia mangium, Acacia auriculiformis*, or hybrid (*Acacia mangium × Acacia auriculiformis*), intercropped with cassava in the early years after tree planting until acacia canopy is closed	North East, Red River Delta, South Central Coast, Mekong River Delta
Rubber (*Hevea brasiliensis*)-based	20.5	Usually intercropped with cassava or maize in the early years after tree planting until rubber canopy is closed	North West, North Central Coast, Central Highlands
Arabica coffee (*Coffea arabica*)-based	10.5	*Leucaena leucocephala*, longan (*Dimocarpus longan*), mango (*Mangifera indica*), plum (*Prunus salicina*) as shade trees	North West
Cashew (*Anacardium occidentale*)-based	10.4	Intercropped with maize, black pepper, or Robusta coffee	Central Highlands, South East
Tea (*Camellia sinensis*)-based	9.5	*Acacia mangium* or hybrid, *Cassia siamea*, or *Illicum verum* as shade trees	North East, North Central Coast
Other systems	79.8	Various fruit- or timber tree-based systems with relatively small areas	Spread across regions

* Ordered by total area; ** Total area in the country based on the Spatially Characterized Agroforestry (SCAF) database.

Figure 2. *Cont.*

Figure 2. Examples of key common agroforestry systems in Vietnam: (**a**) Arabica and leucaena in the North West region, (**b**) robusta, cassia, and black pepper in the Central Highlands, (**c**) tea and cassia in the North Central Coast, (**d**) acacia and cassava in the North East, (**e**) rubber, potato, and maize in the North Central Coast, (**f**) cashew, robusta, and black pepper in the South East, (**g**) melaleuca and rice in the Mekong River Delta, and (**h**) Rhizophora and shrimp farming in the Mekong River Delta. Source of photos, (**a–g**) (SCAF), (**h**) (https://nongnghiep.vn/tom---rung-voi-tang-truong-xanh-d233998.html).

2.1.2. Step 2: Estimate Sequestered Carbon in Existing Areas of Agroforestry

The existing areas of agroforestry in Vietnam refer to the 820,000 ha occupied by the eight key agroforestry systems. Due to a lack of information on crop ages, the total AGC sequestered in these areas was estimated using time-average AGCs obtained from the literature (Table A1). For the six key agroforestry systems beside the two in wetlands, we generally used the BGC/AGC biomass partitioning factor for agroforestry from [14] to estimate the BGC, and the response ratio from [10] to estimate the time-average SOC assuming all areas of agroforestry were converted from logged-over forests. For the two systems in wetlands, species-specific BGC and SOC are available from the literature (Table A1). The aggregate of AGC, BGC, and SOC from the eight key agroforestry systems constitutes the carbon contribution from existing areas of agroforestry. We assumed the total area and sequestered carbon in the existing agroforestry systems are constant until 2030.

2.1.3. Step 3: Determine Potential Areas for Agroforestry Expansion

Due to the lack of a recent national land cover map from a reliable institution in Vietnam, we used Vietnam's 2018 land cover map from [15] to determine potential areas for agroforestry expansion. Among other land cover types in the map, we selected croplands as potential areas for agroforestry expansion. The croplands are defined by [15] as "lands with herbaceous and shrubby crops followed by harvest and bare soil period". They exclude orchards, annual crops with trees, forest plantations, and wet and low-land paddy fields. We excluded wet paddy lands because they are the main source of staple food for Vietnam, forest plantations and other tree-based systems because they are high-biomass land uses, and barren forest lands because they can be used for forest restoration. Croplands are spread across the country and have a total area of about 3.6 million ha (Figure 3a). Due to the absence of a spatial boundary around existing agroforestry areas, and because croplands exclude all systems with trees, we assumed the existing areas of agroforestry and croplands (hereafter called the expansion domain) are thoroughly separated.

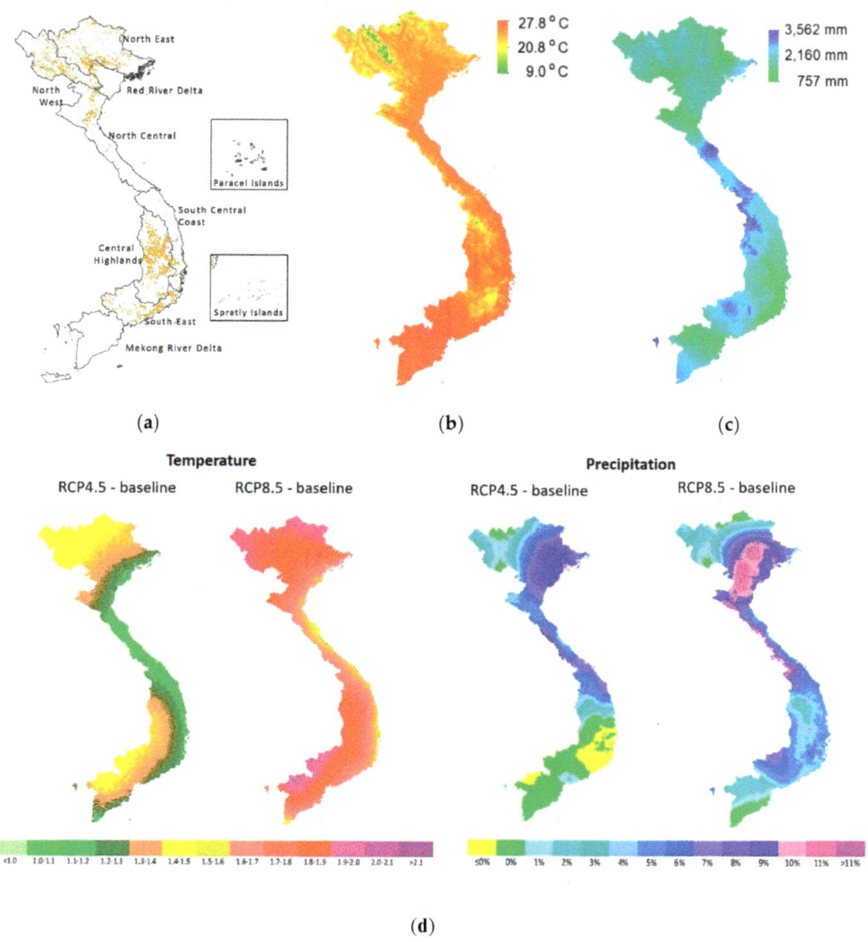

Figure 3. (**a**) Potential expansion domain of agroforestry in Vietnam; (**b**) baseline average annual temperature; (**c**) baseline average annual precipitation; (**d**) changes in temperature and precipitation under representative concentration pathway (RCP) 4.5 and RCP 8.5.

2.1.4. Step 4: Select Agroforestry Systems for Expansion and Land Suitability Analysis

We selected five out of the eight key agroforestry systems for expansion, excluding rubber-, Rhizophora-, and Melaleuca-based systems. For rubber, the Vietnam's Master Plan on Agricultural Production Development to 2020 vision to 2030 focuses on strengthening the processing industry instead of area expansion. This orientation also applies to robusta coffee and tea; however, coffee- and tea-based agroforestry provide diverse products such as timber, nuts, or fruits, for which the country is still unable to meet national demand. We excluded Rhizophora- and Melaleuca-based systems because considerable further research is necessary. Spatial data for determining their potential expansion areas is scarce and we lack information such as inundation frequency, water salinity, and tide intensity, that is crucial for land suitability analysis of wetland systems e.g., see [16]. Moreover, we could not find facts about the suitable growing conditions of these two systems from a reliable institution in the country. The most popular acacia-based system in Vietnam is the short-rotation (3–5 years) type for pulp and paper with annual crops such as cassava in the first or second year after tree planting [17]. However, the system for expansion is long-rotation (8–12 years) for timber purposes to help minimize Vietnam's dependence on timber importation. For the cashew, the selected system for expansion is alley cropping with annual crops such as maize, not the perennial shade system with coffee. In the alley cropping system, cashew plants have a higher density to maximize production for national and export markets.

Based on the guideline of land evaluation from [18,19] and the available spatial data, the land suitability analysis for each agroforestry system for expansion considered the topographical, soil, and climatic conditions within the expansion domain. We used slope as an indicator for topographical conditions, soil depth and type for soil conditions, and average annual temperature and precipitation for climatic conditions. The suitability of areas within the expansion domain depends on the indicator values. The assignment of suitability levels involved two steps: first, the indicator values within the expansion domain were compared with thresholds of the growing conditions of the agroforestry systems. We utilized thresholds from a three-tiered ranking of growing conditions reported by reliable institutions in Vietnam (Table 2). Ranking s2 implies little to no limitation of enabling factors for sustaining crop productivity; s1 indicates moderate to severe limitation of such factors, requiring modest or substantial additional inputs or plot management practices for sustaining crop productivity; and s0 describes conditions in which the specified crop cannot grow even with any additional inputs or plot management practices. Subsequently, each area was classified as "highly suitable" if all its indicators met the s2 growing condition, "not suitable" if all indicators met the s0 growing condition, and "less suitable" if all indicators met the s1 growing condition, or if not all indicators met the s2 or s0 condition. For simplicity, we only assessed the suitability of the main perennial crop species in each system, e.g., tea for the tea-based system.

Table 2. Growing condition of the main crop species of selected agroforestry systems for expansion.

	Acacia *	Cashew	Robusta	Arabica	Tea
Reference	[20,21]	NIAPP ** (data unpublished)	NIAPP (data unpublished)	NIAPP (data unpublished)	[19]
s2					
Soil type ***	Ac, Fl, RhFe, Gl	RhFe, Fe, Ac	RhFe, XaFe, Fe, Ac	RhFe, Fe	RhFe, Fe
Soil depth (m)	>1	>1	>1	>1	>1
Slope (°)	<8	<8	<8	<8	<8
Annual rainfall (mm)	1500–2500	2100–2500	1600–2000	1000–2000	>1800
Annual temperature (°C)	23–26	22–25	22–24	18–22	>22–25
s1					
Soil type	Lu, Fe	Ac	Ac, Fl	Ac	Ac, HuFe
Soil depth (m)	0.5–1	0.5–1	0.5–1	0.5–1	0.5–1
Slope (°)	8 to 35	8–<25	8–20	8–20	8 to 20
Annual rainfall (mm)	800–<1500; >2500–3500	1300–<2100; >2500	1200–1600; >2000	800–1000; >2000	1000–1800
Annual temperature (°C)	20–<23; >26	18–<22; >25	–8–22, >24	14–<18; >22–24	15–22; >25–35
s0					
Soil type	Others	Others	Others	Others	Others
Soil depth (m)	<0.5	<0.5	<0.5	<0.5	<0.5
Slope (°)	>35	>=25	>20	>20	>20
Annual rainfall (mm)	<800; >3500	<1300	<1200	<800	<1000
Annual temperature (°C)	<20	<18	<18	<14; >24	<15; >35

* Common thresholds for *Acacia mangium, Acacia auriculiformis, Acacia crassicarpa,* or acacia hybrid, ** Vietnam National Institute of Agricultural Planning and Projection (NIAPP), *** Ac: acrisols; Fe: ferrasols; RhFe: rhodic ferralsols; XaFe: xantic ferralsols; Fl: fluvisols; Gl: gleysols; HuFe: humic ferralsols.

We used the 1960–1990 climate data from WorldClim 1.4 [22] as the baseline climate (Table 3). The World Meteorological Organization recommended 1960–1990 as the baseline period [23]. To investigate the impact of future climate on suitable areas for agroforestry expansion, we selected the CNRM-CM5 climate model under RCP 4.5 and RCP 8.5 for the period 2041–2060, also from the WorldClim 1.4, which is available at 30 arc seconds resolution. WorldClim 2.0 provides climate data for the period 2021–2040 but at coarser resolution, and a stronger impact of climate change on the area for expansion and carbon sequestration potential from agroforestry can be expected using the 2041–2060 data. CNRM-CM5 is one of the best climate models for South East Asia [24]. Because we only considered the main perennial crop component of each agroforestry system for expansion in the land suitability analysis, the impact of future climate was also assessed for the main crop component only to represent the impact on each agroforestry system, neglecting the potential of agroforestry to modify micro-climates through an integration of multiple plant components that can reduce the intensity of climate change impact. Due to the absence of a spatial boundary to existing areas of agroforestry, we only investigated the impact of future climate on the potential expansion areas. All spatial inputs were standardized into one arc second, considered as a suitable resolution for country analysis.

Table 3. Input maps used for the land suitability analysis.

Input Maps	Resolution	Coordinate System	Date	Source
Land cover	1 arc second	Lat/Long World Geodetic System (WGS) 84	2018	[15]
Digital Elevation Model (DEM)	1 arc second	Lat/Long WGS 84	2019	Advanced Land Observing Satellite (ALOS) Global Digital Surface Model (AW3D30) version 2.2
Slope	1 arc second	Lat/Long WGS 84	2019	Generated from DEM
Soil type	1: 1,000,000	WGS 84 UTM Zone 48N	2010	NIAPP
Soil depth	1: 1,000,000	WGS 84 UTM Zone 48N	2010	NIAPP
Baseline average annual precipitation and temperature	30 arc seconds	Lat/Long WGS 84	1960–1990	WorldClim 1.4
Future average annual precipitation and temperature	30 arc seconds	Lat/Long WGS 84	2041–2060	WorldClim 1.4

Based on the baseline climate data from WorldClim 1.4, the northern part of the country generally has lower average annual temperatures and precipitation than the central and southern parts (Figure 3b,c). A similar pattern was reported by, e.g., [13]. This pattern will potentially change in the future, especially under the RCP 8.5 scenario, due to substantial increases in temperature and precipitation in the northern part of the country (Figure 3d).

2.1.5. Step 5: Estimate Sequestered Carbon in Potential Expansion Areas

We estimated the carbon sequestered through agroforestry expansion by subtracting the AGC and BGC stored in croplands from the total carbon accumulated over ten years in areas suitable for agroforestry expansion. We used the average AGC of croplands of 5 tC ha^{-1} for "cropland containing annual crops" according to the Intergovernmental Panel on Climate Change (IPCC) [25] and a BGC/AGC partitioning factor of 0.05–0.2 for "permanent cropland" [14]. The SOC is 53–158 tC ha^{-1} for "permanent cropland" [14]. We assumed that the conversion of croplands into agroforestry retained the SOC in the soil, but we excluded this SOC from the estimation of accumulated SOC in the ten-year agroforestry expansion. All input AGC, BGC, and SOC data for the five agroforestry systems for expansion are given in Table A1.

There were five scenarios of agroforestry expansion, one for each agroforestry system. In each scenario, the system expands into highly suitable areas only, or into 10% and, at most, 25% of the less suitable areas. We assumed that agroforestry systems that have expanded into less suitable areas accumulate AGC, BGC, and SOC at the same rates as in highly suitable areas.

The highly and less suitable areas for expansion were gradually converted into agroforestry across the ten years at a constant conversion rate. Therefore, in the first year, only 10% of suitable areas was converted into agroforestry, to reach the total suitable area in the tenth year. For simplification, we applied a constant carbon accumulation rate across the years for AGC, BGC, and SOC.

To assess the advantages of using agroforestry to offset the GHG emissions from the agriculture sector, we compared the accumulated carbon levels in the expansion domains from agroforestry and from sole crop plantations. We used three commodities in the comparison, namely arabica coffee, robusta coffee, and tea. Acacia and cashew crops were not considered because the sole plantation of the two commodities are forestry, not agricultural systems. The comparison was made for highly suitable areas and AGC only because the BGC and SOC input data are not species-specific. All estimated carbon values were converted into CO_2e with a factor of 3.67.

2.1.6. Step 6: Estimate the Cost Efficiency of Agroforestry Expansion

We defined cost efficiency as the investment cost required to sequester one ton of CO_2 equivalent. We calculated cost efficiency using the total investment cost to establish and maintain agroforestry systems for ten years under the baseline climate conditions. The cost efficiency will not change under the future climate conditions because the change in investment cost is proportionate to that in the potential expansion area. We estimated all investment costs according to their monetary worth in 2019 (Table 4) by using the annual inflation rate of Vietnam. We compared cost efficiency among agroforestry expansion scenarios, and between agroforestry and sole plantations of arabica, robusta, and tea.

Table 4. Input investment costs for agroforestry and sole crop expansion.

System	Investment Cost (USD ha^{-1} year^{-1})	Source
Acacia agroforestry	173 ± 4.6	[26] for the case of South Central Coast
Arabica agroforestry	2587	SCAF database for the case of North West
Cashew agroforestry	213	SCAF database for the case of South East
Robusta agroforestry	2124 ± 574	SCAF for the case of Central Highlands
Tea agroforestry	2806	SCAF database for the case of North East
Arabica sole plantation	1835	[27] for the case of North Central Coast
Robusta sole plantation	941 ± 24	[28] for the case of Central Highlands
Tea sole plantation	2642	[29] for the case of Central Highlands

3. Results

3.1. Sequestered Carbon in Existing Areas of Agroforestry

The total (TOC) of sequestered AGC, BGC, and SOC in the existing areas of agroforestry, from the eight key systems in Vietnam, reaches 1346 ± 92 mil tCO_2e. The two systems in wetlands, namely Rhizophora- and Melaleuca-based, contribute about 82% to the TOC (Figure 4) thanks to their high carbon storage per hectare and large areas. About 52% of the TOC is SOC, 27% is BGC, and 21% is AGC. The share of TOC from BGC is higher than from AGC due to a large contribution from the root biomass of the Rhizophora-based system. The AGC, BGC, and SOC sequestered by agroforestry systems are shown in Table A2.

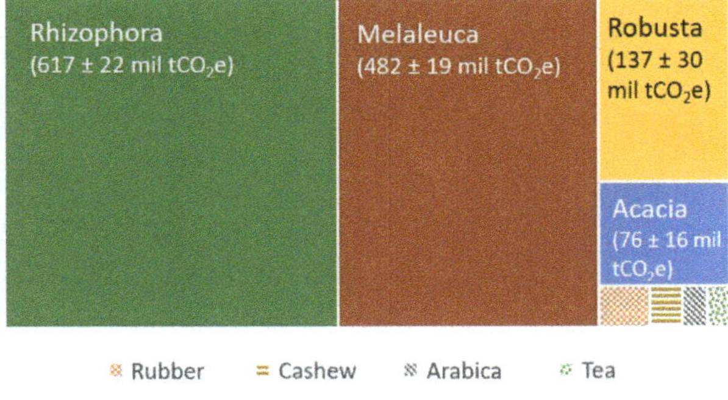

Figure 4. Main contributions to the total sequestered carbon in existing areas of agroforestry.

3.2. Suitable Areas for Agroforestry Expansion

Highly suitable areas for the five agroforestry systems for expansion range from 24 to 419 thousand ha nationwide under the baseline climate (Table 5). Cashew has the most limited area, while acacia has the largest. However, in terms of aggregate of highly and less suitable areas, the most limited is arabica. Neglecting the capacity of agroforestry system to modify micro-climate that can reduce the impact of climate change especially warming climate, the highly suitable areas for agroforestry expansion were substantially affected by the potential change in climate (Table 5). For example, large areas across the country are projected to have an increase in average annual temperature by 1.8–2 °C by 2050 under RCP 8.5, and this reduces the highly suitable area of the arabica coffee-based system by 89% compared to baseline. For most of the agroforestry systems, the change in climate transformed the highly suitable areas into less suitable areas for expansion. The two coffee-based systems are the most severely affected, while acacia is relatively resistant to the climate change. On average, for all agroforestry systems, the highly suitable areas are potentially reduced by 34% and 48% in 2050 under RCP 4.5 and RCP 8.5, respectively.

Table 5. Suitable areas for agroforestry expansion under baseline and future climate conditions.

	Acacia	Arabica	Robusta	Cashew	Tea
	Highly suitable (thousand ha)				
Baseline	419	54	189	24	181
RCP 4.5	435	15	36	21	170
RCP 8.5	389	6	32	17	127
	Combined (highly and less) suitable (thousand ha)				
Baseline	2407	937	1985	1789	1787
RCP 4.5	2447	362	1980	1761	1784
RCP 8.5	2468	268	1997	1827	1797
	Impact of climate change on highly suitable area (%)				
RCP 4.5	4%	−72%	−81%	−14%	−6%
RCP 8.5	−7%	−89%	−83%	−32%	−30%

Among the eight regions, Central Highlands contains the largest area that is highly suitable for agroforestry expansion (Figure 5a) due to favorable soil and climate conditions for crop cultivation. The South Central Coast and Mekong River Delta have the smallest suitable areas due to higher temperatures, lower precipitation, or unsuitable soil types for agroforestry systems' expansion. These two regions, along with the Red River Delta, also have the smallest suitable area for agroforestry expansion when both highly and less suitable areas are combined (Figure 5b). Suitable areas for agroforestry systems organized by region under baseline and future climate conditions are given in Table A3.

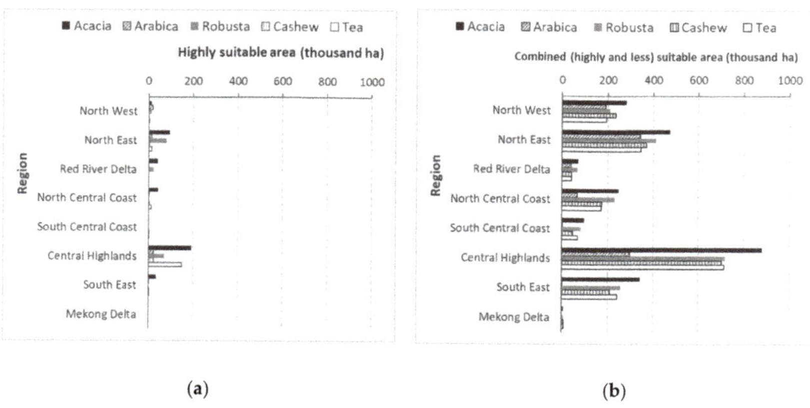

Figure 5. (**a**) Highly and (**b**) combined (highly and less suitable) areas by species and region.

3.3. Sequestered Carbon in the Agroforestry Expansion Areas

Among the five scenarios, expansion using arabica- and cashew-based agroforestry in highly suitable areas accumulated the smallest TOC over the ten-year expansion period under baseline climate conditions (Figure 6). Acacia-based AF accumulated the highest TOC that reaches 44 ± 4.5 mil tCO_2e by 2030. The inclusion of 10% or 25% of the less suitable areas substantially increased the TOC on average by a factor of 3.3 and 6.8, respectively (Figure 6b,c), compared to the TOC from highly suitable areas only. The sequestered TOC values under baseline and future climate conditions with or without the inclusion of less suitable areas are provided in Table A4.

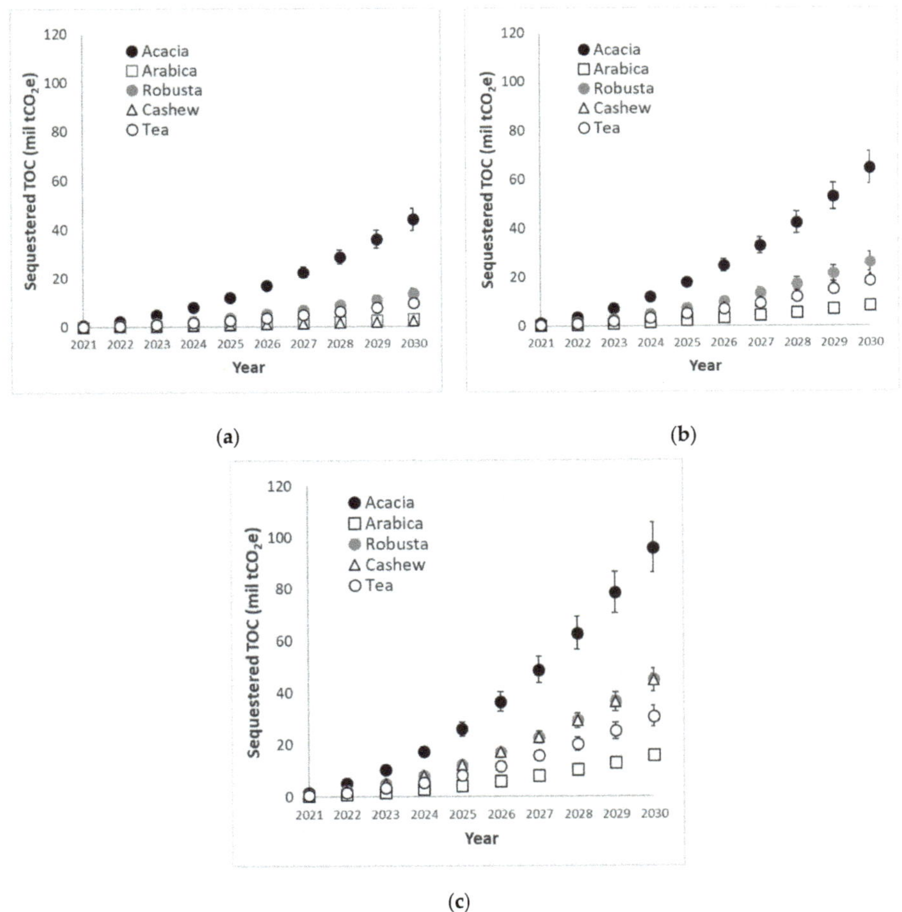

Figure 6. Sequestered total carbon (TOC) over ten-year agroforestry expansion in (**a**) highly suitable and (**b**) highly suitable and 10% or (**c**) 25% of less suitable areas under baseline climate conditions.

3.4. Cost Efficiency of Agroforestry Expansion

The investment cost for agroforestry expansion in the highly suitable areas under baseline climate conditions ranges from USD 28 to 2790 million for the ten-year period (Figure 7a). The cost doubled when the combined suitable areas were included. Among the five agroforestry systems, the expansion of tea-based systems into the total highly suitable area of 180,000 ha requires the largest investment cost. For the purpose of sequestering carbon, neglecting potential economic returns, acacia- and cashew-based systems are the most cost efficient among the five agroforestry systems for expansion. The cost efficiency of the two systems is USD 8–12 per tCO$_2$e. The cost efficiencies of sole crop plantations are USD 9121, 1041, and 934 per tCO$_2$e for arabica, robusta, and tea, respectively. Agroforestry expansion is 1.3–17 times more cost-efficient for sequestering carbon than sole crop plantations.

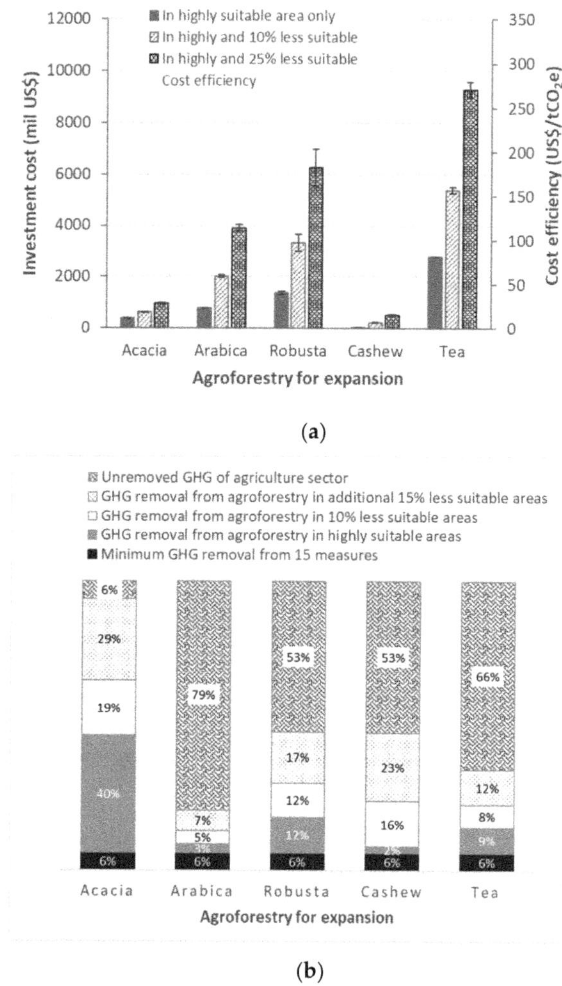

(a)

(b)

Figure 7. (a) Total investment cost and cost efficiency of the five agroforestry expansion scenarios, (b) mitigation contribution from agroforestry expansion for removing greenhouse gas (GHG) emission of Vietnam's agriculture sector.

3.5. Mitigation Contribution to Agriculture Sector

The sequestered TOC in the existing areas of agroforestry that reaches 1346 ± 92 mil tCO$_2$e can thoroughly offset the projected GHG emissions of the agriculture sector by 2030. However, if only the carbon contribution from post-2020 programs is considered, the sequestered TOC in agroforestry expansion areas under baseline climate can remove 15–88% of the total GHG emissions of the agriculture sector compared to if 25% of the less suitable areas is included in the carbon assessment (Figure 7b). The acacia-based expansion provides the largest contribution, while the arabica-based provides the smallest. If the 15 mitigation measures in Vietnam's first NDC provide the minimum contribution of

6%, about 6–79% of projected emissions by 2030 would remain unremoved. A higher contribution from agroforestry can be expected if more than 25% of the less suitable areas were considered in the assessment.

4. Discussion

4.1. Agroforestry in Vietnam's 2020 NDCs

In Vietnam's 2020 NDC[1], agroforestry is mentioned in Section 2.4.3 as part of measures for the LULUCF sector. Agroforestry and forest protection and restoration are most likely placed as part of the same sector due to their potential usefulness for carbon sequestration and land conservation. Agroforestry-related activities and targets will likely be elaborated in the NDC action plan that is still under development by the government at this time.

To offset sectoral emissions, it is, however, more effective to include agroforestry as part of the agriculture sector. By 2010, in Vietnam, forests and other land uses, excluding agriculture, such as grasslands had generated a net negative emission of 19.2 mil tCO_2e with a projected emission of −42.5 and −45.3 mil tCO_2e by 2020 and 2030, respectively [11]. Moreover, the country has committed to the UNFCCC's Koronivia Joint Work on Agriculture (KJWA) to promote and enhance investment in climate-smart agriculture such as agroforestry [30]. Enhancing terrestrial and soil carbon is among the priorities of the KJWA [31], along with improved nutrient and water management for food security and resilience to climate change, for which agroforestry can also generate relevant benefits.

As in its first NDC, Vietnam's 2020 NDC focused on improving the efficiency of inputs, plot management practices, and waste treatment as mitigation measures for the agriculture sector.

4.2. Agroforestry Systems for Expansion and Impact of Climate Change

Several studies in the literatures [32–35] also projected a strong impact of climate change on suitable area and production of arabica and robusta coffee. Globally, climate change potentially reduces the suitable area for coffee by 50% across climate scenarios, and Vietnam is one among several coffee producing countries that will be severely affected [32,34]. The change in climate, especially warming temperature beyond the optimal threshold for growth, affects the coffee biological process that results in reduced photosynthesis, a slower or halted ripening process, or flower abortion [36]. In addition, the change in temperature can trigger pest and disease outbreak. The impact of climate change on suitable area and production of other species, such as tea, has also been reported in the literature [37–39].

Our study only assessed the impact of climate change on the main perennial crop species of each agroforestry system for expansion. However, the projected impact does not necessarily represent impact on agroforestry that uses that species as the main crop. The presence of other plant components in agroforestry can potentially modify the micro-climate, which reduces the intensity of climate change impact. For example, shading trees or ground cover crops in coffee agroforestry systems can keep soil moisture high and soil temperature low, and these benefits are absent in sole coffee plantation [40–42]. The role of coffee agroforestry systems for mitigating the impact of a warming climate has been demonstrated in the southeast Brazil [33]. The suitable area for sole coffee plantation is projected to decrease by 60% in 2050, driven by a 1.7 °C ± 0.3 increase in the average annual temperature. However, coffee agroforestry systems with a 50% shade cover can reduce the average annual temperatures within the systems, and 75% of the total area of coffee agroforestry in the region will still be suitable for production by 2050. Therefore, the strong impact of the warming climate on robusta and arabica coffee projected in our study clearly suggests the need for prioritizing coffee agroforestry rather than sole

---

1 https://www4.unfccc.int/sites/ndcstaging/PublishedDocuments/Viet%20Nam%20First/Viet%20Nam_NDC_2020_Eng.pdf submitted on September 2020.

coffee plantation for expansion in Vietnam and promoting a gradual conversion of existing sole coffee plantations into agroforestry. The projected impact of the warming climate on other species such as tea, as shown in our study, also suggests the need for modifying the micro-climate through agroforestry.

Vietnam's Master Plan on agricultural production development to 2020, vision to 2030, specifies the Central Highlands, South East, and North Central Coast as the main production regions for robusta coffee. Our land suitability analysis shows that the three regions have suitable, both highly and less suitable combined, areas for robusta coffee, each above 200,000 ha, and thus supports the specification. The other regions such as the North West and North East also have large suitable areas for robusta coffee. However, the Central Highlands has been the production house of robusta coffee in Vietnam for decades, with strong market access, role of private sectors, and supporting infrastructure, including the processing industry [43]. Therefore, the Plan is likely focusing the production of robusta coffee in regions around the Central Highlands. The Plan also specifies main production regions for tea, namely the North East, North West, and Central Highlands. Our land suitability analysis suggests that the South East and North Central Coast are also suitable for tea expansion. Therefore, aside from existing infrastructures and market access for tea that are currently concentrated in the North East, North West, and Central Highlands [44], expansion of tea agroforestry in the other two regions can become an option. For cashew, the Plan specifies the Central Highlands, South East, and South Central Coast. As for robusta coffee and tea, the specification likely considers existing infrastructures and access to market as main factors. Putting those factors aside, our analysis shows that other regions such as North West, North East, and North Central Coast have larger suitable areas for cashew than that of the South Central Coast. Another factor that drives the Plan to consider the South Central Coast is likely its territorial bordering with the Central Highlands. The Plan does not specify regions for arabica coffee and acacia. According to our analysis, three regions, namely the North West, North East, and Central Highlands, have suitable areas for arabica coffee, each above 180,000 ha, and according to [43], the main arabica cultivations in Vietnam can be found in the North West and Central Highlands. Therefore, the expansion of arabica-based agroforestry can target the North West and Central Highlands, with the North East as an option. Acacia can grow in a wide range of terrain, soil, and climate conditions and is relatively easy to manage [45]. In addition, it is a nitrogen-fixing species that can help restore soil fertility. Our analysis shows that acacia is suitable in all regions except the Mekong River Delta.

Rhizophora-based agroforestry should be considered as an agroforestry system for expansion, although our study could not assess its potential area for expansion and sensitivity to climate change due to lack of input data. Rhizophora-based agroforestry can potentially reconcile mangrove restoration and livelihood improvement through combining mangrove plantation and shrimp farming. The area of mangrove forests in Vietnam, especially in the Mekong River Delta, has decreased by 35,000 ha, or 32% of their initial area, during the past three decades (1988–2018) [46], mainly due to anthropogenic activities. About 7300 ha of degraded mangrove forests have been regenerated from 2013 to 2018, but stronger efforts for mangrove restoration in Vietnam are necessary to counter the escalated impact of sea water intrusion amidst livelihood pressure from surrounding populations. By 2050, about 13% of existing rice agricultural areas in the Mekong River Delta are likely to be converted to shrimp farms due to the high economic benefit and impact of sea water intrusion [47]. An effective land share between mangrove plantations and shrimp farms within Rhizophora-based agroforestry systems has been suggested by several studies, e.g., [48,49], for the purpose of optimizing economic return from the shrimp farming and ecological contribution from the mangrove. A recent study [49] claims that mangrove density has no ecological impacts on shrimp farming, and the recommended mangrove coverage for shrimp farming is about 60%. Rhizophora-based agroforestry systems are also a solution for restoring mangrove biodiversity and coastal food chains [49], apart from their substantial contribution to carbon sequestration for GHG removal.

4.3. Advantages of Using Agroforestry Rather than Sole Crop Plantation for NDCs

Our study shows that agroforestry is more cost efficient in sequestering carbon than sole crop plantation (see also [12]). Consequently, the cost required to remove GHG emissions is lower when using agroforestry than sole crop plantation. For example, expansion of robusta-based agroforestry in Vietnam can potentially remove 41% of total GHG emissions of the country's agriculture sector by 2030 (see Figure 7b), equivalent to about 45 mil tCO_2e. The expansion requires an investment cost of about USD 6.3 billion (see Figure 7a). Removing a similar amount of GHG emissions using sole robusta plantations will require an additional cost of about USD 41 billion. For the case of arabica coffee and tea, the required additional costs are about USD 145 billion and 20 billion, respectively.

Agroforestry can offer more climate change mitigation and adaptation co-benefits as compared to sole crop plantations. For example, robusta-based agroforestry systems in the Central Highlands region were reported to sequester, on average, 0.16 tCO_2e per ton coffee produced thanks to sequestered carbon in shade trees and contribution from nitrogen-fixing trees to reduce chemical inputs. On the other hand, sole coffee plantations were net GHG emitters with, on average, 0.37 tCO_2e per ton coffee produced [50]. In terms of economic return, crop diversification in agroforestry can increase and stabilize farmers' incomes [7]. For example, in the Central Highlands, the net income of a robusta-macadamia (*Macadamia integrifolia*) system reached USD 2500 ha^{-1} $year^{-1}$ compared to USD 1793 ha^{-1} $year^{-1}$ for sole robusta plantations [29]. For tea, the net annual income from a sole tea plantation in the Central Highlands was USD 720 ha^{-1} $year^{-1}$ [29], while a tea-*Acacia mangium* system in the North East generated USD 1688 ha^{-1} $year^{-1}$, as reported in the SCAF database. There is also evidence of agroforestry's role in stabilizing farmers' annual incomes from fruit tree-based systems in the North West [51] and acacia-based systems in the North Central Coast [26,52]. Higher and more stable incomes from agroforestry mean that farmers can reinvest in improved adaptation strategies.

4.4. Caveats in the Carbon and Cost Assessment for Agroforestry

In our study, due to limited information from the literature, we assumed that the sequestered carbon per ha of agroforestry in highly and less suitable areas, when used as inputs for the national-scale estimation of carbon potential from agroforestry expansion, are comparable. Likewise, the input investment costs per ha for agroforestry expansion were assumed to be comparable for all suitable areas. To overcome the scarcity of input data, future studies can consider the use of tools for simulating soil–climate–crop interaction in agroforestry—e.g., see [53]—to obtain projection of plot-level sequestered carbon and investment cost in different growing conditions, for example, in highly and less suitable areas. The projected carbon and cost per ha can be used to better estimate carbon sequestration potential and associated cost of agroforestry expansion. For example, if the projected carbon per ha of agroforestry in less suitable areas is reduced by 10% compared to that in highly suitable areas, agroforestry expansion under baseline climate can remove 13–83% instead of 15–88% of the total GHG emissions of the agriculture sector by 2030 if 25% of the less suitable areas were included in the calculation. Future studies on the return on investment from agroforestry that include sensitivity analyses of climate change impacts are also necessary. For such assessment, the studies can use the projected impact of future climate on suitable areas for agroforestry expansion as described in the current study and the projected plot-level investment cost, including level of crop production from the simulation tools for highly and less suitable areas of agroforestry.

4.5. Ways Forward to Foster Agroforestry in Vietnam's NDC

Vietnam's 2020 NDC outline national climate programs spanning from 2021 to 2030, with an update to the UNFCCC required by 2025. Among efforts needed to enhance the role of agroforestry in climate mitigation and adaptation informed by the revised NDC, including the inclusion of agroforestry as part of measures for the agriculture sector, are continuous improvement of the database on agroforestry practices, further investigation on their potential mitigation and adaptation benefits, and the development of reliable monitoring, reporting, and verification systems. Vietnam's 2020 NDCs also emphasize that proposed measures need to demonstrate synergy and co-benefits among climate change adaptation, mitigation, and sustainable development goals, including the promotion of gender equality. For the agriculture sector, Section 3.1.4 of the country's NDC underlines that "women's decision-making power within the family is often limited which constrains them to apply their experience and knowledge to selecting varieties and cultivation techniques". Additional scientific evidence is needed to demonstrate advantages of agroforestry over other agricultural practices in contributing to such synergies, including empowering women in decision making. Several studies have shown certain gendered preferences when selecting tree species for agroforestry, e.g., [54–56]. Similar studies for investigating such preference in Vietnamese farmers are necessary. Understanding the reasoning and factors that drive these preferences can have long-term impacts on interventions for climate change mitigation and adaptation, as well as enabling progress on gender equality.

The inclusion of agroforestry in mechanisms that can reward carbon sequestration can also foster agroforestry in Vietnam's NDC. For example, some studies claimed that agroforestry can be a direct or indirect target of Reducing Emissions from Deforestation and Forest Degradation (REDD+) [57], which, to date, remains as the UNFCCC's sole framework that suggests a carbon-reward mechanism. Globally, about half of the 73 developing countries that have REDD+ strategies cited agroforestry as a potential measure for reducing forest degradation and deforestation [58]. Vietnam's REDD+ action plan for 2030 does not, however, explicitly mention agroforestry. Agroforestry is likely relevant for measure 4.1.2 on "promoting sustainable and deforestation-free agriculture"[2], but no elaboration of that measure is provided in the action plan, which creates uncertainty on the relevancy. Recently, Vietnam's national policy on payments for forest ecosystem services is under amendment to include forest carbon in addition to water services. The policy applies a mandatory payment to beneficiaries of forest water service and, very much in the same way, the proposed scheme of payment for carbon services suggests large GHG emitters such as the cement industry and coal-generated power plants would have to pay forest communities and landowners to support forest protection and expansion [59]. Some types of agroforestry, such as acacia-based systems for timber purposes with temporary intercrops, probably classify under the category of forestry land uses that can receive such payments. A verification from relevant authorities is, however, necessary for timber-based systems with permanent intercrops. Furthermore, a feasible monitoring, reporting, and verification system to track progress towards agroforestry mitigation and adaptation targets should be developed to foster agroforestry in Vietnam's NDC. In the current study, the variety and areas of existing agroforestry practices in Vietnam were known from the SCAF database, relying on provincial partners as the principal data sources. Technical, financial, and institutional challenges for developing a monitoring, reporting, and verification system for agroforestry in Vietnam should be identified and properly addressed.

5. Conclusions

The suitable area of each agroforestry system for expansion at the national and sub-national scales and the estimated mitigation potential under baseline and future climate conditions described in this study can support Vietnam in specifying agroforestry activities, mitigation targets, and associated costs for developing the action plan of its 2020 NDC. Among the crop species assessed for = agroforestry expansion, the two varieties of coffee, namely robusta and arabica, will be the most severely affected by climate conditions in 2050. To reduce the impact of climate change, future agricultural expansion in Vietnam should consider these crop species in agroforestry, which has potential to modify micro-climates and has other co-benefits such as being more cost efficient in sequestering carbon compared to sole crop plantation. Thanks to these benefits, agroforestry will bring less investment risks than sole crop plantation.

---

[2]   http://vietnam-redd.org/Upload/CMS/Content/Library-GovernmentDocuments/419%20NRAP%202030%20En.pdf

The 2020 NDC for Vietnam includes agroforestry as part of measures for the LULUCF sector. However, for the purpose of offsetting sectoral GHG emissions, it is more effective to include agroforestry as part of agriculture sector. Excluding the sequestration contribution from the existing areas of agroforestry, the total AGC, BGC, and SOC sequestered in the potential areas for agroforestry expansion can remove 15–88% of the total GHG emissions of the agriculture sector. This is achieved when both highly suitable and 25% of less suitable areas for agroforestry expansion were included in the carbon assessment.

Efforts to foster the implementation of Vietnam's NDC through agroforestry would benefit from continuous data provision for estimating agroforestry's mitigation and adaptation benefits, capacity to achieve synergy and co-benefits between climate change adaptation, mitigation and sustainable development goals, and to promote gender equality. In addition, a reliable monitoring, reporting, and verification system is necessary to track progress towards agroforestry's mitigation and adaptation targets. If all these further efforts generate at least preliminary outputs accessible to relevant authorities within the next 3–4 years, we can expect agroforestry to have a broader role in climate mitigation and adaptation, informed by Vietnam's revised NDC by 2025.

Author Contributions: Conceptualization, R.M., M.P.N., P.S. and T.R.; data curation, R.M., D.D.N., M.P.N., P.S., V.T.P., H.A.L., T.R. and E.S.; formal analysis, D.D.N., M.P.N. and P.S.; funding acquisition, R.M. and T.R.; investigation, R.M., D.D.N., M.P.N., P.S. and V.T.P.; methodology, R.M., D.D.N., M.P.N., P.S., V.T.P. and E.S.; software, D.D.N.; validation, R.M., D.D.N., M.P.N., P.S., H.A.L., T.R. and E.S.; visualization, D.D.N. and V.T.P.; writing—original draft preparation, R.M.; writing—review and editing, R.M., D.D.N., M.P.N., P.S., V.T.P., H.A.L., T.R. and E.S. All authors have read and agreed to the published version of the manuscript.

Funding: This research was funded by the CGIAR Research Program on Climate Change, Agriculture and Food Security (CCAFS), which is carried out with support from the CGIAR Trust Fund and through bilateral funding agreements. We specifically thank the United States Agency for International Development (USAID) for the bilateral funding. The views expressed in this document cannot be taken to reflect the official opinions of these organizations. The APC was funded by CCAFS.

Acknowledgments: This work was implemented as part of the CGIAR Research Program on Climate Change, Agriculture and Food Security (CCAFS), which is carried out with support from the CGIAR Trust Fund and through bilateral funding agreements. We specifically thank USAID for funding this work. For details, please visit https://ccafs.cgiar.org/donors. The views expressed in this document cannot be taken to reflect the official opinions of these organizations. We sincerely thank Delia Catacutan, Nguyen Quang Tan, Vu Tan Phuong, Chu Van Chuong, Bui My Binh, Nguyen Phu Hung, and Meine van Noordwijk for numerous and valuable inputs to the study. We also thank valuable comments from three anonymous reviewers.

Conflicts of Interest: The authors declare no conflict of interest. The funders had no role in the design of the study; in the collection, analyses, or interpretation of data; in the writing of the manuscript, or in the decision to publish the results.

Appendix A

Table A1. Input data for estimation of sequestered carbon in agroforestry systems.

Agroforestry System	Tree Density (trees ha^{-1}) *	AGC **	BGC	SOC ***	Source
Tea-based	Agroforestry (AF): 17,000 (tea), 150–200 (shade trees) Sole: 18,000–25,000	TA: 13.3 tC ha^{-1} SR AF: 1.7 ± 0.18 tC ha^{-1} year^{-1} SR sole: 1.4 ± 0.14 tC ha^{-1} year^{-1}	BGC/AGC: 0.10–0.34	RR-TA: 1.0 ± 0.2 SR: 1.59 ± 0.4 tC ha^{-1} year^{-1}	AGC-TA and AGC-SR AF: [60] for the case in North East Vietnam. AGC-SR sole: calculated from [61]. BGC/AGC for AF [14]. RR-TA from 'forest to shade perennial' for tropical region [10]. SOC-SR from 'croplands to multistrata system' for tropical region [10]
Coffee robusta-based	AF: 500–1100 (coffee), 85–150 (shade trees) Sole: 750–1500	TA: 13.4 ± 0.3 tC ha^{-1} SR AF: 2.63 ± 0.55 tC ha^{-1} year^{-1} SR sole: 0.75 ± 0.01 tC ha^{-1} year^{-1}	BGC/AGC: 0.10–0.34	RR-TA: 1.0 ± 0.2 SR: 1.59 ± 0.4 tC ha^{-1} year^{-1}	AGC-TA [50]. AGC-SR AF: calculated from [62]. AGC-SR sole and calculated using allometric equation from [63]. All for the case in Central Highlands. BGC/AGC for AF [14]. RR-TA and SOC-SR similar as for tea.
Coffee arabica-based	AF: 3000–5000 (coffee), 100–200 (shade trees) Sole: 4000–8000	TA: 13.5 ± 5.5 tC ha^{-1} SR AF: 1.9 ± 0.24 tC ha^{-1} year^{-1} SR sole: 0.5 ± 0.06 tC ha^{-1} year^{-1}	BGC/AGC: 0.10–0.34	RR-TA: 1.0 ± 0.2 SR: 1.59 ± 0.4 tC ha^{-1} year^{-1}	AGC-TA and AGC-SR AF: [64] for the case in North West. AGC-SR sole: calculated using allometric equation from [65] for the case in North West. RR-TA and SOC-SR similar as for robusta.
Cashew-based	100–200	TA: 54.2 tC ha^{-1} SR: 4.2 ± 0.35 tC ha^{-1} year^{-1}	TA: 9.2 tC ha^{-1} SR: 0.7 ± 0.06 tC ha^{-1} year^{-1}	RR-TA: 0.95 SR: 0.84 ± 0.26 tC ha^{-1} year^{-1}	AGC-TA, AGC-SR, BGC-TA, BGC-SR: stem diameter from [66] and allometry equation from [67]. RR-TA from forest to silvoarable for tropical region [10]. SOC-SR from 'croplands to silvoarable' for tropical region [10].
Rubber-based	500	TA: 25.3 ± 2.76 tC ha^{-1}	TA: 4.5 ± 0.3 tC ha^{-1}	RR-TA: 0.95	AGC-TA: stem diameter [68], allometric equation: [69]. BGC-TA: using AGC-BGC relation from [69]. RR-TA is similar as for cashew [10]
Acacia spp.-based #	Short-rotation (3–5 years): 4500–8000 Long-rotation (8–12 years): 600–1200	TA: 25.3 ± 6.2 tC ha^{-1} SR: 4.0 ± 0.37 tC ha^{-1} year^{-1}	BGC/AGC: 0.10–0.33	RR-TA: 0.95 SR: 0.84 ± 0.26 tC ha^{-1} year^{-1}	AGC-TA and AGC-SR: [26] for the case in South Central Coast. BGC/AGC for tree plantation [14]. RR-TA is similar as for rubber [10]. SOC-SR similar as for cashew.
Rhizophora-based	8000–10,000	TA: 156.4 tC ha^{-1}	TA: 568.4 tC ha^{-1}	TA: 386 tC ha^{-1}	AGC-TA and BGC-TA calculated based on [70] for the case in Mekong River Delta. SOC-TA [71]
Melaleuca-based	5500–6500	TA: 178.4 ± 14.6 tC ha^{-1}	TA: 44.6 ± 4.3 tC ha^{-1}	TA: 312.2 ± 25.4 tC ha^{-1}	AGC-TA, BGC-TA, SOC-TA: calculated based on [72] for the case in Mekong River Delta

* Common tree density, not specifically those reported in the cited literature, ** TA: time-average, SR-AF: sequestration rate of AF system, SR-sole: sequestration rate of sole crop system. In the C estimation, for the AGC inputs without standard error, we applied a 10% standard error from the mean. AGC input without standard error *** RR-TA: response ratio to estimate SOC of the eight key existing AFs. The land cover before conversion into those AFs was assumed as logged over forest with SOC ranges from 68–205 tC ha^{-1} [14]. # AGC-TA is for short-rotation (3–5 years) as the current most popular acacia systems in Vietnam. AGC-SR is for long-rotation (8–12 years), for expansion.

Table A2. Sequestered carbon in the existing areas of agroforestry.

Agroforestry System *	Area (10^3 ha)	AGC (mil tCO_2e) Total	SE	BGC (mil tCO_2e) Total	SE	SOC (mil tCO_2e) Total	SE	TOC ** (mil tCO_2e) Total	SE
Melaleuca-based	246	161	6.19	40	1.83	281	10.8	482	18.8
Robusta-based	245	12	0.12	2.4	0.57	123	29.1	137	29.5
Rhizophora-based	149	86	4.04	320	8.93	211	9.97	617	22.4
Acacia-based	130	12	1.40	2.4	0.57	62	14.6	76.1	15.6
Rubber-based	21	1.9	0.10	0.4	0.09	9.8	2.3	12.0	2.4
Arabica-based	11	0.5	0.06	0.1	0.02	5.3	1.25	5.9	1.3
Cashew-based	10	2.1	0.10	0.4	0.03	4.9	1.17	7.4	1.25
Tea-based	10	0.5	0.04	0.1	0.02	4.8	1.13	5.4	1.17
All systems	820	275	12.0	366	12.1	701	70.4	1343	92.4

* ordered by the total area in the country, ** total of AGC, BGC and SOC.

Table A3. Suitable areas for agroforestry expansion by species and region under baseline and future climate.

ER *	Acacia Baseline	RCP 4.5	RCP 8.5	Arabica Baseline	RCP 4.5	RCP 8.5	Robusta Baseline	RCP 4.5	RCP 8.5	Cashew Baseline	RCP 4.5	RCP 8.5	Tea Baseline	RCP 4.5	RCP 8.5
						Highly suitable area (thousand ha)									
NW	12	22	23	19	6	3	11	8	6			0.01	4	8	6
NE	94	112	109	13	3	1	81	9	5	0.01	0.01	0.01	13	38	32
RRD	41	42	42				23	0.01						8	1
NCC	42	55	47	0.01			3			0.01	0.01	0.01	12	1	1
SCC	2	0.01	1				0.01	0.01	0.01	0.01	0.01	0.01	0.01	0.01	0.01
CH	192	202	166	23	6	2	72	19	21	23	20	16	150	114	86
SE	35	1	1	0.01			0.01		0.01	1	0.01		2	0.01	
MRD	0.01														
Total	419	435	389	54	15	6	189	36	32	24	21	17	181	170	127
						Combined (highly and less) suitable area (thousand ha)									
NW	282	307	309	190	141	120	210	210	210	236	236	236	194	194	194
NE	472	491	493	342	102	80	411	413	414	369	373	373	345	345	345
RRD	72	72	72	39			68	68	68	41	41	41	40	40	40
NCC	250	250	250	66	5	2	232	232	232	175	176	176	173	173	173
SCC	99	99	99	1	0	0	82	77	84	48	42	52	69	68	72
CH	881	884	885	298	114	65	718	718	718	704	679	723	715	715	715
SE	344	337	353	1	0	0	257	254	263	209	207	218	243	241	251
MRD	8	8	8				8	8	8	8	8	8	7	7	7
Total	419	435	389	54	15	6	189	36	32	24	21	17	181	170	127

* North West (NW), North East (NE), Red River Delta (RRD), North central Coast (NCC), South Central Coast (SCC), Central Highlands (CH), South East (SE) and Mekong River Delta (MRD).

Table A4. Sequestered TOC in highly suitable with/out partial less-suitable areas under baseline and future climate.

| Scenario ** | Sequestered TOC (mil tCO$_2$e) by 2030 * | | | | | | | | |
| | Highly Suitable | | | Highly and 10% Less Suitable | | | Highly and 25% Less Suitable | | |
	Baseline	RCP4.5	RCP8.5	Baseline	RCP4.5	RCP8.5	Baseline	RCP4.5	RCP8.5
1	44 ± 4.5	46 ± 4.5	40 ± 4.1	65 ± 7.1	67 ± 7.2	62 ± 6.6	96 ± 11	98 ± 11	95 ± 11
2	3.2 ± 0.4	0.9 ± 0.1	0.4 ± 0.1	8.3 ± 1.1	3.0 ± 0.4	1.8 ± 0.2	16 ± 2.1	6.0 ± 0.8	4.0 ± 0.5
3	13 ± 2.1	2.5 ± 0.4	2.3 ± 0.4	26 ± 1.9	16 ± 1.8	16 ± 1.9	45 ± 4.5	37 ± 4.4	37 ± 4.5
4	2.3 ± 0.2	1.9 ± 0.2	1.6 ± 0.2	19 ± 4.2	19 ± 2.5	19 ± 2.5	45 ± 7.4	44 ± 5.7	45 ± 5.7
5	9.6 ± 1.2	8.9 ± 1.1	6.7 ± 0.9	18 ± 2.4	18 ± 2.3	16 ± 2.1	31 ± 4.2	31 ± 4.1	29 ± 3.8

* Average and standard error ** 1: expansion of acacia-, 2: arabica-, 3: robusta-, 4: cashew-, and 5: tea-based AF over the country.

References

1. UNFCCC. *Aggregate Effect of the Intended Nationally Determined Contributions: An Update*; UNFCCC: Marrakech, Morocco, 2016.
2. UNFCCC. Parties. Available online: https://unfccc.int/process/parties-non-party-stakeholders/parties-convention-and-observer-states?field_national_communications_target_id%5B514%5D=514 (accessed on 15 September 2020).
3. FAO. *The Agriculture Sectors in the Intended Nationally Determined Contribution: Analysis*; FAO: Italy, Rome, 2016.
4. Feliciano, D.; Ledo, A.; Hillier, J.; Nayak, D.R. Which agroforestry options give the greatest soil and above ground carbon benefits in different world regions? *Agric. Ecosyst. Environ.* **2018**, *254*, 117–129. [CrossRef]
5. Zomer, R.J.; Neufeldt, H.; Xu, J.; Ahrends, A.; Bossio, D.; Trabucco, A.; Van Noordwijk, M.; Wang, M. Global tree cover and biomass carbon on agricultural land: The contribution of agroforestry to global and national carbon budgets. *Sci. Rep.* **2016**, *6*, 29987. [CrossRef] [PubMed]
6. Lasco, R.D.; Delfino, R.J.P.; Espaldon, M.L.O. Agroforestry systems: Helping smallholders adapt to climate risks while mitigating climate change. *Wiley Interdiscip. Rev. Clim. Chang.* **2014**, *5*, 825–833. [CrossRef]
7. Catacutan, D.C.; van Noordwijk, M.; Nguyen, T.H.; Öborn, I.; Mercado, A.R. *Agroforestry: Contribution to Food Security and Climate-Change Adaptation and Mitigation in Southeast Asia*; World Agroforestry Centre (ICRAF) Southeast Asia Regional Program; ASEAN-Swiss Partnership on Social Forestry and Climate Change: Bogor, Indonesia, 2017.
8. van Noordwijk, M.; Bayala, J.; Hairiah, K.; Lusiana, B.; Muthuri, C.; Khasanah, M.; Mulia, R. Agroforestry solutions for buffering climate variability and adapting to change. In *Climate Change Impact and Adaptation in Agricultural Systems*; Fuhrer, J., Gregory, P.J., Eds.; CAB International: Wallingford, UK, 2014; pp. 216–232.
9. Simelton, E.; Dam, B.V.; Catacutan, D. Trees and agroforestry for coping with extreme weather events: Experiences from northern and central Vietnam. *Agrofor. Syst.* **2015**, *89*, 1065–1082. [CrossRef]
10. Cardinael, R.; Umulisa, V.; Toudert, A.; Olivier, A.; Bockel, L.; Bernoux, M. Revisiting IPCC Tier 1 coefficients for soil organic and biomass carbon storage in agroforestry systems. *Environ. Res. Lett.* **2018**, *13*, 124020. [CrossRef]
11. MONRE. *The Initial Biennial Updated Report of Vietnam to the United Nations Framework Convention on Climate Change*; MONRE: Hanoi, Vietnam, 2014.
12. Escobar-Carbonari, D.; Grosjean, G.; Läderach, P.; Nghia, T.D.; Sander, B.O.; McKinley, J.; Sebastian, L.; Tapasco, J. Reviewing Vietnam's Nationally Determined Contribution: A new perspective using the marginal cost of abatement. *Front. Sustain. Food Syst.* **2019**, *3*, 14. [CrossRef]
13. Vu, T.P.; Nguyen, T.M.L.; Nguyen, N.L.; Do, D.S.; Nguyen, X.Q.; Tran, V.L.; Ngo, D.Q.; Tran, V.C.; Nguyen, D.K.; Lai, V.C.; et al. *Final Report on Forest Ecological Stratification in Vietnam*; Research Centre for Forest Ecology and Environment: Hanoi, Vietnam, 2011.
14. Ziegler, A.D.; Phelps, J.; Yuen, J.Q.; Webb, E.L.; Lawrence, D.; Fox, J.M.; Bruun, T.B.; Leisz, S.J.; Ryan, C.M.; Dressler, W.; et al. Carbon outcomes of major land-cover transitions in SE Asia: Great uncertainties and REDD+ policy implications. *Glob. Chang. Biol.* **2012**, *18*, 3087–3099. [CrossRef]
15. Saah, D.; Tenneson, K.; Poortinga, A.; Nguyen, Q.; Chishtie, F.; Aung, K.S.; Markert, K.N.; Clinton, N.; Anderson, E.R.; Cutter, P.; et al. Primitives as building blocks for constructing land cover maps. *Int. J. Appl. Earth Obs. Geoinf.* **2020**, *85*, 101979. [CrossRef]
16. Shih, S.S. Spatial habitat suitability models of mangroves with *Kandelia obovata*. *Forests* **2020**, *11*, 477. [CrossRef]
17. Pistorius, T.; Ho, D.T.H.; Tennigkeit, T.; Merger, E.; Wittmann, N.; Conway, D. *Business Models for the Restoration of Short-Rotation Acacia Plantations in Vietnam*; UNIQUE Forestry and Land Use GmbH: Hanoi, Vietnam, 2016.
18. FAO. *A Framework for Land Evaluation*; Food and Agriculture Organization of the United Nations: Rome, Italy, 1976.
19. MARD. *National Guideline Number TCVN 8409:2010 on Land Evaluation Procedure for Land Use Planning at District Level*; MARD: Hanoi, Vietnam, 2010.
20. Nguyen, V.L. *Use of GIS Modelling in Assessment of Foresty Land's Potential in Thua Thien Hue Province of Central Vietnam*; Georg-August-University: Gottingen, Germany, 2008.
21. VAFS. *Determine Appropriate Regional Planting of A. Mangium × A. Auriculiformis for Timber Supply in North Central*; VAFS: Hanoi, Vietnam, 2010.

22. Hijmans, R.J.; Cameron, S.E.; Parra, J.L.; Jones, P.G.; Jarvis, A. Very high resolution interpolated climate surfaces for global land areas. *Int. J. Climatol.* **2005**, *25*, 1965–1978. [CrossRef]

23. IPCC-TGCIA. *General Guidelines on the Use of Scenario Data for Climate Impact and Adaptation Assessment*; Finnish Environment Institute: Helsinki, Finland, 2007.

24. Kamworapan, S.; Surussavadee, C. Performance of CMIP5 global climate models for climate simulation in Southeast Asia. In Proceedings of the IEEE Region 10 Annual International Conference, Proceedings/TENCON, Penang, Malaysia, 5–8 November 2017; pp. 718–722.

25. Saah, D.; Battles, J.; Gunn, J.; Buchholz, T.; Schmidt, D.; Roller, G.; Romsos, S. *Technical Improvements to the Greenhouse Gas (GHG) Inventory for California Forests and Other Lands*; Spatial Informatics Group: Pleasanton, CA, USA, 2015.

26. Mulia, R.; Khasanah, N.; Catacutan, D.C. Alternative forest plantation systems for Southcentral Coast of Vietnam: Projections of growth and production using the WaNuLCAS model. In *Towards Low-Emission Landscapes in Vietnam*; Mulia, R., Simelton, E., Eds.; World Agroforestry (ICRAF) Vietnam, World Agroforestry (ICRAF) Southeast Asia Regional Program: Hanoi, Vietnam, 2018; pp. 45–60.

27. Kuit, M.; Jansen, D.M.; Nguyen, V.T. *Manual for Arabica Cultivation*; Tan Lam Agricultural Product Joint Stock Company: Quang Tri, Vietnam, 2004.

28. Catacutan, D.C.; Huy, B.; Pham, T.V.; Mulia, R. *Gender, Land Use and Land Use Change in Vietnam: Coffee as Commodity Driver of Land Use Change in the Central Highlands of Vietnam*; World Agroforestry (ICRAF): Hanoi, Vietnam, 2015.

29. Mulia, R.; Do, H.T.; Pham, T.V.; Nguyen, Q.K.; Nguyen, T.K.D. *Green Growth Action Plan for Lam Dong Province for the Period of 2021–2030, Vision to 2050*; World Agroforestry (ICRAF): Hanoi, Vietnam, 2019.

30. UNFCCC. *Submission from the Socialist Republic of Vietnam in Response to Decision 4/CP.23*; UNFCCC: New York, NY, USA, 2018.

31. Chiriacò, M.V.; Perugini, L.; Bombelli, A.; Bernoux, M.; Gordes, A.; Kaugure, L. *Koronivia Joint Work on Agriculture: Analysis of Submissions on Topic 2(a)–Modalities for Implementation of the Outcomes of the Five In-Session Workshops*; Food Agriculture Organization of United Nations: Rome, Italy, 2018.

32. Bunn, C.; Läderach, P.; Ovalle Rivera, O.; Kirschke, D. A bitter cup: Climate change profile of global production of Arabica and Robusta coffee. *Clim. Chang.* **2015**, *129*, 89–101. [CrossRef]

33. Gomes, L.C.; Bianchi, F.J.J.A.; Cardoso, I.M.; Fernandes, R.B.A.; Filho, E.I.F.; Schulte, R.P.O. Agroforestry systems can mitigate the impacts of climate change on coffee production: A spatially explicit assessment in Brazil. *Agric. Ecosyst. Environ.* **2020**, *294*, 106858. [CrossRef]

34. Ovalle-Rivera, O.; Läderach, P.; Bunn, C.; Obersteiner, M.; Schroth, G. Projected Shifts in *Coffea arabica* Suitability among Major Global Producing Regions Due to Climate Change. *PLoS ONE* **2015**, *10*, e0124155. [CrossRef] [PubMed]

35. Pham, Y.; Reardon-Smith, K.; Mushtaq, S.; Cockfield, G. The impact of climate change and variability on coffee production: A systematic review. *Clim. Chang.* **2019**, *156*, 609–630. [CrossRef]

36. UTZ. *Climate Change and Vietnamese Coffee Production-Manual on Climate Change Mitigation and Adaptation in the Coffee Sector for Local Trainers and Coffee Farmers*; UTZ: Amsterdam, The Netherlands, 2016.

37. Duncan, J.M.A.; Saikia, S.D.; Gupta, N.; Biggs, E.M. Observing climate impacts on tea yield in Assam, India. *Appl. Geogr.* **2016**, *77*, 64–71. [CrossRef]

38. Han, W.; Li, X.; Yan, P.; Zhang, L.; Ahammed, G. Tea cultivation under changing climatic conditions. In *Global Tea Science: Current Status and Future Needs*; Sharma, V.S., Gunasekare, M.T.K., Eds.; Burleigh Dodds Science Publishing: Cambridge, UK, 2018; pp. 1–18.

39. Marx, W.; Haunschild, R.; Bornmann, L. Global Warming and Tea Production—The Bibliometric View on a Newly Emerging Research Topic. *Climate* **2017**, *5*, 46. [CrossRef]

40. Lin, B.B. The role of agroforestry in reducing water loss through soil evaporation and crop transpiration in coffee agroecosystems. *Agric. For. Meteorol.* **2010**, *150*, 510–518. [CrossRef]

41. Moreira, S.L.S.; Pires, C.V.; Marcatti, G.E.; Santos, R.H.S.; Imbuzeiro, H.M.A.; Fernandes, R.B.A. Intercropping of coffee with the palm tree, macauba, can mitigate climate change effects. *Agric. For. Meteorol.* **2018**, *256–257*, 379–390. [CrossRef]

42. de Souza, H.N.; de Goede, R.G.M.; Brussaard, L.; Cardoso, I.M.; Duarte, E.M.G.; Fernandes, R.B.A.; Gomes, L.C.; Pulleman, M.M. Protective shade, tree diversity and soil properties in coffee agroforestry systems in the Atlantic Rainforest biome. *Agric. Ecosyst. Environ.* **2012**, *146*, 179–196. [CrossRef]

43. ICC. *Country Coffee Profile: Vietnam*; ICC: Nairobi, Kenya, 2019.

44. Hopkins, G. Vietnamese Tea: Origin, Geography, Cultivars, Current Affairs. Available online: https://specialtyteaalliance. org/world-of-tea/vietnamese-tea-origin/#:~{}:text=Themaincentresforoolongbyvolumeofteaproduced (accessed on 12 October 2020).

45. Nambiar, E.S.; Harwood, C.E.; Kien, N.D. Acacia plantations in Vietnam: Research and knowledge application to secure a sustainable future. *South. For.* **2015**, *77*, 1–10. [CrossRef]

46. Mokievsky, V.O.; Son, T.; Dobrynin, D.V. The Dynamics of Mangroves in the Mekong Delta (Vietnam): From Degradation to Restoration. *Dokl. Earth Sci.* **2020**, *494*, 745–747. [CrossRef]

47. Zhu, T.; Trinh, M. Climate Change Impacts on Agriculture in Vietnam. Available online: https://www. slideshare.net/tingjuzhu/climate-change-impacts-on-agric (accessed on 5 September 2020).

48. Binh, C.T.; Phillips, M.J.; Demaine, H. Integrated shrimp-mangrove farming systems in the Mekong delta of Vietnam. *Aquac. Res.* **1997**, *28*, 599–610. [CrossRef]

49. Truong, T.D.; Do, L.H. Mangrove forests and aquaculture in the Mekong river delta. *Land Use Policy* **2018**, *73*, 20–28. [CrossRef]

50. Kuit, M.; Guinée, L.; Jansen, D.C.S. *The Carbon Footprint of Vietnam Robusta Coffee Source or Sink?* IDH: Utrecht, The Netherlands, 2019.

51. Do, V.H.; La, N.; Mulia, R.; Bergkvist, G.; Dahlin, A.S.; Nguyen, V.T.; Pham, H.T.; Öborn, I. Fruit Tree-Based Agroforestry Systems for Smallholder Farmers in Northwest Vietnam—A Quantitative and Qualitative Assessment. *Land* **2020**, *9*, 451. [CrossRef]

52. Catacutan, D.C.; Do, T.H.; Simelton, E.; Hanh, V.T.; Hoang, T.L.; Patton, I.; Hairiah, K.; Le, T.T.; van Noordwijk, M.; Nguyen, M.P.; et al. *Assessment of the Livelihoods and Ecological Conditions of Bufferzone Communes in Song Thanh National Reserve, Quang Nam Province, and Phong Dien Natural Reserve, Thua Thien Hue Province*; World Agroforestry (ICRAF): Hanoi, Vietnam, 2017.

53. Luedeling, E.; Smethurst, P.J.; Baudron, F.; Bayala, J.; Huth, N.I.; van Noordwijk, M.; Ong, C.K.; Mulia, R.; Lusiana, B.; Muthuri, C.; et al. Field-scale modeling of tree-crop interactions: Challenges and development needs. *Agric. Syst.* **2016**, *142*, 51–69. [CrossRef]

54. Mulyoutami, E.; Roshetko, J.M.; Martini, E.; Awalina, D. Janudianto Gender roles and knowledge in plant species selection and domestication: A case study in South and Southeast Sulawesi. *Int. For. Rev.* **2015**, *17*, 99–111. [CrossRef]

55. Sari, R.; Saputra, D.; Hairiah, K.; Rozendaal, D.; Roshetko, J.; van Noordwijk, M. Gendered species preferences link tree diversity and carbon stocks in cacao agroforest in Southeast Sulawesi, Indonesia. *Land* **2020**, *9*, 108. [CrossRef]

56. Ureta, J.U.; Evangelista, K.P.A.; Habito, C.M.; Lasco, R. Exploring gender preferences in farming system and tree species selection: Perspectives of smallholder farmers in southern Philippines. *J. Environ. Sci. Manag.* **2016**, *19*, 56–73.

57. Minang, P.A.; Duguma, L.A.; Bernard, F.; Mertz, O.; van Noordwijk, M. Prospects for agroforestry in REDD+ landscapes in Africa. *Curr. Opin. Environ. Sustain.* **2014**, *6*, 78–82. [CrossRef]

58. Rosenstock, T.S.; Wilkes, A.; Jallo, C.; Namoi, N.; Bulusu, M.; Suber, M.; Mboi, D.; Mulia, R.; Simelton, E.; Richards, M.; et al. Making trees count: Measurement and reporting of agroforestry in UNFCCC national communications of non-Annex I countries. *Agric. Ecosyst. Environ.* **2019**, *284*, 106569. [CrossRef]

59. Clouse, C. *Vietnam's New Conservation Plan Prioritizes Trees and People. Emissions? Not So Much*; Mongabay: Brooklyn, NY, USA, 2020.

60. Mulia, R.; Nguyen, M.P.; Do, H.T. Forest and crop-land intensification in the four agro-ecological regions of Vietnam: Impact assessment with the FALLOW model. In *Towards Low-Emission Landscapes in Vietnam*; Mulia, R., Simelton, E., Eds.; World Agroforestry (ICRAF) Vietnam, World Agroforestry (ICRAF) Southeast Asia Regional Program: Hanoi, Vietnam, 2018; pp. 89–108.

61. Kalita, R.; Kalita, R.M.; Das, A.K.; Nath, A.J. Carbon stock and sequestration potential in biomass of tea agroforestry system in Barak Valley, Assam, North East India. *Int. J. Ecol. Environ. Sci.* **2017**, *42*, 107–114.

62. Phan, V.T.; Huynh, V.Q.; Dang, T. Determination of raw biomass, yield and income from intercropping models of coffee (*Coffea canephora* Pierre) compare with perennial trees in tropical climate. *AJCS* **2020**, *14*, 1835–2707. [CrossRef]

63. Guillemot, J.; le Maire, G.; Munishamappa, M.; Charbonnier, F.; Vaast, P. Native coffee agroforestry in the Western Ghats of India maintains higher carbon storage and tree diversity compared to exotic agroforestry. *Agric. Ecosyst. Environ.* **2018**, *265*, 461–469. [CrossRef]

64. Nguyen, M.P.; Mulia, R.; Nguyen, Q.T. *Fruit Tree Agroforestry in Northwest Vietnam: Sample Cases of Current Practices and Benefits*; World Agroforestry (ICRAF): Hanoi, Vietnam, 2020.

65. Segura, M.; Kanninen, M.; Suárez, D. Allometric models for estimating aboveground biomass of shade trees and coffee bushes grown together. *Agrofor. Syst.* **2006**, *68*, 143–150. [CrossRef]

66. Avtar, R.; Suzuki, R.; Sawada, H. Natural forest biomass estimation based on plantation information using PALSAR data. *PLoS ONE* **2014**, *9*, e86121. [CrossRef]

67. Mlagalila, H.E. *Assessment of Volume, Biomass, and Carbon Stock of Cashew Nuts Trees in Liwale District, Tanzania*; Sokoine University of Agriculture: Morogoro, Tanzania, 2016.

68. Vietnam Rubber Association. *Vietnam's Rubber Industry on International Integration*; Vietnam Rubber Association: Ho Chi Minh City, Vietnam, 2015.

69. Brahma, B.; Sileshi, G.W.; Nath, A.J.; Das, A.K. Development and evaluation of robust tree biomass equations for rubber tree (*Hevea brasiliensis*) plantations in India. *For. Ecosyst.* **2017**, *4*, 14. [CrossRef]

70. Dang, T. *Biomass of Rhizophora Apiculata Forest*; Minh Hai Centre for Mangrove Forest Technique Application: Ho Chi Minh City, Vietnam, 2011.

71. Bridgham, S.D.; Megonigal, J.P.; Keller, J.K.; Bliss, N.B.; Trettin, C. The carbon balance of North American wetlands. *Wetlands* **2006**, *26*, 889–916. [CrossRef]

72. Phan, T.; Subasinghe, S.M.C.U. An assessment of the carbon stocks of Melaleuca forests in the lower U Minh National Park in Ca Mau province of Southern Vietnam. *Am. J. Eng. Res.* **2018**, *7*, 305–315.

Publisher's Note: MDPI stays neutral with regard to jurisdictional claims in published maps and institutional affiliations.

 *land*

Article

Agroforestry as Policy Option for Forest-Zone Oil Palm Production in Indonesia

Edi Purwanto [1,*], Hery Santoso [2], Idsert Jelsma [1], Atiek Widayati [1], Hunggul Y. S. H. Nugroho [3] and Meine van Noordwijk [4,5]

[1] Tropenbos Indonesia, Bogor 16163, Indonesia; ijelsma@gmail.com (I.J.); atiekwidayati@tropenbos-indonesia.org (A.W.)
[2] Java Learning Centre (JAVLEC), Klidon, Yogyakarta 55581, Indonesia; herys@javlec.org
[3] Ministry of Environment and Forestry, Makassar 90243, Indonesia; hunggul@balithutmakassar.org
[4] World Agroforestry (ICRAF), Bogor 16155, Indonesia; m.vannoordwijk@cgiar.org
[5] Plant Production Systems, Wageningen University and Research, 6708 PB Wageningen, The Netherlands
* Correspondence: edipurwanto@tropenbos-indonesia.org

Received: 31 October 2020; Accepted: 16 December 2020; Published: 18 December 2020

Abstract: With 15–20% of Indonesian oil palms located, without a legal basis and permits, within the forest zone ('Kawasan hutan'), international concerns regarding deforestation affect the totality of Indonesian palm oil export. 'Forest zone oil palm' (FZ-OP) is a substantive issue that requires analysis and policy change. While spatial details of FZ-OP remain contested, we review literature on (1) the legal basis of the forest zone and its conversion, (2) social stratification in oil palm production (large-scale, plasma and independent growers), and (3) environmental consequences of forest conversion to FZ-OP, before discussing policy options in a range of social and ecological contexts. Policy options range from full regularization (as FZ-OP stands could meet international forest definitions), to conditional acceptance of diversified smallholder plantings in 'agroforestry concessions', to gradually phasing out FZ-OP and eviction/destruction. A nuanced and differentiated approach to FZ-OP is needed, as certification of legality along supply chains is vulnerable to illegal levies and corruption. Corporate actors trading internationally can avoid use of uncertified raw materials, effectively shifting blame and depressing farmgate prices for domestic-market palm oil, but this will not return forest conditions or stop further forest conversion. We discuss an agenda for follow-up policy research.

Keywords: certification; deforestation; palm oil; forest classification; Jambi; legality; independent smallholders; agroforestry concessions; Sumatra; West Kalimantan

1. Introduction

A reader cannot safely assume to understand what the word 'forest' means in any new context, as the ecological (vegetation-based) and social (institutional, rule-based) meanings of the word 'forest' only partially overlap [1,2]. There can be 'trees outside a forest' [3] and 'forests without trees' [4]. Conversion of natural forest to plantation forestry is not considered to be deforestation, an issue that sparked debate when the UNFCCC climate change convention considered forest-specific policies [5]. What is excluded from the 'tree' concept, at the heart of common forest definitions, also matters: rubber (*Hevea brasiliensis*) can be considered to be a forest tree when grown for timber, or an agricultural commodity when tapped for its latex [6]. A number of global studies, however, have included the conversion of rubber monoculture (regardless of its primary purpose) to an oil palm monoculture under global 'deforestation' statistics [7]. The debate whether or not palms are included under the 'tree' concept has direct implications for whether or not the conversion of natural forest to an oil palm monoculture can be called 'deforestation' [8] and for how much a 'deforestation-free' commitment of the palm oil industry means [9]. National definitions of the tree and forest concepts can differ from the global ones [10,11]. At the social-ecological system scale of a landscape, the concept of 'agroforest' describes tree-based vegetation managed by farmers, who often see labeling this as a forest as a threat [12,13]. Institutionally such land use may be legalized under 'community-based forest management' rules [14], but these arrangements may maintain forest authorities in the 'landlord' role, expecting a share in any harvestable goods or sellable services that the land may generate [15]. The absence, at least until recently, of formal recognition for agroforestry as a valid form of land use intermediate to 'forest' and 'agriculture' has not prevented the existence and spread of such land uses that defy the rules [16] (Figure 1).

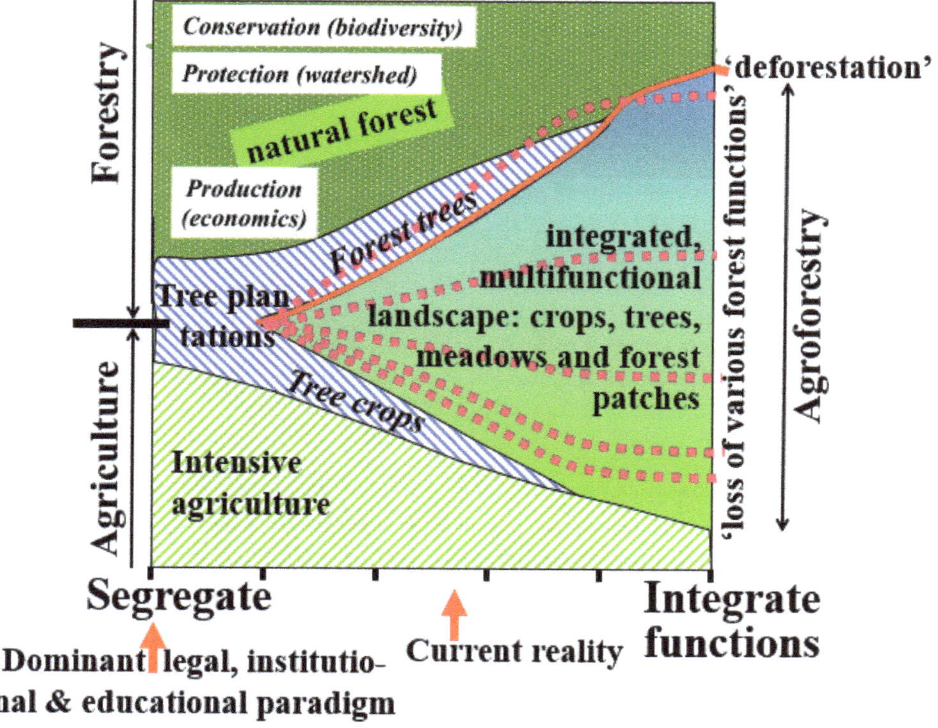

Figure 1. Forest-Agriculture interface with the current segregated policy interpretation distinct from the more fluid 'integrated' reality in the landscape, with consequences for the meaning of 'deforestation' and 'loss of forest functions' [17].

In the international debate over consequences of tropical commodity production, 'forest' is primarily seen as a vegetation type or ecosystem, observable through remote sensing [18], with a rather arbitrary tree cover threshold used in a dichotomous classification. Global data sets based on a continuum interpretation with tree cover as metric are available and have been used to monitor changes in tree cover in agricultural lands [19]. Concerns about deforestation, and the concomitant loss of biodiversity and net C emissions, have led to institutional support in the climate change convention for government action to support 'reforestation' [20] and 'reducing emission from deforestation and forest degradation' [21,22]. In parallel, citizens in importing countries expressing concern about the footprint of their consumption have induced corporate actors to dissociate themselves from deforestation-linked supply chains [23]. Palm oil has been a primary target for the emergence of voluntary private standards [24,25]. The Roundtable on Sustainable Palm Oil (RSPO) and related institutions have triggered responses in Malaysia and Indonesia (jointly around 85% of global palm oil production) that saw their national sovereignty at risk in international markets [26]. A study of RSPO impacts in Kalimantan suggested that it increased forest conversion in agricultural lands but decreased such within the forest estate [27].

Recent estimates, discussed below, show that, based on remote sensing, 15–20% of Indonesian oil palms are located within the 'forest zone' (Kawasan hutan), a primarily institutional indication of land status. We indicate these as Forest Zone Oil Palm (FZ-OP). The FZ-OP percentage can be interpreted in multiple ways. To an international audience that has come to believe that most, if not all, of Indonesian palm oil is linked to deforestation, the number may seem to be small. To the Indonesian palm oil sector, it may give the impression of some bad apples that need to be dealt with, as they spoil the good name of the vast majority of those falsely accused of deforestation. To the forest authorities, the numbers are a clear sign of illegality, as current rules do not allow oil palm to be planted within the forest zone as FZ-OP. Actually, oil palms are the most readily recognizable of the four major tree crops (coffee, cocoa, rubber and oil palm) that occur both outside and inside the forest zone [28], but part of the smallholder oil palm may be part of a finer-grained mosaic and not as easily noticeable. Key questions regarding FZ-OP are: (1) What came first: local land use or designation of the land as part of the forest zone?, (2) How does the land use with tree crops relate to the primary designated forest function in (A) production forest (income generation), (B) protection forest (watershed protection) and (C) Conservation areas (Biodiversity), with recent additional interests in restricting drainage of peat soils in the production forest zone, (3) How do the various stakeholders along the palm-oil value chain respond to the lack of legality of forest-based oil palm production?

Issues related to oil palm (and its product palm oil) require an understanding of the whole value chain, as concerns of end-users, often based on limited or biased information they have access to, inform the product manufacturers, who want to be able to claim a 'clean' product, and their suppliers, the partly refined products derived from Crude Palm Oil (CPO), processed from Fresh Fruit Bunches (FFB) transported from farmgate to mill, and grown by smallholders or large-scale plantations. Current leverage on 'illegal' palm oil may primarily come from making mills responsible for their source areas. However, it is not clear whether these efforts mostly lead to 'shifting blame' [29,30] or whether they actually contribute to changing land-use and effectively protecting forests.

While uncertainty over the exact numbers and locations on FZ-OP continues, we set out to formulate lessons from earlier interactions between forest authorities and the other tree crops (coffee, cocoa, rubber) and to contextualize and analyze policy options and how these might be applied to the various contexts that contribute to the overall issue. We will review the available data and literature on:

- The history of forest legality and oil palm expansion in Indonesia as context for current forest-related issues of local land use versus designation of the land as part of the forest zone;
- Spatial analysis of FZ-OP at the intersection of forests and tree crops for Indonesia as a whole and zoomed in on two provinces with higher-resolution data;
- Social and economic concerns on oil palm expansion and the role of smallholders in FZ-OP;
- Environmental concerns on tree crops in relation to the primary designated forest function in (A) production forest (income generation), (B) protection forest (watershed protection) and (C) Conservation areas (Biodiversity);
- Policy options linked to forest-related contexts, informed by the responses of various stakeholders along the palm-oil value chain to FZ-OP.

We conclude by formulating more specific policy research questions as a follow-up.

2. Methods

2.1. Literature Review

Starting from several recent reviews of oil palm in Indonesia, we used a 'snowball' method to identify more recent papers citing these sources and followed up on the citation network thus established to document a synthetic view on the various sub-topics.

2.2. Maps

For a more detailed analysis of the different forest designations within the forest zone and for the time frames of expansion within the forest zone, we compiled data for Jambi, a province in Sumatra around the Equator, and West-Kalimantan. The two provinces were selected to represent the two islands where oil palm is major (Sumatra and Kalimantan) while they had different historical settings on traditional land use of agroforest and its dynamics. Availability of the data also determined the choice of the two provinces. We traced the development of oil palm and agroforestry areas in the 2000–2020 period in each of the Forest-zone categories and sought the trajectories of changes and observed interlinks between oil palm, forest and agroforestry. Where the data were available, we incorporated the other major agroforestry tree-crops; otherwise, we utilized the common class of 'mixed agricultural tree-crops' mapped in most datasets. The following dataset was utilized for the analyses (Table 1).

Table 1. Maps utilized for case studies.

No	Map Title	Year (s)	Extracted Class (es)	Source
1	Peta Penutupan Lahan Indonesia (Indonesia Land Cover Map)	2000, 2009, 2018	Agroforest	[31]
2	Ecological Vegetation Map of West Kalimantan	2019	Agroforest Rubber plantations Oil palm plantations	[32]
3	Global map of smallholder and industrial closed canopy oil palm plantations	2019	Oil palm plantations Independent smallholder oil palm	[33]
4	National Main Commodity Maps in Indonesia	2019	Oil palm Rubber	[28]
5	Kawasan hutan Provinsi Kalimantan Barat (Forest-zone Lands of West Kalimantan Province)	2014		[34]
6	Kawasan hutan Provinsi Jambi (Forest-zone Lands of Jambi Province)	2014		[35]

2.3. Public Consultation

In a webinar in September 2020, attended by 374 participants, the issue of overlap between Indonesian forest estate and oil palm was highlighted and a number of the basic data presented; the questions and discussions formed the start of this manuscript. Specific aspects discussed are listed in Appendix A at the end of this manuscript.

3. Indonesian Context

3.1. Forest Legality

While efforts to agree on a forestry law during the Colonial period stranded [36], the post-independence (1945) constitution clarified that all the country's resources were for the benefit of the Indonesian people—with subsequent debate on the degree to which the Government of Indonesia has to respect prior claims of customary communities ('Masyarakat Hukum Adat') [37]. The Basic Agrarian Law of 1960 that recognizes colonial period documents as a basis for valid rights to land and the Forestry Law of 1967 was considered to be complementary but had gaps between them [38]. The 1999 revision of the

Forestry Law specified that state ownership claims within the designated 'forest zone' ('Kawasan Hutan') required completion of a gazettement procedure (verifying that there are no valid pre-existing claims), which has been very slow to implement [39], leaving doubts on the legality of government forestry policies that treat the Forest-zone as government-owned.

The total area of forest-zone lands in Indonesia (63% of total land area) is subdivided by primary function, as 'Production Forest' (68.8 M ha; 36.1% of the country), 'Protection Forest' (29.7 M ha; 15.6% of the country) and 'Conservation Areas' (22.1 M ha; 1.6% of the country) [40]. Logging is only allowed in the production forests, with further restrictions in part of the zone, and conversion to monocultural plantations for the pulp and paper industry in other parts. Although logging-based economic interests prevail in this zone, the production forests still have a recognized role in biodiversity conservation, carbon storage and maintenance of watershed functions. Most of the Protection Forest (the Indonesian term could also indicate '(watershed) protecting' forests) are on mineral soils and have been defined through a scoring system emphasizing slope and similar criteria for soil and water protection. For peatland areas, there are separate criteria for determining a Protection Function ('Fungsi Lindung') indication, based on peat depth and the peatland moratorium policy [41]. The total area of Protection Function in Indonesia's peatlands is 12.3 M ha [42].

Building on the specific example of the damar agroforests in Krui (West Lampung, Sumatra) where government ownership claims to forest-like vegetation could not be substantiated [43,44], it was realized that 'agroforests', of various specific histories and intensities of management, were included in the Forest-zone, either with Production or Protection Forest designations, but also claimed by local communities as existing before the Forestry law was formalized [45]. Since then, a government commitment to 'community-based forest management' and 'village forests' has allowed some conflict zones to be resolved, but much remains to be done [46,47]. Meanwhile, calls to redefine the Indonesian forest estate and get the boundaries of its Forestry Regulatory Framework and Agrarian rules right [12,48–51] were largely unheeded. Global interest in and country-level expectations of, new financing mechanisms through Reducing Emissions from Deforestation and (forest) Degradation (REDD+) created a new dynamic where issues of 'forest legality' obtained new relevance [52,53]. Indonesia's challenge to balance the new global environmental agenda on forest conservation, with the existing development deficits is part of a wider global pattern [54].

3.2. Complexity of Frontier Situation

In Southeast Asia's long engagement with global markets, cocoa, coffee, fast-growing trees, oil palm, rubber and shrimp have all had periods of rapid expansion or 'booms' when the region proved to be a low-cost producer in which land, labor, know-how, and market access could be rapidly mobilized [55]. The 'land grab' literature tries to answer key questions on who seeks to exercise control over land to grow export-oriented crops under boom conditions, how would-be producers navigate regulatory powers and market forces to gain control over land, and how booms differentially affect areas with secure and insecure land control relations.

Since the 19th century, commodities such as rubber [56], coffee [57] and cocoa [58] have boomed in specific parts of Indonesia, creating new 'forest frontier' conditions, involving local people, attracting investors to come in, but also stimulating people from other location to move in, spontaneously and/or in government-sponsored transmigration [59]. The relationships between spontaneous migrants, large-scale plantation companies, local communities and various branches of government have become complex in many locations [60–63]. Due to its location, boom crop production is closely associated with the issue of deforestation and environmental degradation [64].

Indigenous people and local communities (jointly described here as IP/LC) in forest frontier areas have had a long tradition of a 'dual economy' [65] where food security continued to depend on swiddening, while cash-crop agroforestry (rubber, coffee, resin trees and now oil palm) provided a financial basis for their livelihoods [66]. Where the terms of trade were sufficiently favorable, staple food production could be 'outsourced' [67]. Accordingly, most of the oil palm grown by IP/LC, just not to say smallholders, has not been a direct cause of deforestation as they developed in a village environment of orchards, old rubber groves and swidden fallows. Smallholders also tend to use fertilizer sparingly, as they are unable to access subsidized fertilizer supplies that are monopolized by larger plantations [68].

While global policies and markets are often held responsible for accelerated deforestation in the tropics [69], local knowledge is generally assumed to lead to overall positive outcomes in matters of conservation [70,71]. Even in matters of primary forest conversion, local communities are usually presented as the best managers to reconcile conservation and development [69]. Increased attention for forest protection in the context of REDD+ may have increased pressure on converting agroforests outside the forest zone to become oil palm monocultures [72].

The Indonesian forest frontier is home to approximately 26 thousand villages (often defined as 'forest villages') and more than 37 M peoples, with some 18% still struggling to escape from poverty based on the national standard [73]. Most of the forest villages are also not formally registered yet, as nationally only 30% of villages are already registered, mostly in Java. They have no clear and defined boundary and are dominated by state forest land, with formal restrictions to use for crops to generate food or income. In such landscapes, boom crop production, such as oil palm, has taken place. Expansion of agriculture into the forest area, often following onto a logging phase that brought people and road access (however poor) into the forest zone, is considered because of these complexities, that for decades have remained unresolved.

3.3. Total Oil Palm Area

According to the most recent official figure, the oil palm area in Indonesia was 14.7 M ha in 2019 [74]. Several organizations have also mapped oil palm areas for different objectives and with a range of methods (Table 2). Bahktiar et al. [75] claimed a substantially larger area (16.8 M ha), by including areas cleared for oil palm but without stands recognizable by remote sensing yet. The lowest recent estimate was 11,530,000 ha [33], in a study that only included fully-developed stands. Time-series data of oil palm for Borneo were analyzed for forest loss or deforestation [76] and biodiversity impacts [77]. The breakdowns of oil palm areas varied, with one study differentiating 'industrial' and 'smallholder' oil palm [33], while another study [75] focused on oil palm areas within forest zone lands. Condro et al. [28] mapped major commodities including coffee and cacao, and obtained area estimates close to the data from the Ministry of Agriculture. Across Indonesia, the largest oil palm areas are in Sumatra and the second largest in Kalimantan, covering respectively 6.4 M ha and 4.9 M ha in 2018 [74], while another study [75] mentioned 10.5 M ha and 5.7 M ha, respectively, for these two islands. Oil palm is expanding in W and SE Sulawesi, and in Papua, but current areas remain relatively small.

Table 2. Published estimates of the total area of oil palm in Indonesia.

No	Oil Palm Hectarage	Year	Approach and Methods	Notes	Reference
1	16,800,000	2015–2017	Multi-data analyses	Including unplanted area for industrial oil palm within forest zone	[75]
2	14,896,964	2019	Automatic classification using multi-data in GEE	RS methods; Oil palm and other major commodities	[28]
3	14,724,420	2019	Using trade statistics and yield data as a basis	Official Oil Palm statistics; industrial (private and state) and smallholders	[74]
4	11,530,000	2019	Automatic classification of sentinel imageries with NN	RS methods; Differentiation of industrial and smallholders	[33]
5	11,100,000	2015–2017	Visual interpretation of Landsat imageries	Oil palm and deforestation	[78]

The distribution of oil palm in Indonesia is uneven and strongly related to climatic conditions. In North Sumatra (Figure 2A), the province with the most even rainfall distribution and virtually no dry season oil palm can reach up to 50% of the area at district (Kabupaten) level; elsewhere in lowland Sumatra and the southern half of Kalimantan it can reach up to 30% of the area, but in most of Indonesia, oil palm forms less than 10% of the landscape or is virtually absent. For Indonesia as a whole oil palm covers about 8% of the total land area. The fraction of oil palm under smallholder ownership (see below for the distinctions within this category) is approaching 50% for Sumatra, but half of that in most of Kalimantan (Figure 2B). The recent expansion of oil palm, indicated by relatively young stands, is evident outside of the areas that already have a large oil palm fraction (Figure 2C).

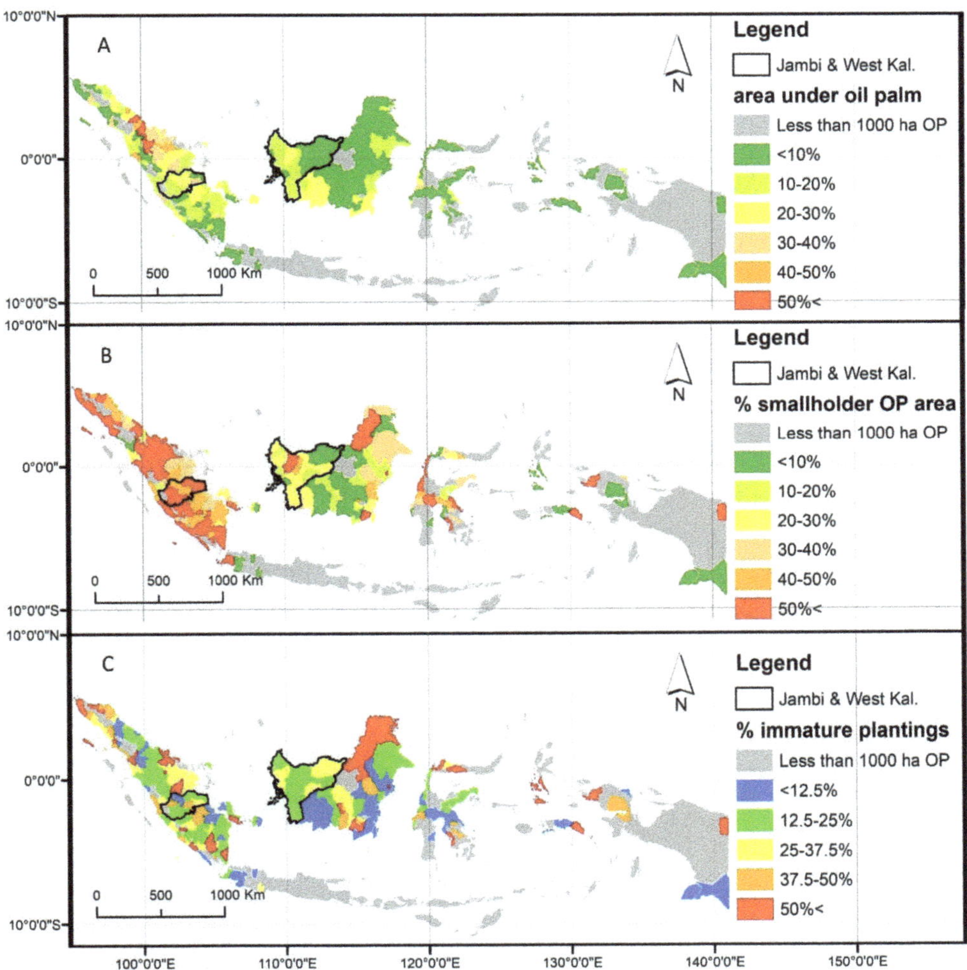

Figure 2. Maps on 2017 oil palm distribution at the district level in Indonesia, highlighting Jambi and West Kalimantan provinces; (**A**) Area fraction under oil palm; (**B**) Share of oil palm cultivated by smallholders; (**C**) Share of oil palm that does not yield yet as it is too young (usually less than three years old); Sources: Own maps, oil palm data from [79]; district boundary and area data from [80].

Several authors have quantified oil palm development within the forest zone in Indonesia [76,78]. Estimates vary from 2.5 [76] to 3.4 M ha [75]. This implies that 15–20% of Indonesian oil palms grow within the designated forest zone. The dataset of [75] allows some further exploration of the geographical patterns of Indonesia's Forest-zone oil palm (FZ-OP). The relative share in Indonesia's FZ-OP at the province level does not match shares in total oil palm production. With 31.5% of

Indonesian oil palm, the two provinces of Riau + C Kalimantan have 66.8% of its FZ-OP (Riau 20.7% OP, 42.1% of FZ-OP; C. Kalimantan 10.9% of OP and 24.7% of FZ-OP). With 41.2% of Indonesian OP, the rest of Sumatra has 24.0% of its FZ-OP; with 23.7% of Indonesian OP, the rest of Kalimantan has 7.7% of its FZ-OP and with 3.6% of Indonesian OP, Eastern Indonesia + Java have 1.4% of its FZ-OP. Across Indonesia, 2.96% of FZ-OP is designated as conservation area, 4.71% as (watershed) protection forest, 15.15% as production forest with restrictions, 45.4% as regular production forest, and 34.8% as production forest indicated for conversion. The total area of FZ-OP is 2.88 times the amount of production forest indicated as 'conversion forest' (planned, legal deforestation).

For further exploration of that pattern as well as to showcase the development of oil palm gardens as part of agroforestry pathways, we focused on two provinces, one each in Sumatra (Jambi ranked #7 in oil palm production and #5 in FZ-OP) and Kalimantan (W. Kalimantan, ranked #3 in oil palm production and #6 in FZ-OP). Both are outside the historical oil palm core area in N. Sumatra, but they were part of the expansion since the 1990s; the district-level presence of oil palm within these provinces varies from 0–30%. The two provinces have harmonized their Kawasan Hutan areas with the provincial planning (RTRWP) with ministerial decrees that legalized the status of 'Forest-zone lands.' Therefore, issues of agroforest and oil palm practices in these two provinces are not burdened by different interpretations of forest legality between the central government (MoEF) and the provincial governments.

4. Oil Palm in the Forest Zones of West Kalimantan and Jambi

4.1. Data by Province and Forest Category

The total area of Forest-zone lands, encompassing all Conservation Areas (National Park and forest reserves), Protection Forest and all Production Forest status (HP, HPT, HPK) in West Kalimantan is 8.1 M ha or 56% of the province area, while for Jambi it is 2 M ha or 43% of the province.

For West Kalimantan, agroforest area occupied 38% of all Production Forest status and 16% of Protection Forest areas [31], while according to [32], agroforest covered <10% (Table 3). Some of the agroforests in [32] are mapped by [31] as disturbed forest, and that was the main differentiating factor of forest and agroforest areas in the two maps. Major commodities of oil palm and rubber are negligible in West Kalimantan's Forest-zone lands, the highest being oil palm in Production Forest areas (4%) mapped by [28].

Table 3. Areas of agroforest and major tree crops in West Kalimantan and Jambi in 2018–2019.

Agroforest and Major Tree Crops	Source	Production Forest (ha)	Protection Forest (ha)	Conservation Areas (ha)
West Kalimantan, total area		**4,444,111**	**2,295,424**	**1,423,567**
Agroforest	[32] +#	318,521 (7%)	66,631 (3%)	10,576 (1%)
	[31] +	1,676,606 (38%)	363,714 (16%)	35,053 (2%)
Rubber plantations	[32]	7889 *		14 *
	[28]	8157 *	764 *	264 *
Oil palm plantations	[32]	80,771 (2%)	6663 *	1852 *
	[28]	186,037 (4%)	29,753 (1%)	4610 *
	[33]	29,382 *	3017 *	59 *
Independent smallholder oil palm	[33]	8267 *	906 *	53 *
Jambi, total area		**1,235,639**	**180,170**	**634,431**
Agroforest	[31] +#	231,112 (19%)	12,463 (7%)	35,594 (6%)
Rubber plantations	[28]	735 *		
Oil palm plantations	[28]	99,848 (8%)	2964 (2%)	20,065 (3%)
	[33]	40,399 (3%)	337 *	902 *
Independent smallholder oil palm	[33]	38,164 (3%)	1032 (1%)	1915 *

+ processed/reclassified for this paper, # utilized for further analyses, * percentage lower than 1% of the respective forest-zone land category.

In Jambi, agroforest occupied 19% of the land with Production Forest status, and <10% for Protection Forest and Conservation Areas [31] (Table 3). Rubber and oil palm monocultures were <10% in all Forest-zone land categories in Jambi [28,33]. A breakdown of FZ-OP by elevation and forest category (Figure A1) shows most are below 100 m above sea level.

Taking the total amount of palm oil produced in each of the provinces as point of references, the likely source areas differ between the two provinces, and according to three data sources specified in Table 4, with 6–17% derived from the forest zone. Within the forest zone, the fractions derived from production forest are highest (even after correction for the total area), followed by protection forest and conservation areas.

Table 4. Likely source area of palm oil produced in West Kalimantan and Jambi according to three spatial data sources, expressed as %, assuming homogenous productivity per unit land.

| Province | West Kalimantan | | | Jambi | |
Data Source	[75]	[33]	[28]	[33]	[28]
Non-forest land	93.65	96.30	89.93	89.05	83.05
Production forest	5.70	3.30	8.44	10.40	13.78
Protection forest	0.47	0.34	1.35	0.18	0.41
Conservation areas	0.13	0.01	0.20	0.37	2.77

4.2. Data by Time Period—Land Cover Changes Involving Agroforest and Smallholder Oil Palm

Agroforest in West Kalimantan was mostly as rubber gardens and *tembawang* (Dayak traditional land-use system) having been practiced for generations [81,82]. Inside the Forest-zone lands in West Kalimantan, the majority (76%) of 300,000 ha agroforest areas in 2019 was already agroforest in 2000, and only 11% was forest (Figure 3)

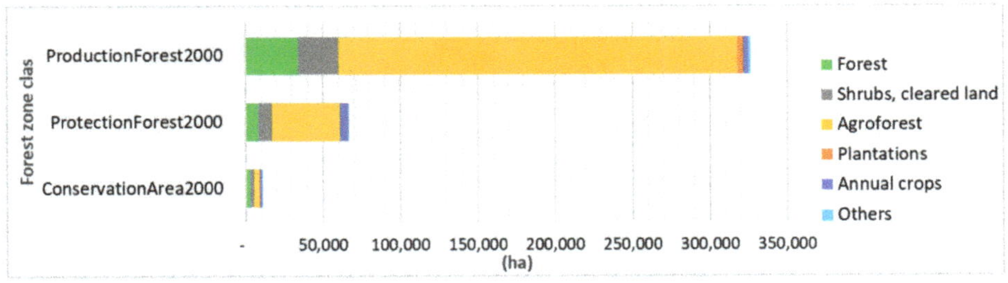

Figure 3. Land cover trajectory (2000) of 2019 agroforest in each Forest-zone class in West Kalimantan (analyzed from [31,32]).

The development of independent smallholder oil palm was closely linked to the conversions of the locally managed agroforest areas due to the lucrative oil palm market in the vicinity of the villages and the decreasing popularity of rubber due to its volatile price [82]. That trend is observed in the land cover trajectory of independent smallholder areas in West Kalimantan, i.e., 57% of which was agroforest in 2000 and 54% in 2009 (Figure 4). Only 13% and 6% of independent oil palm was a forest in 2000 and in 2009, respectively.

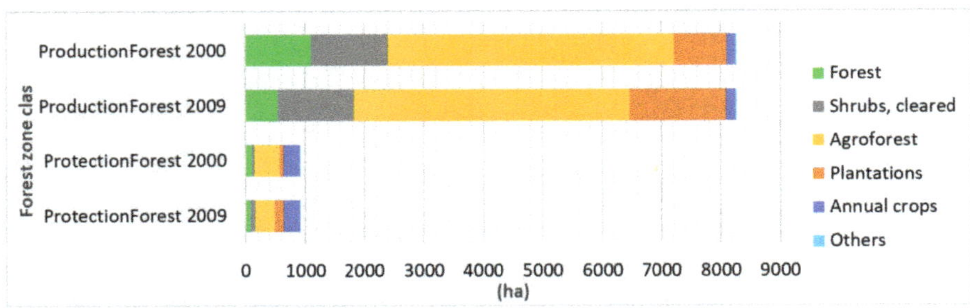

Figure 4. Land cover trajectories (2000 and 2009) of 2019 independent smallholder oil palm in each Forest-zone class in West Kalimantan (based on [31,33]).

For Jambi, agroforest was originally developed by local communities with rubber as the major crop introduced in the beginning of the 1900s during the Dutch period [56,82,83]. It dominated private lands (APL), but its growth reached the areas designated as Forest-zone lands. The 2018 agroforest in Jambi's Forest-zone lands was equally agroforest (44%) and forest (44%) in 2000 (Figure 5), proving the historical presence of this locally managed land use in state's lands.

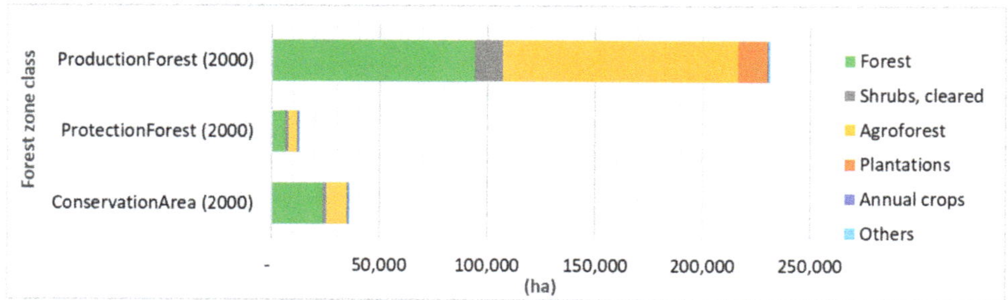

Figure 5. Past land cover types of agroforest in Forest-zone lands in Jambi (analyzed from [31]).

Expectations of high profitability of conversion to oil palm attracted many agroforest rubber farmers in Jambi, also signifying the changes from food sufficiency to cash-cropping [84], in addition to the volatile price mentioned earlier. Independent smallholder oil palm in Jambi's Forest-zone lands was 39% agroforest and 15% plantations in 2009, while in 2000 it was 11% agroforest and 21% plantations (Figure 6). These demonstrate the development of independent smallholder oil palm in Forest zone lands involving agroforest and plantations, mostly rubber. Growth of oil palm also reached degraded areas, shown by 15% shrubs in 2000 and 34% in 2009 (Figure 6), which included peatlands in the lowland areas [85,86].

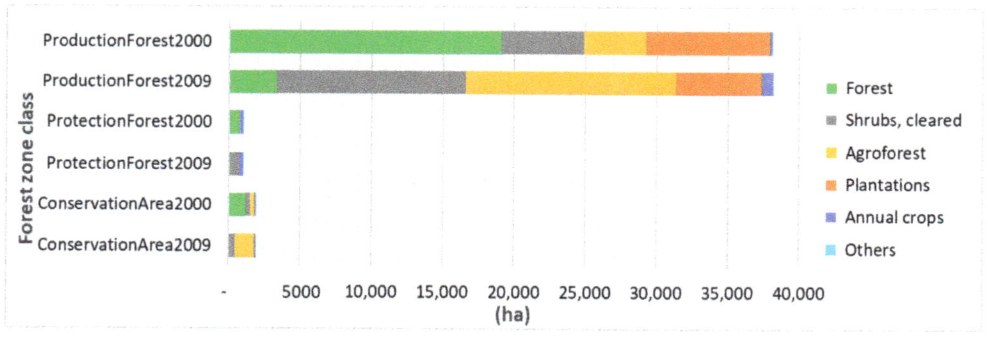

Figure 6. Land cover trajectories (2000 and 2009) of 2019 independent smallholder oil palm in each Forest-zone class in Jambi (based on [31,33]).

Agroforest that has long been managed by local communities has evolved in Forest-zone lands, with the highest proportions and areas in Production Forest status (Figures 4 and 6). With the long history, this land use had faced unresolved forest zone issues since before the 1999 Forestry Law (see Section 3.1) until now, under the *Kawasan Hutan* officiated in 2014 for the two provinces. Since the past decade, oil palm has emerged to become part of the dynamics of this local farming, including in Forest-zone lands. The data show that these dynamics mostly occupied the Production Forest status (Table 3), the class allocated to support various production functions. Looking at the 'predecessor' cash-crop of rubber as part of the ever-evolving mixed-crop agroforest land use [56,81,82], independent oil palm might grow similarly, spatially and or temporally, as part of community-based production functions in land with Production Forest status.

5. Social Dimensions

5.1. Overview Indonesian Oil Palm Sector

The palm oil value chain has a distinct hourglass shape (Figure 7). Upstream there are vast conglomerates that own plantations covering hundreds of thousands of hectares and may be involved in all stages of the value chain. At the other extreme are millions of smallholders that own only a small plot and sell their produce at the farmgate. In between are many shades of grey, ranging from smaller groups, independent large-scale plantations, investors owning a hundred hectares, farmer groups and independent farmers who accumulated multiple small plots over time. Their produce is eventually processed in just over 1000 mills [87], subsequently sold to a few dozen refinery companies, which subsequently supply a few trading companies, that sell their products to a vast number of stakeholders as consumer goods companies and retailers [88].

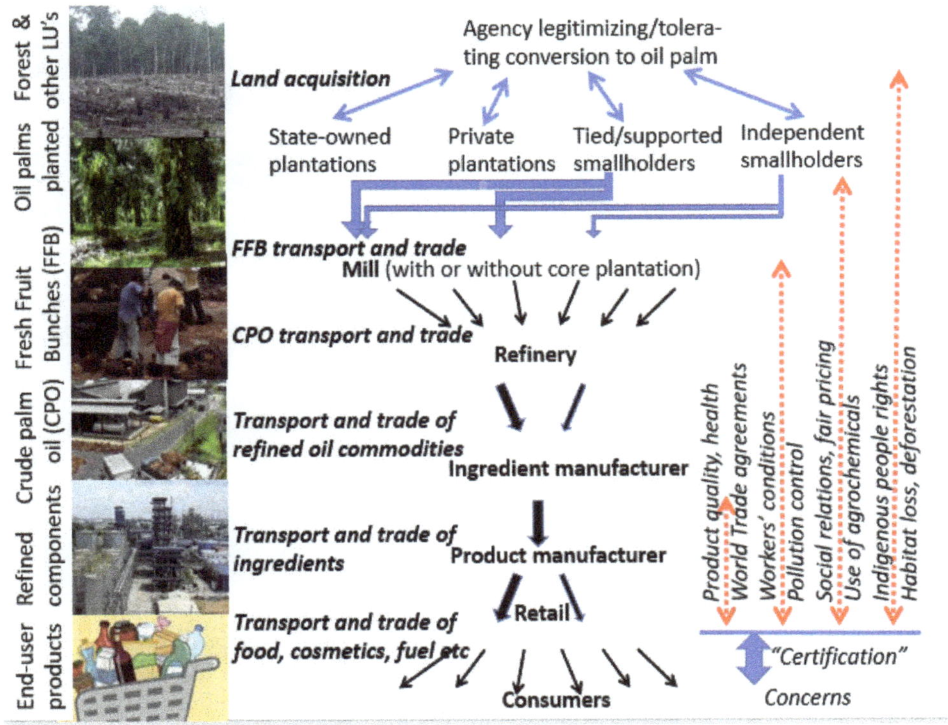

Figure 7. Schematic structure of the oil palm value chain [26].

5.2. Heterogeneity and Expansion amongst Oil Palm Smallholders

Starting on Sumatra's east coast plantation belt, oil palm has been cultivated commercially in Indonesia since 1911 [89]. Smallholders, however, were not involved in oil palm cultivation till the late 1970s, when the New Order regime assigned State-owned plantation companies to develop oil palm plantations that included smallholder farmers. This policy was directed to support the socio-economic development of new settlers and local people in Indonesia's outer islands[1] and develop export commodities that reduce dependency on mineral oil exports. Whereas smallholders first appeared

---

[1]   Outer islands usually refer to islands beyond densely populated Java, Madura and Bali islands.

in oil palm statistics in 1979, covering a mere 3125 ha [90], by 2018 their area covered an estimated 5.8 million hectares, which is equivalent to 40.6% of Indonesia's palm oil area [79]. Whereas oil palm smallholders are often categorized in national statistics as a single entity it is increasingly clear that this vast sub-sector is highly heterogeneous and there is a need for targeted policy measures to improve the environmental and socio-economic performance of the different types of smallholders [91,92]. A key differentiation amongst oil palm smallholders is the scheme vs. independent smallholder dichotomy [93,94].

Scheme smallholder plantations, often also referred to as plasma or '*kemitraan*', are smallholder plantations that have been developed by a plantation company and usually involve company credit for plantation establishment and other agro-inputs, knowledge sharing, and a take-off agreement between smallholders and partner company. Smallholders are usually allocated 2–3 hectares [95]. However, partnership agreements can be highly diverse depending on the regulatory framework present at the time of partnership development, and the negotiations between stakeholders [95,96]. For example, early schemes allocated 60–80% of the plantation to smallholders, but under the influence of World Bank policies and to accommodate private sector investments, this shifted to 40% for smallholders in the late 1980s and 1990s [97,98]. In 2007 the new plantation law stipulated that companies needed to allocate 20% of their concession to smallholders within their concession boundary, and in 2013 this was changed to 20% for smallholders with land possibly outside the company concession [68]. Many companies do not reach this target, however, and this requirement may well be deleted in the new 'omnibus' law that Indonesia is currently preparing [99].

Whilst in some partnerships smallholders are strongly involved in plantation management decisions, participate in plantation labor, and may even outperform their partner company in terms of yields [100], in other arrangements smallholder plantations are fully managed by the company. This last category is often referred to as one-roof management, in which smallholders are effectively mere shareholders [96,101]. Whereas production levels in schemes are generally close to company plantations and thus relatively high for smallholder producers, such schemes frequently suffer transparency issues and especially beneficial to companies [68,102]. Over time government policy regarding oil palm smallholders has increasingly shifted from being a poverty reduction strategy towards one that strengthens industry interests and requires minimal government investments.

Independent oil palm smallholders quickly emerged once the benefits of oil palm cultivation by scheme farmers became obvious. Mills started buying produce outside their plasma and own plantations, and basic infrastructure developed by logging, plantation, and mineral oil companies opened up lands that were previously uneconomic to develop [103,104]. Findings from Jambi indicate that direct economic profits of independent oil palm plantations were significantly higher than forests [105]. Also, oil palm demands less labor compared to competitor cash-crops as rubber, thus allowing for more land to be cultivated or freeing time to engage in other economic activities [106,107]. Independent smallholders generally have no direct links to mills and do not receive extension services from the government or companies. Subsequently, independent smallholders generally have limited knowledge of good agricultural practices, use poor planting material, apply minimal and unbalanced fertilizer regimes, receive low prices for their produce, and suffer yields well below companies and plasma farmers [96,108,109].

The independent oil palm smallholder sector is highly heterogeneous, with many different layers in society engaging in smallholder oil palm development. The independent oil palm smallholders sector includes small local farmers with a few dozen oil palms, migrants attracted by cheap land, oil palm company employees investing in nearby land, local government officials and shop owners looking for investment opportunities and purchasing 5–10 ha, local or urban elites that engage in semi-corporate oil palm plantations and everything in between [93,101]. Multiple typologies on independent oil palm smallholders in Indonesia have been developed [93,101,110], showing presence and ratios between types of smallholders differ in different landscapes. For example, remote peat frontiers often involve relatively large investors whilst transmigration areas or traditionally relatively densely populated areas have a relatively large share of smaller farmers that converted their traditional land uses [93,101].

Whereas land rights and land ownership documentation are generally well arranged in plasma, this is often not the case amongst independent oil palm smallholders [93]. Krishna et al. [111] suggested that especially indigenous populations can obtain forested land and develop oil palm or cash crops in the forestry domain, but sales of such land are limited as it is undervalued due to lack of marketability. Although migrants often do not have nationally recognized land ownership, they require more security in land titles and pay higher prices for land. These authors [111] therefore warn that land titling programs may well lead to the increasing value of land, indigenous people selling their land, claiming new land, and thereby triggering new deforestation [112]. Purnomo et al. [113] provide a detailed analysis of the stakeholders involved in the conversion of land into independent smallholder oil palm plantations in Riau and highlight the considerable profits that accrue to those involved at different stages of the land conversion, including the land mafia. It is increasingly clear that the diversity amongst smallholders and the different landscapes in which they operate needs to be acknowledged for developing adequate policies that foster more sustainable landscapes.

Although reliable current data on plasma vs. independent smallholder are not available, smallholders in plasma schemes were numerically overtaken in 2005 by 'independent' smallholders (including those in various partnership schemes) [114]. Since then some existing plasma plantations have transformed into 'independent' smallholder plantations and oil palm sector growth rates between 2015 and 2018 show that private sector large-scale oil palm plantations grew by 1.91 million ha and that smallholder oil palm area increased by 1.28 million ha [79]. If companies meet their plasma obligations, this would mean a 382,345 ha increase in plasma area and a 901,143 ha increase in independent smallholder area, highlighting that it is especially the independent smallholder sector that is expanding. Whereas companies and associated plasma smallholdings are relatively easy to identify, monitor, and sanction due to the size of their plantations and relative ease of targeting managements, this is not the case with independent smallholders. These independent smallholders are huge in the number of management units that occupy relatively small areas, making it highly complex to manage independent smallholder oil palm expansion. With limited external support and little monitoring, the independent smallholder oil palm has developed into a cheap buffer for the industry; this relates to setting sustainability standards as well as organizing a supply chain.

6. Ecological Dimensions

The economic benefits of oil palm production reviewed above have been discussed in tradeoff with negative impacts on ecosystem services [25,115–117]. However, specific issues of FZ-OP have not been discussed as such in the literature we reviewed. Oil palm cultivation, but especially its continued expansion, has led to severe negative effects on environmental quality such as biodiversity depletion [118,119], loss of hydrological function [118,120–123], increased carbon emissions [85,124], and reduction of water and soil quality [116,118,121,125,126]. The magnitude, and sometimes the sign, of the changes in ecosystem service levels, depend on what land cover is compared to oil palm, and to a smaller extent on the specific way oil palm is grown. Where conversion of natural forest to oil palm represents a drastic change in conditions, comparison of oil palm with the monocultures of fast-growing timber for the pulp- and paper industry, a legal change within production forest, may come out in favor of oil palm in several ecological dimensions. Although the Indonesian rules for logging in the production forest were supposed to secure sustainability across multiple harvest and recovery cycles, there are very few examples where such has been actually achieved in the reality of social-ecological systems.

6.1. Biodiversity

The conversion of natural forests to oil palm plantations threatens biodiversity according to many authors [115,127–129]. Forest fragmentation and large-scale homogenization landscape, whether for the development of plantation forestry or oil palm, pose a serious threat to tropical biodiversity [118,130,131], reducing species richness [128,132,133] and leading to a strongly declined functional diversity [134,135]. However, the response of biodiversity to land-cover change depends upon the extent to which natural habitat features are replicated while the sensitivity of species to change varies [130].

Several global studies have highlighted that past and predicted expansion of oil palm plantations into forest areas threatens mammal [136] and bird species with extinction [134,137,138]. In Indonesia, both Sumatran and Bornean orangutans (*Pongo abelii, P. pygmaeus*) and proboscis monkeys (*Nasalis larvatus*) population trends are decreasing due to the expansion of large-scale oil palm plantations [139–144]. Herds of elephants, tigers and rhinos are reported to be critically threatened due to this expansion [137,145]. In peat swamp forests, the expansion of oil palm monocultures is likely to have negative effects on, among others, macrofungal biodiversity [146].

A study [131] quantifying the impact on the herpetofauna of the Pacific lowlands of Costa Rica found that total species richness of amphibians and reptiles was reduced to 45–49% compared to forest area due to the almost complete absence of leaf litter, understory vegetation, and woody debris and the more open canopy. However, different research [147] conducted in Sumatra (Indonesia) showed different results. Rare amphibians were much more abundant in riparian forests and common amphibians were more prevalent in oil palm plantations. Indeed, reptile richness and abundance were higher in oil palm plantations than in all other habitats. Surprisingly, a meta-analysis [137] found that average invertebrate species richness did not differ significantly between oil palm and forest sites when published studies were compiled. It is commonly found that different aspects of biodiversity are influenced in opposite ways during land-use change [148].

Protection of any forested habitats and enhancing understory vegetation can help improve opportunities for some species [131]. Forest patches, even when small, fragmented and degraded, i.e., riparian sites, doubtlessly are required to sustain the species in human-transformed landscapes [139,146]. Substantial biodiversity loss can only be avoided if future oil palm expansion is managed by avoiding deforestation and cultivating non-forested and abandoned areas for sustainable oil palm cultivation [78,143,149]. The small-scale oil palm production due to greater vegetation heterogeneity is also likely to provide greater ecosystem services provision and should be considered as one of the strategies for achieving sustainable oil palm production [150]. By reducing deforestation and ecosystem degradation, stopping land clearing techniques by land burning, and stopping the development of oil palm plantations on unsuitable land, sustainable palm oil management can be achieved [120].

6.2. Watershed Functions

Oil Palm plantations are known to alter the hydrology of the subwatersheds in which they develop, by increasing the risk of flooding, increasing soil erosion (especially in the planting phase) and nutrient leaching, polluting ground- and surface water [124]. Once established, oil palm-dominated landscapes are 'water greedy' [122], responsible for decreasing local water tables and water supplies as well as changes in streamflow levels and water quality [118,122,151]. Several studies reported that the conversion of forests to young plantations initially decreased evapotranspiration (ET), but also decreased infiltration rates [122,124]. Researchers in Brazil [152] reported that water shortages have occurred more often since oil palm cultivation has become the dominant land use and large-scale deforestation has taken place. A study of land-use change impact on flooding frequency in Batanghari Watershed, Indonesia, showed that land cover change from forest to rubber and oil palm plantation contribute to the higher flooding frequency [153].

Though water availability, air, and water quality were perceived to be the most heavily impacted ecosystem services by oil palms [151], there is a knowledge gap about the magnitude of these ecohydrological changes and their variations over the oil palms lifetime [118,123]. It is unclear whether young and mature plantations have similar or different ET rates when compared to native forests [118,123]. The impacts of oil palm on the major components of the hydrological cycle (e.g., total water yields, dry season baseflow, streamflow dynamics, evaporative return to the atmosphere) are still at an early stage [118,121,152,154,155].

The constraint is that in general, countries with large oil palm plantations are developing countries with limited resources to assess the impact of plantations throughout the growing period of oil palm on hydrological functions [119]. Moreover, oil palm plantations are generally homogeneous monoculture stands of various age classes, with varying water use characteristics [154]. Most of the hydrological studies in oil palm plantations are carried out at the plot scale, while oil palm plantations can reach thousands of hectares across several watersheds [121]. Indeed, research on the hydrologic impacts of oil palm plantations at the scale of watersheds is rare [153].

Quantitative data from field measurement on the relationship between water availability at each stage and water yield are still limited as well as the actual water use of oil palm at the field level, and the minimum amount of water needed to optimize oil palm productivity [156]. Research on the impact of oil palm cultivation on hydrology is still focused mainly on the study of plot-scale ecohydrological fluxes, such as the impact of tree age on canopy rainfall interception [157,158] and transpiration rates in palms with varying ages and grown on different slopes [152,159].

However, in general, large scale conversions of forests to intensively managed plantations, result in significant changes in the hydrological cycle including periodic water scarcity [122,160]. In natural ecosystems, such as forests, most of the rainfall water is absorbed by the soil and plays a role in increasing plant transpiration and replenishing groundwater. In oil palm monoculture plantations, water does not penetrate the eroded and compacted soil properly [122]. Decreased infiltration reduces water storage [124] and increases surface runoff [121] potentially reducing the access to usable water and increasing the risk of flooding [161].

It is well known that highly productive monoculture stands are dependent on an abundant supply of water over time [150]. Indeed, drainage on oil palm plantations can lead to a substantial reduction in streamflow during dry seasons or droughts, soil subsidence, and potentially increasing future flood risks [162]. Thus, the expansion of oil palm plantations will affect water consumption which, in the long term, could affect local water resources [118]. Rather than to high water use of oil palms per se, local water scarcity seems connected to the modified redistribution of water after precipitation at the landscape scale, reducing effective buffering [122].

There may be a potential trade-off between water use and management intensity of oil palm plantations [154]. Optimizing water use and oil palm production and increasing cultivated varieties can result in more efficient land use and reduced conversion of natural landscapes [153]. The practices of retaining old palms during crop rotation and cultivating ground cover crops can mitigate some of the impacts [118]

Retaining a riparian buffer zone helped reduce the negative impacts of oil palm plantations on streams [163,164]. A good riparian zone with dense native vegetation of complex structure adjacent to the streams that flow through the plantations can help restrict the transport of soil, sediment from the oil palm land, and in preventing chemicals pollutant from reaching waterways. They can also serve as valuable natural habitats for riparian and terrestrial species, as well as carbon storage areas. Appropriate riparian zone management, combined with effective monitoring, is essential to maintain or enhance the ecological function and biodiversity of stream ecosystems [165].

6.3. Greenhouse Gas Emissions

Forest conversion to oil palm is responsible for significant net greenhouse gas (GHG) emissions [116,124,128,139,166]. However, information of net carbon emissions throughout the life of the oil palm plantation, especially on peat soils, is still very limited [167]. Aboveground C stocks in oil palm plantation are, averaged over a plantation life-cycle, around 40 t C ha^{-1}, which is lower than most predecessor vegetation (unless this is 'grassland'), leading to a 'carbon debt' in the first cycle [168,169]. On mineral soils, good-practice management can maintain soil carbon levels over a production cycle (compensating for early losses) [170]. Intensive use of fertilizer to increase yields adds greenhouse gas emissions as well, and the environmental optimum, minimizing the footprint per unit of palm oil, depends on context [171]. A synthesis of peat-oxidation emission values for tropical peatland by the Intergovernmental Panel on Climate Change [172], suggested default values of 51 Mg CO_2 ha^{-1} y^{-1} for smallholder systems, 55 Mg CO_2 ha^{-1} y^{-1} for commercial plantations (oil palm, industrial timber), and 10 Mg CO_2 ha^{-1} y^{-1} for disturbed secondary forest. A recent study in smallholder oil palm on peat soils in Jambi suggested values may be higher [173] than that. While the 'moratorium' has restricted large-scale oil palm development on peat soils, smallholder expansion continues and becomes a more prominent part of the overall problem.

7. Policy Responses and Options

7.1. FZ-OP as a Policy Issue

From the evidence so far, we conclude that FZ-OP is a substantive issue that cannot be denied or ignored at the policy level. Credible spatial data sets indicate that 15–20% of Indonesian oil palms are growing within the forest zone, with FZ-OP identified in all provinces in which oil palm is grown. FZ-OP is, however, unevenly distributed in Indonesia, with the provinces of Riau and Central Kalimantan, where agreement between provincial and national authorities over the boundaries of the forest zone was only reached in 2014, as the main contributor. But even if these two provinces are seen as a 'policy anomaly', the case studies in the neighboring provinces of Jambi and West Kalimantan showed that FZ-OP is real, but of mixed history and characteristics that need to be understood before appropriate policies can be designed. FZ-OP is most common within the 'production forest' category, but also includes 'watershed protection forest' and 'conservation areas'. The probability that a unit of land in the elevation and climate zone suitable for oil palm has been converted to oil palm varies from more than 10% to around 1% in conservation areas (Figure A1). Taking palm oil production as a point of reference for the two case-study provinces, up to 14% may derive from 14% production forests and up to 3% from protection forests and conservation areas. As there is no legal basis for oil palm presence within the forest zone, any FZ-OP indicates a lack of legality. Whether it indicates a loss of forest functions or not depends on the comparator land cover—which in the case of production forest can be an industrial timber plantation, with properties similar to oil palm, but should be naturally established in the protection and conservation forests, with substantially higher diversity. Meanwhile, three types of oil palm owners (large-scale plantations, tied or vertically integrated smallholders, and independent smallholders) represent different social contexts and likely require differentiated policy responses. Such responses can focus on rights and permits in land use and water management, or on the transport and processing stages of the value chain.

7.2. Institutional Responses

Given the importance of palm oil exports for the Indonesian economy (around 7% of total export value), the initial 'denial' phase of environmental issues around its production was promoted by the industry and its supporters in the Ministry of Economic Affairs. Yet, with 15–20% of the production area (and a probably lower share of the harvested produce) potentially spoiling the national reputation of all palm oil, the phase of 'shooting the messengers' ended, and communication efforts shifted to demonstrating active responses, while emphasizing the social dimensions of smallholder interests in escaping from rural poverty [26].

7.2.1. RSPO—Market Segmentation ('Shifting Blame')

The Round Table for Sustainable Palm Oil (RSPO) is a private sector-led initiative set up in 2004 to counter environmental and social concerns surrounding the rapid expansion of oil palm. It is often regarded as the most credible sustainability scheme in the oil palm sector and has among the most stringent and explicit principles and criteria [174,175]. Still, more critical analyses emphasize that RSPO certification merely provides a technical managerial solution to satisfy mainly Northern sustainability concerns but does not touch upon the underlying social relations in the production of the commodity that needs to be improved for truly sustainable palm oil production [176]. Examples of this are land and labor rights issues in Indonesia, unequal distribution of profits, but also the fact that even RSPO-certified oil palm monocultures present a huge loss of biodiversity compared to the forests, whether primary, secondary, or agro-forestry systems [171]. Moreover, the RSPO is voluntary and in 2015 covered an estimated 15% of Indonesia's oil palm area. With plenty of 'brown' supply chains to accommodate plantations that do not join RSPO, it is clear that this initiative cannot change the sector as a whole [88,177]. Santika et al. [178] also note that RSPO certification has not improved the wellbeing of local communities. They associate this with the considerably larger size of RSPO-certified plantation compared to non-certified plantations, which leaves little space for, especially forest-dependent, communities to maintain their livelihoods. Although the RSPO acknowledges that the inclusion of small and medium-sized growers is a key target [179], these small and medium actors are struggling with the costs, knowledge and institutional requirements needed for certification. With RSPO definitions of sustainability, there appears a clear risk that especially large corporate actors are classified as 'sustainable', as these are better able to fulfill RSPO criteria. Smaller actors on the other hand run the risk of being marginalized and blamed for the broader ills as land management and poor rural development strategies, for which oil palm is only one of many accelerants.

7.2.2. ISPO—National Sovereignty

Transnational business initiatives as the RSPO do not include governments. The Government of Indonesia, as well as the Indonesian Palm Oil Association (IPOA/GAPKI), regarded such initiatives as primarily reflecting Northern concerns, hollowing out the role of the State, and threatening the socio-economic development within Indonesia [180,181]. Therefore, in 2011, the Indonesian government launched the Indonesian Sustainable Palm Oil (ISPO) initiative to regain momentum and authority in sustainability discussions and to develop a sustainability framework that bypasses hard-to-comply-with sustainability criteria that are not demanded in key Southern markets such as China, India, and Indonesia itself [182]. However, ISPO requires compliance with existing legislation and whereas at first it is only mandatory for company palm oil plantations, with the latest update in 2020 it will obligatory for smallholders in five years, as well. The large area of smallholder SF-OP cannot be certified by ISPO under current rules and will prevent the target of full ISPO coverage to be achieved—unless the rules change and existing SF-OP can become legal. This will require the Ministry of Environment and Forestry and the Ministry of Agriculture to reach an agreement on how this can be done.

From the start, the ISPO organization suffered a lack of capacity to implement its mandate [183], and its principles and criteria are often deemed insufficient for guaranteeing sustainable development [169,184]. Although ISPO has been revised and supposedly strengthened in 2015 and 2020, the latest update of ISPO still poorly addresses human rights issues, lacks protection for non-primary forests, and lacks clear definitions, procedures and independent monitoring [185]. Again, as with RSPO, current smallholder certification only reaches a fraction of the total number of smallholders and smallholder exclusion looms.

7.2.3. Deregulation and Crisis Responses

Recently the Indonesian parliament adopted an 'omnibus' law that, to facilitate business development and job creation, removes legal obligations for environmental impact assessment, simplifies procedures for obtaining permits, and abolishes the requirement for plantations to support smallholder producers as part of their land concessions. The law has been critiqued as both environmental and social (worker rights) concerns appear to lose existing safeguards. Its consequences for the SF-OP issue are not yet clear.

7.3. Current Policy Options Based on Land and Water Management

Several policy options have been discussed in the public debate so far, summarized in Table 5. Just as 'stop the bleeding' is the first line of medical defense in dealing with injuries, the first efforts made, through a 'moratorium' are efforts to stop a further expansion of tree crops within the forest zone. Once some opportunity has been created for evaluating longer-term options, these can be linked to the diversity of contexts, in a 'policy options by context' table (Table 5), that requires an evaluation from the full range of relevant stakeholders before negotiations can progress and decisions with a chance of implementation success can be made. The options can target the land-base for oil palm production and/or the transport and entry into the mills and further value chains (compare Figure 7).

7.3.1. Legalize by Including Oil Palm in Forest Definition

The simplest solution to any illegality issue, at least from a formal governance perspective, is to legalize it. As the internationally used forest definition is ambiguous as to the 'tree' concept that underpins quantitative criteria about canopy cover and spatial scale, an argument can be made that palms, including *Elaeis guineensis*, are included in the woody perennial category, while quantitative criteria on potential tree height, the density of the canopy cover and longevity are easily within the range of what are commonly understood to be 'forestry trees'. Such an approach would avoid complexity, but not address any of the environmental and social concerns of the status quo. It would, however, put the pulp and paper industry, supported by Forestry authorities as core economic activity, and the oil palm industry on a level playing field. It will make all oil palm production deforestation-free by definition and greatly help in achieving the target that all of the Indonesian palm oil can be ISPO-certified, without major changes on the ground.

However, as simple and attractive as the option may seem in the short term, it would severely undermine Indonesia's international standing as a country balancing development and environmental concerns. It would also open the door for a much further expansion of oil palm within forests, without legal means to control such expansion.

7.3.2. Grandfathering

As part of an overall policy package, a grandfathering approach here refers to defining a cut-off date that tolerates (although not fully legalizes) oil palm cultivation in the forest zone if started before an agreed date. It is a common practice where 'new' regulations emerge, such as the RSPO standards that refer to the time standards were published. It would normally be restricted to the current plantation cycle and not allow replanting of oil palm.

A 'grandfather' approach could be a policy option that shares responsibility for the status quo and its path dependency where government agencies have been involved in the facilitation of illegal activities. Where smallholders from the local or indigenous communities shifted to oil palm from past rubber involvement in their swidden/fallow rotations they may have done so on land that is classified as 'production forest' but has not been formally gazetted. In the two provinces with the highest fraction of forest-zone oil palm, the designation of forest zone has remained contested between provincial and national forest authorities until 2014, creating a grey zone in which law enforcement was a low priority.

A case could also be made for migrants who moved to the frontier area for income generation, not for investment and capital accumulation (as 'white-collar' farmers). Part of these migrants (or their parents) were mobilized by the government through the "transmigration program" and looked for opportunities to increase their land, acquiring land from informal local land markets that disrespect forest zone categorization. Where this happened in the past, a 'generic pardon' or grandfather rule could close the books on it—but at risk of setting precedents that current rules can still be breached.

Table 5. Policy options to deal with oil palm within the forest zone, by applicability domain.

Policy Option	Applicability Domain				Expected Consequences		
	Conservation Areas	Protection/Production Forest on Peat	Protection Forest on Mineral Soil	Production Forest on Mineral Soil	Social	Economic	Environmental
Land-focused:							
1.Legalize	NA	NA	NA	NA	++	++	−
2. Grandfathering	NA	NA	A	A	+	+	+/-
3. Evict	NA	NA	NA	NA	−	-	+/-
4. Charge	A	NA	A	A	+/-	+/-	+/-
5. Agroforestry concessions	NA	A	A	A	+	+/-	+/-
6. Focus on high-value locations	A	A	A	A	+/-	+/-	+
7. Rewet peatlands	NA	A	NA	NA	+/-	+/-	+
Value-chain based:							
8. Mill certification	A	NA	A	A	-	-	+/-
9.Transport permits	A	NA	A	A	-	-	+/-
10.Segment markets	A	NA	A	A	-	+/-	+/-

Remarks: A (applicable), NA (Not applicable) from a forestry institutional perspective.

7.3.3. Evict Farmers, Destroy the Crops

On the other side of the spectrum from a blanket legalization of all forest-zone oil palm, there are voices for strict law enforcement. Evicting farmers and destroying their crops, to prevent their return, has been implemented before, e.g., in the 1990s in coffee encroachment areas in Sumatra. However, this sparked further conflicts, while the uprooting may have aggravated environmental issues in the short term, as did the type of reforestation with fast-growing exotic tree species in the longer term from an environmental perspective, as documented for Sumberjaya in Lampung (Sumatra) [185] and Manggarai (Flores) [186].

During the 'New Order' regime (1965–1998) military enforcement of government-set rules was common, but not free from corruption where economic interests were involved. After the regime change in 1998, reliance on court procedures, prisons and fines has shifted to the role of local governments in illegal procedures. However, attempts to sue provincial authorities in C Kalimantan and Riau for transgressing forest-zone rules backfired when the court found the forest zone in these provinces had not been gazette according to the prevailing Forestry Law [187].

In areas where production forests have been largely transformed into oil palm plantations for many years, such as in parts of Riau [91], it is unlikely that evicting small farmers, local upper-middle-class families, elites and companies with vested interests is feasible. They are not going to accept the destruction of their investment, and therefore this option may well be politically unachievable and risky. Already there have been cases where 'farmer groups' have held captive inspectors from the national authorities [188]. As highlighted elsewhere [111], the land mafia is making large profits with land conversion and in certain frontiers, this goes well beyond poor farmers looking for a few hectares of land [91]. Furthermore, the mere destruction of crops will not necessarily lead to increased environmental performance as oil palm plantations capture considerably more carbon than the degraded lands and *Imperata* (alang-alang) fields that destroyed plantations are likely to turn into. Heavily degraded lands without clear ownership and proper management are prone to fire and a key source of repeatedly occurring mass forest fires in Indonesia, with their detrimental effect on GHG emissions and human health [111,189].

7.3.4. Charge Land-Owner Benefit Shares to Pay for Forest Management Elsewhere

In Indonesia 'share-cropping' has a long tradition of arrangements where a land owner allows others to use farmland, based on a share of the crop yield obtained. Such arrangements are often preferred by both parties over hired labor or land rent contracts; the land owners share of the yield, the payment for and accounting of agricultural inputs tend to vary with local circumstances and the supply–demand balance for contracts. Often patron–client relations are involved with social dimensions beyond economic rationality. Rules can apply to trees, as well as to land as a production factor [190]. In the tradition of the Java Social Forestry program, forest authorities allowed farmers to grow annual crops, perennial fodder grass, or low-stature tree crops such as coffee under similar rules of a yield-share for the forest authorities [15]. Reference to such a system could lead to a financial charge on oil palm in the forest zone, for example within production forest lands where gazettement has been completed and the state is the legal owner.

Currently, forest management, from protected areas to production forests that are in the recovery phase, is severely restricted by lack of funding. Charging a land owner's share of palm oil produced within the production forest zone could provide funding for forest management elsewhere.

Of the 3.4 M ha oil-palm (OP) plantation in state forest areas, 700,000 ha is smallholder OP [75]. Rules may have to be differentiated by scale, for example, distinguishing between oil palm farm sizes of >100, 25–100, 5–25, and 0–5 ha, as only the last category is likely to be a family farm without external labor. Such differentiation could be part of the Social Forestry (SF) and Agrarian Reform (AR)/UUCK schemes. A legal basis for such charges could be found in Law 32/2009 on Environment protection and management.

If ways can be found to implement this, it will be attractive to local forest authorities—possibly too attractive and providing incentives for further oil palm expansion within the forest zone.

7.3.5. Agroforestry Concession

Peru, dealing with similar issues of past undocumented and illegal agricultural encroachment into forest zones, has created a legal category of an 'agroforestry concession' [191]. These schemes allow for agreements that re-establish government authority over forest lands but allow current tree-based land uses to continue, within agreed conditions. They resolve the issue that the current illegality of the land use is a bottleneck for government support of any type in the area, including a lack of extension services that restrain the development of socially and environmentally desirable practices. Similar issues have been noted in Riau, where support for companies and especially oil palm farmers in the forestry domain has been minimal to non-existent as illegal activities cannot formally be supported [91,92].

Agroforestry concessions as a legal instrument would complement current 'community-based forest management' and 'village forest' schemes, without claims that oil palm plantations are considered 'forests'. They would also allow investment in upgrading different types of production systems, systems that are more acceptable than mere oil palm monocultures. Current research interests in testing diversified oil palm agroforestry systems are opening new perspectives that might match farm economies [192–194].

7.3.6. Swaps with High-Value Legal Deforestation Locations

Side-by-side with the 3.6 M of illegal oil palm in the forest zone, there is still a substantial area (1.2 M ha by recent estimates [195]) of old-growth forest that has been legally transferred (PKH/release) to large scale oil palm development (as a land bank), but has not yet been converted into oil-palm plantations. Earlier efforts under a 'land swap' umbrella to exchange land for companies with such concession rights have not had much impact [196], but in combination with a substantial landowners' charge on existing illegal forest-zone oil palm, it may be feasible to recover high-value forest while giving up (from a forestry perspective) production forest land already converted. The challenge is, however, that such swaps usually involve locations in different districts or even provinces. Other high-value locations that deserve priority in resolving existing forest-zone oil palm are riparian zones and ecological corridors.

Specific ideas have been formulated on how such swaps could, in the context of the Oil Palm moratorium, return existing non-converted OP concessions to government control. It could be implemented by buying out rightsholders using funds obtained by charging FZ-OP units of more than 25 ha based on Law 32/2009 on Environmental protection and management, charging land-rents to 'big smallholder OP' (5–25 ha), while 'real smallholder OP' (<5 ha) could be exonerated in a Social Forestry (SF) and Agrarian Reform (AR)/UUCK perspective.

What has yet to be explored, however, is how the records on spatial boundaries of forest zone would be adjusted to 'legalize' the FZ-OP for the longer term. Moreover, it is unclear how such swaps would be understood internationally, and whether the smallholder OP can, after such swaps, enter international markets within a 'deforestation-free' label. One would have to explain that the oil palms were planted on illegally deforested land, but became legalized because elsewhere deforestation had been legalized, did not happen and its rights were withdrawn. This will be quite a mouthful of legalese, that only has a chance if the Government that implements such schemes has a record of transparency beyond what currently exists.

7.3.7. Rewet Peatlands

Specific solutions are needed for peatland forests converted to oil palm. The current situation of illegality makes it hard for government agencies to engage. For example, oil palm plantations located in Riau peatlands that were part of the Forestry Domain suffered severe water table management issues, with larger farmers just digging canals without coordination, leading to drought and flooding, leaving the area prone to peat fires [91]. Such undesirable practices, which are relevant in other landscapes as well [92], may be countered by (partial, temporary or full) legalization of their plantations, charging farmers for their activities in the forestry domain and using these proceeds to finance and coordinate proper water table management. Such an approach may lead to landscape-wide improved water table management and associated reductions in draught, fire, flooding and associated GHG emissions. The government rules about maintaining groundwater tables in peatlands no deeper than 40 cm below the surface are not incompatible with growing oil palm—at least not under smallholder, non-mechanized management.

7.4. Policy Options Based on Transport and Processing

7.4.1. Impose Legality Checks at the Mills

Current sustainability initiatives as RSPO and ISPO heavily depend on legality checks at the mill, restricting certification to mills with adequate traceability of the fresh fruit bunches they obtain. However, supply chain characteristic for especially independent smallholders, who usually sell to middlemen that mix and may resell their produce to other traders, are unable to fully trace all produce that enters the mills. This system appears to benefit especially the larger actors, who have the most integrated value chains, shifting blame and costs to others. As noted elsewhere [29,30] certification may solve problems of guilt for downstream users who do not want to be part of value chains with negative social and environmental consequences, but it may primarily shift blame, rather than transforming other means of production, unless economic signals, accounting for increased transaction costs, are sending a clear message.

7.4.2. Apply a Transport Permit System

Forestry has a long history of trying to control illegal logging by applying transport systems and road checks. As documented for charcoal trade in Africa [197], it can increase transaction costs by legal and illegal levies, with little impact other than reducing farmgate profitability at the sites of production. This approach thus suffers similar drawbacks as legality checks at the mills. It is prone to corruption. In areas where there are a lot of roads, this becomes cumbersome. However, in some contexts where there is less infrastructure, it may work to some extent.

7.4.3. Segment Markets

Legality checks at the mills and chain of custody rules [93], potentially supported by transport permit systems, can support the bifurcations into 'green' and 'brown' supply chains, targeting different market segments (e.g., international and domestic). It will not solve the issue of illegal forest-zone oil palm, but it may affect the profitability of the lower-grade production system if transaction costs of the certification are controlled or subsidized. The new EU anti-deforestation policy and RSPO are heading in this direction, recognizing that importing countries have to respect the sovereignty of producing countries within world trade rules. Keep your own street clean and shift blame for existing problems to others. The EU with its deforestation-free supply chains is picking the low-hanging fruits, with a 'grandfather' cut-off date for deforestation before 2008, which can easily cover EU demands, whilst leaving the tougher issues to other parties.

7.5. Follow-Up Policy Research

Table 5 provides an early assessment of ten policy options for four subcategories of FZ-OP, from three overarching perspectives. The tentative evaluation of options could be the start of a further process of stakeholder consultations, that may modify perspectives on expected effectiveness and acceptability by main stakeholder groups. As some of the options proposed, especially the 'agroforestry concession' idea would be new to the Indonesian forest management framework, deeper analysis of legal opportunities and consequences will be needed before specific recommendations can be made.

8. Conclusions

Up to one-fifth of oil palms in Indonesia are located within the 'forest zone' (Kawasan hutan), especially in lands indicated as production forest, with smaller fractions in protection forest and conservation areas and substantial variation between the provinces and islands. Two (out of 34) provinces are responsible for two-thirds of the forest-zone oil palm, while harboring nearly one-third of Indonesian oil palms. While state-owned and private large-scale plantations dominated in the early expansion phase of oil palm, smallholders have become prominent, especially in Sumatra, where 60% of oil palm capacity is located, especially those without a contract with mills ('independents'). Part of the smallholder oil palm within the forest zone has been derived from earlier agroforests (often rubber-based).

As there is no legal basis for oil palm within the forest zone, one-fifth of Indonesia's oil palms cannot meet current criteria for sustainability certification, whether the voluntary international RSPO standard or the domestic ISPO standard. Solutions for the ensuing policy problem will have to differentiate between the primary forest functions indicated for different parts of the forest zone, with the smallest gap between desired function and current reality in the production forest zone, where most of the oil palms are located, and largest gap for the small fraction (<1%) of oil palms in conservation areas. Policies will have to differentiate between the scale of the production unit, the opportunities to charge a 'land owner's' share by forest authorities, the history of the conversion process, and the expected impact through discouraging further expansion.

Policy options for dealing with forest-zone oil palm can focus on the land where oil palms are grown and/or on the transport and mills that are processing the fresh fruit bunches produced. Where mill-based regulation can be effective in segregating the market and ensuring that part of the supply chain is 'deforestation-free', it is unlikely to close marketing channels for forest-zone oil palm products. Land-based options will have to explore a middle ground as neither blanket legalization of oil palm as a forest commodity, nor evictions and destroying of the crops are realistic in the institutional context. Combinations of financial charges and permits for agroforestry development may be the most feasible basis of a negotiated policy package. Further policy research and stakeholder consultations will be needed before specific policy recommendations can be made, adjusted to the legal environment and responsibilities of national, provincial and district-level government authorities, as well as likely responses of international trade and domestic industry.

Author Contributions: Conceptualization, E.P., H.S., I.J., A.W. and M.v.N.; Investigation, E.P., H.S., I.J., A.W., H.Y.S.H.N. and M.v.N.; Methodology, E.P., A.W. and M.v.N.; Writing—original draft, E.P., H.S., I.J., A.W., H.Y.S.H.N. and M.v.N.; Writing—review & editing, E.P., H.S., I.J., A.W. and M.v.N. All authors have read and agreed to the published version of the manuscript.

Funding: This research is financially supported by the Forested Landscape for Equity (GLA 1.0) project, Component: Sustainable Agrocommodity Landscape (No. Project 04400407) managed by Tropenbos Indonesia.

Acknowledgments: The authors acknowledge the availability of spatial data sets through Yves Laumonier (CIFOR), Lilik Prasetyo and Aryo Condro. The National Main Commodity Maps in Indonesia (Condro et al. (2020)) were produced under collaboration between the Faculty of Forestry, IPB, and the United Nations Development Programme (UNDP). We acknowledge questions and inputs from speakers and participants of the on-line seminar on 26 September 2020.

Conflicts of Interest: The authors declare no conflict of interest.

Appendix A

Appendix A.1. Details of the Initial Stakeholder Consultation

The webinar "The Future of Agro-Commodities in Forest Areas" on 26 September 2020 had 373 registered participants. They attended the on-line event geographically spread over 33 (out of 34) provinces; institutionally, 30% represented government agencies (local, national; forestry, non-forestry),

3% research organizations, 44% universities, 11% private sector, 13% NGO's. Demographically 35% was female and 27% below 25 years of age.

As first speaker the Director General of Natural Resources and Ecosystem Conservation (KSDAE), Ministry of Environment and Forestry (KLHK), Ir. Wiratno, MSc, marked the shift from a timber focus in the past to a much wider range of economic commodities as potentially derived from and grown in forests, as long as the societal functions of forests can be secured. Speakers from NGO's reviewed data on forest-associated poverty levels (around 21% of poor people in Indonesia are forest villagers) and presence of tree crops, with most specific debate about oil palm.

The discussion focused on the reliability of data of various sources and on opportunities to resolve issues through implementation of Law No.34/2014 on Soil and Water Conservation and Government Regulation No.46/2017 on Environmental Economic Instruments. Panelists emphasized a need for the involvement of government agencies in intensive assistance for large-scale plantations and smallholders in forest areas, and for law enforcement for deliberate occupation in forest areas and resolving tenure issues. The various strands of discussion informed the current manuscript.

Appendix A.2. Analysis of FZ-OP by Elevation Zone and Forest Class

The spatial data on FZ-OP and oil palm on non-forest lands are differentiated by forest class and elevation in Figure A1.

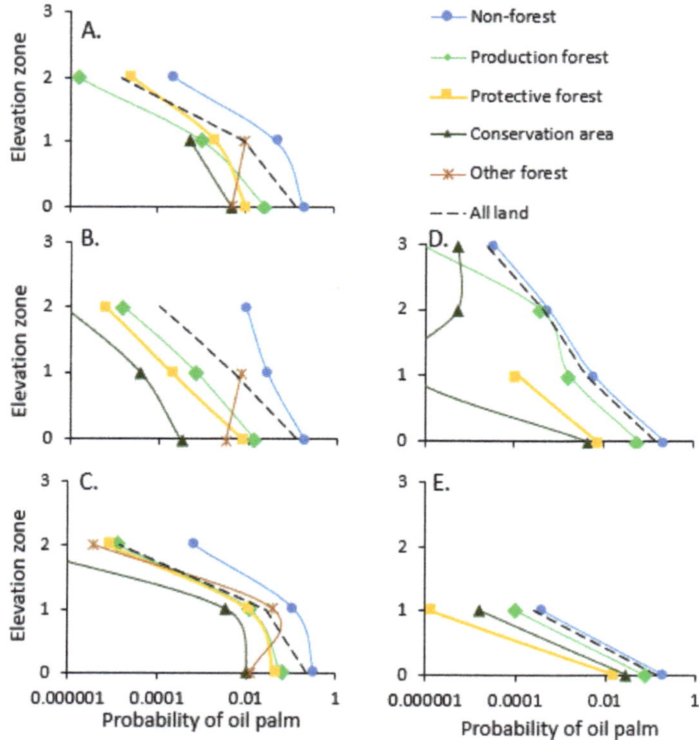

Figure A1. Probability of land having oil palm as land cover at four elevation zones (0 = 0–100; 1 = 100–300; 2 = 300–1000; 3 = >1000 m a.s.l.) in two provinces ((**A–C**), West Kalimantan; (**D,E**) Jambi), for specified land use designations outside and inside the forest-zone, according to three data sources ((**A**) [75]; (**B,D**) [33]; (**C,E**) [28]); (NB log scale on the X-axis).

References

1. Chazdon, R.L.; Brancalion, P.H.; Laestadius, L.; Bennett-Curry, A.; Buckingham, K.; Kumar, C.; Moll-Rocek, J.; Vieira, I.C.G.; Wilson, S.J. When is a forest a forest? Forest concepts and definitions in the era of forest and landscape restoration. *Ambio* **2016**, *45*, 538–550. [CrossRef]
2. van Noordwijk, M.; Minang, P.A. *If We Cannot Define It, We Cannot Save It*; ASB Policy Brief 15; ASB Partnership for the Tropical Forest Margins: Nairobi, Kenya, 2009; p. 4. Available online: http://www.asb.cgiar.org/pdfwebdocs/ASBPB15.pdf (accessed on 24 October 2020).
3. de Foresta, H.; Temu, A.; Boulanger, D.; Feuilly, H.; Gauthier, M. *Towards the Assessment of Trees Outside Forests: A Thematic Report Prepared in the Framework of the Global Forest Resources Assessment 2010*; Food and Agriculture Organization of the United Nations: Rome, Italy, 2013.
4. van Noordwijk, M.; Suyamto, D.A.; Lusiana, B.; Ekadinata, A.; Hairiah, K. Facilitating agroforestation of landscapes for sustainable benefits: Tradeoffs between carbon stocks and local development benefits in Indonesia according to the FALLOW model. *Agric. Ecosyst. Environ.* **2008**, *126*, 98–112. [CrossRef]
5. Romijn, E.; Ainembabazi, J.H.; Wijaya, A.; Herold, M.; Angelsen, A.; Verchot, L.; Murdiyarso, D. Exploring different forest definitions and their impact on developing REDD+ reference emission levels: A case study for Indonesia. *Environ. Sci. Policy* **2013**, *33*, 246–259. [CrossRef]
6. FAO. *Global Forest Resources Assessment 2010*; FAO Forestry Paper No. 163; UN Food and Agriculture Organization: Rome, Italy, 2010.
7. Keenan, R.J.; Reams, G.A.; Achard, F.; de Freitas, J.V.; Grainger, A.; Lindquist, E. Dynamics of global forest area: Results from the FAO Global Forest Resources Assessment 2015. *For. Ecol. Manag.* **2015**, *352*, 9–20. [CrossRef]
8. van Dijk, K.; Savenije, H. *Oil Palm or Forests? More Than a Question of Definition*; Policy Brief; Tropenbos International: Wageningen, The Netherlands, 2010.
9. van Noordwijk, M.; Dewi, S.; Minang, P.; Simons, T. 1.2 Deforestation-free claims: Scams or substance. In *Zero Deforestation: A Commitment to Change*; No. 58; Tropenbos International: Wageningen, The Netherlands, 2017; pp. 11–16.
10. Sasaki, N.; Putz, F.E. Critical need for new definitions of "forest" and "forest degradation" in global climate change agreements. *Conserv. Lett.* **2009**, *2*, 226–232. [CrossRef]
11. Zomer, R.J.; Trabucco, A.; Verchot, L.V.; Muys, B. Land area eligible for afforestation and reforestation within the Clean Development Mechanism: A global analysis of the impact of forest definition. *Mitig. Adapt. Strateg. Glob. Chang.* **2008**, *13*, 219–239. [CrossRef]
12. Fay, C.; Michon, G. Redressing forestry hegemony when a forestry regulatory framework is best replaced by an agrarian one. *For. Trees Livelihoods* **2005**, *15*, 193–209. [CrossRef]
13. Silvianingsih, Y.A.; Hairiah, K.; Suprayogo, D.; van Noordwijk, M. Agroforests, swiddening and livelihoods between restored peat domes and river: Effects of the 2015 fire ban in Central Kalimantan (Indonesia). *Int. For. Rev.* **2020**, *22*, 382–396. [CrossRef]
14. Akiefnawati, R.; Villamor, G.B.; Zulfikar, F.; Budisetiawan, I.; Mulyoutami, E.; Ayat, A.; van Noordwijk, M. Stewardship agreement to reduce emissions from deforestation and degradation (REDD): Case study from Lubuk Beringin's Hutan Desa, Jambi Province, Sumatra, Indonesia. *Int. For. Rev.* **2010**, *12*, 349–360. [CrossRef]
15. Cahyono, E.D.; Fairuzzana, S.; Willianto, D.; Pradesti, E.; McNamara, N.P.; Rowe, R.L.; van Noordwijk, M. Agroforestry Innovation through Planned Farmer Behavior: Trimming in Pine–Coffee Systems. *Land* **2020**, *9*, 363. [CrossRef]
16. Van Noordwijk, M.; Zomer, R.J.; Xu, J.; Bayala, J.; Dewi, S.; Miccolis, A.; Cornelius, J.P.; Robiglio, V.; Nayak, D.; Rizvi, J. Agroforestry options, issues and progress in pantropical contexts. In *Sustainable Development through Trees on Farms: Agroforestry in Its Fifth Decade*; Van Noordwijk, M., Ed.; World Agroforestry (ICRAF): Bogor, Indonesia, 2019.
17. Van Noordwijk, M.; Williams, S.; Verbist, B. Towards integrated natural resource management in forest margins of the humid tropics: Local action and global concerns. In *ASB Lecture Notes*; ICRAF: Bogor, Indonesia, 2001.
18. Hansen, M.C.; Potapov, P.V.; Moore, R.; Hancher, M.; Turubanova, S.A.; Tyukavina, A.; Thau, D.; Stehman, S.V.; Goetz, S.J.; Loveland, T.R.; et al. High-resolution global maps of 21st-century forest cover change. *Science* **2013**, *342*, 850–853. [CrossRef] [PubMed]

19. Zomer, R.J.; Neufeldt, H.; Xu, J.C.; Ahrends, A.; Bossio, D.; Trabucco, A.; van Noordwijk, M.; Wang, M. Global tree cover and biomass carbon on agricultural land: The contribution of agroforestry to global and national carbon budgets. *Sci. Rep.* **2016**, *6*, 29987. [CrossRef] [PubMed]

20. Thomas, S.; Dargusch, P.; Harrison, S.; Herbohn, J. Why are there so few afforestation and reforestation Clean Development Mechanism projects? *Land Use Policy* **2010**, *27*, 880–887. [CrossRef]

21. Corbera, E.; Schroeder, H. Governing and implementing REDD+. *Environ. Sci. Policy* **2011**, *14*, 89–99. [CrossRef]

22. Minang, P.A.; Van Noordwijk, M.; Duguma, L.A.; Alemagi, D.; Do, T.H.; Bernard, F.; Agung, P.; Robiglio, V.; Catacutan, D.; Suyanto, S.; et al. REDD+ Readiness progress across countries: Time for reconsideration. *Clim. Policy* **2014**, *14*, 685–708. [CrossRef]

23. Garrett, R.D.; Levy, S.; Carlson, K.M.; Gardner, T.A.; Godar, J.; Clapp, J.; Dauvergne, P.; Heilmayr, R.; de Waroux, Y.L.P.; Ayre, B.; et al. Criteria for effective zero-deforestation commitments. *Glob. Environ. Chang.* **2019**, *54*, 135–147. [CrossRef]

24. Sheil, D.; Casson, A.; Meijaard, E.; Van Noordwijk, M.; Gaskell, J.; Sunderland-Groves, J.; Wertz, K.; Kanninen, M. *The Impacts and Opportunities of Oil Palm in Southeast Asia: What do We Know and What Do We Need to Know*; Center for International Forestry Research: Bogor, Indonesia, 2009. [CrossRef]

25. Sayer, J.; Ghazoul, J.; Nelson, P.; Boedhihartono, A.K. Oil palm expansion transforms tropical landscapes and livelihoods. *Glob. Food Secur.* **2012**, *1*, 114–119. [CrossRef]

26. van Noordwijk, M.; Pacheco, P.; Slingerland, M.; Dewi, S.; Khasanah, N. *Palm Oil Expansion in Tropical Forest Margins or Sustainability of Production? Focal Issues of Regulations and Private Standards*; World Agroforestry (ICRAF): Bogor, Indonesia, 2017. [CrossRef]

27. Heilmayr, R.; Carlson, K.M.; Benedict, J.J. Deforestation spillovers from oil palm sustainability certification. *Environ. Res. Lett.* **2020**, *15*, 075002. [CrossRef]

28. Condro, A.A.; Setiawan, Y.; Prasetyo, L.B.; Pramulya, R.; Siahaan, L. Retrieving the National Main Commodity Maps in Indonesia Based on High-Resolution Remotely Sensed Data Using Cloud Computing Platform. *Land* **2020**, *9*, 377. [CrossRef]

29. Mithöfer, D.; van Noordwijk, M.; Leimona, B.; Cerutti, P.O. Certify and shift blame, or resolve issues? Environmentally and socially responsible global trade and production of timber and tree crops. *Int. J. Biodivers. Sci. Ecosyst. Serv. Manag.* **2017**, *13*, 72–85.

30. Leimona, B.; van Noordwijk, M.; Mithöfer, D.; Cerutti, P.O. Certifying Environmental Social Responsibility. Environmentally and socially responsible global production and trade of timber and tree crop commodities: Certification as a transient issue-attention cycle response to ecological and social issues. *Int. J. Biodivers. Sci. Ecosyst. Serv. Manag.* **2018**, *13*, 497–502. [CrossRef]

31. KLHK. *Peta Penutupan Lahan Indonesia, 2000, 2010, 2018*; Kementrian Lingkunan Hidup dan Kehutanan: Jakarta, Indonesia, 2018; Available online: http://geoportal.menlhk.go.id/arcgis/rest/services/ (accessed on 9 October 2020).

32. Pribadi, U.A.; Setiabudi Suryadi, I.; Laumonier, Y. West Kalimantan Ecological Vegetation Map 1:50000. 2020. Available online: https://data.cifor.org/dataset.xhtml?persistentId=doi:10.17528/CIFOR/DATA.00203 (accessed on 15 October 2020). [CrossRef]

33. Descals, A.; Wich, S.; Meijaard, E.; Gaveau, D.L.A.; Peedell, S.; Szantoi, Z. High-resolution global map of smallholder and industrial closed-canopy oil palm plantations. *Earth Syst. Sci. Data Discuss* **2020**. [CrossRef]

34. SK Menteri Kehutanan no 733/2014. Available online: https://fdokumen.com/document/sk-no733-tahun-2014-penunjukan-kalbar.html (accessed on 30 October 2020).

35. SK Menteri Kehutanan no 863/2014. Available online: https://fdokumen.com/document/surat-keputusan-menteri-kehutanan-nomor-sk-863menhut-ii2014-tentang-kawasan.html (accessed on 30 October 2020).

36. Galudra, G.; Sirait, M. A discourse on Dutch colonial forest policy and science in Indonesia at the beginning of the 20th century. *Int. For. Rev.* **2009**, *11*, 524–533. [CrossRef]

37. Rachman, N.F.; Siscawati, M. Forestry Law, Masyarakat Adat and Struggles for Inclusive Citizenship in Indonesia. *Routledge Handb. Asian Law* **2016**, 224–249. [CrossRef]

38. Bakker, L.; Moniaga, S. The space between: Land claims and the law in Indonesia. *Asian J. Soc. Sci.* **2010**, *38*, 187–203. [CrossRef]

39. Hermosilla, A.C.; Fay, C. *Strengthening Forest Management in Indonesia through Land Tenure Reform: Issues and Framework for Action*; The World Bank: Washington, DC, USA, 2005; p. 55.

40. Ministry of Environment and Forestry. *The State of Indonesia's Forest*; Ministry of Environment and Forestry, Republic of Indonesia: Jakarta, Indonesia, 2018.

41. Suwarno, A.; van Noordwijk, M.; Weikard, H.P.; Suyamto, D. Indonesia's forest conversion moratorium assessed with an agent-based model of Land-Use Change and Ecosystem Services (LUCES). *Mitig. Adapt. Strateg. Glob. Chang.* **2018**, *23*, 211–229. [CrossRef] [PubMed]

42. Directorate of Peat Degradation Control. Penetapan Fungsi Ekosistem Gambut (Peatland Ecosystem Function Establishment). 2020. Available online: http://pkgppkl.menlhk.go.id/v0/penetapan-fungsi-ekosistem-gambut/ (accessed on 15 October 2020).

43. Decree of Minister of Forestry No. 47/Kpts-II/1998 Concerning the Designation of +/– 29,000 Hectares of the Protective Forest and Limited Production Forest Zone, From the Pesisir Forest Grouping, in West Lampung District, Province of Lampung, Which Are Covered by Damar Agroforests ('Repong Damar') Managed by Communities under Customary Law (Masyarakat Hukum Adat), as a Zone with Distinct Purpose (Kawasan Dengan Tujuan Istimewa). Available online: https://peraturan.huma.or.id/pub/surat-keputusan-menteri-kehutanan-nomor-47-kpts-ii-1998 (accessed on 30 November 2020).

44. Kusters, K.; De Foresta, H.; Ekadinata, A.; van Noordwijk, M. Towards solutions for state vs. local community conflicts over forestland: The impact of formal recognition of user rights in Krui, Sumatra, Indonesia. *Hum. Ecol.* **2007**, *35*, 427–438. [CrossRef]

45. Michon, G. *Domesticating Forests—How Farmers Manage Forest Resources*; CIFOR: Bogor, Indonesia, 2005.

46. De Royer, S.; Van Noordwijk, M.; Roshetko, J.M. Does community-based forest management in Indonesia devolve social justice or social costs? *Int. For. Rev.* **2018**, *20*, 167–180. [CrossRef]

47. Rakatama, A.; Pandit, R. Reviewing social forestry schemes in Indonesia: Opportunities and challenges. *For. Policy Econ.* **2020**, *111*, 102052. [CrossRef]

48. Kelly, A.B.; Peluso, N.L. Frontiers of commodification: State lands and their formalization. *Soc. Nat. Resour.* **2015**, *28*, 473–495. [CrossRef]

49. Brown, D.W. *Addicted to Rent: Corporate and Spatial Distribution of Forest Resources in Indonesia: Implications for Forest Sustainability and Government Policy*; Indonesia-UK Tropical Forest Management Programme, Provincial Forest Management Programme: Jakarta, Indonesia, 1999.

50. Resosudarmo, B.P. (Ed.) *The Politics and Economics of Indonesia's Natural Resources*; Institute of Southeast Asian Studies: Singapore, 2005. [CrossRef]

51. Brown, T.; Leitmann, J.; Boccucci, M.; Jurgens, E. *Sustaining Economic Growth, Rural Livelihoods, and Environmental Benefits: Strategic Options for Forest Assistance in Indonesia*; Worldbank: Jakarta, Indonesia, 2006; Available online: http://documents1.worldbank.org/curated/en/986501468049447840/pdf/392450REVISED0IDWBForestOptions.pdf (accessed on 24 October 2020).

52. Mulyani, M.; Jepson, P. REDD+ and forest governance in Indonesia: A multistakeholder study of perceived challenges and opportunities. *J. Environ. Dev.* **2013**, *22*, 261–283. [CrossRef]

53. Van Noordwijk, M.; Agus, F.; Dewi, S.; Purnomo, H. Reducing emissions from land use in Indonesia: Motivation, policy instruments and expected funding streams. *Mitig. Adapt. Strateg. Glob. Chang.* **2014**, *19*, 677–692. [CrossRef]

54. Singer, B.; Giessen, L. Towards a donut regime? Domestic actors, climatization, and the hollowing-out of the international forests regime in the Anthropocene. *For. Policy Econ.* **2017**, *79*, 69–79. [CrossRef]

55. Hall, D. Land grabs, land control, and Southeast Asian crop booms. *J. Peasant Stud.* **2011**, *38*, 837–857. [CrossRef]

56. Gouyon, A.; De Foresta, H.; Levang, P. Does 'jungle rubber' deserve its name? An analysis of rubber agroforestry systems in southeast Sumatra. *Agrofor. Syst.* **1993**, *22*, 181–206. [CrossRef]

57. Verbist, B.; Putra, A.E.D.; Budidarsono, S. Factors driving land use change: Effects on watershed functions in a coffee agroforestry system in Lampung, Sumatra. *Agric. Syst.* **2005**, *85*, 254–270. [CrossRef]

58. Ruf, F.; Ehret, P. Smallholder cocoa in Indonesia: Why a cocoa boom in Sulawesi? In *Cocoa Pioneer Fronts Since 1800*; Palgrave Macmillan: London, UK, 1996; pp. 212–231. [CrossRef]

59. Li, T.M. Practices of Assemblage and Community Forest Management. *Econ. Soc.* **2007**, *36*, 264–294. [CrossRef]

60. Galudra, G.; van Noordwijk, M.; Agung, P.; Suyanto, S.; Pradhan, U. Migrants, land markets and carbon emissions in Jambi, Indonesia: Land tenure change and the prospect of emission reduction. *Mitig. Adapt. Strateg. Glob. Chang.* **2014**, *19*, 715–731. [CrossRef]

61. Pelzer, K.J. Pioneer Settlement in the Asiatic Tropics, Studies in land utilisation and agricultural colonization in southeastern Asia. In *Pioneer Settlement in the Asiatic Tropics. Special Publication No. 29*; American Geographical Society: New York, NY, USA, 1945.

62. De Koninck, R.; McTaggart, W.D. Land settlement processes in Southeast Asia: Historical foundations, discontinuities, and problems. *Asian Res. Serv.* **1987**, *15*, 341–356.

63. Angelsen, A. *Forest Cover Change in Space and Time: Combining the von Thunen and Forest Transition Theories*; The World Bank: Washington, DC, USA, 2007.

64. De Koninck, R. The challenges of the agrarian transition in Southeast Asia. *Labour Cap. Soc.* **2004**, *37*, 285–288.

65. Dove, M. *The Banana Tree at the Gate: A History of Marginal Peoples and Global Markets in Borneo*; Yale University Press: New Haven, CT, USA, 2011.

66. Geertz, C. *Agricultural Involution: The Processes of Ecological Change in Indonesia*; Univ. of California Press: Berkeley, CA, USA, 1963.

67. van Noordwijk, M.; Bizard, V.; Wangpakapattanawong, P.; Tata, H.L.; Villamor, G.B.; Leimona, B. Tree cover transitions and food security in Southeast Asia. *Glob. Food Secur.* **2014**, *3*, 200–208. [CrossRef]

68. Potter, L. How can the people's sovereignty be achieved in the oil palm sector? Is the plantation model shifting in favour of smallholders. In *Land and Development in Indonesia: Searching for the People's Sovereignty*; McCarthy, J.F., Robinson, K., Eds.; ISEAS-Yusof Ishak Institute: Singapore, 2016; pp. 315–342.

69. Colfer, C.J.P.; Resosudarmo, I.A.P. Which way forward. In *People, Forests and Policymaking in Indonesia*; Resources for the Future: Washington, DC, USA, 2002.

70. Berkes, F.; Colding, J.; Folke, C. Rediscovery of traditional ecological knowledge as adaptive management. *Ecol. Appl.* **2000**, *10*, 1251–1262. [CrossRef]

71. Gadgil, M.; Olsson, P.; Berkes, F.; Folke, C. Exploring the role of local ecological knowledge in ecosystem management: Three case studies. In *Navigating Social-Ecological Systems: Building Resilience for Complexity and Change*; Berkes, F., Colding, J., Folkes, C., Eds.; Cambridge University Press: Cambridge, UK, 2003. [CrossRef]

72. Villamor, G.B.; Pontius, R.G.; van Noordwijk, M. Agroforest's growing role in reducing carbon losses from Jambi (Sumatra), Indonesia. *Reg. Environ. Chang.* **2014**, *14*, 825–834. [CrossRef]

73. Santika, T.; Wilson, K.A.; Budiharta, S.; Kusworo, A.; Meijaard, E.; Law, E.A.; Friedman, R.; Hutabarat, J.A.; Indrawan, T.P.; St John, F.A.; et al. Heterogeneous impacts of community forestry on forest conservation and poverty alleviation: Evidence from Indonesia. *People Nat.* **2019**, *1*, 204–219. [CrossRef]

74. Directorate General of Estate Crops. *Tree Crop Estate Statistics of Indonesia 2018–2020—Palm Oil*; Directorate General of Estate, Ministry of Agriculture: Jakarta, Indonesia, 2019.

75. Bakhtiar, I.; Suradiredja, D.; Santoso, H.; Saputra, W. *Hutan Kita Bersawit–Gagasan Penyelesaian untuk Perkebunan Kelpa Sawit dalam Kawasan Hutan*; KEHATI: Jakarta, Indonesia, 2019.

76. Gaveau, D.; Sheil, D.; Husnayaen Salim, M.A.; Arjakusuma, S.; Ancrenaz, M.; Pacheco, P.; Meijaard, E. Rapid conversions and avoided deforestation: Examining four decades of industrial plantation expansion in Borneo. *Sci. Rep.* **2016**, *6*, 3201. [CrossRef]

77. Meijaard, E.; Garcia-Ulloa, J.; Sheil, D.; Wich, S.A.; Carlson, K.M.; Juffe-Bignoli, D.; Brooks, T.M. (Eds.) *Oil Palm and Biodiversity. A Situation Analysis by the IUCN Oil Palm Task Force*; IUCN Oil Palm Task Force: Gland, Switzerland, 2018; p. 116. [CrossRef]

78. Austin, K.G.; Mosnier, A.; Pirker, J.; McCallum, I.; Fritz, S.; Kasibhatla, P.S. Shifting patterns of oil palm driven deforestation in Indonesia and implications for zero-deforestation commitments. *Land Use Policy* **2017**, *69*, 41–48. [CrossRef]

79. DJP. *Statistik Perkebunan Indonesia; Kelapa Sawit 2018–2020*; Direktorat Jenderal Perkebunan: Jakarta, Indonesia, 2020; p. 68.

80. Badan Pusat Statistik, Sistem Informasi Administrasi Kependudukan, Kemendagri, 2019. Indonesia Administrative Level 0–4 Boundaries, Indonesia (IDN) Administrative Boundary Common Operational Database (COD-AB), Update 23 December 2019, Badan Pusat Statistik, Sistem Informasi Administrasi Kependudukan. Kemendagri (Ministry of Home Affairs). Available online: https://data.humdata.org/dataset/indonesia-administrative-level-0-4-boundariesx (accessed on 16 May 2020).

81. Potter, L. Where Are the Swidden Fallows Now? An Overview of Oil Palm and Dayak Agriculture across Kalimantan, with Case Studies from Sanggau, in West Kalimantan. In *Shifting Cultivation and Environmental Change: Indigenous People, Agriculture and Forest Conservation*; Routledge: Abingdon, UK, 2015.

82. Penot, E.; Chambon, B.; Wibawa, G. History of rubber agroforestry systems development in Indonesia and Thailand as alternatives for sustainable agriculture and income stability. *Int. Proc. IRC* **2017**, *1*, 497–532. [CrossRef]

83. Feintrenie, L.; Levang, P. Sumatra's rubber agroforests: Advent, rise and fall of a sustainable cropping system. *Small-Scale For.* **2009**, *8*, 323–335. [CrossRef]

84. Therville, C.; Feintrenie, L.; Levang, P. Farmers' perspectives about agroforests conversion to plantations in Sumatra. Lessons learnt from Bungo District (Jambi, Indonesia). *For. Trees Livelihoods* **2011**, *20*, 15–33. [CrossRef]

85. Koh, L.P.; Miettinen, J.; Liew, S.C.; Ghazoul, J. Remotely sensed evidence of tropical peatland conversion to oil palm. *Proc. Natl. Acad. Sci. USA* **2011**, *108*, 5127–5132. [CrossRef]

86. Afriyanti, D.; Hein, L.; Kroeze, C.; Zuhdi, M.; Saad, A. Scenarios for withdrawal of oil palm plantations from peatlands in Jambi Province, Sumatra, Indonesia. *Reg. Environ. Chang.* **2019**, *19*, 1201–1215. [CrossRef]

87. Global Forest Watch. Universal Mill List. 2020. Available online: http://data.globalforestwatch.org/datasets/universal-mill-list/data (accessed on 21 March 2020).

88. ten Kate, A.; Kuepper, B.; Piotrowski, M. NDPE Policies Cover 83% of Palm Oil Refineries; Implementation at 78%. Available online: https://chainreactionresearch.com/report/ndpe-policies-cover-83-of-palm-oil-refineries-implementation-at-75/ (accessed on 25 October 2020).

89. Budidarsono, S.; Susanti, A.; Zoomers, A. Oil palm plantations in Indonesia: The implications for migration, settlement/resettlement and local economic development. In *Biofuels—Economy, Environment and Sustainability*; INTECH: London, UK, 2013; pp. 173–193. [CrossRef]

90. DJP. *Statistik Perkebunan Indonesia*; Kelapa Sawit; Direktorat Jenderal Perkebunan: Jakarta, Indonesia, 1990.

91. Jelsma, I.; Schoneveld, G.C.; Zoomers, A.; van Westen, A.C.M. Unpacking Indonesia's independent oil palm smallholders: An actor-disaggregated approach to identifying environmental and social performance challenges. *Land Use Policy* **2017**, *69*, 281–297. [CrossRef]

92. Schoneveld, G.C.; van der Haar, S.; Ekowati, D.; Andrianto, A.; Komarudin, H.; Okarda, B.; Jelsma, I.; Pacheco, P. Certification, good agricultural practice and smallholder heterogeneity: Differentiated pathways for resolving compliance gaps in the Indonesian oil palm sector. *Glob. Environ. Chang.* **2019**, *57*, 101933. [CrossRef]

93. Hidayat, K.N.; Glasbergen, P.; Offermans, A. Sustainability Certification and Palm Oil Smallholders' Livelihood: A comparison between scheme smallholders and independent smallholders in Indonesia. *Int. Food Agribus. Manag. Rev.* **2015**, *18*, 25–48. [CrossRef]

94. Euler, M.; Schwarze, S.; Siregar, H.; Qaim, M. Oil Palm Expansion among Smallholder Farmers in Sumatra, Indonesia. *J. Agric. Econ.* **2016**, *67*, 658–676. [CrossRef]

95. IFC. *Diagnostic Study on Indonesian Palm Oil Smallholders: Developing a Better Understanding of Their Performane and Potential*; International Finance Corporation: Washington, DC, USA, 2013; p. 96.

96. Gillespie, P. How Does Legislation Affect Oil Palm Smallholders in the Sanggau District of Kalimantan, Indonesia? *Australas. J. Nat. Resour. Law Policy* **2011**, *14*, 1–37.

97. World Bank. Indonesia: Strategies for Sustained Development of Tree Crops. In *Agriculture Operations Division*; Asia Regional Office, Ed.; World Bank: Washington, DC, USA, 1989.

98. Badrun, M. *Milestone of Change: Developing a Nation through Oil Palm "PIR"*; Directorate General of Estate Crops: Jakarta, Indonesia, 2011; p. 229.

99. Mongabay. Indonesia Moves to End Smallholder Guarantee Meant to Empower Palm Oil Farmers. 2020. Available online: https://news.mongabay.com/2020/05/indonesia-palm-oil-plasma-plantation-farmers-smallholders/ (accessed on 30 November 2020).

100. Jelsma, I.; Slingerland, M.; Giller, K.E.; Bijman, J. Collective action in a smallholder oil palm production system in Indonesia: The key to sustainable and inclusive smallholder palm oil? *J. Rural Stud.* **2017**, *54*, 198–210. [CrossRef]

101. Daemeter. *Overview of Indonesian Smallholder Farmers: A Typology of Organizational Models, Needs, and Investment Opportunities*; Daemeter: Bogor, Indonesia, 2015.

102. Cramb, R.A. Palmed Off: Incentive Problems with Joint-Venture Schemes for Oil Palm Development on Customary Land. *World Dev.* **2013**, *43*, 84–99. [CrossRef]

103. Susanti, A.; Maryudi, A. Development narratives, notions of forest crisis, and boom of oil palm plantations in Indonesia. *For. Policy Econ.* **2016**, *73*, 130–139. [CrossRef]

104. Bissonnette, J.-F.; De Koninck, R. The return of the plantation? Historical and contemporary trends in the relation between plantations and smallholdings in Southeast Asia. *J. Peasant Stud.* **2017**, *44*, 918–938. [CrossRef]

105. Grass, I.; Kubitza, C.; Krishna, V.V.; Corre, M.D.; Mußhoff, O.; Pütz, P.; Drescher, J.; Rembold, K.; Ariyanti, E.S.; Barnes, A.D.; et al. Trade-offs between multifunctionality and profit in tropical smallholder landscapes. *Nat. Commun.* **2020**, *11*, 1186. [CrossRef] [PubMed]

106. Feintrenie, L.; Chong, W.; Levang, P. Why do Farmers Prefer Oil Palm? Lessons Learnt from Bungo District, Indonesia. *Small-Scale For.* **2010**, *9*, 379–396. [CrossRef]

107. Kubitza, C.; Krishna, V.V.; Alamsyah, Z.; Qaim, M. The economics behind an ecological crisis: Livelihood effects of oil palm expansion in Sumatra, Indonesia. *Hum. Ecol.* **2018**, *46*, 107–116. [CrossRef]

108. Jelsma, I.; Woittiez, L.S.; Ollivier, J.; Dharmawan, A.H. Do wealthy farmers implement better agricultural practices? An assessment of implementation of Good Agricultural Practices among different types of independent oil palm smallholders in Riau, Indonesia. *Agric. Syst.* **2019**, *170*, 63–76. [CrossRef]

109. Woittiez, L.S.; Slingerland, M.; Rafik, R.; Giller, K.E. Nutritional imbalance in smallholder oil palm plantations in Indonesia. *Nutr. Cycl. Agroecosyst.* **2018**, *111*, 73–86. [CrossRef]

110. McCarthy, J.F.; Zen, Z. Agribusiness, agrarian change, and the fate of oil palm smallholders in Jambi. In *The Oil Palm Complex: Smallholders, Agribusiness and the State in Indonesia and Malaysia*; Cramb, R., McCarthy, J.F., Eds.; NUS Press: Singapore, 2016; pp. 109–155. ISBN 978-981-4722-06-3.

111. Krishna, V.; Euler, M.; Siregar, H.; Qaim, M. Differential livelihood impacts of oil palm expansion in Indonesia. *Agric. Econ.* **2017**, *48*, 639–653. [CrossRef]

112. Krishna, V.V.; Kubitza, C.; Pascual, U.; Qaim, M. Land markets, Property rights, and Deforestation: Insights from Indonesia. *World Dev.* **2017**, *99*, 335–349. [CrossRef]

113. Purnomo, H.; Shantiko, B.; Sitorus, S.; Gunawan, H.; Achdiawan, R.; Kartodihardjo, H.; Dewayani, A.A. Fire economy and actor network of forest and land fires in Indonesia. *For. Policy Econ.* **2017**, *78*, 21–31. [CrossRef]

114. Zen, Z.; Barlow, C.; Gondowarsito, R.; McCarthy, J.F. Interventions to promote smallholder oil palm and socio-economic improvement in Indonesia. In *The Oil Palm Complex: Smallholders, Agribusiness and the State in Indonesia and Malaysia*; Cramb, R., McCarthy, J.F., Eds.; NUS Press: Singapore, 2016; pp. 78–108. ISBN 978-981-4722-06-3.

115. Aulia, A.; Sandhu, H.; Millington, A. Quantifying the Economic Value of Ecosystem Services in Oil Palm Dominated Landscapes in Riau Province in Sumatra, Indonesia. *Land* **2020**, *9*, 194. [CrossRef]

116. Ayompe, L.M.; Schaafsma, M.; Egoh, B.N. Towards sustainable palm oil production: The positive and negative impacts on ecosystem services and human wellbeing. *J. Clean. Prod.* **2021**, *278*, 123914. [CrossRef]

117. Manoli, G.; Meijide, A.; Huth, N.; Knohl, A.; Kosugi, Y.; Burlando, P.; Fatichi, S. Ecohydrological changes after tropical forest conversion to oil palm. *Environ. Res. Lett.* **2018**, *13*, 064035. [CrossRef]

118. Fitzherbert, E.B.; Struebig, M.J.; Morel, A.; Danielsen, F.; Brühl, C.A.; Donald, P.F.; Phalan, B. How will oil palm expansion affect biodiversity? *Trends Ecol. Evol.* **2008**, *23*, 538–545. [CrossRef] [PubMed]

119. Koh, L.P.; Wilcove, D.S. Is oil palm agriculture really destroying tropical biodiversity? *Conserv. Lett.* **2008**, *1*, 60–64. [CrossRef]

120. Booth, S.L. The Relationship between Carbon Emissions, Land Use Change and the Oil Palm Industry within Southeast Asia. Master's Thesis, University of San Fransisco, San Fransisco, CA, USA, 2017. Available online: https://repository.usfca.edu/capstone/562 (accessed on 30 November 2020).

121. Comte, I.; Colin, F.; Whalen, J.; Olivier, G.; Caliman, J.-P. Agricultural practices in oil palm plantations and their impact on hydrological changes, nutrient fluxes and water quality in Indonesia. A review. *Adv. Agron.* **2012**, *116*, 71–122. [CrossRef]

122. Merten, J.; Röll, A.; Guillaume, T.; Meijide, A.; Tarigan, S.; Agusta, H.; Hölscher, D. Water scarcity and oil palm expansion: Social views and environmental processes. *Ecol. Soc.* **2016**, *21*, 21. [CrossRef]

123. Obidzinski, K.; Andriani, R.; Komarudin, H.; Andrianto, A. Environmental and Social Impacts of Oil Palm Plantations and their Implications for Biofuel Production in Indonesia. *Ecol. Soc.* **2012**, *17*. [CrossRef]

124. Dislich, C.; Keyel, A.C.; Salecker, J.; Kisel, Y.; Meyer, K.M.; Auliya, M.; Wiegand, K. A review of the ecosystem functions in oil palm plantations, using forests as a reference system. *Biol. Rev. Camb. Philos. Soc.* **2017**, *92*, 1539–1569. [CrossRef]

125. Morana, L.S. *Oil Palm Plantations: Threats and Opportunities for Tropical Ecosystems. Thematic Focus: Ecosystem Management and Resource Efficiency*; United Nation environment Programme Global Environmental Alert Service (GEAS): Geneva, Switzerland, 2011.

126. Safitri, L.; Sastrohartono, H.; Purboseno, S.; Kautsar, V.; Saptomo, S.; Kurniawan, A. Water Footprint and Crop Water Usage of Oil Palm (*Eleasis guineensis*) in Central Kalimantan: Environmental Sustainability Indicators for Different Crop Age and Soil Conditions. *Water* **2018**, *11*, 35. [CrossRef]

127. Barthel, M.; Jennings, S.; Schreiber, W.; Sheane, R.; Royston, S.; Fry, J.; McGill, J. *Study on the Environmental Impact of Palm Oil Consumption and on Existing Sustainability Standards*; (07.0201/2016/743217/ETU/ENV.F3); European Commission: Brussels, Belgium, 2018; Available online: https://ec.europa.eu/environment/forests/pdf/palm_oil_study_kh0218208enn_new.pdf (accessed on 30 November 2020).

128. Azhar, B.; Lindenmayer, D.B.; Wood, J.; Fischer, J.; Manning, A.; McElhinny, C.; Zakaria, M. The conservation value of oil palm plantation estates, smallholdings and logged peat swamp forest for birds. *For. Ecol. Manag.* **2011**, *262*, 2306–2315. [CrossRef]

129. Petrenko, C.; Paltseva, J.; Searle, S. *Ecological Impacts of Palm Oil Expansion in Indonesia*; White Paper; International Council on Clean Transportation: Washington, DC, USA, 2016.

130. Fischer, J.; Lindenmayer, D. Landscape modification and habitat fragmentation: A synthesis. *Glob. Ecol. Biogeogr.* **2007**, *16*, 265–280. [CrossRef]

131. Gallmetzer, N.; Schulze, C.H. Impact of oil palm agriculture on understory amphibians and reptiles: A Mesoamerican perspective. *Glob. Ecol. Conserv.* **2015**, *4*, 95–109. [CrossRef]

132. Murdiyarso, D.; Van Noordwijk, M.; Wasrin, U.R.; Tomich, T.P.; Gillison, A.N. Environmental benefits and sustainable land-use options in the Jambi transect, Sumatra. *J. Veg. Sci.* **2002**, *13*, 429–438. [CrossRef]

133. Aratrakorn, S.; Thunhikorn, S. Changes in bird communities following conversion of lowland forest to oil palm and rubber plantations in Thailand. *Bird Conserv. Int.* **2006**, *16*, 71–82. [CrossRef]

134. Edwards, F.A.; Edwards, D.P.; Hamer, K.C.; Davies, R.G. Impacts of logging and conversion of rainforest to oil palm on the functional diversity of birds in Sundaland. *IBIS* **2013**, *155*, 313–326. [CrossRef]

135. Konopik, O.; Gray, C.L.; Grafe, T.U.; Steffan-Dewenter, I.; Fayle, T.M. From rainforest to oil palm plantations: Shifts in predator population and prey communities, but resistant interactions. *Glob. Ecol. Conserv.* **2014**, *2*, 385–394. [CrossRef]

136. Payán, E.; Boron, V. The Future of Wild Mammals in Oil Palm Landscapes in the Neotropics. *Front. For. Glob. Chang.* **2019**, *2*, 61. [CrossRef]

137. Danielsen, F.; Beukema, H.; Burgess, N.; Parish, F.; Brühl, C.; Donald, P.; Fitzherbert, E. Biofuel plantations on forested lands: Double jeopardy for biodiversity and climate. *IOP Conf. Ser. Earth Environ. Sci.* **2009**, *6*, 242014. [CrossRef]

138. Vijay, V.; Pimm, S.L.; Jenkins, C.N.; Smith, S.J. The Impacts of Oil Palm on Recent Deforestation and Biodiversity Loss. *PLoS ONE* **2016**, *11*, e0159668. [CrossRef]

139. Ancrenaz, M.; Oram, F.; Ambu, L.; Lackman, I.; Ahmad, E.; Elahan, H.; Meijaard, E. Of Pongo, palms and perceptions: A multidisciplinary assessment of Bornean orang-utans Pongo pygmaeus in an oil palm context. *Oryx* **2014**, *49*, 465–472. [CrossRef]

140. Baraniuk, C. Orangutans Are Hanging on in the Same Palm Oil Plantations That Displace Them, Conservationists Still Need to Work to Minimize Conflict between the Endangered Apes and Humans. *Sci. Am.* 2020. Available online: https://www.scientificamerican.com/article/orangutans-are-hanging-on-in-the-same-palm-oil-plantations-that-displace-them/ (accessed on 30 November 2020).

141. Jonas, H.; Abram, N.K.; Ancrenaz, M. *Addressing the Impact of Large-Scale Oil Palm Plantations on Orangutan Conservation in Borneo: A Spatial, Legal and Political Economy Analysis*; IIED: London, UK, 2017.

142. Nellemann, C.; Miles, L.; Kaltenborn, B.; Virtue, M.; Ahlenius, H. *Last Stand of the Orang-Utan*; A UNEP Rapid Response Assessment: Nairobi, Kenya, 2007.

143. Swarna Nantha, H.; Tisdell, C. The orangutan–oil palm conflict: Economic constraints and opportunities for conservation. *Biodivers. Conserv.* **2009**, *18*, 487–502. [CrossRef]

144. Suba, R.B.; van de Ploeg, J.; Zelfde, M.; Lau, Y.W.; Wissingh, T.F.; Kustiawan, W.; de Iongh, H.H. Rapid expansion of oil palm is leading to human–elephant conflicts in North Kalimantan province of Indonesia. *Trop. Conserv. Sci.* **2017**, *10*, 1940082917703508. [CrossRef]

145. Shuhada, S.N.; Salim, S.; Nobilly, F.; Lechner, A.M.; Azhar, B. Conversion of peat swamp forest to oil palm cultivation reduces the diversity and abundance of macrofungi. *Glob. Ecol. Conserv.* **2020**, *23*, e01122. [CrossRef]

146. Paoletti, A.; Darras, K.; Jayanto, H.; Grass, I.; Kusrini, M.; Tscharntke, T. Amphibian and reptile communities of upland and riparian sites across Indonesian oil palm, rubber and forest. *Glob. Ecol. Conserv.* **2018**, *16*, e00492. [CrossRef]

147. Gillison, A.N.; Bignell, D.E.; Brewer, K.R.; Fernandes, E.C.; Jones, D.T.; Sheil, D.; May, P.H.; Watt, A.D.; Constantino, R.; Couto, E.G.; et al. Plant functional types and traits as biodiversity indicators for tropical forests: Two biogeographically separated case studies including birds, mammals and termites. *Biodivers. Conserv.* **2013**, *22*, 1909–1930. [CrossRef]

148. Unjan, R.; Nissapa, A.; Phitthayaphinant, P. An Identification of Impacts of Area Expansion Policy of Oil Palm in Southern Thailand: A Case Study in Phatthalung and Nakhon Si Thammarat Provinces. *Procedia Soc. Behav. Sci.* **2013**, *91*, 489–496. [CrossRef]

149. Anamulai, S.; Sanusi, R.; Zubaid, A.; Lechner, A.M.; Ashton-Butt, A.; Azhar, B. Land use conversion from peat swamp forest to oil palm agriculture greatly modifies microclimate and soil conditions. *PeerJ* **2019**, *7*, e7656. [CrossRef]

150. Larsen, R.K.; Jiwan, N.; Rompas, A.; Jenito, J.; Osbeck, M.; Tarigan, A. Towards 'hybrid accountability' in EU biofuels policy? Community grievances and competing water claims in the Central Kalimantan oil palm sector. *Geoforum* **2014**, *54*, 295–305. [CrossRef]

151. Córdoba, D.; Juen, L.; Selfa, T.; Peredo, A.M.; Montag, L.F.d.A.; Sombra, D.; Santos, M.P.D. Understanding local perceptions of the impacts of large-scale oil palm plantations on ecosystem services in the Brazilian Amazon. *For. Policy Econ.* **2019**, *109*, 102007. [CrossRef]

152. Tarigan, S.D. Land Cover Change and its Impact on Flooding Frequency of Batanghari Watershed, Jambi Province, Indonesia. *Procedia Environ. Sci.* **2016**, *33*, 386–392. [CrossRef]

153. Heidari, A.; Mayer, A.; Watkins, D.; Castillo, M.M. Hydrologic impacts and trade-offs associated with developing oil palm for bioenergy in Tabasco, Mexico. *J. Hydrol. Reg. Stud.* **2020**, *31*, 100722. [CrossRef]

154. Röll, A.; Niu, F.; Meijide, A.; Hardanto, A.; Hendrayanto, H.; Knohl, A.; Hölscher, D. Transpiration in an oil palm landscape: Effects of palm age. *Biogeosciences* **2015**, *12*, 5619–5633. [CrossRef]

155. Carr, M. The water relations and irrigation requirements of oil palm (*Elaeis guineensis*): A review. *Exp. Agric.* **2011**, *47*, 629–652. [CrossRef]

156. Chong, S.Y.; Teh, C.B.S.; Ainuddin, A.N.; Philip, E. Simple net rainfall partitioning equations for nearly closed to fully closed canopy stands. *Pertanika J. Trop. Agric. Sci.* **2018**, *41*, 81–100.

157. Farmanta, Y.; Dedi, S. Rainfall Interception by Palm Plant Canopy. 2016. Available online: http://repository.unib.ac.id/11344/1/046%20Yong%20F%20%26%20Dedi%20S.pdf (accessed on 30 November 2020).

158. Hardanto, A.; Röll, A.; Niu, F.; Meijide, A.; Hendrayanto; Hölscher, D. Oil Palm and Rubber Tree Water Use Patterns: Effects of Topography and Flooding. *Front. Plant Sci.* **2017**, *8*, 452. [CrossRef]

159. Bruijnzeel, L.A. Hydrological functions of tropical forests: Not seeing the soil for the trees? *Agric. Ecosyst. Environ.* **2004**, *104*, 185–228. [CrossRef]

160. Tarigan, S.; Stiegler, C.; Wiegand, K.; Knohl, A.; Murtilaksono, K. Relative contribution of evapotranspiration and soil compaction to the fluctuation of catchment discharge: Case study from a plantation landscape. *Hydrol. Sci. J.* **2020**, *65*, 1239–1248. [CrossRef]

161. Sumarga, E.; Hein, L.; Hooijer, A.; Vernimmen, R. Hydrological and economic effects of oil palm cultivation in Indonesian peatlands. *Ecol. Soc.* **2016**, *21*, 19. [CrossRef]

162. Chellaiah, D.; Yule, C.M. Effect of riparian management on stream morphometry and water quality in oil palm plantations in Borneo. *Limnologica* **2018**, *69*, 72–80. [CrossRef]

163. Juen, L.; Cunha, E.J.; Carvalho, F.G.; Ferreira, M.C.; Begot, T.O.; Andrade, A.L. Effects of oil palm plantations on the habitat structure and biota of streams in eastern Amazon. *River Res. Appl.* **2016**, *32*, 2081–2094. [CrossRef]

164. Sheaves, M.; Johnston, R.; Miller, K.; Nelson, P.N. Impact of oil palm development on the integrity of riparian vegetation of a tropical coastal landscape. *Agric. Ecosyst. Environ.* **2018**, *262*, 1–10. [CrossRef]

165. Jamaludin, N.F.; Muis, Z.A.; Hashim, H. An integrated carbon footprint accounting and sustainability index for palm oil mills. *J. Clean. Prod.* **2019**, *225*, 496–509. [CrossRef]

166. Lewis, K.; Rumpang, E.; Kho, L.K.; McCalmont, J.; Teh, Y.A.; Gallego-Sala, A.; Hill, T.C. An assessment of oil palm plantation aboveground biomass stocks on tropical peat using destructive and non-destructive methods. *Sci. Rep.* **2020**, *10*, 2230. [CrossRef] [PubMed]

167. Khasanah, N.M.; van Noordwijk, M.; Ningsih, H. Aboveground carbon stocks in oil palm plantations and the threshold for carbon-neutral vegetation conversion on mineral soils. *Cogent Environ. Sci.* **2015**, *1*, 1119964. [CrossRef]

168. Kotowska, M.M.; Leuschner, C.; Triadiati, T.; Meriem, S.; Hertel, D. Quantifying above- and belowground biomass carbon loss with forest conversion in tropical lowlands of Sumatra (Indonesia). *Glob. Chang. Biol.* **2015**, *21*, 3620–3634. [CrossRef] [PubMed]

169. Paterson, R.R.M.; Lima, N. Climate change affecting oil palm agronomy, and oil palm cultivation increasing climate change, require amelioration. *Ecol. Evol.* **2018**, *8*, 452–461. [CrossRef]

170. Khasanah, N.M.; van Noordwijk, M.; Ningsih, H.; Rahayu, S. Carbon neutral? No change in mineral soil carbon stock under oil palm plantations derived from forest or non-forest in Indonesia. *Agric. Ecosyst. Environ.* **2015**, *211*, 195–206. [CrossRef]

171. van Noordwijk, M.; Khasanah, N.M.; Dewi, S. Can intensification reduce emission intensity of biofuel through optimized fertilizer use? Theory and the case of oil palm in Indonesia. *GCB Bioenergy* **2017**, *9*, 940–952. [CrossRef]

172. IPCC. 2013 Supplement to the 2006 IPCC guidelines for national greenhouse gas inventories: Wetlands. In *Intergovernmental Panel on Climate Change (IPCC)*; Hiraishi, T., Krug, T., Tanabe, K., Srivastava, N., Baasansuren, J., Fukuda, M., Troxler, T.G., Eds.; IPCC: Geneva, Switzerland, 2014.

173. Khasanah, N.M.; van Noordwijk, M. Subsidence and carbon dioxide emissions in a smallholder peatland mosaic in Sumatra, Indonesia. *Mitig. Adapt. Strateg. Glob. Chang.* **2019**, *24*, 147–163. [CrossRef]

174. Rival, A.; Montet, D.; Pioch, D. Certification, labelling and traceability of palm oil: Can we build confidence from trustworthy standards? *OCL* **2016**, *23*, D609. [CrossRef]

175. Tinhout, B.; van de Hombergh, H. *Setting the Biodiversity Bar for Palm Oil Certification; Assessing the Rigor of Biodiversity and Assurance Requirements of Palm Oil Standards*; IUCN: Amsterdam, The Netherlands, 2019.

176. Pye, O. Commodifying sustainability: Development, nature and politics in the palm oil industry. *World Dev.* **2019**, *121*, 218–228. [CrossRef]

177. Wiggs, C.; Kuepper, B.; Piotr, M. Spot Market Purchases Allow Deforestation-Linked Palm Oil to Enter NDPE Supply Chains. Chain Reaction Research. Available online: https://chainreactionresearch.com/report/spot-market-purchases-allow-deforestation-linked-palm-oil-to-enter-ndpe-supply-chains/ (accessed on 30 October 2020).

178. Santika, T.; Law, E.; Wilson, K.A.; John, F.A.S.; Carlson, K.; Gibbs, H.; Morgans, C.L.; Ancrenaz, M.; Meijaard, E. Impact of Palm Oil Sustainability Certification on Village Well-Being and Poverty in Indonesia. 2020. Available online: https://osf.io/preprints/socarxiv/5qk67/ (accessed on 30 November 2020).

179. RSPO. Impact Update 2019. In *Round Table on Sustainable Palm Oil*; RSPO: Kuala Lumpur, Malaysia, 2019; p. 52.

180. Astari, A.J.; Lovett, J.C. Does the rise of transnational governance 'hollow-out' the state? Discourse analysis of the mandatory Indonesian sustainable palm oil policy. *World Dev.* **2019**, *117*, 1–12. [CrossRef]

181. Schouten, G.; Bitzer, V. The emergence of Southern standards in agricultural value chains: A new trend in sustainability governance? *Ecol. Econ.* **2015**, *120*, 175–184. [CrossRef]

182. Higgins, V.; Richards, C. Framing sustainability: Alternative standards schemes for sustainable palm oil and South-South trade. *J. Rural Stud.* **2019**, *65*, 126–134. [CrossRef]

183. Hidayat, N.K.; Offermans, A.; Glasbergen, P. Sustainable palm oil as a public responsibility? On the governance capacity of Indonesian Standard for Sustainable Palm Oil (ISPO). *Agric. Hum. Values* **2018**, *35*, 223–242. [CrossRef]

184. Kaoem, T. A False Hope? In *An Analysis of the Draft New Indonesia Sustainable Palm Oil (ISPO) Regulation*; EIA: London, UK, 2020; p. 6.

185. Van Noordwijk, M.; Leimona, B.; Amaruzaman, S. Sumber Jaya from conflict to source of wealth in Indonesia: Reconciling coffee agroforestry and watershed functions. In *Sustainable Development through Trees on Farms: Agroforestry in Its Fifth Decade*; Van Noordwijk, M., Ed.; World Agroforestry (ICRAF): Bogor, Indonesia, 2019; pp. 177–192.

186. Erb, M.; Anggal, W. Conflict and the growth of democracy in Manggarai District. In *Deepening Democracy in Indonesia.*; Erb, M., Sulistiyanto, P., Eds.; Institute of Southeast Asian Studies: Singapore, 2009; pp. 283–302.

187. Setiawan, E.N.; Maryudi, A.; Purwanto, R.H.; Lele, G. Opposing interests in the legalization of non-procedural forest conversion to oil palm in Central Kalimantan, Indonesia. *Land Use Policy* **2016**, *58*, 472–481. [CrossRef]

188. TEMPO English. *Held Captive at Rokan Hulu*; Tempo English: Jakarta, Indoenisa, 2016; pp. 36–37.

189. Gaveau, D.L.A.; Salim, M.A.; Hergoualc'h, K.; Locatelli, B.; Sloan, S.; Wooster, M.; Marlier, M.E.; Molidena, E.; Yaen, H.; DeFries, R.; et al. Major atmospheric emissions from peat fires in Southeast Asia during non-drought years: Evidence from the 2013 Sumatran fires. *Sci. Rep.* **2014**, *4*, 6112. [CrossRef] [PubMed]

190. Suryanata, K. Fruit trees under contract: Tenure and land use change in upland Java, Indonesia. *World Dev.* **1994**, *22*, 1567–1578. [CrossRef]

191. Robiglio, V.; Reyes, M. Restoration through formalization? Assessing the potential of Peru's Agroforestry Concessions scheme to contribute to restoration in agricultural frontiers in the Amazon region. *World Dev. Perspect.* **2016**, *3*, 42–46. [CrossRef]

192. Slingerland, M.A.; Khasanah, N.M.; van Noordwijk, M.; Susanti, A.; Meilantina, M. Improving smallholder inclusivity through integrating oil palm with crops. In *Exploring Inclusive Palm Oil Production*; ETFRN and Tropenbos International: Wageningen, The Netherland, 2019; pp. 147–154.

193. Khasanah, N.; Van Noordwijk, M.; Slingerland, M.; Sofiyudin, M.; Stomph, D.; Migeon, A.F.; Hairiah, K. Oil Palm Agroforestry Can Achieve Economic and Environmental Gains as Indicated by Multifunctional Land Equivalent Ratios. *Front. Sustain. Food Syst.* **2020**, *3*, 122. [CrossRef]

194. Woittiez, L.S.; van Wijk, M.T.; Slingerland, M.; van Noordwijk, M.; Giller, K.E. Yield gaps in oil palm: A quantitative review of contributing factors. *Eur. J. Agron.* **2017**, *83*, 57–77. [CrossRef]

195. Kehati. *Solusi Sawitt Salam Kawasn Hutan [Oilmpalm Solutins in the Forest Zone]. SymSPOSia, Episode 10*; Kehati: Jakarta, Indonesia, 2020; Available online: https://www.youtube.com/watch?v=0fsEfK5jsGU&feature=youtu.be (accessed on 29 November 2020).

196. Nurrochmat, D.R.; Boer, R.; Ardiansyah, M.; Immanuel, G.; Purwawangsa, H. Policy forum: Reconciling palm oil targets and reduced deforestation: Landswap and agrarian reform in Indonesia. *For. Policy Econ.* **2020**, *119*, 102291. [CrossRef]

197. Iiyama, M.; Neufeldt, H.; Dobie, P.; Hagen, R.; Njenga, M.; Ndegwa, G.; Mowo, J.G.; Kisoyan, P.; Jamnadass, R. Opportunities and challenges of landscape approaches for sustainable charcoal production and use. In *Climate-Smart Landscapes: Multifunctionality in Practice*; World Agroforestry Centre (ICRAF): Nairobi, Kenya, 2015; pp. 195–209.

Publisher's Note: MDPI stays neutral with regard to jurisdictional claims in published maps and institutional affiliations.

 *land*

Article

People-Centric Nature-Based Land Restoration through Agroforestry: A Typology

Meine van Noordwijk [1,2,3,]*[id], Vincent Gitz [4], Peter A. Minang [5], Sonya Dewi [1], Beria Leimona [1], Lalisa Duguma [5], Nathanaël Pingault [4] and Alexandre Meybeck [4]

[1] World Agroforestry (ICRAF), Bogor 16155, Indonesia; S.Dewi@cgir.org (S.D.); L.Beria@cgiar.org (B.L.)
[2] Plant Production Systems, Wageningen University & Research, 6700 AK Wageningen, The Netherlands
[3] Agroforestry Research Group, Brawijaya University, Malang 65145, Indonesia
[4] Centre for International Forestry Research (CIFOR), Bogor 16115, Indonesia; V.Gitz@cgiar.org (V.G.); nathanael.pingault@gmail.com (N.P.); A.Meybeck@cgiar.org (A.M.)
[5] World Agroforestry (ICRAF), Nairobi 00100, Kenya; A.Minang@cgiar.org (P.A.M.); L.Duguma@cgiar.org (L.D.)
*   Correspondence: m.vannoordwijk@cgiar.org

Received: 30 June 2020; Accepted: 28 July 2020; Published: 29 July 2020

Abstract: Restoration depends on purpose and context. At the core it entails innovation to halt ongoing and reverse past degradation. It aims for increased functionality, not necessarily recovering past system states. Location-specific interventions in social-ecological systems reducing proximate pressures, need to synergize with transforming generic drivers of unsustainable land use. After reviewing pantropical international research on forests, trees, and agroforestry, we developed an options-by-context typology. Four intensities of land restoration interact: R.I. Ecological intensification within a land use system, R.II. Recovery/regeneration, within a local social-ecological system, R.III. Reparation/recuperation, requiring a national policy context, R.IV. Remediation, requiring international support and investment. Relevant interventions start from core values of human identity while addressing five potential bottlenecks: Rights, Know-how, Markets (inputs, outputs, credit), Local Ecosystem Services (including water, agrobiodiversity, micro/mesoclimate) and Teleconnections (global climate change, biodiversity). Six stages of forest transition (from closed old-growth forest to open-field agriculture and re-treed (peri)urban landscapes) can contextualize interventions, with six special places: water towers, riparian zone and wetlands, peat landscapes, small islands and mangroves, transport infrastructure, and mining scars. The typology can help to link knowledge with action in people-centric restoration in which external stakeholders coinvest, reflecting shared responsibility for historical degradation and benefits from environmental stewardship.

Keywords: assisted natural regeneration (ANR); co-investment; ecosystem services; environmental stewardship; equity; forest and landscape restoration (FLR); landscape approach; rights-based approach; tree planting; water

---

1. Introduction

With the Bonn Challenge (2011) [1], the New York Declaration on Forests (2014) [2] and the UN Decade on ecosystem restoration [3] launched in March 2019, forest and landscape restoration (FLR) is gaining traction on the global political agenda. Within the goal of reversing centuries of damage to forests, wetlands, and other ecosystems, getting it right will be key to putting the planet back on a sustainable course. However, despite the high level of political engagement; despite the number and diversity of actors and institutions involved, from public and private sectors, civil society and local communities, research, and academia, at all levels, from local to global; and beyond some success stories, restoration is not happening at scale. As agriculture is a major driver of degradation but remains a primary source of rural livelihoods, international agricultural research can be part of the solution, contributing to the design of successful restoration approaches to be implemented at scale in the coming years.

As part of the international agricultural research consortium (CGIAR) as food security-oriented international agricultural research body [4], the partners in the Forests, Trees and Agroforestry (FTA) program, building on many decades of research, identified the need to uncover the diverse understandings and perspectives about 'restoration' and to construct a typology that can help to clarify contrasts, similarities and possible synergies across the many starting points for targeted interventions that are currently propagated and implemented under the common heading of 'restoration' of forests, landscapes and/or land. The aim of such typology is to better describe links between evolving knowledge, stakeholder-driven action, and achievement of Sustainable Development Goals.

As a first step toward such a typology of restoration as part of international agricultural research, three Common Research Programs (CRPs) of the CGIAR—Forests, Trees and Agroforestry (FTA), Policies, Institutions and Markets (PIM) and Water, Land and Ecosystems (WLE)—conducted a joint stocktaking of CGIAR work on forest and landscape restoration [5]. They covered a wide range of field projects and case studies, decision making supporting tools, modeling, mapping, conceptual approaches, and frameworks across the geographical area of interest of the CGIAR, i.e., the tropics and sub-tropics in Africa, Asia, and Latin America. They covered a broad range of issues directly or indirectly related to restoration, including sustainable land and water management; seed supply systems and genetic diversity; climate change adaptation and mitigation: land tenure security and land governance reform. They showed how restoration efforts could contribute to SDG2 by supporting smallholder farmers' ability to increase food production, while also addressing SDG15 "protect, restore and promote sustainable use of terrestrial ecosystems" and "halt and reverse land degradation", contingent on SDG16 "peace, justice and strong institutions", assist with SDG6 "availability and sustainable management of water and sanitation for al", and most, if not all the 17 sustainable development goals [6]. They unveiled the many drivers of land degradation, not only biophysical, but also socio-economic, political, and institutional and the variety of actors involved. They suggest different ways to categorize restoration interventions.

Building on the results of this survey, the main objective of this paper is to elaborate a typology of restoration options by context applicable in a wide range of situations across the tropics and sub-tropics. We adopt here a people-centric nature-based perspective focusing on land restoration through agroforestry. For that purpose, Section 2 discusses the underlying concepts and definitions and presents our approach. Section 3 focuses on land degradation, its symptoms, drivers, and indicators. Section 4 suggests and discusses a possible typology of restoration options by context.

2. Underlying Concepts, Definitions and Approach

2.1. Beyond Tree Planting, the Various Aspects of Restoration

Tree planting as a way to restore local ecosystems appeals to many farmers, local communities, national policymakers, and private companies, for a variety of reasons. Tree planting ceremonies as a symbol of peace and commitment to stability and prosperity have an important place in the diplomatic world and national policy agendas for post-disaster contexts [7]. The 'tree planter hero' portrayal has strong emotional appeal [8–10]. Claims of the millions, billions [11,12] or trillions [13] of trees planted capture the public imagination and get news headlines. More than seven billion humans share the planet with approximately three trillion trees [14] 46% less trees than at the start of human history. Approximately 1.36 trillion of these trees exist in tropical and subtropical regions 0.84 trillion in temperate regions and 0.84 trillion in the boreal region; overall nearly one-third are outside forests [15]. Tree diversity in agroforestry landscapes varies over three orders of magnitude, from 1–1000 [16].

Enthusiasm for tree planting only partially aligns with the focus of thousands of experts around the world who have dedicated their professional lives to the protection and restoration of ecosystems [17]. Guidelines for restoration include recommendations to first consider and deal with root causes of degradation, to work with nature, rather than against it with technical means, and to work with people. The guidelines suggest that experts can provide advice on where and how tree planting can be helpful. The idea, however, that planting trees is at the core of restoration ignores the expertise of millions of agroforesters (practitioners) around the world who learned that people-centric and nature-based land restoration through agroforestry can be as simple as 'assisted natural regeneration', with selective retention of the right trees growing in the right places [18,19]. While the practice may be as old as agriculture, agroforestry as a branch of applied science started four decades ago. 'Restoration' was and still is one of its main motivators [20–23]. Where it is ecologically and socially feasible [24–27], approaches such as assisted natural regeneration need to be upfront part of a 'restoration typology' message and list of options to be considered for any given context. In a more comprehensive 'options by context' typology for restoration a wider spectrum of activities beyond 'tree planting' is needed, as we will explore in this contribution.

The growing consensus on the relevance of a social-ecological systems perspective has seen restoration ecology [28–30] and forestry-oriented implementation guidelines [31–33] evolving to restoration science [34–37]. This meant a stronger social orientation [38–40], with attention to success factors for community forestry [41,42]. Yet agendas on food security [43] and public health [44] remain poorly connected to the dominant restoration discourse. A gap still exists between Land Restoration from an agricultural perspective, as perceived in CGIAR efforts [9], and the main ideas in Forest and Landscape Restoration. The wording of principles of Forest and Landscape Restoration [45], such as "Engage stakeholders and support participatory governance", "Taylor to the local context using a variety of approaches", "Manage adaptively for long-term resilience" suggest a genuine attempt to connect with bottom-up farmer perspectives, but also serious challenges to actually achieve that. Principles such as "Restore multiple functions for multiple benefits" may be redundant if local stakeholders have a real say in what happens on the ground. A list of common governance challenges for Forest and Landscape Restoration in a recent review [46] included (1) Poor alignment across levels and sectors of government, (2) Environmental and social heterogeneity, (3) Lack of enabling conditions and implementation capacity. A non-involved reader might wonder whose agenda such type of restoration actually is. Thus, a recent review of FLR practice concluded that "Existing guidelines and best practices documents do not satisfy, at present, the need for guiding the implementation of Forest and Landscape Restoration (FLR) based on core principles. Given the wide range of FLR practices and the varied spectrum of actors involved, a single working framework is unlikely to be effective, but tailored working frameworks can be co-created based on a common conceptual framework" [47]. While FLR is supposed to support sustainable agricultural production, there is very little discussion on how this can be achieved. When farmers are asked about their own decision making with regards to landscape restoration, responses may be surprising. Recent efforts to obtain a deeper understanding of farmer decision making for landscape restoration in Malawi revealed that the expectations that 'planting more trees will attract reliable rains' figured prominently in local perspectives, before expected benefits for soil fertility or beekeeping [48]. One of the milestones of restoration science is the "reference ecosystem", specifying the desired successional stage of recovery, the species (or group of species) that are the target of rehabilitation and the expected time of recovery of the degraded ecosystems after initial treatment (change in management). Follow-up questions are whether the time involved is socially accepted, economically feasible and ecologically reasonable.

One of the key propositions of the paper is to advance restoration science to a more complex set of objectives and functions to be restored and to put local land users at the center of the process. The confrontation, managing synergies, and trade-offs between such functions, and at nested scales, is fundamental to the challenge, and in line with decades of analysis of landscape multifunctionality and integrated natural resource management in international agricultural research [49,50]. Rather than a static reference ecosystem continued change and agility will be needed as a vision for multi-functional restoration [51]. Therefore, we propose to extend the concept of "reference ecosystem" to the "reference social-ecological system" and resulting functions.

Innovation, development, intensification, adaptation, rejuvenation, and restoration appear to refer to different actions, with emphasis on the new, the existing or the past. In practice, however, similarities exceed the differences as all need to deal with motivation, rights, know-how, markets, environment from local environmental effects to global effects through bio-geochemical and hydrological cycles and their global teleconnections [52–54]. This means that all efforts need to match options for interventions to context at the nested scales of farms, landscapes, nation-states, and the changing global context. This means involving a multiplicity of actors: farmers, communities, private sector, public sector, and global investors [55–59]. Interventions also range from projects and programs to wider policies at national or even international (regional) level [60]. Unless these are people-centric, however, their chance of sustainable success is small [61,62]. Land restoration, as we present it and contrarily to what the word often means in other contexts, is not backward-looking but forward-looking. Innovative restoration (or restorative innovation) reconciles historical path-dependency of the degraded status quo with forward-looking theories of induced change that are empirically grounded, rather than wish-lists of over-optimistic planners. It requires science-based and across-scales diagnosis of the underlying causes ('driving forces') that shaped current context, mobilizing a wide range of conceptual frameworks to understand social-institutional constraints, drivers of change and sustainers of long-term action. It then relates that context to options for interventions. Common interventions in land restoration focus on modifying land cover or structural land surface properties but are aimed at improvement of land use and functionality in support of multiple goals.

Land cover change can be achieved through natural regeneration (with various degrees of human assistance and farmer management), tree and grass planting (with its dependence on seed supply and nursery value chains), or remediating management of soil and water. Land-use change depends on who is allowed to be a user, for what purpose and for what use, their motivation, preferences, restrictions, the know-how of managing land in local contexts, market opportunities, concerns about local environmental impacts and external co-investment in land stewardship. Across all modes of regeneration there are tree genetic resources issues [63–65], agronomic options in context [66,67],value chains [68–70], hydrology [71], global teleconnections through climate [72,73], and biodiversity [74–76], as well as policy reform of rights [77], cross-scale incentive systems for stewardship [78–80], and distributional and process concerns over equity and inclusiveness [81]. Key constituencies (including policy-makers, funders, local stakeholders, scientists) need commonly understood metrics to achieve progress [82,83]. Restoration is commonly differentiated by geographic contexts, such as China [84], Southeast Asian fallow to forest transitions [85], East and South African drylands [86], Horn of Africa [87], or Brazil [88]. Still, a more incisive way of describing similarities and differences is needed.

2.2. Definitions

The Society for Ecological Restoration (SER) [89], in line with the Intergovernmental Science-Policy Platform on Biodiversity and Ecosystem Services (IPBES) [90] defines ecological restoration as "any activity with the goal of achieving substantial ecosystem recovery relative to an appropriate reference model, regardless of the time required to achieve recovery. Reference models used for ecological restoration projects are informed by native ecosystems, including many traditional cultural ecosystems". Both the SER and the IPBES distinguish ecological restoration from rehabilitation. The latter refers to restoration activities that aim at "reinstating a level of ecosystem functioning for renewed and ongoing provision of ecosystem services potentially derived from nonnative ecosystems as well" [79] but may "fall short of fully restoring the biotic community to its pre-degradation state" [80]. Both see restoration activities as a continuum, from rehabilitation to ecological restoration, aiming at initiating or accelerating the recovery of an ecosystem from a degraded state. These various acceptances of the term 'restoration' reflect the various perspectives, motivations and behaviors of the different actors involved. While ecologists, scientists, and activists might focus more on ecological restoration as the return to a pristine state of natural ecosystems, governments and economists may be more interested in a discourse focusing on regaining ecological functionality in degraded landscapes in order to enhance food security and livelihoods, reduce poverty and contribute to sustainable development. In line with this latter perspective, in the context of the Bonn Challenge, the International Union for Conservation of Nature (IUCN) and other partners adopted the following definition [91]: "Forest landscape restoration (FLR) is the ongoing process of regaining ecological functionality and enhancing human well-being across deforested or degraded forest landscapes. FLR is more than just planting trees–it is restoring a whole landscape to meet present and future needs and to offer multiple benefits and land uses over time".

For the purpose of this paper, the term 'restoration' encompasses the whole continuum of restoration activities, from rehabilitation to ecological restoration, covers any kind of ecosystems (natural forest, agroforest or agricultural landscapes) and gives a central place to the concept of 'ecological functions' (Figure 1), i.e., the functions that allow ecosystems to generate various regulating, supporting, provisioning and cultural benefits to people or 'services' [92], including those generating direct economic value.

Building on these considerations, this paper adopts the following definitions:

- Degradation: Loss of functionality of e.g., land or forests, usually from a specific human perspective, linked to a change in land cover with consequences for (at least one category of) ecosystem services,
- Degraded lands: Lands that have lost functionality beyond what can be recovered by natural processes and existing land use practices in a defined, policy-relevant time frame,
- Syndrome: Set of concurrent diagnostical indicators or symptoms that can be the result of different and often interacting causes or drivers,
- Restoration: Efforts to halt ongoing and reverse past degradation, by aiming for increased functionality of ecosystems supporting land use (not necessarily recovering past system states)

Restoration covers a broad range of changes (innovations) relative to the current system state—from land use practices and land cover changes to physical infrastructure and institutional changes. As highlighted above, the objective here is to regain ecological functionality and enhance human well-being, not necessarily to go back to the initial ecosystem state or function. That may simply be impossible in some places because of the change in local demographic conditions. Living with the current 7 billion people on the planet would not allow that. Moreover, the final 'restored' state of the ecosystem shall be self-sustaining. This means that in a particular context, given the set of ecological functions to 'restore', restoration interventions need to lead to social, economic, and ecological benefits lasting in the long-term. While in other aspects of human life binary classifications have been recognized as being problematic, past distinctions between 'Nature' (ecosystems, wilderness) and human endeavors (agro-ecosystems, plantations, (peri)urban systems) remain evident in the way 'ecological restoration' is distinguished from 'forest and landscape restoration'. Under the heading 'land restoration' we aim to bridge this divide. Therefore, we propose here a broader umbrella to clarify the full perspective of 'stopping degradation *plus* recovering damage' and addressing in a sustainable way the underlying drivers, including those related to the production function of the lands and to the needs of people living in it and from it. Reconciling these perspectives is critical to the sustained, long-term success of land restoration and especially relevant in the pan-tropical domain. Dealing with the generic drivers of degradation, rather than area-specific pressures, builds on decades of policy development, although progress on Aichi 2020 target of the Convention of Biological Diversity on pollution control is less than that on target 11 achieving an increase in protected areas. [93].

2.3. The Social-Ecological Cascade Framework

The distinctions between land cover (as structure) and land use (as a set of services derived) are compatible with a Social-Ecological System (SES) cascade (Figure 1A) [94]. Mainstream 'forest' definitions combine aspects of scale (minimum area), structure (tree cover, tree size), function (primary designated purpose) and institutional control by forestry agencies, segregating trees into 'forestry' and 'agricultural' ones [95,96]. Important parts of the structure are also the condition of the topsoil (including its soil organic matter [97], protective litter layer, macroporosity [98] and soil biota), while the related functions include processes such as rainfall infiltration and (absence of) surface runoff and erosion [99].

Feedback options from 'stakeholders' based on the values at stake for them to 'actors' in the landscape make the cascade flow into a social-ecological system with self-adjustment or learning ability. These feedbacks link 'bio-geo-physical' units, social actors and institutions, across scales. Many changes in the landscape increase some and decrease other functions and values, and as such they are either degradation or restoration depending on the weight given to various functions by stakeholders, or the strength of their voice in public discourse. Post-logging forest management to increase the growth of the most desirable species, may imply degradation from an ecological perspective, for example. Draining swamps for improved public health, implies ecological degradation, as a second example. The feedback loops that aim to shift ongoing degradation toward restoration require specific ways of linking knowledge (on options in context) with action (getting societal traction on issues and agreement on goals) (Figure 1B). Five issue cycle steps depend on and strengthen the knowledge-action linkages: A. Agenda setting, B. Better and shared understanding of what is at stake, C. Commitment to common principles, often based on coalitions, D. Devolved details of design and delivery, dealing with trade-offs, and E. Efforts to evaluate, and where necessary restart [50,100].

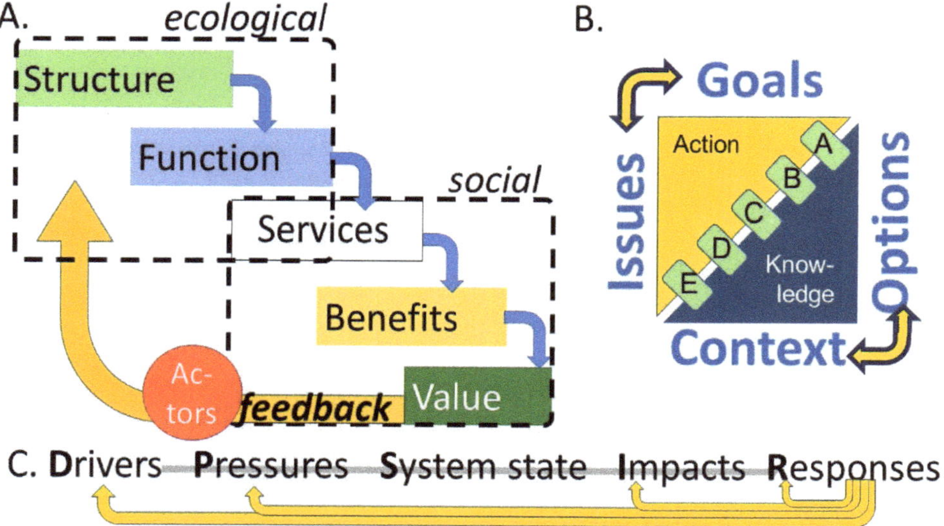

Figure 1. (**A**) Social-ecological system (SES) as a cascade with feedbacks via actors (modified from [101]), (**B**) Multiple links between knowledge and action along the intervention cycle, with two-way interactions between issues and goals, as well as between options and context (modified from [50]), (**C**). Responses can target drivers, pressures, impacts or the emergence of responses itself.

The SES cascade framework and the above-mentioned knowledge-action linkages are conceptually close to a third commonly used framework, which depicts how Drivers, Pressures, State change, Impact and Responses (DPSIR) interact in a feedback loop, with responses addressing the immediate pressures and/or the underlying causes (drivers) (Figure 1C). Often Drivers, Pressures and Impacts operate in nested scales, necessitating Responses (including 'restoration') to do the same to be effective.

2.4. Our Approach

Interventions to support the restoration of the functionality of lands, forests and landscapes can be many folds, and operate at different scales (local, national, global), involve various combinations of private and public initiatives and investment, relate to the rights and authority of local actors and stakeholders in multiple ways (ranging from eviction to full consultation and respect of Free and Prior Informed Consent (FPIC) or public and private support for local initiatives). These interventions seek various entry points into Social-Ecological Systems other than directly addressing land cover (e.g., by tree planting), such as modified rights, enhanced know-how, supported markets, or incentive systems. These are always related to specific contexts that are in turn very diverse depending on the social-ecological system in place and its historical path-dependency.

The dual purpose of our proposed typology is:

1. to allow a more effective exchange of knowledge and experience between settings where restoration (of any kind) is initiated, planned or in process, based on recognized similarities and contextualized differences, enabling a better confrontation and/or integration of the corresponding evidence-base, and
2. to assist planning and priority setting, especially where scarce public resources are involved.

The first purpose is to reflect an actor-centric (bottom-up) perspective, the second a planner (top-down) one.

As typologies of goals and issues can be derived from other frameworks, such as the set of Sustainable Development Goals (SDG) targets and indicators (or any other national or local level framework), we focus here on the knowledge side of the interaction and thus on a combined typology of intervention options and contexts, within the main steps of issue cycles, as illustrated in Figure 1B [102].

Restoration interventions are chosen out of a wide array of options, supposedly finetuned to the local context. Therefore, to elaborate a typology of restoration options by context, we cross two different typologies:

- a typology of contexts/situations, focused on types and levels of ecological degradation in their social contexts, at the system state level of DPSIR, but with attention to the pressures, drivers and impacts as levels of analysis;
- a typology of options for restoration interventions that address symptoms of ecological degradation and/or its drivers, and/or support the social conditions that sustain further improvement.

The resulting 'restoration' typology positions (combinations of) interventions in a specific context, expecting considerable 'endogeneity' in what is attempted and has success where. Endogeneity is one of the main obstacles to interpreting observable patterns of associations (e.g., forest cover and human well-being) in terms of replicable mechanisms and generic theories of change. Real learning from track records of any intervention elsewhere requires contextualization of its initiation and operationalization.

A technical perspective on restoration takes 'ecosystem structure' as the direct target for interventions, triggering the cascade (Figure 1A) to function, services and human benefits, but a social-ecological perspective starts from the Response part of the DPSIR cycle (Figure 1C) and identifies leverage points (preferably at Driver level), leading to land use change that leads to changes in ecosystem structure (Figure 2).

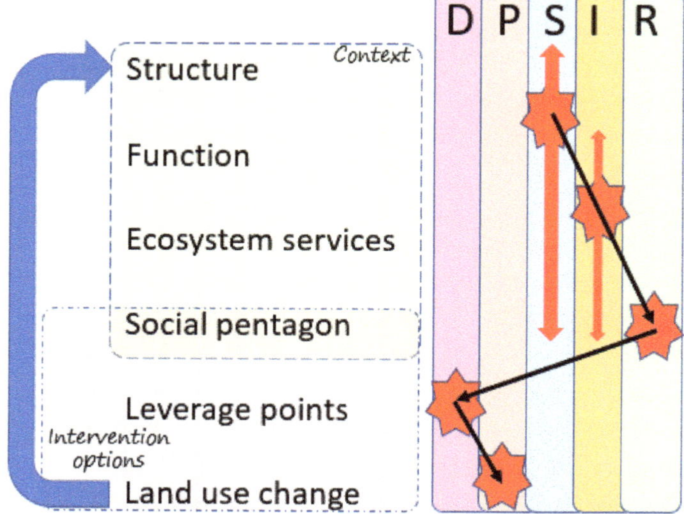

Figure 2. Relating the cascade and DPSIR concepts of Figures 1A and 1C, respectively, in social-ecological systems that shift from degradation to restoration; the social pentagon is shorthand for rights, know-how, markets, local ES issues and teleconnections.

3. Land Degradation: Symptoms, Drivers, and Indicators

3.1. Symptoms, Syndromes and Diagnostics

In a social-ecological system interpretation, the importance of breaking current trends of degradation (loss of functionality) emerges in many different contexts, with a wide range of 'entry points' that get issues inscribed into an agenda for action. These starting points can be compared with 'symptoms' in the medical tradition, signs that something is amiss with system health, but requiring diagnosis. As in the medical tradition a determined set of symptoms can appear concurrently, a syndrome, giving further indications for diagnosis. Diagnosis aims to identify the location-specific pressures (Table 1) and their underlying generic drivers as targets for interventions beyond the symptom level.

Table 1. Examples of diagnostic links between symptoms along the SES cascade and the underlying pressures (that themselves respond to generic 'drivers' outside the targeted landscape).

Starting Point	Symptom	Example of Contributing Pressures
Structure	Loss of land cover	Uncontrolled conversion
	Loss of perennials	Overgrazing
	Loss of tree cover	Overlogging, overharvest
	Loss of plant diversity	Specialization, markets
	Skewed tree age distribution	Lack of rejuvenation investment
	Loss of soil structure, carbon	Over-cropping
Function	Reduced primary production	Local climate change, soil fertility loss
	Disturbed hydrology: quantity, quality, timing of river flow; freshwater retention	Local climate change, loss of filter and buffer functions by vegetation
	Loss of soil retention, downstream sedimentation, landslides	Loss of effective land cover, increased rainfall intensity
	Spread of uncontrolled fire	Loss of functional fire-breaks
	Increase of pest, disease, weed pressures	Disturbed ecology and food-webs
Services	Loss of harvestable and marketable production ('provisioning services')	Loss of productivity, consumer trust in responsible production
	Decline in usable water, increased floods	Disturbed hydrology
	Loss of net greenhouse gas sequestration	Loss of soil functions, vegetation
	Loss of human health	Loss of healthy ecosystems
	Loss of cultural and spiritual value	Loss of respect and recognition
Benefits	Loss of local livelihoods	Loss of harvestable/marketable production
	Loss of secure and trusted value chains	Production costs, market prices, loss of trust
	Loss of existence value of global biodiversity	Awareness of existence and threats
Value	Increased resource use conflicts	Lack of rights, lack of law implementation
	Increased inequity and gender inequity	Lack of voice in decisions
	Loss of options for young people	Land shortage, lack of livelihood options
	Loss of local ecological knowledge, rules	Lack of attention, respect and rejuvenation

Behind the location-specific pressures that lead to degradation in Table 1, there are generic underlying causes (such as population growth, economic growth ambitions, globalization, urbanization and changes in diets and lifestyles) that require generic responses [103]. Such ultimate causes of land degradation cannot be ignored by restoration scientists and practitioners. However, a long history of claims that intensifying agriculture, especially in the tropical forest margins, would by itself reduce environmental impact (known as the 'Borlaug' hypothesis) [104,105] was rejected based on evidence from the field. Agricultural land abandonment, providing space for restoration, may be expected in less-favorable conditions where agricultural labor moves to cities [106], rather than as a direct consequence of intensified agriculture elsewhere. Population policies, both the migration and birth rate side of them are closely linked to national identity issues, and hardly modifiable by directly environmental concerns or agricultural research [4]. SDG 12 "responsible production and consumption" may be the closest approximation among the SDGs to deal with market-related drivers of degradation and restoration, but the goal is largely aspirational, with weak operationalization [107].

3.2. Structural Indicators: Tree Cover Linked to Ecological Functions

Restoration issues can start with any of the five elements of the SES cascade of Figure 1: structure, function, services, benefits, or value, and need to identify where 'lack of' function is caused by 'loss of' function. At the highest level of this cascade is the category Land or Land Health, as vegetation and surface soil are parts of the structure—which also is most readily observed. Remote sensing has so far had a dominating role in informing top-down priority setting for restoration interventions, but it does not always allow separating anthropogenic change from natural patterns of variation. Anthropogenic changes (over time) in land cover and surface soil structure such as tree cover transitions (also known as forest transition (FT)) interact with two other dimensions of tree cover variation: latitude and topography, jointly shaping climate and soils. The natural spatial distribution of vegetation types responds to climate and soil, with temperature and rainfall varying by both latitude (from tropics to boreal) and topography and water availability in any given climate by topographical variation in water acquisition and drainage (Figure 3).

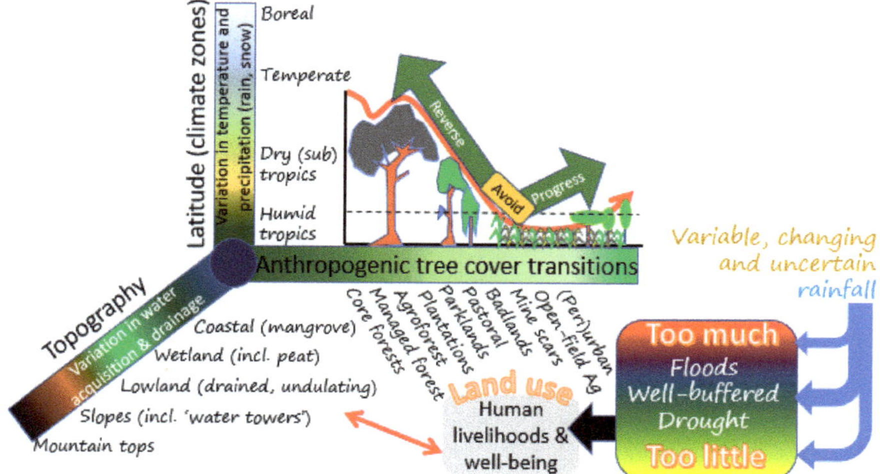

Figure 3. Three major sources of spatial variation in tree cover: latitude, topography and place-based anthropogenic tree cover transition in interaction with rainfall regimes as the cause of water excess, shortage or well-buffered conditions (modified from [54]).

Tree cover in itself cannot be a universal proxy for land degradation or restoration, or for assessing the human influence in the SES cascade. For example, vegetation with 20% tree cover could be the natural vegetation in semi-arid conditions or on mountain slopes but indicates human influence elsewhere (e.g., increased fire frequency in anthropogenic savanna conditions). With several of the worst effects of 'degradation' related to disturbed hydrology (e.g., floods and droughts), one needs to unpack anthropogenic effects from natural background variation.

The forest transitions (FT) typology [108], illustrated in Figure 3 and presented in more detail in Section 4.4, clarified that the highest human population densities (FT6) in a pantropical study were associated with around 30% tree cover at sub-watershed scale (Figure 4A), while areas with less tree cover (e.g., open-field agriculture) are associated with lower population densities (Figure 4B). Within the humid and (per)humid tropics there is a strong negative relationship of the 'more people, less forest' type (Figure 4D), but in arid and semiarid an opposite trend (more people, more trees) can be noted (Figure 4C).

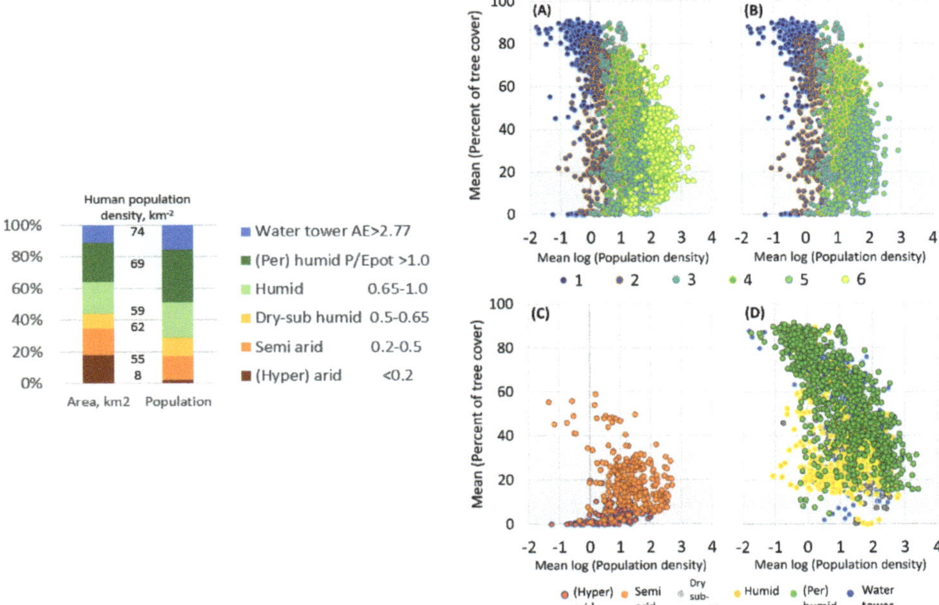

Figure 4. (**Left panel**): area and human population size in the tropics in five climatic zones (defined on the basis of the ratio of precipitation P and potential evapotranspiration E_{pot} and the water tower configuration as defined in [108]); (**Right panel**): pantropical relationships at sub-watershed level between human population density and tree cover, for six identified stages of forest transition FT and the six climate zones as defined in [108]; (**A**) for all sub-watersheds differentiated by forest-transition stages; (**B**) only for sub-watersheds of stages 1–5; (**C**) for sub-watersheds within (hyper-) arid and semi-arid zones; (**D**) for sub-watersheds in the remaining agro-ecological zones.

Reductions, rather than the expected increases, in streamflow are a common result of tree planting and forest restoration [109] that do not have to come as a surprise. Depending on the environmental conditions an intermediate tree density can be optimal from a groundwater recharge perspective [110], while desirable effects on flood-causing peak flows can be stronger than undesirable reductions in annual water yield [111,112]. Flood control may depend on using trees as part of river restoration [113] rather than blanket reforestation of landscapes. Increasing attention to the rainfall-generating effects of high evapotranspiration [114,115] justifies renewed attention to the savanna-forest hysteresis theory of local climate influences [116,117].

3.3. Social Indicators: Services, Benefits And Values

At the lower part of Table 1 we see closely connected, primarily socially defined, indicators of services, benefits, and value. While there are many ways to unpack the complexity further (e.g., the 'five capital' framing), we found five aspects that operate at the landscape scale and capture important dimensions of human well-being. These five aspects illustrated by the 'social pentagon' (Figure 5), are discussed in more detail in the next section.

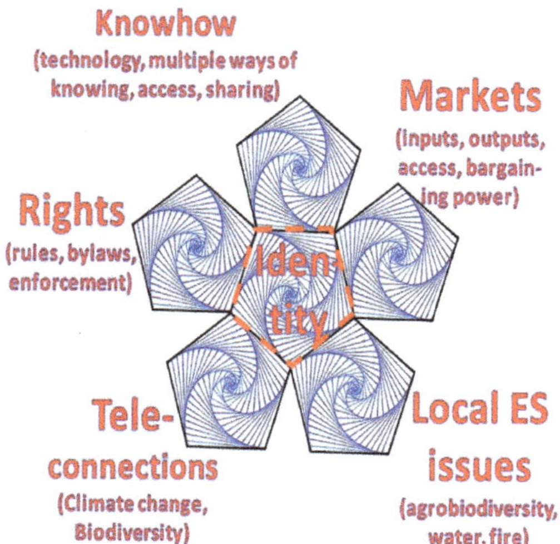

Figure 5. Representation of six key attractors of attention in Social-Ecological Systems that relate to both degradation and restoration phases.

4. Typology of Interventions

4.1. The Social Pentagon as a Starting Point for a Typology of Restoration Options

At the heart of the social pentagon (Figure 5) a lost or modified sense of identity may be the worst impact of degradation and has to be on the basis of any success in restoration. Human identity (self-image) relates to local institutions (that include education, religion, cultural values, collective action), motivation, usually stratified by age- and gender, as well as to other dimensions such as social stratifiers of wealth and influence (e.g., caste, social class, patronage, entrepreneurship, or other aspects of political ecology).

Identity and the human capacity to adapt to gradual degradation processes can be the main obstacle to transformative change or to tackling the underlying degradation drivers, often requiring an external event such as a disaster to trigger or catalyze the needed transformative changes. As noted in a review for Southeast Asia, many examples of locally led restoration have a dearly paid locally learned lesson of a disaster (including landslides, floods, drought, fire) as their turning point in local history, generating the energy needed to overcome vested interests and status quo degradation [118].

Identity interacts with rights (as defined at national scale in laws governing forests, tenure and inheritance rules, and land-use planning, or in local bylaws clarifying stewardship and collective action), know-how (accumulated in local knowledge and interacting with externally supported ways of knowing), markets (for both inputs such as planting materials and products), local ecosystem service issues (often with water, microclimate, agrobiodiversity and fire as focal points of concern) and global teleconnections, interactions and feedback loops (especially those regarding climate change and biodiversity conservation).

A substantial body of case studies and action research engagements in Africa and Asia tested several ecological, economic, social, and governance propositions on the way 'Payments for Ecosystem Services (PES)' may have to be renamed 'Co-investment in Environmental Stewardship' to understand sensitivities and misunderstandings that arise from the most commonly used terminology [119]. The risk of 'crowding' social motivation for pro-environmental behavior by a focus on financial transactions was demonstrated to exist experimentally [120] but depends on the communication of programs and opportunities for local re-interpretation of terms [121].

The social pentagon interacting with identity defines not only the ultimate effect of degradation, it also forms the starting point of interventions for change (Figure 6), challenging the neat two-way options by context typology by emphasizing that it needs to be considered along an issues-cycle of awareness, motivation for change and steps toward a break with the past.

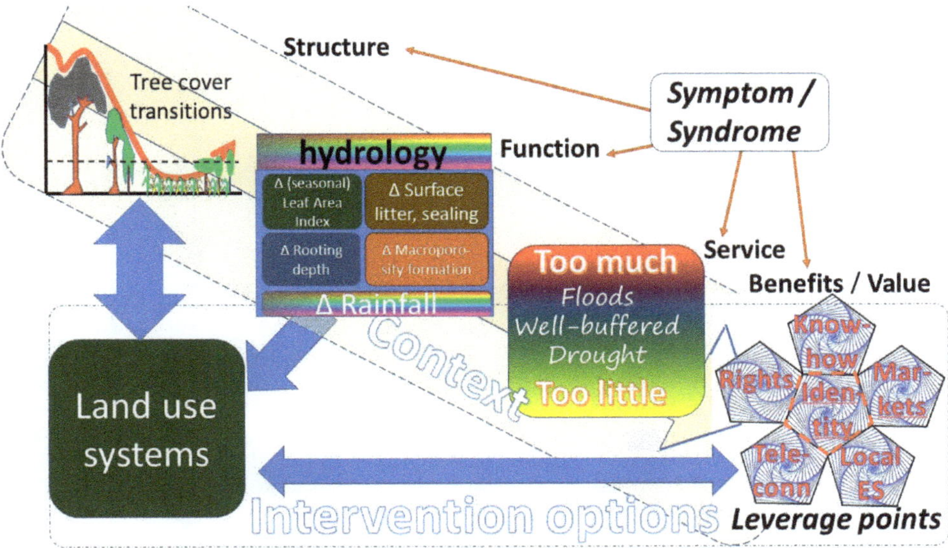

Figure 6. Connecting typologies of degradation/restoration context with intervention options through leverage points in the social pentagon and core identities.

While specific 'development' organizations take rights-, knowledge-, or market-based approaches as their starting point for restoration interventions in local social-ecological systems, trying to nudge current degradation into a restoration and innovation trajectory, 'conservation' organizations have focused on globally important teleconnections (climate, biodiversity, footprints of commodity trade) and/or local ecosystem services and associated local livelihoods. Regardless of the starting point, however, the relevance of an integrative livelihood orientation that relates to all five aspects has become clear to all actors. For example, lessons learned in the past 'Integrated Conservation and Development' projects were (partially) learned in designs for subsequent 'Reducing Emissions from Deforestation and forest Degradation' (REDD+) efforts [122], and lessons learned in REDD+ pilots can inform current Forest Landscape Restoration efforts [123]. Policy reform in community forestry was more difficult to achieve then was expected, as the balance between visionary 'prophets', a practical profit orientation for the main stakeholders and the transparency requirements of 'prove it' agencies (within and outside of formal government, at local, national and international scales) is hard to achieve [124]. Long-term efforts to get operational programs to reduce emissions from deforestation and forest degradation (REDD+) have had much less success than expected, revealing complexities in implementing ideas that at first had large appeal [125].

Engagement in the Sumberjaya landscape in Indonesia, a hot spot of degradation and conflict around 2000 and an inspiration for watershed restoration and resolution of similar conflicts elsewhere went through three phases: first addressing tenure conflicts in the forest margins, then providing incentives for 'river-care' efforts to reduce sediment loads by engagement with the hydro-power company, and thirdly support for marketing environment-friendly products (especially coffee) through more rewarding channels [126].

A case study on restoring traditional water harvesting structures in India showed that groundwater recharge could indeed be enhanced, facilitating an extra crop and fruit tree production, while reducing the need for seasonal migration to a nearby urban center, but negative impacts on water capture by a downstream dam suggest that tradeoffs across scales are complex [127].

Some of the appearances of integrated approaches, however, may be informed by opportunities to tap into financial incentive streams that focus on specific entry points that represent current donor/investor priorities. Although many projects claim to be 'people-centric', their trust in specific 'theories of change' and publicly declared targets in terms of area or number of trees planted can be at odds with adaptive management and local control over process and speed.

4.2. Land-Use Change as the Target of Interventions

While land cover refers to the (bio)physical cover observed on the Earth's surface (FAO, 2005), land use is characterized by all "the arrangements, activities and inputs people undertake in a certain land cover type to produce, change or maintain it" [128–130]. As an important interface between 'actors' and 'land cover', the concept of 'land use' integrates social, economic and ecological aspects. It thus forms the target of restoration interventions. Only where land use is functional from a local perspective, restoration efforts, including land cover changes and changes in the ecosystem structure, will have a chance to be sustainable. Typologies of land use systems have to deal with life-cycle accounting, e.g., in swidden/fallow cycles, or rotational plantation or grazing management, and environmental impacts that are time-dependent.

Based on the level of land degradation, and the intended impact of restoration on land use and land-use changes, we suggest distinguishing here the following four intensities/levels of restoration:

- R.I. Ecological intensification: where improvements to the resource base are possible within existing land use by combining provisioning, regulating and regenerative aspects of agro-ecosystem functioning, within a context of supportive input and output markets. It may include a re-integration of livestock on farms that specialized into arable-only types of farming, as 'leys' as part of a rotation can be both productive and support the recovery of desirable soil properties.
- R.II Recovery/regeneration: where forms of fallow, resting land, exclosures from grazing, fire control and assisted natural regeneration can bring back conditions within which ecological intensification is possible. This level often entails a change in land use, at least temporarily.
- R.III Reparation/recuperation: where more intense action than recovery/regeneration is performed (e.g., tree planting), with additional external support, e.g., by creating access to nurseries for diversified germplasm, knowledge not locally available, inputs (including soil amendments) not currently used, supporting local institutions (and bridging social capital with institutions outside the landscape) not currently effective and/or changing tenurial relations with the state or private sector.
- R IV. Remediation: where past activities such as long-term unsustainable land use, mining, soil pollution or deep drainage have substantially or completely destroyed the ecosystem, preventing its natural functioning or its sustainable exploitation for forestry or agricultural production. This level requires intense specific, typically externally supported, and financed efforts and economic reparation of past damage, e.g., by those who benefited from the unsustainable resource exploitation.

The overarching goal of restoration is to progress across this restoration intensity scale down to the first level. In other words, the aim is to disrupt existing degradation spirals and transform lives and landscapes to bring them progressively back into the domain were 'ecological intensification' (R.I) becomes possible again.

The degree of needed intervention/support is likely to increase across the four levels, and so does the perimeter of the system and the reach of main institutions to be mobilized. Ecological intensification (R I) is generally applicable within the current local land use system at the farm or landscape scale. Recovery/regeneration (R II) within a local social-ecological system; reparation/recuperation (R III) within a broader national policy context; while remediation (R IV) usually requires a stronger external, or even international support and investment (Figure 7).

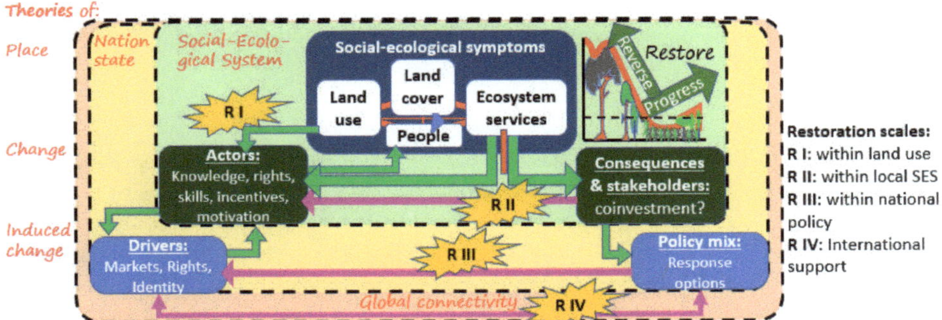

Figure 7. Four scales of restoration interventions and enabling actions in a nested social-ecological system perspective.

Enabling actions can be identified across these four levels that relate land use to current farm-gate profitability, additional local efforts to internalize externalities, changes in the national policy mix that influence the profitability of alternative land uses and allow sustainable land use planning and integrated landscape management, and/or global co-investment and efforts to address global challenges and strengthen global value chains. Responses at one scale (for example local SES) may help to 'scale up' to another scale, via pressures to national policies or international support, or by the cumulative effect to global drivers.

4.3. Reconciling Bottom-Up and Top-Down Restoration Initiatives

The strength of local motivation for change is probably the switch for any 'restoration' success, but often external support is needed to make change sustainable. Where 'restoration' is to be managed as a program or project, it requires 'metrics' as markers of progress and clarity on targets. To do so it needs to link bottom-up local initiatives and drives for return/increase of ecological functionality with the understanding of local social-ecological systems that are needed to define the elements of project designs that can attract funding and investment.

Some of the interventions targeting the surface soil structure (such as stone-rows, Zai pits and terracing in Sahelian agriculture [131]) are aimed at modifying lateral flows of soil and water in the landscape. Their effectiveness is likely to depend on scale and position on a topo-sequence, as they combine water harvesting source and sink zones. Physical interventions in surface soil structure tend to be labor demanding, with limited opportunities for mechanization, triggering a search for low-labor alternatives such as the naturally vegetated strips (started by not-plowing contour strips in intensively used slopes in the Philippines) as an alternative to tree planting and hedgerow pruning.

Local preferences are also an important aspect of obtaining the right mix between three primary ways of obtaining a change in land cover:

- Leave alone: rely on seed banks and seed rains (depending on the dispersal mode of (tree) seeds, distances to 'mother' trees, and presence of flying or walking animals as dispersal agents) as the basis of a diverse, locally adapted vegetation,
- Assisted/managed natural regeneration of vegetation or of land infrastructure: dealing with fire, grazing and other pressures and selective retention of desirable trees that can still derive from seedbanks and seed rain,
- Tree and grass planting, land infrastructure building: taking full control of the vegetation and infrastructures that will form the next land cover.

Especially a choice for the latter has important implications for local input markets, in terms of tree seeds and/or local nurseries that can provide diverse and good quality planting materials at an affordable cost. The widespread tendency to provide cost-free externally produced planting materials is now seen as a way to achieve short-term success (and project deliverables) at the cost of the long-term sustainability of solutions. Reconciling such with a gradual shift to lower intensities of restoration (toward the RI level) is important.

4.4. Typology of Contexts

Our typology of contexts ('theory of place') is based on the forest transition concept as applied to a pantropical dataset at the subwatershed level by Dewi et al. [108]. The classification relied primarily on (i) forest fraction, (ii) forest configuration (core, edge, and mosaic forest) and (iii) the logarithm of human population density. Stage 1 represents sub-basins where core forest covers around 80% of the total area of the sub-basins and population density is below 1 km^{-2}. As population density increased, the fraction of core natural forest tended to decrease while the fraction of non-forest increased, and the ratio of planted to natural tree cover (both classified as forest in national statistics) increased. While some may envision a 'transition' to have distinct stages, in a large data set these are markers along a continuum of gradual change. In a parallel paper [132] the typology is applied to thirteen pantropical landscapes (from up-river Suriname in FT1 to densely polluted East Java in Indonesia in FT6, with key issues in the forest-water-people nexus identified for each landscape but changing in character. Generic aspects of degradation that occur across a range of forest transition stages are discussed elsewhere for a range of landscapes in Southeast Asia [118]: Forest classification conflicts in FT2-4, Over-intensified monocropping in FT3-5, Degraded hillslopes in FT3-6 and Fire-climax coarse-grass lands in FT3-6.

Beyond this general typology of contexts, six 'special places' have so far been identified that deserve specific attention in the analysis of pressures, drivers, and restoration options, because of their specific importance in the interactions between ecosystem functions and human activities:

Water towers—areas that generate river flow for neighboring landscapes but tend to have an above-average human population density and opportunities to supply local markets with vegetables and other commodities, as well as providing the highest quality types of coffee to global markets [108]. These are prominent in East and West Africa and various parts of Asia, often with substantial downstream areas depending on the rivers that originate in such water towers. They can include 'cloud forests' that capture moisture beyond what precipitation gauges register.

Riparian zone & wetlands—As riparian zones often have fertile soils, sedimented from uphill erosion over long periods, offer easy ways of transport and access to water for irrigation and human use, as well as fishing as a complementary source of food security and nutrition, they have been the parts of the landscape with the longest settlement history in many parts of the world. Exceptions, with a ridge-based settlement pattern can be related to either prevalence of human disease vectors, or invading human enemies using the same river. Controlling disease vectors has often been a primary reason for draining wetlands, further allowing for increases in human population density. Many of the problem-solving interventions, however, displace pressures, such as the increase in downriver flood frequency if local flooding risk is reduced by increased drainage or removal of riparian vegetation that slowed down river flow. Restoring upstream water storage capacity and flow buffering is one of the primary targets of watershed restoration, but typically requires new upstream-downstream coordination of land use patterns and redistribution of economic benefits ('payments for ecosystem services').

Peat landscapes—Limited in area but now recognized in Congo and Amazon basin beyond their better-studied examples in Southeast Asia, they are disproportionately important in terrestrial carbon storage. Peat domes and lowland peat areas developed where drainage was restricted, and a year-round level of water saturation reduced organic matter decomposition to rates below the annual above- and belowground inputs [133]. Current understanding is that restoration focus should be on the peatland hydrological units (from dome to the river) essential for the continued function of peat domes including riparian zones that are not classified as peat soils themselves, spanning all land from river to river across the dome [134,135]. Beyond a 'zoning' perspective, focused on conservation of the dome, restoration should target the landscape, as human livelihoods in the riparian zone are both part of the problem (as the starting point for exploitation of the peat) and at the core of any livelihoods-focused solution.

Small islands, and mangroves in coastal zones—Small islands are miniature universes, where sectors of society are not as divergent as in larger on main-lands, even when national structures promote segregation. Small islands often have limited sources and supplies of freshwater, which is particularly challenging when tourism discovers the attraction of coastal zones and possibly adjacent coral reefs [136]. Developing tourism increases pressure on natural resources but may also provide a financial basis for restorative innovation, as seen in pioneering mangrove restoration on the tourist island Bali in Indonesia. Similar issues and opportunities to link terrestrial and marine ecosystems exist in the mangroves and other coastal vegetation of larger islands and continents. Attractive economic returns on mangrove destruction, e.g., for shrimp farming, as drivers of degradation affect the disproportionately high concentrations of people in coastal areas, exposing them to sea level rise, and surges due to typhoons and tsunami's [137–139].

Mining scars—areas where economic interests sparked particularly destructive change, especially where open-cast mining is used, leaving scars in the landscape where all vegetation and soil is disturbed to such a degree that natural recovery tends to be very slow. As there are frequently high metal concentrations in mine spoils affecting downstream water quality, remediating action is urgent [140]. While large-scale mining permits, and international pressure on transnational mining companies have 'internalized' such externalities by obligations to leave landscapes behind in a multifunctional condition, past mining and small and medium scale enterprises are not effectively bound by the same rules. External involvement in cleaning up the mess left by rogue mining companies can have a 'moral hazard' aspect, as the costs should morally be borne by those who benefited from the destruction.

Transport infrastructure—High investments in roads, canals and power lines make such, typically linear, landscape elements specifically vulnerable to floods, landslides, and similar disasters—to which they often contributed by disturbing hydrology and cutting into mountain slopes. As for the mines, engineers and constructors have a specific responsibility to avoid and mitigate such effects, but some of the past damage may require external 'restoration' support.

4.5. A Typology of Restoration Intervention Options by Context

Table 2 combines the two typologies developed in the previous sections (typology of contexts in Section 4.4; and typology of restoration options in Section 4.1), giving further specification at cell-level (possible options for a specific context), and identifying some among the special "hotspots".

Further descriptors of context that are likely to be relevant for restoration are the climatic zones (Figure 4A) and soil properties. Soil quality is a key context parameter for restoration: it is determined by soil type, texture and depth (as it relates to hydrology and erosion), topography (slope angle and length) and indicators of current soil condition relative to what could be expected for soil in a given location under undisturbed conditions, such as the ratios of current soil carbon and bulk density, to their reference values [141–143]. Such characteristics can be complemented by a characterization of the social, institutional, and economic context that often conditions both the drivers of degradation and the possibility to overcome them, often linking to goals. As the relations between tree cover, soil quality, and human population density within a given climatic zone are fairly strong, we can similarly identify areas where tree cover is above or below what would be expected for the same demographic condition (e.g., with a 10% bandwidth around the expected value). Further classification within the six columns for options are needed and can build on existing assessment methods and typologies for (gender-specific) rights, knowledge, and market access, and for ecosystem services and associated co-investment prototypes.

Table 2. Proposed typology of restoration options by social-ecological context (with further distinctions by climate zone, human development index) and options for inducing change, with examples of intervention targets.

Context	Motivation, Local Insti-Tutions	Know-how Tree	Soil Management	Rights	Markets Inputs	Markets Outputs	Local ES, Human Water Health		Teleconnections: Climate Biodiversity	
SDG-links	**2,3,4,5,10,16**	**5,10,16**		**4,5,8,17**	**1,2,8,9,12**		**3, 6,7,11**		**13,17**	**14,15,17**
FT-stage1: Core forests, low population density, swiddening		FGR				EcoT			REDD+	PA
FT-stage2: e.g., Logged over forests, swiddening		FGR/RIL /CBFM				Logs NTFP			REDD+	PA
FT-stage3: e.g., Mosaic of agriculture, secondary forests, agroforests	Local identity and initiatives, empowered and supported by participation and co-investment by external stakeholders	CBFM/ AF/TGR		National law & local bylaw reform, forest classification and land use rights planning	Nurseries, Tree seed sources	AFP, Plant	Disease vector control, clean water supply, sanitation	Water quantity, regularity of flow (floods/droughts), quality	REDD+/ NDC	PA
FT-stage4: Mosaic of agriculture, secondary forests and plantations		AF/TGR	ISFM			AFP, Plant			NDC	PA
FT-stage5: Open-field agriculture		AF/TGR	ISFM			AFP			NDC	PA
FT-stage6: Peri-urban, tree cover higher than FT5		AF/TGR				AFP			NDC	PA
Special places: Water towers				Water rights						
Riparian zone & wetlands				RC						
Peat		Palud	Rewet	Drainage rights RC						RC
Small islands, Mangroves				Hotel permits						
Transport infrastructure			Land-slides	Development contracts						
Mining scars				Permits, RC						

Acronyms used: AFP (agroforestry product), CBFM (Community-based forest management), EcoT (Ecotourism), ES (Ecosystem services), FGR (Forest genetic resources), ISFM (Integrated spoil fertility management), Log (Logging), NDC (Nationally determined contribution), NTFP (non-timber forest product), PA (Protected area), Palud (Paludiculture), RC (Restoration concession), REDD$^+$ (Reducing emissions from deforestation and forest degradation), RIL (Reduced impact logging), TGR (Tree genetic resources).

As the option-by-context cells are mostly of a target 'land use' nature (see Figure 6 and Section 4.2), existing efforts to achieve generic classifications of land-use systems and intensity of land use, such as used and further developed in the ASB matrix studies can help [144,145]. Participatory land use planning methods and the LUMENS (land use for multiple environmental services) procedures are relevant here [83,146].

4.6. Discussion: Linking the Options in Context Typology to Issues and Goals Across Scales

With a basic option-by-context typology (Table 2), we can revisit the links with an issue-and-goals level typology (as indicated in Figure 1B). Figure 8 suggests how the six critical enabling or resulting dimensions/aspects of degradation and restoration (earlier indicated as a social pentagon) relate across scales to the 17 sustainable development goals (SDGs). Most, if not all, goals relate to conditioning factors for restoration success (e.g., SDG16 on governance and rights, SDG4 on education or SDG5 and 10 on gender and generic equity), but also as policy domains that can benefit from the successful restoration of the land and livelihoods base of national economies. The challenge of coordinated approaches, however, exists within international organizations, as much as it does within national governments [147].

Figure 8. Cross-scale (local, national, global) linkage of the five determinants of local livelihoods, and the local institutional core needed to achieve the 17 sustainable development goals (SDGs).

Boundary work to relate the spheres of knowledge to the arenas of action [148] is needed to take steps from the 'symptoms' (Table 1) to potentially effective restoration actions (Table 2). Figure 9 suggests key researchable questions that can support a shared understanding among stakeholders of existing land users and use (who?, where?, how?), its consequences (so what?), stakeholders (who cares?) and the underlying drivers and pressures that need to be tackled (Why?), across scales (compare Figure 8).

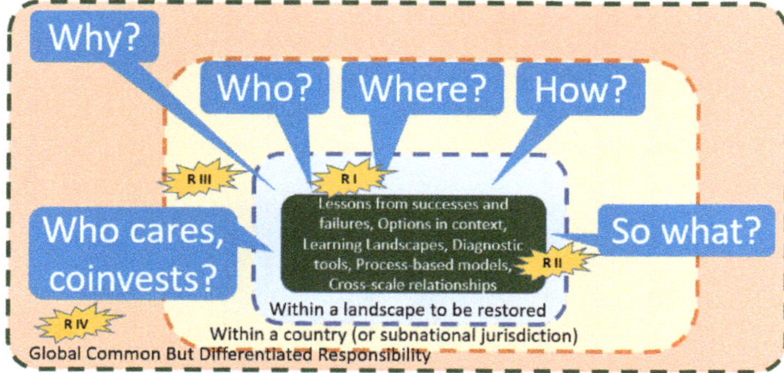

Figure 9. Questions that can drive efforts to understand theories of place and change for local social -ecological systems and their connections with higher-level pressures and drivers.

The boundary work to link existing and emerging knowledge to desirable action to make a difference on the ground has to recognize the different pace and dynamic of the five steps indicated in Figure 1B: A. Agenda setting (bridging between existing global and national 'restoration', 'climate change adaptation' or other 'sustainable development' initiatives), B. Better and shared understanding of what is at stake locally, but also in a wider (e.g., regional) perspective and how it interconnects with others, C. Commitment to common principles (e.g., specific targets within the SDG agenda and/or national development strategies, Nationally Determined Contributions to the Paris agreement on climate change, Aichi targets in the Convention on Biological Diversity (CBD)), D. Devolved details of design and delivery, dealing with trade-offs (what institutional change is needed to provide the essential context for change on the ground), and E. Efforts to evaluate and provide a basis for adjustments. Ten-point progress markers have been proposed for these steps [50].

Development of integrated policy responses [149] in the face of trade-offs [150],has to effectively deal with anticipating actor choices in response to proposed policies [151] and the diversity of opinions and interpretations of current system state, trends and leverage factors [152], including the balance between what can be locally achieved versus what is determined at national scale [153].

Finally, efforts can be presented as 'innovative restoration', strengthening the livelihoods dimension of existing restoration efforts, or as 'restorative innovation', supporting the integrative aspects of SDGs, pursuing restoration as an enabler (and co-benefit) of other key objectives. These are two sides of the same coin. International public-funded research programs such as FTA can support knowledge and implementation gaps in both aspects. Lessons learned from past research for development work in the restoration area suggests that a priority for future research investments is to co-develop with stakeholders system-level knowledge, combining pattern and process type understanding, to provide real solutions to actors on the ground, and the agility to answer to new, emerging issues.

5. Conclusions

Restoration is increasingly an object of interest for a multitude of institutional actors and groups of interests with specific objectives and perspectives as well as of diverse scientific disciplines and approaches. Each category of actors, each scientific approach has its own definition of restoration modeled by its specific perspective. This multiplicity of definitions, of ideal visions of what restoration should be is a significant impediment to collective, long-term engagement.

The typology presented in its paper can support a broader and more precise understanding of the very notion of restoration, in its diversity, generated by the diversity of contexts, objectives and solutions. Moreover, because it is precisely grounded on the interests, objectives and perspectives of the actors engaged in restoration it can help the diverse categories of actors to be involved understand the diversity of their objectives and find common ground uniquely adapted to the specificity of the situation.

The following are some of the key points arising from the assessment and analysis presented in the study.

- There is no single generalizable approach to restoration despite the general notion underlying all restoration interventions is improving or rehabilitating the functionality of the system. Restoration involves multiple entry points most of which arise from either the state of the social-ecological systems or the choices of the people who depend on it.
- Despite the goal similarity (see point 1 above), the entry points to restoration also vary widely. The entry points are defined based on the specific contexts (both the place and its social-ecological aspects) of the system to be restored. It is thus important to understand how the identified entry point could lead to the desired state of the landscape or location and what benefits (e.g., ecosystem services) should be generated from the process of restoration for the effort to be characterized as being impactful.
- Progress along the recovery trajectory in a given landscape or location needs to embrace structural, functional, and social attributes all of which are required to form an agile system with its own improved structural, functional, and social 'identity' or characteristics.
- The typology proposed in this paper is a useful one in view of the lack of such structured interpretations of the typologies. It can support further studies on the effectiveness of interventions in different contexts.
- The framework can facilitate extrapolations from a single case, combining an intervention in a context, to other potential combinations. Such a framework can facilitate the establishment of sound research findings, public goods applicable in a diversity of cases, from a reduced number of well- studied cases in the long-term. This multiplication factor is particularly important given the length of restoration projects that increase both the time to observe results and the risks of non or bad intervention.
- Such a typology, grounded on the two axes of contexts and intervention options, can also orient comparative studies in order to support decision making. In particular, it could facilitate estimations ex-ante of costs and benefits, whether marketable or not, of a specific intervention in a specific context by comparison with past comparable interventions in comparable contexts. Such evidence-based estimations are necessary for actors to engage in long-term actions that are often disrupting established interests and to attract long-term investments of external actors, public and private.
- It is also important to move the restoration beyond the forest and agricultural systems and include the 'special' places of high social and ecological values. Such places may include water towers, riparian zone and wetlands, peat landscapes, small islands, and mangroves in coastal zones, mining scars and transport infrastructure. The global emphasis on greenhouse gases and food security has given generic forest and agricultural land issues high visibility in claims of tens or hundreds of Mha of restoration, while other critical ecosystems are not getting sufficient attention for their specific needs, which may imply higher costs per unit of land for greater societal benefits.

Author Contributions: Conceptualization, M.v.N., V.G. and P.A.M.; Investigation, M.v.N., S.D., B.L., L.D., N.P. and A.M.; Writing—original draft, M.v.N.; Writing—review & editing, V.G., P.A.M., S.D., B.L., L.D., N.P. and A.M. All authors have read and agreed to the published version of the manuscript.

Funding: This research was funded by the CGIAR fund through the Forests, Trees and Agroforestry (FTA) program.

Acknowledgments: We acknowledge the ideas, reflections and inputs of many participants in Forests, Trees and Agroforestry (FTA) workshops held in August 2018 in Nairobi and January 2019 in Rome.

Conflicts of Interest: The authors declare no conflict of interest.

References

1. Available online: http://www.bonnchallenge.org (accessed on 18 July 2020).
2. Available online: http://forestdeclaration.org (accessed on 18 July 2020).

3. Available online: https://www.decadeonrestoration.org/ (accessed on 18 July 2020).

4. Tomich, T.P.; Lidder, P.; Coley, M.; Gollin, D.; Meinzen-Dick, R.; Webb, P.; Carberry, P. Food and agricultural innovation pathways for prosperity. *Agric. Syst.* **2019**, *172*, 1–15. [CrossRef]

5. Gitz, V.; Place, F.; Koziell, I.; Pingault, N.; van Noordwijk, M.; Meybeck, A.; Minang, P. *A Joint Stocktaking of CGIAR Work on Forest and Landscape Restoration*; Working Paper 4; The CGIAR Research Program on Forests, Trees and Agroforestry (FTA): Bogor, Indonesia, 2020.

6. Katila, P.; Colfer, C.J.P.; de Jong, W.; Galloway, G.; Pacheco, P.; Winkel, G. (Eds.) *Sustainable Development Goals: Their Impacts on Forests and People*; Cambridge University Press: Cambridge, UK, 2019; p. 653.

7. Tidball, K.G. Seeing the forest for the trees: Hybridity and social-ecological symbols, rituals and resilience in postdisaster contexts. *Ecol. Soc.* **2014**, *19*, 25. [CrossRef]

8. Giono, J. *The Man Who Planted Trees*; Chelsea Green Publishing Co.: Chelsea, VT, USA, 1985.

9. Robbins, J. *The Man Who Planted Trees: Lost Groves, Champion Trees, and an Urgent Plan to Save the Planet*; Spiegel & Grau Random House Group: New York, NY, USA, 2012.

10. McLauchlan, L. A multispecies collective planting trees: Tending to life and making meaning outside of the conservation heroic. *Cult. Stud. Rev.* **2019**, *25*, 135–153. [CrossRef]

11. Available online: https://www.nature.org/en-us/get-involved/how-to-help/plant-a-billion/ (accessed on 15 July 2020).

12. Available online: https://www.bbc.com/news/world-africa-50813726 (accessed on 15 July 2020).

13. Available online: https://en.wikipedia.org/wiki/Trillion_Tree_Campaign (accessed on 15 July 2020).

14. Crowther, T.W.; Glick, H.B.; Covey, K.R.; Bettigole, C.; Maynard, D.; Thomas, S.M.; Smith, J.R.; Hintler, G.; Duguid, M.C.; Amatulli, G.; et al. Mapping tree density at a global scale. *Nature* **2015**, *525*, 201. [CrossRef]

15. Creed. I.F.; van Noordwijk, M. Forests, trees and water on a changing planet: A contemporary scientific perspective. In *Forest and Water on a Changing Planet: Vulnerability, Adaptation and Governance Opportunities: A Global Assessment Report (No. 38)*; Creed, I.F., van Noordwijk, M., Eds.; International Union of Forest Research Organizations (IUFRO): Vienna, Austria, 2018; pp. 13–24.

16. van Noordwijk, M.; Rahayu, S.; Gebrekirstos, A.; Kindt, R.; Tata, H.L.; Muchugi, A.; Ordonez, J.C.; Xu, J. Tree diversity as basis of agroforestry. In *Sustainable Development Through Trees on Farms: Agroforestry in its Fifth Decade*; van Noordwijk, M., Ed.; World Agroforestry (ICRAF): Bogor, Indonesia, 2019; pp. 17–44.

17. Available online: https://www.decadeonrestoration.org/Interactive/tree-planting-and-ecosystem-restoration-crash-course (accessed on 29 July 2020).

18. Garrity, D.P.; Bayala, J. Zinder: Farmer-managed natural regeneration of Sahelian parklands in Niger. In *Sustainable Development Through Trees on Farms: Agroforestry in its Fifth Decade*; van Noordwijk, M., Ed.; World Agroforestry (ICRAF): Bogor, Indonesia, 2019; pp. 151–172.

19. Duguma, L.A.; Minang, P.A.; Kimaro, A.A.; Otsyina, R.; Mpanda, M. Shinyanga: Blending old and new agroforestry to integrate development, climate change mitigation and adaptation in Tanzania. In *Sustainable Development Through Trees on Farms: Agroforestry in its Fifth Decade*; van Noordwijk, M., Ed.; World Agroforestry (ICRAF): Bogor, Indonesia, 2019; pp. 121–131.

20. van Noordwijk, M. (Ed.) *Sustainable Development Through Trees on Farms: Agroforestry in its Fifth Decade*; World Agroforestry (ICRAF): Bogor, Indonesia, 2019.

21. Miccolis, A.; Peneireiro, F.M.; Marques, H.R.; Vieira, D.L.M.; Arcoverde, M.F.; Hoffmann, M.R.; Rehder, T.; Pereira, A.V.B. *Agroforestry Systems for Ecological Restoration: How to Reconcile Conservation and Production. Options for Brazil's Cerrado and Caatinga Biomes*; Instituto Sociedade, População e Natureza—ISPN/World Agroforestry Centre (ICRAF): Brasília, Brasil, 2016; Available online: https://www.worldagroforestry.org/publication/agroforestry-systems-ecological-restoration-how-reconcile-conservation-and-production (accessed on 29 July 2020).

22. Cornelius, J.P.; Miccolis, A. Can market-based agroforestry germplasm supply systems meet the needs of forest landscape restoration? *New For.* **2018**, *49*, 457–469. [CrossRef]

23. HLPE. Agroecological and other innovative approaches for sustainable agriculture and food systems that enhance food security and nutrition. In *A Report by the High Level Panel of Experts on Food Security and Nutrition of the Committee on World Food Security*; FAO: Rome, Italy, 2019.

24. Shono, K.; Cadaweng, E.A.; Durst, P.B. Application of assisted natural regeneration to restore degraded tropical forestlands. *Restor. Ecol.* **2007**, *15*, 620–626. [CrossRef]

25. Friday, K.S. *Imperata Grassland Rehabilitation Using Agroforestry and Assisted Natural Regeneration*; World Agroforestry Centre: Bogor, Indonesia, 1999.

26. Reij, C.; Garrity, D. Scaling up farmer-managed natural regeneration in Africa to restore degraded landscapes. *Biotropica* **2016**, *48*, 834–843. [CrossRef]

27. Chazdon, R.L.; Guariguata, M.R. Natural regeneration as a tool for large-scale forest restoration in the tropics: Prospects and challenges. *Biotropica* **2016**, *48*, 716–730. [CrossRef]

28. Perrow, M.R.; Davy, A.J. (Eds.) *Handbook of Ecological Restoration*; Cambridge University Press: Cambridge, UK, 2002; Volume 2.

29. Van Andel, J.; Aronson, J. *Restoration Ecology*; Blackwell Publishing: Malden, MA, USA, 2006.

30. Vaughn, K.J.; Porensky, L.M.; Wilkerson, M.L.; Balachowski, J.; Peffer, E.; Riginos, C.; Young, T.P. Restoration ecology. *Nat. Educ. Knowl.* **2010**, *3*, 66.

31. Clewell, A.; Rieger, J.; Munro, J. *Guidelines for Developing and Managing Ecological Restoration Projects*, 2nd ed.; Society for Ecological Restoration: Tucson, AZ, USA, 2005; Available online: http://c.ymcdn.com/sites/www.ser.org/resource/resmgr/custompages/publications/ser_publications/Dev_and_Mng_Eco_Rest_Proj.pdf (accessed on 29 July 2020).

32. Buckingham, K.; Weber, S. *Assessing the ITTO Guidelines for the Restoration, Management, and Rehabilitation of Degraded Secondary Tropical Forests—Case Studies of Ghana, Indonesia, and Mexico*; International Tropical Tree Organization (ITTO) Consultancy with the World Resources Institute (WRI): Yokahama, Japan, 2015; Available online: http://www.itto.int/direct/topics/topics_pdf_download/topics_id=4632&no=1 (accessed on 29 July 2020).

33. Stanturf, J.; Mansourian, S.; Kleine, M. (Eds.) *Implementing Forest Landscape Restoration, A Practitioner's Guide*; International Union of Forest Research Organizations, Special Programme for Development of Capacities (IUFRO-SPDC): Vienna, Austria, 2017; p. 128.

34. Chazdon, R.L.; Brancalion, P.H.; Lamb, D.; Laestadius, L.; Calmon, M.; Kumar, C. A policy-driven knowledge agenda for global forest and landscape restoration. *Conserv. Lett.* **2017**, *10*, 125–132. [CrossRef]

35. Clewell, A.F.; Aronson, J. *Ecological Restoration: Principles, Values, and Structure of an Emerging Profession*; Island Press: Washington, DC, USA, 2013.

36. Duguma, L.A.; Minang, P.A.; Mpanda, M.; Kimaro, A.; Alemagi, D. Landscape restoration from a social-ecological system perspective? In *Climate-Smart Landscapes: Multifunctionality in Practice*; World Agroforestry Centre (ICRAF): Nairobi, Kenya, 2015; pp. 63–72.

37. Erbaugh, J.T.; Oldekop, J.A. Forest landscape restoration for livelihoods and well-being. *Curr. Opin. Environ. Sustain.* **2018**, *32*, 76–83. [CrossRef]

38. Kerr, J.; Verbist, B.; Suyanto; Pender, J. Placement of a payment for watershed services program in indonesia: Social and ecological factors. In *Co-Investment in Ecosystem Services: Global Lessons from Payment and Incentive Schemes*; Namirembe, S., Leimona, B., van Noordwijk, M., Minang, P., Eds.; World Agroforestry Centre (ICRAF): Nairobi, Kenya, 2017.

39. IUCN/WRI. *A Guide to the Restoration Opportunities Assessment Methodology (ROAM): Assessing Forest Landscape Restoration Opportunities at the National or Sub-National Level*; Working Paper (Road-test edition); IUCN: Gland, Switzerland, 2014; p. 125. Available online: http://www.bonnchallenge.org/content/restoration-opportunities-assessment-methodology-roam (accessed on 29 July 2020).

40. Ghazoul, J.; Chazdon, R. Degradation and recovery in changing forest landscapes: A multiscale conceptual framework. *Annu. Rev. Environ. Resour.* **2017**, *42*, 161–188. [CrossRef]

41. de Souza, S.E.F.; Vidal, E.; Chagas, G.D.F.; Elgar, A.T.; Brancalion, P.H. Ecological outcomes and livelihood benefits of community-managed agroforests and second growth forests in Southeast Brazil. *Biotropica* **2016**, *48*, 868–881. [CrossRef]

42. Baynes, J.; Herbohn, J.; Smith, C.; Fisher, R.; Bray, D. Key factors which influence the success of community forestry in developing countries. *Glob. Environ. Chang.* **2015**, *35*, 226–238. [CrossRef]

43. HLPE. *Sustainable Forestry for Food Security and Nutrition: A Report by the High-Level Panel of Experts on Food Security and Nutrition of the Committee on World Food Security*; FAO: Rome, Italy, 2017; Available online: http://www.fao.org/3/a-i7395e.pdf (accessed on 29 July 2020).

44. IRP. *Land Restoration for Achieving the Sustainable Development Goals: An International Resource Panel Think Piece*; Herrick, J.E., Abrahamse, T., Abhilash, P.C., Ali, S.H., Alvarez-Torres, P., Barau, A.S., Branquinho, C., Chhatre, A., Chotte, J.L., Von Maltitz, G.P., Eds.; United Nations Environment Programme: Nairobi, Kenya, 2019.

45. Besseau, P.; Graham, S.; Christophersen, T. *Restoring Forests and Landscapes: The Key to a Sustainable Future*; IUFRO On Behalf of the Global Partnership on Forest and Landscape Restoration: Vienna, Austria, 2018.

46. Chazdon, R.L.; Wilson, S.J.; Brondizio, E.; Guariguata, M.R.; Herbohn, J. Key challenges for governing forest and landscape restoration across different contexts. *Land Use Policy* **2020**, 104854. [CrossRef]

47. Chazdon, R.L.; Gutierrez, V.; Brancalion, P.H.; Laestadius, L.; Guariguata, M.R. Co-creating conceptual and working frameworks for implementing forest and landscape restoration based on core principles. *Forests* **2020**, *11*, 706. [CrossRef]

48. Djenontin, I.N.S.; Zulu, L.C.; Ligmann-Zielinska, A. Improving representation of decision rules in LUCC-ABM: An example with an elicitation of farmers' decision making for landscape restoration in central malawi. *Sustainability* **2020**, *12*, 5380. [CrossRef]

49. Lee, D.R.; Barrett, C.B. (Eds.) *Tradeoffs or Synergies? Agricultural Intensification, Economic Development, and the Environment*; CABI: Wallingford, UK, 2000; p. 538.

50. van Noordwijk, M. Integrated natural resource management as pathway to poverty reduction: Innovating practices, institutions and policies. *Agric. Syst.* **2019**, *172*, 60–71. [CrossRef]

51. Jackson, L.; van Noordwijk, M.; Bengtsson, J.; Foster, W.; Lipper, L.; Pulleman, M.; Said, M.; Snaddon, J.; Vodouhe, R. Biodiversity and agricultural sustainagility: From assessment to adaptive management. *Curr. Opin. Environ. Sustain.* **2010**, *2*, 80–87. [CrossRef]

52. Rockström, J.; Steffen, W.; Noone, K.; Persson, Å.; Chapin, F.S.; Lambin, E.; Lenton, T.M.; Scheffer, M.; Folke, C.; Schellnhuber, H.J. Planetary boundaries: Exploring the safe operating space for humanity. *Ecol. Soc.* **2009**, *14*, 1–33. [CrossRef]

53. Rockström, J.; Falkenmark, M.; Allan, T.; Folke, C.; Gordon, L.; Jägerskog, A.; Kummu, M.; Lannerstad, M. The unfolding water drama in the Anthropocene: Towards a resilience-based perspective on water for global sustainability. *Ecohydrology* **2014**, *7*, 1249–1261. [CrossRef]

54. van Noordwijk, M.; Bargues-Tobella, A.; Muthuri, C.W.; Gebrekirstos, A.; Maimbo, M.; Leimona, B.; Bayala, J.; Ma, X.; Lasco, R.; Xu, J.; et al. Agroforestry as part of nature-based water management. In *Sustainable Development through Trees on Farms: Agroforestry in its Fifth Decade*; van Noordwijk, M., Ed.; World Agroforestry (ICRAF): Bogor, Indonesia, 2019; pp. 305–334.

55. Sayer, J.; Sunderland, T.; Ghazoul, J.; Pfund, J.-L.; Sheil, D.; Meijaard, E.; Venter, M.; Boedhihartono, A.K.; Day, M.; Garcia, C.; et al. Ten principles for a landscape approach to reconciling agriculture, conservation, and other competing land uses. *Proc. Natl. Acad. Sci. USA* **2013**, *110*, 8349–8356. [CrossRef] [PubMed]

56. Freeman, O.E.; Duguma, L.A.; Minang, P.A. Operationalizing the integrated landscape approach in practice. *Ecol. Soc.* **2015**, *20*, 24. Available online: http://www.ecologyandsociety.org/vol20/iss1/art24/ (accessed on 29 July 2020). [CrossRef]

57. Minang, P.A.; van Noordwijk, M.; Freeman, O.E.; Mbow, C.; de Leeuw, J.; Catacutan, D. (Eds.) *Climate-Smart Landscapes: Multifunctionality In Practice*; World Agroforestry Centre (ICRAF): Nairobi, Kenya, 2015; p. 444. ISBN 978-92-9059-375-1. Available online: http://asb.cgiar.org/climate-smart-landscapes/ (accessed on 29 July 2020).

58. Reed, J.; van Vianen, J.; Deakin, E.L.; Barlow, J.; Sunderland, T. Integrated landscape approaches to managing social and environmental issues in the tropics: Learning from the past to guide the future. *Glob. Chang. Biol.* **2016**, *22*, 2540–2554. [CrossRef]

59. Sayer, J.; Margules, C.; Boedhihartono, A.K.; Sunderland, T.; Langston, J.D.; Reed, J.; Riggs, R.A.; Buck, L.E.; Campbell, B.M.; Kusters, K.; et al. Measuring the effectiveness of landscape approaches to conservation and development. *Sustain. Sci.* **2016**, *12*, 465–476. [CrossRef]

60. Wilson, S.J.; Cagalanan, D. Governing restoration: Strategies, adaptations and innovations for tomorrow's forest landscapes. *World Dev. Perspect.* **2016**, *4*, 11–15. [CrossRef]

61. Evans, K.A.; Guariguata, M.R. Success from the ground up: Participatory monitoring and forest restoration. In *Occasional Paper 159*; Center for International Forestry Research (CIFOR): Bogor, Indonesia, 2016. Available online: http://www.cifor.org/publications/pdf_fies/OccPapers/OP-159.pdf (accessed on 20 July 2020).

62. FAO. *Global Guidelines for the Restoration of Degraded Forests and Landscapes in Drylands: Building Resilience and Benefiting Livelihoods*; FAO Forestry Paper 175; Food and Agricultural Organization of the United Nations: Rome, Italy, 2015; Available online: http://www.fao.org/3/a-i5036e.pdf (accessed on 20 July 2020).

63. Bozzano, M.; Jalone, R.; Thomas, E.; Boschier, D.; Gallo, L.; Cavers, S.; Bordacs, S.; Smith, P.; Loo, J. (Eds.) *Genetic Considerations in Ecosystem Restoration Using Native Tree Species*; FAO and Bioversity International: Rome, Italy, 2014; Available online: http://www.fao.org/3/a-i3938e.pdf (accessed on 29 July 2020).

64. Dawson, I.K.; Leakey, R.; Clement, C.R.; Weber, J.C.; Cornelius, J.P.; Roshetko, J.M.; Vinceti, B.; Kalinganire, A.; Tchoundjeu, Z.; Masters, E.; et al. The management of tree genetic resources and the livelihoods of rural communities in the tropics: Non-timber forest products, smallholder agroforestry practices and tree commodity crops. *For. Ecol. Manag.* **2014**, *333*, 9–21. [CrossRef]

65. Jalonen, R.; Valette, M.; Boshier, D.; Duminil, J.; Thomas, E. Forest and landscape restoration severely constrained by a lack of attention to the quantity and quality of tree seed: Insights from a global survey. *Conserv. Lett.* **2018**, *11*, e12424. [CrossRef]

66. Sinclair, F.L.; Coe, R. The options by context approach: A paradigm shift in agronomy. *Exp. Agric.* **2019**, *55*, 1–13. [CrossRef]

67. van Noordwijk, M.; Coe, R. Methods in agroforestry research across its three paradigms. In *Sustainable Development Through Trees on Farms: Agroforestry in Its Fifth Decade*; van Noordwijk, M., Ed.; World Agroforestry (ICRAF): Bogor, Indonesia, 2019; pp. 379–402.

68. Mithöfer, D.; van Noordwijk, M.; Leimona, B.; Cerutti, P.O. Certify and shift blame, or resolve issues? Environmentally and socially responsible global trade and production of timber and tree crops. *Int. J. Biodivers. Sci. Ecosyst. Serv. Manag.* **2017**, *13*, 72–85. [CrossRef]

69. Leimona, B.; van Noordwijk, M.; Mithöfer, D.; Cerutti, P.O. Environmentally and socially responsible global production and trade of timber and tree crop commodities: Certification as a transient issue-attention cycle response to ecological and social issues. *Int. J. Biodivers. Sci. Ecosyst. Serv. Manag.* **2018**, *13*, 497–502. [CrossRef]

70. van Noordwijk, M.; Dewi, S.; Minang, P.A.; Simons, A.J. Deforestation-free claims: Scams or substance? In *Zero Deforestation: A Commitment to Change*; Pasiecznik, N., Savenije, H., Eds.; Tropenbos International: Wageningen, The Netherlands, 2017; pp. 11–16.

71. Creed, I.F.; van Noordwijk, M. (Eds.) *Forest and Water on a Changing Planet: Vulnerability, Adaptation and Governance Opportunities: A Global Assessment Report (No. 38)*; International Union of Forest Research Organizations (IUFRO): Vienna, Austria, 2018.

72. Harris, J.A.; Hobbs, R.J.; Higgs, E.; Aronson, J. Ecological restoration and global climate change. *Restor. Ecol.* **2006**, *14*, 170–176. [CrossRef]

73. Meli, P.; Holl, K.D.; Benayas, J.M.R.; Jones, H.P.; Jones, P.C.; Montoya, D.; Mateos, D.M. A global review of past land use, climate, and active vs. passive restoration effects on forest recovery. *PLoS ONE* **2017**, *12*, e0171368. [CrossRef] [PubMed]

74. Mansourian, S.; Dudley, N.; Vallauri, D. Forest landscape restoration: Progress in the last decade and remaining challenges. *Ecol. Restor.* **2017**, *35*, 281–288. [CrossRef]

75. Swinfield, T.; Afriandi, R.; Antoni, F.; Harrison, R.D. Accelerating tropical forest restoration through the selective removal of pioneer species. *For. Ecol. Manag.* **2016**, *381*, 209–216. [CrossRef]

76. Harrison, R.D.; Tan, S.; Plotkin, J.B.; Slik, F.; Detto, M.; Brenes, T.; Itoh, A.; Davies, S.J. Consequences of defaunation for a tropical tree community. *Ecol. Lett.* **2013**, *16*, 687–694. [CrossRef]

77. de Royer, S.; van Noordwijk, M.; Roshetko, J. Does community-based forest management in Indonesia devolve social justice or social costs? *Int. For. Rev.* **2018**, *20*, 167–180. [CrossRef]

78. van Noordwijk, M.; Leimona, B. Principles for fairness and efficiency in enhancing environmental services in Asia: Payments, compensation, or co-investment? *Ecol. Soc.* **2010**, *15*, 17. [CrossRef]

79. Leimona, B.; van Noordwijk, M.; de Groot, R.; Leemans, R. Fairly efficient, efficiently fair: Lessons from designing and testing payment schemes for ecosystem services in Asia. *Ecosyst. Serv.* **2015**, *12*, 16–28. [CrossRef]

80. Namirembe, S.; Leimona, B.; van Noordwijk, M.; Minang, P.A. *Co-Investment in Ecosystem Services: Global Lessons from Payment and Incentive Schemes*; World Agroforestry (ICRAF): Nairobi, Kenya, 2017; Chapter 1; Available online: https://www.worldagroforestry.org/sites/default/files/u884/Ch1_IntroCoinvest_ebook.pdf (accessed on 20 July 2020).

81. Meinzen-Dick, R.; Quisumbing, A.; Doss, C.; Theis, S. Women's land rights as a pathway to poverty reduction: Framework and review of available evidence. *Agric. Syst.* **2017**, *172*, 72–82. [CrossRef]

82. Kusters, K.; Buck, L.; de Graaf, M.; Minang, P.A.; van Oosten, C.; Zagt, R. Participatory planning, monitoring and evaluation of multi-stakeholder platforms in integrated landscape initiatives. *Environ. Manag.* **2017**, *2*, 170–181. [CrossRef] [PubMed]

83. van Noordwijk, M.; Kim, Y.S.; Leimona, B.; Hairiah, K.; Fisher, L.A. Metrics of water security, adaptive capacity and agroforestry in Indonesia. *Curr. Opin. Environ. Sustain.* **2016**, *21*, 1–8. [CrossRef]

84. Ahrends, A.; Hollingsworth, P.M.; Beckschäfer, P.; Chen, H.; Zomer, R.J.; Zhang, L.; Wang, M.; Xu, J. China's fight to halt tree cover loss. *Proc. R. Soc. B* **2017**, *284*. [CrossRef]

85. Wangpakapattanawong, P.; Kavinchan, N.; Vaidhayakarn, C.; Schmidt-Vogt, D.; Elliott, S. Fallow to forest: Applying indigenous and scientific knowledge of swidden cultivation to tropical forest restoration. *For. Ecol. Manag.* **2010**, *260*, 1399–1406. [CrossRef]

86. Chirwa, P.W.; Mahamane, L. Overview of restoration and management practices in the degraded landscapes of the Sahelian and dryland forests and woodlands of East and southern Africa. *South. For. J. For. Sci.* **2017**, *79*, 87–94. [CrossRef]

87. Mokria, M.; Tolera, M.; Sterck, F.J.; Gebrekirstos, A.; Bongers, F.; Decuyper, M.; Sass-Klaassen, U. The frankincense tree Boswellia neglecta reveals high potential for restoration of woodlands in the Horn of Africa. *For. Ecol. Manag.* **2017**, *385*, 16–24. [CrossRef]

88. Miccolis, A. Restoration through agroforestry: Options for reconciling livelihoods with conservation in the cerrado and caatinga biomes in Brazil. *Exp. Agric.* **2017**, *55*, 208–225. [CrossRef]

89. Gann, G.D.; McDonald, T.; Walder, B.; Aronson, J.; Nelson, C.R.; Jonson, J.; Hallett, J.G.; Eisenberg, C.; Guariguata, M.R.; Liu, J.; et al. International principles and standards for the practice of ecological restoration. Second edition. *Restor. Ecol.* **2019**, *27*, S1–S46. [CrossRef]

90. IPBES. *The IPBES Assessment Report on Land Degradation And Restoration*; Montanarella, L., Scholes, R., Brainich, A., Eds.; Secretariat of the Intergovernmental Science-Policy Platform on Biodiversity and Ecosystem Services: Bonn, Germany, 2018; p. 44. Available online: https://ipbes.net/sites/default/files/2018_ldr_full_report_book_v4_pages.pdf (accessed on 29 July 2020).

91. Available online: https://www.iucn.org/theme/forests/our-work/forest-landscape-restoration (accessed on 29 July 2020).

92. Millennium Ecosystem Assessment. *Ecosystems and Human Well-Being: Current State and Trends*; Island Press: Washington, DC, USA, 2005; Volume 5.

93. Available online: https://www.iucn.org/news/protected-areas/201905/enhancing-progress-towards-aichi-target-11#:~{}:text=As%20reported%20in%20the%20latest,and%2010%25%20of%20the%20sea.&text=In%20an%20effort%20to%20ramp,Partnership%20on%20Aichi%20Target%2011 (accessed on 18 July 2020).

94. Potschin-Young, M.; Haines-Young, R.; Görg, C.; Heink, U.; Jax, K.; Schleyer, C. Understanding the role of conceptual frameworks: Reading the ecosystem service cascade. *Ecosyst. Serv.* **2018**, *29*, 428–440. [CrossRef]

95. Van Noordwijk, M.; Minang, P.A. *If We Cannot Define It, We Cannot Save It: Forest Definitions and REDD*; Tropenbos International: Wageningen, The Netherlands, 2009.

96. Chazdon, R.L.; Brancalion, P.H.; Laestadius, L.; Bennett-Curry, A.; Buckingham, K.; Kumar, C.; Wilson, S.J. When is a forest a forest? Forest concepts and definitions in the era of forest and landscape restoration. *Ambio* **2016**, *45*, 538–550. [CrossRef]

97. Dignac, M.F.; Derrien, D.; Barré, P.; Barot, S.; Cécillon, L.; Chenu, C.; Chevallier, T.; Freschet, G.T.; Garnier, P.; Guenet, B.; et al. Increasing soil carbon storage: Mechanisms, effects of agricultural practices and proxies. A review. *Agron. Sustain. Dev.* **2017**, *37*, 14. [CrossRef]

98. Suprayogo, D.; van Noordwijk, M.; Hairiah, K.; Meilasari, N.; Rabbani, A.L.; Ishaq, R.M.; Widianto, W. Infiltration-Friendly Agroforestry Land Uses on Volcanic Slopes in the Rejoso Watershed, East Java, Indonesia. *Land* **2020**, *9*, 240. [CrossRef]

99. Shepherd, K.D.; Shepherd, G.; Walsh, M.G. Land health surveillance and response: A framework for evidence-informed land management. *Agric. Syst.* **2015**, *132*, 93–106. [CrossRef]

100. Tomich, T.P.; Chomitz, K.; Francisco, H.; Izac, A.M.N.; Murdiyarso, D.; Ratner, B.D.; Thomas, D.E.; van Noordwijk, M. Policy analysis and environmental problems at different scales: Asking the right questions. *Agric. Ecosyst. Environ.* **2004**, *104*, 5–18. [CrossRef]

101. Namirembe, S.; Leimona, B.; van Noordwijk, M.; Minang, P.A. *Co-Investment in Ecosystem Services: Global Lessons from Payment and Incentive Schemes*; World Agroforestry (ICRAF): Nairobi, Kenya, 2019.

102. van Noordwijk, M.; Duguma, L.A.; Dewi, S.; Leimona, B.; Catacutan, D.M.; Lusiana, B.; Öborn, I.; Hairiah, K.; Minang, P.A. SDG synergy between agriculture and forestry in the food, energy, water and income nexus: Reinventing agroforestry? *Curr. Opin. Environ. Sustain.* **2018**, *34*, 33–42. [CrossRef]

103. Minang, P.A.; van Noordwijk, M.; Duguma, L.A. Policies for ecosystem services enhancement. In *Sustainable Development Through Trees on Farms: Agroforestry in its Fifth Decade*; van Noordwijk, M., Ed.; World Agroforestry (ICRAF): Bogor, Indonesia, 2019; pp. 361–376.

104. Rudel, T.K. Did a green revolution restore the forests of the American South. In *Agricultural Technologies and Tropical Deforestation*; Angelsen, A., Kaimowitz, D., Eds.; CABi: Wallingford, UK, 2001; pp. 53–68.

105. Minang, P.A.; van Noordwijk, M.; Kahurani, E. *Partnership in the Tropical Forest Margins: A 20-Year Journey in Search of Alternatives to Slash-and-Burn*; World Agroforestry: Nairobi, Kenya, 2014; p. 241.

106. Li, S.; Li, X. Global understanding of farmland abandonment: A review and prospects. *J. Geogr. Sci.* **2017**, *27*, 1123–1150. [CrossRef]

107. Schröder, P.; Antonarakis, A.S.; Brauer, J.; Conteh, A.; Kohsaka, R.; Uchiyama, Y.; Pacheco, P. SDG 12: Responsible consumption and production–Potential Benefits and impacts on forests and livelihoods. In *Sustainable Development Goals: Their Impacts on Forests and People*; Katila, P., Colfer, C.J.P., de Jong, W., Galloway, G., Pacheco, P., Winkel, G., Eds.; Cambridge University Press: Cambridge, UK, 2019; pp. 386–418.

108. Dewi, S.; van Noordwijk, M.; Zulkarnain, M.T.; Dwiputra, A.; Prabhu, R.; Hyman, G.; Gitz, V.; Nasi, R. Tropical forest-transition landscapes: A portfolio for studying people, tree crops and agro-ecological change in context. *Int. J. Biodivers. Sci. Ecosyst. Serv. Manag.* **2017**, *13*, 312–329. [CrossRef]

109. Filoso, S.; Bezerra, M.O.; Weiss, K.C.; Palmer, M.A. Impacts of forest restoration on water yield: A systematic review. *PLoS ONE* **2017**, *12*, e0183210. [CrossRef]

110. Ilstedt, U.; Tobella, A.B.; Bazié, H.R.; Bayala, J.; Verbeeten, E.; Nyberg, G.; Sanou, J.; Benegas, L.; Murdiyarso, D.; Laudon, H.; et al. Intermediate tree cover can maximize groundwater recharge in the seasonally dry tropics. *Sci. Rep.* **2016**, *6*. [CrossRef]

111. van Noordwijk, M.; Tanika, L.; Lusiana, B. Flood risk reduction and flow buffering as ecosystem services—Part I: Theory on a flow persistence indicator. *Hydrol. Earth Syst. Sci.* **2017**, *21*, 2321–2340. [CrossRef]

112. van Noordwijk, M.; Tanika, L.; Lusiana, B. Flood risk reduction and flow buffering as ecosystem services—Part II: Land use and rainfall intensity effects in Southeast Asia. *Hydrol. Earth Syst. Sci.* **2017**, *21*, 2341–2360. [CrossRef]

113. Dufour, S.; Piégay, H. From the myth of a lost paradise to targeted river restoration: Forget natural references and focus on human benefits. *River Res. Appl.* **2009**, *25*, 568–581. [CrossRef]

114. Ellison, D.; Morris, C.; Locatelli, B.; Sheil, D.; Cohen, J.D.M.; Murdiyarso, D.; Gutierrez, V.; Van Noordwijk, M.; Creed, I.F.; Pokorny, J.; et al. Trees, forests and water: Cool insights for a hot world. *Glob. Environ. Chang.* **2017**, *43*, 51–61. [CrossRef]

115. Bonnesoeur, V.; Locatelli, B.; Guariguata, M.R.; Ochoa-Tocachi, B.F.; Vanacker, V.; Mao, Z.; Stokes, A.; Mathez-Stiefel, S.L. Impacts of forests and forestation on hydrological services in the Andes: A systematic review. *For. Ecol. Manag.* **2019**, *433*, 569–584. [CrossRef]

116. Sternberg, L.D. Savanna-forest hysteresis in the tropics. *Glob. Ecol. Biogeogr.* **2001**, *10*, 369–378. [CrossRef]

117. Hirota, M.; Holmgren, M.; Van Nes, E.H.; Scheffer, M. Global resilience of tropical forest and savanna to critical transitions. *Science* **2011**, *334*, 232–235. [CrossRef]

118. van Noordwijk, M.; Ekadinata, A.; Leimona, B.; Catacutan, D.C.; Martini, E.; Tata, H.L.; Öborn, I.; Hairiah, K.; Wangpakapattanawong, P.; Mulia, R.; et al. *Agroforestry Options for Degraded Landscapes in Southeast Asia*; Dagar, J.C., Gupta, S.R., Teketay, D., Eds.; Agroforestry for Degraded Landscapes Springer: Singapore, 2020; in press.

119. Leimona, B.; van Noordwijk, M.; Kennedy, S.; Namirembe, S.; Minang, P.A. Synthesis and lessons on ecological, economic, social and governance propositions. In *Co-Investment in Ecosystem Services: Global Lessons from Payment and Incentive Schemes*; Namirembe, S., Leimona, B., van Noordwijk, M., Minang, P.A., Eds.; World Agroforestry Centre (ICRAF): Nairobi, Kenya, 2019; Chapter 38.

120. Van Noordwijk, M.; Leimona, B.; Jindal, R.; Villamor, G.B.; Vardhan, M.; Namirembe, S.; Catacutan, D.; Kerr, J.; Minang, P.A.; Tomich, T.P. Payments for Environmental Services: Evolution toward efficient and fair incentives for multifunctional landscapes. *Annu. Rev. Environ. Resour.* **2012**, *37*, 389–420. [CrossRef]

121. Chapman, M.; Satterfield, T.; Wittman, H.; Chan, K.M. A payment by any other name: Is Costa Rica's PES a payment for services or a support for stewards? *World Dev.* **2020**, *129*, 104900. [CrossRef]

122. Minang, P.A.; van Noordwijk, M. Design challenges for achieving reduced emissions from deforestation and forest degradation through conservation: Leveraging multiple paradigms at the tropical forest margins. *Land Use Policy* **2013**, *31*, 61–70. [CrossRef]

123. Barr, C.M.; Sayer, J.A. The political economy of reforestation and forest restoration in Asia–Pacific: Critical issues for REDD+. *Biol. Cons.* **2012**, *154*, 9–19. [CrossRef]

124. van Noordwijk, M. Prophets, profits, prove it: Social forestry under pressure. *One Earth* **2020**, *2*, 394–397. [CrossRef]

125. Minang, P.A.; Van Noordwijk, M.; Duguma, L.; Alemagi, D.; Do, T.H.; Bernard, F.; Agung, P.; Robiglio, V.; Catacutan, D.; Suyanto, S.; et al. REDD+ Readiness progress across countries: Time for reconsideration. *Clim. Policy* **2014**, *14*, 685–708. [CrossRef]

126. van Noordwijk, M.; Leimona, B.; Amaruzaman, S. Sumber Jaya from conflict to source of wealth in Indonesia: Reconciling coffee agroforestry and watershed functions. In *Sustainable Development Through Trees on Farms: Agroforestry in its Fifth Decade*; van Noordwijk, M., Ed.; World Agroforestry (ICRAF): Bogor, Indonesia, 2019; pp. 177–192.

127. Singh, R.; van Noordwijk, M.; Chaturvedi, O.P.; Garg, K.K.; Dev, I.; Wani, S.P.; Rizvi, J. Bundelkhand: Public co-investment in groundwater recharge in India. In *Sustainable Development Through Trees on Farms: Agroforestry in its Fifth Decade*; van Noordwijk, M., Ed.; World Agroforestry (ICRAF): Bogor, Indonesia, 2019; pp. 201–208.

128. FAO/UNEP. *Terminology for Integrated Resources Planning and Management*; United Nations Environmental Programme: Kenya, Nairobi; FAO: Rome, Italy, 1999.

129. FAO. *Land Cover Classification System Classification Concepts and User Manual Software Version (2)*; Di Gregorio, A., Jansen, L.J.M., Eds.; FAO: Rome, Italy, 2005; Available online: http://www.fao.org/3/y7220e/y7220e00.htm (accessed on 20 July 2020).

130. FAO/ITPS. *Status of the World's Soil Resources (SWSR)—Main Report*; Food and Agriculture Organization of the United Nations and Intergovernmental Technical Panel on Soils: Rome, Italy, 2015; p. 650. Available online: http://www.fao.org/3/a-i5199e.pdf (accessed on 20 July 2020).

131. Bayala, J.; Sanou, J.; Teklehaimanot, Z.; Ouedraogo, S.J.; Kalinganire, A.; Coe, R.; Van Noordwijk, M. Advances in knowledge of processes in soil–tree–crop interactions in parkland systems in the West African Sahel: A review. *Agric. Ecosyst. Environ.* **2015**, *205*, 25–35. [CrossRef]

132. van Noordwijk, M.; Speelman, E.; Hofstede, G.J.; Farida, A.; Abdurrahim, A.Y.; Miccolis, A.; Hakim, A.L.; Wamucii, C.N.; Lagneaux, E.; Andreotti, F.; et al. Sustainable agroforestry management: Changing the game. *Land* **2020**, *9*, 243. [CrossRef]

133. Saragi-Sasmito, M.F.; Murdiyarso, D.; June, T.; Sasmito, S.D. Carbon stocks, emissions, and aboveground productivity in restored secondary tropical peat swamp forests. *Mitig. Adapt. Strat. Glob. Chang.* **2018**, *24*, 521–533. [CrossRef]

134. Widayati, A.; Tata, H.L.; van Noordwijk, M. Agroforestry in peatlands: Combining productive and protective functions as part of restoration. In *Agroforestry Options for ASEAN Series No. 4*; World Agroforestry Centre (ICRAF): Bogor, Indonesia, 2016.

135. van Noordwijk, M.; Matthews, R.B.; Agus, F.; Farmer, J.; Verchot, L.; Hergoualc'h, K.; Persch, S.; Tata, H.L.; Lusiana, B.; Widayati, A.; et al. Mud, muddle and models in the knowledge value- chain to action on tropical peatland issues. *Mitig. Adapt. Strat. Glod. Chang.* **2014**, *19*, 863–885.

136. van Noordwijk, M. Small-island agroforestry in an era of climate change and sustainable development goals. In *Sustainable Development Through Trees on Farms: Agroforestry in its Fifth Decade*; van Noordwijk, M., Ed.; World Agroforestry (ICRAF): Bogor, Indonesia, 2019; pp. 233–248.

137. Chan, H.T.; Baba, S. *Manual on Guidelines for Rehabilitation of Coastal Forests Damaged by Natural Hazards in the Asia-Pacific Region*; International Society for Mangrove Ecosystems and International Tropical Timber Organization (ITTO): Yokahama, Japan, 2009; Available online: http://www.preventionweb.net/fies/13225_ISMEManualoncoastalforestrehabilita.pdf (accessed on 20 July 2020).

138. Global Nature Fund. *Mangrove Restoration Guide, Best Practices and Lessons Learned from a Community-Based Conservation Project*; Global Nature Fund: Radolfzell, Germany, 2015; Available online: https://www.globalnature.org/bausteine.net/f/8281/GNF_Mangrove_Handbook_2015.pdf?fd=0 (accessed on 20 July 2020).

139. Sasmito, S.D.; Murdiyarso, D.; Friess, D.A.; Kurnianto, S. Can mangroves keep pace with contemporary sea level rise? A global data review. *Wetl. Ecol. Manag.* **2016**, *24*, 263–278. [CrossRef]

140. Wong, M.H. Ecological restoration of mine degraded soils, with emphasis on metal contaminated soils. *Chemosphere* **2003**, *50*, 775–780. [CrossRef]

141. Hairiah, K.; van Noordwijk, M.; Sari, R.R.; Saputra, D.D.; Widianto; Suprayogo, D.; Kurniawan, S.; Prayogo, C.; Gusli, S. Soil carbon stocks in Indonesian (agro)forest transitions: Compaction conceals lower carbon concentrations in standard accounting. *Agric. Ecosyst. Environ.* **2020**, *294*, 106879. [CrossRef]

142. van Noordwijk, M.; Goverse, T.; Ballabio, C.; Banwart, S.; Bhattacharyya, T.; Goldhaber, M.; Nikolaidis, N.; Noellemeyer, E.; Zhao, Y.; Mline, E. Soil carbon transition curves: Reversal of land degradation through management of soil organic matter for multiple benefits. In *Soil Carbon: Science, Management and Policy for Multiple Benefits*; Banwart, S.A., Noellemeyer, E., Milne, E., Eds.; CAB International: Harpenden, UK, 2015; pp. 26–46.

143. Minasny, B.; Malone, B.; McBratney, A.B.; Angers, D.; Arrouays, D.; Chambers, A.; Chaplot, V.; Chen, Z.-S.; Cheng, K.; Das, B.S.; et al. Soil carbon 4 per mille. *Geoderma* **2017**, *292*, 59–86. [CrossRef]

144. Tomich, T.P.; van Noordwijk, M.; Budidarsono, S.; Gillison, A.N.; Kusumanto, Y.; Murdiyarso, D.; Stolle, F.; Fagi, A.M. Agricultural intensification, deforestation, and the environment: Assessing tradeoffs in Sumatra, Indonesia. In *Tradeoffs or Synergies? Agricultural Intensification, Economic Development and the Environment*; Lee, D.R., Barrett, C.B., Eds.; CAB-International: Wallingford, UK, 2001; pp. 221–244.

145. Tomich, T.P.; van Noordwijk, M. Trade-off matrix between private and public benefits of land-use systems (ASB Matrix). In *Negotiation-Support Toolkit for Learning Landscapes*; van Noordwijk, M., Lusiana, B., Leimona, B., Dewi, S., Wulandari, D., Eds.; World Agroforestry Centre (ICRAF): Bogor, Indonesia, 2013; pp. 118–119.

146. Dewi, S.; Ekadinata, A.; Galudra, G.; Agung, P.; Johana, F. LUWES: Land use planning for low emission development strategy. *World Agrofor. Cent. Bogor. Indones* **2011**. 47p. Available online: http://www.worldagroforestry.org/output/land-use-planning-low-emission-development-strategy-luwes (accessed on 20 July 2020).

147. Neely, C.; Bourne, M.; Chesterman, S.; Kouplevatskaya-Buttoud, I.; Bojic, D.; Vallée, D. *Implementing 2030 Agenda for Food and Agriculture: Accelerating Impact through Cross-Sectoral Coordination*; World Agroforestry Centre (ICRAF) and Food and Agricultural Organization of The United Nations (FAO): Nairobi, Kenya; Rome, Italy, 2017; Available online: http://www.fao.org/3/a-i7749e.pdf (accessed on 20 July 2020).

148. Clark, W.C.; Tomich, T.P.; Van Noordwijk, M.; Guston, D.; Catacutan, D.C.; Dickson, N.M.; McNie, E. Boundary work for sustainable development: Natural resource management at the Consultative Group on International Agricultural Research (CGIAR). *Proc. Natl. Acad. Sci. USA* **2016**, *113*, 4615–4622. [CrossRef]

149. Robiglio, V.; Reyes, M. Restoration through formalization? Assessing the potential of Peru's Agroforestry Concessions scheme to contribute to restoration in agricultural frontiers in the Amazon region. *World Dev. Perspect.* **2016**, *3*, 42–46. [CrossRef]

150. Villamor, G.B.; van Noordwijk, M.; Leimona, B.; Duguma, L. Tradeoffs. In *Co-Investment in Ecosystem Services: Global Lessons from Payment and Incentive Schemes*; Namirembe, S., Leimona, B., van Noordwijk, M., Minang, P.A., Eds.; ICRAF: Nairobi, Kenya, 2017; Chapter 3; Available online: http://www.worldagroforestry.org/sites/default/files/chapters/Ch3%20Trade-offs_ebookB-DONE2.pdf (accessed on 20 July 2020).

151. Suwarno, A.; van Noordwijk, M.; Weikard, H.P.; Suyamto, D. Indonesia's forest conversion moratorium assessed with an agent-based model of Land-Use Change and Ecosystem Services (LUCES). *Mitig. Adapt. Strat. Glob. Chang.* **2018**, *23*, 211–229. [CrossRef]

152. Amaruzaman, S.; Leimona, B.; van Noordwijk, M.; Lusiana, B. Discourses on the performance gap of agriculture in a green economy: A Q-methodology study in Indonesia. *Int. J. Biodivers. Sci. Ecosyst. Serv. Manag.* **2017**, *13*, 233–247. [CrossRef]

153. Langston, J.; McIntyre, R.; Falconer, K.; Sunderland, T.J.C.; van Noordwijk, M.; Boedihartono, A.K. Discourses mapped by Q-method show governance constraints motivate landscape approaches in Indonesia. *PLoS ONE* **2019**, *14*, e0211221. [CrossRef]

 *land*

Article

Cost-Benefit Analysis of Landscape Restoration: A Stocktake

Priscilla Wainaina *[iD], Peter A. Minang, Eunice Gituku and Lalisa Duguma

World Agroforestry Centre, UN Avenue, Gigiri, Nairobi 00100, Kenya; p.minang@cgiar.org (P.A.M.);
E.Gituku@cgiar.org (E.G.); l.duguma@cgiar.org (L.D.)
* Correspondence: P.Wainaina@cgiar.org; Tel.: +254-20-722-4182

Received: 29 October 2020; Accepted: 18 November 2020; Published: 19 November 2020

Abstract: With the increase in demand for landscape restoration and the limited resources available, there is need for economic analysis of landscape restoration to help prioritize investment of the resources. Cost-benefit analysis (CBA) is a commonly applied tool in the economic analysis of landscape restoration, yet its application seems limited and varied. We undertake a review of CBA applications to understand the breadth, depth, and gaps. Of the 2056 studies identified in literature search, only 31 met our predefined criteria. Three studies offered a global perspective, while more than half were conducted in Africa. Only six countries benefit from at least 2 CBA studies, including Brazil, Ethiopia, Kenya, Vietnam, South Africa, and Tanzania. About 60% focus on agroforestry, afforestation, reforestation, and assisted natural regeneration practices. Only 16% covered all cost categories, with opportunity costs being the least covered. Eighty-four percent apply direct use values, while only 16% captured the non-use values. Similarly, lack of reliable data due to predictions and assumptions involved in data generation influenced CBA results. The limited number of eligible studies and the weaknesses identified hereinabove suggest strong need for improvements in both the quantity and quality of CBA to better inform planning, policies, and investments in landscape restoration.

Keywords: cost-benefit analysis; landscape restoration; global; stocktake

1. Introduction

There is a growing demand for restoration globally aimed at stopping further degradation and reversing degradation. Forest degradation, soil erosion, peatland, wetland drainage, and salinization have been the leading causes of land degradation globally over the past 50 years. According to the Intergovernmental Science-Policy Platform on Biodiversity and Ecosystem Services [1], 75% of the global land surface is significantly degraded, 66% of the ocean area is experiencing increasing cumulative impacts, and over 85% of wetlands have been lost. This has affected almost a quarter of the world's total land area, and the damage is felt through the loss of ecosystem goods and services. The damage costs the world an estimated USD 6.3 trillion a year in lost ecosystem service value, which includes climate regulation, clean air, recreational opportunities, freshwater, and fertile soils [2]. Besides, due to land degradation, the livelihoods of about half a billion people, especially poor people who depend on agricultural and forestlands, are jeopardized. Decreasing land productivity threatens water and food security, destabilizes sustainable development, and results in civil conflicts and human migration.

Global efforts have been put in place to stop further degradation and reverse degradation through restoration of forest and landscapes ecosystems. Over the last decade, there have been global restoration initiatives, including 'The Bonn Challenge' and 'The New York Declaration on Forests (NYDF)'. The Bonn Challenge is a global effort to bring 150 million hectares of the world's deforested and degraded land into restoration by 2020, and 350 million hectares by 2030[2], while the NYDF is a political declaration among governments, companies, indigenous peoples, and civil society to take action to halve the loss of natural forests by 2020 and halt it by 2030[1]. Within Africa, the African Forest Landscape Restoration Initiative, AFR100 aims to restore 100 million hectares of land in Africa by 2030[3]. Specific countries have also set definite restoration goals; for example, Kenya has committed to restoring 5.1 million hectares of degraded land[4], and Malawi committed to about 4.5 million hectares of land by 2030 [3].

Restoration is generally a costly undertaking, partly because it often begins after the environmental degradation is well-advanced and expensive to reverse, and is often labor and resource-intensive [4]. Funding sources for investment in landscape restoration include (1) Private finance, where capital is managed mainly to earn a financial return for the investor. (2) Public finance where funding comes from the government bodies. Public finance can further be divided into international donor support and domestic public expenditure. In public finance, public investments are largely made to generate economic, environmental and social benefits for the public. However, there may be a return to the government. (3) Philanthropic finance, which is charitable giving by individuals or organizations, typically with no intention of earning a financial return [5].

Numerous environmental, economic and social benefits are generated when degraded lands are restored. These benefits may range from conservation of biodiversity, creation of jobs, improvement in agricultural productivity, and so on. Despite the numerous benefits accruing from restoration, funding for landscape restoration falls short by about USD 300 billion a year [2]. Investment is inadequate for several key reasons, among them: (1) many of the benefits are public goods, which are difficult to monetize; (2) the long-term nature of investments does not match investors' desire for liquidity, i.e., restoration usually requires large investments upfront and has long lags before generating benefits, and (3) landscape projects are perceived to be risky. Overall, landscape restoration activities are often misunderstood as involving high up-front costs and low rates of return, and these ideas persist because few evaluations of restoration activities include a comprehensive and objective accounting of restoration's ecological and economic impacts. Economic analysis can encourage investment in restoration by clearly laying out the benefits and costs of restoration projects and their distribution among stakeholders. It is also instrumental in prioritizing scarce resources accordingly.

Cost-benefit analysis (CBA) is the commonly applied approach in the economic analysis of landscape restoration. However, compared to the significant number of landscape restoration projects and studies globally, relatively few CBA studies have been conducted on restoration projects. Reference [6] conducted a review of the rehabilitation of degraded dryland ecosystems. They found that, while numerous studies have been undertaken on restoration and rehabilitation of degraded ecosystems, there remains a gap in cost-benefit analysis of these interventions. They largely attributed this to missing data on the benefits and costs accruing from the restoration projects. Similarly, in a review of more than 2000 restoration case studies, the "The Economics of Ecosystems and Biodiversity (TEEB)[5] "2009 study found that less than 5% of the case studies provided meaningful cost data, and none provided an analysis of both costs and benefits [7].

While CBA continues to be a primary approach of economic assessment, adequate use of the tool requires a clear understanding of its limitations and pitfalls. Thus, the paper seeks to understand the breadth and depth of existing CBA studies. It also highlights gaps in the existing CBA studies on landscape restoration and how further studies can address these gaps.

---

[2] www.nydfglobalplatform.org
[1] www.bonnchallenge.org
[3] http://afr100.org/
[4] https://afr100.org/content/kenya
[5] http://teebweb.org/

2. Background

2.1. Brief Background on Landscape Restoration

Ecological restoration is the process of assisting the recovery of an ecosystem that has been degraded, damaged, or destroyed [8]. The goal is to repair the ecosystem to its integrity and its health and to reverse land degradation, increase the resilience of biodiversity, and deliver important ecosystem services [9,10]. Ecological restoration also includes forest and landscapes restoration, which is a process that aims to regain ecological integrity and enhance human well-being in deforested or degraded forest landscapes. A key feature of forest and landscapes restoration is that a combination of forest and non-forest ecosystems, land uses, and restoration approaches can be accommodated within a landscape to achieve sustainable food production, ecosystem services provisioning, and biodiversity conservation [11].

Landscape restoration is not just a matter of planting trees but also involves assisting the recovery of a damaged or destroyed ecosystem. The restoration activities range from small local activities carried out by individuals or community groups through to regional, country, and even global-scale activities involving multiple agencies and large numbers of people [12]. Restoration can either be achieved through natural regeneration or active restoration. For example, natural forest regeneration is the spontaneous recovery of native tree species that colonize and establish in abandoned fields or natural disturbances. This process can also be assisted natural regeneration, whereby it is assisted through human interventions, such as fencing to control livestock grazing, weed control, and fire protection [13]. Active restoration, on the other hand, involves human involvement in a range of ways and involves a considerable cost in terms of labor and time [14]. It may require planting of nursery-grown seedlings, direct seeding, and/or the manipulation of disturbance regimes (for example, thinning and burning) to speed up the recovery process. This is often at a high cost to establish the vegetation structure, reassemble local species composition, and/or catalyze ecological succession.

Effective restoration should aim at the reestablishment of fully functioning ecosystems. To ensure effective restoration and achieve sustainability and resilience into the future, Reference [15] advocates for four principles of restoration; (1) restoration should increase ecological integrity (2) restoration should be sustainable in the long term (3) restoration ought to be informed by the past and future, and (4) the restoration benefits and engages society through direct participation. Ecological integrity has been defined as the "ability of an ecosystem to support and maintain a balanced, adaptive community of organisms having a species composition, diversity, and functional organization comparable to that of natural habitat [16].

Several factors ought to be considered while deciding on the restoration strategy to be employed; the specific ecosystem resilience, the land-use history, the landscape context, the goal for the restoration, and available resources [14]. In addition, depending on the level of land degradation, and the intended impact of restoration on land use and land-use changes [17] distinguishes four intensities/levels of restoration; ecological intensification, recovery/regeneration, reparation/recuperation, and remediation. Table 1 presents a summary of some of the commonly practiced landscape restoration options/strategies in different land use or ecosystem types adapted from Reference [18,19]. A detailed explanation of most of these landscape restoration practices and their restoration goals is provided by Reference [19].

Table 1. Landscape restoration options for different land-use types.

Land Use/Ecosystem Type	Landscape Restoration Options/Strategies
Forest land	- Afforestation and reforestation - Planted forests and woodlots - Natural regeneration - Silviculture - Assisted natural regeneration/Reclamation/Rehabilitation

Table 1. *Cont.*

Land Use/Ecosystem Type	Landscape Restoration Options/Strategies
Agricultural land	- Agroforestry - Integrated soil fertility management - Climate-smart agriculture - Improved fallow - Extended rotations in plantations - Farmer managed natural regeneration (FMNR)
Protective land and buffers	- Mangrove restoration - Watershed protection - Erosion control - Bamboo planting along water bodies and wetlands
Urban areas	- Green and blue infrastructure in urban areas
Wetlands	- Wetlands restoration and conservation
Freshwater (Rivers/lakes)	- River and lake restoration - Sediment management - Pound restoration - Integrated watershed management
Grasslands and Shrublands	- Assisted natural regeneration

Adapted from Reference [18,19].

2.2. Economic Cost-Benefit Analysis of Landscape Restoration

Generally, cost-benefit analysis can be either financial or economic; there are similarities and differences in both. In both analyses, the net benefits of a project investment are estimated, and the estimation is based on the difference between with-project and without-project situations. While conducting both analyses, the assumption of constant prices is made and, the techniques of evaluating costs and benefits through the discounting method remain the same. However, in financial CBA of projects, benefits and costs are compared to the enterprise, while, in economic CBA, benefits and costs are compared to the whole economy.

The true value that a project holds for society is highly considered in economic analysis. It considers all members of society and measures the project's positive and negative impacts in terms of willingness to pay (WTP) for units of increased consumption and to accept compensation for foregone units of consumption. Importantly, the economic analysis covers even the costs and benefits of goods and services which have no market price. In financial analysis, the project's sustainability and balance of investment is checked using market prices. In economic analysis, the legitimacy of using national resources to a certain project is measured using economic price, which has been converted from the market price by excluding tax, profit, subsidy, etc. In financial analysis, the taxes and subsidies included in the price of goods and services are integral parts of financial prices, but they are treated differently in economic analysis.

There is also a significant difference between financial and economic analysis in the way they treat their external effects (costs and benefits). Such externalities, health effects and non-technical losses tend to be valued in economic analysis. Both financial and economic analyses are supposed to include such externalities (side effects). In addition, economic and financial returns in both analyses do not converge. This is because what counts as a benefit or a cost to the project operator does not necessarily count as a benefit or cost to the economy. When restoration is viewed through a financial accounting lens that ignores public values and the inter-generational nature of restoration, the conclusions that are drawn tend to favor investing in less restoration than society would prefer [20]. For this review, we consider those studies that conducted an economic CBA.

There are nine steps for conducting an economic cost-benefit analysis for landscape restoration as outlined by Reference [21]: (1) Specify the set of restoration transitions: Define which degraded land uses will be restored and the activities that will be used to restore them, (2) Define the stakeholders who will be impacted by restoration, (3) Catalogue the impacts and define how they will be measured: Which impacts matter most to the stakeholders, who will be impacted by restoration and what units of measurement are most useful for measuring them? (4) Predict the impacts quantitatively over the time horizon of the project: Use ecosystem service models, household surveys, stakeholder engagement, and other estimation methods to quantify the expected impacts of restoration activities, (5) Monetize all of the impacts: Use appropriate direct and indirect methods to value the estimated impacts, (6) Discount benefits and costs to obtain present values: Select appropriate discount rates to make streams of future benefits and costs comparable at the present moment, (7) Calculate the Net Present Value (NPV) of each alternative: Subtract the discounted stream of implementation, transaction, and opportunity costs from the discounted stream of benefits as shown in the equation.

$$Net\ Present\ Value = \sum_{t=0}^{n} \frac{B_t}{(1+r)^t} - \sum_{t=0}^{n} \frac{C_t}{(1+r)^t},$$

where B_t is the benefits from restoration at time t; r is the discount rate, and C_t is the total restoration costs at time t.

(8) Perform sensitivity analysis: The results of the CBA depend on assumptions, and the sensitivity of the results to changes in the underlying assumptions should be evaluated, (9) Make policy recommendations: From a Pareto-efficiency perspective, the restoration activities with the largest NPV should be recommended.

Overall, several categories of costs and benefits ought to be accounted for in a comprehensive CBA. The total cost of landscape restoration depends on how degraded a site is and how difficult it is to restore. Additionally, costs vary according to geography, degradation category, the objectives and contexts of specific restoration activities, and the types of restoration methods that are used. There are three investment phases involved in forest and landscape restoration: (1) Phase 1 is the initial readiness investment or up-front investment. During this phase, investments flow towards designing projects, planning, stakeholder engagement and participation, developing safeguards and capacity building. (2) Phase 2 is the investment for actual implementation. During this phase, it may involve policy reforms, implementation of the restoration of degraded lands, educational activities, land-use zoning and strengthening of capacities. (3) Phase 3 focuses on sustained financing for landscape ecosystem services and product services, for self-sustaining funding of the project's long-term running costs [5,18].

Similarly, these costs can be broadly categorized into three: (1) Implementation costs (usually very high); these costs include costs of raw materials, such as tree seedling, fencing, labor costs, transport costs, and other costs. In addition, this includes costs incurred in capacity building and training the local stakeholders. They are mostly the direct costs incurred in the project. Land users usually incur these costs, or they can be covered by the project. (2) Opportunity cost: this represents the cost of foregone opportunities and, represents the tangible goods and services that were foregone to make restoration possible. To capture the opportunity cost, it is necessary to conduct a baseline before implementing the project. (3) Transaction costs; transaction costs represent the cost for landowners and implementing agencies to identify viable land to restore and negotiate over terms that ensure restoration meets both local and national priorities [21]. These may include monitoring costs, as well.

Additionally, for a comprehensive economic CBA on restoration, the benefits to be considered ought to include both use and non-use values, as well as private and public benefits arising from the restoration, i.e., the total economic value of the restoration. The total economic value (TEV) of an investment attempts to estimate and monetize all economic impacts of the investment [22]. TEV recognizes that benefits and costs radiate far beyond the landowner or investor to the global effects. The TEV approach accounts for both the use and non-use benefits values [22].

The use-values are categorized into:

(a) Direct use values: These relate to the benefits obtained from the direct use of an ecosystem. Most of the direct products have market values, such as timber, poles, charcoal, gum arabica, and medicine, as well as other non-timber forest products (NTFPs), such as wild fruits, honey, fodder, crop harvests, recreation value, and others. They are the most straightforward benefit category to capture and account for since the data on quantities and value is available for most of the restoration projects.

(b) Indirect use values: These are usually associated with regulating services. These may include services, such as carbon sequestration, water treatment and regulation, soil erosion control, pollination, and so on.

(c) Option values: These include valuing ecosystem services for the option of future use, such as medicinal purposes.

Non-use values, on the hand, are categorized into:

(a) Bequest value: This captures the value arising from the satisfaction of knowing that future generations will access nature's benefits. This value is concerned with intergenerational equity.

(b) Altruist value: This value concerns intragenerational equity, i.e., the satisfaction of knowing that other people can also access nature's benefits.

(c) Existence value: This value is derived from the satisfaction of knowing a certain species exists. For example, indigenous trees, endangered species, and medical trees.

Direct use values are relatively easy to identify and value since they are tangible and usually have market values. On the other hand, indirect use and non-use value pose a challenge in valuation, and the valuation methods employed are often time and resource intensive. Several studies, such as Reference [23–25], detail how to adequately value these benefits through a wide range of valuation methods, including: production methods, choice experiments, contingent valuation methods (CVM), hedonic pricing, travel cost methods, cost-based approaches, benefit transfer, and mean-variance analysis.

3. Materials and Methods

Reviews aim to identify the most reliable research on a given question in a manner that minimizes selection biases in the literature search and screening process [26]. A comprehensive review of studies was conducted on cost-benefit analysis for landscape restoration by following the Preferred Reporting Items for Systematic Reviews and Meta-Analyses (PRISMA)[6] guidelines. Figure 1 shows the process of the review followed. The first step was a keyword search using the "publish or perish"[7] software. We searched for a combination of three phrases (1) "Cost-benefit analysis" and "Landscape restoration" (2) "Economics" and "Landscape restoration" (3) "Cost-benefit analysis" and "Land restoration". The search for "Cost-benefit analysis" and "Landscape restoration" generated 522 studies, while the search for "Economics" and "Landscape restoration" generated 922 studies and that of "Cost-benefit analysis" and "Land restoration" generated 612 studies.

For step two, the total 2056 publications were filtered down by titles and abstracts, resulting in 102 relevant publications. For the 102 publications, after reading through the full text, 31 publications were found that either entirely focused on CBA of landscape restoration or had a component of CBA of landscape restoration. The 31 publications are the ones that have been considered in the review, and a summary of these studies is presented in the Appendix A (Table A1). The summary is based on restoration options/strategies, country of focus, the year the study was conducted, data sources, the time considered in the CBA analysis, benefits and costs components, sensitivity analysis conducted, and the reported NPV. These variables are discussed in detail in Section 4 under results.

---

6 http://www.prisma-statement.org/
7 https://harzing.com/resources/publish-or-perish/windows

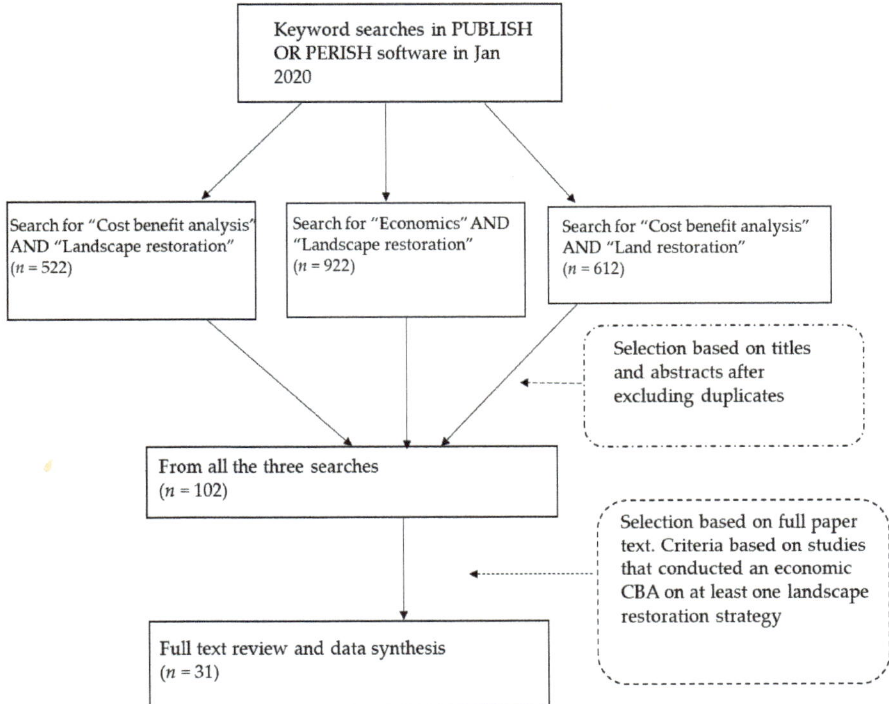

Figure 1. Review process of the studies on cost-benefit analysis (CBA) of landscape restoration.

4. Results

From the review process explained in the materials and method section, we found 31 publications of CBA on land restoration. In this chapter, we discuss these publications based on several attributes.

4.1. Country of Focus and Study Year

The 31 studies under review were conducted in about 20 countries distributed across five regions: Africa (10), Europe (2), N. America (2), S. America (3), and Asia & Middle East (3). Almost all the 31 studies reviewed, were conducted in a single country and only five focused on multiple countries. Of the five, three studies focused on a global perspective [18,27,28], while one focused on multiple African countries[8] [29], and one focused on several countries in Latin America[9] [7]. As shown in Figure 2, countries that have had the most CBA studies conducted there over the years had, on average, three studies and include South Africa, Brazil, Tanzania. Ethiopia, Kenya, and Vietnam, where each have had two CBA studies conducted.

---

[8] Angola, Benin, Botswana, Burkina Faso, Burundi, Cameroon, Central African Republic, Chad, Congo, Côte D'Ivoire, Djibouti, DR Congo, Egypt, Eritrea, Ethiopia, Gabon, Ghana, Guinea, Kenya, Lesotho, Liberia, Madagascar, Malawi, Mali, Mauritania, Morocco, Mozambique, Namibia, Niger, Nigeria, Rwanda, Senegal, Sierra Leone, South Africa, Sudan, Swaziland, Togo, Tunisia, Uganda, UR of Tanzania, Zambia, Zimbabwe.

[9] The focus Latin America countries include: Mexico, Argentina, and Chile.

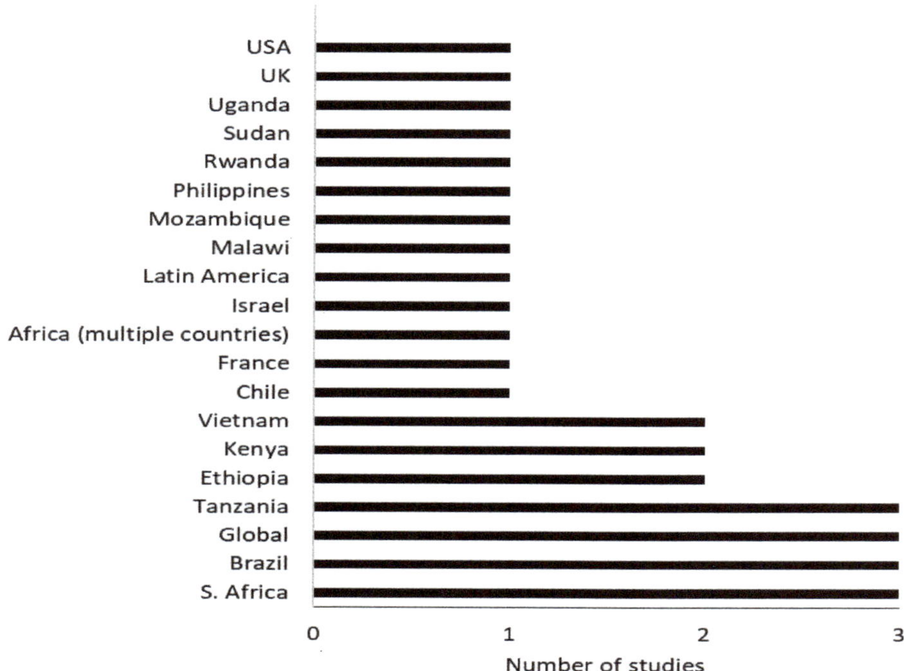

Figure 2. Distribution of CBA studies across various countries.

Figure 3 presents a trend of the year studies on restoration under review were conducted. The oldest study we reviewed was conducted in 1997 in France and focused on landscape restoration using hedgerows [30]. The was no study under review conducted between 1997 and 2005. The most recent studies were conducted in Kenya by Reference [31] and in Brazil by Reference [32].

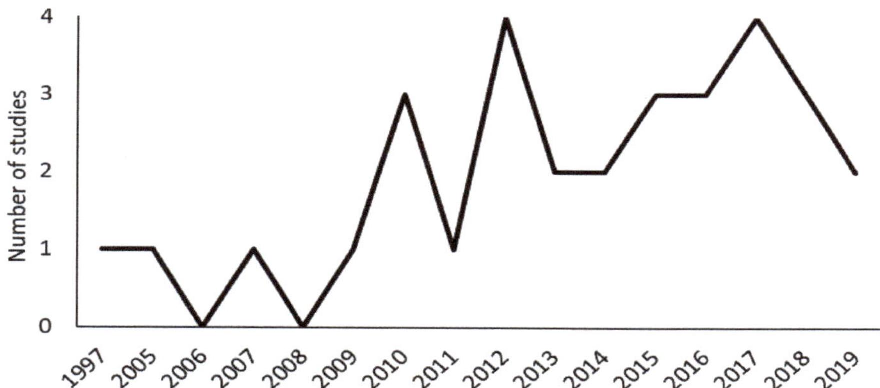

Figure 3. Year the CBA studies were conducted.

4.2. Landscape Restoration Options

As shown in Table 2, the CBA studies we reviewed were conducted for different types of restoration options/strategies, including: reforestation & afforestation, agroforestry, biofuel agroforestry, participatory forest management, woodlots establishment, sustainable land management practices, natural regeneration, assisted natural regeneration, mangrove restoration, clearing of invasive alien species, urban area restoration, and buffer areas restoration. The specific studies are presented in the Appendix A in Table A1. Some studies focused on only one type of restoration, while others conducted a comparison of different restoration options depending on the land use type. One of the most comprehensive CBA study we reviewed was conducted in Kenya by Reference [31] and compared returns for several landscape restoration strategies, including: afforestation or reforestation of degraded natural forests, rehabilitation of degraded natural forests, agroforestry in cropland, commercial tree and bamboo growing on potentially marginal cropland and un-stocked forest plantation forests, tree-based buffer zones along water bodies and wetlands, tree-based buffer zones along roads, and restoration of degraded rangelands. Overall, the most popular landscape restoration options for which CBA studies were conducted include: reforestation and afforestation (8), agroforestry (7), farmer-managed natural regeneration/Assisted natural regeneration (5), soil and water conservation practices (5), and establishment of woodlots (4). Further still, Reference [20] assessed the net present value of the Bonn Challenge, which is a global effort to restore 350 million hectares of degraded forest landscape.

Table 2. Landscape restoration options considered in the CBA studies reviewed.

Landscape Restoration Options	Number of Studies	Countries of Focus
Reforestation and Afforestation	8	Brazil, Chile, Ethiopia, Kenya, Uganda, USA, Chile, Tanzania
Agroforestry	7	Brazil, Ethiopia, Kenya, Sudan, Tanzania, Uganda, Malawi
Participatory forest management/FMNR/ANR	5	Brazil, Ethiopia, Malawi, Vietnam, Tanzania
Soil and water conservation measures and SLM practices, e.g., bunds, terracing, zero tillage,	5	Ethiopia, Kenya, Malawi, Vietnam, multiple countries in Africa
Establishment of woodlots	4	Ethiopia, Malawi, Uganda, Tanzania, Rwanda
Mangrove restoration (protective and planting)	3	Mozambique, Philippines, Vietnam
Natural regeneration	2	Uganda, Rwanda
River restoration/habitat restoration for river catchment	2	Israel, UK
Dryland forest restoration	2	Latin America, Chile
Alien vegetation clearing for water yield and tourism	2	South Africa (2)
Biofuels agroforestry or biofuel for energy, thus reducing deforestation	1	Tanzania,
Landscape restoration using hedgerows	1	France
Urban area restoration (green and blue infrastructure in urban areas)	1	Multiple countries
Tree-based buffer zones along water bodies and wetlands	1	Kenya
Tree-based buffer zones along roads and riparian land	1	Kenya
Restoration of degraded rangelands	1	Kenya
Subtropical thicket restoration	1	South Africa

Figure 4 below shows reported NPV (positive or negative) by the various CBA studies for the different landscape restoration options. For some of the restoration options, all the studies conducted reported positive NPV; agroforestry (8 studies), soil and water conservation (5), mangrove restoration (3), and alien vegetation clearing (3). However, for some of the restoration strategies, some of the studies reported negative NPV; for reforestation and afforestation, the number of studies that reported positive NPV (4) was equal to those that reported negative NPV (4). For other restoration options—FMNR/ANR and woodlot establishment, the number of studies that reported positive NPV was higher than those which reported negative NPV. None of the restoration strategies had more studies that reported negative NPV compared to those that reported positive NPV.

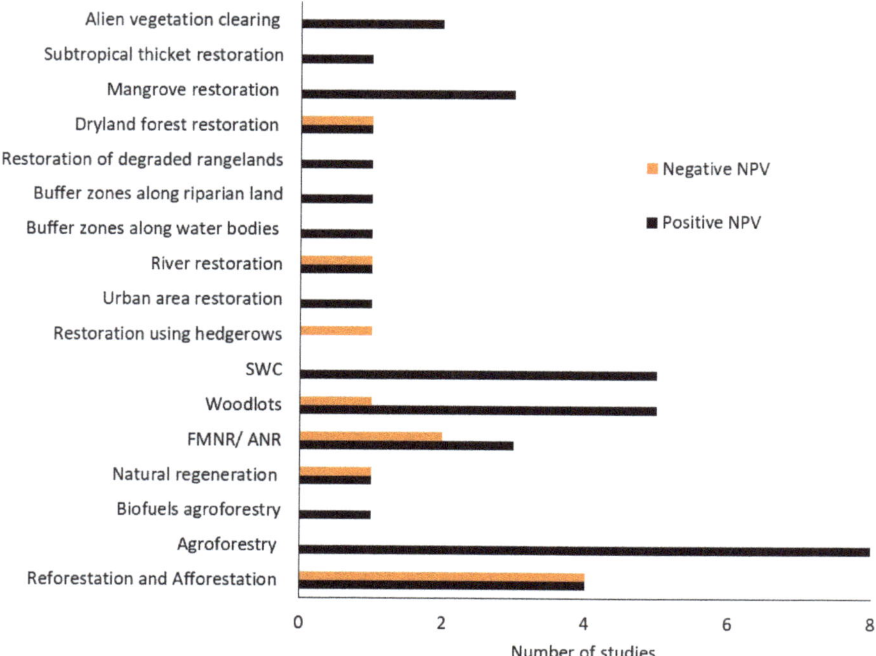

Figure 4. Reported Net Present Value (NPV) for the various landscape restoration type by the reviewed studies.

4.3. Age of Restoration in the CBA Studies

Figure 5 shows the time in years for which different studies under review considered. A higher number of CBA studies, (8) covered restoration benefits and costs for 16–20 years. Another six studies covered between 21–25 years. The maximum duration that was considered was 100 years by two studies, Reference [33] in Mozambique and [30] in France. A further three CBA studies, Reference [34–36], covered restoration benefits and costs for 50 years.

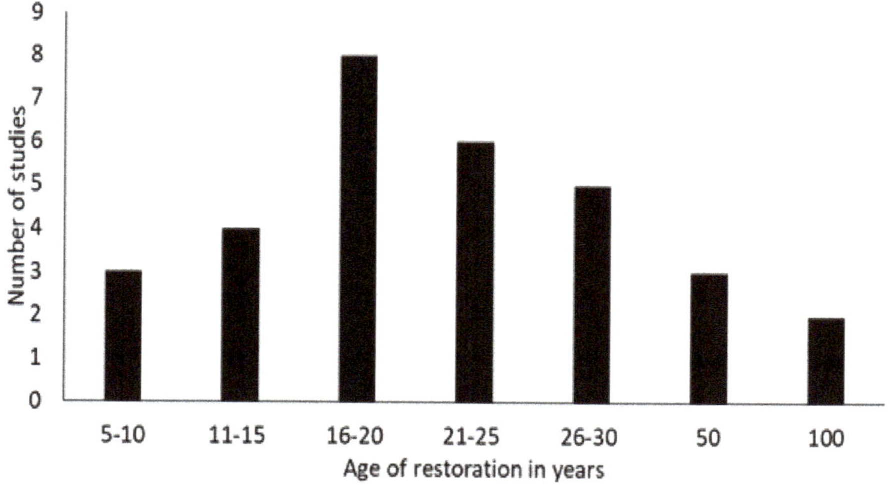

Figure 5. Restoration project timeline considered in the CBA studies.

4.4. Data Source

The CBA studies we reviewed sourced data from different sources; 12 studies used primary data, 11 applied secondary data, and the remaining 8 used both primary and secondary data. For primary data, various studies applied different data collection methods and techniques, as shown in Figure 6. Approximately seven studies used expert discussions, key informants' interviews (KIIs) or focus group discussions (FGDs). Others employed other data collection techniques, including: surveys, reviewing budgets, spatial analysis, and field observations.

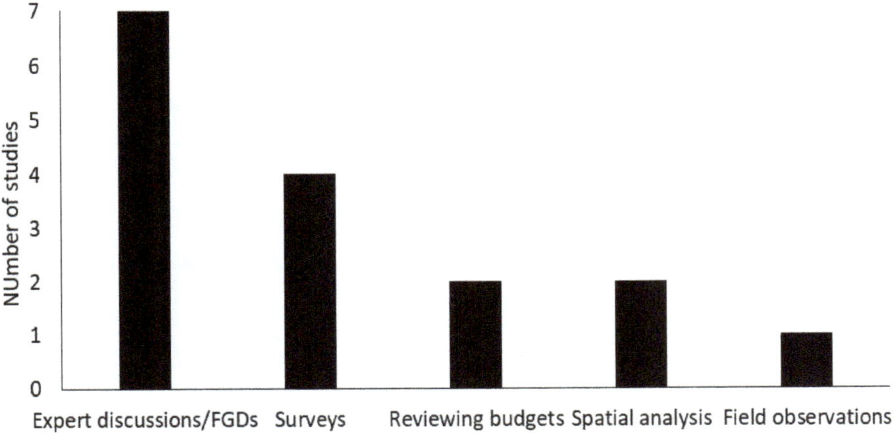

Figure 6. Primary data collection methods employed in the restoration CBA studies.

For those studies that considered indirect use and/or non-use benefits in their analysis, different approaches were applied in valuing these benefits since they mostly do not have a market value. Three studies applied the contingent valuation method, and at least one used either hedonic analysis, travel cost method, or benefit transfer method.

4.5. Benefits and Cost Components in the CBA Studies

Figure 7 shows the proportion of studies that considered different benefits and cost categories. Most of the studies accounted for the use-values only (either direct use, indirect use, or both) and only around 16% accounted for the total economic value of the project (both use and non-use values). About half the studies accounted for both direct and indirect use-values. Of these, the indirect use benefit that was mostly considered in these restoration studies was carbon sequestration. Other indirect use benefits that were accounted for include: erosion control, stormwater control, air regulation, temperature regulation, recreational value, avoided nutrient loss, nitrogen fixation, soil fertility improvement, and aquifer recharge. Non-use values that the reviewed studies accounted for include: aesthetic value, bequest (inheritance) values, and existence value.

The studies applied different valuation methods to value the respective benefits, as shown in Table 3. A more detailed explanation of these benefits by the specific study is provided in Appendix A, Table A1. The provisioning services were considered in almost all the studies. They were commonly assessed by collecting primary/secondary data on the quantities of these goods and valuing them at the current market price. Carbon sequestration was valued by either spatial analysis (e.g., Reference [7,37]) or sourcing the quantities from secondary data and valuing them at the market price of carbon. Contingent valuation method was applied by several studies in valuing benefits, including: air pollution & regulation [38], public benefits [30], cultural, aesthetic & recreation values [35], and inheritance and bequest values [39].

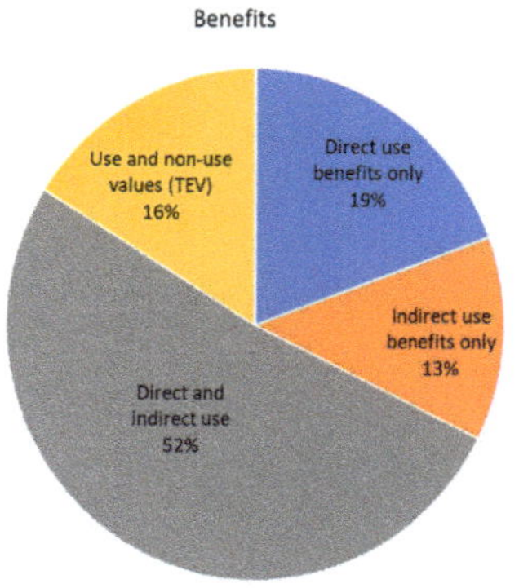

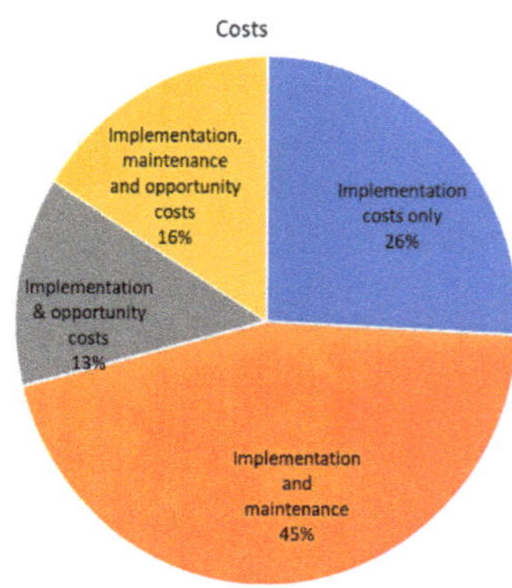

Figure 7. Benefits and cost categories considered in the CBA studies reviewed (*n* = 31).

Table 3. Valuation methods applied across various benefits.

Benefits	Valuation Methods
Timber, wood fuel, non-timber forest products (NTFPs), livestock, yield benefits	- Primary data valued at existing market price
	- Expert opinion
Tourism	- Value of tickets paid at the entrance
	- Travel cost method
	- Spatial analysis
Carbon sequestration	- Secondary data valued at the market price of carbon
Public benefits	- Contingent valuation methods, i.e., assessing willingness to pay for these benefits
	- Hedonic method
Air pollution and air regulation	- Contingent valuation method
	- Benefit transfer
Stormwater control/reduction	- Hedonic methods
	- Benefit transfer
Increased soil fertility	- Replacement cost method
	- Benefit transfer
Soil erosion control	- Avoided cost
	- Benefit transfer
Temperature regulation	- Benefit transfer
Cultural, aesthetic and recreation	- Contingent valuation method; assessing willingness to pay
Positive health effects	- Benefit transfer
Biodiversity recovery	- Secondary data with simulations
	- Benefit transfer
Inheritance and bequest values	- Contingent valuation methods
Total economic value for global studies	- Secondary data and benefit transfer particularly from the TEEB valuation database

Benefit transfer method was frequently applied in valuing many indirect use benefits, such as air pollution & regulation, biodiversity improvement, soil fertility, soil erosion control, temperature control, positive health effects, recreation, and stormwater control. A comprehensive example of the use of benefit transfer is in Reference [27], where they used data collected from 94 studies, from which they created a database of benefits and costs for conducting CBA in seven biomes. In addition, Reference [18] used secondary data and the comprehensive TEEB database due to Reference [40] in valuing benefits while assessing the CBA of forest landscape restoration within the Bonn Challenge for six different biomes. In the absence of site-specific valuation information, benefit transfer is an alternative to estimating non-existing values. It adapts existing valuation information to a new context (location or time), and it is principally useful when there are budget and time constraints with the collection of primary data [24]. However, there is a need to ensure that the ecological conditions are the same; otherwise, the value may be overstated or understated.

Hedonic pricing method was the least frequently applied valuation method; only one study used the approach in valuing stormwater control and air pollution control [34]. The study employed the hedonic models through controlling for the price of housing in different locations. The reason the hedonic method is rarely applied is that the data required can be quite intensive. Similarly, this method works well if markets can pick up quality differentials, which may not be the case for agricultural and forest land, due to the non-observability of some attributes [23]. Further still, Reference [31] applied the replacement cost approach to value the soil fertility and the avoided loss approach to value soil erosion control.

Three cost categories were considered in the studies we reviewed, as shown in Figure 8: implementation costs, maintenance costs (mostly annual costs of maintaining the restoration infrastructure), and opportunity costs (cost of foregone opportunities). All the CBA studies we reviewed considered implementation cost since this is the most direct cost in restoration and is easily captured. Approximately 25% considered only the implementation costs, and a further 45% covered both implementation and maintenance costs—only 16% of the reviewed studies included all the three cost categories. Implementation and maintenance costs were mostly sourced from the costs incurred in the projects. Implementation costs included the investment costs, such as seedlings, materials, and other inputs labor, and training costs, among others. Maintenance cost includes monitoring and transaction costs. For opportunity cost, on the other hand, the studies had to conduct baseline assessments to ascertain the value of foregone opportunities. For example, Reference [41] assessed the baseline situation of agriculture and grazing, which was translated into the opportunity cost of the land.

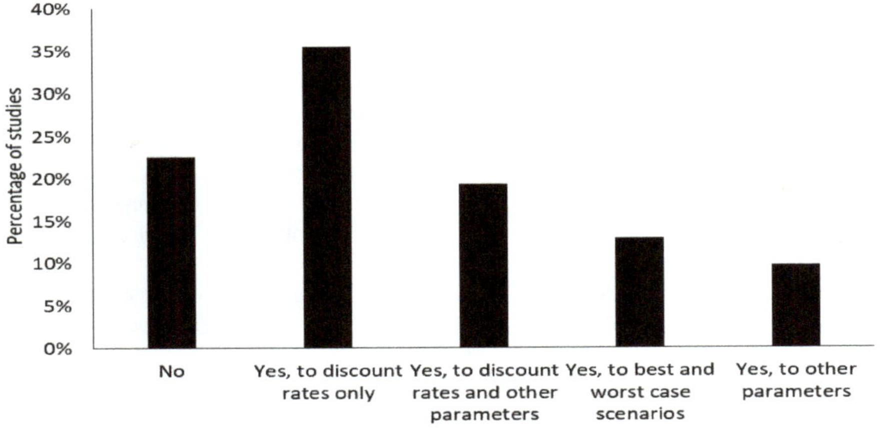

Figure 8. Sensitivity analysis across the CBA studies under review (*n* = 31).

4.6. Sensitivity Analysis

The uncertainties in the entire CBA process, such as fluctuating prices, discount rates, and unseen events within the restoration lifetime, can affect the estimated results especially when the CBA is conducted ex-ante. Hence, after conducting a CBA, it is necessary to conduct a sensitivity analysis by altering various parameters of the estimation, such as the discount rate or prices. Alternatively, one may conduct a Monte Carlo simulation. A Monte Carlo simulation is similar to sensitivity analysis in that it demonstrates how a project's profitability varies. However, instead of altering one input variable and analyzing how that changes in that variable affects the project viability, a Monte Carlo simulation attempts to model uncertainty across multiple inputs assumptions. The model is run thousands of times to understand different possible outcomes and the likelihood of them occurring [19]. Hence, it is more rigorous and robust compared to regular sensitivity analysis.

Figure 8 shows the proportion of studies under review that conducted a sensitivity analysis to test the validity of their results. Of the 31 studies we reviewed, none of the studies applied the rigorous Monte Carlo simulation. Approximately 23% of the studies did not conduct any form of sensitivity analysis. The majority of the studies (about 35%) conducted a sensitivity analysis by varying the discount rates only. Additionally, a further 19% varied the discount rates and other parameters, such as carbon prices, products prices, maintenance costs, and so on. Approximately 13% tested the validity of the CBA results by varying the best- and worst-case scenarios by assuming a very optimistic scenario where most of the assumptions hold and a very pessimistic scenario where most of the assumptions do not hold. Still, 10% of the reviewed studies only varied other parameters without varying the discount rate.

5. Discussion

We present a discussion of the results with respect to the quantity and quality of CBA for landscape restoration and the implications for moving forward with restoration in terms of planning, policies, and investments. There is a need to grow the quantity of evidence in specific areas, as well as economic evidence to make restoration attractive to investors. A number of areas for improvement emerge from our analysis. These include capturing all costs categories; going beyond direct use values; capturing public benefits; sensitivity analysis; and the need for standardization and or guidance. We briefly discuss each below.

5.1. Capturing All Costs Categories

Most of the studies do not account for all the costs associated with restoration, thus overstating the profitability of restoration. All the reviewed studies accounted for implementation costs, but relatively few covered all the cost categories (16%). The least accounted for cost category is the opportunity cost; probably because it is often difficult to estimate this cost since it is not a direct cost. Estimating the opportunity cost requires a baseline assessment to identify the foregone uses of the land. For example, Reference [41] assessed the baseline situation of agriculture and grazing, which was translated into the opportunity cost of the land. Most of the studies do not conduct baseline assessments since this involves committing more resources and time, making it challenging to provide an estimate of the opportunity cost. In addition, for some land uses, the opportunity cost may be negligible, especially if the land is highly degraded. However, degraded lands with high surface runoff can be used for water harvesting, providing a positive externality that may get lost if infiltration and vegetation water use increase as a result of landscape restoration [42].

In addition, there is a need to include maintenance and monitoring costs in accounting for the total economic costs. Most restoration projects fail to account for maintenance and monitoring costs since they view restoration as a one-time cost activity as opposed to a continuous activity—for example, tree planting as opposed to tree growing [43]. Tree planting is a one-time cost activity where only the implementation cost will be significant. On the other hand, tree-growing is a continuous activity, implying that maintenance and monitoring costs are significant and accountable, as well. Hence, all the three cost categories ought to be accounted for, in order for a cost-benefit analysis to reflect the actual economic viability of a restoration project.

5.2. Going Beyond Direct Use Values

A major challenge in conducting an economic CBA is that it is difficult and controversial to monetize social and environmental benefits. Environmental benefits are valued differently by different stakeholders, which disputes the findings of the analysis. For instance, greenhouse gases (GHG) mitigation, which is a global ecosystem service, maybe prioritized over reducing soil erosion which is a local benefit. Additionally, the value of some environmental benefits, such as supporting biodiversity, are often excluded because monetizing them is challenging [19]. However, it is still possible to include non-monetized values, such as biodiversity, in the decision-making.

Based on the stocktake, a significant proportion of existing CBA studies do not account for indirect use benefits, and an even more substantial proportion do not account for non-use benefits (84%). This is mostly because most of these benefits are "invisible" and do not have a market value, thus making valuing them quite challenging. Some of the studies that included the indirect and non-use values applied various methods for evaluating them, including: benefit transfer, contingent valuation method, travel cost method, hedonic pricing models, replacement cost, avoided loss, and spatial analysis. Benefit transfer and contingent valuation methods were frequently applied compared to the other methods. This is probably because, these methods can be applied in the valuation of almost all the benefits, such as air pollution control, biodiversity, inheritance and bequest value, cultural and aesthetic values, and so on [23,24]. Similarly, benefit transfer is generally cost-effective and is commonly applied when there is limited budgetary allocation.

On the other hand, some of the other methods (e.g., hedonic models, avoided loss, replacement costs) are data-intensive and, cannot be universally applied for most of the benefits. For example, only one study, Reference [34], employed the hedonic models in valuing stormwater control and air pollution control through controlling for the price of housing in different locations. This method works well if markets can pick up quality differentials, which may not be the case for agricultural and forest land, due to the non-observability of some attributes [23]. Hence, there is a need for understanding the methods that can be used to reasonably value the specific benefits arising from the project depending on the nature of the benefit/ecosystem services data availability, time and cost constraint, and so on. Conducting a comprehensive economic CBA for restoration is costly and time-consuming; thus, there is a need for restoration projects to budget for this.

5.3. Capturing Public Benefits and Implications for Government Investments in Restoration

In addition,, some of these benefits are public benefits attributable to other stakeholders beyond those directly targeted by the restoration projects. A comprehensive economic CBA ought to account for all these benefits. Otherwise, the estimated NPVs for these restoration projects are undervalued. Hence, to present a true picture of the profitability of restoration projects, future CBA studies should aim to capture all the benefits arising from the restoration projects—use and non-use benefits, as well as private and public benefits. This particularly useful for large-scale restoration projects where the benefits accrue to the broader public beyond the targeted stakeholders. Reference [20] also found that when the value of public goods and services are accounted for in the cost-benefit analysis, the benefits of large-scale restoration outweigh the costs and targets like the Bonn Challenge can be met efficiently.

There is also a need for information on public benefits to drive government investments in restoration. Investors require good information on costs and benefits for investment proofing and decision-making. To this end, there is a need for a cost-benefit analysis (CBA) database compiling existing data on landscape restoration costs and benefits more so information on indirect and public benefits [18].

5.4. Improving Sensitivity Analysis

CBA of restoration attempts to model or estimate the future; therefore, a certain degree of uncertainty is involved. For example, unforeseen events and climate change may affect productivity in ways that are difficult to predict, and they should be considered. Similarly, applying a discount rate is also an inherently subjective decision, but it is important for prioritizing near-term benefits versus long-term benefits [19]. Thus, CBA is based on certain assumptions that vary in their degree and level of confidence. The assumptions made during CBA include political and/or social assumptions that may not necessarily hold.

This calls for the need for sensitivity analysis in the CBA process to provide a robustness check for the results. While conducting sensitivity analysis, almost all the existing studies we reviewed conducted a direct sensitivity analysis by varying only one or just a few variables, mostly the discount rate, e.g., Reference [7,27,32] or carbon prices, e.g., Reference [35]. None of the studies applied a more rigorous sensitivity test, such as the Monte Carlo simulation approach. For more robust and comparable results, future CBA on restoration should consider more rigorous approaches for sensitivity analysis, such as the Monte Carlo simulation.

5.5. The Need for Standardization and Guidelines

During CBA of land restoration, it is difficult to monetize social or political considerations. Restoration options selected during CBA should produce maximum benefits for all, but this is usually not the case. Due to political reasons, benefits to one group may be valued more than the benefits of another group and such is not usually included in the CBA [19]. Data collection to be used for CBA of restoration can be time-consuming and expensive since the impacts of restoration transitions are felt over long-time periods. One requirement for a comprehensive CBA is to quantify all the impacts for each land use (degraded and restored) for the relevant time horizon of the project [21]. Predictions about the levels of inputs (i.e., costs) and the production of ecosystem services must be made for each year and each land use in a restoration transition. This can be the most challenging aspect of CBA because there is not always a complete scientific understanding of how complex natural systems work, especially when significant changes to their structure are made.

Closely related, lack of reliable data owing to poor data-keeping during the restoration period also affects the CBA results. It takes time to realize the actual profitability of these restoration investments since returns to landscape restoration projects are not immediate. For example, in the studies we reviewed, the restoration age considered was even up to 100 years for some projects, with the minimum being seven years. This requires data over several years, and most projects do not keep a record of this data. Hence, even for ex-post CBA evaluations, a lot of predictions and assumptions are involved in data generation. Thus, there is need to adopt standardized methods of data prediction if the results are to be comparable across different restoration projects in deciding the allocation of funds. Similarly, the Food and Agriculture Organization (FAO)"The Economics of Ecosystem Restoration" (TEER), points to need for a comprehensive tool on costs and benefits of Ecosystem Restoration (and FLR) interventions[10]. In an on-going project, TEER aims to "offer a reference point for the estimation of costs and benefits of future ecosystem restoration projects in all major biomes, based on information from comparable projects on which data are collected through a standardized framework". Of course, the question of whether a standardized approach is feasible remains, considering the diversity in landscapes and land-use practices. But such an initiative is a good starting point for providing the missing database on all costs and benefits categories, particularly the indirect use and non-use benefits, as well as maintenance and opportunity costs categories, which are rarely captured in CBA. It can also provide guidelines to be used broadly by supporting donors, investors, and a wider range of stakeholders10.

5.6. Economic Attractiveness of Restoration and Implications for Private and Impact Investments

From the analysis, proportionately, more studies reported positive NPVs for most of the restoration strategies. In fact, for some restoration options, all the studies conducted reported positive NPV: agroforestry, soil and water conservation, mangrove restoration, and alien vegetation clearing. None of the restoration strategies had more studies that reported negative NPV compared to those that reported positive NPV. Positive NPV and economic viability, as confirmed in these studies, is a good starting point for promoting investments and financing. Investors will only be attracted to landscape restoration if their risks are covered, or at least mitigated to an acceptable level [18]. CBA is a first step in documenting the economic viability of landscape restoration and providing empirical evidence that in the long-term, benefits accruing from restoration outweigh the vast investment costs associated with landscape restoration.

---

[10] https://www.vi-med.forestweek.org/sites/default/files/presentations/docs/c5-teer-garavaglia.pdf

5.7. Insufficient CBA Evidence on Landscape Restoration

Compared to the relatively large number of restoration projects and studies, few have conducted comprehensive CBA. These studies were skewed towards some regions and some restoration options. For example, almost half of the existing studies were conducted in Africa. Similarly, substantially more studies focused on reforestation & afforestation and agroforestry; this is probably because these are among the common landscape restoration options. However, there remains a gap in CBA studies for other popular restoration options, including soil and water conservation practices and establishment of woodlots.

Nonetheless, some of the studies took a global focus, and some assessed and compared CBA results over many restoration strategies. For example, one of the most comprehensive CBA study we reviewed was conducted in Kenya by Reference [31] and compared returns for several landscape restoration strategies in different landscapes. In addition, Reference [20] took a global focus by assessing the net present value of the Bonn Challenge. Such studies form a good starting point for building a comprehensive CBA database upon which resource allocation in restoration can be based.

Overall, a major reason for the relatively few CBA studies is because most restoration projects do not budget for a CBA study. This may be due to an assumption that such projects always yield positive gains which may not necessarily hold. However, owing to scarce resources and the growing global demand for restoration, CBA studies can provide empirical evidence of restoration options with good returns on investment under different landscapes. Conducting a comprehensive economic CBA for landscape restoration is costly, data-intensive, and time-consuming; thus, there is a need for restoration projects to budget for this adequately.

6. Conclusions

This study set out to understand the breadth and depth of current CBA applications in landscape restoration, in a bid to find ways of improving its usefulness in planning, investments, and policies related to land restoration. Thirty-one out of 2056 studies were found to meet the CBA study selection criteria, i.e., they had conducted an economic CBA on at least one landscape restoration strategy. Three of these studies were of global character, while more than half covered African countries, with about two each covering Europe, Asia, Latin America, and the Middle East, respectively. Agroforestry, afforestation, reforestation, and assisted natural regeneration seem to be the most studied with at least five studies each. Other forms of land restoration are lagging. Most studies show a positive NPV for at least one restoration option, pointing to and confirming that restoration can be a viable private investment. Because most studies do not capture public benefits, evidence for public investments remains thin and could potentially hamper prioritization of government investments where resources are scarce. The study also identifies a number of areas for improvement in CBA from the stocktake. These include capturing all costs categories, including opportunity costs and maintenance and monitoring costs; going beyond direct use values; capturing public benefits; conducting thorough sensitivity analysis; the need for standardization and or guidance; and the insufficient CBA evidence on landscape restoration. Overall, the limited extent and depth in landscape restoration CBA studies suggest a great need to improve both quantity and quality in order to better inform planning, policies and investments in landscape restoration.

Author Contributions: Conceptualization, P.W., P.A.M., and L.D.; methodology, P.W. and E.G.; validation, P.W., P.A.M., and E.G.; formal analysis, P.W. and E.G.; investigation, E.G.; resources, P.W.; data curation E.G. and P.W.; writing—original draft preparation, P.W.; writing—review and editing, P.W., P.A.M, E.G., and L.D.: supervision, P.A.M.; project administration, P.M.; funding acquisition, P.A.M. and L.D. All authors have read and agreed to the published version of the manuscript.

Funding: This work was funded by the Consultative Group on International Agricultural Research (CGIAR) Research Program on Forests Trees and Agroforestry (FTA).

Appendix A

Table A1. A summary of studies conducted to assess CBA of Landscape restoration following a systematic review.

No.	Paper	Country	Type of Restoration	Data Source	Years	Benefits	Costs	Net Present Value	Sensitivity Analysis	Private or Communal
1	Baig et al. (2016) [44]	Philippines	Ecosystem-based adaptation using mangrove protection and planting	Secondary data	20	Total economic values (direct use, non-direct use values and non-use values	Implementation costs,	Positive	Yes, to discount rate	Communal
2	Becker et al. (2018) [39]	Israel	Full and partial river restoration	Primary data through CVM and travel cost methods	varies	Total economic values; Use values (direct and optional use values) and non-use values (inheritance (bequest) and existence values)	Restoration costs (fixed value and the yearly value of maintenance)	Positive	No	Communal
3	Birch et al. (2010) [7]	Latin America	Dryland forest restoration	Primary data through spatial analysis	20	Carbon sequestration, NTFPs, timber, tourism and livestock products (benefit between restoration and BAU scenario)	Implementation costs (fencing and fire suppression), opportunity costs (cost foregone from livestock production from forest expansion	Positive	Yes, to discount rates and market price of carbon	Private and communal
4	Bonnieux and Le Goffe (1997) [30]	France	Landscape restoration using hedgerows	WTP (to assess public benefits	100	Firewood, timber and public benefits	Planting and regenerating costs, maintenance costs	Negative	No	Communal
5	Chadourne et al. (2012) [34]	USA (Tennessee)	Forest landscape restoration	Primary data-hedonic models were used to	50	Indirect use values (air pollution mitigation and stormwater control)	Explicit costs (land acquisition, labor, seedlings, materials) and amenity value	Positive	No	Private and communal
6	De Groot et al. (2013) [27]	Global	Restoration of 9 different biomes	Secondary data Reviews of 94 studies	20	Total Economic value of all services	Implementation costs, Maintenance costs	Positive for the various restoration types considered	Yes, to discount rate and to worst-case and best-case scenarios	Communal
7	ELD Initiative and UNEP (2015) [29]	Africa (FDjibouti,	Sustainable land management against soil erosion	Secondary data mostly from FAO and world bank data	15	Avoided crop damages from erosion control	SLM establishment cost and SLM maintenance costs	Positive	Yes, by varying discount rates, prices of cereals, capital and maintenance costs.	Private and communal

Table A1. *Cont.*

No.	Paper	Country	Type of Restoration	Data Source	Years	Benefits	Costs	Net Present Value	Sensitivity Analysis	Private or Communal
8	Elmqvist et al. (2015) [28]	Global	Urban areas restoration (Green and blue infrastructure in urban areas)	Secondary data (review)-Benefit transfer	20	Ecosystem services (pollution and air regulation, carbon sequestration, stormwater reduction, temperature regulation, recreation, positive health effects)	Costs for planning, preparation, modest soil restoration, plant propagation, and planting both for grasslands and woodlands	Positive	Yes, to discount rates and max/min benefits and costs	Communal
9	Verdone and Seidl (2017) [20]	Global	CBA of FLR within the Bonn Challenge within six different biomes	Primary and secondary data from TEEB	varies	Total Economic value both direct and indirect benefits	Bonn challenge cost of restoration	Positive	No	Private and communal
10	Gasparinetti et al. (2019) [32]	Brazil (South Amazon)	FLR through agroforestry with cocoa, coffee and Guarana	Primary data	30	Direct outputs and ecosystem services	Maintenance, fencing, labor costs, machine costs	Positive	Yes, to discount rates	Private
11	Hofer et al. (2010) [45]	Brazil (Amazon)	Reforestation (land use from pastures to forest for carbon sequestration (carbon for credits)	Secondary data	20	Carbon sequestration	Opportunity, implementation, and transaction costs	Negative	Yes, to different carbon prices	Communal
12	Narayan et al. (2017) [33]	Mozambique	Mangrove restoration to shelter against storms and flooding	Secondary data from an adaptation project conducted in 2013	100	Reduction in storm damages to houses, fish production, aquaculture, apiculture, carbon sequestered by growing mangroves	Costs of buying the seedlings; labor for planting, maintenance, and support staff; and hydrological restoration	Positive	Yes, to different carbon prices and discount rates	Private and communal
13	Newton et al. (2012) [35]	UK	Habitat restoration for river catchment	Primary data	10 and 50 years	Marginal value of benefits- carbon, timber, crops, livestock and recreational, aesthetic and cultural values	Initial capital investment and annual maintenance costs	Negative	Yes, to different carbon prices and discount rates	Communal
14	Pistorius et al. (2017) [37]	Ethiopia	FLR-(1) Afforestation/reforestation (2) participatory forest management (3) sustainable woodland management (4) restoration of afro-alpine or sub-afro-alpine (5) establishment of woodlots	Primary data-spatial analysis and expert opinion	20	Provisioning services (timber & NTFPs), Carbon sequestration	Investment costs and labor costs	Positive except for the afro-alpine slope restoration	No	Communal

Table A1. *Cont.*

No.	Paper	Country	Type of Restoration	Data Source	Years	Benefits	Costs	Net Present Value	Sensitivity Analysis	Private or Communal
15	Mills et al. (2007) [36]	South Africa	Restoration of natural capital through Subtropical Thicket Restoration	Secondary data with simulations	50	livestock and game production, harvesting plant products (assuming natural recovery of biodiversity), and carbon sequestration	Transaction costs (including costs of verification of carbon stocks), labor costs, opportunity costs	Positive	Yes, to various parameters including biomass growth rate,	Communal
16	Holmes et al. (2007) [46]	South Africa	Restoring Natural Capital Following Alien Plant Invasions in Fynbos Ecosystems	Projections from secondary data	30	Direct and indirect use benefits	Clearing costs, installation costs	Positive	Yes, to discount rates	Communal
17	Rizzetti et al. (2018) [47]	Vietnam	FLR through ANR, extended acacia rotation, native species rotation, SWC	Projections from secondary data	23, 30, 2	Crop income, income from timber	Labor costs, seedling cost	Positive	No	Private and communal
18	Schiappacasse et al. (2012) [38]	Chile	Dryland forest restoration thru reforestation using native trees	Primary data using contingent valuation method	25	WTP for forest restoration for the entire pollution of the city	Implementation costs, operating costs,	Negative	Yes, to discount rate	Communal
19	Currie et al. (2009) [48]	South Africa	Alien vegetation clearing for water yield and tourism	Primary data and projections	15	water and tourism benefits (tourism benefits involved revenue from the sale of tickets)	Costs of alien invasive plant removal, gully-erosion repair and reseeding with indigenous plants	Positive	Yes, to discount rates and with realistic and pessimistic scenarios	Communal
20	Silva and Nunes (2017) [49]	Brazil Amazon	Forest restoration through sustainable forest management (legal logging) and agroforestry	Secondary	11	Timber from logging, financial benefits of AFS (timber and NTFPs)	Implementation costs, transaction costs, opportunity costs (loss from agriculture and livestock)	Negative	Yes, to discount rates and different scenarios	Communal
21	Tuan and Tinh (2013) [50]	Vietnam	Mangrove restoration	Secondary and primary using CVM to value non-use values and market methods to value use values	22	Direct use values, indirect use values and non-use values. WTP for non-use values,	Maintenance and protection costs, mangrove restoration	Positive	Yes, to discount rates	Private and communal
22	Monela (2005) [51]	Tanzania	FLR through agroforestry and silviculture (Ngitili)	Primary data through expert evaluation and literature review	20	Direct use values (timber and other NTFPs), time saved in collecting firewood and water,	Total project cost for restoration	Positive	Yes, to discount rates	Private and communal

Table A1. *Cont.*

No.	Paper	Country	Type of Restoration	Data Source	Years	Benefits	Costs	Net Present Value	Sensitivity Analysis	Private or Communal
23	Cheboiwo et al. (2019) [31]	Kenya	Afforestation or reforestation of degraded natural forests, Rehabilitation of degraded natural forests, Agroforestry in cropland, Commercial tree and bamboo growing on potentially marginal cropland and un-stocked forest plantation forests, Tree-based buffer zones along water bodies and wetlands, Tree-based buffer zones along roads and restoration of degraded rangelands)	Expert discussions, activity restoration budgets and extensive review of various land use literature. Benefits and opportunity costs were valued using market prices, avoided cost/replacement cost and benefit transfer approaches	30	Direct (crop harvests, timbers and NTFPs) and indirect use values (carbon sequestration, soil erosion control and increased soil fertility)	Implementation costs, opportunity costs, monitoring and maintenance costs	Positive for the various restoration types considered	Yes, to discount rates	Private and communal
24	Ministry of Natural resources, energy and mining-Malawi (2017) [3]	Malawi	Conservation agriculture, agroforestry, FMNR, Community plantations and private woodlots, Natural forest management	Primary data	20	Direct and indirect use benefits	Implementation costs and opportunity costs	Positive for the various restoration types considered	Yes, to discount rates	Private and communal
25	Ministry of Water & Env-Uganda (2016) [52]	Uganda	Reforestation and afforestation, woodlots, Agroforestry, Natural regeneration	Budgets, expert discussions and secondary sources	30	Direct and indirect use benefits	Implementation costs	Positive for the various restoration types	Yes, to discount rates	Private and communal
26	FAO and UNHCR (2018) [53]	Tanzania	Wood-energy rehabilitation (Afforestation and reforestation), Agroforestry, Rehabilitation of degraded native forests	Primary data (field observations, KIIs, FGDs) secondary data	10	Direct benefits (wood fuel) and indirect use benefits (carbon sequestration)	Implementation, operational and opportunity costs	Positive for wood energy plantations and agroforestry but negative for the rest	Yes, to discount rates and wood prices	Private and communal
27	Aymeric et al. (2014) [54]	Sudan	SLM through *A. senegal* Agroforestry	Secondary data	25	Direct use benefits (fuelwood and Gum arabica) and indirect use benefits (N fixation, avoided nutrient loss, aquifer recharge, carbon sequestration)	Implementation and maintenance costs	Positive	Yes, to discount rates	Private

Table A1. *Cont.*

No.	Paper	Country	Type of Restoration	Data Source	Years	Benefits	Costs	Net Present Value	Sensitivity Analysis	Private or Communal
28	Ministry of Natural resources-Rwanda (2014) [55]	Rwanda	Agroforestry, Well managed woodlots, Natural forest regeneration, protective forests	Primary and secondary data through simulations and predictions	20–30	Direct use benefits (crops, wood) and indirect use benefits (carbon sequestration and erosion control	Implementation, operational and monitoring costs		Yes, to discount rates	Private and communal
29	Tesfaye et al. (2016) [56]	Ethiopia	Soil conservation measures (soil bunds, stone bunds, Fanya juu bunds)	Primary data	27	Yield increment from implementation of bunds	Investment and maintenance costs	Positive	Yes, to investment and maintenance costs and the market price of yield	Private
30	Wiskerke et al. (2010) [41]	Tanzania	A small-scale forestation project for carbon sequestration, a short rotation woodlot and a Jatropha plantation	Primary data (expert opinions and field survey) and secondary data	7	Direct benefits (wood fuel, electricity from jatropha, etc.) and indirect benefits (avoided deforestation, improved health, indirect economic benefits)	Production costs and opportunity costs	Positive for woodlots and jatropha for electrification and soap production, negative for forestation for C credits	No	Private and communal
31	Onduru and Muchena (2011) [57]	Kenya	SWC practices, such as mulching, zero tillage, stone lines, contour ridges, micro catchments with bananas, terracing, and others	Primary data	15	Incremental yield benefits from the adoption of SWC practices	Investment and maintenance costs	Positive	Yes, to discount rates	Private

References

1. Intergovernmental Platform on Biodiversity and Ecosystem Services (IPBES). *Summary for Policymakers of the Global Assessment Report on Biodiversity and Ecosystem Services of the Intergovernmental Science-Policy Platform on Biodiversity and Ecosystem Services*; IPBES Secretariat: Bonn, Germany, 2019.
2. Ding, H.; Altamirano, J.C.; Anchondo, A.; Faruqi, S.; Verdone, M.; Wu, A.; Ortega, A.A.; Zamora Cristales, R.; Chazdon, R.; Vergara, W. *Roots of Prosperity: The Economics and Finance of Restoring Land*; World Resources Institute: Washington, DC, USA, 2017.
3. Ministry of Natural Resources; Energy and Mining—Malawi. *Forest Landscape Restoration Opportunities Assessment for Malawi*; NFLRA: Lilongwe, Malawi; IUCN: Gland, Switzerland; WRI: Washington, DC, USA, 2017.
4. Crookes, D.J.; Blignaut, J.N.; De Wit, M.P.; Esler, K.J.; Le Maitre, D.C.; Milton, S.J.; Mitchell, S.A.; Cloete, J.; De Abreu, P.; Fourie, H.; et al. System dynamic modelling to assess economic viability and risk trade-offs for ecological restoration in South Africa. *J. Environ. Manag.* **2013**, *120*, 138–147. [CrossRef]
5. Gichuki, L.; Brouwer, R.; Davies, J.; Vidal, A.; Kuzee, M.; Magero, C.; Walter, S.; Lara, P.; Oragbade, C.; Gilbey, B. *Reviving Land and Restoring Landscapes: Policy Convergence between Forest Landscape Restoration and Land Degradation Neutrality*; IUCN: Gland, Switzerland, 2019.
6. Yirdaw, E.; Tigabu, M.; Monge, A. Rehabilitation of degraded dryland ecosystems—Review. *Silva Fenn.* **2017**, *51*, 1–32. [CrossRef]
7. Birch, J.C.; Newton, A.C.; Aquino, C.A.; Cantarello, E.; Echeverría, C.; Kitzberger, T.; Schiappacasse, I.; Garavito, N.T. Cost-effectiveness of dryland forest restoration evaluated by spatial analysis of ecosystem services. *Proc. Natl. Acad. Sci. USA* **2010**, *107*, 21925–21930. [CrossRef]
8. Society for Ecological Restoration International Science & Policy Working Group. *The SER International Primer on Ecological Restoration*; Society for Ecological Restoration International: Washington, DC, USA, 2004; Volume 2, pp. 206–207.
9. Ciccarese, L.; Mattsson, A.F.; Pettenella, D. Ecosystem services from forest restoration: Thinking ahead. *New For.* **2012**, *43*, 543–560. [CrossRef]
10. Wortley, L.; Hero, J.-M.; Howes, M.J. Evaluating Ecological Restoration Success: A Review of the Literature. *Restor. Ecol.* **2013**, *21*, 537–543. [CrossRef]
11. Chazdon, R.L.; Brancalion, P.H.S.; Lamb, D.; Laestadius, L.; Calmon, M.; Kumar, C. A Policy-Driven Knowledge Agenda for Global Forest and Landscape Restoration. *Conserv. Lett.* **2016**, *10*, 125–132. [CrossRef]
12. Menz, M.H.M.; Dixon, K.W.; Hobbs, R.J. Hurdles and Opportunities for Landscape-Scale Restoration. *Science* **2013**, *339*, 526–527. [CrossRef] [PubMed]
13. Shono, K.; Cadaweng, E.A.; Durst, P.B. Application of Assisted Natural Regeneration to Restore Degraded Tropical Forestlands. *Restor. Ecol.* **2007**, *15*, 620–626. [CrossRef]
14. Holl, K.; Aide, T. When and where to actively restore ecosystems? *For. Ecol. Manag.* **2011**, *261*, 1558–1563. [CrossRef]
15. Suding, K.N.; Higgs, E.; Palmer, M.; Callicott, J.B.; Anderson, C.B.; Baker, M.; Gutrich, J.J.; Hondula, K.L.; LaFevor, M.C.; Larson, B.M.H.; et al. Committing to ecological restoration. *Science* **2015**, *348*, 638–640. [CrossRef]
16. Dellasala, D.A.; Martin, A.; Spivak, R.; Schulke, T.; Bird, B.; Criley, M.; Van Daalen, C.; Kreilick, J.; Brown, R.; Aplet, G. A Citizen's Call for Ecological Forest Restoration: Forest Restoration Principles and Criteria. *Ecol. Restor.* **2003**, *21*, 14–23. [CrossRef]
17. Van Noordwijk, M.; Gitz, V.; Minang, P.A.; Dewi, S.; Leimona, B.; Duguma, L.; Pingault, N.; Meybeck, A. People-Centric Nature-Based Land Restoration Through Agroforestry: A Typology. *Land* **2020**, *9*, 251. [CrossRef]
18. FAO; Global Mechanism of the UNCCD. *Sustainable Financing for Forest and Landscape Restoration*; Discussion paper; Food and Agriculture Organization: Rome, Italy, 2015.
19. Gromko, D.; Pistorius, T.; Seebauer, M.; Braun, A.; Meier, E. *Economics of Forest Landscape Restoration. Estimating impacts, costs and benefits from ecosystem services*; Unique Forestry and Land Use: Freiburg, Germany, 2019.
20. Verdone, M.; Seidl, A. Time, space, place, and the Bonn Challenge global forest restoration target. *Restor. Ecol.* **2017**, *25*, 903–911. [CrossRef]

21. Verdone, M. *A Cost-Benefit Framework for Analyzing Forest Landscape Restoration Decisions*; IUCN: Gland, Switzerland, 2015.

22. Pascual, U.; Muradian, R.; Brander, L.; Gómez-Baggethun, E.; Martín-López, B.; Verma, M.; Armsworth, P.; Christie, M.; Cornelissen, H.; Eppink, F.; et al. The economics of valuing ecosystem services and biodiversity. In *The Economics of Ecosystems and Biodiversity: Ecological and Economic Foundations*; UNEP: Hertfordshire, UK, 2010; pp. 183–256.

23. Gundimeda, H.; Markandya, A.; Bassi, A.M. TEEBAgriFood methodology: An overview of evaluation and valuation methods and tools. In *TEEB for Agriculture & Food: Scientific and Economic Foundations*; UN Environment: Geneva, Switzerland, 2018.

24. Chiputwa, B.; Ihli, H.J.; Wainaina, P.; Gassner, A. Accounting for the invisible value of trees on farms through valuation of ecosystem services. In *The Role of Ecosystem Services in Sustainable Food Systems*; Elsevier BV: Amsterdam, The Netherlands, 2020; pp. 229–261.

25. Ring, I.; Hansjürgens, B.; Elmqvist, T.; Wittmer, H.; Sukhdev, P. Challenges in framing the economics of ecosystems and biodiversity: The TEEB initiative. *Curr. Opin. Environ. Sustain.* **2010**, *2*, 15–26. [CrossRef]

26. Malkamäki, A.; D'Amato, D.; Hogarth, N.J.; Kanninen, M.; Pirard, R.; Toppinen, A.; Zhou, W. A systematic review of the socio-economic impacts of large-scale tree plantations, worldwide. *Glob. Environ. Chang.* **2018**, *53*, 90–103. [CrossRef]

27. De Groot, R.S.; Blignaut, J.N.; Van Der Ploeg, S.; Aronson, J.; Elmqvist, T.; Farley, J. Benefits of Investing in Ecosystem Restoration. *Conserv. Biol.* **2013**, *27*, 1286–1293. [CrossRef]

28. Elmqvist, T.; Setälä, H.; Handel, S.; Van Der Ploeg, S.; Aronson, J.K.; Blignaut, J.N.; Gomez-Baggethun, E.; Nowak, D.J.; Kronenberg, J.; De Groot, R. Benefits of restoring ecosystem services in urban areas. *Curr. Opin. Environ. Sustain.* **2015**, *14*, 101–108. [CrossRef]

29. ELD Initiative; UNEP. The Economics of Land Degradation in Africa: Benefits of Action Outweigh the Costs. A Complementary Report to the ELD Initiative. 2015. Available online: www.eld-initiative.org (accessed on 20 January 2020).

30. Bonnieux, F.; Le Goff, P. Valuing the Benefits of Landscape Restoration: A Case Study of the Cotentin in Lower-Normandy, France. *J. Environ. Manag.* **1997**, *50*, 321–333. [CrossRef]

31. Cheboiwo, J.; Langat, D.; Muga, M.; Kiprop, J. *Economic Analysis of Forest Land Restoration Options in Kenya*; Ministry of Environment and Forestry: Nairobi, Kenya, 2019.

32. Gasparinetti, P.; Brandão, D.O.; Araujo, V.; Araujo, N. *Economic Feasibility Study for Forest Landscape Restoration Banking Models: Cases from Southern Amazonas State*; CSF-Brazil and WWF-Brazil: Brasilia, Brazil, 2019.

33. Narayan, T.; Foley, L.; Haskell, J.; Cooley, D.; Hyman, E. *Cost-Benefit Analysis of Mangrove Restoration for Coastal Protection and an Earthen Dike Alternative in Mozambique*; USAID: Washington, DC, USA, 2017.

34. Chadourne, M.H.; Cho, S.-H.; Roberts, R.K. Identifying Priority Areas for Forest Landscape Restoration to Protect Ridgelines and Hillsides: A Cost-Benefit Analysis. *Can. J. Agric. Econ. Can. D'agroeconomie* **2012**, *60*, 275–294. [CrossRef]

35. Newton, A.C.; Hodder, K.; Cantarello, E.; Perrella, L.; Birch, J.C.; Robins, J.M.; Douglas, S.J.; Moody, C.E.; Cordingley, J. Cost-benefit analysis of ecological networks assessed through spatial analysis of ecosystem services. *J. Appl. Ecol.* **2012**, *49*, 571–580. [CrossRef]

36. Mills, A.J.; Turpie, J.K.; Cowling, R.M.; Marais, C.; Kerley, G.I.; Lechmere-Oertel, R.G.; Sigwela, A.M.; Powell, M. Assessing costs, benefits, and feasibility of restoring natural capital in subtropical thicket in South Africa. In *Restoring Natural Capital, Science, Business, and Practice*; Island Press: Washington, DC, USA, 2007; pp. 179–187.

37. Pistorius, T.; Carodenuto, S.; Wathum, G. Implementing Forest Landscape Restoration in Ethiopia. *Forests* **2017**, *8*, 61. [CrossRef]

38. Schiappacasse, I.; Nahuelhual, L.; Vásquez, F.; Echeverría, C. Assessing the benefits and costs of dryland forest restoration in central Chile. *J. Environ. Manag.* **2012**, *97*, 38–45. [CrossRef] [PubMed]

39. Becker, N.; Greenfeld, A.; Shamir, S.Z. Cost–benefit analysis of full and partial river restoration: The Kishon River in Israel. *Int. J. Water Resour. Dev.* **2018**, *35*, 871–890. [CrossRef]

40. Van der Ploeg, S.; de Groot, R.S. *The TEEB Valuation Database–A Searchable Database of 1310 Estimates of Monetary Values of Ecosystem Services*; Foundation for Sustainable Development: Wageningen, The Netherlands, 2010.

41. Wiskerke, W.; Dornburg, V.; Rubanza, C.; Malimbwi, R.; Faaij, A.P. Cost/benefit analysis of biomass energy supply options for rural smallholders in the semi-arid eastern part of Shinyanga Region in Tanzania. *Renew. Sustain. Energy Rev.* **2010**, *14*, 148–165. [CrossRef]

42. Singh, R.; van Noordwijk, M.; Chaturvedi, O.; Garg, K.K.; Dev, I.; Wani, S.P.; Rizvi, J. Public co-investment in groundwater recharge in Bundelkhand, Uttar Pradesh, India. In *Sustainable Development through Trees on Farms: Agroforestry in Its Fifth Decade*; van Noordwijk, M., Ed.; World Agroforestry (ICRAF) Southeast Asia Regional Program: Bogor, Indonesia, 2019.

43. Duguma, L.; Centre, W.A.; Minang, P.; Aynekulu, E.; Carsan, S.; Nzyoka, J.; Bah, A.; Jamnadass, R. *From Tree Planting to Tree Growing: Rethinking Ecosystem Restoration Through Trees*; ICRAF: Nairobi, Kenya, 2020.

44. Baig, S.P.; Rizvi, A.; Josella, M.; Palanca-Tan, R. *Cost and Benefits of Ecosystem Based Adaptation: The Case of the Philippines*; IUCN: Gland, Switzerland, 2015.

45. Hofer, C.T. *Carbon Finance & Cattle Externalities in the Brazilian Amazon: Pricing Reforestation in Terms of Restoration Ecology*; University of Connecticut: Storrs, CT, USA, 2010.

46. Holmes, P.M.; Richardson, D.M.; Marais, C. Cost and benefits of restoring natural capital following alien plant invasions in Fynbos ecosystems in South Africa. In *Restoring Natural Capital, Science, Business, and Practice*; Island press: Washington, DC, USA, 2007; pp. 188–197.

47. Rizzeti, D.; Swaans, K.; Holden, J.; Brunner, J.; Le, T.; Nguyen, T. *Assessing Opportunities in Forest Landscape Restoration in Quang Tri, Vietnam*; IUCN: Gland, Switzerland, 2018; Volume 46.

48. Currie, B.; Milton, S.J.; Steenkamp, J. Cost–benefit analysis of alien vegetation clearing for water yield and tourism in a mountain catchment in the Western Cape of South Africa. *Ecol. Econ.* **2009**, *68*, 2574–2579. [CrossRef]

49. Silva, D.; Nunes, S. *Evaluation and Economic Modelling of Forest Restoration in the State of Pará, Eastern Brazilian Amazon*; Imazon: Pará, Brasil, 2017.

50. Tuan, T.H.; Tinh, B.D. *Cost-Benefit Analysis of Mangrove Restoration in Thi Nai Lagoon, Quy Nhon City, Vietnam*; IIED: London, UK, 2013.

51. Monela, G.C.; Chamshama, S.A.O.; Mwaipopo, R.; Gamassa, D.M. *A Study on the Social, Economic and Environmental Impacts of Forest Landscape Restoration in Shinyanga Region, Tanzania*; Final Report to the Ministry of Natural Resources and Tourism and IUCN; IUCN: Gland, Switzerland, 2005.

52. Ministry of Water and Environment—Uganda. *Forest Landscape Restoration Opportunity Assessment Report for Uganda*; IUCN: Gland, Switzerland, 2016.

53. Gianvenuti, A.; Vyamana., V.G. *Cost–Benefit Analysis of Forestry Interventions for Supplying Woodfuel in a Refugee Situation in the United Republic of Tanzania*; Food and Agriculture Organization of the United Nations (FAO): Rome, Italy; United Nations High Commissioner for Refugees (UNHCR): Geneva, Switzerland, 2018.

54. Aymeric, R.; Myint., M.M.; Westerberg, V. An economic valuation of sustainable land management through agroforestry in eastern Sudan. In *Report for the Economics of Land Degradation Initiative by the International Union for Conservation of Nature*; ELD: Nairobi, Kenya, 2014; Available online: www.eld-initiative.org (accessed on 17 January 2020).

55. Ministry of Natural Resources—Rwanda. *Forest Landscape Restoration Opportunity Assessment for Rwanda*. MINIRENA: Kigali Kigali, Rwanda; IUCN: Gland, Switzerland; WRI: Washington, DC, USA, 2014.

56. Tesfaye, A.; Brouwer, R.; Van Der Zaag, P.; Negatu, W. Assessing the costs and benefits of improved land management practices in three watershed areas in Ethiopia. *Int. Soil Water Conserv. Res.* **2016**, *4*, 20–29. [CrossRef]

57. Onduru, D.D.; Muchena, F.N. *Cost-Benefit Analysis of Land Management Options in the Upper Tana, Kenya*; Green water credits Reports 15; ISRIC—World Soil Information: Wageningen, The Netherlands, 2011.

Publisher's Note: MDPI stays neutral with regard to jurisdictional claims in published maps and institutional affiliations.

Article

Sustainable Agroforestry Landscape Management: Changing the Game

Meine van Noordwijk [1,2,3,*], Erika Speelman [4], Gert Jan Hofstede [5,6], Ai Farida [7], Ali Yansyah Abdurrahim [8,9], Andrew Miccolis [10,11], Arief Lukman Hakim [3,11,12], Charles Nduhiu Wamucii [13], Elisabeth Lagneaux [1,14], Federico Andreotti [4,15], George Kimbowa [16,17], Gildas Geraud Comlan Assogba [1], Lisa Best [4,18], Lisa Tanika [11], Margaret Githinji [5], Paulina Rosero [5], Rika Ratna Sari [1,3], Usha Satnarain [19], Soeryo Adiwibowo [9], Arend Ligtenberg [4], Catherine Muthuri [20], Marielos Peña-Claros [11], Edi Purwanto [21], Pieter van Oel [16], Danaë Rozendaal [1,22], Didik Suprayogo [3] and Adriaan J. Teuling [13]

1. Plant Production Systems, Wageningen University and Research, 6700 AK Wageningen, The Netherlands; elisabeth.lagneaux@wur.nl (E.L.); gildas.assogba@wur.nl (G.G.C.A.); rika.sari@wur.nl (R.R.S.); danae.rozendaal@wur.nl (D.R.)
2. World Agroforestry (ICRAF), Bogor 16155, Indonesia
3. Agroforestry Research Group, Faculty of Agriculture, Brawijaya University, Jl. Veteran no 1, Malang 65145, Indonesia; arief.hakim@wur.nl (A.L.H.); suprayogo@ub.ac.id (D.S.)
4. Laboratory of Geo-Information Science and Remote Sensing, Environmental Sciences, Wageningen University & Research, 6708 PB Wageningen, The Netherlands; Erika.speelman@wur.nl (E.S.); federico.andreotti@wur.nl (F.A.); lisa.best@wur.nl (L.B.); arend.ligtenberg@wur.nl (A.L.)
5. Information Technology Group, Social Sciences, Wageningen University & Research, 6706 KN Wageningen, The Netherlands; Gertjan.hofstede@wur.nl (G.J.H.); Margaret.githinji@wur.nl (M.G.); paulina.roseroanazco@wur.nl (P.R.)
6. Centre for Applied Risk Management (UARM), North-West University, Potchefstroom 2520, South Africa
7. Applied Climatology, IPB University, Bogor 16680, Indonesia; farida_ai@apps.ipb.ac.id
8. Research Centre for Population, Indonesian Institute of Sciences (LIPI), Jakarta 12190, Indonesia; aliy002@lipi.go.id
9. Rural Sociology, IPB University, Bogor 16680, Indonesia; adibowo3006@gmail.com
10. World Agroforestry (ICRAF), 66095-903 Belém/PA, Brazil; a.miccolis@cgiar.org
11. Forest Ecology and Forest Management Group, Wageningen University & Research, 6708 PB Wageningen, The Netherlands; lisa.tanika@wur.nl (L.T.); marielos.penaclaros@wur.nl (M.P.-C.)
12. Faculty of Science and Technology, Universitas Islam Raden Rahmat, Malang 65163, Indonesia
13. Hydrology and Quantitative Water Management Group, Wageningen University & Research, 6700 AA Wageningen, The Netherlands; charles.wamucii@wur.nl (C.N.W.); ryan.teuling@wur.nl (A.J.T.)
14. iES Landau, Institute for Environmental Sciences, University of Koblenz-Landau, 76829 Landau, Germany
15. CIRAD, UPR GREEN, F-34398 Montpellier, France
16. Water Resources Management, Wageningen University & Research, 6700AA Wageningen, The Netherlands; George.kimbowa@wur.nl (G.K.); Pieter.vanoel@wur.nl (P.v.O.)
17. Faculty of Engineering, Busitema University, 236 Tororo, Uganda
18. Tropenbos Suriname, University Campus (CELOS Building), Paramaribo, Suriname
19. Environmental Policy, Department of Social Sciences, Wageningen University & Research, 6706 KN Wageningen, The Netherlands; usha.satnarain@wur.nl
20. World Agroforestry (ICRAF), 00100 Nairobi, Kenya; C.Muthuri@cgiar.org
21. Tropenbos Indonesia, Bogor 16163, Indonesia; edipurwanto@tropenbos-indonesia.org
22. Centre for Crop Systems Analysis, Wageningen University & Research, 6700 AK Wageningen, The Netherlands
* Correspondence: m.vanNoordwijk@cgiar.org

Received: 28 June 2020; Accepted: 22 July 2020; Published: 24 July 2020

Abstract: Location-specific forms of agroforestry management can reduce problems in the forest–water–people nexus, by balancing upstream and downstream interests, but social and ecological finetuning is needed. New ways of achieving shared understanding of the underlying ecological and social-ecological relations is needed to adapt and contextualize generic solutions. Addressing these challenges between thirteen cases of tropical agroforestry scenario development across three continents requires exploration of generic aspects of issues, knowledge and participative approaches. Participative projects with local stakeholders increasingly use 'serious gaming'. Although helpful, serious games so far (1) appear to be ad hoc, case dependent, with poorly defined extrapolation domains, (2) require heavy research investment, (3) have untested cultural limitations and (4) lack clarity on where and how they can be used in policy making. We classify the main forest–water–people nexus issues and the types of land-use solutions that shape local discourses and that are to be brought to life in the games. Four 'prototype' games will be further used to test hypotheses about the four problems identified constraining game use. The resulting generic forest–water–people games will be the outcome of the project "Scenario evaluation for sustainable agroforestry management through forest-water-people games" (SESAM), for which this article provides a preview.

Keywords: boundary work; ecohydrology; forest–water–people nexus; landscape approach; participatory methods; scenario evaluation; social-ecological systems; tropical forests

---

1. Introduction

1.1. Agroforestry and the Forest–Water–People Nexus

Current understanding of the term agroforestry links plot, landscape and policy aspects of the ways farmers interact with trees [1]. We present early results and planned next steps of a new interdisciplinary research program on scenario evaluation for sustainable agroforestry management, focusses on the (agro) forest–water–people nexus and exploring how serious games can help bridge the science–policy divide, as adequate supplies of clean water are a key development challenge while the roles of trees and forest in modifying ecohydrology are context specific and often contested. A top-down policy perspective needs to be reconciled with a bottom-up farmer understanding, while social, cultural aspects may be as important as ecological, technical ones. To unpack the context dependence of past research results, a comprehensive diagnosis has to be the basis for any comparative study. We restrict ourselves here to a pan-tropical perspective.

The 2030 agenda for Sustainable Development, adopted by all United Nations member states in 2015, provides a shared blueprint for peace and prosperity for people and the planet, now and into the future [2]. As desirable as this sounds, there was a long way to go to achieve these goals even before the emergence of the SARS-CoV-2 virus showed the fragility [3] of the increasingly globalized and interdependent world. As part of the Sustainable Development agenda, land use, often a mix of forests, agroforestry, agriculture and built-up areas, has to meet many and partly conflicting needs, including the provision of food, energy, water as well as environmental protection from floods, droughts and biological extinctions. The interconnected global social-ecological system needs to be understood as a dynamic feedback system before 'leverage points' [4] can be used wisely to shift development to more sustainable trajectories understanding the forces working in opposite directions. By understanding the landscape as a social-ecological system [5], the ecological pattern of 'land cover', its hydrological consequences and the social overlay of 'land use' and 'water use' can be analyzed as part of multiple feedback loops [6].

The nexus of forest–water–people interactions is central to the landscape as social-ecological system; it is not restricted to institutional forest definitions, but relates to actual tree cover of different quantity, quality, age and spatial pattern [7]. Many authors have noted that current drivers and pressures lead to partial-interest decisions, suboptimal and often contested consequences, including disturbances of hydrological cycle and flow regimes [8,9]. Diagnosing the issue, from plot-level soil effects to continental hydro-climates [10], is a relevant first step, but unless it contributes to decisions and actions that correct the system, it remains an academic exercise. Socially, the ecological impacts can increase conflict, aggravating land mismanagement (Figure 1). Rather than trying to redress the symptoms one by one, a more incisive analysis is needed of the social-ecological system as a whole. Change will require a concurrence of local, national and international decisions translated into actions, and as such will be helped by a shared understanding of issues, commitments to achievable goals and action at appropriate temporal (immediate improvement plus long-term results) and spatial scales.

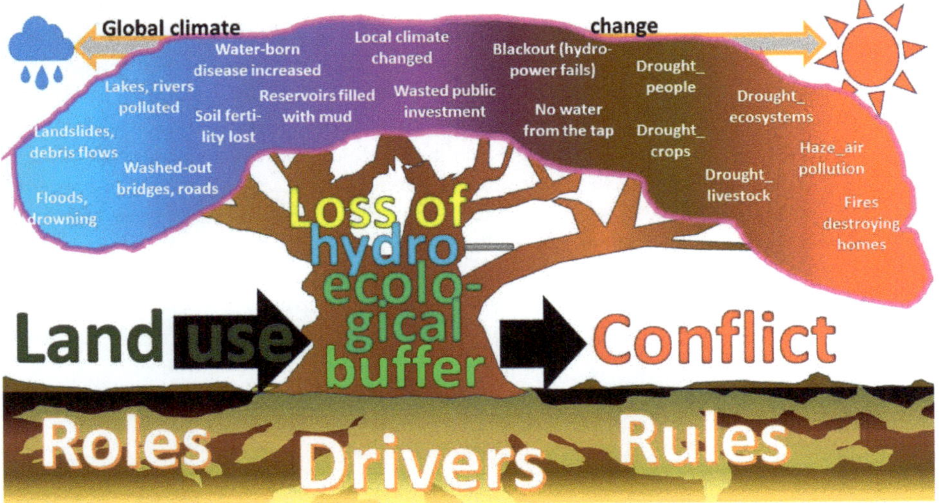

Figure 1. Problem tree of symptoms of disturbed hydrology (ranging from 'too wet' to 'too dry') and its underlying causes internal to the landscape, and external (including climate change).

Regional planning for resilient landscape forest-water management that can adapt to change and that can incorporate a diversity of knowledge has been shown to be more effective when collective approaches for problem-solving are employed [11]. However, while active stakeholder participation has become a main approach in problem solving and solution exploration [12–15], methods are less commonly used for supporting stakeholders to explore and evaluate alternative scenarios and facilitate (social) learning [16].

1.2. Use of Serious Games: Issues Arising

An innovative participatory approach to learn about, discuss and explore the complexity of the various dimensions of complex contested landscapes to facilitate social learning among stakeholders can be best described by the keyword 'serious gaming' [17]. Serious games are emerging as a valid possible intervention tool [18–20]. In multi-stakeholder settings, games function as (social) learning tools and boundary objects to discuss local voices and concerns. A game-based approach can stimulate participants, in a safe space, to explore system behavior through scenario evaluation and to support negotiations in local contexts. Shared experience and jointly acquired knowledge in scenario evaluation games may help in the emergence of coalitions for change in the real world. In this context, playing games has shown to provide information that can support better-informed decision making [21,22].

While the use of serious games has become increasingly popular in research and development projects, there still are relevant methodological questions that remain unanswered [23].

The development and use of these games appear to suffer from a number of drawbacks that will be addressed in this paper:

(1). Games are commonly ad hoc, case dependent with poorly defined extrapolation domains for responsible use, and therefore less relevant of applicable in other contexts;

(2). Games often require heavy research investment from intervention experts to be constructed in ways that are relevant for important local discussions;

(3). Games have untested cultural limitations in where and how they can be used [24];

(4). Game users lack clarity on where and how games relate to policy making in local and/or global issue cycles, negotiations and reforms of governance instruments.

Here, we review these four identified drawbacks of serious gaming as background to the Scenario Evaluation for Sustainable Agro-forestry Management (SESAM) program (2019–2025), which includes 15 PhD research projects, geared at credible, salient and legitimate action research (Figure 2). The overall SESAM program and the individual PhD projects within it are designed to (1) be systematic in their coverage of the pantropical forest–water–people nexus in its main manifestations and issues, using generic forest and tree cover transitions as continuum description rather than forest–agriculture dichotomies; (2) use well-established hydrological, ecological, social and economic concepts to complement local empirical knowledge; (3) be cognizant of the main dimensions in which cultural contexts of inter-human and human–nature relations vary to guide responsible game use; and (4) be explicitly adapted (or adaptable) to different stages of local and global issue (policy) cycles [25], where issues become part of a political agenda, get debated and (partly) resolved (often sowing seeds for the next issue to emerge).

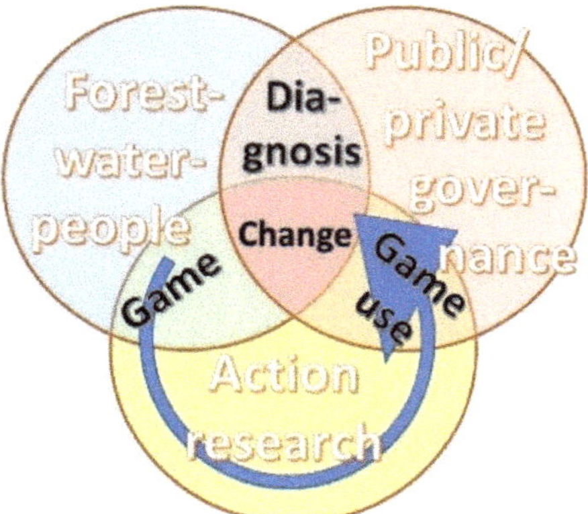

Figure 2. The three spheres in which SESAM will operate to ensure credible (based on current understanding), salient (actionable in terms of public and/or private governance) and legitimate (aligned with stakeholder interests) games, used appropriately.

These four targets are highly ambitious when taken one by one but tackling them jointly may prove to tap into synergies. In this early stage of confronting reality and our ambitions, we will describe (A) Steps to make games less 'ad hoc' by a selection of existing frameworks that can be used in a systemic understanding of forest–water–people nexus issues, human decision making in cultural contexts, and constructing games, (B) Link generic understanding of pantropical variation in forest–water–people configurations to the current set of landscapes/sub-watersheds that form the primary contexts and a list of emerging issues for more efficient linking of scientific, local and policy-oriented knowledge, (C) Place the set of landscapes in the known geography of cultural variation as background for exploring cultural limitations in game re-use, (D) Present a set of game prototypes that can match stages of issue cycles in policy making in local to global issue cycles, negotiation support and reforms of governance instruments.

2. Frameworks for Understanding Social-Ecological System Change

A central issue in todays' Sustainability Science is how knowledge, power and values interact in decision making and how long-term impacts and planetary responsibility can be woven into existing institutions. While there are hard trade-offs that lead to difficult choices in prioritizing one goal over another, existing space for multifunctionality is missed due to incomplete understanding of positive and negative consequences of land and water management decisions, oversimplified ideas of a forest vs. agriculture dichotomy (missing out on intermediate land uses with trees), 'free and prior informed consent' and inadequate stakeholder involvement. Therefore, 'game-changing' ways to involving diverse types of knowledge, plural values, multiple voices and heterogeneous stakes are needed. Science has always contributed to 'environmental issue cycles', but current concepts of "Boundary work' across the science–policy interface include roles in agenda setting, understanding patterns and processes, exploring options and scenarios, commitment to principles and implementation decisions in a complex reality of issues and concerns. Boundary work aims for salient, credible and legitimate information and understanding [26], avoiding being 'normative' in presenting insights and results. Three research traditions and their tools may need to be combined for effective boundary work (Figure 3):

1. The environmental science tradition of analyzing Drivers, Pressures, System states, Impacts and Responses (DPSIR) as multi-scale phenomena that can be studied one by one, but need to be understood jointly [27–29].
2. The natural resource governance 'issue cycle' concept [30,31] in five boundary work steps that clarify the R of DPSIR: (a) Agenda setting, (b) Better and widely shared understanding of what is at stake, (c) Commitment to principles, (d) Details of operation, devolved to (newly created or existing) formal institutions that handle implementation and associated budgets, and (e) Efforts to monitor and evaluate effects ('outcomes'); it thus relates to the 'Responses' part of DPSIR, and
3. The social side of human decision making based on Groups, Rituals, Affiliation, Status, Power (GRASP) [32], as will be further discussed in Section 5.

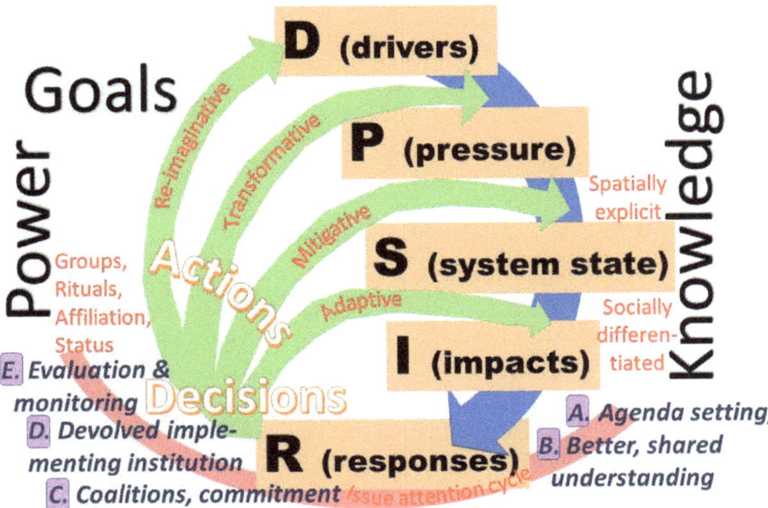

Figure 3. Drivers, pressures, system state, impacts and responses (also known as DPSIR) of landscape level change, with a decision cycle (**A–E**) closing various feedback loops interacting with knowledge, power and human sociality and its GRASP determinants.

With the 'system state' (S) of land cover/use as starting point, one can analyze the 'pressures' (P) that the landscape actors respond to (e.g., food production, income generation, health, water and energy requirements, protection from disasters), or look at underlying drivers (D) that lead to the pressures (e.g., demographic trends, international markets, national policies). One can also focus on the impacts (I) that the system state has on social and ecological well-being, or on the responses (R) that are triggered. These responses can lead to a change in the S→I relationship, the often spatially explicit P→S relationship, the often-generic D→P relationships, or challenge the drivers themselves. Such decisions can be labelled as adaptive, mitigative, transformative and re-imaginative, requiring increasingly drastic change in the existing social-ecological system (deeper roots of the problem tree in Figure 1). While the 'adaptive' decisions can be taken individually or in small groups, the others depend on collective action (especially the mitigative ones) and policy-level institutional change.

Basic questions in any landscape approach [33] are the 'Who?', 'What?' and 'Where?' of current land use, in a temporal perspective ('When?'), as a start for exploring 'Why?' questions of drivers and pressures. Impacts ('So what?') are the entry point to 'Who cares?' stakeholder analysis, and opportunities to close the loop if those who care can directly change the who/what/where of land use, or at more fundamental level the why of drivers and pressures. These questions lead to a further elaboration of the DPSIR framework defining the breadth of the type of drivers and pressures that we may need to understand on the human side of the eco-hydrological-social system that deals with the forest–water–people nexus issues. (Figure 4).

Figure 4. State variables and processes relevant for change in eco-hydrological-social systems in the forest–water–people nexus grouped in a DPSIR (drivers, pressures, system state, impacts, responses) framework.

The System-state variables (at the heart of DPSIR) deal with ecosystem structure and function on the (hydro-) ecological side, and with identity, (collective) culture, and economic indicators on the social side (Figure 4, center of the diagram). On the Impacts side, the most relevant indicators can be grouped as livelihoods and local ES, Markets and global ES. Among the pressures, migration (rather than birth rates), rights and know-how, relate to underlying drivers that include legal frameworks, demography and knowledge systems. Responses can include land and water use plans (linked to rights within a legal framework), trade reform and climate action, emphasizing relations between the local system and global change. Part of the latter are attempts to make incentive structures at the land-user level more aligned with 'downstream' impacts, via Payments for Ecosystem Services (PES) [34]. The recent PRIME (Productivity, Rights, complementary Investments, Market access and Ecosystem services) framework of the World Bank that has helped conceptualize forests' contribution to poverty reduction and guide intervention design [35] is compatible with the middle hexagon in Figure 4. Further conceptual framing will be presented after the portfolio of landscapes are briefly introduced.

3. Representativeness and Diagnostic DPSIR Analysis of the SESAM Landscape Portfolio

Brief descriptions of the 13 landscapes in which SESAM will diagnose issues, develop games and test their relevance for local decision making are provided in Appendix A. In this section, we will consider the degree to which the portfolio can be expected to include major variation in pantropical manifestation of forest–water–people nexus issues and apply a DPSIR analysis to identify the major issues at stake.

A pantropical typology of forest–people interactions [36] has proposed six recognizable stages of a 'forest transition', starting with landscapes with >80% of core forest and human population densities below 1 to over 1000 km^{-2}. All forest transition stages can be found in any of five ecoclimatic zones, that link to a forest water analysis, and are based on the ratio of rainfall (precipitation, P) and potential evapotranspiration (E_{pot}). This typology separates drylands and semi-arid zones (P/E_{pot} < 0.5; 35% of tropical area, 20% of people) from a dry–sub-humid zone (0.5 < P/E_{pot} < 0.65; 10% of area, 11% of people). On the wetter side, we can distinguish (relatively) high 'water towers' with P/E_{pot} > 0.65 and generating streamflow to lower, drier parts of the same watershed (11% of area; 15% of people), a lowland (non-water tower) humid forest zone (0.65 < P/E_{pot} < 0.9; 19% of area, 22% of people) and a per-humid lowland forest zone (P/E_{pot} > 0.9) with 25% of pantropical land area and 32% of its human population. Human life in the drier zones depends on rivers (or groundwater flows) that originate in wetter areas, especially in the 'water tower' configuration where the wetter part is higher in the landscape. All eco-climatic zones can contain all stages of the forest transition, although in the drier parts of the tropics climate and human agency are not easily disentangled as causes of low tree cover vegetation. Forest transitions are defined by a phase of declining cover of (natural) forest, an inflection point and a phase of increasing forest cover (secondary forest in landscapes with rural land abandonment and urbanization, and, more commonly, planted forests). The pantropical classification at sub-watershed level [35] forms a basis for judging pantropical representativeness of our SESAM landscapes. Summary statistics of the various SESAM landscapes (Table 1) show that the landscapes include all five ecoclimatic zones (dryland/semiarid; dry-sub-humid; water tower; humid per-humid lowland) and all six forest transition stages. On further analysis, the forest–water–people nexus issues in each of the landscapes involve plot-level aspects that relate to lack or presence of trees, hillslope and watershed hydrological relations (and associated water rights), and policies for agriculture, forestry, land-use rights (Table 2). These three issues match the three current agroforestry paradigms at plot/farm, landscape and policy scales, within a common definition of agroforestry as the interaction of agriculture and trees, including the agricultural use of trees [37].

Table 1. Basic properties of SESAM landscapes; W = water-shed, R = Region, C = coastal zone.

Context	Location	Coordinates	Hydro-Climate	Mean Annual Rainfall, mm	Human Population Density, km^{-2}	Forest Cover, %	Forest Transi-tion Stage	Scale:	Area of Focus, km^2
Core forests, upriver	1. Suriname upriver	3–4°N, 54–56° W	Per-humid	2700	<1	>70	1	W, R	7860
	2. Madre de Dios, Peru	12°36′ S, 69°11′ W	Per-humid	2221	1.3	95	1	W	85300
Mangrove coast	3A. Nickerie, Suriname	5°51′N, 55°12′W	Per-humid	1800	6.4	90	1	C, District	5353
	3B. Paramaribo, Suriname	5°56′N, 57°01′W	Per-humid	2210	1297	19	3	C, District	182
Agroforestry mosaic	4. Tomé Açu, Pará, Brazil	2°25′S, 48°09′W	Per-humid	2371	10	37	2	Municipality	5145
Coastal peatland	5. Ketapang, Indonesia	1°27′–2°0′S, 110°4′–110°8 E	Per-humid	3169	20	34	2	Peatland Hydrolo-gical Units	948 and 1048
Mountain lakes	6. Singkarak, Indonesia	0°30′–°45′N 100°20′–100°43′E;	Water-tower	1700–3200	338	16	3	W	1135
Watertower/Semi-arid gradient	7. Mount Elgon, Uganda	01°07′ 06″ N, 34°31′ 30″ E	Water-tower	1600	355 (up),606 (md),870 (lw)	25 (up), 63(md) 36(lw)	4	W	4200
	8. Ewaso Ng′iro, Mt Kenya	0°15′S–1°00′N, 36° 30′– 37°45′ E	Water-tower	600 (lw), 1600 (up)	150 (up), 12 (lw)	18	4 or 6	W	15,200
Mountain farms	9. Andes, Peru	15°50′S, 70°01′W	Dry-sub humid	700	5.1	0.1	5	W	8490
Dryland farming	10. Mossi plateau, Burkina Faso	12°45′– 13° 06′N, 0°.99–1°33′W	Semi-arid	400–700	148	10	5 or 6	Village territory	
Agroforestry mosaic	11. Kali Konto, Indonesia	7°45′–7°57′S, 112°19′–112°29′E	Water tower	2995–4422	453	20	6	Sub-watershed	240
Water towers under pressure	12. Rejoso, Indonesia	7°32′–7°57′S, 112°34′–113°06′E	Water tower	2776	414 (up), 693 (md), 1925 (lw)	11	6	W	628
	13. Upper Brantas, Indonesia	7°44′–8° 26′S, 112°17′–112°57′E	Water tower	875–3000	1042	24	6	Sub-watershed	180

Table 2. Drivers, pressures/system state and impacts across the SESAM landscapes, based on descriptions in Appendix A.

Location	Drivers of Land-use Change	Pressure/System State: Plot-Level Land Use	Pressure/System State: Landscapes, Watersheds	Pressure/System State: Policy Interactions	Impacts on Forest–Water–People Nexus
1. Suriname upriver	Pressures: Shifting cultivation, logging, road infrastructure, encroaching gold mining. Drivers: income generation, weak law enforcement.	Agriculture in the landscape is basically all shifting cultivation. Traditionally within the plot there are more crops and only a few trees (usually palm-fruit trees).	A multifunctional landscape with agriculture, forestry, nature-based tourism Good water quality, availability of drinking water and maintaining water levels in the streams and rivers for agriculture.	Forestry is the better regulated sector than agriculture sector No sector-coordinated policy. Low capacity in district government institutions.	High vulnerability of rain-fed agriculture and water security, forest degradation due to logging and shifting cultivation, increasing deforestation in the Guiana Shield.
2. Madre de Dios, Peru	Illegal gold mining and informal agricultural expansion along the recently paved Inter-Oceanic Highway.	Diverse agroforestry systems are expected to increase hydraulic redistribution, soil macroporosity and in general to represent a sustainable alternative to gold mining.	Degraded or deforested areas, when dedicated to farming, can be agro(re-)forested to increase the tree cover at landscape level while improving socioecological resilience.	Regional governments are supporting market-oriented low-diversity agroforestry systems as an alternative to the illegal gold mining and slash-and-burn farming. Land-use planning does not take the impact on watershed functions into account, and vulnerability to droughts (fires).	Increase in drought and flooding episodes, drought-related fires, mercury contamination of rivers resulting from gold-mining activities.
3. Mangrove coast, Suriname	Drivers: Population growth (urbanization), income generation (for different actors such as fishermen, tour operators), poor law enforcement.	The mangrove forest along the coast prevent erosion and its ecosystems also serves as a potential water source for the nearby agriculture land mainly used for rice cultivation.	Unsustainable use of the mangrove ecosystem services and the removal of mangrove trees for various purposes (building infra-structure, housing, etc.) resulting in saltwater intrusion, reducing water quality land inwards.	There is poorly integrated coordination among the different stakeholders (direct and indirect users) to foster mangrove conservation.	Mangrove forest degradation affecting coastal resilience. Excessive and unsustainable use of mangrove ecosystem service
4. Tomé Açu, Pará, Brazil	Poor environmental governance and enforcement, migration, rural poverty.	Logging, extensive grazing, slash and burn farming (cassava and annual crops), monocrop oil palm.	Oil palm drives contamination of waterways through chemical fertilizer/pesticide use and mill effluents. Fire, low-yielding extensive grazing and land degradation influence water flow regimes.	Lack of coordination between various levels of government (municipal, state and federal) and lack of landscape-level land-use planning.	Oil palm drives contamination of waterways through chemical fertilizer/pesticide use and mill effluents. Different land-use types have different implications on water flows at the plot and landscape scale.

Table 2. *Cont.*

Location	Drivers of Land-use Change	Pressure/System State: Plot-Level Land Use	Pressure/System State: Landscapes, Watersheds	Pressure/System State: Policy Interactions	Impacts on Forest–Water–People Nexus
5. Ketapang, Indonesia	Increase in number of migrant communities and expansion of plantations in peat-swamp forest that led to massive canal construction.	Maintain soil infiltration for water level in the agricultural area in the peatland. Control drainage (smart 'canal blocking') for more constant water levels.	Converting the burnt areas surrounding protected forests into agro-forestry system will restore the functions as well as improving local peoples' livelihood. Saltwater intrusion in coastal zone when peat domes are drained.	Managing the land-use planning in peatland area based on peat depth and characteristics: Areas with peat depth > 3 m have protected forest status; with peat depth < 3 m plantation or agroforestry systems are allowed; with peat depth < 0.5 m and sapric peat (open-field) agriculture is allowed.	Haze episodes from the forest fire of the degraded peatland ecosystem, as an impact of decreasing groundwater level during dry season.
6. Singkarak, Indonesia	Highly intensified agriculture, population increment, urbanization.	Soil erosion occurs in the agricultural land (highly intensive horticulture) which is located in the upstream and hillslope area.	Land-use change into highly intensive agri-culture and residential in upper basin impact on water quantity and quality of the lake.	There is no single integrated authority for watershed and lake management.	Water resource, forest and land-use change impact on the lake (water quantity, quality and biodiversity), impact of climate variability (dry years) on river basin.
7. Mount Elgon, Uganda	High population, High poverty levels (low income generating activities), Land fragmentation (land tenure system), Favorable Climate, Urban extension, food gap, Conflicting policies.	Due to the topography of the region, soil erosion is common; Planting shade trees in existing coffee fields/systems controls soil erosion and boosts coffee production hence enhancing and sustaining crop yield and food security.	Fragmented forests due to population growth and increased agricultural activities; Subsistence farmers cultivating wooded areas and practicing agroforestry (with other crops and coffee); The degraded soil/land needs to be rehabilitated in order to promote ecosystem services of the mountainous forests (East Africa's water tower).	Empowering the local agroforestry communities and cooperatives to plant more trees; Supporting payments to local communities to avoid deforestation and restore forest inside the park; Joint environment policy Implementation—as community motivation to encroach into park forest is dependent on policy, commodity prices, law enforcement and political interests; Enforcement of forest/environment bylaws, and resource use agreements.	LULC changes—High deforestation levels (urban area extension and agricultural land expansion); Upstream–downstream conflicts (decreasing rivers base flow); Human encroachment (national park, riparian zones of riverbanks and swamps); Riverbank degradation and Land degradation (soil erosion and declining soil fertility); Seasonal downstream floods and landslides

Table 2. *Cont.*

Location	Drivers of Land-use Change	Pressure/System State: Plot-Level Land Use	Pressure/System State: Landscapes, Watersheds	Pressure/System State: Policy Interactions	Impacts on Forest–Water–People Nexus
8. Ewaso Ng'iro, Mt Kenya	Increasing irrigation water demands, increasing human population, changing weather patterns (erratic rainfall, prolonged droughts), poor governance, political interference.	Sustainable Land Management practices that reduces soil erosion and increase water infiltration. The area under water-demanding crops affects irrigation de-mands. Increasing rainwater harvesting and storage at plot level would reduce water demands.	Land cover/use changes at watershed level affects water retention/water yields Climate change and variability (P, PET) affects distribution of water resources Excessive abstraction of river water upstream affects river flows downstream.	Watershed planning and governance needs to be improved by capacity building community level structures such as Water Resources Users Association (WRUA) in Kenya. Irrigation water management efficiency is constrained by lack of knowledge, understanding and technology access for farmers.	Uncontrolled abstraction of water, human encroachment (e.g., farming in riverbank protected/riparian areas, limited downstream flows (dry riverbeds); Poor enforcement of policies (metering, riparian corridors), local politics versus national government interests, climate change/variability
9. Andes, Peru	Agricultural boom: quinoa production for export supported by cooperatives and NGOs.	Highland agroecological zone with good soil organic matter. Depending on the on exposure, soils and drainage, agricultural activity remains in the high altitude.	A multifunctional landscape with agroecological practices quinoa with their crop wild relatives and diverse activities as silvo-pastoral systems.	This landscape is recognized from the United Nations Organization as Globally Important Agricultural Heritage Systems (GIAHS) valorizing the ancestral systems of cultivation of quinoa.	Increased drought, soil fertility loss, loss of cultivated diversity—the export market only includes few quinoa varieties.
10. Mossi plateau, Burkina Faso	Demography, climate variability and land degradation	Onset of rainy season, soil quality, infiltration, crop water use, yield formation.	Overland flow capture, Village level transfers of biomass, (fodder), farmer groups.	Land and water use rights, local institutions for collective action, grazing management.	Low and unstable biomass production.
11. Kali Konto, Indonesia	Agricultural and dairy products demand (market), poor land management, and population increment.	Intensive agricultural farming with minimum tree cover in hillslope increase soil surface exposure, increase on livestock.	Land-use change from forested area to open field (including grassland for fodder), reservoir siltation due to high sedimentation which affect water quality and quantity.	There is lack of integrated coordination among stake holder (government, farmers, and water beneficiary) for achieving sustainable watershed conservation effort.	Increase on horticultural area, soil fertility lost, landslide (debris flows), reservoir filled with muds.

Table 2. *Cont.*

Location	Drivers of Land-use Change	Pressure/System State: Plot-Level Land Use	Pressure/System State: Landscapes, Watersheds	Pressure/System State: Policy Interactions	Impacts on Forest–Water–People Nexus
12. Rejoso, Indonesia	Population growth (all), changes in forest and agricultural crop commodity prices (all), deforestation for horticulture (upland), rock and sand mining (midland), groundwater exploitation (lowland)	Upper zone erosive Andisols used for intensive vegetable production with low tree cover. Local communities are not allowed to use forest resources or use forest areas for farming and raising livestock; Community only carries out agricultural activities outside the state forest area.	Local communities are involved in planting trees allowed to farm among tree stands (intercropping) in production forests with the rules of cooperation for the results. Negative impacts of forest conversion to agri-culture; increasing tourism, stone mining, uncontrolled water use for paddy rice	The implementation of recent Community Based Forest Management Programs Collaboration between stakeholders A PES (Payment for Ecosystem Services) scheme began to be implemented	Land conversion and commodity change without planning, reduced flowing springs, drought in the dry season, flooded and landslides in the rainy season, an explosion of agricultural pests almost every year, overexploiting downstream use of bore wells and potential conflicts, environmental damage due to central rock mining and potential conflicts.
13. Upper Brantas, Indonesia	Population growth (0.95%/year), urbanization; upland vegetable markets, tourism industry.	Upland vegetable area, degraded state-owned forest soil erosion and sedimentation, intensive use of fertilizer and chemical pesticides.	Gradient land-use change from grand forest protected area, production forest, upland vegetable, settlement and tourism are. Deficit water balance in dry season triggering water conflict.	There is no coordination and comprehend Policy of Land and (Ground) Water Resource.	Land-use change, ground water and deep well extraction, water user conflict, flood, water supply and demand.

4. Unpacking the Forest–Water–People Nexus

A large number of 'issues' in the case studies relate to water and need to be understood in their landscape context (Figure 5). They all relate to the way water flows from mountains ultimately back to oceans, unless it follows the atmospheric 'short cycle' downwind route back to rainfall over land. Along the streamflow pathway, water changes in quality (sediment load, pollution), quantity (annual water yield to reservoirs or lakes), and flow regime (regularity of flow, flood risks), affecting settlements, agriculture, fisheries and human health. A major theme is the upstream versus downstream (with the 'water tower' configuration of freshwater sources for lowlands specifically challenging), complemented by upwind versus downwind for rainfall recycling, and specific features such as riparian wetlands and forests, peat swamp forests and mangroves protecting the coastal zone, while supporting marine fisheries. The scientific debate on the mechanisms and patterns of 'biological rainfall infrastructure' is undecided [6,38]. While the 'biotic pump' theory suggests forests cause the flow of moist air towards them, observations of windspeeds show that rainforests are associated with relatively low wind speeds that imply the 'short cycle' can remain relatively local, with transport distances during the mean atmospheric residence time of 8 days of hundreds rather than thousands of kilometers [39].

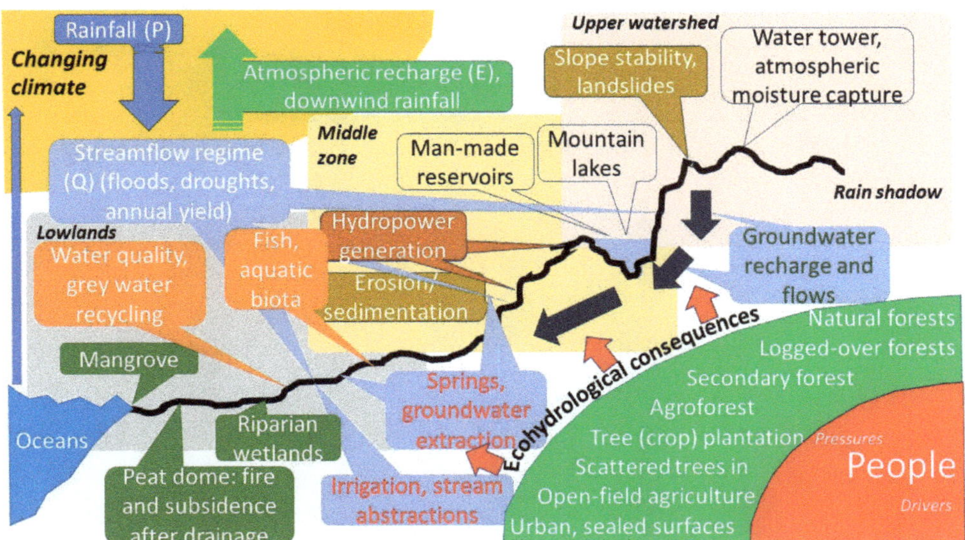

Figure 5. A first synthesis of eco-hydrological relations across the various landscapes in Table 2.

Where issues relate to forests and/or trees outside the forest, the cascade of processes that start with rainfall, are followed by canopy interception, infiltration into the soil or overland flow and contribution to local water storage and/or direct response of streams and rivers (Figure 6). Ecosystem structure interacts with hydro-ecological processes, jointly shaping a set of 'ecosystem services' that relate to benefits humans derive from well-functioning systems. A typology of nine such 'watershed functions' are indicated in Figure 6 and described in Table 3. These functions (or 'services') can be used to analyze site-specific differences between land cover types, relating the actions of roots in soil to landscape-level streamflow [40].

The principal concept in all of ecohydrology is that of the water balance, where input (precipitation P or snowmelt in cooler parts of the world) is related to two main pathways out of the system: back to the atmosphere via evapotranspiration (E) and streamflow Q (surface or groundwater flows) and to changes in the 'buffer' of water held inside the system (Figure 7). At timescales of a year or more this buffering can be ignored (except for interannual climate variability), but at the timescale of a rainstorm it is key to reduce downstream flooding risk [41,42].

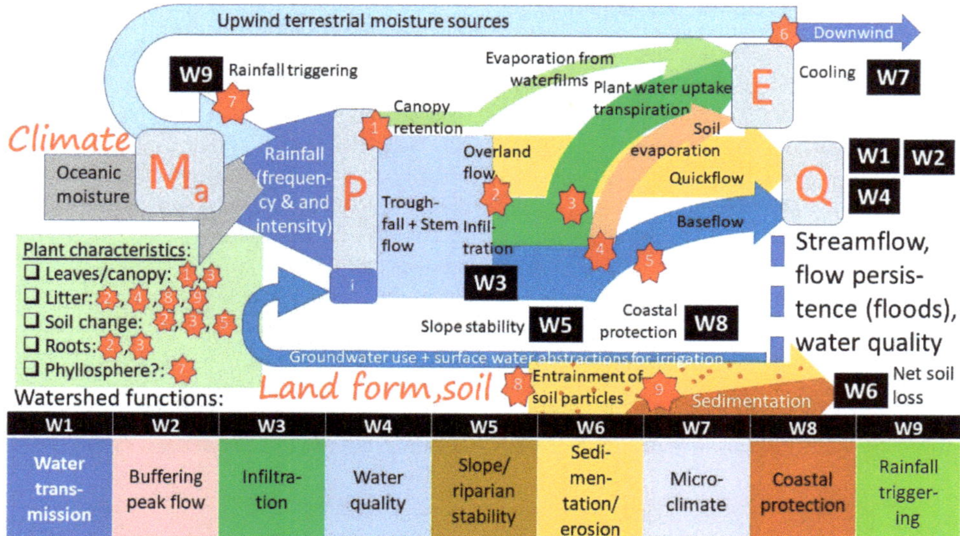

W1	W2	W3	W4	W5	W6	W7	W8	W9
Water trans-mission	Buffering peak flow	Infiltra-tion	Water quality	Slope/ riparian stability	Sedi-men-tation/ erosion	Micro-climate	Coastal protection	Rainfall trigger-ing

Figure 6. Ecosystem structure (landform, soil, plant characteristics), processes (**1–9**), and 'watershed functions' (**W1–W9**) based on water balance: P = precipitation, Q = stream flow, E = evapotranspiration, M_a = atmospheric moisture, i = irrigation.

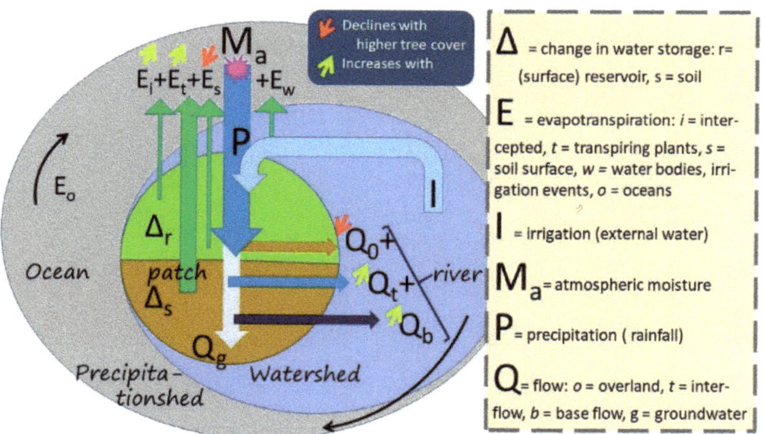

Figure 7. Basic elements of a water balance at patch, watershed and precipitationshed scale levels, that can form a biophysical basis for forest–water–people nexus games, with expected changes in terms with increasing/decreasing forest cover.

We can now relate a typology of 'watershed functions' (**W1–W9**) to an understanding of ecosystem structure and function, across the various landscapes (Table 3). Decisions on adding landscapes to the SESAM portfolio were made with this representation in mind. Probably the most contested in this list is W9, the influence of vegetation on rainfall regimes at landscape and continental scales [38,43–46]. Three of the Latin America landscapes are at the start of the Amazon rainfall recycling system, one is at the receiving end, and one in the much drier Andes range that depends on left-over atmospheric moisture. The two East African water towers are part of the complex hydro-climatic system, influencing rainfall further North (including to some extent that in the Burkina Faso landscape, along with water coming in from the Atlantic Ocean).

Table 3. Reference (V, major, or v, relevant) to nine 'watershed issues' across the various landscapes [47,48].

Locations	Watershed Functions								
	W1	W2	W3	W4	W5	W6	W7	W8	W9
	Water Trans-mission	Buffering Peak Flow	Infiltra-tion	Water Quality	Slope/ Riparian Stability	Sedi-men-tation/ Erosion	Micro-Climate	Coastal Protec-tion	Rainfall Trigger-ing
Upper Suri-name River	v		V			v	V		v
Madre de Dios, Peru	v		V	V			V		V
Mangrove coast, Suriname				V			V	V	
Para, Brazil		v		V		v	V		v
Ketapang, peatland	v	V	V	V			V	v	
Singkarak Lake	v	V		V	V	v			
Mount Elgon		V		V	V	V			
Mount Kenya	v	V	V			v	V		
Peruvian Andes	v		V			v			
Mossi plateau			V	V			V		
Kali Konto		v	V	V		v	V		
Rejoso Watershed	v	V	V	V	V	v			
Upper Brantas		V	V	V	V	V			

Beyond the 'provisioning' and 'regulating' services captured in W1 … W9, there also are 'cultural' services based on 'relational values' between Humans and Nature. The two interact when visits by domestic (or international) tourists to the mangroves of Nickerie district in Suriname or the mountain resorts in East Java increase pressure on freshwater resources. At the global scale the 'Ecosystem Service' (ES) language is currently embedded as 'instrumental values' in a wider 'relational values' paradigm on 'Nature's Contributions to People' (NCP) [49]. The way cultural relations with a landscape and its forest and water aspects is expressed varies with history and religion. The springs in the upper Brantas still are sacred places, where mountain spirits are brought offerings to secure the gift of fresh water continues. On densely populated Java, the *wayang* tradition re-tells stories of the past. All stories start with the *gunungan* or mountain symbol (Figure 8), a tree of life connecting forest animals and creatures to people's homes and lives. The flip-side, shown occasionally for dramatic effect, shows demons and fire: social conflict and mismanaging human–nature relations destroys human livelihoods. Throughout human history, perspectives on spirits, deities, personified nature or a single Almighty have been described in metaphors of words that also describe human–human relations in terms of (A) family, such as ancestors, (grand-) parents (Mother Nature), siblings, offspring, partners, in-laws), (B) neighbors, friends, business partners, (C) adversaries, competitors, armed attackers and defenders, (D) servants or (E) educators. A subset of these relations (the providing mother of NCPs and ES servants) can be interpreted as 'instrumental', directly supporting human goals and objectives, but even those imply that there will have to be a two-way (rather than unidirectional) relationship to maintain or support what is relevant to people because it cannot be taken for granted [50].

Across the various landscapes we can now understand that debates about increasing/decreasing forest cover and/or tree cover in agroforestry or other land uses, can have both instrumental and wider relational aspects. Technical solutions to local issues may increase problems if they do not lead to shared understanding and negotiated trade-offs. Such effects will have to be represented in applicable games at a reasonable level of detail and accuracy, and in their cultural context.

Figure 8. Instrumental as part of relational values of nature from a human perspective as exemplified in the Javanese *gunungan* representation with its idyllic harmony and fearful disaster side, and the UN Sustainable Development Goals; NCP = Nature's contributions to people, PCN = People's contributions to nature.

The immediate issues of public concern to which landscape management contributes (or can help solve) are very diverse among the SESAM portfolio, even though they all relate to the same water balance. Upstream on the Suriname river, increasing deforestation (gold mining) and forest degradation (logging) in the Guiana Shield will need to be managed in a better way. Increasing food security, and water security during the long dry season, depends on maintaining and restoring the flow buffering aspect of the river, along with sustained biodiversity as a basis for other forest ecosystem services beneficiaries depend on for subsistence and income. In Madre de Dios (Peru) focus is on reduced water shortages for agriculture and household use (including fruit production) in the dry season, but also on fire prevention during the dry season. In Pará State, water recycling is reduced by deforestation and land degradation and water quality undermined by chemical inputs used on large-scale monocrop plantations, pointing to the need for restorative and climate-smart agroecological practices including agroforestry. In the Ketapang peatlands (Indonesia), reducing public health impacts of haze episodes due to peat-swamp forest fire may align with an interest in reducing agricultural damage due to flooding during the rainy season by restoring the water storage capacity of peat domes and reducing the speed of drainage. Restoring the mangrove forest in Suriname will protect the coastal zone from saltwater intrusion and floods, while the mangrove ecosystem provides services to coastal fisheries, beyond its direct products.

Water resources management in water towers of Kenya and Uganda involves better crop selection, caution in increasing middle-zone tree densities and attention to water distribution for downstream users. It also involves reducing landslides, minimizing population displacement and deaths, loss of fertile land and famine. Reducing flooding in the downstream area during the rainy season can go hand in hand with reducing drought (severe dry Spells).

In the densely populated Brantas river basin in Indonesia, the targets are water security and water quality for domestic use and tourism, and irrigation water for dry-season paddy rice production (three or even four crops per year). Similarly, around Lake Singkarak there's a need for increasing water availability for irrigation and hydropower during the dry season, but also a need for maintaining the local endemic fish habitat due to maintained water quality. In the Rejoso Watershed, beyond these points, reducing water use in lowland rice paddies may be needed to secure groundwater for a nearby metropole. Before using these insights, however, the cultural context needs to be considered as it may call for different ways of constructing games, despite hydro-ecological similarity.

5. Cultural Diversity in Response to Forest–Water–People Nexus Issues

The response options chosen in any society and any point in time are under the influence of culture, as a layer between generic human nature and individual personality. Research on geographic variation in culture at any point in time (Figure 9A) and cultural change as part of economic development can be reconciled in identifying at least two (but up to six) main axes of variation [51]. In the simplest portrayal (Figure 9B) a collectivism–individualism axis aligns with distance to the equator, and a monumentalism-flexibility axis on which Latin America and East Asia are the bookends.

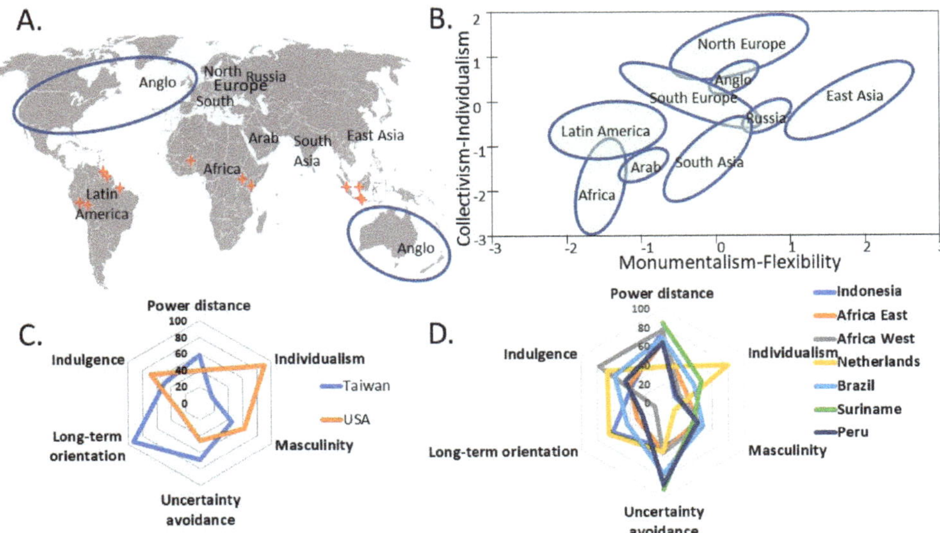

Figure 9. Geography of cultural dimensions: (**A**) Global map with red signs indicating landscapes where SESAM is active; (**B**) Two main axes of cultural variation and the relative position of countries (based on national surveys) and country groupings [52]; (**C**) Comparison of six Hofstede dimensions for two contexts in which the 'so long sucker' game was played; (**D**) Hofstede dimensions for the countries (or where such data are lacking regions) involved in SESAM [53].

People follow any of four directions in addressing any (new) issue, with preferences depending on culture [54]: hierarchy (clarity on power in decision making), unleashing private initiative (aiming for efficiency and public benefits via an 'invisible hand'), (perceived) fairness of social outcomes and (public) transparency (accountability, anticorruption). Forestry issues follow this general pattern around the globe. The high degrees of private (agro)forest ownership in Scandinavian countries reflect a low sense of hierarchy and belief in private initiative as opposed to a history of centralized control in much of the (formerly colonized) tropics. Recent responses to the COVID-19 pandemic may confirm basic cultural patterns of relying on central authority versus citizen responsibility, and in orientation on long-term goals versus immediate gratification.

A simulation game of an agroforestry landscape is a model of a socio-ecological system, clarifying components, actors, interactions, roles and rules. A run of such a game puts that system into action. Some of the components will have been designed, while others are implicitly embodied in the participants, making the game dependent on the players. This mixed composition is crucial for how the game unfolds. In doing so, we may need to refer to the six culture dimensions identified in [52,55], although the axes are only partially independent of each other in a statistical sense.

As an illustration of the way this matters in how a single game can be interpreted differently, consider some intercultural experiences with the game 'So Long Sucker' [56]. This board game allowed four players to form coalitions, rapidly eliminating one another until one remained. It had been designed in the USA to show how an incentive structure can lead to selfish behavior. When a class of Taiwanese came for a visit and groups from this class played the game, they managed to turn the dynamics around, using exactly the same rules to create a sustained group of participants that played for hours until the facilitator intervened. The reason for this is that the Taiwanese used different unwritten rules, implicit in their culture, than the US students were using, or the US designers of the game had anticipated. This story reminds us that one game may not fit all cultures [57]. Let us look at these cultural differences, since they may hold a message for SESAM. The USA and Taiwan have cultures that are widely apart (Figure 9C) on the six dimensions of culture [52].

For the US players, of individualistic and short-term-oriented mindset, the spirit of the game was "each one for himself and the devil take the hindmost". There was no moral penalty on kicking out your fellow players. For the Taiwanese partners, excluding someone from the game would be morally wrong and also imply a destroyed future of the game. This made the difference in behaviors of both types of players rational from their perspective. For SESAM we need to distinguish the inherent properties of a game from the multiple ways it can be used (re-interpreted) in various contexts.

A further example of the way games can be understood in different ways depending on cultural context emerged when researchers adopted a game originally developed for university students in Sudan to understand land degradation due to interaction of grazing pressure and erratic rainfall to farmers in N. Ghana [58]. The game centers on cattle, grass and watering points, but missed out on an important consideration of managing grazing pressure in Ghana: Guinea fowl like short grass, and especially female farmers found that the lack of these animals in the game missed opportunities for their land management ideas to emerge. The game became a 'boundary object' for debating gender balance in the local context, as much as discussing climate change adaptation.

6. Action Orientation: Game Typology and Prototypes

The game development process is conventionally divided in four major steps: Step 1—Baseline study of the local landscape and context, diagnosing key issues, Step 2—Game development, Step 3—Game implementation, Step 4—Game impact assessment. In each of the SESAM landscapes, research will be strongly embedded in participatory action research and executed in close collaboration with (key)stakeholders. SESAM will facilitate and guide research in the four steps by providing a toolbox of relevant methods for each of the research steps. The toolbox will be filled with a collection of existing tools and methods as well as potentially newly developed methods. Inspiration will be taken from relevant well-established gaming communities, namely companion modelling network [59] and the International Simulation and Gaming Association [60].

In step 1, a thorough understanding of the complexity and context of the studied landscape will be developed by identifying and exploring the current social-ecological system. Proposed methods for this include stakeholder assessment [61], (P)ARDI [62], and fuzzy cognitive mapping [63]. Through these methods a conceptual understanding of the system will be developed. In addition, the Q methodology is recommended to identify the current perceptions of various stakeholder groups. In Step 2, the game will be developed based on the conceptual understanding of the systems developed in Step 1. Key actors, elements, of the system and their interactions will be represented in the game to allow for relevant dynamics and patterns to be reproduced. From the initial stakeholder assessment, stakeholders will be selected to be actively involved in the development and implementation of the game. Step 3 will be part of an initial learning loop with Step 2 in the target area, although it is common to pre-test games on other target audiences (e.g., students...). If games are re-used beyond their initial place of origin, Step 3 can follow directly from Step 1. Step 4, the impact of using serious games on (social) learning and stakeholder opinions will be assessed through qualitative, and semi-quantitative analysis. Some of the existing innovative methods, e.g., Q-method, to distinguish between concurrent discourses and ways of explaining observed phenomena [64,65] are currently being explored in the context of game impact (before vs. after) assessment. Existing natural resource management frameworks [29] are also relevant in this context as a way to assess stakeholder understanding, perception, and willingness to act and adopt improved management options.

Reflecting on the types of games that we may need to develop and test, we may need at least four types of games (Figure 10): A. games that share the discovery process of a diagnostic stage (Table 4), B. games that focus on land-use decisions (Table 5), C. games that add hydrological consequences (with their human impacts) to land-use decisions (Table 6), and D. games that also include responses, where stakeholders outside the landscape try to influence land-use decisions (Table 7).

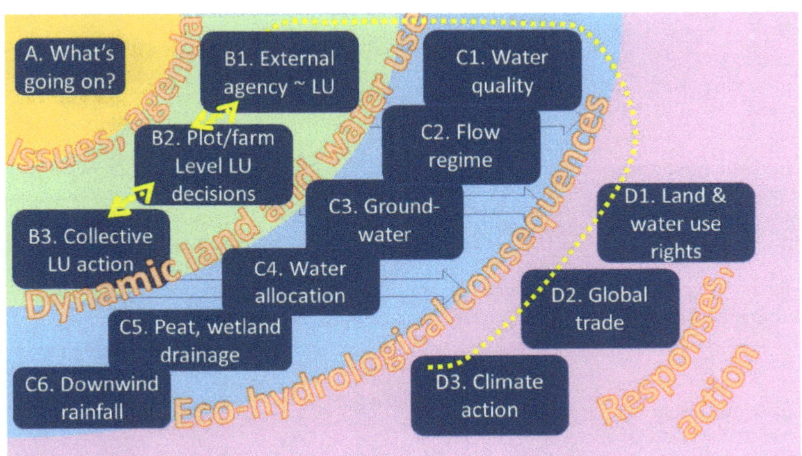

Figure 10. Initial game typology based on scope (A for agenda setting, B for better understanding land-use decisions, C for consequences of land-use decisions for water flows, D. for driver level responses by stakeholders external to the landscape considered in B).

Table 4. Prototype of 'What's going on' agenda-setting game (journalist/detective quest), with targeted/expected endpoints as hypothesis for further corroboration; type A in Figure 10.

Starting Point	Dynamic	Targeted/Expected End Point
An issue of public concern that involves water and people, and in which forests and trees may play a role.	One or more journalist/detective teams are formed and have opportunity to interact with stakeholders.	The various pieces of the puzzle come together and start to give an 'emergent' perspective on what's going on.
Multiple stakeholders of the issue have diverse interpretations of what is at stake, how it works, what are (alternative) facts.	Stakeholder groups have their own interpretation, e.g., deforestation, climate change, technical failures, water grabs, of underlying causes of the issue and possible solutions.	A first 'agenda setting' conclusion may well be that the issue is indeed an important one, that it is 'wicked' (no easy way out), requiring deeper analysis.
There is no consensus on 'what's going on', tensions may be rising, conflicts emerging.	If the journalists/detectives interact appropriately with a stakeholder group, they may get 'a piece of the puzzle'.	Depending on how the process is managed, an overarching 'framing' of the issue may emerge that is shared by all.

Table 5. Prototype of a who? what? where? land and water use game, with targeted/expected endpoints as hypothesis for further corroboration; type B in Figure 10

Starting Point	Dynamic	Targeted/Expected End Point
A locally recognizable functional terminology of land uses along the forest transition curve.	Land users (farmers, communities) make choices with direct consequences for their livelihoods, leading to an emerging 'land-use mosaic'.	Patterns of change in land (and water) use mosaic that are made visible along with the multiple 'causes' that were at play.
A spatial representation of topography, soils and water flows as interaction 'arena'.	B1: Focus on plot- and farm-level decisions, including trees and tree diversity.	Reported experience of players in various roles, (partially) achieving their goals.
Characterization of local livelihoods, on-farm and off-farm, leading to 'land use'.	B2: Focus on external agency, pulling and pushing local land-use decisions according to various agendas.	Clarity, within the game, on what are 'externalities' for the various actors and how this contributes to an overall result.
Identification of external agents, influences and pressures that shape land-use decisions.	B3: Focus on collective action in land and/or water use and the decision making that can enhance synergy.	Depending on physical landscape context, a better understanding of its role in shaping land use.

Table 6. Prototype of a so what? who cares? game with eco-hydrological consequences, with targeted/expected endpoints as hypothesis for further corroboration; type C in Figure 10.

Starting Point	Dynamic	Targeted/Expected End Point
A climate plus topography description of abiotic context.	C1: Focus on water quality ('pollution'), consequences for health, sedimentation.	As follow-up to games A and B, clarify the consequences for a wider range of 'stakeholders.
Pre-human vegetation interacting with abiotic context.	C2: Focus on water quantity and flow regime: water yield, floods, droughts.	C1–C5: Identify downstream people influenced by decisions made upstream: 'who cares?'
Human land use modifying vegetation, soils, drainage pat-terns (as shaped in games B).	C3: Focus on (blue) water availability and its allocation to (appropriation by) competing users.	C6: Identify downwind people influenced by decisions made upwind: 'who cares?'
Awareness (based on game A) of the 'down-stream' issues land-use change can influence.	C4: Focus on groundwater recharge and availability through springs and wells.	Identify vulnerability to climate change (trend, increased variability).
A technical water balance model that may stay in the background, but provides 'ballpark' rules for the game.	C5: Focus on wetland and peat drainage and its consequences for subsidence and/or fire risk.	Identify the contributors to 'buffering' that reduce impacts of external variability.
Climate variability and climate change scenarios that provide challenges to existing land use.	C6: Focus on atmospheric moisture recycling, downwind effects on rainfall.	Reflect on the 'wicked' nature of the underlying issue (game A).

Table 7. Prototype of a 'closing-the-loop' game, with targeted/expected endpoints as hypothesis for further corroboration; type D in Figure 10.

Starting Point	Dynamic	Targeted/Expected End Point
Current land (and water) use is a direct cause of problems downstream/downwind.	D1: Land-use planning and water use rights negotiations modify future land-use change and incentives (incl. PES?).	Unexpected winners and losers of various 'feedback loops', deepening the sense of 'wicked' problems.
Current land use is a resultant of local + external forces that expect to benefit from their choices, but don't take 'externalities' into account.	D2: Global trade as driver of land-use change becomes aware of its social and ecological 'footprint' and starts to take responsibility, e.g., by standards and 'certification'.	Deeper understanding of 'common but differentiated responsibility' in resolving issues at landscape, national and/or global scale.
Those affected by 'externalities can take action, depending on power relations, political and cultural context.	D3: Global climate action expands from its current carbon focus to concerns over water cycles and downwind effects of tree cover change.	Need to balance *'efficiency'* and *'fairness'* in interacting with social-ecological systems in a given cultural context.

7. Discussion: the Four Challenges to Use of Serious Games

Returning to the four issues restricting use of 'serious games' in the context of forest–water–people nexus issues, we need to take stock of how the SESAM program, in our current preview, will address the critique that games (1) are ad hoc with poorly defined extrapolation domains; (2) require heavy research investment from intervention experts; (3) have untested cultural limitations in where and how they can be used; and (4) lack clarity on where and how they can be used in policy making in local or global issue cycles. In doing so, we will need to return to the perspective of Figure 2, articulating it further in Figure 11. The figure shows (A) real-world social-ecological systems and their forest–water–people issues, (B) games as simplified representations of such systems, focused on specific aspects, and (C) use of these games in real-world contexts (potentially beyond where they were initially constructed). The match can be viewed from an ecological, a social or policy-oriented perspective.

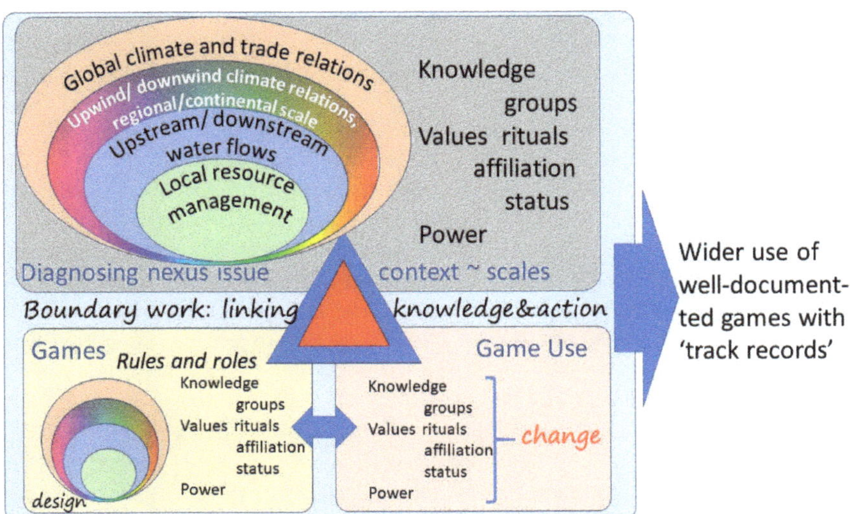

Figure 11. Real-world forest–water–people nexus issues across scale levels, games that reflect these, and game use that can modify the trajectory of real-world issues, where the same game can be used in different settings, in different ways, yielding a track record that can be documented.

7.1. Providing a Scaffold for Scenario Evaluation Games in the Forest–Water–People Nexus

In order to address the current rather ad hoc nature of serious gaming studies in agroforestry, SESAM aims to develop a scaffold for scenario evaluation games in the forest–water–people nexus that could form an example in other fields in which games have been increasingly used. We will build up a library of games, describing games based on scope, content, format, track record (experiences) and identify games (e.g., as subtitle) by generic issue-in-context (e.g., tropical mountain lake). By creating this overview of games, starting from the games developed within SESAM, but not limited to the set of SESAM games, the program aims to be able to draw grounded conclusions on the interplay at the core of the forest–water–people nexus and the relevance and contribution of a serious gaming approach in this context. Through a cross-case study comparative analysis, we aim to identify how different interpretations of forest–water interaction influences the understanding, awareness and action related to landscape change and management at different levels of governance in different locations on a pantropical forest transition curve.

SESAM will perform a comparative analysis of all games developed and their impact on actor learning in different contexts. This will provide insights into larger methodological gaming-related questions contributing to the wider application of serious games as a generic approach for decision making under high levels of complexity and uncertainty. SESAM will match case-by-case games with other landscapes with (a) similar issue, (b) similar ecosystem and social characteristics, (c) same country or region, while starting from a 'User demand' point of view ('what is the question?' rather than 'supply' 'we want to disseminate our research results'). The portfolio of games developed within the SESAM program should enable a more in depth understanding of (1) The relative value of different type of games, e.g., simple vs. complex, or fully defined vs. open, (2) game comfort zones linked to cultural context, age, gender and other social stratifiers, and, (3) learning effects (through before–after comparison) of game session participants who enter with different levels of knowledge. Classifying games and assessing the relative learning impact of gaming participants experienced will allow comparing learning from different game sessions with different games in distinct systems and contexts around the world. Based on this, recommendations will be developed to facilitate a wider use of well-documented games with track records.

7.2. Optimizing Research Investment in Game Development

Games currently rely on long-term, expensive prior research involvement. Optimizing research investment is key to support the broader use of serious games in participatory social-ecological systems research on the forest–water–people nexus and development of agroforested landscapes. SESAM will contribute towards optimizing research investments by (i) streamlining the serious gaming process from start to finish, and (ii) exploring options for the re-using games.

In order to streamline the serious gaming process, SESAM will develop guidelines for the development, implementation and analysis of games. These guidelines will consist of a standardized step-wise approach based on a number of complementary methods (see Section 6). By developing these SESAM guidelines and implementing them in our SESAM case studies, we aim to offer a streamlined game development process as well as to facilitate communication about games and between research teams in the serious gaming community. By doing so, SESAM aims to provide a scaffold to build upon existing knowledge and experience in the field of serious gaming in scenario evaluation and allow for research teams to connect and share experiences.

In addition, SESAM will address the question of transposability of games by exploring the use of games in contexts and settings beyond the original case study they have been developed in and assessing the impact of co-designed vs. off-the-shelf games. Some authors have already been successfully developing and exploring the implementation of games beyond their place or origin [66,67]. The development of such games that allow for more generic applications, requires the concepts used to be 'valid' and 'robust' in a wider range of circumstances, without claiming high accuracy in any specific location. Within SESAM we will cross-test games between case study areas within the comprehensive coverage of the forest–water–people nexus among the sites. SESAM will develop means of communication to assist in finding existing games and finding out whether they could be of use in a specific (new) situation. In addition, SESAM will describe how 'long-term' adaptability goals can be reconciled with 'short-term' match with data (e.g., by shifting from data-driven heuristics to first-principles models underlying the game).

SESAM will also develop Agent-Based Models (ABM) in parallel and interacting with game development, as this provides valuable insights in the adequacy of current process-level understanding of the social-ecological systems represented in the game [68,69].

7.3. Culture-Sensitive Gaming

Scenario games are artefacts that embody certain perceptions and values, but also allow various usages, e.g., through selection of player groups, setting of incentives, or usage of group discussions surrounding game rounds. The larger cultural setting influences these issues [52]. In SESAM we will have to create games that involve the future of agroforestry and the sustained livelihood of people. Based on existing national-level data on culture [51], there is a considerable 'band-width' for each of the six culture dimensions across the SESAM countries (Figure 9D), especially if the Netherlands is included as well. Within countries, there will undoubtedly be further variation, for example between the mountain and lowland forest people of Peru, the upriver and coastal zone people in Surinam and the islands (Sumatra, Kalimantan, Java) of Indonesia, within the SESAM portfolio.

The knowledge that unwritten rules of culture play a role implies that SESAM participants will be playing similar games across sites and looking for potential differences in game dynamics even if the pre-designed rules and incentives are identical. At design time, cultural differences between the designers and the intended users can cause blind spots [52]. We will be careful to design games that are meaningful and acceptable to participants. The local knowledge of our PhD candidates, as well as intensive contacts with stakeholders, should guarantee this. Dimensions of culture that we expect to be important for game design and dynamics are individualism and power distance—mentioned in one breath because they tend to be strongly correlated with tropical countries typically more hierarchical and collectivist than Western ones; and long-term orientation, which varies across countries in both the Tropics and the West. In conclusion, while a game is just an artefact and has no cultural awareness, its design and its usage can be done in culture-aware ways. SESAM intends to do this and document the results.

7.4. Game Relevance in the Policy Domain

One of the most confounding elements of policy issues is that there is a can of worms (and other parts of belowground biodiversity) of actors and interests, and nobody has an overview. This is precisely where games are strong: they provide a shared system boundary and show how actions of stakeholders impact the overall behavior of the system. In doing so, they also allow emotional responses to events, as well as joint (collective) action. In terms of system patterns, games allow to experience patterns such as the Tragedy of the Commons, or a Fix that Fails.

A review of 43 serious games and gamified applications related to water [70], covering a diversity of serious games, noted the still unsettled terminology in the research area of gamification and serious gaming and discussed how existing games could benefit early steps of decision making by problem structuring, stakeholder analysis, defining objectives, and exploring alternatives. Behavioral games on common pool resources may be used to facilitate self-governance [71], with groundwater a particularly challenging common pool resource to govern due to its low visibility, resulting in resource depletion in many areas. Serious games can promote values that transcend self-interest (transcendental values), based on the contributions of social psychology, but to do so, their design should incorporate the many value conflicts that are faced in real life water management and promote learning by having players reflect on the reasoning behind value priorities across water management situations [72,73].

This implies that potentially, not only playing games but also designing or adapting them can be very valuable for stakeholders in a policy-setting context [74]. A scenario game of one of the types in Figure 10 can be contextualized for and by local stakeholders; this exercise will generate relevant discussion and important learning. Being inclusive in which actors to select, as well as having support from important local persons, is essential [75]. Games also fit in a movement towards more plural and participatory approaches to 'valuation' of natural capital and ecosystem services [76].

All SESAM games will build on the boundary work tradition of taking the three 'ways of knowing' (local ecological knowledge, public/policy knowledge and science-based knowledge) as potentially complementing each other [32,77], with scope for new solutions to emerge at interfaces.

How can game use trigger policy change? From the five steps mentioned in Figure 3, it is relatively easy to see how games (including those of type 1) can be used for 'Agenda setting' and raising awareness of issues. Games are also a good vehicle for the 'Better understanding' part, as they offer insights not only into what happens, but how actor decisions contribute to outcomes of the game.

Hypothesis 1. *The impact of serious games on 'agenda setting' and 'better understanding' parts of issue cycles is reflected in increased consensus about what is important to do and what changes can be expected to result from actions.*

However, more is needed to nudge decisions locally and/or globally into desirable transformative and re-imaginative (game-changing) directions: commitment to aspirational goals.

Hypothesis 2. *Embedding serious games in a stakeholder negotiation process, can contribute to commitment to aspirational goals for addressing underlying causes ('drivers'), while addressing immediate symptoms.*

Real progress depends, beyond ambitious policy language, on action [6]. Most 'serious games', however, will have limited precision and use an oversimplified problem description. Scenario analysis at 'implementation' level will often require more detailed spatial analysis of tradeoffs.

Hypothesis 3. *Games are not a safe basis for operational decisions due to their limited specificity and (spatial) precision.*

We expect that the SESAM studies will provide further evidence to judge the contexts in which 'boundary work' in the forest–water–people nexus can be supported through serious games to support game-changing transformations at 'driver' level.

8. Conclusions

Four challenges have been raised to the use of serious games in addressing issues such as those in the forest–water–people nexus: games so far (1) appear to be ad hoc, case dependent, with poorly defined extrapolation domains, (2) require heavy research investment, (3) have untested cultural limitations and (4) lack clarity on where and how they can be used in policy making. Reviewing the literature and considering a set of case study landscapes, we conclude that these challenges can be addressed at the design, testing and communication stages of games to be shared in a wider community.

The SESAM program of networked PhD research programs will be geared at credible, salient and legitimate action research designing, testing and using 'serious games', that are meant to (1) be systematic in their coverage of the pantropical forest–water–people nexus in its main manifestations and issues, using generic forest and tree cover transitions as continuum description rather than forest–agriculture dichotomies, supporting easy 'localization' of games to match local contexts, (2) use 'basic hydrological, ecological, social and economic concepts, (3) be cognizant of the main dimensions in which cultural contexts of inter-human and human–nature relations vary to guide responsible game use, and (4) be explicitly adapted (or adaptable) to different stages of local and global issue cycles. The 13 landscapes provide a wide diversity of forest–water–people nexus aspects, as they range from low (<1 km^{-2}) to very high (>1000 km^{-2}) human population densities, high (>90%) to low (<5%) forest cover, five ecoclimatic zones (from drylands, sub-humid, via 'water tower' to humid and per-humid), and have focal issues that range from concerns over groundwater and streamflow depletion, flood-and-drought flow regimes, erosion and water quality to atmospheric moisture recycling, all (supposedly) influenced by quantity, quality and spatial pattern of tree cover ('agroforestry'). Game prototypes are described for (1) a diagnostic phase where multiple explanations for identified local issues are explored, (2) a deeper understanding of individual and collective land-use decisions, (3) explicit consequences for a range of ecohydrological landscape functions of such decisions and (4) societal feedback to land users based on landscape-level consequences, through land and water use rights ('planning'), global trade and climate action. Games are expected (hypothesized) to help in agenda setting phases, in achieving a common understanding of what is at stake and in political commitment to solutions—but will need more specific information to guide decisions on actual solutions. These hypotheses will be tested in the coming years.

Author Contributions: Conceptualization, M.v.N., E.S. and G.J.H.; Writing—original draft, M.v.N., E.S., G.J.H., A.F., A.Y.A., A.M., A.L.H., C.N.W., E.L., F.A., G.K., G.G.C.A., L.B., L.T., M.G., P.R., R.R.S. and U.S.; Writing—review and editing, S.A., A.L., P.v.O., C.M., M.P.-C., E.P., D.R., D.S. and A.J.T. All authors have read and agreed to the published version of the manuscript.

Funding: This research was funded by Wageningen University as one its INREF programs, with further support from twelve participating institutions.

Acknowledgments: The authors acknowledge assistance by Toha Zulkarnain for interpreting Forest Transition stages of the sites.

Conflicts of Interest: The authors declare no conflict of interest.

Appendix A. Brief Descriptions of the SESAM Landscapes

Appendix A.1. Upstream Remote Forests of the Suriname River Basin

Suriname, a country on the Guiana Shield in Latin America, is rich in rainforests and freshwater resources [78,79]. Its average annual rainfall in the area of 2700 mm [80] is an important source of evaporation that precipitates further inland in South America [41]. The Upper Suriname River Basin (USRB) is still mostly covered by primary and secondary forests [81]. The USRB is since several centuries inhabited by the Saamaka afro-descendent groups who, now around 20,000 people [82], live in 62 villages along the river. The Saamaka do not have legal land tenure rights, permanent electricity or running water, but live off ecosystem services provided by the forest (e.g., food, water, medicines, materials for housing and boats) [77]. The main income sources are rain-fed shifting cultivation and non-timber forest products, especially for the women. However, nature-based tourism has also become important. Some villages have community forest concessions, exploited by third parties. Competing interests and the community's increasing participation in market economies lead to land-use conflicts and sustainability challenges. There is a clear need for better coordination and improved capacities to manage the USRB in a sustainable way, benefitting rights and stakeholders. Government decentralization was only partly successful as district governments have limited capacity and budget, and coordination between stakeholders remains poor. There have been recent improvements in forestry policies, but enforcement is a challenge. Forest, water and people in the USRB are inextricably linked, yet knowledge and understanding are lacking about these interactions and how different land uses affect them. Sustainably managing forests for their vital role in watersheds, precipitation sheds [83] and global climate is crucial, especially for vulnerable communities such as the Saamaka. Our study aims to improve shared understanding on the multi-scale effects of land-use changes on the forest–water–people nexus and apply gaming as a tool to enhance coordinated management of the landscape.

Appendix A.2. Upstream Forests of the Amazon in Madre de Dios, Peru

The Madre de Dios region, in the western part of the Amazon basin, is located in the south-eastern part of Peru and shares borders both with Brazil and Bolivia. The region is almost entirely covered (95%) by forest [84] and its mean annual precipitation is 2200 mm [85]. The region's name originates from the Madre de Dios river that forms part of the vast Amazon River watershed [86]. The population of about 150,000 inhabitants is growing at a 2.6% yearly rate, partly due to immigration [87]. This migration originates mainly in the highlands and is driven by the economic perspectives offered to rural workers in the sectors of (illegal) gold mining and informal agriculture. Both these sectors have benefited of the recent pavement of the Inter-oceanic Highway, along which all recent deforestation hotspots are located [88]. In order to provide economic alternatives to mining, the regional government has started investing more resources in the agricultural sector. This strategy is mainly focused on cash crops, i.e., cacao (native to the Amazon) with a cultivated surface that has risen about 100% between 2010 and 2017 [87]. This agricultural intensification is likely to affect most smallholder farmers in the region and the high agrobiodiversity that is traditionally managed on their land. While small-scale farming remains little studied in the region, there is a growing need for transdisciplinary research that, by giving a voice to the farmers, allows them to become part of the conversation about land use. This case-study aims to provide a better understanding of smallholder farmer's decision-making processes and how it is influenced by their systemic perception, i.e., the forest–water–people nexus. In the longer run, this will provide NGOs and the regional government with tools to better adapt their development programs to the local farmer's needs and visions [89].

Appendix A.3. Mangrove Coasts of Suriname

The majority of the population of Suriname resides in the coastal area where also 90% of the economic activities is concentrated. The 370-km-long coastline harboring the largest and pristine mangrove forest of the Guianan Ecoregion serves the entire nation with its numerous ecosystem services. The Districts Nickerie and Paramaribo represent rural and urban parts of this coastal zone. The Suriname climate is tropical humid, with an average air humidity in the coastal area of 80-90% and a north-eastern (land inward) wind direction. Poor mangrove forest management and its ultimate destruction threaten the livelihoods of coastal communities, but also removes flood protection for the whole coastal zone. The root cause of the problem is lack of awareness about the mangrove ecosystem, its services and the effects of stakeholder activities on mangrove ecosystem services. Mangrove management can be achieved with the current legislation; however, the legislation is very much fragmented and sectoral in its orientation. In 2019, a national mangrove strategy has been developed for the ministry of Spatial Planning, Land and Forest Management. This study will apply serious games to increase awareness as well as to stimulate an effective decision-making process among stakeholders.

Appendix A.4. Amazonian Agroforestry Mosaics in Para State, Brazil

Pará State, which comprises a large portion of the Eastern Amazon basin in Brazil, has the highest level of deforestation of any subnational area in the tropical world [90] and is deemed a critical hotspot to contain the Amazon dieback [91] (breakdown of rainfall recycling). It thus has become a priority landscape for corporations to promote deforestation-free supply chains [92]. Nestled along the Acará river in the mesoregion of Northeastern Pará, the municipality of Tomé-Açu, has a population density of 10 km^{-2} and an area of roughly 5000 km^2. It is currently dominated by oil palm, pastures and perennial crops, particularly cocoa, cupuaçu (Theobroma grandiflorum), black pepper, and açaí (Euterpe oleracea), a native palm, as well as slash and burn agriculture for cassava production, a mainstay of local livelihoods, and other annual crops [93]. Tomé Açu has seen 56% of its original forest cleared, mostly dating back to the 1970s and 1980s, whereas some neighboring municipalities still face high deforestation rates to this day. Since 2010, the municipality has been a hotspot of oil palm expansion against a backdrop of logging and extensive grazing juxtaposed by clusters of agroforestry innovation. Plummeting prices and fusarium disease that ravaged monocrop black pepper plantations in the 1960s and 1970s led Japanese settlers [94] to diversify their production systems to reduce such risks, developing what became known as the Tomé Açu Agroforestry Systems (SAFTAS) supported by a fruit-processing cooperative (CAMTA) [95]. These commercially oriented systems have become a key example widely followed by family farmers in Tomé Açu and far beyond in the Brazilian Amazon. This case study aims to shed light on the factors underlying agroecological intensification and agroforestry transition pathways in the Tomé Açu landscape, focusing on the constraints and levers for scaling agroforestry, with and without oil palm, and trade-offs between different agroforestry systems and other land-use options.

Appendix A.5. Tropical Peatland Restoration in Indonesia

Tropical peatlands form where drainage is limited, and organic matter decomposition cannot keep up with its production—often in lowland interfluvial locations. In the Ketapang Peatland in the coastal zone of West Kalimantan, (Indonesia) two Peatland Hydrological Unit (PHU): PHU Tolak river—Pawan river (948 km^2) and PHU Pawan river—Pesaguan river (1048 km^2) were selected to explore options for hydrological restoration. Human population density in these areas is low by Indonesian standards (20 people km^{-2} in 2017) and is dominated by people of Malay background, along with migrants from Java and Bali. Average annual rainfall is 3168 mm [96], but the June—October period is relatively dry in most years and, especially in El Niño years serious droughts develop. The area is dominated by logged-over peatland forest and oil palm plantations, with some rubber plantations and paddy rice fields. Expansion of plantations and agricultural areas has led to massive canal construction [97]. Besides their functions for log transport and human access, these canals also drain water from the peatlands so that oil-palm, rubber and crops can grow. As a result, however, the peatland becomes drier during the dry season and more vulnerable to forest fires, with haze directly affecting human health and well-being. Drainage can also cause saline seawater intrusion into the coastal zone. Peatland restoration aims to restore the hydrological function in the peatland area by increasing and stabilizing the groundwater levels in the peat dome, especially in the dry season, to reduce fire risk. Understanding the relationship between the environment and people is the key to build commitment among all stakeholders, so they can engage and cooperate to address the environmental issues. The objective of this study is to develop communication tools that can use to increase the level of understanding and to facilitate the communication process among stakeholders.

Appendix A.6. A Tropical Mountain Lake: Singkarak in Sumatra, Indonesia

Lake Singkarak is one of the highland lakes in the Bukit Barisan mountain range that runs the length of Sumatra. Inflow to the lake derives from an area of 1135 km^2 through a number of streams and rivers, some with smaller natural lakes that provide some buffering. The watershed is home to 440,000 people with intensive horticulture in the upstream areas and extensive paddy cultivation before inflow to the lake. Located across two districts (Solok and Tanah Datar) the lake provides opportunities for fishing (an endemic fish, overfished due to high demand), year-round water supply for surrounding villages and tourism. The outflow of the lake provides irrigation water for downstream farmers and hydropower for West Sumatra Province. With annual rainfall ranging from 1700–3200 mm there is enough water for the hydropower plant in average years, but in years with long dry seasons there is a shortfall, as dropping the water level in the lake disturbs local livelihoods. The W side of the lake is dominated by forest, mixed gardens (agroforests) and agricultural fields, but the E Side is drier and dependent on the natural outflow from the lake that was disturbed by the construction in the 1990's of the Singkarak Hydro Electric Power Plant [98]. There have been a number of reforestation project implemented in the area to increase amount of forest cover [99] and the local village surrounding the lake tried to protect the water quality and endemic fish by setting up their village regulation. Understanding the water allocation management (between water flows in the original riverbed flowing East), supporting traditional rice farmers, and the use for hydropower and irrigation schemes west of the mountain range [96]. The objective of this study to have a tool to support and enhance social learning and action by actors involved in multi-level decision-making processes around the nexus, and to explore how participatory decision-making on water and (agro)forest landscape management can be improved as part of climate change adaptation.

Appendix A.7. Water Tower for Adjacent Drylands: Mount Elgon, Uganda

Overlapping the international boundary between Uganda and Kenya, Mount Elgon is the 7th highest mountain in Africa rising to 4320 m a.s.l. It is approximately 100 km North-east of Lake Victoria. The mean annual rainfall is 1600 mm [100]. Forests in the Mt. Elgon ecosystem have become restricted to the protected upper slopes (23% forest cover in 2016). With 1000 people km^{-2} and a 3.4% annual increase the mid-slope is a densely populated agricultural landscape [101]. The region is a highly productive agricultural zone, growing arabica coffee and horticultural crops. The watershed contributes to Lake Victoria (and thus to its outflow, the Nile river), and lake Turkana) [102]. However, the ecosystem functionality and integrity has been compromised and impacted by climate change. The population increase has directly raised demand and competition for natural resources including land and water [103,104]. Due to land inheritance, land fragmentation is common and is expected to worsen with population increment mounting pressures on resources [105,106]. Besides, the declining land productivity has led to reduction in food produce thus deficit in food supplies to the continuously fast-growing human population. The resultant food gap has sparked encroachment into the national park and on riparian zones of riverbanks, swamps and steep slopes leading to soil erosion, siltation and flooding. There is overgrazing, destruction of forest for urban extension and high levels of conflict between the park authorities and locals. Landscape actors with divergent values/interests include resource users, local government, national conservation agencies, international donors and NGOs, local politicians, etc. To strengthen local natural resource governance, participatory scenario evaluation games for supporting and enhancing joint planning and social learning by actors involved in multi-level decision-making processes around the forest–water–people nexus will be developed and implemented.

Appendix A.8. Water Tower for Adjacent Drylands: the Ewaso Ng'iro River NW of Mount Kenya

Mount Kenya is located on the equator and rises to 5199 m a.s.l, 180 km north of Nairobi. The mountain has different zones of influence. The eastern climatic gradient is relatively humid (windward) and the western climatic gradient (leeward) stretches from humid (upstream) to semi-arid areas in the downstream. The mountain contributes about 50% of the entire flow of the Tana River, the largest river basin in Kenya providing water supply to 50% of the Kenyan population and 70% of the country's hydroelectric power. Both rainfed and irrigated agriculture are major economic activities in the upstream. Further downstream many others rely on the river flows for irrigation, energy generation, pastoralism and hotels/tourism. Problems are related to over-abstraction of water [107], farm encroachment to fragile lands and deforestation [108]. Over-abstraction of river water in upper reaches and unwillingness to be held accountable through metering has been a major threat to meeting downstream flows. Different water management institutions exist, but activities are uncoordinated, leading to gaps in water resources management. The national government reacts by closing down water intakes in desperate attempts to resolve downstream water crises. Such orders affect schools, hospitals and thousands of households in the downstream. The local politicians strongly oppose the national government and defend the community water projects (most of them are unmetered). Consequently, an endless cycle of water crises characterizes the Mt. Kenya region [109]. Climate change and land-use changes continue to exert pressure on limited water resources. There is a need for increased knowledge and understanding among stakeholders for sustainable solutions. This study will explore the influence of climate change and land-use changes on downstream hydrology and apply serious games to explore decision making processes among stakeholders. This will provide a better understanding and knowledge for sustainable development.

Appendix A.9. Mountain Farming in the Andes, Peru

The Peruvian region of Puno is located in the south-eastern Peru on the shore of Lake Titicaca close to the border with Bolivia. The region is poorly covered (0.1%) by forest [84] and the mean annual rainfall averages 700 mm [110]. The watershed of Lake Titicaca extends for 8490 km^2. The highlands—over 3000 m high—region of Lake Titicaca supports a population density of 5.1 km^{-2} [87]. In the region of Puno of the Altiplano, in Peru, farmers with their knowledge and practices are growing the highest quinoa diversity hotspot in the world [111]. The potential of quinoa was promoted by the United Nations during the International Year of Quinoa (IYQ) in 2013 [112]. However, IYQ-2013 did not cover aspects of the worldwide spread of and commercial interests in quinoa, and the unbalanced competition between producers from the Andes and producers from North America and Europe [113]. Collective governance instruments as collective trademark (CT) are used to defend property rights on trading products and to recognize their anteriority and origin. Co-developing an Andean quinoa CT rise our research objective: explore the process of co-constructing a CT for recognize and promote quinoa in the global market. The region of Puno in Peru (the highest quinoa diversity hotspot in the world) is selected as case study. Participatory games and agent-based models will be developed and explored to assess the gap between local and regional farmers, and higher level including Andean quinoa farmers for developing an Andean CT for quinoa. The proposed research will explore the role CT can play to preserve the Andean generic hotspot of quinoa as well as to provide a potentially relevant governance tool for other neglected crops that suddenly get global consumer attention.

Appendix A.10. Farming Drylands on the Mossi Plateau, Burkina Faso

The Mossi Plateau in the central part of Burkina-Faso in West-Africa spans the Sudano–Sahelian climate gradient, with unimodal rainfall from June to September, with an annual average ranging from 400-700 mm [114]. Two villages (Yilou and Tansin) represent this gradient, with sandy clay loam soils, low in soil organic carbon (0.2 and 0.4, respectively) [115]. Agriculture in this area is challenged by low and unstable biomass production, limiting farmers' resilience. Water, nutrients, biomass, labor, information and money move in the landscape, while many studies focused on biomass production at farm scale only considering single households [116–118], without their interactions. Biomass management through farmers' decisions on biomass allocation [119], include use of crop residues as livestock feed [120], indirectly influencing manure availability for crop production. At village scale, biomass production is determined by farmers' organization, the spatial and temporal interactions amongst farmers, and between farmers and their biophysical environment (soil fertility), as farmers can organize themselves in "labor-groups" to till each other's fields and thus be able to cultivate more land. As rainfall tends to vary spatially, risk management depends on the collective use of space. A spatio-temporal modelling approach [121,122] needs to take the various spatial and social interactions into account. We aim to achieve this by co-designing tailored and realistic biomass management and organization options to improve farmers' production and livelihoods through participatory modelling.

Appendix A.11. Upland Agroforestry Mosaics in Kali Konto, East Java, Indonesia

The Kali Konto sub-watershed in East Java contributes to the Brantas river, and covers an area of approximately 240 km^2 with elevation ranging from 600-2800 m a.s.l. The Kali Konto river is approximately 40 km long and empties into Selorejo reservoir. The average daily temperature is 20–22 °C with a mean annual rainfall ranging from 2995–4422 mm [123]. Forest area has significantly decreased in the last two decades (from 30.5% of forest cover in 1990 to 20.4% in 2005), following the increased demand for timber, firewood, fodder and other agriculture (and dairy) products [124,125]. On the other hand, annual crops area increased by 26%. In the last three decades, the population increased by approximately 20% from 1990 adding more pressure to the system [126]. Intensive agricultural farming with minimum tree cover in hill slope leads to the increase of soil surface exposure, which is followed by higher soil erosion and sedimentation (top 'fertile' soil washed away) to the river [127] and impacts on crop production [128]. The polluted reservoir brings negative consequences for both the state water management corporation that is responsible for managing water quality and quantity in the region and downstream 'water' beneficiary (HIPAM–drinking water user association) [129]. Efforts have been made by PJT to encourage soil conservation practices. However, it has not yet accommodated farmers' choice on the type of land management in the program. It is essential to understand farmers' preferences and decision making along with the direct consequences for livelihoods and environmental impact. It benefits on exploring environmentally friendly management options based on their preference and increase the awareness on the importance of achieving sustainable watershed conservation through collaborative action among stakeholders.

Appendix A.12. Water Tower for a Metropole: Rejoso, East Java, Indonesia

The Rejoso watershed stretches from the Mount Bromo volcano crater (summit at 2329 m a.s.l.) to the Pasuruan coastal area (0 m a.s.l.) covering an area of 62,773 ha. Land in the watershed is used in a variety of ways: a national park (conservation forests), protected forests (natural jungle forests), monoculture production forests, agroforestry (production forest + agricultural + livestock), irrigated lowland agriculture and built-up areas. The watershed covers 14 sub-districts in Pasuruan Regency and three sub-districts in Pasuruan City with a total population in 2018 of 838,313 people [130,131], with 100,497 people living in the upstream part, 271,908 in the middle zone and 465,728 in the downstream area. Most of the population, except residents of Pasuruan City, depend for their livelihoods on their agricultural and agroforestry systems, supported by the ecosystem services provided by land and water resources, including ecotourism [132–134]. They worked collaboratively (but there are conflicts as well) with multi-stakeholder institutions in charge of managing natural resources. Since 2003, for example, residents in all parts who are members of the forest village community (LMDH) collaborate with Perhutani (a state-owned forestry company) in 'the Community Joint Forest Management Program (PHBM). They are allowed to plant (intercropping) various agricultural commodities (horticulture, food crops, and fruit trees) and fodder in between the stands of production forest trees. They also get wages and/or profit-sharing after contributing to maintaining the main trees (teak, pine, mahogany, eucalyptus) which are managed together with Perhutani. There are at least 114 village communities that use the Perhutani area of 11,713 ha [135]. Before the PHBM was implemented, conflicts often occurred between communities and Perhutani. In the upstream part, communities collaborate with national park managers (Ministry of Environment and Forestry (MoEF)), natural resource conservation offices (MoEF), and tourism agencies authorities. Another collaboration is 'the Rejoso-Kita Program' which implements an integrated watershed management model. Hydrological analysis suggests that the tree cover required for 'infiltration friendly' land uses depends on the altitudinal zone [136]. The program involved various stakeholders: communities, NGOs, private sectors, and multi-level governments [137–139]. Although, there have been collaborations, conflicts have also been reported, including communities and the private sector (mining industry in the middle and water industry in the downstream). The study to be carried out at this location aims to design participatory collective action games to strengthen sustainability community–private–government collaboration in agroforestry management.

Appendix A.13. Rehabilitating a Water Tower Under Pressure: Brantas, East Java, Indonesia

The Upper Brantas sub watershed is the source area of the Brantas river that starts from the southwestern mountainous slope of active volcanoes of the Arjuna-Anjasmara mountain complex (a protected forest area). Originating at the Sumberbrantas spring, the river flows southward through the cities of Batu and Malang, before bending to the West and then North to the Provincial capital Surabaya. Total area of the sub watershed is 180 km^2. Annual rainfall ranges from 875 to 3000 mm [140]. Its forest fraction is 23.8%. The Upper Brantas sub-watershed population size in 2018 was 207,490, with a density 1092 km^{-2} and an annual population growth of 0.95 per year [141]. Forest encroachment was a serious environmental problem in the 1998–2000 era. Conflicts over state forest land became visible in 1998 during the political reformation, when many local farmers occupied state forest land [142]. In this period power was decentralized to provincial and district levels. This period meant the end of timber forest management. Over 2003–2007, the settlement area increased by 9%, plantation and farms increased by 7%, and the forest area decreased by 6%. The watershed response can be observed through the increasing of run-off coefficient from 0.59 to 0.67 [143] and an increase in peak discharge (at the outflow of the sub-watershed) from 96.8 m^3 s^{-1} in 2003 to 189.2 m^3 s^{-1} in 2007 [144]. Meanwhile the springs were affected (data to be further collected) and groundwater recharge was probably reduced. Policy level response to the forest encroachment was a number of new programs. In 2001 the state forest agency introduced a co-management program, specially focused on forest resource management. In 2007 the co-management concept became relabeled as community-based forest management. The objective of the case study is to develop games and simulations of land-use change of recharge area management for sustainable groundwater supply and demand. There are many volcanic slope watersheds in Indonesia similar to the upper Brantas sub-watershed.

References

1. van Noordwijk, M.; Coe, R.; Sinclair, F.L. Agroforestry paradigms. In *Sustainable Development through Trees on Farms: Agroforestry in its Fifth Decade*; van Noordwijk, M., Ed.; World Agroforestry (ICRAF): Bogor, Indonesia, 2019; pp. 1–12. Available online: http://www.worldagroforestry.org/trees-on-farms (accessed on 20 July 2020).
2. Sustainable Developent Goals. Available online: https://sustainabledevelopment.un.org/sdgs (accessed on 21 June 2020).
3. Taleb, N.N.; Douady, R. Mathematical definition, mapping, and detection of (anti) fragility. *Quant. Financ.* **2013**, *13*, 1677–1689. [CrossRef]
4. Meadows, D. Leverage Points: Places to Intervene in a System. 1999. Available online: http://donellameadows.org/archives/leverage-points-places-to-intervene-in-a-system/ (accessed on 21 June 2020).
5. Minang, P.A.; van Noordwijk, M.; Freeman, O.E.; Mbow, C.; de Leeuw, J.; Catacutan, D. (Eds.) *Climate-Smart Landscapes: Multifunctionality in Practice*; World Agroforestry Centre (ICRAF): Nairobi, Kenya, 2015; p. 405.
6. Climate-Smart Landscapes: Multifunctionality in Practice. Available online: http://www.worldagroforestry.org/publication/climate-smart-landscapes-multifunctionality-practice (accessed on 20 July 2020).
7. van Noordwijk, M.; Lusiana, B.; Villamor, G.B.; Purnomo, H.; Dewi, S. Feedback loops added to four conceptual models linking land change with driving forces and actors. *Ecol. Soc.* **2011**, *16*. Available online: http://www.ecologyandsociety.org/vol16/iss1/resp1/ (accessed on 20 July 2020). [CrossRef]
8. van Noordwijk, M. Prophets, Profits, Prove It: Social Forestry under Pressure. *One Earth* **2020**, *2*, 394–397. [CrossRef]
9. Creed, I.F.; van Noordwijk, M. *Forest and Water on a Changing Planet: Vulnerability, Adaptation and Governance Opportunities: A Global Assessment Report (No. 38)*; International Union of Forest Research Organizations (IUFRO): Vienna, Austria, 2018; p. 188. Available online: https://www.iufro.org/fileadmin/material/publications/iufro-series/ws38/ws38.pdf (accessed on 20 July 2020).
10. Creed, I.F.; Jones, J.A.; Archer, E.; Claassen, M.; Ellison, D.; McNulty, S.G.; van Noordwijk, M.; Vira, B.; Wei, X.; Bishop, K.; et al. Managing Forests for both Downstream and Downwind Water. *Front. For. Glob. Chang.* **2019**. [CrossRef]
11. Nobre, C.A. Land-use and climate change risks in the Amazon and the need of a novel sustainable development paradigm. *Proc. Natl. Acad. Sci. USA* **2016**, *113*, 10759–10768. [CrossRef] [PubMed]
12. Page, S.E. *The Difference: How the Power of Diversity Creates Better Groups, Firms, Schools, and Societies (New Edition)*; Princeton University Press: Princeton, NJ, USA, 2008; p. 448.
13. Walker, B. Resilience management in social-ecological systems: A working hypothesis for a participatory approach. *Conserv. Ecol.* **2002**, *6*, 14. Available online: http://www.consecol.org/vol6/iss1/art14 (accessed on 20 July 2020). [CrossRef]
14. Reed, M.S. Stakeholder participation for environmental management: A literature review. *Biol. Conserv.* **2008**, *141*, 2417–2431. [CrossRef]

15. Scholz, G. An Analytical Framework of Social Learning Facilitated by Participatory Methods. *Syst. Prac. Action Res.* **2013**, *27*, 575–591. [CrossRef]

16. Angelstam, P.; Andersson, K.; Annerstedt, M.; Axelsson, R.; Elbakidze, M.; Garrido, P.; Grahn, P.; Jönsson, K.I.; Pedersen, S.; Schlyter, P.; et al. Solving problems in social–ecological systems: Definition, practice and barriers of transdisciplinary research. *Ambio* **2013**, *42*, 254–265. [CrossRef]

17. Tschakert, P. Anticipatory Learning for Climate Change Adaptation and Resilience. *Ecol. Soc.* **2010**, *15*, 11. Available online: http://www.ecologyandsociety.org/vol15/iss2/art11 (accessed on 20 July 2020). [CrossRef]

18. Rodela, R.; Ligtenberg, A.; Bosma, R. Conceptualizing Serious Games as a learning-Based Intervention in the Context of Natural Resources and Environmental Governance. *Water* **2019**, *11*, 245. [CrossRef]

19. García-Barrios, L.E.; Speelman, E.N.; Pim, M. An educational simulation tool for negotiating sustainable natural resource management strategies among stakeholders with conflicting interests. *Ecol. Model.* **2008**, *210*, 115–126. [CrossRef]

20. Villamor, G.B.; van Noordwijk, M. Social role-play games vs individual perceptions of conservation and PES agreements for maintaining rubber agroforests in Jambi (Sumatra), Indonesia. *Ecol. Soc.* **2011**, *16*, 27. [CrossRef]

21. Speelman, E.N.; García-Barrios, L.E.; Groot, J.C.J.; Tittonell, P. Gaming for smallholder participation in the design of more sustainable agricultural landscapes. *Agric. Syst.* **2014**, *126*, 62–75. [CrossRef]

22. Speelman, E.N.; van Noordwijk, M.; Garcia, C. Gaming to better manage complex natural resource landscapes. In *Co-investment in Ecosystem Services: Global Lessons from Payment and Incentive Schemes*; Namirembe, S., Leimona, B., van Noordwijk, M., Minang, P., Eds.; World Agroforestry Centre (ICRAF): Nairobi, Kenya, 2017; pp. 480–489.

23. García-Barrios, L.; Cruz-Morales, J.; Vandermeer, J.; Perfecto, I. The Azteca Chess experience: Learning how to share concepts of ecological complexity with small coffee farmers. *Ecol. Soc.* **2017**, *22*. [CrossRef]

24. Mayer, I.; Bekebrede, G.; Harteveld, C.; Warmelink, H.; Zhou, Q.; Van Ruijven, T.; Lo, J.; Kortmann, R.; Wenzler, I. The research and evaluation of serious games: Toward a comprehensive methodology. *Br. J. Educ. Technol.* **2014**, *45*, 502–527. [CrossRef]

25. Hofstede, G.J. One game does not fit all cultures. In *Organizing and Leaning through Gaming and Simulation, Proceedings of ISAGA, Nijmegen, The Netherlands, 9–13 July 2007*; Mayer, I., Mastik, H., Eds.; Eburon Uitgeverij BV: Nijmegen, The Netherlands, 2007; pp. 103–110.

26. Jann, W.; Wegrich, K. Theories of the policy cycle. In *Handbook of Public Policy Analysis: Theory, Politics, and Methods*; CRC Press: Boca Raton, FL, USA, 2007; Volume 125, pp. 43–62.

27. Clark, W.C.; Tomich, T.P.; van Noordwijk, M.; Guston, D.; Catacutan, D.; Dickson, N.M.; McNie, E. Boundary work for sustainable development: Natural resource management at the Consultative Group on International Agricultural Research (CGIAR). *Proc. Natl. Acad. Sci. USA* **2016**, *113*, 4615–4622. [CrossRef]

28. Kristensen, P. The DPSIR framework. In *Workshop on a Comprehensive/Detailed Assessment of the Vulnerability of Water Resources to Environmental Change in Africa Using River Basin Approach*; UNEP Headquarters: Nairobi, Kenya, 2004; p. 10. Available online: http://enviro.lclark.edu:8002/rid%3D1145949501662_742777852_522/DPSIR%20Overview.pdf (accessed on 20 July 2020).

29. Svarstad, H.; Petersen, L.K.; Rothman, D.; Siepel, H.; Wätzold, F. Discursive biases of the environmental research framework DPSIR. *Land Use Policy* **2008**, *25*, 116–125. [CrossRef]

30. Diaz, S.; Settele, J.; Brondízio, E.; Ngo, H.; Guèze, M.; Agard, J.; Arneth, A.; Balvanera, P.; Brauman, K.; Butchart, S.; et al. *Summary for Policymakers of the Global Assessment Report on Biodiversity and Ecosystem Services of the Intergovernmental Science-Policy Platform on Biodiversity and Ecosystem Services*; IPBES: Bonn, Germany, 2020; p. 56.

31. van Noordwijk, M. Integrated natural resource management as pathway to poverty reduction: Innovating practices, institutions and policies. *Agric. Syst.* **2019**, *172*, 60–71. [CrossRef]

32. Mithöfer, D.; van Noordwijk, M.; Leimona, B.; Cerutti, P.O. Certify and shift blame, or resolve issues? Environmentally and socially responsible global trade and production of timber and tree crops. *Int. J. Biodivers. Sci. Ecosyst. Serv. Manag.* **2017**, *13*, 72–85. [CrossRef]

33. Hofstede, G.J. GRASP agents: Social first, intelligent later. *Ai Soc.* **2019**, *34*, 535–543. [CrossRef]

34. van Noordwijk, M.; Dewi, S.; Leimona, B.; Lusiana, B.; Wulandari, D. *Negotiation-Support Toolkit for Learning Landscapes*; World Agroforestry Centre: Bogor, Indonesia, 2013; p. 285. Available online: http://www.worldagroforestry.org/publication/negotiation-support-toolkit-learning-landscapes (accessed on 20 July 2020).

35. van Noordwijk, M.; Leimona, B.; Jindal, R.; Villamor, G.B.; Vardhan, M.; Namirembe, S.; Catacutan, D.; Kerr, J.; Minang, P.A.; Tomich, T.P. Payments for Environmental Services: Evolution towards efficient and fair incentives for multifunctional landscapes. *Annu. Rev. Environ. Resour.* **2012**, *37*, 389–420. [CrossRef]

36. Shyamsundar, P.; Ahlroth, S.; Kristjanson, P.; Onder, S. Supporting pathways to prosperity in forest landscapes–A PRIME framework. *World Dev.* **2020**, *125*, 104622. [CrossRef]

37. Dewi, S.; van Noordwijk, M.; Zulkarnain, M.T.; Dwiputra, A.; Hyman, G.; Prabhu, R.; Gitz, V.; Nasi, R. Tropical forest-transition landscapes: A portfolio for studying people, tree crops and agro-ecological change in context. *Int. J. Biodivers. Sci. Ecosyst. Serv. Manag.* **2017**, *13*, 312–329. [CrossRef]

38. van Noordwijk, M.; Bruijnzeel, S.; Ellison, D.; Sheil, D.; Morris, C.E.; Sands, D.; Gutierrez, V.; Cohen, J.; Sullivan, C.A.; Verbist, B.; et al. Ecological rainfall infrastructure: Investment in trees for sustainable development. In *ASB Policy Brief 47*; World Agroforestry Centre: Nairobi, Kenya, 2015; p. 6. Available online: http://www.asb.cgiar.org/Publications%202015/PBs/ASB_PB47.p (accessed on 20 July 2020).

39. Ellison, D.; Morris, C.E.; Locatelli, B.; Sheil, D.; Cohen, J.; Murdiyarso, D.; Gutierrez, V.; van Noordwijk, M.; Creed, I.F.; Pokorny, J.; et al. Trees, forests and water: Cool insights for a hot world. *Glob. Environ. Chang.* **2017**, *43*, 51–61. [CrossRef]

40. van Noordwijk, M.; Creed, I.F.; Jones, J.A.; Wei, X.; Gush, M.; Blanco, J.A.; Sullivan, C.A.; Bishop, K.; Murdiyarso, D.; Xu, J.; et al. Climate-forest-water-people relations: Seven system delineations. In *Forest and Water on a Changing Planet: Vulnerability, Adaptation and Governance Opportunities: A Global Assessment Report (No. 38)*; International Union of Forest Research Organizations (IUFRO): Vienna, Austria, 2018; pp. 27–58.

41. van Noordwijk, M.; Tanika, L.; Lusiana, B. Flood risk reduction and flow buffering as ecosystem services: I. Theory on a flow persistence indicator. *Hydrol. Earth Syst. Sci.* **2017**, *21*, 2321–2340. [CrossRef]

42. van Noordwijk, M.; Tanika, L.; Lusiana, B. Flood risk reduction and flow buffering as ecosystem services II. Land use and rainfall intensity effects in Southeast Asia. *Hydrol. Earth Syst. Sci.* **2017**, *21*, 2341–2352. [CrossRef]

43. van der Ent, R.J.; Savenije, H.H.G.; Schaefli, B.; Steele-Dunne, S.C. Origin and fate of atmospheric moisture over continents. *Water Resour. Res.* **2010**, *46*, W09525. [CrossRef]

44. Ellison, D.; Wang-Erlandsson, L.; van der Ent, R.; van Noordwijk, M. Upwind forests: Managing moisture recycling for nature-based resilience. *Unasylva* **2019**, *251*, 14–26.

45. Teuling, A.J.; Taylor, C.M.; Meirink, J.F.; Melsen, L.A.; Miralles, D.G.; Van Heerwaarden, C.C.; Vautard, R.; Stegehuis, A.I.; Nabuurs, G.J.; de Arellano, J.V.G. Observational evidence for cloud cover enhancement over western European forests. *Nat. Commun.* **2017**, *8*, 14065. [CrossRef]

46. Tuinenburg, O.A.; Staal, A. Tracking the global flows of atmospheric moisture and associated uncertainties. *Hydrol. Earth Syst. Sci.* **2020**, *24*, 2419–2435. Available online: https://www.hydrol-earth-syst-sci.net/24/2419/2020/ (accessed on 17 July 2020). [CrossRef]

47. van Noordwijk, M.; Bargues-Tobella, A.; Muthuri, C.W.; Gebrekirstos, A.; Maimbo, M.; Leimona, B.; Bayala, J.; Ma, X.; Lasco, R.; Xu, J.; et al. Agroforestry as part of nature-based water management. In *Sustainable Development through Trees on Farms: Agroforestry in its Fifth Decade*; van Noordwijk, M., Ed.; World Agroforestry (ICRAF): Bogor, Indonesia, 2019; pp. 261–287.

48. Guswa, A.J.; Tetzlaff, D.; Selker, J.S.; Carlyle-Moses, D.E.; Boyer, E.W.; Bruen, M.; Cayuela, C.; Creed, I.F.; van de Giesen, N.; Grasso, D.; et al. Advancing ecohydrology in the 21st century: A convergence of opportunities. *Ecohydrology* **2020**, e2208. [CrossRef]

49. Kadykalo, A.N.; López-Rodriguez, M.D.; Ainscough, J.; Droste, N.; Ryu, H.; Ávila-Flores, G.; Le Clec'h, S. Disentangling 'ecosystem services' and 'nature's contributions to people'. *Ecosyst. People* **2019**, *15*, 269–287. [CrossRef]

50. van Noordwijk, M. Did Mother Nature consent in the human 'appropriation' of 'gifts' so sweet, or would a #MeToo hashtag get flooded if she could tweet? *Sci. E-Lett.* **2019**, *359*, 270.

51. Beugelsdijk, S.; Welzel, C. Dimensions and dynamics of national culture: Synthesizing Hofstede with Inglehart. *J. Cross-Cult. Psychol.* **2018**, *49*, 1469–1505. [CrossRef] [PubMed]

52. Minkov, M. A revision of Hofstede's model of national culture: Old evidence and new data from 56 countries. *Cross Cult. Strateg. Manag.* **2018**, *25*, 231–256. [CrossRef]
53. Dimension Data Matrix. Available online: https://geerthofstede.com/research-and-vsm/dimension-data-matrix/ (accessed on 20 July 2020).
54. Hofstede, G.; Hofstede, G.J.; Minkov, M. *Cultures and Organizations: Software of the Mind*, 3rd ed.; McGraw-Hill: New York, NY, USA, 2010; p. 561.
55. Hofstede, G.J. Culture's causes: The next challenge (Distinguished scholar essay). *Cross Cult. Manag.* **2015**, *22*, 545–569. [CrossRef]
56. Hofstede, G.J.; Murff, E.J. Repurposing an Old Game for an International World. *Simul. Gaming* **2012**, *43*, 34–50. [CrossRef]
57. Hofstede, G.J.; de Caluwé, L.; Peters, V. Why Simulation Games Work-In Search of the Active Substance: A Synthesis. *Simul. Gaming* **2010**, *41*, 824–843. [CrossRef]
58. Villamor, G.B.; Badmos, B.K. Grazing game: A learning tool for adaptive management in response to climate variability in semiarid areas of Ghana. *Ecol. Soc.* **2016**, *21*. [CrossRef]
59. Barreteau, O. Our companion modelling approach. *J. Artif. Soc. Soc. Simul.* **2003**, *6*, 1. Available online: http://jasss.soc.surrey.ac.uk/6/2/1.html (accessed on 20 July 2020).
60. International Simulation and Gaming Association. Available online: https://www.isaga.com/ (accessed on 20 July 2020).
61. Reed, M.S.; Graves, A.; Dandy, N.; Posthumus, H.; Hubacek, K.; Morris, J.; Prell, C.; Quinn, C.H.; Stringer, L.C. Who's in and why? A typology of stakeholder analysis methods for natural resource management. *J. Environ. Manag.* **2009**, *90*, 1933–1949. [CrossRef] [PubMed]
62. Etienne, M.; Du Toit, D.R.; Pollard, S. ARDI: A co-construction method for participatory modeling in natural resources management. *Ecol. Soc.* **2011**, *16*. Available online: http://www.ecologyandsociety.org/vol16/iss1/art44/ (accessed on 20 July 2020). [CrossRef]
63. Vanwindekens, F.M.; Baret, P.V.; Stilmant, D. A new approach for comparing and categorizing farmers' systems of practice based on cognitive mapping and graph theory indicators. *Ecol. Model.* **2014**, *274*, 1–11. [CrossRef]
64. Langston, J.; McIntyre, R.; Falconer, K.; Sunderland, T.; Van Noordwijk, M.; Boedhihartono, A.K. Discourses mapped by Q-method show governance constraints motivate landscape approaches in Indonesia. *PLoS ONE* **2019**, *14*, e0211221. [CrossRef] [PubMed]
65. Amaruzaman, S.; Leimona, B.; van Noordwijk, M.; Lusiana, B. Discourses on the performance gap of agriculture in a green economy: A Q-methodology study in Indonesia. *Int. J. Biodivers. Sci. Ecosyst. Serv. Manag.* **2017**, *13*, 233–247. [CrossRef]
66. Page, C.L.; Dray, A.; Perez, P.; Garcia, C. Exploring how knowledge and communication influence natural resources management with ReHab. *Simul. Gaming* **2016**, *47*, 257–284. [CrossRef]
67. Andreotti, F.; Speelman, E.N.; Van den Meersche, K.; Allinne, C. Combining participatory games and backcasting to support collective scenario evaluation: An action research approach for sustainable agroforestry landscape management. *Sustain. Sci.* **2020**. [CrossRef]
68. Ligtenberg, A.; van Lammeren, R.J.A.; Bregt, A.K.; Beulens, A.J.M. Validation of an agent-based model for spatial planning: A role-playing approach. *Comput. Environ. Urban Syst.* **2010**, *34*, 424–434. [CrossRef]
69. Van Oel, P.R.; Mulatu, D.W.; Odongo, V.O.; Willy, D.K.; Van der Veen, A. Using data on social influence and collective action for parameterizing a geographically-explicit agent-based model for the diffusion of soil conservation efforts. *Environ. Model. Assess.* **2019**, *24*, 1–19. [CrossRef]
70. Aubert, A.H.; Bauer, R.; Lienert, J. A review of water-related serious games to specify use in environmental Multi-Criteria Decision Analysis. *Environ. Model. Softw.* **2018**, *105*, 64–78. [CrossRef]
71. Meinzen-Dick, R.; Janssen, M.A.; Kandikuppa, S.; Chaturvedi, R.; Rao, K.; Theis, S. Playing games to save water: Collective action games for groundwater management in Andhra Pradesh, India. *World Dev.* **2018**, *107*, 40–53. [CrossRef]
72. Marini, D.; Medema, W.; Adamowski, J.; Veissière, S.P.; Mayer, I.; Wals, A.E. Socio-psychological perspectives on the potential for serious games to promote transcendental values in IWRM decision-making. *Water* **2018**, *10*, 1097. [CrossRef]
73. Buyinza, J.; Nuberg, I.K.; Muthuri, C.W.; Denton, M.D. Psychological Factors Influencing Farmers' Intention to Adopt Agroforestry: A Structural Equation Modeling Approach. *J. Sustain. For.* **2020**. [CrossRef]

74. Guyot, P.; Honiden, S. Agent-based participatory simulations: Merging multi-agent systems and role-playing games. *J. Artif. Soc. Soc. Simul.* **2006**, *9*, 8. Available online: http://jasss.soc.surrey.ac.uk/9/4/8.html (accessed on 20 July 2020).

75. Speelman, E.N.; Rodela, R.; Doddema, M.; Ligtenberg, A. Serious gaming as a tool to facilitate inclusive business; a review of untapped potential. *Curr. Opin. Environ. Sustain.* **2019**, *41*, 31–37. [CrossRef]

76. Zafra-Calvo, N.; Balvanera, P.; Pascual, U.; Merçon, J.; Martin-Lopez, B.; van Noordwijk, M.; Heita Mwampamba, T.; Lele, S.; Ifejika Speranza, C.; Arias-Arévalok, P.; et al. Plural valuation of nature for equity and sustainability: Insights from the Global South. *Glob. Environ. Chang.* **2020**, *63*, 102115. [CrossRef]

77. Leimona, B.; Lusiana, B.; van Noordwijk, M.; Mulyoutami, E.; Ekadinata, A.; Amaruzaman, S. Boundary work: Knowledge co-production for negotiating payment for watershed services in Indonesia. *Ecosyst. Serv.* **2015**, *15*, 45–62. [CrossRef]

78. Ouboter, P.E. (Ed.) *The Fresh Water Ecosystems of Suriname*; Kluwer Academic Publishers: Dordrecht, The Netherlands, 1993; p. 315.

79. Misiedjan, D.J.E. Towards a Sustainable Human Right to Water: Supporting Vulnerable People and Protecting Water Resources with Suriname as a Case Study. Ph.D. Thesis, Utrecht University, Utrecht, The Netherlands, 2017; p. 220.

80. Nurmohamed, R.; Naipal, S.; Becker, C. Changes and variation in the discharge regime of the Upper Suriname River Basin and its relationship with the tropical Pacific and Atlantic SST anomalies. *Hydrol. Process.* **2008**, *22*, 1650–1659. [CrossRef]

81. Ramirez, S.O.I.; Verweij, P.; Best, L.; van Kanten, R.; Rambaldi, G.; Zagt, R. P3DM as a socially engaging and user-useful approach in ecosystem service assessment among marginalized communities. *Appl. Geogr.* **2017**, *83*, 63–77. [CrossRef]

82. *Census Statistics: Population by Resort*; Statistics Bureau Suriname: Paramaribo, Suriname, 2012.

83. Keys, P.W.; van der Ent, R.J.; Gordon, L.J.; Hoff, H.; Nikoli, R.; Savenije, H.H.G. Analyzing precipitationsheds to understand the vulnerability of rainfall dependent regions. *Biogeosciences* **2012**, *9*, 733–746. [CrossRef]

84. Global Forest Watch. Dashboard by Country and Region. Available online: https://www.globalforestwatch.org/dashboards/country/PER (accessed on 20 July 2020).

85. Climate-data, Puerto Maldonado Climate. Available online: https://en.climate-data.org/south-america/peru/madre-de-dios/puerto-maldonado-27856/ (accessed on 20 July 2020).

86. Madre de Dios Web Portal, Ríos y Quebradas. Available online: https://madrededios.com.pe/rios-y-quebradas.html (accessed on 20 June 2020).

87. *Madre de Dios, Compendio Estadistico*; Instituto Nacional de Estadística e Informática (INEI): Puerto Maldonado, Peru, 2018.

88. Finer, M.; Mamani, N.; García, R.; Novoa, S. MAAP: 78: Deforestation Hotspots in the Peruvian Amazon. 2017. Available online: http://maaproject.org/2018/hotspots-peru2017/ (accessed on 20 July 2020).

89. Rojas, R. *Vulnerabilidad al cambio climático en el sector agropecuario de la región Madre de Dios*; Proyecto Especial Madre de Dios; Consorcio Madre de Dios: La Molina, Peru, 2014.

90. Stickler, C.; Duchelle, A.E.; Nepstad, D.; Ardila, J.P. Subnational jurisdictional approaches. In *Transforming REDD*; CIFOR: Bogor, Indonesia, 2018; pp. 145–159.

91. Lovejoy, T.E.; Nobre, C. Amazon tipping point. *Sci. Adv.* **2018**, *4*, eaat2340. [CrossRef] [PubMed]

92. Tropical Forest Alliance (TFA). Available online: https://www.tropicalforestalliance.org/ (accessed on 20 July 2020).

93. *Censo Demográfico*; Instituto Brasileiro de Geografia e Estatística (IBGE): Brasilia, Brasil, 2010.

94. Smith, N.J.; Falesi, I.C.; Alvim, P.D.T.; Serrão, E.A.S. Agroforestry trajectories among smallholders in the Brazilian Amazon: Innovation and resiliency in pioneer and older settled areas. *Ecol. Econ.* **1996**, *18*, 15–27. [CrossRef]

95. Saes, M.S.M.; Silva, V.L.; Nunes, R.; Gomes, T.M. Partnerships, learning, and adaptation: A cooperative founded by Japanese immigrants in the Amazon rainforest. *Int. J. Bus. Soc. Sci.* **2014**, *5*, 131–141.

96. Salafsky, N. Drought in the rain forest: Effects of the 1991 El Niño-Southern Oscillation event on a rural economy in West Kalimantan, Indonesia. *Clim. Chang.* **1994**, *27*, 373–396. [CrossRef]

97. Carlson, K.M.; Curran, L.M.; Ratnasari, D.; Pittman, A.M.; Soares-Filho, B.S.; Asner, G.P.; Trigg, S.N.; Gaveau, D.A.; Lawrence, D.; Rodrigues, H.O. Committed carbon emissions, deforestation, and community land conversion from oil palm plantation expansion in West Kalimantan, Indonesia. *Proc. Natl. Acad. Sci. USA* **2012**, *109*, 7559–7564. [CrossRef] [PubMed]

98. Farida, A.; Jeanes, K.; Kurniasari, D.; Widayati, A.; Ekadinata, A.; Hadi, D.P.; Joshi, L.; Suyamto, D.; van Noordwijk, M. *Rapid Hydrological Appraisal (RHA) of Singkarak Lake in the Context of Rewarding Upland Poor for Environmental Services (RUPES)*; World Agroforestry Centre: Bogor, Indonesia, 2005; p. 119.

99. Burgers, P.; Farida, A. Community management for agro-reforestation under a voluntary carbon market scheme in West Sumatra. In *Co-Investment in Ecosystem Services: Global Lessons from Payment and Incentive Schemes*; Namirembe, S., Leimona, B., van Noordwijk, M., Minang, P., Eds.; World Agroforestry Centre (ICRAF): Nairobi, Kenya, 2017; Available online: http://www.worldagroforestry.org/sites/default/files/chapters/Ch29%20Collective%20action_ebookB-DONE2.pdf (accessed on 20 July 2020).

100. Musau, J.; Sang, J.; Gathenya, J.; Luedeling, E. Hydrological responses to climate change in Mt. Elgon watersheds. *J. Hydrol. Reg. Stud.* **2015**, *3*, 233–246. [CrossRef]

101. Shames, S.; Heiner, K.; Kapukha, M.; Kiguli, L.; Masiga, M.; Kalunda, P.N.; Ssempala, A.; Recha, J.; Wekesa, A. Building local institutional capacity to implement agricultural carbon projects: Participatory action research with Vi Agroforestry in Kenya and ECOTRUST in Uganda. *Agric. Food Secur.* **2016**, *5*, 13. [CrossRef]

102. Muhweezi, A.B.; Sikoyo, G.M.; Chemonges, M. Introducing a Transboundary Ecosystem Management Approach in the Mount Elgon Region. *Mt. Res. Dev.* **2007**, *27*, 215–219. [CrossRef]

103. Mugagga, F. The Effect of Land Use on Carbon Stocks and Implications for Climate Variability on the Slopes of Mount Elgon, Eastern Uganda. *Int. J. Reg. Dev.* **2015**, *2*, 58–75. [CrossRef]

104. Nsubuga, F.N.W.; Namutebi, E.N.; Nsubuga-Ssenfuma, M. Water Resources of Uganda: An Assessment and Review. *J. Water Resour. Prot.* **2014**, *6*, 1297–1315. [CrossRef]

105. Mugagga, F.; Buyinza, M. Land tenure and soil conservation practices on the slopes of Mt Elgon National Park, Eastern Uganda. *J. Geogr. Reg. Plan.* **2013**, *6*, 255–262. [CrossRef]

106. Sassen, M. Conservation in a Crowded Place: Forest and people on Mount Elgon Uganda. Ph.D. Thesis, Wageningen University, Wageningen, The Netherlands, 2014. Available online: https://edepot.wur.nl/293853 (accessed on 20 July 2020).

107. Mutiga, J.K.; Mavengano, S.T.; Zhongbo, S.; Woldai, T.; Becht, R. Water allocation as a planning tool to minimise water use conflicts in the Upper Ewaso Ng'iro North Basin, Kenya. *Water Resour. Manag.* **2010**, *24*, 3939–3959. [CrossRef]

108. Ngigi, S.N.; Savenije, H.H.; Gichuki, F.N. Land use changes and hydrological impacts related to up-scaling of rainwater harvesting and management in upper Ewaso Ng'iro river basin, Kenya. *Land Use Policy* **2007**, *24*, 129–140. [CrossRef]

109. Chepyegon, C.; Kamiya, D. Challenges faced by the Kenya water sector management in improving water supply coverage. *J. Water Resour. Prot.* **2018**, *10*, 85–105. [CrossRef]

110. Climate-Data, Puno Climate. Available online: https://en.climate-data.org/search/?q=puno (accessed on 20 July 2020).

111. Fagandini, R.F. Distribution des parents sauvages du quinoa cultivé en lien avec les pratiques et usages des communautés andines dans la région de Puno au Pérou. Ph.D. Thesis, AgroParisTec, Paris, France, 2019.

112. International Year of Quinoa 2013. Available online: http://www.fao.org/quinoa-2013/iyq/en/?no_mobile=1 (accessed on 25 June 2020).

113. Winkel, T.; Álvarez-Flores, R.; Bertero, D.; Cruz, P.; Castillo, C.D.; Joffre, R.; Parada, S.P.; Tonacca, L.S. Calling for a reappraisal of the impact of quinoa expansion on agricultural sustainability in the Andean highlands. *Idesia* **2014**, *32*, 95–100. [CrossRef]

114. Diarisso, T.; Corbeels, M.; Andrieu, N.; Djamen, P.; Douzet, J.-M.; Tittonell, P. Soil variability and crop yield gaps in two village landscapes of Burkina Faso. *Nutr. Cycl. Agroecosyst.* **2016**, *105*, 199–216. [CrossRef]

115. Hengl, T.; de Jesus, J.M.; Heuvelink, G.B.; Gonzalez, M.R.; Kilibarda, M.; Blagotić, A.; Shangguan, W.; Wright, M.N.; Geng, X.; Bauer-Marschallinger, B. SoilGrids250m: Global gridded soil information based on machine learning. *PLoS ONE* **2017**, *12*, e0169748. [CrossRef]

116. Chikowo, R.; Zingore, S.; Snapp, S.; Johnston, A. Farm typologies, soil fertility variability and nutrient management in smallholder farming in Sub-Saharan Africa. *Nutr. Cycl. Agroecosyst.* **2014**, *100*, 1–18. [CrossRef]

117. Tittonell, P.; van Wijk, M.T.; Herrero, M.; Rufino, M.C.; de Ridder, N.; Giller, K.E. Beyond resource constraints–Exploring the biophysical feasibility of options for the intensification of smallholder crop-livestock systems in Vihiga district, Kenya. *Agric. Syst.* **2009**, *101*, 1–19. [CrossRef]

118. van Wijk, M.T.; Tittonell, P.; Rufino, M.C.; Herrero, M.; Pacini, C.; de Ridder, N.; Giller, K.E. Identifying key entry-points for strategic management of smallholder farming systems in sub-Saharan Africa using the dynamic farm-scale simulation model NUANCES-FARMSIM. *Agric. Syst.* **2009**, *102*, 89–101. [CrossRef]

119. Grillot, M.; Guerrin, F.; Gaudou, B.; Masse, D.; Vayssières, J. Multi-level analysis of nutrient cycling within agro-sylvo-pastoral landscapes in West Africa using an agent-based model. *Environ. Model. Softw.* **2018**, *107*, 267–280. [CrossRef]

120. Rufino, M.C.; Dury, J.; Tittonell, P.; van Wijk, M.T.; Herrero, M.; Zingore, S.; Mapfumo, P.; Giller, K.E. Competing use of organic resources, village-level interactions between farm types and climate variability in a communal area of NE Zimbabwe. *Agric. Syst.* **2011**, *104*, 175–190. [CrossRef]

121. Andrieu, N.; Vayssières, J.; Corbeels, M.; Blanchard, M.; Vall, E.; Tittonell, P. From farm scale synergies to village scale trade-offs: Cereal crop residues use in an agro-pastoral system of the Sudanian zone of Burkina Faso. *Agric. Syst.* **2015**, *134*, 84–96. [CrossRef]

122. Baudron, F.; Delmotte, S.; Corbeels, M.; Herrera, J.M.; Tittonell, P. Multi-scale trade-off analysis of cereal residue use for livestock feeding vs. soil mulching in the Mid-Zambezi Valley, Zimbabwe. *Agric. Syst.* **2015**, *134*, 97–106. [CrossRef]

123. Meteorological, Climatological, and Geophysical Agency of Karangploso. *Rainfall Data of Karangploso Station Year 2013–2017*; BMKG Karangploso: Malang, East Java, Indonesia, 2018.

124. Kurniawan, S.; Prayogo, C.; Widianto, M.; Zulkarnain, M.T.; Lestari, N.D.; Aini, F.K.; Hairiah, K. *Estimasi Karbon Tersimpan di Lahan-lahan Pertanian di DAS Konto, Jawa Timur: RACSA (Rapid Carbon Stock Appraisal).* *Working Paper no.120*; World Agroforestry Centre: Bogor, Indonesia, 2010; p. 59.

125. Lusiana, B.; van Noordwijk, M.; Cadisch, G. Land sparing or sharing? Exploring livestock fodder options in combination with land use zoning and consequences for livelihoods and net carbon stocks using the FALLOW model. *Agric. Ecosyst. Environ.* **2012**, *159*, 145–160. [CrossRef]

126. BPS (Central Bureau of Statistics). *Malang Regency in Figures 2018*; BPS-Statistic of Malang Regency: Malang, East Java, Indonesia, 2018.

127. Andriyanto, C.; Sudarto, S.; Suprayogo, D. Estimation of soil erosion for a sustainable land use planning: RUSLE model validation by remote sensing data utilization in the Kalikonto watershed. *J. Degrad. Min. Lands Manag.* **2015**, *3*, 459–468. [CrossRef]

128. Jackson, L.; Bawa, K.; Pascual, U.; Perrings, C. *Agro-Biodiversity–A New Science Agenda for Biodiversity in Support of Sustainable Agroecosystems*; Diversitas: Paris, France, 2005; p. 40. Available online: http://www.diversitasinternational.org/cross_agriculture.html (accessed on 20 July 2020).

129. Rijsdijk, A.; Bruijnzeel, L.S.; Sutoto, C.K. Runoff and sediment yield from rural roads, trails and settlements in the upper Konto catchment, East Java, Indonesia. *Geomorphology* **2007**, *87*, 28–37. [CrossRef]

130. *Pasuruan Regency in Figures*; BPS-Statistics of Pasuruan Regency: Pasuruan, Indonesia, 2019.

131. *Pasuruan City in Figures*; BPS-Statistics of Pasuruan City: Pasuruan, Indonesia, 2019.

132. Ross, H.; Adhuri, D.S.; Abdurrahim, A.Y.; Phelan, A. Opportunities in community-government cooperation to maintain marine ecosystem services in the Asia-Pacific and Oceania. *Ecosyst. Serv.* **2019**, *38*, 100969. [CrossRef]

133. Charles, A.; Loucks, L.; Berkes, F.; Armitage, D. Community science: A typology and its implications for governance of social-ecological systems. *Environ. Sci. Policy* **2020**, *106*, 77–86. [CrossRef]

134. van Noordwijk, M.; Tomich, T.P.; Verbist, B. Negotiation support models for integrated natural resource management in tropical forest margins. *Conserv. Ecol.* **2001**, *5*, 21. Available online: http://www.consecol.org/vol5/iss2/art21 (accessed on 17 July 2020). [CrossRef]

135. Perhutani. *KPH Pasuruan*; Jakarta, Indonesia. Available online: https://perhutani.co.id/tentang-kami/struktur-organisasi-perum-perhutani/divisi-regional/jatim/kph-pasuruan/ (accessed on 20 July 2020).

136. Suprayogo, D.; van Noordwijk, M.; Widianto; Hairiah, K.; Meilasari, N.; Rabbani, A.L.; Ishaq, R.M. Infiltration-Friendly Land Uses for Climate Resilience on Volcanic Slopes in the Rejoso Watershed, East Java, Indonesia. *Land* **2020**, *9*, 240. [CrossRef]

137. Amaruzaman, S.; Khasanah, N.; Tanika, L.; Dwiyanti, E.; Lusiana, B.; Leimona, B. *Landscape Characteristics of Rejoso Watershed: Land Cover Dynamics, Farming Systems, and Community Strategies*; World Agroforestry Centre (ICRAF) Southeast Asia Regional Program: Bogor, Indonesia, 2018.

138. Leimona, B.; Khasanah, N.; Lusiana, B.; Amaruzaman, S.; Tanika, L.; Hairiah, K.; Suprayogo, D.; Pambudi, S.; Negoro, F.S. *A Business Case: Co-Investing for Ecosystem Service provIsions and Local Livelihoods in Rejoso Watershed*; World Agroforestry Centre (ICRAF) Southeast Asia Regional Program: Bogor, Indonesia, 2018.

139. Tanika, L.; Khasanah, N.; Leimona, B. *Simulasi Dampak Perubahan Tutupan Lahan Terhadap Neraca Air di DAS dan Sub-DAS Rejoso Menggunakan Model GenRiver*; World Agroforestry Centre (ICRAF) Southeast Asia Regional Program: Bogor, Indonesia, 2018.

140. Mahzum, M.M. *Analisis Ketersediaan Sumber Daya Air dan Upaya Konservasi SUb DAS Brantas Hulu Wilayah Kota Batu*; Civil Engineering and Planning Facultry, Institut Tehnologi Sepuluh November: Magister, Surabaya, 2015; p. 187.

141. *Batu Municipality in Figures*; BPS_Statistical Bureau: Jakarta, Indonesia, 2020; p. 151.

142. Maryudi, A.; Citraningtyas, E.R.; Purwanto, R.H.; Sadono, R.; Suryanto, P.; Riyanto, S.; Siswoko, B.D. The emerging power of peasant farmers in the tenurial conflicts over the uses of state forestland in Central Java, Indonesia. *For. Policy Econ.* **2016**, *67*, 70–75. [CrossRef]

143. Witjaksono, A.; Surjono, S.; Suharyanto, A.; Muhammad, B. Spatial analysis of land use in Bumiaji subdistrict, Batu city, East Java, Indonesia. *Int. J. GEOMATE* **2018**, *15*, 139–144. [CrossRef]

144. Nurrizqi, E.; Suyono, S. Pengaruh Perubahan Penggunaan Lahan Terhadap Perubahan Debit Puncak Banjir Di Sub DAS Brantas Hulu. *Jurnal Bumi Indonesia* **2012**, *1*, 363.

Perspective

Agroforestry-Based Ecosystem Services: Reconciling Values of Humans and Nature in Sustainable Development

Meine van Noordwijk [1,2,3]

1   World Agroforestry (ICRAF), Bogor 16155, Indonesia; M.vannoordwijk@cgiar.org
2   Plant Production Systems, Wageningen University & Research, 6700 AK Wageningen, The Netherlands
3   Study Group Agroforestry, Faculty of Agriculture, Brawijaya University, Jl Veteran, Malang 65145, Indonesia

Abstract: Agroforestry as active area of multi-, inter-, and transdisciplinary research aims to bridge several artificial divides that have respectable historical roots but hinder progress toward sustainable development goals. These include: (1) The segregation of "forestry trees" and "agricultural crops", ignoring the continuity in functional properties and functions; the farm-scale "Agroforestry-1" concept seeks to reconnect perennial and annual, woody and nonwoody plants across the forest–agriculture divide to markets for inputs and outputs. (2) The identification of agriculture with provisioning services and the assumed monopoly of forests on other ecosystem services (including hydrology, carbon storage, biodiversity conservation) in the landscape, challenged by the opportunity of "integrated" solutions at landscape scale as the "Agroforestry-2" concept explores. (3) The gaps among local knowledge of farmers/agroforesters as landscape managers, the contributions of social and ecological sciences, the path-dependency of forestry, environmental or agricultural institutions, and emerging policy responses to "issue attention cycles" in the public debate, as is the focus of the "Agroforestry-3" concept. Progress in understanding social–ecological–economic systems at the practitioners–science–policy interface requires that both instrumental and relational values of nature are appreciated, as they complement critical steps in progressing issue cycles at the three scales. A set of hypotheses can guide further research.

Keywords: coinvestment; instrumental values; landscape; relational values; restoration; social–ecological systems; stewardship; sustainable development goals (SDGs); trees; water

Citation: van Noordwijk, M. Agroforestry-Based Ecosystem Services: Reconciling Values of Humans and Nature in Sustainable Development. *Land* **2021**, *10*, 699. https://doi.org/10.3390/land10070699

Academic Editor: Daniel S. Mendham

Received: 28 May 2021
Accepted: 1 July 2021
Published: 2 July 2021

Publisher's Note: MDPI stays neutral with regard to jurisdictional claims in published maps and institutional affiliations.

1. Introduction

Agroforestry-based ecosystem services, the title of the special issue this perspective is part of, refers to an active arena of international agricultural research connected with global sustainability science at the science–policy interface [1]. This research deals with "theories of place" (how and why do social–ecological contexts differ from each other as part of multiscale spatial patterns?), theories of change (how, when, and why does the process of change happen?), and theories of induced change (how can change processes in their existing context be nudged into a direction that is deemed to be desirable?) [2]. It also deals with value plurality as depending on context and stakeholders' expressions of value that express "instrumentality" complement those that emphasize "relations" [3]. "Ecosystem services" (ES), "agroforestry" (AF), and "value plurality" all refer to connections. Their combination may thus need to clarify the components of all three terms as well as their interfaces. Ecosystem services is commonly defined as the benefits people derive in various ways from functioning ecosystems (including agro-ecosystems and forest ecosystems) and that can be at risk due to human activities from local to global scales. It connects "service providers" (nature and its guardians) and "service beneficiaries" (e.g., people, companies, cities, nations). While terminologies and metaphors have been proposed other than that of a servant, the alternative terms referring to "nature" and "contributions" (voluntary? appropriated?) may have similar semantic challenges [4]. Ecosystem services as ongoing benefit flows have been promoted as reason to appreciate the underlying "natural capital", but it may have become (too) strongly associated with economic valuations of these services. Economic estimates of costs to society of loss of natural ecosystems [5,6] are astronomical and call for a realignment of economic priorities in achieving progress in human well-being under the umbrella of sustainable development. Quantitative economic language, however, has been interpreted as underrating qualitative social dimensions [4]. Current interest in "relational values" can be understood as emphasizing social ("in-group") aspects over economic (costs, benefits) ones [7].

Agroforestry, the word constructed by combining agro- and forestry, has from its start been concerned with bridging between concepts that appeared to contradict each other [2]. It is currently understood to refer to three nested scales: as "AF1" to the scale of trees, management practices, plot-level technology, and farm level decisions; as "AF2" to landscapes with trees and forest (patches) in which productivity ("provisioning services") interacts with other ES ("regulating" and "cultural" services); and as "AF3" to the reconciliation of agriculture and forestry as separate policy domains that interact at "land use" and natural resource management levels, connecting local knowledge, sciences and policy framing in "issue cycles" [8]. As partial tree cover in what is classified as agricultural lands is common (more than a third of such lands has at least 10% tree cover) and appears to be increasing [9], agroforestry is part of mainstream land use.

As a dominant theory of inducible change, the Sustainable Development Goals have tried to define a "safe space for humanity" that links local to global concerns and vice versa. Following the Doughnut model of Raworth [10], these goals can be summarized as centrifugal expansion in an inner circle of reducing development deficits (essentially the Millennium Development Goals that have not yet been achieved), with simultaneous centripetal movement avoiding overshoot of planetary boundaries and restoring functions that were lost from "degraded" parts of the planet (Figure 1). The Earth is too small to achieve all SDGs in silos, with separate land allocations for each goal. Provision of water (of acceptable quality and regularity of flow), food, fiber, and energy may have to be achieved on the same units of land that also provide jobs and contribute to public health. Multifunctional landscapes that simultaneously contribute to multiple goals are in demand, that combine fairness (ensuring that nobody is left behind in reducing development deficits that contribute to poverty) and efficiency of avoiding planetary overshoot [11]. Rather than the "productivity gaps" that still are a major focus of agronomy [12], the broader concept of "multifunctionality gaps" has been the focus of agroforestry in its first four decades [13,14]. To avoid further land degradation and promote land restoration, multifunctional use of land is needed within the boundaries of the soil–water system; new business models in robust economic systems are needed based on environmental systems thinking integrating environmental, social, and economic interests [15].

This leads to the three questions that frame this perspective as a review of relevant literature: 1. how are value concepts of agroforestry evolving in relation to ES discourse and sustainability concerns at the multifunctional landscape scale?; 2. how can the interaction among the three agroforestry scales and concepts (AF1, AF2, and AF3) be understood?; and 3. what roles can research play in connecting theories of place and change to policies that aim for applicable theories of induced change, in support of SDGs? The rest of this paper will review these three questions and then formulate some hypotheses for follow-up research.

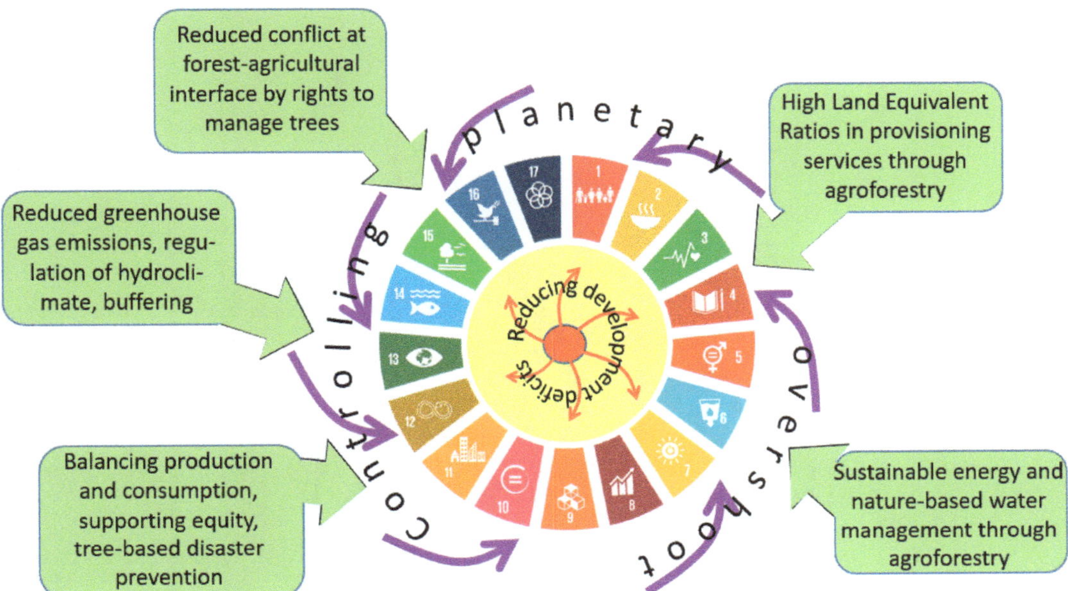

Figure 1. Agroforestry contributes to the various Sustainable Development Goals as safe space for humanity, between "development deficits" and "planetary boundary overshoot".

2. Roots of Sustainability Concepts Relevant to Agroforestry

2.1. Three Changes of Theory Foundational to Sustainable Development

Changes in the way human society effectively relates to and values nature, and nature-derived land uses such as agroforestry, require changes in theory, and thus a theory of change of theory [16]. IPBES reports record the ongoing loss of biodiversity and identify market-based production responding to ever-growing demand as major drivers of destruction and a principal theory of change [17]. Three "changes of theory" that originated in the 18th, 19th, and 20th century, respectively "the blind watchmaker" (a later metaphor for the ideas of Charles Darwin), "the invisible hand" (Adam Smith), and "the not-so-tragic commons" (Elinor Ostrom), have shaped how sustainability scientists perceive humanity (Figure 2). All three theories connected basic mechanisms at the level of the individual to impacts at scales (time, space, systems) many orders of magnitude beyond the location and lifespan of the individual. Previous theories lacked mechanisms for change in organisms after their creation, which lived in autocratically governed static economies where neither individuals nor local communities counted. Although all three changes of theory are now mainstream, the radical break with preceding theories of change can hardly be overstated.

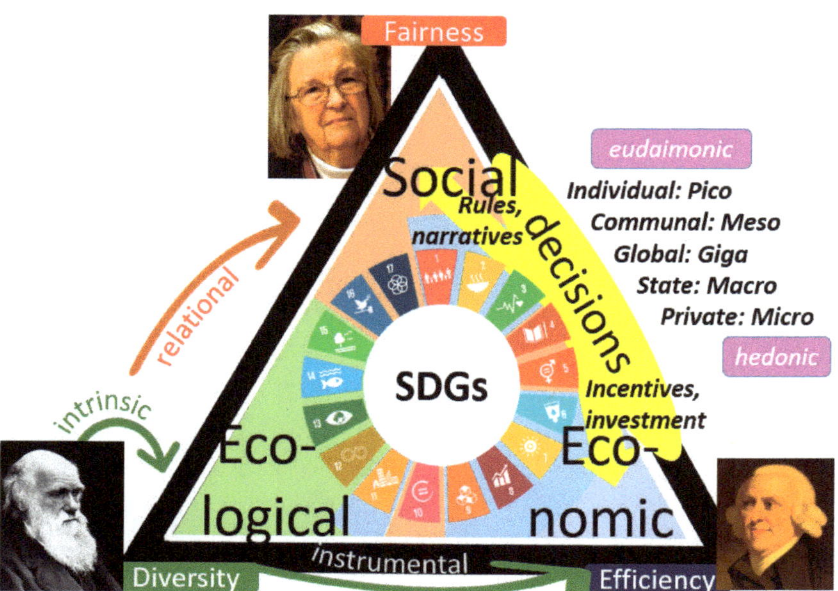

Figure 2. The triple bottom line of People–Profit–Planet or social, economic, and ecological aspects of value (importance, exchangeability, and long-term survival) for sustainability is reflected in the current set of 17 Sustainable Development Goals (SDGs) and connects the intellectual breakthroughs of the invisible hand, blind watchmakers, and not-so-tragic commons, attributed to Adam Smith [18]., Charles Darwin [19], and Elinor Ostrom [20], respectively; human decision making across the social–economic spectrum depends on the scale (see Figure 3 for the pico–giga terminology) and reflects different types of values: the intrinsic, instrumental, and relational values of nature and its diversity, the efficiency of using scarce resources, and the hedonic and eudaimonic improvements of human quality of life (see below for a further discussion).

A first powerful cross-scale mechanism was found in the invisible hand that efficiently links supply and demand in markets through the self-interest of individuals [21]. Adam Smith also explored [22] how moral sentiments in a society constrain (or should constrain) this self-interest, but this part of his heritage received less attention than his ode to market forces. The second mechanism revolutionized biology and its application in medical agricultural and biotechnological sciences, with the current rate of evolution of a specific virus remaining headline news. Moderate rates of errors in replication of genetic code, in combination with sexual reproduction involving recombination, and selective phenotype survival with feedback to gene frequencies were understood to be the blind watchmaker's [23] recipe for the overwhelming biodiversity of our planet. The diversification mechanism interacts with factors such as climate, water, carbon and nutrient cycles and geomorphology. It also implies coevolution in plant–herbivore, predator–prey, disease–host, symbionts, pollinator–flower, and seed dispersal–dispersant relations providing positive feedback loops to ever-increasing diversity levels. These relations between organisms translate to differential survival, but also lead to interconnected sensitivity to collapse. Plate tectonics as the source of continental drift supported (was a driver of) further speciation, but also set up the current sensitivity to "invasive" species overcoming geographic isolation. Any seed-dispersing animal modifies vegetation in a direction of fruits it likes to eat, but our species has learned over the past 10,000 years to coevolve and further work with, rather than against nature, in domesticating the major sources of food, gradually adding visioning, design, planning, and rationality to the blind watchmaker. Agriculture, with the option of storing, hoarding, and transporting food, is widely seen as the start of social stratification and power structures [24,25]. The popular Jenga game, where a fair number of blocks of wood can be extracted from a stack before an inevitable collapse follows, provides a visualization of this risk [26] when the invisible hand is not constrained in its resource extraction.

Once governments learned to work with, rather than against, market forces, economies blossomed. They did so well that the invisible hand gradually, and initially imperceptibly to the blind watchmaker, started to destroy its work. This is the theory of change the IPBES reports indicate as a major driver behind ongoing trends: market-driven expansion of the human sphere of influence, destroying habitat. However, the emergence of SARS-Cov-2 at the end of 2019 showed that the opposite, results of the blind watchmakers' activity destroying those of the invisible hand, can happen as well [27]. A modified theory of change has negative outcomes on both sides of the interaction and urges a coevolutionary path of humans and the rest of the planet's life forms. The mutual dependence of nature (N) and humans (H) has gradually dawned on mainstream political systems with current financial understanding of "nature risk" [28]. The main religions of the world have assigned H a special, privileged position that, however, comes with obligations for stewardship and adaptive social–ecological governance [29]. Prioritizing N has also been discussed (and objected to) as a "dark-green religion" [30].

The third intellectual breakthrough, the revision of what is tragic about the commons, is the most recent of the three foundations of sustainability science, and it is still in the process of mainstreaming. To some extent it fills the place left by the failure of a third theory of change rooted in social concerns over inequity and in the expectation that political power will (or at least can) shift to (what used to be) the proletariat. In practice new autocracies emerged where the ideas of Karl Marx were applied [31]. Yet, at the smaller scale of "commons", bottom-up institutions have historically and to this day been able to manage resources, avoiding the free-for-all degradation expected. Where the lack of private property rights (especially the right of exclusion) had been portrayed as a tragedy of the commons and as the primary bottleneck to environmental management benefitting the well-being of local communities, a counter-narrative emerged of the comedy of the commons [32], to be better understood and heard. Two thousand years ago, Roman society, the legal thinking of which greatly influenced later European law, was sufficiently interested in public property (ius publicum) to distinguish various categories within the concept, as applied, for example, to issues of water at various scales. Debate on the relative merits of "common law" revolved about a bundle of rights and the roles of community-based institutions at a scale intermediate between the state and the individual. The "institutional economics" analyses by Elinor Ostrom [33] brought back respect for the ability of local communities to develop their own rules, incentives, and narratives to manage resources, including land, forest, and inland and coastal fisheries. With the concept of commons broadening from grazing land to various types of land and water use, and more recently the atmosphere and global climate system, the scale at which resources can be managed needs to be reflected in the institutions managing it. Private, communal, national, and global scales all have a role to play. The triple bottom line of these ideas is that working with, rather than against, nature, markets, and communities is essential to transcend conflicts and move towards diverse, efficient, and fair societies. The challenge is where to start.

Research on human well-being or qualities of lives has explored two general perspectives [34] that date back to Greek philosophers more than 2000 years ago: the hedonic (pleasure-oriented) approach, which focuses on happiness and defines well-being in terms of pleasure attainment and pain avoidance; and the eudaimonic approach, which focuses on meaning and self-realization and defines well-being in terms of the degree to which a person is fully functioning [35]. It refers to a perception of social harmony, reflected in societal concepts as adat in Indonesian cultures [36], satoyama landscapes in Japan [37], yin-yang in Chinese philosophy [38], or religious concepts such as kosher or halal [39]. Eudaimonic well-being may well be an acquired taste during an individual's development, achieved when aspirations, rationalizations, and actions match [40,41]. Self-actualization, personal expressiveness, and perceived vitality are indicators at the interface of individuals and the social groups in which they function. In Bahasa Indonesia, there are two words for "we": an "in-group" we (kami) and a "we" (kita) that includes those spoken to, regardless of the relationship. Adat applies to the first, with a dotted line that is fluid and negotiable; internalization may strive for widening the kami concept. Grossly simplifying matters: the heritage of Adam Smith is mostly invoked in market-based efficiency, pursuit of hedonic values, and defending individual freedom versus "the state", the ideas of Elinor Ostrom in balancing rights and responsibilities, eudaimonic values of well-functioning community-based resource management, and appreciating the institutional dynamics multilayered governance where fairness concepts are not to be ignored in the search for efficiency.

2.2. Theories of Induced Change

In connection with an analysis of drivers, pressure, system state, impacts, and responses (DPSIR) change processes can be classified as (1) responsive—connecting (undesirable) impacts to a (corrective) response, (2) adaptive—optimizing impacts given a changed system state, (3) mitigative—given existing pressures reducing (spatially explicit) negative changes in the state of the system, (4) transformative—given the drivers, reduce (generic) pressures, or (5) reimaginative—challenge the drivers in view of the overarching goals of safe human well-being on the only planet we have [42].

Sustainable development that connects local to global scales and vice versa requires cross-scale analysis. Theories of induced change with mechanisms at pico (brain synapses, genes), micro (self-interest), and meso (rules) levels can have impacts at macro (national) and (giga) planetary scales [43] (Figure 3).

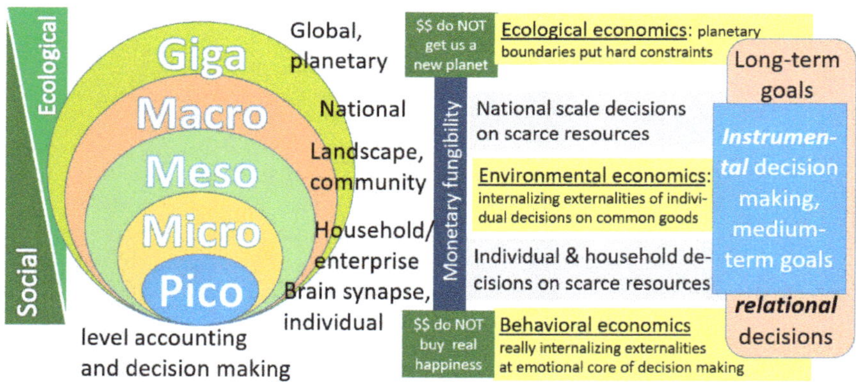

Figure 3. Linking five scales of economic analysis of social–ecological systems to two ways of decision making instrumental (medium-term goal oriented) and relational [44].

Economic analysis of decision making has long been differentiated by scale. Beyond the micro (household, enterprise) and macro (state) accounting stance of economics as a discipline that aims to understand human decisions facing scarcity, three further scales have been explored in recent decades: meso, giga, and pico [44] (Figure 3). The contrast between the meso and giga scales has been discussed in the debate of "environmental" versus "ecological" economics. The first aims for a meso-scale perspective where "market failures" that are the consequence of explicit valuation of commons (including ecosystem services) can be addressed ("internalizing externalities") by facilitating the emergence of new (regulated) markets that try to get "prices" more aligned with "values at stake", influencing household-level decisions and choices both in the consumptive and investment spheres.

Ecological economics (giga), in contrast, is not content with fitting environmental issues into a mainstream economics perspective but starts from planetary boundary perspectives on the types of changes needed to fit humans into what a living planet can deal with, aiming to stop short of a number of precipices, of which global climate change is now probably the best understood for its risks of self-propelling changes. The fifth scale, pico, refers to synapses in human brains where actual decisions are made, as explored in behavioral economics, interfacing psychology, neurology, game theory, and choice experiments. The relevance of branding, influencing perceptions, and priming human decisions had been long understood in marketing research. More formal understanding and academic recognition within economics only arose in the past three decades. Its application of "nudging" [45] as a way of less visibly (obnoxiously?) influencing individual- and household-level decisions aligned with public governance values has been practiced for less than two decades.

Thus, modification of prices is only one of several ways to internalize externalities (Figure 4), as human decision-making responds to both rational and emotional clues [46], and human sociality [47] analyzes decisions in terms that are understood by "influencers" and social media but have not yet been integrated in standard economic theory.

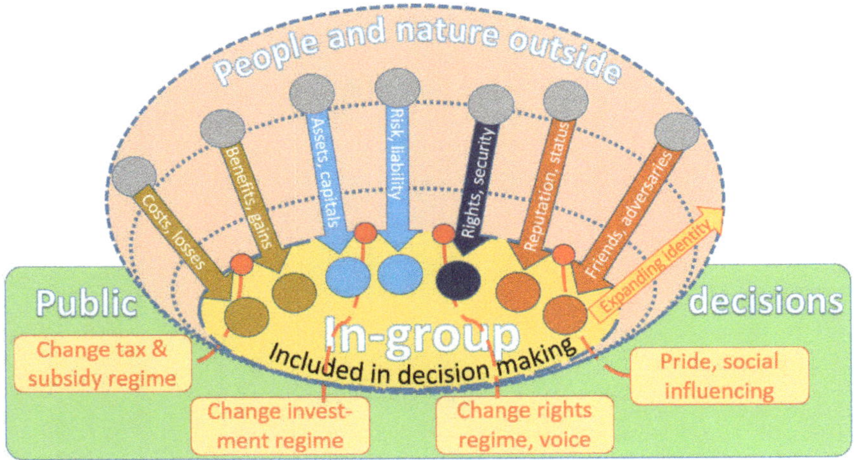

Figure 4. Visualization of the various approaches to internalize externalities in human decision making, which tends to consider consequences (costs and benefits) for an in-group but not for people and nature beyond this emotional inner circle unless efforts to expand the inner circle are effective.

2.3. Teleconnections and Distractors

Thirty years ago, the world was, at a conference in Noordwijk, as close as it has ever been to binding international agreement on combatting climate change [48], but the fossil fuel lobby successfully obstructed agreements at government level in 1989 [49,50] and ever since through disinformation and fear campaigns that suggested citizen self-interest would be better off ignoring the signals of planetary overshoot. Small-scale effects with global consequences became known as the butterfly effect as part of chaos theory, which term refers to the positive feedback loops, metaphorically, of a butterfly clapping its wings triggering hurricanes far away [51].

A stronger, more current story of teleconnections, although other theories are around as well, is that caged wild animals (illegally traded) in the Wuhan market were in close enough contact with people to allow a virus to jump the species barrier. The rest is rewriting history books as we speak, demonstrating that cross-scale theories of change are rightly known as "going viral", in the world of memes as well as genes. A single schoolgirl who decided one day to skip school and sit in front of her parliament with a "Skolstrejk för klimatet" poster, started a movement now known as the Extinction Rebellion, and, once she got noticed, was strengthened by her rebuttal of denial and conspiracy theories and became a role model not only for her own generation but for the world [52]. Her urge to action focused on the climate change issue the world had been talking about since she was born (and before). Planetary anxiety may trigger what technical, solution-oriented talks to rein in market forces failed to achieve: consumer restraint to reduce footprints to what is affordable within planetary boundaries. Consumer power may achieve what governments, subject to its interaction with industry, cannot.

The metaphors used here, the invisible hand that drives global environmental destruction in the name of human welfare and that ignores its ticking wristwatch, made by a blind watchmaker, while understanding the commons, can now be appreciated as a social–ecological system in need of reimaginative as well as transformative changes beyond the tinkering of adaptive and mitigative responses. The typology of "leverage points" that was formulated on the basis of early whole-Earth system models [53] still applies; the most effective leverage points relate to rules and incentives (#5) and goals (#3) on the instrumental side and self-organization (#4) and mindsets (#2) on the relational value side. However, the most powerful leverage point of all, the power to transcend the existing system (#1), may refer to ways to go beyond the limits of the current valuation debate and the way it has been framed.

2.4. Value Types

Two main categories of "values" of nature to humans are currently distinguished: instrumental (goal-oriented, potentially substitutable means to achieve goals) and relational (based on a two-way interaction that establishes affinity and is not easily substitutable) [7,54] (Figure 5).

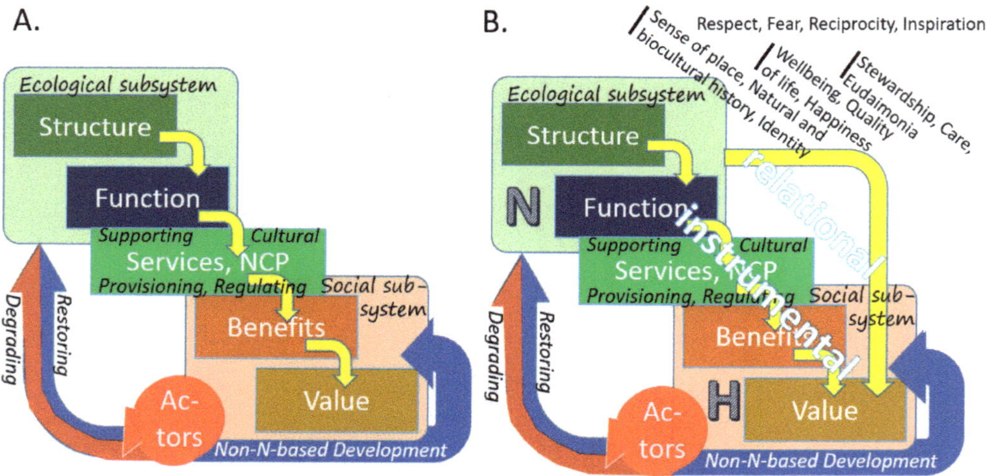

Figure 5. (**A**) The ecosystem cascade concept of ecological–social systems, with ecosystem services (alternatively described as nature's contributions to people or NCPs) at the interface of the two subsystems and feedback by human actors using, degrading, and/or restoring the ecological subsystem; (**B**) Distinction between instrumental (goal-oriented) values of nature and relational values that are not dependent on material benefits.

When positioned in a feedback loop between human and nature (Figure 6), relational and instrumental values may have complementary roles with respect to decisions to improve qualities of human lives. In this interpretation, the degree of substitutability may not be most important distinction between instrumental and relational values: the first is counted in terms of benefits and contribution by nature, the second by the investment in effort. The latter matches the long-term experience that getting school children to care for plants or animals shapes their values on environmental issues [55]. Stewardship as a central concept of relational value has emerged in many human societies [56]. This view on relational values as based on direct engagement may also help to further analyze the noted gender differences in appreciating landscape elements and land use decisions [57].

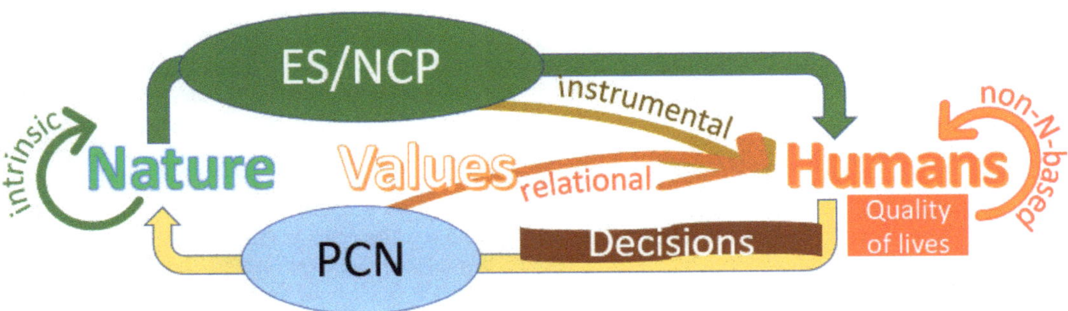

Figure 6. Interpretation of the complementarity of goal-oriented, instrumental and engagement-based, relational values of nature to humans, with respect to ecosystem services (ES) (or nature's contributions to people, NCP) and people's contributions to nature (PCN).

A substantial share of the words and metaphors that can be used to describe relational values of nature to humans, are derived from the way relations between humans are understood, within and outside an in-group or family concept. Tigers may be described as "uncles" living in Asian forest, not to be disturbed but fair in their retaliatory actions against human trespassers [58]; nature may be described as a "mother". The relations themselves grade from fear and respect, competition, enemies, friends, reciprocity, love and care, or stewardship. Elaborating on a diagrammatic representation in [42], instrumental values can be interpreted as part of an overarching relational value concept (with servant or contributor as a one-direction oriented relation) (Figure 7).

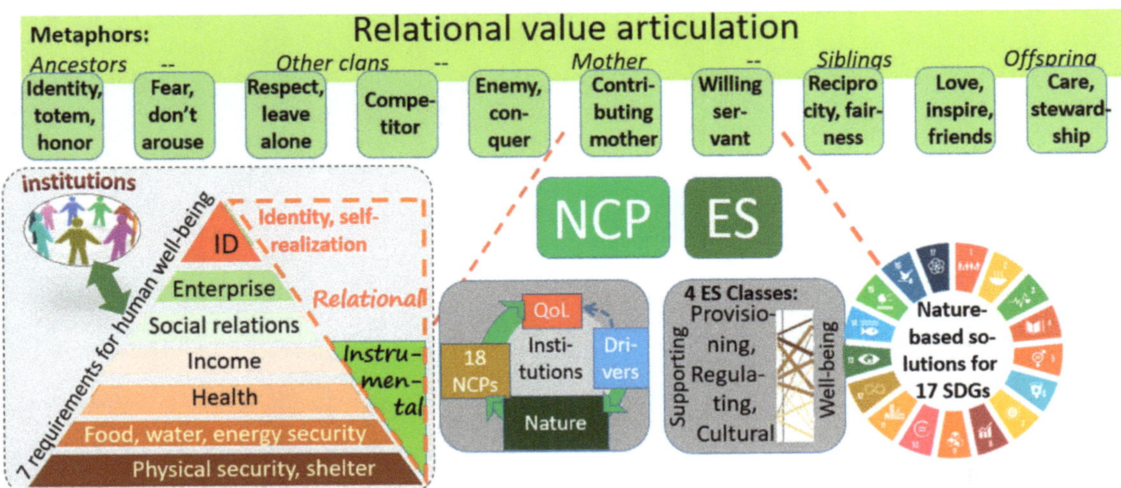

Figure 7. Portrayal of instrumental as a specific subset of relational values [7], including the material values involved in a modified Maslow pyramid of human wellbeing [59] and nature-based solutions [60] to the various SDGs.

In agroforestry the primary relational targets are the tree, the forest and the farmer. Trees play many symbolic roles, ranging from the "tree of life" to the peace-making rituals of tree planting as joint relational effort to (symbolically) transcend conflicts [61]. The popularity of tree planting as activity that responds to global climate change and is expected to protect people from harm has had various instrumental rationalizations but has relational roots [62,63]. Seeing both trees and forests at the same time is a recognized challenge. The derivation of the term forest is on the exclusion of local communities of resources claimed by the central state power, and despite the interest in community-based forest management or social forestry, it is still hard to bridge the differences in relational value expressed by the term forest [64]. Agroforests, as farmer-managed forests, often with complex bundles of rights for individuals and local in-group customary community members, have both instrumental and relational values that need to be appreciated.

3. Nested-Scale Agroforestry Concepts in Relation to Ecosystem Services

3.1. Three Agroforestry Concepts

The three agroforestry concepts that relate patch to landscape- and policy-level analysis (Table 1) relate to both instrumental and relational values—but to different degrees. Instrumental values of nature to humans have been the basis of (meso- to micro-)economic analysis that connect the ideas of Adam Smith to those of Charles Darwin; they are relevant at all three AF concept scales. Relational values may be primarily located at the picoeconomic scale of behavioral economics, bringing in aspects of collective action and commons and thus the ideas of Elinor Ostrom.

Table 1. Agroforestry (AF) concepts with their social and ecological aspects across scales, in relation to ecosystem services issues and underlying science.

AF Concept—Scale	Social Aspects	Ecological Aspects	ES Aspects	Underlying Science
AF1—patch/farm	Farmer resources, knowledge, targets, management choices (~ gender, ~ age, ~ context); input and output markets; value chain relations	Tree cover, tree diversity; tree–soil–crop interactions in spatial and climatic contexts; response to farmer management; land equivalent ratio (LER) for productivity	Generate, Influence, Manage	Knowledge types; tree growth (architecture and functioning); water, nutrient, light, and carbon capture and cycles; options in context evaluation; farm economics
AF2—landscape	Ecosystem (dis)service perceptions across stakeholders at local to global scales; instrumental and relational value of nature as concepts	Lateral flows (water, organisms, fire, nutrients, soil, etc.), buffers and filters; biodiversity change; land equivalent ratio for multifunctionality (land shparing index) [1]	Express, Interact, Manage	Quantifying scale relations; instrumental and relational values influencing decision making in issue cycles; self-regulation of industry, ~ certification
AF3—policy	Sustainable Development Goals (SDGs); reconciling rights and incentives in agricultural and forestry institutional traditions	Planetary boundaries linked to land and water use: climate change, biodiversity loss, pollution, land degradation	Recognize, Regulate, Reward	Issue cycles; subsidiarity (devolution of governance); transparency; environmental and intergenerational justice

[1] The term shparing indicates that the land sparing versus land sharing debate may have been a false dichotomy, where land sparing can be achieved through land sharing with land equivalent ratios above 1.0 (as discussed below).

Early analysis of the interactions between trees and crops in agroforestry has built on the land equivalent ratio concept that emerged from the intercropping literature [65]. Rather than expecting benefits for each component as such, when compared to dedicated specialized ways of producing any single component, LER provides a metric for how much land can be spared by combining multiple elements on a single piece of land (Figure 8). If all relations with tree cover are linear, the LER may be 1.0, but values below 1.0 are possible when convex relations prevail, and values above 1.0 if concave relations dominate. The LER index obviously depends on the functions that are included—in the original concept, the key function was production per unit area, while more recent interpretations include many aspects of multifunctionality [66].

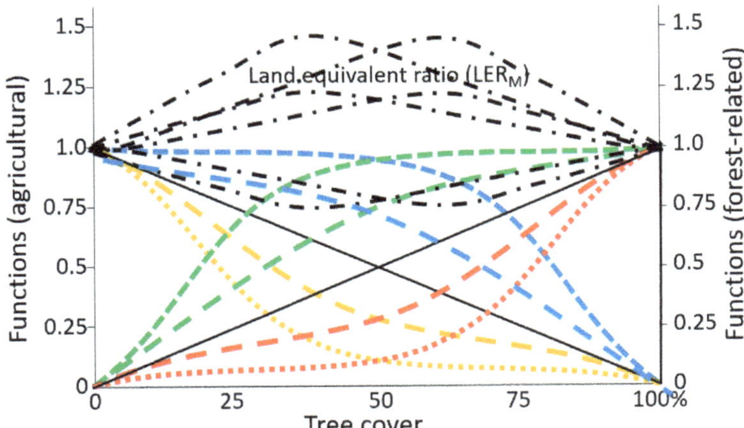

Figure 8. Schematic relationships of various land functions, either agricultural or forest-related, in relation to the percentage tree cover on a unit of land [67]; depending on the weights assigned to the various functions, the LER sum across functions can be convex, with an intermediate tree-density optimum, or concave, with an advantage for segregating.

3.2. Agroforestry as Part of a Landscape Mosaic

At the AF2 and AF3 scales, ecosystem services of agroforestry are not standalone entities that can be expressed per unit area and multiplied with the area under agroforestry for a total contribution to human well-being. Rather, agroforestry at these scales needs to be understood and appreciated in a spatial context: it may be the recipient of ecosystem services from "upstream", providing benefits to people in the agroforestry systems themselves, as well as passing on (modified) services to those further "downstream" (Figure 9). Landscape mosaics can be characterized by the relative proportions of the components and the spatial pattern or grain size of the mosaic, with edge length per unit area and types of neighbors as distinguishing features. The components are here presented as four categories: forest, agroforestry (here understood as half-open vegetation with trees, intermediate between open-field agriculture and closed-canopy forest), agricultural, and urban environments. Finer-grained mosaics have been associated with land sharing views on multifunctionality, coarse-grained mosaics with economic specialization and trade globalization.

In this view there are ten types ecosystem services: 1, 2, 3, and 4 provided by the local environment to people living in forest, agroforestry, agricultural, and urban environments, respectively; 5, 6, and 8 provided by forests (and the people living there) to human beneficiaries in agroforestry, agricultural, and urban environments, respectively; 7 and 9 derived from agroforestry and provided to agricultural and urban parts of the landscape; and 10 derived from agricultural parts of the landscape and provided to urban people. Agroforestry thus plays a role in the provisioning services of (domesticated) forest products [68–70] and the regulating services on water flows [71] or carbon storage [71–73]. Part of these agroforestry services are supported by the presence of forests in the landscape (arrow 5) and contribute (arrow 7) to the services agriculture in turn provides (arrow 10).

Similar to a value added tax concept that applies to economic value chains, each landscape component could be held accountable for the net balance of incoming and outgoing services. There are some building blocks for such analysis, that include the local forest typology of Karen in Thailand with a specific word for "forests protecting rice-fields" [74] and the recent analysis of watersheds where agricultural over-use of ground- or surface water competes with urban beneficiaries [75]. However, a full analysis along such lines does not yet exist. Few studies, if any, have effectively dissected these interactions and clarified the degrees of substitutability between them. Partial compensation of forest-based ES by agroforestry (including provisioning of medicinal plants and animals, forest fauna and flora, supporting human food security) has been documented in several landscapes [76–78]. If landscape management is based on local knowledge and experience, such explicit dissection may not be needed, but where national policies and international incentive systems and policy support continue to make distinctions between forest and agriculture without space for intermediate agroforestry that can match both definitions [79], evidence from well-quantified case studies can contribute to policy change, potentially rediscovering, reinventing, or reimagining agroforestry [42]. An example where the mosaic concept has been applied is in the flow persistence metric that translates the plot-level pathways of peak rainfall events, to the temporal dynamic of rivers and floods [80]. Similar analyses of the net balance of erosion and sedimentation in landscapes with the existing spatial filter and buffer elements [74,81,82] defy the simplicity of expressing ES values as entities per unit area, with a total derived by multiplying area fractions with value per unit area. What needs to be valued is the mosaic with its structure and functions.

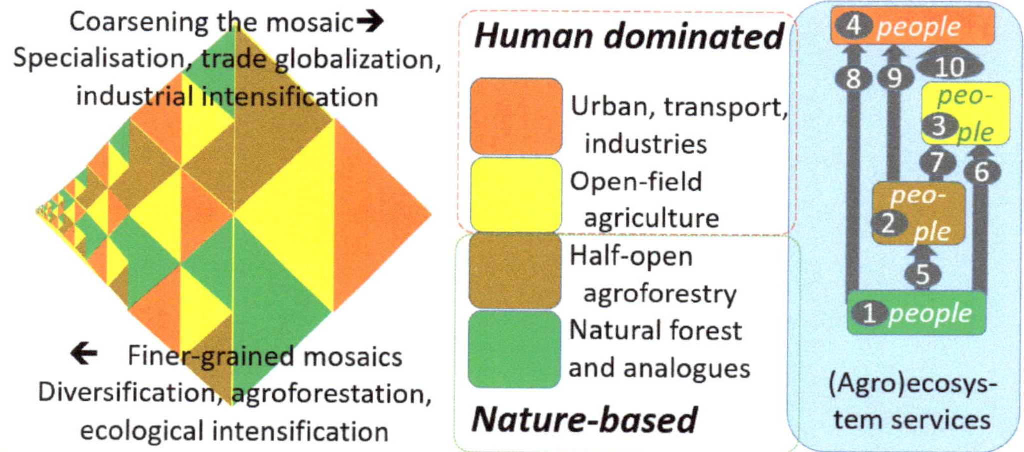

Figure 9. Relative composition (fractions of various colors) and spatial pattern (grain size of the mosaic) as two aspects of land cover at the forestry–agricultural–urban interfaces that jointly determine a set of six (agro-)ecosystem services via direct and indirect benefits people obtain (Source: [27]).

3.3. Agroforestry and Metrics for Multifunctionality

A central metric for progress towards multifunctionality across all relevant goals and efficient use of the land available could be that the land equivalent ratio for multifunctionality, LER_M, exceeds 1.0 [66]. LER_M is the sum of the areas of land managed specifically to achieve any of the various functions that match what multifunctional land use provides jointly. Examples include the provisioning of food, building materials, firewood, groundwater recharge, C storage, and belowground biodiversity conservation [66]. A subset of this index, the land equivalent ratio for productivity (LER_P), is commonly found to be in the range 1.1–1.3 for legume and cereal systems, for example; this may indicate negative yield gaps when current practice monocultures are taken as point of reference for the productivity of each component of a mixture. Forms of agricultural/agroforestry land use that do not maximize productivity as such may achieve high LER_M values through a range of ecosystem services and thus have land sparing properties through their land sharing [83]. Such land uses contradict the usually inferred contrast between sparing (by minimizing yield gaps) and sharing as two ways to reconcile production and other ecosystem services [84]. As the land sparing argument refers to the proportions of land required for agriculture and land sharing to the mosaic pattern (Figure 9), the sparing versus sharing debate may be based on a false dichotomy. Sparing through sharing (or "shparing") is feasible when LER_M exceeds 1.0.

A recent contribution to the debate on whether or not the promotion of agroforestry in tropical landscapes is a sensible policy stated [85], "Agroforestry is widely promoted as a potential solution to address multiple UN Sustainable Development Goals, including Zero Hunger, Responsible Consumption and Production, Climate Action, and Life on Land. Nonetheless, agroforests in the tropics often result from direct forest conversions, displacing rapidly vanishing and highly biodiverse forests with large carbon stocks, causing undesirable trade-offs." These authors concluded that forest-derived agroforestry supports higher biodiversity than open-land-derived agroforestry but essentially represents a degradation of forest, whereas open-land-derived agroforestry rehabilitates formerly forested open land. Ironically, expansion of the first type (forest derived), although associated with higher biodiversity indicators than the second (restoration based), might be rated as less biodiversity friendly than the second. Policies that promote the second might need to be combined with conserving remaining natural forest and maintaining tree diversity in forest-derived agroforests [86].

3.4. Agroforestry and the Half Earth Debate

Beyond the relational biophilia argument [87], current proposals to formulate targets for the post-2020 global biodiversity agenda that reserve half of the Earth for Nature blend a strong intrinsic (moral, ethical) values with reference to the planetary boundaries as existential risk for all life on the planet. The proposal builds on the relative success in achieving the Aichi target for increasing protected areas relative to the lower degrees of progress on other targets [88]. As the initial formulation and support came mainly from the "fortress conservation" lobbies, groups concerned with the rights, perspectives, and livelihoods of the people currently living in the areas that would get a primarily "conservation" designation responded with alarm [89,90]. Some follow-up proposals referred to 30% of the earth as conservation area, plus 20% under indigenous people's management, but the discussion continues. An aspect that is relatively absent from the debate so far is the spatial consequences and the alternative ways of achieving such targets, e.g., through spatial segregation of "production" and conservation areas (Figure 10A) or in a spatially integrated mosaic (Figure 10B) that may have a (fractal) scale distribution.

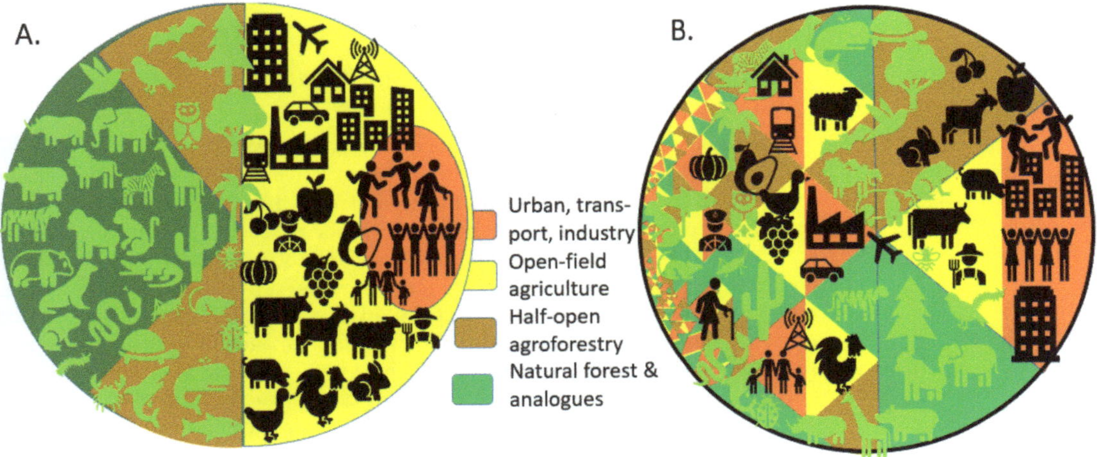

Figure 10. Visualization of the half earth debate and two contrasting spatial ways of achieving it: (**A**). through spatial segregation minimizing the nature–human interface and maximizing space for species with low compatibility with humans, or (**B**). through spatial integration with more intensive nature–human contact and proximity.

The segregated option, with large, protected areas segregated from intensive agriculture and urban domains, may achieve the best outcomes for conservation of threatened wilderness and associated species (especially those with large home ranges and low tolerance of human presence); in the spatially integrated option, more people would experience nature, or at least natural landscape elements, in their neighborhoods and potentially benefit from them in various ways. In terms of instrumental and relational values, the segregated and integrated options [91] thus have different consequences for change relative to the current mosaic of urban, open-field agriculture (including monoculture tree crops and short-rotation plantation forestry), half-open agroforestry (including trees outside forest), and natural vegetation (with levels of management that match sustainable use criteria). Many landscapes still are coarsening the grain size of the mosaic [92,93], influencing the environmental and social interactions between nature and people. The land sparing versus land sharing debate of the past decades [94,95] has focused on either closing mono-cultural yield gaps (through conventional intensification) or exploiting land equivalent ratios (through "ecological intensification"; [52]) as approaches to reconciling production and conservation needs of society.

A recent economic analysis [96] suggested that half Earth scenarios could be economically viable if technological progress towards land sparing continues and if the recreational needs of an increasingly urban and affluent population provide incentives and employment outside the urban and primary production areas. The authors may not have realized (and the debate has not picked up) that these arguments refer to a Figure 10B rather than Figure 10A scenarios. Value arguments have suggested that people move towards nature if nature does not move towards people [97]. In follow-up to the COVID pandemic, however, the spatial integration of people and nature will have to be reconciled with relational/instrumental concerns about public health [27]; a main objection to the half Earth concept is that the half of the planet where people dominate still needs elements of nature to function [98]. In this debate, the relational and instrumental value categories are fluid and part of a continuum.

The scope for a more connective agroforestry landscape concept has never been larger than now, as a key challenge for implementing the landscape approach is that political processes and conservation initiatives still operate in "silos", largely disconnected from farmers and local key agents responsible for tree governance [99]. The recent "making peace with nature" report [100] urges for a coherent approach to climate change, loss of biodiversity, and pollution as part of reimagining and transforming the ways in which the values of humans and nature are reconciled. An ambitious vision of the way agroforestry can be part of the solution needs to connect local to global scales and vice versa.

4. Roles for and Contributions by Agroforestry Research

4.1. Issue Cycle Stages

The recent "The Future is Now" UN sustainability science report [101] identified four levers that are essential in opening transformative pathways toward sustainable development: (1) governance (public decision making, rules); (2) economy and finance (incentives and investment); (3) individual and collective action (motivation and social feedback); and (4) science and technology (knowledge and understanding). The boundary work research tradition has long analyzed how the latter can be more effectively linked to the first three [1,102]. These levers match, but in different order, the various knowledge-to-action chains identified along an issue cycle [8]: (a) agenda setting; (b) better and more widely shared understanding of what is at stake and how it can be monitored; (c) commitment to common principles and moving on from denial and conspiracy theories; (d) differentiated responsibility in practice through means of implementation devolved to (newly created or existing) formal institutions; and (e) efforts to monitor and evaluate effects, which can be the start of a new issue cycle while allowing for genuine innovation. Progress markers along these five chains are available [8] (Table 2); bottlenecks in any of the chains can slow progress in others, including in the way knowledge of values and decision making progress.

Table 2. Progress markers (from 1 to 10) for four interrelated knowledge-to-action chains in the issue (decisions) cycle [8] classified by instrumental [　], relational [　] or mixed [　] value articulations.

Chains A: Agenda Setting and B: Science-Based Understanding of Ongoing Change and Emerging Issues	Chain C: Societal Willingness to Act—From Denial to Responsiveness, Common Goals, and Responsibility	Chain D: Governability Pathways to Change from Blame Games to Taking Responsibility and Means of Implementation	Chain E: Technological, Institutional Innovation: Real-Life Solutions and Learning
ab1. Initial guesstimates of seriousness of impacts of "emerging issues" based on current understanding of "systems"	c1. Steps from "ignoring" to "denial", based on conflicting evidence from "best" and "worst" cases in public discourse	d1. Identification of current rules, incentives, and motivational instruments as contributors/aggravators of the issue at stake, and options to reform them	e1. Adequate grounding of potential innovators in existing knowledge and theories to explore new applications, and in lists of "unresolved questions" for society at large
b2. Operational definitions of the entities and processes associated with the "issue" (potentially reframing, splitting and lumping of issues based on increased understanding of causation and/or effects)	c2. Steps from denial to accepting issues as part of the concurrent "agenda", requiring debate in a multiple stakeholder context with multiple knowledge claims	d2. Reflection on an "at least do no harm" precautionary principle in the face of remaining uncertainty and existing communication pathways with the wider stakeholder community	e2. Safe spaces for innovators, in terms of resources (finances, facilities) needed and protection from micromanagers
b3. Cause–effect mechanisms, feedback loops and system dynamics associated with the issue	c3. Steps from "blaming others" and "victim roles" to facing complex reality and taking shared responsibility	d3. Path dependency of the issue and opportunities to deal with the established context and its spatial variation	e3. Support for functional diversity of pathways explored and delayed, stepwise selection of increased support for "likely winners", within clear societal goals and criteria
b4. Agreed methods with known biases to allow replicable research and mapping	c4. Initial estimates of differential (by geographic and social strata) vulnerability	d4. Relevance of and steps towards legal change in rights and responsibilities in the existing constitutional framing	e4. Risk awareness and compliance with agreed safeguards by all innovators, but especially publicly supported ones
b5. Studies of spatial extent and temporal change of key aspects of the issue, its drivers, and its consequences	c5. Initial estimates of differential contribution to "causes" and likely need to change behavior and/or pay for damage done	d5. Economic (efficiency) dimensions of proposed pathways for dealing with the issue (at cause and/or consequence level)	e5. Early awareness of scale relations (in applicability, undesired/unexpected consequences) of emerging innovations
b6. Articulation of the planetary boundaries associated with the issue	c6. Initial estimates of differential opportunities to adapt to consequences and reduce contributions to "causation"	d6. Motivational and social (fairness) dimensions of proposed pathways for dealing with the issue (at driver and/or consequence level)	e6. Effective two-way feedback where existing theory ("first principles") appears to contrast with emerging practices (Pasteur quadrant)
b7. Using understanding of nonlinearity and feedback loops, proposition of thresholds for "safe operating space"	c7. Articulation of culture- and religion-based motivation to act in solidarity or direct self-interest	d7. Intersectoral integration across all relevant aspects of current agendas (i.e., beyond the focal issue)	e7. Early feedback from potential users and stakeholders of potential consequences that are to be avoided
b8. Agreed monitoring, reporting, and verification tools for collective action at relevant scales (local to global)	c8. Dynamic coalitions for change in the face of tradeoffs and synergy with other issues in various stages of their own "cycle"	d8. Polycentric governance dimensions of rights and responsibilities across institutional scales	e8. Opportunities to evaluate likely wider consequences in scenario tools that are sufficiently robust to extrapolate beyond known empirics

Table 2. Cont.

Chains A: Agenda Setting and B: Science-Based Understanding of Ongoing Change and Emerging Issues	Chain C: Societal Willingness to Act—From Denial to Responsiveness, Common Goals, and Responsibility	Chain D: Governability Pathways to Change from Blame Games to Taking Responsibility and Means of Implementation	Chain E: Technological, Institutional Innovation: Real-Life Solutions and Learning
b9. Scenario-evaluation tools to judge likely effectiveness of proposed and emerging innovations in their multidimensional characteristics (incl. tradeoffs and synergy)	c9. Prioritization among concurrent issues and negotiated trade-offs between agendas of multiple negotiating parties	d9. Opportunities for new public–private partnerships (covenants, phased change, clarity on long-term goals and standards)	e9. Stepwise empirical tests at relevant scales for "promising candidates", with clarity on standards to be applied for societal risk management
b10. Regular reassessment and recalibration of simplified proxies used for monitoring compliance and progress in dealing with the issue	c10. Sufficiently ambitious goals and adequate governance instruments (incl. monitoring compliance and effectiveness, sanctions) at all relevant scales in agreements and plans of action, with "common but differentiated responsibility"	d10. Where necessary, adjusting governance instruments based on litigation by specific stakeholder groups	e10. Adequate recognition (remuneration, influence) for past success (recognizing its limited predictive skill for future successes)

Taking the values of others, beyond one's in-group, into account is an important dimension of "governability" [103]. This process can be based on moral or ethical values and (relational) choices, but it can also derive from pragmatic (instrumental) considerations and experience that the "others" may have the power to disrupt and that they need incentives to not do so. In the analysis of social systems and their decision making, (reference) groups, rituals, affiliation, status, and power have been identified as aspects that need to be understood in their interaction [104,105]. Typically, public decisions consist of a declaration of principles followed by the creation of (or "outsourcing" to existing) institutions mandated to implement agreed goals, within delineated powers to enforce [106,107]. Communication and nudging [108] are likely to be central to such a theory of change of theory at the scale of human societies.

4.2. Methods and Interdisciplinarity of Agroforestry Research

The methods sections of the papers in the special issue on agroforestry-based ecosystem services demonstrate that agroforestry research has derived and embraced methods from many research traditions, ranging from agronomic trials, ecological, soil science, economics, social sciences and policy analysis. This reflects the trends in a recent overview of agroforestry research methods [109]. Further progress on the appreciation of agroforestry-based ecosystem services will require, for example,

1. Enhanced quantification of trees-outside-forests [9,110,111] in relation for "forest function" thresholds such as those for rainfall [112] and infiltration [113];
2. Distinctions between forest-derived and restoration-based agroforestry practices [15,85,86] and the expected societal costs and benefits [114];
3. Reconciling local, science-based, and policy-oriented ecological knowledge [115,116];
4. Process-level understanding of tree–soil–crop interactions used in bio-economic models [117] and the dependence on tree cover (Figure 8) of specific ecosystem services in each context [118–120];
5. Market-oriented domestication of local fruit trees [121], regularization of smallholder cash-crop production within the supposedly permanent forest estate [122];
6. Further landscape-scale studies of biodiversity impacts beyond plot-level effects on average farms [123–125];

7. Location specificity of trade-off management in the food–energy–water nexus [42,66,126] and in biodiversity [85,127–129], with increased interest in disease risks outside and within agroforestry [130,131];
8. Balancing the two sides of the doughnut in challenging any siloed interpretation of SDGs ([8,10,66,132]; Figure 1);
9. Realistic and critical impact assessments of agroforestry-enhancing projects in local contexts [133–135], in view of livelihood strategies and tactics that include migration [136] and grasping new economic opportunities [137];
10. Use of participatory methods in scenario development and reimagining desirable futures a local scale [42,138], reconciled with understanding agent behavior through "sociality" research [139].

4.3. Research Roles in Theories of Induced Change

In line with the stronger impact orientation of most of the funding sources for applied or use-oriented research, it is important that "theories of induced change" are explicit in how research can contribute to the progress of policy issue cycles (Table 3).

Table 3. Achievable goals for researchers interacting with policy issue cycles stages [109].

Policy Cycle Stage	Researcher Goals	Impact Looks Like …
Problem alert	Spotting new social and environmental problems or phenomena that (someone believes) limit progress to development goals such as SDGs	Raised interest and concern among researchers (and others? Activists?)
Problem scope and basis	Understanding: (A) extent of the problem (areas, people affected), (B) drivers and mechanisms, (C) connections to current or new theory	Either increasing numbers of people aware of and understanding nature of the problem and why it matters, or (if it turns out to be an unimportant problem) efforts redirected to areas with more potential effect
Potential solutions and interventions	Showing that there are actions that will alleviate the problem and policies that will promote those actions	Pilot projects that excite people, increase demands, generate more nuanced research
Political agenda setting	Getting relevant policy makers interested and pushing towards policy change	Convincing demonstrations that the problem impacts things policy makers care about and that policies proposed will help
Policy formulation	Systematic investigation of a problem and thoughtful assessment of options and alternatives	Convincing policy options formulated
Selection Process	Prioritization (decision-making) of available options given costs/benefits and compromises across diverse stakeholder interests	New policies adopted and followed
Implementing	Introduce actions based on policy aimed at changing the problem	Change in state of problem
Evaluation and monitoring	Confirm that the problem is under control (or tracking in the right direction) and remains so	Problem is solved—extent of "fix" and role of the policy.

4.4. Hypotheses for Further Research

In each of the three questions, several hypotheses emerged from our exploration of concepts:

    i.   The way plural value concepts of agroforestry evolve in relation to ES discourse and sustainability concerns at the multifunctional landscape scale.

Hypothesis 1. *The broader landscape context influences the roles trees and agroforestry play in value addition in forest-to-urban land use gradients, with the suggested ten-point classification of beneficiaries and service providers (Figure 9) largely untested.*

Hypothesis 2. *Critical tree cover thresholds for ecohydroclimatic functions in the agroforestry–forest continuum (Figure 8), including rainfall triggering and infiltration, depend on terrain properties and position relative to global atmospheric circulation patterns.*

    ii.   The interaction between the three agroforestry scales (AF1, AF2, and AF3) and value-of-nature concepts.

Hypothesis 3. *Instrumental, goal-oriented values of agroforestry land uses can be more effectively expressed in national policy debates in relation to nature-based solutions to a suite of SDGs (such as those linked to food, health, water, energy, jobs, sustainable cities, and climate change), rather than purely in economic value (SDG1 only),*

Hypothesis 4. *Relational values of nature to humans play an under-appreciated role in the communication of agroforestry options and the concerns in producer–consumer teleconnections of tropical commodity supply and value chains.*

    iii.   The roles research can play in connecting theories of place and change to policies that aim for applicable theories of induced change, in support of SDGs.

Hypothesis 5. *The complementarity of relational and instrumental value concepts in overcoming the successive hurdles that hinder progress of public policy issue cycles calls for boundary work teams with broad sets of analytical and communicative skills.*

Hypothesis 6. *A healthy research portfolio requires intellectual and financial space to invest in "high-risk, high potential impact" early stages of emerging issues in which agroforestry plays a role as cause and/or solution, beyond the more easily plannable later stages of issue cycles.*

5. Conclusions and Outlook

This review of a wide-ranging literature suggested that value concepts of agroforestry are evolving as part of the wider ES discourse and sustainability concerns at and beyond the multifunctional landscape scale. Basic understanding of the existing and evolving diversity of living organisms, the strengths and limitations of profit-oriented individual actions that drive markets, and the relevance and history of local community-scale institutions are essential now that it is the global commons of reconciling planetary boundaries with remaining development deficits that drives the agenda at the highest level. Both instrumental (goal-orientation, rationality) and relational (social in-group orientation) values are used in communication and broadening the political platforms for theories of induced change at landscape and regional scales.

The three agroforestry scales and concepts (farm, landscape, and policy integration) interact with the way these values can be articulated and, where relevant, quantified. In terms of the efficient use of land argument, land sparing can be achieved by land sharing if the latter meets the quantitative requirement that the land equivalent ratio, across relevant functions, is above 1.0. To evaluate this metric in practice, it is important to understand how intermediate (partial) tree cover affects a wide range of ecological functions that can contribute to services from a human perspective.

Current research on agroforestry and the ecosystem services it provides at local, national and global scales commonly embraces a social–ecological system framing. Two-way feedbacks between social and ecological subsystems across scales are at the center of research interest. Most research has, however, been framed in an instrumental ecosystem services perspective, where the ecological subsystem serves the social system, rather than as a two-way relational one, where stewardship enhances and broadens relational values.

Many contemporary studies are both multi- and interdisciplinary in nature, and they can become transdisciplinary if they explicitly compare farmers' ecological knowledge with science-based perspectives, farmers' choices with regard to what economic analysis suggests as ways to reconcile multiple goals, and/or policy ambitions with the social–ecological reality of landscapes connected to global trade. Relevant, use-oriented agroforestry research can play active roles in connecting theories of place and change to policies that aim for applicable theories of induced change in support of SDGs. The substantial number of hurdles that need to be taken to link knowledge with action in environmental issue cycles require both instrumental and relational value arguments to play out, without suggestions that they are anything but complementary perspectives rather than separate "values" that would need to be added up to attain a "total value". The potential emotional appeal of trees (and especially tree planting) will continue to be both an opportunity and a challenge for promoting agroforestry unless the relational and instrumental values of agroforestry are more widely understood as complementary articulations of a rich and complex reality in plural societies.

Funding: This research received no external funding.

Institutional Review Board Statement: Not applicable.

Informed Consent Statement: Not applicable.

Data Availability Statement: Not applicable.

Acknowledgments: The authors acknowledge that the ideas presented here are the tip of an iceberg, build on discussions with many colleagues, who are acknowledged collectively. We thank anonymous reviewers for helpful comments on an earlier draft.

Conflicts of Interest: The author declares no conflict of interest. No funders had a role in the design of the study; in the collection, analyses, or interpretation of data; in the writing of the manuscript, or in the decision to publish the results.

References

1. Clark, W.C.; Tomich, T.P.; Van Noordwijk, M.; Guston, D.; Catacutan, D.; Dickson, N.M.; McNie, E. Boundary work for sustainable development: Natural resource management at the Consultative Group on International Agricultural Research (CGIAR). *Proc. Natl. Acad. Sci. USA* **2016**, *113*, 4615–4622. [CrossRef] [PubMed]
2. Van Noordwijk, M.; Coe, R.; Sinclair, F.L. Agroforestry paradigms. In *Sustainable Development through Trees on Farms: Agroforestry in Its Fifth Decade*; World Agroforestry (ICRAF): Bogor, Indonesia, 2019; pp. 1–12. Available online: https://worldagroforestry.org/trees-on-farms (accessed on 2 July 2021).
3. Zafra-Calvo, N.; Balvanera, P.; Pascual, U.; Merçon, J.; Martín-López, B.; van Noordwijk, M.; Mwampamba, T.H.; Lele, S.; Speranza, C.I.; Arias-Arévalo, P.; et al. Plural valuation of nature for equity and sustainability: Insights from the Global South. *Glob. Environ. Chang.* **2020**, *63*, 102115. [CrossRef]
4. Kadykalo, A.N.; López-Rodriguez, M.D.; Ainscough, J.; Droste, N.; Ryu, H.; Ávila-Flores, G.; Le Clec'h, S.; Muñoz, M.C.; Nilsson, L.; Rana, S.; et al. Disentangling 'ecosystem services' and 'nature's contributions to people'. *Ecosyst. People* **2019**, *15*, 269–287. [CrossRef]
5. Costanza, R.; D'Arge, R.; de Groot, R.; Farber, S.; Grasso, M.; Hannon, B.; Limburg, K.; Naeem, S.; O'neill, R.V.; Paruelo, J.; et al. The value of the world's ecosystem services and natural capital. *Nature* **1997**, *387*, 253–260. [CrossRef]
6. Costanza, R.; de Groot, R.; Braat, L.; Kubiszewski, I.; Fioramonti, L.; Sutton, P.; Farber, S.; Grasso, M. Twenty years of ecosystem services: How far have we come and how far do we still need to go? *Ecosyst. Serv.* **2017**, *28*, 1–16. [CrossRef]
7. Himes, A.; Muraca, B. Relational values: The key to pluralistic valuation of ecosystem services. *Curr. Opin. Environ. Sustain.* **2018**, *35*, 1–7. [CrossRef]
8. Van Noordwijk, M. Integrated natural resource management as pathway to poverty reduction: Innovating practices, institutions and policies. *Agric. Syst.* **2019**, *172*, 60–71. [CrossRef]

9.   Zomer, R.J.; Neufeldt, H.; Xu, J.; Ahrends, A.; Bossio, D.; Trabucco, A.; Van Noordwijk, M.; Wang, M. Global Tree Cover and Biomass Carbon on Agricultural Land: The contribution of agroforestry to global and national carbon budgets. *Sci. Rep.* **2016**, *6*, 1–12. [CrossRef]

10.   Raworth, K. A safe and just space for humanity: Can we live within the doughnut. *Oxfam Policy Pract. Clim. Chang. Resil.* **2012**, *8*, 1–26.

11.   Van Noordwijk, M.; Leimona, B. Principles for fairness and efficiency in enhancing environmental services in Asia: Payments, compensation, or co-investment? *Ecol. Soc.* **2010**, *15*, 17. Available online: http://www.ecologyandsociety.org/vol15/iss4/art17/ (accessed on 2 July 2021). [CrossRef]

12.   Woittiez, L.S.; van Wijk, M.T.; Slingerland, M.; van Noordwijk, M.; Giller, K.E. Yield gaps in oil palm: A quantitative review of contributing factors. *Eur. J. Agron.* **2017**, *83*, 57–77. [CrossRef]

13.   Van Noordwijk, M.; Brussaard, L. Minimizing the ecological footprint of food: Closing yield and efficiency gaps simultaneously? *Curr. Opin. Environ. Sustain.* **2014**, *8*, 62–70. [CrossRef]

14.   Minang, P.A.; van Noordwijk, M.; Duguma, L. Policies for ecosystem services enhancement. In *Sustainable Development through Trees on Farms: Agroforestry in Its Fifth Decade*; World Agroforestry (ICRAF): Bogor, Indonesia, 2019; pp. 361–375. Available online: https://worldagroforestry.org/trees-on-farms (accessed on 2 July 2021).

15.   Keesstra, S.; Mol, G.; De Leeuw, J.; Okx, J.; Molenaar, C.; De Cleen, M.; Visser, S. Soil-Related Sustainable Development Goals: Four Concepts to Make Land Degradation Neutrality and Restoration Work. *Land* **2018**, *7*, 133. [CrossRef]

16.   Van Noordwijk, M. *Theories of Place, Change and Induced Change for Tree-Crop-Based Agroforestry*; World Agroforestry (ICRAF): Bogor, Indonesia, 2021.

17.   Díaz, S.; Settele, J.; Brondízio, E.S.; Ngo, H.T.; Guèze, M.; Agard, J.; Arneth, A.; Balvanera, P.; Brauman, K.; Butchart, S.H.; et al. *Summary for Policymakers of the Global Assessment Report on Biodiversity and Ecosystem Services of the Intergovernmental Science-Policy Platform on Biodiversity and Ecosystem Services*; IPBES: Bonn, Germany, 2019.

18.   Available online: https://commons.wikimedia.org/wiki/File:Charles_Darwin_01.jpg (accessed on 25 June 2021).

19.   Available online: https://commons.wikimedia.org/wiki/File:Adam_Smith,_1723_-_1790._Political_economist_-_Google_Art_Project.jpg (accessed on 25 June 2021).

20.   Available online: https://creativecommons.org/2012/06/13/honoring-elinor-ostrom/ (accessed on 25 June 2021).

21.   Smith, A. *An Inquiry into the Nature and Causes of the Wealth of Nations*; W. Strahan and T. Cadell: London, UK, 1776.

22.   Smith, A. *The Theory of Moral Sentiments*, 2nd ed.; A. Millar; A. Kincaid and J. Bell: Strand: Edinburgh, UK, 1761; Available online: https://books.google.co.id/books?id=bZhZAAAAcAAJ&q=editions:u_L0P5LRqXkC&pg=PP3&redir_esc=y#v=onepage&q=editions%3Au_L0P5LRqXkC&f=false (accessed on 2 July 2021).

23.   Dawkins, R. *The Blind Watchmaker: Why the Evidence of Evolution Reveals a Universe without Design*; WW Norton & Company: New York, NY, USA, 1996.

24.   Mazoyer, M.; Roudart, L. *A History of World Agriculture: From the Neolithic Age to the Current Crisis*; Monthly Review Press: New York, NY, USA, 2006.

25.   Diamond, J. *Collapse: How Societies Choose to Fail or Survive*; Lane: London, UK, 2005.

26.   De Ruiter, P.C.; Wolters, V.; Moore, J.C.; Winemiller, K.O. Food web ecology: Playing Jenga and beyond. *Science* **2005**, *309*, 68–71. [CrossRef] [PubMed]

27.   Duguma, L.A.; van Noordwijk, M.; Minang, P.A.; Muthee, K. COVID-19 pandemic and agroecosystem resilience: Early insights for building better futures. *Sustainability* **2021**, *13*, 1278. [CrossRef]

28.   Nature-Risk-Rising. Available online: https://www.weforum.org/reports/nature-risk-rising-why-the-crisis-engulfing-nature-matters-for-business-and-the-economy (accessed on 30 March 2021).

29.   Folke, C.; Hahn, T.; Olsson, P.; Norberg, J. Adaptive governance of social-ecological systems. *Annu. Rev. Environ. Resour.* **2005**, *30*, 441–473. [CrossRef]

30.   Taylor, B.R. *Dark Green Religion: Nature Spirituality and the Planetary Future*; University of California Press: Berkeley, CA, USA, 2010.

31.   Rosenau, J.N. *Turbulence in World Politics: A Theory of Change and Continuity*; Princeton University Press: Princeton, NJ, USA, 2018.

32.   Rose, C. The comedy of the commons: Custom, commerce, and inherently public property. *Univ. Chic. Law Rev.* **1986**, *53*, 711–781. Available online: https://digitalcommons.law.yale.edu/fss_papers/1828/ (accessed on 25 March 2021). [CrossRef]

33.   Ostrom, E. *Governing the Commons: The Evolution of Institutions for Collective Action*; Cambridge University Press: Cambridge, UK, 1990.

34.   Ryan, R.M.; Deci, E.L. On happiness and human potentials: A review of research on hedonic and eudaimonic well-being. *Annu. Rev. Psychol.* **2001**, *52*, 141–166. [CrossRef]

35.   Bauer, J.J.; McAdams, D.P.; Pals, J.L. Narrative identity and eudaimonic well-being. *J. Happiness Stud.* **2008**, *9*, 81–104. [CrossRef]

36.   Davidson, J.; Henley, D. (Eds.) *The Revival of Tradition in Indonesian Politics: The Deployment of Adat from Colonialism to Indigenism*; Routledge: London, UK; New York, NY, USA, 2007.

37.   Takeuchi, K. Rebuilding the relationship between people and nature: The Satoyama Initiative. *Ecol. Res.* **2010**, *25*, 891–897. [CrossRef]

38.   Fang, T. Yin Yang: A new perspective on culture. *Manag. Organ. Rev.* **2012**, *8*, 25–50. [CrossRef]

39.   Ahmad, J.; Kadir, A.; Bahari, M.A. Disequilibrium and divinity salience as invariant structures in the halal executives' experience of eudaimonia. *Adv. Bus. Res. Int. J.* **2019**, *5*, 1–15. [CrossRef]

40. Steger, M.F.; Kashdan, T.B.; Oishi, S. Being good by doing good: Daily eudaimonic activity and well-being. *J. Res. Personal.* **2008**, *42*, 22–42. [CrossRef]

41. Ryff, C.D.; Singer, B.H. Know thyself and become what you are: A eudaimonic approach to psychological well-being. *J. Happiness Stud.* **2008**, *9*, 13–39. [CrossRef]

42. Van Noordwijk, M.; Speelman, E.; Hofstede, G.J.; Farida, A.; Abdurrahim, A.Y.; Miccolis, A.; Hakim, A.L.; Wamucii, C.N.; Lagneaux, E.; Andreotti, F.; et al. Sustainable Agroforestry Landscape Management: Changing the Game. *Land* **2020**, *9*, 243. [CrossRef]

43. Minang, P.A.; Duguma, L.A.; Alemagi, D.; van Noordwijk, M. Scale considerations in landscape approaches. In *Climate-Smart Landscapes: Multifunctionality in Practice*; Minang, P.A., van Noordwijk, M., Freeman, O.E., Mbow, C., de Leeuw, J., Catacutan, D., Eds.; World Agroforestry Centre (ICRAF): Nairobi, Kenya, 2015; pp. 121–133.

44. Van Noordwijk, M.; Leimona, B.; Jindal, R.; Villamor, G.B.; Vardhan, M.; Namirembe, S.; Catacutan, D.; Kerr, J.; Minang, P.A.; Tomich, T.P. Payments for environmental services: Evolution toward efficient and fair incentives for multifunctional landscapes. *Annu. Rev. Environ. Resour.* **2012**, *37*, 389–420. [CrossRef]

45. Thaler, R.H.; Ganser, L.J. *Misbehaving: The Making of Behavioral Economics*; WW Norton: New York, NY, USA, 2015.

46. Kahneman, D. *Thinking, Fast and Slow*; Macmillan: London, UK, 2011.

47. Fiske, A.P. The four elementary forms of sociality: Framework for a unified theory of social relations. *Psychol. Rev.* **1992**, *99*, 689. [CrossRef]

48. Gupta, J.; Matthews, R.; Minang, P.; van Noordwijk, M.; Kuik, O.; van der Grijp, N. Climate change and forests: From the Noordwijk Declaration to REDD. In *Climate Change, Forests and REDD. Lessons for Institutional Design*; Routledge: London, UK, 2013; pp. 1–24.

49. The Noordwijk Ministerial Declaration on Climate Change. Available online: https://unfccc.int/resource/ccsites/senegal/fact/fs218.htm (accessed on 15 February 2021).

50. Hoe Het Redden van de Aarde Strandde in Noordwijk aan Zee. Available online: https://www.youtube.com/watch?v=LqjZStt9quE (accessed on 15 February 2021).

51. Lorenz, E. The butterfly effect. *World Sci. Ser. Nonlinear Sci. Ser. A* **2000**, *39*, 91–94.

52. Thunberg, G. *No One Is Too Small to Make a Difference: Illustrated Edition*; Allen Lane, Penguin: London, UK, 2019; ISBN 9780241453445.

53. Meadows, D. Leverage points: Places to intervene in a system. *Solut. A Sustain. Desirable Future* **1999**, *1*, 41–49.

54. Stålhammar, S.; Thorén, H. Three perspectives on relational values of nature. *Sustain. Sci.* **2019**, *14*, 1201–1212. [CrossRef]

55. Van Noordwijk-Van Veen, J.C. *Biologieonderwijs en Milieueducatie op de Basisschool [Biology Education and Environmental Education at Primary Schools]*; Van Gorcum: Assen, The Netherlands, 1973.

56. Ross, A.; Sherman, R.; Snodgrass, J.G.; Delcore, H.D. *Indigenous Peoples and the Collaborative Stewardship of Nature: Knowledge Binds and Institutional Conflicts*; Left Coast Press: Walnut Creek, CA, USA, 2011.

57. Villamor, G.B.; Chiong-Javier, E.; Djanibekov, U.; Catacutan, D.C.; van Noordwijk, M. Gender differences in land-use decisions: Shaping multifunctional landscapes? *Curr. Opin. Environ. Sustain.* **2014**, *6*, 128–133. [CrossRef]

58. Nyhus, P.; Tilson, R. Agroforestry, elephants, and tigers: Balancing conservation theory and practice in human-dominated landscapes of Southeast Asia. *Agric. Ecosyst. Environ.* **2004**, *104*, 87–97. [CrossRef]

59. Namirembe, S.; Leimona, B.; van Noordwijk, M.; Minang, P.A. (Eds.) *Co-Investment in Ecosystem Services: Global Lessons from Payment and Incentive Schemes*; World Agroforestry Centre (ICRAF): Nairobi, Kenya, 2018.

60. Cohen-Shacham, E.; Walters, G.; Janzen, C.; Maginnis, S. (Eds.) *Nature-Based Solutions to Address Global Societal Challenges*; IUCN: Gland, Switzerland, 2016.

61. Rival, L. (Ed.) *The Social Life of Trees: Anthropological Perspectives on Tree Symbolism*; Routledge: London, UK, 2021.

62. Van Noordwijk, M.; Hoang, M.H.; Neufeldt, H.; Öborn, I.; Yatich, T. *How Trees and People Can Co-Adapt to Climate Change: Reducing Vulnerability in Multifunctional Landscapes*; World Agroforestry Centre: Nairobi, Kenya, 2011.

63. Van Noordwijk, M.; Coe, R.; Sinclair, F.L.; Luedeling, E.; Bayala, J.; Muthuri, C.W.; Cooper, P.; Kindt, R.; Duguma, L.; Lamanna, C.; et al. Climate change adaptation in and through agroforestry: Four decades of research initiated by Peter Huxley. *Mitig. Adapt. Strateg. Glob. Chang.* **2021**, *26*, 1–33. [CrossRef]

64. Van Noordwijk, M. Prophets, profits, prove it: Social forestry under pressure. *One Earth* **2020**, *2*, 394–397. [CrossRef]

65. Mead, R.; Willey, R. The concept of a 'land equivalent ratio' and advantages in yields from intercropping. *Exp. Agric.* **1980**, *16*, 217–228. [CrossRef]

66. Van Noordwijk, M.; Duguma, L.A.; Dewi, S.; Leimona, B.; Catacutan, D.C.; Lusiana, B.; Öborn, I.; Hairiah, K.; Minang, P.A. SDG synergy between agriculture and forestry in the food, energy, water and income nexus: Reinventing agroforestry? *Curr. Opin. Environ. Sustain.* **2018**, *34*, 33–42. [CrossRef]

67. Van Noordwijk, M.; Chandler, F.; Tomich, T.P. *An Introduction to the Conceptual Basis of RUPES: Rewarding Upland Poor for the Environmental Services They Provide*; ICRAF-Southeast Asia: Bogor, Indonesia, 2004; p. 41.

68. Michon, G.; De Foresta, H. Agroforests: Pre-domestication of forest trees or true domestication of forest ecosystems? *NJAS Wagening. J. Life Sci.* **1997**, *45*, 451–462. [CrossRef]

69. Wiersum, K.F. Forest gardens as an 'intermediate' land-use system in the nature-culture continuum: Characteristics and future potential. In *New Vistas in Agroforestry*; Springer: Dordrecht, The Netherlands, 2004; pp. 123–134.

70. Van Noordwijk, M.; Rahayu, S.; Gebrekirstos, A.; Kindt, R.; Tata, H.L.; Muchugi, A.; Ordonez, J.C.; Xu, J. Tree diversity as basis of agroforestry. In *Sustainable Development through Trees on Farms: Agroforestry in Its Fifth Decade*; Van Noordwijk, M., Ed.; World Agroforestry (ICRAF): Bogor, Indonesia, 2019; pp. 17–44. Available online: https://worldagroforestry.org/trees-on-farms (accessed on 2 July 2021).

71. Van Noordwijk, M.; Bargues-Tobella, A.; Muthuri, C.W.; Gebrekirstos, A.; Maimbo, M.; Leimona, B.; Bayala, J.; Ma, X.; Lasco, R.; Xu, J.; et al. Agroforestry as part of nature-based water management. In *Sustainable Development through Trees on Farms: Agroforestry in Its Fifth Decade*; Van Noordwijk, M., Ed.; World Agroforestry (ICRAF): Bogor, Indonesia, 2019; pp. 305–334. Available online: https://worldagroforestry.org/trees-on-farms (accessed on 2 July 2021).

72. Rosenstock, T.S.; Wilkes, A.; Jallo, C.; Namoi, N.; Bulusu, M.; Suber, M.; Mboi, D.; Mulia, R.; Simelton, E.; Richards, M.; et al. Making trees count: Measurement and reporting of agroforestry in UNFCCC national communications of non-Annex I countries. *Agric. Ecosyst. Environ.* **2019**, *284*, 106569. [CrossRef]

73. Hairiah, K.; van Noordwijk, M.; Sari, R.R.; Saputra, D.D.; Suprayogo, D.; Kurniawan, S.; Prayogo, C.; Gusli, S. Soil carbon stocks in Indonesian (agro) forest transitions: Compaction conceals lower carbon concentrations in standard accounting. *Agric. Ecosyst. Environ.* **2020**, *294*, 106879. [CrossRef]

74. Van Noordwijk, M.; Poulsen, J.; Ericksen, P. Quantifying off-site effects of land use change: Filters, flows and fallacies. *Agric. Ecosyst. Environ.* **2004**, *104*, 19–34. [CrossRef]

75. Khasanah, N.; Tanika, L.; Pratama, L.D.Y.; Leimona, B.; Prasetiyo, E.; Marulani, F.; Hendriatna, A.; Zulkarnain, M.T.; Toulier, A.; van Noordwijk, M. Groundwater-extracting rice production in the Rejoso watershed (Indonesia) reducing urban water availability: Characterization and intervention priorities. *Land* **2021**, *9*, in press.

76. Pfund, J.L.; Watts, J.D.; Boissiere, M.; Boucard, A.; Bullock, R.M.; Ekadinata, A.; Dewi, S.; Feintrenie, L.; Levang, P.; Rantala, S.; et al. Understanding and integrating local perceptions of trees and forests into incentives for sustainable landscape management. *Environ. Manag.* **2011**, *48*, 334–349. [CrossRef] [PubMed]

77. Vira, B.; Agarwal, B.; Jamnadas, R.; Kleinschmit, D.; McMullin, S.; Mansourian, S.; Neufeldt, H.; Parrotta, J.A.; Sunderland, T.; Wildburger, C. *Introduction: Forests, Trees and Landscapes for Food Security and Nutrition*; IUFRO World Series 33; IUFRO: Vienna, Austria, 2015; pp. 14–23.

78. Van Noordwijk, M.; Bizard, V.; Wangpakapattanawong, P.; Tata, H.L.; Villamor, G.B.; Leimona, B. Tree cover transitions and food security in Southeast Asia. *Glob. Food Secur.* **2014**, *3*, 200–208. [CrossRef]

79. Van Noordwijk, M.; Suyamto, D.; Lusiana, B.; Ekadinata, A.; Hairiah, K. Facilitating agroforestation of landscapes for sustainable benefits: Tradeoffs between carbon stocks and local development benefits in Indonesia according to the FALLOW model. *Agric. Ecosyst. Environ.* **2008**, *126*, 98–112. [CrossRef]

80. Van Noordwijk, M.; Tanika, L.; Lusiana, B. Flood risk reduction and flow buffering as ecosystem services. Part 1: Theory on flow persistence, flashiness and base flow. *Hydrol. Earth Syst. Sci.* **2017**, *21*, 2321–2340. [CrossRef]

81. Van Noordwijk, M.; van Roode, M.; McCallie, E.L.; Cadisch, G. Erosion and sedimentation as multiscale, fractal processes: Implications for models, experiments and the real world. In *Soil Erosion at Multiple Scales, Principles and Methods for Assessing Causes and Impacts*; Penning de Vries, F., Agus, F., Kerr, J., Eds.; CAB International: Wallingford, UK, 1998; pp. 223–253.

82. Verbist, B.; Poesen, J.; van Noordwijk, M.; Widianto, W.; Suprayogo, D.; Agus, F.; Deckers, S. Factors affecting soil loss at plot scale and sediment yield at catchment scale in a tropical volcanic agroforestry landscape. *Catena* **2010**, *80*, 34–46. [CrossRef]

83. Khasanah, N.; van Noordwijk, M.; Slingerland, M.; Sofiyudin, M.; Stomph, D.; Migeon, A.F.; Hairiah, K. Oil Palm Agroforestry Can Achieve Economic and Environmental Gains as Indicated by Multifunctional Land Equivalent Ratios. *Front. Sustain. Food Syst.* **2020**, *3*, 122. [CrossRef]

84. Phalan, B.T. What have we learned from the land sparing-sharing model? *Sustainability* **2018**, *10*, 1760. [CrossRef]

85. Martin, D.A.; Osen, K.; Grass, I.; Hölscher, D.; Tscharntke, T.; Wurz, A.; Kreft, H. Land-use history determines ecosystem services and conservation value in tropical agroforestry. *Conserv. Lett.* **2020**, *13*, e12740. [CrossRef]

86. Van Noordwijk, M.; Gitz, V.; Minang, P.A.; Dewi, S.; Leimona, B.; Duguma, L.; Pingault, N.; Meybeck, A. People-Centric Nature-Based Land Restoration through Agroforestry: A Typology. *Land* **2020**, *9*, 251. [CrossRef]

87. Wilson, E.O. Biophilia and the Conservation Ethic. In *Evolutionary Perspectives on Environmental Problems*; Routledge: London, UK, 2007; pp. 249–257.

88. Secretariat of the Convention on Biological Diversity. *Global Biodiversity Outlook 5*; Secretariat of the Convention on Biological Diversity: Montreal, QC, Canada, 2020; Available online: https://www.cbd.int/gbo/gbo5/publication/gbo-5-en.pdf (accessed on 5 March 2021).

89. Domínguez, L.; Luoma, L. Decolonising conservation policy: How colonial land and conservation ideologies persist and perpetuate indigenous injustices at the expense of the environment. *Land* **2020**, *9*, 65. [CrossRef]

90. Büscher, B.; Fletcher, R.; Brockington, D.; Sandbrook, C.; Adams, W.M.; Campbell, L.; Corson, C.; Dressler, W.; Duffy, R.; Gray, N.; et al. Half-Earth or Whole Earth? Radical ideas for conservation, and their implications. *Oryx* **2017**, *51*, 407–410. [CrossRef]

91. Van Noordwijk, M.; Tomich, T.P.; de Foresta, H.; Michon, G. To segregate-or to integrate? The question of balance between production and biodiversity conservation in complex agroforestry systems. *Agrofor. Today* **1997**, *9*, 6–9.

92. Van Noordwijk, M.; Tata, H.L.; Xu, J.; Dewi, S.; Minang, P.A. Segregate or integrate for multifunctionality and sustained change through rubber-based agroforestry in Indonesia and China. In *Agroforestry-the Future of Global Land Use*; Nair, P.K., Garrity, D.P., Eds.; Springer: Dordrecht, The Netherlands, 2012; pp. 69–104.

93. Asubonteng, K.O.; Pfeffer, K.; Ros-Tonen, M.A.; Baud, I.; Benefoh, D.T. Integration versus segregation: Structural dynamics of a smallholder-dominated mosaic landscape under tree-crop expansion in Ghana. *Appl. Geogr.* **2020**, *108*, 102201. [CrossRef]
94. Phalan, B.; Onial, M.; Balmford, A.; Green, R.E. Reconciling food production and biodiversity conservation: Land sharing and land sparing compared. *Science* **2011**, *333*, 1289–1291. [CrossRef]
95. Ellis, E.C.; Pascual, U.; Mertz, O. Ecosystem services and nature's contribution to people: Negotiating diverse values and trade-offs in land systems. *Curr. Opin. Environ. Sustain.* **2019**, *38*, 86–94. [CrossRef]
96. Dinerstein, E.; Vynne, C.; Sala, E.; Joshi, A.R.; Fernando, S.; Lovejoy, T.E.; Mayorga, J.; Olson, D.; Asner, G.P.; Baillie, J.E.; et al. A global deal for nature: Guiding principles, milestones, and targets. *Sci. Adv.* **2019**, *5*, eaaw2869. [CrossRef]
97. Dolan, R.; Bullock, J.M.; Jones, J.P.G.; Athanasiadis, I.N.; Martinez-Lopez, J.; Willcock, S. The flows of nature to people, and of people to nature: Applying movement concepts to ecosystem services. *Land* **2021**, *10*, 576. [CrossRef]
98. Rolo, V.; Roces-Diaz, J.V.; Torralba, M.; Kay, S.; Fagerholm, N.; Aviron, S.; Burgess, P.; Crous-Duran, J.; Ferreiro-Dominguez, N.; Graves, A.; et al. Mixtures of forest and agroforestry alleviate trade-offs between ecosystem services in European rural landscapes. *Ecosyst. Serv.* **2021**, *50*, 101318. [CrossRef]
99. Zinngrebe, Y.; Borasino, E.; Chiputwa, B.; Dobie, P.; Garcia, E.; Gassner, A.; Kihumuro, P.; Komarudin, H.; Liswanti, N.; Makui, P.; et al. Agroforestry governance for operationalising the landscape approach: Connecting conservation and farming actors. *Sustain. Sci.* **2020**, *15*, 1417–1434. [CrossRef]
100. United Nations Environment Programme. Making Peace with Nature: A Scientific Blueprint to Tackle the Climate, Biodiversity and Pollution Emergencies. Nairobi. 2021. Available online: https://www.unep.org/resources/making-peace-nature (accessed on 1 July 2021).
101. IGS-UN [Independent Group of Scientists appointed by the Secretary-General]. *Global Sustainable Development Report 2019: The Future Is Now—Science for Achieving Sustainable Development*; United Nations: New York, NY, USA, 2019; Available online: https://sustainabledevelopment.un.org/content/documents/24797GSDR_report_2019.pdf (accessed on 5 March 2021).
102. Cash, D.; Clark, W.C.; Alcock, F.; Dickson, N.M.; Eckley, N.; Jäger, J. *Salience, Credibility, Legitimacy and Boundaries: Linking Research, Assessment and Decision Making*; KSG Working Papers Series, RWP02-046; Kennedy School of Governance, Harvard University: Boston, MA, USA, 2003. [CrossRef]
103. Kooiman, J. Exploring the concept of governability. *J. Comp. Policy Anal. Res. Pract.* **2008**, *10*, 171–190. [CrossRef]
104. Hofstede, G.J. GRASP agents: Social first, intelligent later. *Ai Soc.* **2019**, *34*, 535–543. [CrossRef]
105. Kemper, T.D. *Status, Power and Ritual Interaction: A Relational Reading of Durkheim, Goffman and Collins*; Ashgate: Farnham, UK, 2011.
106. Vatn, A. *Institutions and the Environment*; Edward Elgar Publishing: Cheltenham, UK, 2007.
107. Vatn, A. Rationality, institutions and environmental policy. *Ecol. Econ.* **2005**, *55*, 203–217. [CrossRef]
108. Thaler, R.H.; Sunstein, C.R. *Nudge: Improving Decisions about Health, Wealth, and Happiness*; Yale University Press: New Haven, CT, USA, 2008.
109. Van Noordwijk, M.; Coe, R. Methods in agroforestry research across its three paradigms. In *Sustainable Development through Trees on Farms: Agroforestry in Its Fifth Decade*; van Noordwijk, M., Ed.; World Agroforestry (ICRAF): Bogor, Indonesia, 2019; pp. 324–346. Available online: https://worldagroforestry.org/trees-on-farms (accessed on 2 July 2021).
110. Brandt, J.; Stolle, F. A global method to identify trees outside of closed-canopy forests with medium-resolution satellite imagery. *Int. J. Remote Sens.* **2020**, *42*, 1713–1737. [CrossRef]
111. Oloo, F.; Murithi, G.; Jepkoshei, A. Quantifying tree cover loss in urban forests within Nairobi city metropolitan area from Earth observation data. *Environ. Sci. Proc.* **2021**, *3*, 78. [CrossRef]
112. Duku, C.; Hein, L. The impact of deforestation on rainfall in Africa: A data-driven assessment. *Environ. Res. Lett.* **2021**, *16*, 064044. [CrossRef]
113. Suprayogo, D.; van Noordwijk, M.; Hairiah, K.; Meilasari, N.; Rabbani, A.L.; Ishaq, R.M.; Widianto, W. Infiltration-Friendly Agroforestry Land Uses on Volcanic Slopes in the Rejoso Watershed, East Java, Indonesia. *Land* **2020**, *9*, 240. [CrossRef]
114. Wainaina, P.; Minang, P.A.; Gituku, E.; Duguma, L. Cost-Benefit Analysis of Landscape Restoration: A Stocktake. *Land* **2020**, *9*, 465. [CrossRef]
115. Nguyen, M.P.; Vaast, P.; Pagella, T.; Sinclair, F. Local Knowledge about Ecosystem Services Provided by Trees in Coffee Agroforestry Practices in Northwest Vietnam. *Land* **2020**, *9*, 486. [CrossRef]
116. Cahyono, E.D.; Fairuzzana, S.; Willianto, D.; Pradesti, E.; McNamara, N.P.; Rowe, R.L.; van Noordwijk, M. Agroforestry Innovation through Planned Farmer Behavior: Trimming in Pine–Coffee Systems. *Land* **2020**, *9*, 363. [CrossRef]
117. Burgess, P.J.; Graves, A.R.; García de Jalón, S.; Palma, J.H.N.; Dupraz, C.; van Noordwijk, M. Modelling agroforestry systems. In *Agroforestry for Sustainable Agriculture*; Mosquera-Losada, M.R., Prabhu, R., Eds.; Burleigh Dodds Science Publishing Limited: Cambridge, UK, 2019; pp. 209–238.
118. Hairiah, K.; Widianto, W.; Suprayogo, D.; Van Noordwijk, M. Tree Roots Anchoring and Binding Soil: Reducing Landslide Risk in Indonesian Agroforestry. *Land* **2020**, *9*, 256. [CrossRef]
119. Nyberg, Y.; Musee, C.; Wachiye, E.; Jonsson, M.; Wetterlind, J.; Öborn, I. Effects of Agroforestry and Other Sustainable Practices in the Kenya Agricultural Carbon Project (KACP). *Land* **2020**, *9*, 389. [CrossRef]
120. Aynekulu, E.; Suber, M.; van Noordwijk, M.; Arango, J.; Roshetko, J.M.; Rosenstock, T.S. Carbon Storage Potential of Silvopastoral Systems of Colombia. *Land* **2020**, *9*, 309. [CrossRef]

121. Do, V.H.; La, N.; Mulia, R.; Bergkvist, G.; Dahlin, A.S.; Nguyen, V.T.; Pham, H.T.; Öborn, I. Fruit Tree-Based Agroforestry Systems for Smallholder Farmers in Northwest Vietnam—A Quantitative and Qualitative Assessment. *Land* **2020**, *9*, 451. [CrossRef]
122. Purwanto, E.; Santoso, H.; Jelsma, I.; Widayati, A.; Nugroho, H.Y.S.H.; van Noordwijk, M. Agroforestry as policy option for forest-zone oil palm production in Indonesia. *Land* **2020**, *9*, 531. [CrossRef]
123. Cabral, J.P.; Faria, D.; Morante-Filho, J.C. Landscape composition is more important than local vegetation structure for understory birds in cocoa agroforestry systems. *For. Ecol. Manag.* **2020**, *481*, 118704. [CrossRef]
124. Niether, W.; Jacobi, J.; Blaser, W.J.; Andres, C.; Armengot, L. Cocoa agroforestry systems versus monocultures: A multi-dimensional meta-analysis. *Environ. Res. Lett.* **2020**, *15*, 104085. [CrossRef]
125. Kay, S.; Kühn, E.; Albrecht, M.; Sutter, L.; Szerencsits, E.; Herzog, F. Agroforestry can enhance foraging and nesting resources for pollinators with focus on solitary bees at the landscape scale. *Agrofor. Syst.* **2020**, *94*, 379–387. [CrossRef]
126. Elagib, N.A.; Al-Saidi, M. Balancing the benefits from the water–energy–land–food nexus through agroforestry in the Sahel. *Sci. Total Environ.* **2020**, *742*, 140509. [CrossRef]
127. Marconi, L.; Armengot, L. Complex agroforestry systems against biotic homogenization: The case of plants in the herbaceous stratum of cocoa production systems. *Agric. Ecosyst. Environ.* **2020**, *287*, 106664. [CrossRef]
128. Mupepele, A.C.; Keller, M.; Dormann, C.F. European agroforestry is no universal remedy for biodiversity: A time-cumulative meta-analysis. *bioRxiv* **2020**. Available online: https://www.biorxiv.org/content/biorxiv/early/2020/08/28/2020.08.27.269589.full.pdf (accessed on 2 July 2021). [CrossRef]
129. Sari, R.R.; Saputra, D.D.; Hairiah, K.; Rozendaal, D.M.A.; Roshetko, J.M.; van Noordwijk, M. Gendered Species Preferences Link Tree Diversity and Carbon Stocks in Cacao Agroforest in Southeast Sulawesi, Indonesia. *Land* **2020**, *9*, 108. [CrossRef]
130. Cerda, R.; Avelino, J.; Harvey, C.A.; Gary, C.; Tixier, P.; Allinne, C. Coffee agroforestry systems capable of reducing disease-induced yield and economic losses while providing multiple ecosystem services. *Crop Prot.* **2020**, *134*, 105149. [CrossRef]
131. Durand-Bessart, C.; Tixier, P.; Quinteros, A.; Andreotti, F.; Rapidel, B.; Tauvel, C.; Allinne, C. Analysis of interactions amongst shade trees, coffee foliar diseases and coffee yield in multistrata agroforestry systems. *Crop Prot.* **2020**, *133*, 105137. [CrossRef]
132. Mulia, R.; Nguyen, D.D.; Nguyen, M.P.; Steward, P.; Pham, V.T.; Le, H.A.; Rosenstock, T.; Simelton, E. Enhancing Vietnam's Nationally Determined Contribution with Mitigation Targets for Agroforestry: A Technical and Economic Estimate. *Land* **2020**, *9*, 528. [CrossRef]
133. Magaju, C.; Ann Winowiecki, L.; Crossland, M.; Frija, A.; Ouerghemmi, H.; Hagazi, N.; Sola, P.; Ochenje, I.; Kiura, E.; Kuria, A.; et al. Assessing Context-Specific Factors to Increase Tree Survival for Scaling Ecosystem Restoration Efforts in East Africa. *Land* **2020**, *9*, 494. [CrossRef]
134. Hughes, K.; Morgan, S.; Baylis, K.; Oduol, J.; Smith-Dumont, E.; Vågen, T.G.; Kegode, H. Assessing the downstream socioeconomic impacts of agroforestry in Kenya. *World Dev.* **2020**, *128*, 104835. [CrossRef]
135. Sollen-Norrlin, M.; Ghaley, B.B.; Rintoul, N.L.J. Agroforestry benefits and challenges for adoption in Europe and beyond. *Sustainability* **2020**, *12*, 7001. [CrossRef]
136. Mulyoutami, E.; Lusiana, B.; van Noordwijk, M. Gendered migration and agroforestry in Indonesia: Livelihoods, labor, know-how, networks. *Land* **2020**, *9*, 529. [CrossRef]
137. Chizmar, S.; Castillo, M.; Pizarro, D.; Vasquez, H.; Bernal, W.; Rivera, R.; Sills, E.; Abt, R.; Parajuli, R.; Cubbage, F. A Discounted Cash Flow and Capital Budgeting Analysis of Silvopastoral Systems in the Amazonas Region of Peru. *Land* **2020**, *9*, 353. [CrossRef]
138. Andreotti, F.; Speelman, E.N.; Van den Meersche, K.; Allinne, C. Combining participatory games and backcasting to support collective scenario evaluation: An action research approach for sustainable agroforestry landscape management. *Sustain. Sci.* **2020**, *15*, 1383–1399. [CrossRef]
139. Hofstede, G.J.; Frantz, C.; Hoey, J.; Scholz, G.; Schröder, T. Artificial Sociality Manifesto. Review of Artificial Societies and Social Simulation, 8th April 2021. Available online: https://rofasss.org/2021/04/08/artsocmanif/ (accessed on 15 May 2021).

MDPI
St. Alban-Anlage 66
4052 Basel
Switzerland
Tel. +41 61 683 77 34
Fax +41 61 302 89 18
www.mdpi.com

Land Editorial Office
E-mail: land@mdpi.com
www.mdpi.com/journal/land